Lecture Notes in Mathematics   2107

T0213744

For further volumes:
http://www.springer.com/series/304

Lecture Notes in Mathematics

Editors-in-Chief:
J.-M. Morel, Cachan
B. Teissier, Paris

Advisory Board:
Camillo De Lellis, Zürich
Mario di Bernardo, Bristol
Alessio Figalli, Austin
Davar Khoshnevisan, Salt Lake City
Ioannis Kontoyiannis, Athens
Gábor Lugosi, Barcelona
Mark Podolskij, Aarhus
Sylvia Serfaty, Paris and NY
Catharina Stroppel, Bonn
Anna Wienhard, Heidelberg

Lower bound techniques, such as the orthogonal function method of Roth, the Fourier transform technique of Beck, and the integral geometric method of Alexander, have found applications in the study on algorithms and complexity. Upper bound considerations have found applications in quasi Monte Carlo methods and numerical integration. Apart from the pioneering work of Roth, there have been many notable successes, particularly the work of Alexander and Beck on lower bounds and the work of Beck, Chen, and Skriganov on upper bounds.

Four chapters in the book will address these aspects and new developments.

In Chap. 1, William Chen and Maxim Skriganov provide a detailed study of upper bounds, using arguments from diophantine approximation, probability theory, number theory, and Fourier analysis. The chapter is also an introduction to basic concepts and proofs, like probabilistic and deterministic techniques and their comparison, van der Corput sets, Fourier–Walsh analysis, explicit constructions, and orthogonality.

In Chap. 2, Dmitriy Bilyk will present the recent breakthrough by Bilyk, Lacey, and Vagharshakyan on the L^∞ discrepancy problem related to rectangles in dimension $d \geq 3$. This was called the *Great open problem* in the first chapter of the book of Beck and Chen [7]. Until recently, there was only one deep result by Beck [6] in dimension 3. Bilyk will bring the reader to the core of the problem and will show the connections of discrepancy to other areas of mathematics, in particular to the *small ball inequality* which arises in harmonic analysis as well as in the study of the small deviation probabilities for the Brownian sheet.

As we said, a famous problem in the classical geometric theory concerns the study of discrepancies related to discs and more generally to convex bodies; see Schmidt [49], Beck [4, 5], Beck and Chen [8], and Montgomery [37]. Most of these results depend on suitable estimates for the average decay of certain Fourier transforms, a topic which has been recently investigated by a number of authors, among them Brandolini, Colzani, and Travaglini [11], and Brandolini, Hofmann, and Iosevich [12]. In Chap. 3, Luca Brandolini, Giacomo Gigante, and Giancarlo Travaglini will present in a detailed and unified way recent Fourier analysis results and their connections to the above discrepancy problem.

Finding the integer solutions of a Pell equation is equivalent to finding the integer lattice points in a long and narrow tilted hyperbolic region, where the slope is a quadratic irrational. Motivated by this relationship, in Chap. 4, József Beck presents a systematic study of point counting with respect to translated or congruent families of any given long and narrow hyperbolic region. The main results exhibit a fascinating new phenomenon about the extra large discrepancy called *superirregularity* and demonstrate, in a quantitative sense, that in point counting with respect to translated/congruent copies of any long and narrow hyperbolic region, superirregularity is inevitable. The techniques involved depend on ideas from number theory, combinatorics, probability theory, and Fourier analysis.

2. Combinatorial Discrepancy Theory: The basic problem in combinatorial discrepancy theory is to color the nodes of a finite hypergraph with two colors in a way that ideally in every hyperedge the number of nodes in the two colors is the same. The minimum deviation from this optimal situation is called the discrepancy of the hypergraph.

There is a relation between combinatorial discrepancy and classical geometric discrepancy, usually known as the transfer lemma, which allows the transfer of results between geometric discrepancy theory and combinatorial discrepancy theory. On the other hand, for many important problems in combinatorial discrepancy theory the transfer lemma is too weak, and intrinsic combinatorial methods are required.

The foundation of combinatorial discrepancy theory was laid by the work of Beck [3], Beck and Fiala [9], Spencer [52] (*the six-standard-deviation theorem*), and Lovász, Spencer, and Vesztergombi [30]. Combinatorial discrepancies arise in several areas of combinatorics, like Ramsey theory, uni-modular matrices, and extremal set systems. Over the last 10 years a number of new results have appeared, leading to new techniques and sometimes to optimal discrepancy bounds. Among them are the trace bound for the hereditary discrepancy of Chazelle and Lvov [15], the new bounds for geometric set systems of Matoušek, Welzl, and Wernisch [36], Matoušek [31], Chazelle [13], and the resolution of the linear discrepancy conjecture for totally unimodular matrices by Doerr [19]. A new aspect has been the investigation of multicolor discrepancy by Doerr and Srivastav [20], where some unexpected phenomena arise by passing from two colors to several colors.

Among the interesting classes of hypergraphs are certainly those with some arithmetic structures, like the hypergraph of arithmetic progressions in the first n integers (Roth [47]; Matoušek and Spencer [35]) and their generalizations, like products and sums of arithmetic progressions (Doerr, Srivastav and Wehr [21]; Hebbinghaus [27]; Přívětivý [45]) or hyperplanes in finite vector spaces (Hebbinghaus, Schoen and Srivastav [28]).

Quite recently, an efficient randomized algorithm for the construction of a 2-coloring satisfying Spencer's famous six-standard-bound was given by Bansal [1], and in a derandomized version by Bansal and Spencer [2], resolving a long-standing open problem.

In this book, three chapters are concerned with the development of combinatorial discrepancy theory and some of the mentioned directions.

In Chap. 5 on multicolor discrepancy of arithmetic structures, Nils Hebbinghaus and Anand Srivastav present the discrepancy theory for hypergraphs with arithmetic structures, e.g., arithmetic progressions in the first N integer, their various generalizations, like cartesian products, sums of arithmetic progressions, central arithmetic progressions (Bohr sets) in $\mathbb{Z}_p$, and linear hyperplanes in finite vector spaces. At the beginning, the theory of multicolor discrepancy is described, like upper and lower bounds for general hypergraphs and the multicolor generalization of several classical 2-color theorems. It is shown that at several places phenomena not visible in the 2-color theory show up, among them the multicoloring of products of hypergraphs. The focus of the chapter are proofs of lower bounds for the multicolor discrepancy for the hypergraphs mentioned above, where often the application of Fourier analysis or linear algebra techniques is not sufficient and has to be combined with combinatorial arguments.

Chapter 6 by Nikhil Bansal comprises recent breakthrough work on algorithms in combinatorial discrepancy theory. Since 1985 it has been an open problem, whether there is a polynomial-time algorithm which computes for a hypergraph with n nodes and n hyperedges a 2-coloring satisfying Joel Spencer's famous six-standard-deviation bound of $O(\sqrt{n})$ published in the *Transaction of the American Mathematical Society* in 1985. In 2010 N. Bansal [1] solved this problem, with a randomized algorithm based on semi-definite programming and a kind of twofold randomization. In 2011 Bansal and Spencer [2] were able to derandomize the algorithm. Some other exciting developments related to Bansal's work, for example the linear algebra technique of Lovett and Meka [50] and the tightness of the determinant bound for hereditary discrepancy due to Matoušek [34], are discussed as well.

Combinatorics is more and more touched by computational advances in computer science, and practical and efficient algorithms are sought. This is in particular true for discrepancy theory, where we wish to construct point sets or colorings satisfying the best known discrepancy bounds or finding experimental evidence for discrepancy bounds, e.g., the unsolved conjecture of Paul Erdős on the discrepancy of arithmetic progressions in the integers which is part of the *polymath project* initiated by Timothy Gowers in 2009. In Chap. 7, Lasse Kliemann shows how to efficiently compute low-discrepancy colorings using high-performance computing. As a benchmark problem he chooses the hypergraph of arithmetic progressions in the first N integers, for which the optimal discrepancy is $\Theta(N^{1/4})$ up to constants; Roth [47], Matoušek, Spencer [35]. With Bansal's algorithm one can compute, in randomized polynomial time, a coloring with discrepancy $O(N^{1/4}(\log N)^k)$ using semidefinite programs. But as the semidefinite programs grow in the number of hyperedges, the time complexity is too high even for moderately large N. Kliemann devised a new evolutionary algorithm based on estimation of distribution (EDA) on modern multicore computers. The algorithms compute the optimum $\Theta(N^{1/4})$ up to a constant factor for at least up to $N = 250,000$, where we have the astronomical number of $377 \cdot 10^9$ arithmetic progressions.

3. Applications and Constructions: A fundamental problem is the efficient construction of point sets with low geometric discrepancy. Classical constructions are the point sets of Halton [26], Faure [23], and Niederreiter [38]. New constructions are based on the so-called rank-1 lattice rules by Sloan, Kuo and Joe [51], Kuo [29], Dick [17], and Nuyens and Cools [42, 43]. A comprehensive theory of general tractability was developed by Gnewuch and Woźniakowski [24, 25]. The books of Novak and Woźniakowski [39–41] summarize the current state of the art in tractability theory.

Among the prominent applications of discrepancy theory are counting problems in number theory, for example the investigation of the distribution of the solutions of a diophantine equation, or the numerical integration in d-dimensional space, where d is large (e.g., between 30 and 360 in financial mathematics), the so-called quasi Monte Carlo method.

In many experiments the quasi Monte Carlo method is superior to the widely applied Monte Carlo simulation, where random points are used (see Owen [44]). However, the dispute for which applications this observation holds is ongoing and an interesting area of research (see Niederreiter [38], Matoušek [32], Dick and Pillichshammer [18]). On the other hand, a flow of results, particularly by Woźniakowski (1998–2001), determining the complexity of approximation of numerical integration, have shown the worst case limits of approximation. Other results can be found in recent volumes of the *Journal of Complexity*.

In Chap. 8, Ákos Magyar studies the distribution of the solutions of a diophantine equation when projected onto the unit level surface via the dilations, and also when mapped to the flat torus $\mathbf{T}^n$. He obtains quantitative estimates on the rate of equi-distribution in terms of upper bounds on the associated discrepancy. The main technical tool is the Hardy–Littlewood method of exponential sums utilized to obtain asymptotic expansions of the Fourier transform of the solution sets.

Chapter 9 by Josef Dick and Friedrich Pillichshammer is devoted to the application of discrepancy theory to quasi Monte Carlo integration, with an emphasis on explicit error bounds. The chapter is a presentation of the state-of-the-art methods for quasi Monte Carlo integration.

Last, but not least, we return to the basic problem of the geometric discrepancy, where the known bounds and methods have been described in the first chapter. Now we are concerned with efficient constructions of point sets and computation of the discrepancy. In Chap. 10, Carola Doerr, Michael Gnewuch, and Magnus Wahlström present randomized and de-randomized algorithms for the construction of low discrepancy point sets and the calculation of the star discrepancy, prove complexity results, and show interesting and promising connections to integer programming.

Acknowledgements We thank Dr. Volkmar Sauerland for his technical support in preparing the manuscript of this book.

References

1. N. Bansal, Constructive algorithms for discrepancy minimization, in *Proceedings of the 51st Annual IEEE Symposium on Foundations of Computer Science, Las Vegas, Nevada, USA, October 2010 (FOCS 2010)*, 2010, pp. 3–10
2. N. Bansal, J.H. Spencer, Deterministic discrepancy minimization., in *19th annual European symposium, Saarbrücken, Germany, September 5–9, 2011. Proceedings, Algorithms – ESA 2011*, ed. by Demetrescu, Camil (ed.) et al. Lecture Notes in Computer Science, vol. 6942 (Springer, Berlin, 2011), pp. 408–420. doi:10.1007/978-3-642-23719-5_35
3. J. Beck, Roth's estimate of the discrepancy of integer sequences is nearly sharp. Combinatorica **1**, 319–325 (1981). doi:10.1007/BF02579452
4. J. Beck, Irregularities of distribution. I. Acta Mathematica **159**(1–2), 1–49 (1987). doi:10.1007/BF02392553
5. J. Beck, Irregularities of distribution. II. Proc. Lond. Math. Soc. Ser. III **56**(1), 1–50 (1988). doi:10.1112/plms/s3-56.1.1
6. J. Beck, A two-dimensional van Aardenne-Ehrenfest theorem in irregularities of distribution. Compositio Math. **72**(3), 269–339 (1989)
7. J. Beck, W.W.L. Chen, *Irregularities of Distribution*. Cambridge Tracts in Mathematics vol. 89 (Cambridge University Press, Cambridge, 1987)
8. J. Beck, W.W.L. Chen, Note on irregularities of distribution. II. Proc. Lond. Math. Soc. Ser. III. **61**(2), 251–272 (1990). doi:10.1112/plms/s3-61.2.251

9. J. Beck, T. Fiala, "Integer making" theorems. Discrete Appl. Math. **3**(1), 1–8 (1981)
10. J. Beck, V.T. Sós, Discrepancy Theory, in *Handbook of Combinatorics*, ed. by Graham, R. L. and Grötschel, M. and Lovász, László (Elsevier, Amsterdam, 1995), pp. 1405–1446. Chap. 26
11. L. Brandolini, L. Colzani, G. Travaglini, Average decay of Fourier transforms and integer points in polyhedra. Arkiv för Mathematik **35**(2), 253–275 (1997). doi:10.1007/BF02559969
12. L. Brandolini, S. Hofmann, A. Iosevich, Sharp rate of average decay of the Fourier transform of a bounded set. Geomet. Funct. Anal. **13**(4), 671–680 (2003). doi:10.1007/s00039-003-0426-7
13. B. Chazelle, *Discrepancy Bounds for Geometric Set Systems with Square Incidence Matrices*, ed. by B. Chazelle, et al., Advances in discrete and computational geometry. Proceedings of the 1996 AMS-IMS-SIAM joint summer research conference on discrete and computational geometry: ten years later, South Hadley, MA, USA, July 14–18, 1996. Contemp. Math., vol. 223 (American Mathematical Society, Providence, RI, 1999), pp. 103–107.
14. B. Chazelle, *The Discrepancy Method. Randomness and Complexity* (Cambridge University Press, Cambridge, 2000)
15. B. Chazelle, A. Lvov, A trace bound for the hereditary discrepancy. Discrete Comput Geomet. **26**(2), 221–231 (2001). doi:10.1007/s00454-001-0030-2
16. W.W.L. Chen, M.M. Skriganov, Explicit constructions in the classical mean squares problem in irregularities of point distribution. J. für die Reine Angewandte Mathematik **545**, 67–95 (2002). doi:10.1515/crll.2002.037
17. J. Dick, On the convergence rate of the component-by-component construction of good lattice rules. J. Complexity **20**(4), 493–522 (2004). doi:10.1016/j.jco.2003.11.008
18. J. Dick, F. Pillichshammer, *Digital Nets and Sequences. Discrepancy Theory and Quasi-Monte Carlo Integration* (Cambridge University Press, Cambridge, 2010)
19. B. Doerr, Linear discrepancy of totally unimodular matrices, in *Proceedings of the 12th Annual ACM-SIAM Symposium on Discrete Algorithms, Washington, DC, USA, January 2001 (SODA 2001)*, 2001, pp. 119–125
20. B. Doerr, A. Srivastav, Multi-color discrepancies. Combinator. Probab. Comput. **12**, 365–399 (2003). doi:10.1017/S0963548303005662
21. B. Doerr, A. Srivastav, P. Wehr (nee Knieper), Discrepancy of cartesian products of arithmetic progressions. Electron. J. Combinator. **11**(1), 16 (2004)
22. M. Drmota, R.F. Tichy, *Sequences, Discrepancies and Applications*. Lecture Notes in Mathematics (Springer, Berlin, 1997). doi:10.1007/BFb0093404
23. H. Faure, Discrépance de suites associées à un systéme de numération (en dimension un). Bull. de la Société Mathématique de France **109**, 143–182 (1981)
24. M. Gnewuch, H. Woźniakowski, Generalized tractability for multivariate problems. I: Linear tensor product problems and linear information. J. Complexity **23**(2), 262–295 (2007). doi:10.1016/j.jco.2006.06.006
25. M. Gnewuch, H. Woźniakowski, Generalized tractability for multivariate problems. II: Linear tensor product problems, linear information, and unrestricted tractability. Found. Comput. Math. **9**(4), 431–460 (2009). doi:10.1007/s10208-009-9044-6
26. J.H. Halton, On the efficiency of certain quasi-random sequences of points in evaluating multidimensional integrals. Numer. Math. **2**, 84–90 (1960)
27. N. Hebbinghaus, Discrepancy of arithmetic structures, PhD thesis, Christian-Albrechts-Universität Kiel, Technische Fakultät, 2005. http://eldiss.uni-kiel.de/macau/receive/dissertation_diss_00001851
28. N. Hebbinghaus, T. Schoen, A. Srivastav, One-Sided Discrepancy of linear hyperplanes in finite vector spaces, in *Analytic Number Theory, Essays in Honour of Klaus Roth*, ed. by W.W.L. Chen, W.T. Gowers, H. Halberstam, W.M. Schmidt, R.C. Vaughan (Cambridge University Press, Cambridge, 2009), pp. 205–223
29. F.Y. Kuo, Component-by-component constructions achieve the optimal rate of convergence for multivariate integration in weighted Korobov and Sobolev spaces. J. Complexity **19**(3), 301–320 (2003). doi:10.1016/S0885-064X(03)00006-2
30. L. Lovász, J.H. Spencer, K. Vesztergombi, Discrepancies of set-systems and matrices. Eur. J. Combinator. **7**(2), 151–160 (1986)

31. J. Matoušek, Tight upper bounds for the discrepancy of half-spaces. Discrete Comput. Geomet. **13**(3–4), 593–601 (1995). doi:10.1007/BF02574066
32. J. Matoušek, *Geometric Discrepancy* (Springer, Heidelberg, New York, 1999)
33. J. Matoušek, *Geometric Discrepancy* (Springer, Heidelberg, New York, 2010)
34. J. Matoušek, The determinant bound for discrepancy is almost tight. Proc. Am. Math. Soc. **141**(2), 451–460 (2013). doi:10.1090/S0002-9939-2012-11334-6
35. J. Matoušek, J. Spencer, Discrepancy in arithmetic progressions. J. Am. Math. Soc. **9**, 195–204 (1996)
36. J. Matoušek, E. Welzl, L. Wernisch, Discrepancy and approximations for bounded VC–dimension. Combinatorica **13**(4), 455–466 (1993)
37. H.L. Montgomery, *Ten Lectures on the Interface Between Analytic Number Theory and Harmonic Analysis*. Regional Conference Series in Mathematics, vol. 84 (American Mathematical Society, Providence, RI, 1994), p. 220
38. H. Niederreiter, *Random Number Generation and Quasi-Monte Carlo Methods*. SIAM CBMS-NSF Regional Conference Series in Applied Mathematics, vol. 63 (SIAM, Philadelphia, 1992)
39. E. Novak, H. Woźniakowski, *Tractability of Multivariate Problems. Volume I: Linear Information*. EMS Tracts in Mathematics, vol. 6 (European Mathematical Society (EMS), Zürich, 2008)
40. E. Novak, H. Woźniakowski, *Tractability of Multivariate Problems. Volume Ii: Standard Information for Functionals*. EMS Tracts in Mathematics, vol. 12 (European Mathematical Society (EMS), Zürich, 2010)
41. E. Novak, H. Woźniakowski, *Tractability of Multivariate Problems. Volume Iii: Standard Information for Operators*. EMS Tracts in Mathematics, vol. 18 (European Mathematical Society (EMS), Zürich, 2012). doi:10.4171/116
42. D. Nuyens, R. Cools, Fast algorithms for component-by-component construction of rank-1 lattice rules in shift-invariant reproducing kernel Hilbert spaces. Math. Comput. **75**(254), 903–920 (2006). doi:10.1090/S0025-5718-06-01785-6
43. D. Nuyens, R. Cools, Fast component-by-component construction of rank-1 lattice rules with a non-prime number of points. J. Complexity **22**(1), 4–28 (2006). doi:10.1016/j.jco.2005.07.002
44. A.B. Owen, Latin supercube sampling for very high-dimensional simulations. ACM Trans. Model. Comput. Simul. **8**(1), 71–102 (1998). doi:10.1145/272991.273010
45. A. Přívětivý, Discrepancy of sums of three arithmetic progressions. Electron. J. Combinator. **13**(1), 49 (2006)
46. K.F. Roth, On irregularities of distribution. Mathematika **1**, 73–79 (1954). doi:10.1112/S0025579300000541
47. K.F. Roth, Remark concerning integer sequences. Acta Arithmetica **9**, 257–260 (1964)
48. K.F. Roth, On irregularities of distribution. IV. Acta Arithmetica **37**, 67–75 (1980)
49. W.M. Schmidt, *Lectures on Irregularities of Distribution* (Tata Institute of Fundamental Research Lectures on Mathematics and Physics, 56, Tata Institute of Fundamental Research, Bombay, 1977)
50. S. Shachar Lovett, R. Meka, Constructive discrepancy minimization by walking on the edges. CoRR **abs/1203.5747** (2012)
51. I.H. Sloan, F.Y. Kuo, S. Joe, Constructing randomly shifted lattice rules in weighted Sobolev spaces. SIAM J. Numer. Anal. **40**(5), 1650–1665 (2002). doi:10.1137/S0036142901393942
52. J. Spencer, Six standard deviations suffice. Trans. Am. Math. Soc. **289**(2), 679–706 (1985)
53. T. van Aardenne-Ehrenfest, Proof of the impossibility of a just distribution of an infinite sequence of points over an interval. Proc. Nederlandse Akademie van Wetenschappen **48**, 266–271 (1945)
54. T. van Aardenne-Ehrenfest, On the impossibility of a just distribution. Proc. Koninklijke Nederlandse Akademie van Wetenschappen **52**, 734–739 (1949)
55. H. Weyl, Über die Gleichverteilung von Zahlen mod. Eins. Math. Ann. **77**, 313–352 (1916). doi:10.1007/BF01475864

Contents

Part I Classical and Geometric Discrepancy

1 Upper Bounds in Classical Discrepancy Theory 3
William Chen and Maxim Skriganov

**2 Roth's Orthogonal Function Method in Discrepancy
Theory and Some New Connections** 71
Dmitriy Bilyk

**3 Irregularities of Distribution and Average Decay
of Fourier Transforms** 159
Luca Brandolini, Giacomo Gigante, and Giancarlo Travaglini

4 Superirregularity 221
József Beck

Part II Combinatorial Discrepancy

5 Multicolor Discrepancy of Arithmetic Structures 319
Nils Hebbinghaus and Anand Srivastav

6 Algorithmic Aspects of Combinatorial Discrepancy 425
Nikhil Bansal

7 Practical Algorithms for Low-Discrepancy 2-Colorings 459
Lasse Kliemann

Part III Applications and Constructions

8 On the Distribution of Solutions to Diophantine Equations 487
Ákos Magyar

9 Discrepancy Theory and Quasi-Monte Carlo Integration 539
Josef Dick and Friedrich Pillichshammer

10   Calculation of Discrepancy Measures and Applications 621
Carola Doerr, Michael Gnewuch, and Magnus Wahlström

Author Index.. 679

Index.. 685

List of Contributors

Nikhil Bansal Department of Mathematics and Computer Science, Eindhoven University of Technology, Eindhoven, The Netherlands

Jószef Beck Department of Mathematics, Rutgers University, New Brunswick, NJ, USA

Dmitriy Bilyk School of Mathematics, University of Minnesota, Minneapolis, MN, USA

Luca Brandolini Dipartimento di Ingegneria, Università di Bergamo, Bergamo, Italia

William Chen Department of Mathematics, Macquarie University, Sydney, NSW, Australia

Josef Dick School of Mathematics and Statistics, The University of New South Wales, Sydney, NSW, Australia

Carola Doerr Université Pierre et Marie Curie - Paris 6, LIP6, équipe RO, Paris, France *and* Department 1: Algorithms and Complexity, Max-Planck-Institut für Informatik, Saarbrücken, Germany

Giacomo Gigante Dipartimento di Ingegneria, Università di Bergamo, Bergamo, Italia

Michael Gnewuch Mathematisches Seminar, Kiel University, Kiel, Germany

Nils Hebbinghaus Department of Computer Science, Kiel University, Kiel, Germany

Lasse Kliemann Department of Computer Science, Kiel University, Kiel, Germany

Akos Magyar Department of Mathematics, University of British Columbia, Vancouver, BC, Canada

Friedrich Pillichshammer Institute of Financial Mathematics, University of Linz, Linz, Austria

Maxim Skriganov Steklov Mathematical Institute, St. Petersburg, Russia

Anand Srivastav Department of Computer Science, Kiel University, Kiel, Germany

Giancarlo Travaglini Dipartimento di Statistica e Metodi Quantitativi, Università di Milano-Bicocca, Milano, Italia

Magnus Wahlström Department 1: Algorithms and Complexity, Max-Planck-Institut für Informatik, Saarbrücken, Germany

Part I
Classical and Geometric Discrepancy

Chapter 1
Upper Bounds in Classical Discrepancy Theory

William Chen and Maxim Skriganov

Abstract We discuss some of the ideas behind the study of upper bound questions in classical discrepancy theory. The many ideas involved come from diverse areas of mathematics and include diophantine approximation, probability theory, number theory and various forms of Fourier analysis. We illustrate these ideas by largely restricting our discussion to two dimensions.

1.1 Introduction

Classical discrepancy theory, or irregularities of distribution, began as a branch of the theory of uniform distribution but has independent interest. It is often viewed as a quantitative and substantially more precise version of the theory of uniform distribution, in the sense that one seeks to obtain very accurate bounds on various quantities arising from the difference between the discrete and the continuous. Here the discrete concerns the actual point count in a given region, which clearly takes integer values, whereas the continuous refers to the expectation of the point count, which depends on the area or volume of the region concerned and therefore can take non-integer values.

We shall first state the problem in a rather general form. Let $k \geq 2$ be a fixed integer. Our domain U will be a set of unit Lebesgue measure in k-dimensional Euclidean space $\mathbf{R}^k$.

W. Chen (✉)
Department of Mathematics, Macquarie University, Sydney, NSW 2109, Australia
e-mail: william.chen@mq.edu.au

M. Skriganov
Steklov Mathematical Institute, Fontanka 27, St Petersburg 191011, Russia
e-mail: skrig@pdmi.ras.ru

W. Chen et al. (eds.), *A Panorama of Discrepancy Theory*, Lecture Notes in Mathematics 2107, DOI 10.1007/978-3-319-04696-9_1,
© Springer International Publishing Switzerland 2014

Suppose that $\mathscr{A}$ is a set of measurable subsets of U, endowed with an integral geometric measure, normalized so that the total measure is equal to unity. Suppose further that $\mathscr{P}$ is a set of N points in U. For every subset $A \in \mathscr{A}$ of U, let

$$Z[\mathscr{P}; A] = \#(\mathscr{P} \cap A)$$

denote the number of points of $\mathscr{P}$ that fall into A. This is the actual point count of $\mathscr{P}$ in A, with corresponding expectation $N\mu(A)$. By the discrepancy of $\mathscr{P}$ in A, we mean the difference

$$D[\mathscr{P}; A] = Z[\mathscr{P}; A] - N\mu(A).$$

Often, we consider the extreme discrepancy of $\mathscr{P}$ in U, taken to be the L^∞-norm

$$\|D[\mathscr{P}]\|_\infty = \sup_{A \in \mathscr{A}} |D[\mathscr{P}; A]|. \tag{1.1}$$

However, for upper bound considerations, it is far more interesting and challenging to consider the corresponding L^2-norm

$$\|D[\mathscr{P}]\|_2 = \left(\int_{\mathscr{A}} |D[\mathscr{P}; A]|^2 \, \mathrm{d}A \right)^{1/2}, \tag{1.2}$$

as well as the corresponding L^q-norms where $2 < q < \infty$.

For any given choice of U and $\mathscr{A}$, we are interested in studying the growth of the functions (1.1) and (1.2) as functions of N, the number of points of $\mathscr{P}$. It is the cornerstone of discrepancy theory that these quantities become arbitrarily large in many interesting cases, following the early conjecture of van der Corput [37, 38] and the pioneering work of van Aardenne-Ehrenfest [35, 36] and Roth [29]. A lower bound result is thus of the form

$$\|D[\mathscr{P}]\|_\infty > f(N) \quad \text{for all sets } \mathscr{P} \text{ of } N \text{ points in } U,$$

or of the form

$$\|D[\mathscr{P}]\|_2 > f(N) \quad \text{for all sets } \mathscr{P} \text{ of } N \text{ points in } U.$$

For upper bounds, we first make a simple observation. Consider a set $\mathscr{P}$ of N points, where all the points coincide. Then clearly any subset $A \in \mathscr{A}$ of U either contains all points of $\mathscr{P}$ or contains no point of $\mathscr{P}$. In either case, we expect the discrepancy $D[\mathscr{P}; A]$ to have rather large absolute value for many of these sets A. This is an example of an extremely badly distributed point set. Such examples must never be allowed to play a role in upper bound considerations. After all, if the lower bound asserts that all distributions are *bad*, then a complementary upper bound must say that some distributions are *close to as good as they possibly can be*. Hence an upper bound result must be of the form

$$\|D[\mathscr{P}]\|_\infty < g(N) \quad \text{for some sets } \mathscr{P} \text{ of } N \text{ points in } U,$$

or of the form

$$\|D[\mathscr{P}]\|_2 < g(N) \quad \text{for some sets } \mathscr{P} \text{ of } N \text{ points in } U.$$

Our task is therefore to construct such a point set $\mathscr{P}$, or to show that one exists.

Of course, the ultimate task is to establish lower and upper bounds where the two functions $f(N)$ and $g(N)$ have the same order of magnitude. This has been achieved in a few instances, and we shall discuss the upper bound aspects of some of these in some detail in this article.

There are well known choices of U and $\mathscr{A}$ where the quantities (1.1) and (1.2) exceed N^δ for some positive exponent δ. We refer to these as large discrepancy phenomena. On the other hand, there are also well known choices of U and $\mathscr{A}$ where, for suitably chosen point sets $\mathscr{P}$, the quantities (1.1) and (1.2) can be bounded above by $(\log N)^\delta$ for some positive exponent δ. We refer to these as small discrepancy phenomena. As a general rule, upper bound questions are somewhat harder for small discrepancy phenomena, as we shall attempt to illustrate in the course of this article.

Notation. For any complex-valued function f and any positive function g, we write $f = O(g)$ to denote that there exists a positive constant C such that $|f| \le Cg$, and write $f = O_\delta(g)$ if the positive constant C may depend on a parameter δ. We also use the Vinogradov notation, where $f \ll g$ if $f = O(g)$, and $f \ll_\delta g$ if $f = O_\delta(g)$. We also write $f \gg g$ and $f \gg_\delta g$ to denote respectively $g \ll f$ and $g \ll_\delta f$, but here both f and g must be positive functions. The letters $\mathbf{N}$, $\mathbf{Z}$ and $\mathbf{R}$ denote respectively the set of all natural numbers, i.e. positive integers, the set of all integers and the set of all real numbers. We also write $\mathbf{N}_0$ to denote the set of all non-negative integers. For any real number z, we write $e(z) = e^{2\pi i z}$, and write $[z]$ and $\{z\}$ to denote respectively the integer part and the fractional part of z, i.e.

$$[z] = \max\{n \in \mathbf{Z} : n \le z\} \quad \text{and} \quad \{z\} = z - [z].$$

For any finite set $\mathscr{S}$, we denote by $\#\mathscr{S}$ the cardinality of $\mathscr{S}$. For any probabilistic variable ξ, we denote by $\mathbf{E}\xi$ the expected value of ξ.

1.2 Large Discrepancy: Main Results

The work on large discrepancy problems can best be summarized by the following ground-breaking result of Beck [4]. Consider the k-dimensional Euclidean space $\mathbf{R}^k$. We take as our domain U the unit cube $[0, 1]^k$, treated as a torus for simplicity. Let $B \subseteq [0, 1]^k$ be a compact and convex set that satisfies a technical condition

$$r(B) \geq N^{-1/k}, \tag{1.3}$$

where $r(B)$ denotes the radius of the largest inscribed ball in B, and N is the cardinality of the point sets $\mathscr{P}$ under consideration. While this technical condition does not really affect the argument in a serious way, it is nevertheless necessary in order for us to avoid degenerate cases. Let $\mathscr{T}$ denote the group of all orthogonal transformations in $\mathbf{R}^k$, normalized so that the total measure is equal to unity. For any contraction $\lambda \in [0, 1]$, orthogonal transformation $\tau \in \mathscr{T}$ and translation $\mathbf{x} \in [0, 1]^k$, we consider the similar copy

$$B(\lambda, \tau, \mathbf{x}) = \lambda(\tau B) + \mathbf{x}$$

of B. We then consider the collection

$$\mathscr{A} = \{B(\lambda, \tau, \mathbf{x}) : \lambda \in [0, 1], \ \tau \in \mathscr{T}, \ \mathbf{x} \in [0, 1]^k\}$$

of all similar copies of B, where the integral geometric measure is given by a natural combination of the standard Lebesgue measures of λ and $\mathbf{x}$ and the measure of $\mathscr{T}$. More precisely, for any set $\mathscr{P}$ of N points in $[0, 1]^k$, we have the L^2-norm

$$\|D[\mathscr{P}]\|_2 = \left(\int_{[0,1]^k} \int_{\mathscr{T}} \int_0^1 |D[\mathscr{P}; B(\lambda, \tau, \mathbf{x})]|^2 \, d\lambda \, d\tau \, d\mathbf{x} \right)^{1/2}. \tag{1.4}$$

We also have the simpler L^∞-norm

$$\|D[\mathscr{P}]\|_\infty = \sup_{\substack{\lambda \in [0,1] \\ \tau \in \mathscr{T} \\ \mathbf{x} \in [0,1]^k}} |D[\mathscr{P}; B(\lambda, \tau, \mathbf{x})]|. \tag{1.5}$$

The following result is due to Beck [4].

Theorem 1. *Suppose that $B \subseteq [0, 1]^k$ is a compact and convex set that satisfies the condition (1.3). Then for every set $\mathscr{P}$ of N points in $[0, 1]^k$, we have*

$$\|D[\mathscr{P}]\|_2 \gg_B N^{1/2 - 1/2k}. \tag{1.6}$$

This leads immediately to the corresponding statement for the L^∞-norm.

Theorem 2. *Suppose that $B \subseteq [0, 1]^k$ is a compact and convex set that satisfies the condition (1.3). Then for every set $\mathscr{P}$ of N points in $[0, 1]^k$, we have*

$$\|D[\mathscr{P}]\|_\infty \gg_B N^{1/2 - 1/2k}. \tag{1.7}$$

The lower bound (1.6) is essentially best possible, in view of the following result of Beck and Chen which can be established as a simple case of their more general result in [1].

Theorem 3. *Suppose that $B \subseteq [0, 1]^k$ is a compact and convex set. Then for every natural number N, there exists a set $\mathscr{P}$ of N points in $[0, 1]^k$ such that*

$$\|D[\mathscr{P}]\|_2 \ll_B N^{1/2 - 1/2k}. \tag{1.8}$$

The proof of Theorem 3 is an extension of the original ideas needed to establish the following result using ideas in Beck [3]; see also Beck and Chen [5, Section 8.1].

Theorem 4. *Suppose that $B \subseteq [0, 1]^k$ is a compact and convex set. Then for every natural number $N \geq 2$, there exists a set $\mathscr{P}$ of N points in $[0, 1]^k$ such that*

$$\|D[\mathscr{P}]\|_\infty \ll_B N^{1/2 - 1/2k} (\log N)^{1/2}. \tag{1.9}$$

We shall discuss Beck's ideas in Sect. 1.4 and sketch a proof of the special case $k = 2$ of Theorem 4. Most important of all, however, the argument gives us a very good understanding of the exponent in the estimates (1.6)–(1.9).

We shall then sketch a proof of the special case $k = 2$ of Theorem 3 in Sect. 1.5.

1.3 A Seemingly Trivial Argument

We start by making an inadequate attempt to establish the special case $k = 2$ of Theorem 4. Such simple and perhaps naive attempts often play an important role in the study of upper bounds. Remember that we need to find a *good* set of points, and we often start by toying with some specific set of points which we hope will be good. Often it is not, but sometimes it permits us to bring in some stronger techniques at a later stage of the argument.

For simplicity, let us assume that the number of points is a perfect square, so that $N = M^2$ for some natural number M. We may then choose to split the unit square $[0, 1]^2$ in the natural way into a union of $N = M^2$ little squares of side length M^{-1}, and then place a point in the centre of each little square, as shown in Fig. 1.1 below.

Suppose that $A = B(\lambda, \tau, \mathbf{x})$, where $\lambda \in [0, 1]$, $\tau \in \mathscr{T}$ and $\mathbf{x} \in [0, 1]^2$, is a similar copy of a given fixed compact and convex set B. We now attempt to estimate the discrepancy $D[\mathscr{P}; A]$. Let $\mathscr{S}$ denote the collection of the $N = M^2$ little squares S of side length M^{-1}. The additive property of the discrepancy function then gives

$$D[\mathscr{P}; A] = \sum_{S \in \mathscr{S}} D[\mathscr{P}; S \cap A]. \tag{1.10}$$

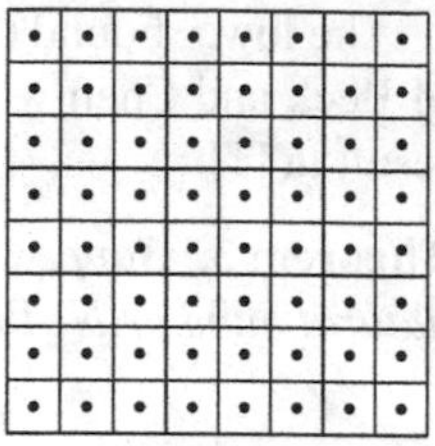

Fig. 1.1 A basic construction
of $N = M^2$ points in the unit
square

Next, we make the simple observation that

$$D[\mathscr{P}; S \cap A] = 0 \quad \text{if } S \subseteq A \text{ or } S \cap A = \emptyset.$$

The identity (1.10) then becomes

$$D[\mathscr{P}; A] = \sum_{\substack{S \in \mathscr{S} \\ S \cap \partial A \neq \emptyset}} D[\mathscr{P}; S \cap A], \tag{1.11}$$

where ∂A denotes the boundary of A. Finally, observe that both $0 \leq Z[\mathscr{P}; S \cap A] \leq 1$ and $0 \leq N\mu(S \cap A) \leq 1$, so that $|D[\mathscr{P}; S \cap A]| \leq 1$, and it follows from (1.11) and the triangle inequality that

$$|D[\mathscr{P}; A]| \leq \#\{S \in \mathscr{S} : S \cap \partial A \neq \emptyset\} \ll M = N^{1/2}. \tag{1.12}$$

This estimate is almost trivial, but very far from the upper bound $N^{1/4}(\log N)^{1/2}$ alluded to in Theorem 4.

We make an important observation here that the term $\#\{S \in \mathscr{S} : S \cap \partial A \neq \emptyset\}$ in (1.12) is intricately related to the length of the boundary curve ∂B of B; note that the set A is a similar copy of the given compact and convex set B. Indeed, in the general case of the problem in k-dimensional space, the corresponding term is intricately related to the $(k-1)$-dimensional volume of the boundary surface ∂B of B. It is worthwhile to record the important role played by boundary surface in large discrepancy problems.

1.4 A Large Deviation Technique

In this section, we continue our study of the special case $k = 2$ of Theorem 4. Again, let us assume that the number of points is a perfect square, so that $N = M^2$ for some natural number M. Again, we choose to split the unit square $[0, 1]^2$ in the natural way into a union of $N = M^2$ little squares of side length M^{-1}. As before, let $\mathscr{S}$ denote the collection of the $N = M^2$ little squares S of side length M^{-1}.

For every little square $S \in \mathscr{S}$, instead of placing a point in the centre of the square, we now associate a random point $\widetilde{\mathbf{p}_S} \in S$, uniformly distributed within the little square S and independent of all the other random points in the other little squares. We thus obtain a random point set

$$\tilde{\mathscr{P}} = \{\widetilde{\mathbf{p}_S} : S \in \mathscr{S}\}. \tag{1.13}$$

Suppose that a fixed compact and convex set $B \subseteq [0, 1]^2$ is given. Let

$$\mathscr{G} = \left\{ B(\lambda, \tau, \mathbf{x}) : \lambda \in \left[0, \frac{11}{10}\right], \tau \in \mathscr{T}, \mathbf{x} \in [0, 1]^2 \right\}.$$

Note that the collection $\mathscr{G}$ contains the collection $\mathscr{A}$ and permits some similar copies of B which are a little bigger than B. Then one can find a subset $\mathscr{H}$ of $\mathscr{G}$ such that

$$\#\mathscr{H} \leq N^{C_1},$$

where C_1 is a positive constant depending at most on B, and such that for every $A \in \mathscr{A}$, there exist $A^-, A^+ \in \mathscr{H}$ such that

$$A^- \subseteq A \subseteq A^+ \quad \text{and} \quad \mu(A^+ \setminus A^-) \leq N^{-1}. \tag{1.14}$$

We comment that such a set $\mathscr{H}$ may not exist if we make the restriction $\mathscr{H} \subseteq \mathscr{A}$ instead of the more generous restriction $\mathscr{H} \subseteq \mathscr{G}$.

Suppose that $A \in \mathscr{H}$ is fixed. Then, analogous to the discrepancy function (1.10), we now consider the discrepancy function

$$D[\tilde{\mathscr{P}}; A] = \sum_{S \in \mathscr{S}} D[\tilde{\mathscr{P}}; S \cap A] = \sum_{\substack{S \in \mathscr{S} \\ S \cap \partial A \neq \emptyset}} D[\tilde{\mathscr{P}}; S \cap A], \tag{1.15}$$

and note as before that

$$\#\{S \in \mathscr{S} : S \cap \partial A \neq \emptyset\} \ll M = N^{1/2}. \tag{1.16}$$

For every $S \in \mathscr{S}$, let

$$\phi_S = \begin{cases} 1, & \text{if } \widetilde{\mathbf{p}_S} \in A, \\ 0, & \text{otherwise.} \end{cases}$$

The observation

$$D[\tilde{\mathscr{P}}; A] = \sum_{S \in \mathscr{S}} (\phi_S - \mathbf{E}\phi_S) = \sum_{\substack{S \in \mathscr{S} \\ S \cap \partial A \neq \emptyset}} (\phi_S - \mathbf{E}\phi_S) \tag{1.17}$$

sets us up to appeal to large deviation type inequalities in probability theory. For instance, we can use the following result attributed to Hoeffding; see, for instance, Pollard [27, Appendix B].

Lemma 5. *Suppose that $\phi_1, \ldots, \phi_m$ are independent random variables that satisfy $0 \le \phi_i \le 1$ for every $i = 1, \ldots, m$. Then for every real number $\gamma > 0$, we have*

$$\text{Prob}\left(\left| \sum_{i=1}^{m} (\phi_i - \mathbf{E}\phi_i) \right| \ge \gamma \right) \le 2e^{-2\gamma^2/m}.$$

In view of (1.17), we now apply Lemma 5 with

$$m = \#\{S \in \mathscr{S} : S \cap \partial A \ne \emptyset\} \le C_2 N^{1/2},$$

where C_2 is a positive constant depending at most on the given set B, and with

$$\gamma = C_3 N^{1/4} (\log N)^{1/2},$$

where C_3 is a sufficiently large positive constant. Indeed,

$$\frac{\gamma^2}{m} \ge \frac{C_3^2}{C_2} \log N,$$

and it follows therefore that

$$2e^{-2\gamma^2/m} \le \frac{1}{2} N^{-C_1} \le \frac{1}{2} (\#\mathscr{H})^{-1}$$

provided that C_3 is chosen sufficiently large in terms of C_1 and C_2. Then

$$\text{Prob}\left(|D[\tilde{\mathscr{P}}; A]| \ge C_3 N^{1/4} (\log N)^{1/2} \right) \le \frac{1}{2} (\#\mathscr{H})^{-1},$$

and so

$$\text{Prob}\left(|D[\tilde{\mathscr{P}}; A]| \ge C_3 N^{1/4} (\log N)^{1/2} \text{ for some } A \in \mathscr{H} \right) \le \frac{1}{2},$$

whence

$$\text{Prob}\left(|D[\tilde{\mathscr{P}}; A]| \le C_3 N^{1/4} (\log N)^{1/2} \text{ for all } A \in \mathscr{H} \right) \ge \frac{1}{2}.$$

In other words, there exists a set $\mathscr{P}^*$ of $N = M^2$ points in $[0, 1]^2$ such that

$$|D[\mathscr{P}^*; A]| \le C_3 N^{1/4} (\log N)^{1/2} \quad \text{for every } A \in \mathscr{H}.$$

Suppose now that $A \in \mathscr{A}$ is given. Then there exist $A^-, A^+ \in \mathscr{H}$ such that (1.14) is satisfied. It is not difficult to show that

$$|D[\mathscr{P}^*; A]| \leq \max\left\{|D[\mathscr{P}^*; A^-]|, |D[\mathscr{P}^*; A^+]|\right\} + N\mu(A^+ \setminus A^-)$$

$$\leq C_3 N^{1/4}(\log N)^{1/2} + 1.$$

Theorem 4 for $k = 2$ in the special case when $N = M^2$ is therefore established.

Finally, we can easily lift the restriction that N is a perfect square. By Lagrange's theorem, every positive integer N can be written as a sum

$$N = M_1^2 + M_2^2 + M_3^2 + M_4^2$$

of the squares of four non-negative integers. We can therefore superimpose up to four point distributions in $[0, 1]^2$ where the number of points in each is a perfect square. This completes the proof of Theorem 4 for $k = 2$.

1.5 An Averaging Argument

In this section, we indicate how the argument in the previous section can be adapted to establish Theorem 3 in the case $k = 2$.

We construct the random point set $\mathscr{P}$, given by (1.13), as before. Suppose that a fixed compact and convex set $B \subseteq [0, 1]^2$ is given. Let $A \in \mathscr{A}$ be fixed. Then (1.15), (1.16) and (1.17) are valid. If we write $\eta_S = \phi_S - \mathbf{E}\phi_S$, then

$$|D[\tilde{\mathscr{P}}; A]|^2 = \sum_{\substack{S_1, S_2 \in \mathscr{S} \\ S_1 \cap \partial A \neq \emptyset \\ S_2 \cap \partial A \neq \emptyset}} \eta_{S_1} \eta_{S_2}.$$

Taking expectation over all the $N = M^2$ random points, we have

$$\mathbf{E}\left(|D[\tilde{\mathscr{P}}; A]|^2\right) = \sum_{\substack{S_1, S_2 \in \mathscr{S} \\ S_1 \cap \partial A \neq \emptyset \\ S_2 \cap \partial A \neq \emptyset}} \mathbf{E}\left(\eta_{S_1} \eta_{S_2}\right). \tag{1.18}$$

The random variables η_S, where $S \in \mathscr{S}$, are independent since the distribution of the random points are independent of each other. If $S_1 \neq S_2$, then

$$\mathbf{E}\left(\eta_{S_1} \eta_{S_2}\right) = \mathbf{E}\left(\eta_{S_1}\right)\mathbf{E}\left(\eta_{S_2}\right) = 0.$$

It follows that the only non-zero contributions to the sum (1.18) come from those terms where $S_1 = S_2$, so that

$$\mathbf{E}\left(|D[\tilde{\mathscr{P}}; A]|^2\right) \leq \#\{S \in \mathscr{S} : S \cap \partial A \neq \emptyset\} \ll_B N^{1/2}.$$

Integrating now over all $A \in \mathscr{A}$ and changing the order of integration, we obtain

$$\mathbf{E}\left(\int_{\mathscr{A}} |D[\tilde{\mathscr{P}}; A]|^2 \, \mathrm{d}A\right) \ll_B N^{1/2}.$$

It follows that there exists a set $\mathscr{P}^*$ of $N = M^2$ points in $[0, 1]^2$ such that

$$\int_{\mathscr{A}} |D[\mathscr{P}^*; A]|^2 \, \mathrm{d}A \ll_B N^{1/2},$$

establishing Theorem 3 for $k = 2$ in the special case when $N = M^2$.

The generalization to all positive integers N follow from Lagrange's theorem as before, and this completes the proof of Theorem 3 for $k = 2$.

We remark that the argument in Sects. 1.3–1.5 can be extended in a reasonably straightforward manner to arbitrary dimensions $k \geq 2$. Also the argument in this section on Theorem 3 can be extended to L^q-norms for all even positive integers q, and hence all positive real numbers q, without too many complications.

1.6 A Comparison of Deterministic and Probabilistic Techniques

In this section, we make a digression and use Fourier transform techniques to try to understand and relate various approaches to upper bounds in large discrepancy problems.

Consider the unit cube $U = [0, 1]^k$, treated as a torus, in Euclidean space $\mathbf{R}^k$. Suppose that a natural number N is given, and that $N = M^k$ for some natural number M. We shall partition U into a union of $N = M^k$ cubes of sidelength M^{-1} in the natural way, and denote by $\mathscr{S}$ the collection of these small cubes. For every cube $S \in \mathscr{S}$, we denote by $\mathbf{p}_S$ the point in the centre of S. Then

$$\mathscr{P} = \{\mathbf{p}_S : S \in \mathscr{S}\} \tag{1.19}$$

is a collection of $N = M^k$ points in $U = [0, 1]^k$.

Let ν be a probabilistic measure on U. For every cube $S \in \mathscr{S}$, let ν_S denote the translation of ν by $\mathbf{p}_S$, so that

$$\int_U f(\mathbf{u}) \, \mathrm{d}\nu_S = \int_U f(\mathbf{u} - \mathbf{p}_S) \, \mathrm{d}\nu$$

for any integrable function f. Furthermore, let $\boldsymbol{\xi}_S$ denote the probabilistic variable associated with the probabilistic measure ν_S. Then

$$\tilde{\mathscr{P}} = \{\boldsymbol{\xi}_S : S \in \mathscr{S}\}$$

is a random set of $N = M^k$ points in $U = [0, 1]^k$.

Let A denote a compact and convex set in $U = [0, 1]^k$. For every $\mathbf{x} \in [0, 1]^k$, let $A(\mathbf{x}) = A + \mathbf{x}$ denote the translation of A by $\mathbf{x}$. Now consider the average

$$D_v^2(N; A) = \int_U \cdots \int_U \left(\int_{[0,1]^k} |D[\tilde{\mathscr{P}}; A(\mathbf{x})]|^2 \, d\mathbf{x} \right) \prod_{S \in \mathscr{S}} dv_S. \tag{1.20}$$

In other words, for every realization of $\tilde{\mathscr{P}}$, we consider the mean square average of the discrepancy function over all translations of A. We then average over all the different realizations of $\tilde{\mathscr{P}}$, with the understanding that the probabilistic measures v_S, where $S \in \mathscr{S}$, are independent.

We can describe $D_v^2(N; A)$ rather precisely in terms of the Fourier transforms of the measure v and of the characteristic function χ_A of the set A.

Proposition 6. *For any natural number $N = M^k$, any compact and convex set A in $U = [0, 1]^k$ and any probabilistic measure v on U, we have*

$$D_v^2(N; A) = N \sum_{0 \neq \mathbf{t} \in \mathbf{Z}^k} |\widehat{\chi_A}(\mathbf{t})|^2 (1 - |\hat{v}(\mathbf{t})|^2) + N^2 \sum_{0 \neq \mathbf{t} \in \mathbf{Z}^k} |\widehat{\chi_A}(M\mathbf{t})|^2 |\hat{v}(M\mathbf{t})|^2. \tag{1.21}$$

Before we proceed to establish this proposition, we shall first of all endeavour to understand the significance of the each of the two terms on the right hand side of (1.21).

Suppose first of all that v is the Dirac measure δ_0 concentrated at the origin. Then the Fourier transform $\hat{v}(\mathbf{t}) = 1$ identically, so the first term on the right hand side of the identity (1.21) vanishes, and we have

$$D_{\delta_0}^2(N; A) = N^2 \sum_{0 \neq \mathbf{t} \in \mathbf{Z}^k} |\widehat{\chi_A}(M\mathbf{t})|^2. \tag{1.22}$$

On the other hand, note that under this measure δ_0, the only realization of the random set $\tilde{\mathscr{P}}$ is the set $\mathscr{P}$ given by (1.19). This represents a deterministic model.

Suppose next that v is the uniform measure λ supported by the cube

$$\left[-\frac{1}{2M}, \frac{1}{2M} \right]^k, \tag{1.23}$$

so that $d\lambda = \lambda(\mathbf{u}) \, d\mathbf{u}$, where

$$\lambda(\mathbf{u}) = N \chi_{[-1/2M,1/2M]^k}(\mathbf{u})$$

denotes the characteristic function of the cube (1.23), suitably normalized. It is well known that for every $\mathbf{t} = (t_1, \ldots, t_k) \in \mathbf{Z}^k$, the Fourier transform

$$\hat{\lambda}(\mathbf{t}) = N \prod_{i=1}^{k} \frac{\sin(\pi M^{-1} t_i)}{\pi t_i},$$

with suitable modification when $t_i = 0$ for some $i = 1, \ldots, k$. Since $\hat{\lambda}(M\mathbf{t}) = 0$ for every non-zero $\mathbf{t} \in \mathbf{Z}^k$, the second term on the right hand side of the identity (1.21) vanishes, and we have

$$D_\lambda^2(N; A) = N \sum_{0 \neq \mathbf{t} \in \mathbf{Z}^k} |\widehat{\chi_A}(\mathbf{t})|^2 (1 - |\hat{\lambda}(\mathbf{t})|^2).$$

On the other hand, note that under this uniform measure λ, each of the probabilistic variables $\boldsymbol{\xi}_S$, where $S \in \mathscr{S}$, represents a random point uniformly distributed within the cube S. This represents a probabilistic model the special case $k = 2$ of which has been described earlier in Sects. 1.4–1.5.

In summary, the two terms on the right hand side of the identity (1.21) may be interpreted as respectively the probabilistic and the deterministic part of the quantity $D_\nu^2(N; A)$.

Proof of Proposition 6. Applying Parseval's identity to the inner integral in (1.20), we obtain

$$D_\nu^2(N; A) = \int_U \cdots \int_U \sum_{0 \neq \mathbf{t} \in \mathbf{Z}^k} |\widehat{\chi_A}(\mathbf{t})|^2 \left| \sum_{X \in \mathscr{S}} e(\mathbf{t} \cdot \boldsymbol{\xi}_X) \right|^2 \prod_{S \in \mathscr{S}} d\nu_S$$

$$= \sum_{0 \neq \mathbf{t} \in \mathbf{Z}^k} |\widehat{\chi_A}(\mathbf{t})|^2 \int_U \cdots \int_U \sum_{X,Y \in \mathscr{S}} e(\mathbf{t} \cdot \boldsymbol{\xi}_X) e(-\mathbf{t} \cdot \boldsymbol{\xi}_Y) \prod_{S \in \mathscr{S}} d\nu_S$$

$$= \sum_{0 \neq \mathbf{t} \in \mathbf{Z}^k} |\widehat{\chi_A}(\mathbf{t})|^2 \sum_{X,Y \in \mathscr{S}} \int_U \cdots \int_U e(\mathbf{t} \cdot \boldsymbol{\xi}_X) e(-\mathbf{t} \cdot \boldsymbol{\xi}_Y) \prod_{S \in \mathscr{S}} d\nu_S$$

$$= \sum_{0 \neq \mathbf{t} \in \mathbf{Z}^k} |\widehat{\chi_A}(\mathbf{t})|^2 \left(N + \sum_{\substack{X,Y \in \mathscr{S} \\ X \neq Y}} \int_U \int_U e(\mathbf{t} \cdot \boldsymbol{\xi}_X) e(-\mathbf{t} \cdot \boldsymbol{\xi}_Y) \, d\nu_X \, d\nu_Y \right).$$

$$(1.24)$$

For $X \neq Y$, we clearly have

$$\int_U \int_U e(\mathbf{t} \cdot \boldsymbol{\xi}_X) e(-\mathbf{t} \cdot \boldsymbol{\xi}_Y) \, d\nu_X \, d\nu_Y = \int_U \int_U e(\mathbf{t} \cdot (\boldsymbol{\xi}_X - \mathbf{p}_X)) e(-\mathbf{t} \cdot (\boldsymbol{\xi}_Y - \mathbf{p}_Y)) \, d\nu \, d\nu$$

$$= e(-\mathbf{t} \cdot \mathbf{p}_X) e(\mathbf{t} \cdot \mathbf{p}_Y) \int_U e(\mathbf{t} \cdot \boldsymbol{\xi}_X) \, d\nu \int_U e(-\mathbf{t} \cdot \boldsymbol{\xi}_Y) \, d\nu = |\hat{\nu}(\mathbf{t})|^2 e(-\mathbf{t} \cdot \mathbf{p}_X) e(\mathbf{t} \cdot \mathbf{p}_Y),$$

and so

$$\sum_{\substack{X,Y \in \mathscr{S} \\ X \neq Y}} \int_U \int_U e(\mathbf{t} \cdot \boldsymbol{\xi}_X) e(-\mathbf{t} \cdot \boldsymbol{\xi}_Y) \, d\nu_X \, d\nu_Y = |\hat{v}(\mathbf{t})|^2 \sum_{\substack{X,Y \in \mathscr{S} \\ X \neq Y}} e(-\mathbf{t} \cdot \mathbf{p}_X) e(\mathbf{t} \cdot \mathbf{p}_Y)$$

$$= |\hat{v}(\mathbf{t})|^2 \left(\sum_{X,Y \in \mathscr{S}} e(-\mathbf{t} \cdot \mathbf{p}_X) e(\mathbf{t} \cdot \mathbf{p}_Y) - N \right)$$

$$= |\hat{v}(\mathbf{t})|^2 \left(\left| \sum_{X \in \mathscr{S}} e(\mathbf{t} \cdot \mathbf{p}_X) \right|^2 - N \right). \tag{1.25}$$

The identity (1.21) follows easily on combining (1.24), (1.25) and the orthogonality relationship

$$\left| \sum_{X \in \mathscr{S}} e(\mathbf{t} \cdot \mathbf{p}_X) \right| = \begin{cases} N, & \text{if } \mathbf{t} \in M\mathbf{Z}^k, \\ 0, & \text{otherwise.} \end{cases}$$

This completes the proof.   □

In the special cases when $N = M^k$ and when we have sufficient knowledge on the Fourier transform of the characteristic function of the given compact and convex set B, we expect to be able to establish the inequality (1.8) in Theorem 3 for the set $\mathscr{P}$ given by (1.19). This will give a deterministic proof of Theorem 3, an alternative to the probabilistic proof briefly described in Sects. 1.4–1.5. However, there is virtually no documentation of results of this kind in the literature, apart from the special case when $N = M^2$ is odd and the set B is a cube, described in Chen [10, Section 3].

Nevertheless, the question arises as to whether a deterministic technique or a probabilistic technique gives a better upper bound. Much of the description in this section arises as a consequence of work done in this direction by Chen and Travaglini [14] for the case when B is a ball of fixed radius, so there is no contraction. Note also that since B is a ball, orthogonal transformation is redundant. Hence there is only translation.

Returning to the beginning of this section, we let A denote a ball in $U = [0, 1]^k$, of fixed radius not exceeding $\frac{1}{2}$. We shall consider translations $A(\mathbf{x}) = A + \mathbf{x}$ of A, where $\mathbf{x} \in [0, 1]^k$. We have the following surprising result.

Proposition 7. *Suppose that $k \not\equiv 1 \bmod 4$.*

a) *If k is sufficiently large, then the inequality $D_\lambda(M^k; A) < D_{\delta_0}(M^k; A)$ holds for all sufficiently large natural numbers M.*

b) *For $k = 2$ and ball A of radius $\frac{1}{4}$, the inequality $D_{\delta_0}(M^k; A) < D_\lambda(M^k; A)$ holds for all sufficiently large natural numbers M.*

Suppose that $k \equiv 1 \bmod 4$.

c) *If* k *is sufficiently large, then the inequality* $D_\lambda(M^k; A) < D_{\delta_0}(M^k; A)$ *holds for infinitely many natural numbers* M.

d) *The inequality* $D_{\delta_0}(M^k; A) < D_\lambda(M^k; A)$ *holds for infinitely many natural numbers* M.

e) *For* $k = 1$, *the inequality* $D_{\delta_0}(M^k; A) < D_\lambda(M^k; A)$ *holds for every natural number* M.

The case $k \not\equiv 1 \bmod 4$ is the standard case, whereas the case $k \equiv 1 \bmod 4$ is the exceptional case. This exceptional case is intimately related to the work of Konyagin, Skriganov and Sobolev [23] concerning the peculiar distribution of lattice points with respect to balls in these dimensions. We shall give a very brief description of the underlying ideas.

It is fairly straightforward to show that for every fixed dimension k, we have

$$D_\lambda^2(M^k; A) \leq \frac{\pi^{k/2} k^{3/2} r^{k-1} M^{k-1}}{2\Gamma(1 + k/2)} \tag{1.26}$$

if M is sufficiently large, where r denotes the radius of the ball A.

To study the term $D_{\delta_0}^2(M^k; A)$, we make use of the identity (1.22). Suppose that A is a ball of radius r centred at the origin. Then the Fourier transform $\widehat{\chi_A}$ can be described in terms of Bessel functions. Roughly speaking, we can write

$$D_{\delta_0}^2(M^k; A) = M^k \sum_{0 \neq \mathbf{t} \in \mathbb{Z}^k} r^k |\mathbf{t}|^{-k} J_{k/2}^2(2\pi r M |\mathbf{t}|), \tag{1.27}$$

where the Bessel function term $J_{k/2}^2(2\pi r M |\mathbf{t}|)$ is dominated by

$$\frac{1}{\pi^2 r M |\mathbf{t}|} \cos^2\left(2\pi r M |\mathbf{t}| - \frac{(k+1)\pi}{4}\right).$$

Suppose that $k \not\equiv 1 \bmod 4$. Then elementary calculation gives

$$\max\left\{\cos^2\left(2\pi r M - \frac{(k+1)\pi}{4}\right), \cos^2\left(4\pi r M - \frac{(k+1)\pi}{4}\right)\right\} > \frac{1}{100},$$

for instance, ensuring a significant contribution to the sum in (1.27) from those $\mathbf{t}$ satisfying $|\mathbf{t}| = 1$ or from those $\mathbf{t}$ satisfying $|\mathbf{t}| = 2$, sufficient for us to show that

$$D_{\delta_0}^2(M^k; A) \geq \frac{k r^{k-1} M^{k-1}}{1000\pi^2 2^k}. \tag{1.28}$$

For sufficiently large k, one has

$$\frac{\pi^{k/2}k^{3/2}}{2\Gamma(1+k/2)} < \frac{k}{1000\pi^2 2^k}.$$

Combining this with (1.26) and (1.28) gives part (a) of Proposition 7.

Suppose that $k \equiv 1 \bmod 4$. Then the Bessel function term $J_{k/2}^2(2\pi rM|\mathbf{t}|)$ in (1.27) is dominated by

$$\frac{1}{\pi^2 rM|\mathbf{t}|} \cos^2\left(2\pi rM|\mathbf{t}| \pm \frac{\pi}{2}\right) = \frac{1}{\pi^2 rM|\mathbf{t}|} \sin^2(2\pi rM|\mathbf{t}|).$$

For many values of M, the terms $\sin^2(2\pi rM|\mathbf{t}|)$ can be simultaneously small for all small $|\mathbf{t}|$, making $D_{\delta_0}(M^k; A)$ unusually small. This goes towards explaining parts (c) and (d) of Proposition 7.

1.7 Small Discrepancy: The Classical Problem

To illustrate the work on small discrepancy problems, we shall consider the pioneering work of Roth [29] on the classical problem in connection with aligned rectangular boxes in the unit cube. Consider the k-dimensional Euclidean space $\mathbf{R}^k$. We take as our domain U the unit cube $[0, 1]^k$. For every $\mathbf{x} = (x_1, \ldots, x_k) \in [0, 1]^k$, we consider the aligned rectangular box

$$B(\mathbf{x}) = [0, x_1) \times \ldots \times [0, x_k),$$

anchored at the origin. Here the condition that the intervals do not include the right hand endpoints is unimportant but a very convenient technical device. On the other hand, the assumption that all such boxes are anchored at the origin is purely historical, but is necessary if one wants to have a deeper understanding of the problem. We then consider the collection

$$\mathscr{A} = \{B(\mathbf{x}) : \mathbf{x} \in [0, 1]^k\}$$

of all such aligned rectangular boxes in U, where the integral geometric measure is given by the natural Lebesgue measure of $\mathbf{x}$. More precisely, for any set $\mathscr{P}$ of N points in $[0, 1]^k$, we have the L^2-norm

$$\|D[\mathscr{P}]\|_2 = \left(\int_{[0,1]^k} |D[\mathscr{P}; B(\mathbf{x})]|^2 \, d\mathbf{x}\right)^{1/2}. \tag{1.29}$$

We also have the simpler L^∞-norm

$$\|D[\mathscr{P}]\|_\infty = \sup_{\mathbf{x}\in[0,1]^k} |D[\mathscr{P}; B(\mathbf{x})]|. \tag{1.30}$$

The following result is due to Roth [29].

Theorem 8. *For every set $\mathscr{P}$ of N points in $[0,1]^k$, we have*

$$\|D[\mathscr{P}]\|_2 \gg_k (\log N)^{(k-1)/2}. \tag{1.31}$$

This leads immediately to the corresponding statement for the l^∞-norm.

Theorem 9. *For every set $\mathscr{P}$ of N points in $[0,1]^k$, we have*

$$\|D[\mathscr{P}]\|_\infty \gg_k (\log N)^{(k-1)/2}. \tag{1.32}$$

It is well known that Theorem 9 is not sharp. In dimension $k = 2$, Schmidt [32] has shown that for every set $\mathscr{P}$ of N points in $[0,1]^2$, we have

$$\|D[\mathscr{P}]\|_\infty \gg \log N. \tag{1.33}$$

An alternative proof of this can be found in Halász [18]. On the other hand, the recent work of Bilyk and Lacey [6] and of Bilyk, Lacey and Vagharshakyan [7] has shown that for every dimension $k \geq 3$, there exists a positive constant $\delta(k)$ such that for every set $\mathscr{P}$ of N points in $[0,1]^k$, we have

$$\|D[\mathscr{P}]\|_\infty \gg (\log N)^{(k-1)/2+\delta(k)}. \tag{1.34}$$

See Chap. 2 in this volume for a detailed discussion of this question.

The lower bound (1.31) is essentially best possible, in view of the following result of Roth [31].

Theorem 10. *For every natural number $N \geq 2$, there exists a set $\mathscr{P}$ of N points in $[0,1]^k$ such that*

$$\|D[\mathscr{P}]\|_2 \ll_k (\log N)^{(k-1)/2}. \tag{1.35}$$

The special cases $k = 2$ and $k = 3$ have been established earlier by Davenport [15] and Roth [30] respectively.

The first proof of Theorem 10 is based on a probabilistic variant of the idea first used to establish the following result of Halton [19].

Theorem 11. *For every natural number $N \geq 2$, there exists a set $\mathscr{P}$ of N points in $[0,1]^k$ such that*

$$\|D[\mathscr{P}]\|_\infty \ll_k (\log N)^{k-1}. \tag{1.36}$$

The special case $k = 2$ has been known for over 100 years through the work of Lerch [24].

Note that in dimension $k = 2$, Theorem 11 shows that Schmidt's lower bound (1.33) is best possible. In dimensions $k \geq 3$, there remains a significant gap between the lower bound (1.34) and the upper bound (1.36). This is sometimes referred to as the *Great Open Problem*.

1.8   Diophantine Approximation and Davenport Reflection

We begin by making a fatally flawed attempt to establish[1] the special case $k = 2$ of Theorem 11.

Again, for simplicity, let us assume that the number of points is a perfect square, so that $N = M^2$ for some natural number M. We may then choose to split the unit square $[0, 1]^2$ in the natural way into a union of $N = M^2$ little squares of sidelength M^{-1}, and then place a point in the centre of each little square. Let $\mathscr{P}$ be the collection of these $N = M^2$ points.

Let ξ be the second coordinate of one of the points of $\mathscr{P}$. Clearly, there are precisely M points in $\mathscr{P}$ sharing this second coordinate. Consider the discrepancy

$$D[\mathscr{P}; B(1, x_2)] \tag{1.37}$$

of the rectangle $B(1, x_2) = [0, 1) \times [0, x_2)$. As x_2 increases from just less than ξ to just more than ξ, the value of (1.37) increases by M. It follows immediately that

$$\|D[\mathscr{P}]\|_\infty \geq \frac{1}{2}M \gg N^{1/2}.$$

Let us make a digression to the work of Hardy and Littlewood [21, 22] on the distribution of lattice points in a right angled triangle. Consider a large right angled triangle T with two sides parallel to the coordinate axes. We are interested in the number of points of the lattice $\mathbf{Z}^2$ that lie in T. For simplicity, the triangle T is placed so that the horizontal side is precisely halfway between two neighbouring rows of $\mathbf{Z}^2$ and the vertical side is precisely halfway between two neighbouring columns of $\mathbf{Z}^2$, as shown in Fig. 1.2.

Note that the lattice $\mathbf{Z}^2$ has precisely one point per unit area, so we can think of the area of T as the expected number of lattice points in T. We therefore wish to understand the difference between the number of lattice points in T and the area of T, and this is the discrepancy of $\mathbf{Z}^2$ in T. The careful placement of the horizontal and

---

[1]It was put to the first author by a rather preposterous engineering colleague many years ago that this could be achieved easily by a square lattice in the obvious way. Not quite the case, as an obvious way would be far from so to this colleague.

Fig. 1.2 Lattice points in a
right angled triangle

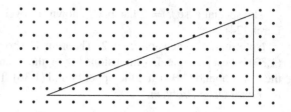

vertical sides of T means that the discrepancy comes solely from the third side of T. In the work of Hardy and Littlewood, it is shown that the size of the discrepancy when T is large is intimately related to the arithmetic properties of the slope of this third side of T. In particular, the discrepancy is essentially smallest when this slope is a badly approximable number.[2]

Returning to our attempt to establish the special case $k = 2$ of Theorem 11, perhaps our approach is not quite fatally flawed as we have thought earlier, in view of our knowledge of the work of Hardy and Littlewood. Suppose that a positive integer $N \geq 2$ is given. The lattice

$$(N^{-1/2}\mathbf{Z})^2 \qquad\qquad (1.38)$$

contains precisely N points per unit area. Inspired by Hardy and Littlewood, we now rotate (1.38) by an angle θ, chosen so that $\tan\theta$ is a badly approximable number. Let us denote the resulting lattice by Λ. Then $\Lambda \cap [0, 1]^2$ has roughly N points. Deleting or adding a few points, we end up with a set $\mathscr{P}$ of precisely N points in $[0, 1]^2$. It can then be shown that $\|D[\mathscr{P}]\|_\infty \ll \log N$, establishing Theorem 11 for $k = 2$. For the details, see the paper of Chen and Travaglini [12].

Indeed, this approach has been known for some time, as Beck and Chen [2] have already used this idea earlier in an alternative proof of Theorem 10 for $k = 2$. In fact, the first proof of this result by Davenport [15] makes essential use of diophantine approximation and badly approximable numbers, but in a slightly different and less obvious way. We now proceed to describe this.

Recall that $U = [0, 1]^2$ in this case. For the sake of convenience, we shall assume that the intervals are closed on the left and open on the right. We are also going to rescale U. Suppose first of all that N is a given even positive integer, with $N = 2M$. We now rescale U in the vertical direction by a factor M to obtain

$$V = [0, 1) \times [0, M).$$

Consider now the infinite lattice Λ_1 on $\mathbf{R}^2$ generated by the two vectors

$$(1, 0) \quad \text{and} \quad (\theta, 1),$$

---

[2]For those readers not familiar with the theory of diophantine approximation, just take any quadratic irrational like $\sqrt{2}$ or $\sqrt{3}$.

where the arithmetic properties of the non-zero number θ will be described later. It is not difficult to see that the set

$$\mathcal{Q}_1 = \Lambda_1 \cap V = \{(\{\theta n\}, n) : n = 0, 1, 2, \ldots, M - 1\}$$

contains precisely M points. We now wish to study the discrepancy properties of the set $\mathcal{Q}_1$ in V. For every aligned rectangle

$$B(x_1, y) = [0, x_1) \times [0, y) \subseteq V,$$

we consider the discrepancy

$$E[\mathcal{Q}_1; B(x_1, y)] = \#(\mathcal{Q}_1 \cap B(x_1, y)) - x_1 y, \tag{1.39}$$

noting that the area of $B(x_1, y)$ is equal to $x_1 y$, and that there is an average of one point of $\mathcal{Q}_1$ per unit area in V. Suppose first of all that y is an integer satisfying $0 < y \leq M$. Then we can write

$$E[\mathcal{Q}_1; B(x_1, y)] = \sum_{0 \leq n < y} (\psi(\theta n - x_1) - \psi(\theta n)),$$

for all but finitely many x_1 satisfying $0 < x_1 \leq 1$, where $\psi(z) = z - [z] - \frac{1}{2}$ for every $z \in \mathbf{R}$. If we relax the condition that y is an integer, then for every real number y satisfying $0 < y \leq M$, we have the approximation

$$E[\mathcal{Q}_1; B(x_1, y)] = \sum_{0 \leq n < y} (\psi(\theta n - x_1) - \psi(\theta n)) + O(1)$$

for all but finitely many x_1 satisfying $0 < x_1 \leq 1$. For simplicity, let us write

$$E[\mathcal{Q}_1; B(x_1, y)] \approx \sum_{0 \leq n < y} (\psi(\theta n - x_1) - \psi(\theta n)).$$

The sawtooth function $\psi(z)$ is periodic, so it is natural to use its Fourier series, and we obtain the estimate

$$E[\mathcal{Q}_1; B(x_1, y)] \approx \sum_{0 \neq m \in \mathbf{Z}} \left(\frac{1 - e(-mx_1)}{2\pi i m} \right) \left(\sum_{0 \leq n < y} e(\theta nm) \right). \tag{1.40}$$

Ideally we would like to square the expression (1.40) and integrate with respect to the variable x_1 over $[0, 1]$. Unfortunately, the term 1 in the numerator on the right hand side, arising from the assumption that the rectangles we consider are anchored at the origin, proves to be more than a nuisance.

To overcome this problem, Davenport's brilliant idea is to introduce a second lattice Λ_2 on $\mathbf{R}^2$ generated by the two vectors

$$(1,0) \quad \text{and} \quad (-\theta, 1).$$

It is not difficult to see that the set

$$\mathscr{Q}_2 = \Lambda_2 \cap V = \{(\{-\theta n\}, n) : n = 0, 1, 2, \ldots, M - 1\}$$

again contains precisely M points. Then the set

$$\mathscr{Q} = \mathscr{Q}_1 \cup \mathscr{Q}_2 = \{(\{\pm\theta n\}, n) : n = 0, 1, 2, \ldots, M - 1\},$$

where the points are counted with multiplicity, contains precisely $2M$ points. Thus analogous to the discrepancy (1.39), we now consider the discrepancy

$$F[\mathscr{Q}; B(x_1, y)] = \#(\mathscr{Q} \cap B(x_1, y)) - 2x_1 y,$$

noting that there is now an average of two points of $\mathscr{Q}$ per unit area in V. The analogue of the estimate (1.40) is now

$$F[\mathscr{Q}; B(x_1, y)] \approx \sum_{0 \neq m \in \mathbf{Z}} \left(\frac{e(mx_1) - e(-mx_1)}{2\pi i m} \right) \left(\sum_{0 \leq n < y} e(\theta nm) \right).$$

Squaring this and integrating with respect to the variable x_1 over $[0, 1]$, we have

$$\int_0^1 |F[\mathscr{Q}; B(x_1, y)]|^2 \, dx_1 \ll \sum_{m=1}^{\infty} \frac{1}{m^2} \left| \sum_{0 \leq n < y} e(\theta nm) \right|^2. \qquad (1.41)$$

To estimate the sum on the right hand side of (1.41), we need to make some assumptions on the arithmetic properties of the number θ. We shall assume that θ is a badly approximable number, so that there is a constant $c = c(\theta)$, depending only on θ, such that the inequality

$$m\|m\theta\| > c > 0 \qquad (1.42)$$

holds for every natural number $m \in \mathbf{N}$, where $\|z\|$ denotes the distance of z to the nearest integer.

Lemma 12. *Suppose that the real number θ is badly approximable. Then*

$$\sum_{m=1}^{\infty} \frac{1}{m^2} \left| \sum_{0 \leq n < y} e(\theta nm) \right|^2 \ll_\theta \log(2y). \qquad (1.43)$$

Proof. It is well known that

$$\left| \sum_{0 \le n < y} e(\theta n m) \right| \ll \min\{y, \|m\theta\|^{-1}\},$$

so that

$$S = \sum_{m=1}^{\infty} \frac{1}{m^2} \left| \sum_{0 \le n < y} e(\theta n m) \right|^2 \ll \sum_{h=1}^{\infty} 2^{-2h} \sum_{2^{h-1} \le m < 2^h} \min\{y^2, \|m\theta\|^{-2}\}.$$

The condition (1.42) implies that if $2^{h-1} \le m < 2^h$, then the inequality

$$\|m\theta\| > c2^{-h}$$

holds. On the other hand, for any pair $h, p \in \mathbf{N}$, there are at most two values of m satisfying $2^{h-1} \le m < 2^h$ and

$$pc2^{-h} \le \|m\theta\| < (p+1)c2^{-h},$$

for otherwise the difference $(m_1 - m_2)$ of two of them would contradict (1.42). It follows that

$$S \ll_\theta \sum_{h=1}^{\infty} \sum_{p=1}^{\infty} \min\{2^{-2h}y^2, p^{-2}\}$$

$$= \sum_{2^h \le y} \sum_{p=1}^{\infty} \min\{2^{-2h}y^2, p^{-2}\} + \sum_{2^h > y} \sum_{p=1}^{\infty} \min\{2^{-2h}y^2, p^{-2}\}$$

$$\ll \sum_{2^h \le y} \sum_{p=1}^{\infty} p^{-2} + \sum_{2^h > y} \left(2^{-2h}y^2 2^h y^{-1} + \sum_{p > 2^h y^{-1}} p^{-2} \right)$$

$$\ll \sum_{2^h \le y} 1 + \sum_{2^h > y} 2^{-h} y \ll \log(2y).$$

This completes the proof.   □

Combining (1.41) and (1.43) and then integrating with respect to the variable y over $[0, M]$, we have

$$\int_0^M \int_0^1 |F[\mathcal{Q}; B(x_1, y)]|^2 \, dx_1 \, dy \ll_\theta M \log(2M).$$

Rescaling in the vertical direction by a factor M^{-1}, we see that the set

$$\mathscr{P} = \{(\{\pm\theta n\}, nM^{-1}) : n = 0, 1, 2, \ldots, M - 1\}$$

of $N = 2M$ points in $[0, 1)^2$ satisfies the conclusion of Theorem 10 for $k = 2$.

Finally, if N is a given odd number, then we can repeat the argument above with $2M = N + 1$ points. Removing one of the points causes an error of at most 1.

1.9 Roth's Probabilistic Technique: A Preview

In this section, we describe an ingenious variation of Davenport's argument by Roth [30]. This is a nice preview of his powerful probabilistic technique, which we shall describe in Sect. 1.11, and which has been generalized in many different ways and applied in many different situations by many other colleagues.

Let us return to the lattice Λ_1 on $\mathbf{R}^2$ generated by the two vectors $(1, 0)$ and $(\theta, 1)$. For any real number $t \in \mathbf{R}$, we consider the translated lattice

$$t(1, 0) + \Lambda_1 = \{t(1, 0) + \mathbf{v} : \mathbf{v} \in \Lambda_1\}.$$

In particular, we are interested in the set

$$\mathscr{Q}_1(t) = (t(1, 0) + \Lambda_1) \cap V = \{(\{t + \theta n\}, n) : n = 0, 1, 2, \ldots, M - 1\}$$

which clearly contains precisely M points. Thus analogous to the discrepancy (1.39), we now consider the discrepancy

$$E[\mathscr{Q}_1(t); B(x_1, y)] = \#(\mathscr{Q}_1(t) \cap B(x_1, y)) - x_1 y,$$

noting that there is now an average of one point of $\mathscr{Q}_1(t)$ per unit area in V. The analogue of the estimate (1.40) is now

$$E[\mathscr{Q}_1(t); B(x_1, y)] \approx \sum_{0 \neq m \in \mathbf{Z}} \left(\frac{1 - e(-mx_1)}{2\pi i m}\right) \left(\sum_{0 \leq n < y} e(\theta n m)\right) e(tm).$$

Squaring this and integrating with respect to the new variable t over $[0, 1]$, we have

$$\int_0^1 |E[\mathscr{Q}_1(t); B(x_1, y)]|^2 \, dt \ll \sum_{m=1}^{\infty} \frac{1}{m^2} \left|\sum_{0 \leq n < y} e(\theta n m)\right|^2. \tag{1.44}$$

Furthermore, if θ is a badly approximable number as in the last section, then integrating (1.44) trivially with respect to the variable x_1 over $[0, 1]$ and with respect to the variable y over $[0, M]$, we have

$$\int_0^1 \int_0^M \int_0^1 |E[\mathscr{Q}_1(t); B(x_1, y)]|^2 \, dx_1 \, dy \, dt \ll_\theta M \log(2M).$$

It follows that there exists $t^* \in [0, 1]$ such that the set $\mathscr{Q}_1(t^*)$ satisfies

$$\int_0^M \int_0^1 |E[\mathscr{Q}_1(t^*); B(x_1, y)]|^2 \, dx_1 \, dy \ll_\theta M \log(2M).$$

Rescaling in the vertical direction by a factor M^{-1}, we see that the set

$$\mathscr{P}(t^*) = \{(\{t^* + \theta n\}, n M^{-1}) : n = 0, 1, 2, \ldots, M - 1\}$$

of $N = M$ points in $[0, 1)^2$ satisfies the requirements of Theorem 10 for $k = 2$.

1.10 Van der Corput Point Sets

In this section, we begin our discussion of those point sets which have been explored in great depth through our study of Theorems 10 and 11.

Our first step is to construct the simplest point sets which will allow us to establish Theorem 11 in the case $k = 2$.

The construction is based on the famous van der Corput sequence $c_0, c_1, c_2, \ldots$ defined as follows. For every non-negative integer $n \in \mathbf{N}_0$, we write

$$n = \sum_{j=1}^{\infty} a_j 2^{j-1} \tag{1.45}$$

as a dyadic expansion. Then we write

$$c_n = \sum_{j=1}^{\infty} a_j 2^{-j}. \tag{1.46}$$

Note that $c_n \in [0, 1)$. Note also that only finitely many digits $a_1, a_2, a_3, \ldots$ are non-zero, so that the sums in (1.45) and (1.46) have only finitely many non-zero terms. For simplicity, we sometimes write

$$n = \ldots a_3 a_2 a_1 \quad \text{and} \quad c_n = 0.a_1 a_2 a_3 \ldots$$

in terms of the digits $a_1, a_2, a_3, \ldots$ of n. The infinite set

$$\mathcal{Q} = \{(c_n, n) : n = 0, 1, 2, \ldots\} \tag{1.47}$$

in $[0, 1) \times [0, \infty)$ is known as the van der Corput point set.

The following is the most crucial property of the van der Corput point set.

Lemma 13. *For all non-negative integers s and ℓ such that $\ell < 2^s$ holds, the set*

$$\{n \in \mathbf{N}_0 : c_n \in [\ell\, 2^{-s}, (\ell + 1)2^{-s})\}$$

contains precisely all the elements of a residue class modulo 2^s in $\mathbf{N}_0$.

Proof. There exist unique integers $b_1, b_2, b_3, \ldots$ such that $\ell\, 2^{-s} = 0.b_1 b_2 b_3 \ldots b_s$. Clearly $c_n = 0.a_1 a_2 a_3 \ldots \in [\ell\, 2^{-s}, (\ell + 1)2^{-s})$ precisely when $0.a_1 a_2 a_3 \ldots a_s = \ell\, 2^{-s}$; in other words, precisely when $a_j = b_j$ for every $j = 1, \ldots, s$. The value of a_j for any $j > s$ is irrelevant. $\qquad\square$

We say that an interval of the form $[\ell\, 2^{-s}, (\ell + 1)2^{-s}) \subseteq [0, 1)$ for some integer ℓ is an elementary dyadic interval of length 2^{-s}. Hence Lemma 13 says that the van der Corput sequence has very good distribution among such elementary dyadic intervals for all non-negative integer values of s.

Lemma 14. *For all non-negative integers s, ℓ and m such that $\ell < 2^s$ holds, the rectangle*

$$[\ell\, 2^{-s}, (\ell + 1)2^{-s}) \times [m2^s, (m + 1)2^s)$$

contains precisely one point of the van der Corput point set $\mathcal{Q}$.

It is clear that there is an average of one point of the van der Corput point set $\mathcal{Q}$ per unit area in $[0, 1) \times [0, \infty)$. For any measurable set A in $[0, 1) \times [0, \infty)$, let

$$E[\mathcal{Q}; A] = \#(\mathcal{Q} \cap A) - \mu(A)$$

denote the discrepancy of $\mathcal{Q}$ in A.

Let $\psi(z) = z - [z] - \frac{1}{2}$ for every $z \in \mathbf{R}$.

Lemma 15. *For all non-negative integers s and ℓ such that $\ell < 2^s$ holds, there exist real numbers α_0, β_0, depending at most on s and ℓ, such that $|\alpha_0| \leq \frac{1}{2}$ and*

$$E[\mathcal{Q}; [\ell\, 2^{-s}, (\ell + 1)2^{-s}) \times [0, y)] = \alpha_0 - \psi(2^{-s}(y - \beta_0)) \tag{1.48}$$

at all points of continuity of the right hand side.

Proof. In view of Lemma 13, the second coordinates of the points of $\mathcal{Q}$ in the region $[\ell\, 2^{-s}, (\ell + 1)2^{-s}) \times [0, \infty)$ fall precisely into a residue class modulo 2^s. Let n_0 be the smallest of these second coordinates. Then $0 \leq n_0 < 2^s$. We now study

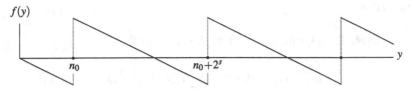

Fig. 1.3 The sawtooth function $E[\mathscr{Q}; [\ell\, 2^{-s}, (\ell + 1)2^{-s}) \times [0, y)]$

$$E[\mathscr{Q}; [\ell\, 2^{-s}, (\ell + 1)2^{-s}) \times [0, y)]$$

as a function of y. For simplicity, denote it by $f(y)$, say. Clearly $f(0) = E[\mathscr{Q}; \emptyset] = 0$. On the other hand, note that

$$\mu([\ell\, 2^{-s}, (\ell + 1)2^{-s}) \times [0, y)) = 2^{-s} y$$

increases with y at the rate 2^{-s}, so that $f(y)$ decreases with y at the rate 2^{-s}, except when y coincides with the second coordinate of one of the points of the set $\mathscr{Q}$ in the region $[\ell\, 2^{-s}, (\ell + 1)2^{-s}) \times [0, \infty)$, in which case $f(y)$ jumps up by 1. The first instance of this jump occurs when $y = n_0$. See Fig. 1.3.

With suitable α_0 and β_0, the right hand side of (1.48) fits all the requirements.

$\square$

We can now prove Theorem 11 for $k = 2$. Let $N \geq 2$ be a given integer. It follows immediately from the definition of $\mathscr{Q}$ that the set

$$\mathscr{Q}_0 = \mathscr{Q} \cap ([0, 1) \times [0, N))$$

contains precisely N points. Let the integer h be determined uniquely by

$$2^{h-1} < N \leq 2^h. \tag{1.49}$$

Consider a rectangle of the form

$$B(x_1, y) = [0, x_1) \times [0, y) \subseteq [0, 1) \times [0, N).$$

Let $x_1^{(0)} = 0$. For every $s = 1, \ldots, h$, let $x_1^{(s)} = 2^{-s}[2^s x_1]$ denote the greatest integer multiple of 2^{-s} not exceeding x_1. Then we can write $[0, x_1)$ as a union of disjoint intervals in the form

$$[0, x_1) = [x_1^{(h)}, x_1) \cup \bigcup_{s=1}^{h} [x_1^{(s-1)}, x_1^{(s)}).$$

It follows that

$$E[\mathcal{Q}_0; [0, x_1) \times [0, y)] = E[\mathcal{Q}; [0, x_1) \times [0, y)]$$

$$= E[\mathcal{Q}; [x_1^{(h)}, x_1) \times [0, y)] + \sum_{s=1}^{h} E[\mathcal{Q}; [x_1^{(s-1)}, x_1^{(s)}) \times [0, y)].$$

$$(1.50)$$

Clearly $[x_1^{(h)}, x_1) \times [0, y) \subseteq [x_1^{(h)}, x_1^{(h)} + 2^{-h}) \times [0, 2^h)$, and the latter rectangle has area 1 and is of the type under discussion in Lemma 14, hence contains precisely one point of $\mathcal{Q}$. It follows that

$$\#(\mathcal{Q} \cap ([x_1^{(h)}, x_1) \times [0, y)) \le 1 \quad \text{and} \quad \mu([x_1^{(h)}, x_1) \times [0, y)) \le 1,$$

and we have the bound

$$|E[\mathcal{Q}; [x_1^{(h)}, x_1) \times [0, y)]| \le 1. \qquad (1.51)$$

On the other hand, for every $s = 1, \ldots, k$, the rectangle

$$[x_1^{(s-1)}, x_1^{(s)}) \times [0, y)$$

either is empty, in which case we have $E[\mathcal{Q}; [x_1^{(s-1)}, x_1^{(s)}) \times [0, y)] = 0$ trivially, or is of the type under discussion in Lemma 15, and we have the bound

$$|E[\mathcal{Q}; [x_1^{(s-1)}, x_1^{(s)}) \times [0, y)]| \le 1. \qquad (1.52)$$

Note that (1.52) still holds in the empty case. Combining (1.49)–(1.52), we arrive at an upper bound

$$|E[\mathcal{Q}_0; [0, x_1) \times [0, y)]| \le 1 + h \ll \log N. \qquad (1.53)$$

For comparison later in Sect. 1.14, let us summarize what we have done. We are approximating the interval $[0, x_1)$ by a subinterval $[0, x_1^{(h)})$, and consequently approximating the rectangle $B(x_1, y)$ by a smaller rectangle $B(x_1^{(h)}, y)$. Then we show that the difference $B(x_1, y) \setminus B(x_1^{(h)}, y)$ is contained in one of the rectangles under discussion in Lemma 14, and inequality (1.51) is the observation that

$$|E[\mathcal{Q}; B(x_1, y)] - E[\mathcal{Q}; B(x_1^{(h)}, y)]| \le 1.$$

To estimate $E[\mathcal{Q}; B(x_1^{(h)}, y)]$, we note that the interval $[0, x_1^{(h)})$ is a union of at most h disjoint elementary dyadic intervals. More precisely, if we write

$$x_1^{(h)} = \sum_{s=1}^{h} b_s 2^{-s}$$

as a dyadic expansion, then $[0, x_i^{(h)})$ can be written as a union of

$$\sum_{s=1}^{h} b_s \leq h$$

elementary dyadic intervals, namely b_1 elementary dyadic intervals of length 2^{-1}, together with b_2 elementary dyadic intervals of length 2^{-2}, and so on. It follows that $B(x_1^{(h)}, y)$ is a disjoint union of at most h rectangles discussed in Lemma 15, each of which satisfies inequality (1.52).

Finally, rescaling the second coordinate of the points of $\mathcal{Q}_0$ by a factor N^{-1}, we obtain a set

$$\mathcal{P} = \{(c_n, N^{-1}n) : n = 0, 1, 2, \ldots, N - 1\} \tag{1.54}$$

of precisely N points in $[0, 1)^2$. For every $\mathbf{x} = (x_1, x_2) \in [0, 1]^2$, we have

$$D[\mathcal{P}; B(\mathbf{x})] = E[\mathcal{Q}_0; [0, x_1) \times [0, Nx_2)] \ll \log N,$$

in view of (1.53) and noting that $0 \leq Nx_2 \leq N$. This now completes the proof of Theorem 11 for $k = 2$.

1.11 Roth's Probabilistic Technique

We now attempt to extend the ideas in the last section to obtain a proof of Theorem 10 for $k = 2$.

Let us first of all consider the special case when $N = 2^h$. Then the set (1.54) used to establish Theorem 11 for $k = 2$ becomes

$$\mathcal{P}(2^h) = \{(c_n, 2^{-h}n) : n = 0, 1, 2, \ldots, 2^h - 1\}$$
$$= \{(0.a_1 a_2 a_3 \ldots a_h, 0.a_h \ldots a_3 a_2 a_1) : a_1, \ldots, a_h \in \{0, 1\}\},$$

$$\tag{1.55}$$

in terms of binary digits. We have the following unhelpful result[3] of Halton and Zaremba [20].

---

[3]In their paper, Halton and Zaremba have an exact expression for the integral under study.

Proposition 16. *For every positive integer h, we have*

$$\int_{[0,1]^2} |D[\mathscr{P}(2^h); B(\mathbf{x})]|^2 \, d\mathbf{x} = 2^{-6}h^2 + O(h). \tag{1.56}$$

Clearly the order of magnitude is $(\log N)^2$, and not $\log N$ as we would have liked. Hence any unmodified van der Corput point set is not sufficient to establish our desired result. To understand the problem, we return to our discussion in the last section. Assume that $N = 2^h$. Consider a rectangle of the form

$$B(x_1, y) = [0, x_1) \times [0, y) \subseteq [0, 1) \times [0, 2^h).$$

For simplicity, let us assume that x_1 is an integer multiple of 2^{-h}, so that $x_1 = x_1^{(h)}$ and (1.50) simplifies to

$$D[\mathscr{P}; B(x_1, 2^{-h}y)] = E[\mathscr{Q}_0; [0, x_1) \times [0, y)] = \sum_{s=1}^{h}{}^* E[\mathscr{Q}; [x_1^{(s-1)}, x_1^{(s)}) \times [0, y)],$$

where the $*$ in the summation sign denotes that the sum includes only those terms where $x_1^{(s-1)} \neq x_1^{(s)}$. Note that when $x_1^{(s-1)} \neq x_1^{(s)}$, we have

$$[x_1^{(s-1)}, x_1^{(s)}) = [\ell \, 2^{-s}, (\ell + 1)2^{-s})$$

for some integer ℓ, so it follows from Lemma 15 that

$$D[\mathscr{P}; B(x_1, 2^{-h}y)] = \sum_{s=1}^{h}{}^* (\alpha_s - \psi(2^{-s}(y - \beta_s))), \tag{1.57}$$

where, for each $s = 1, \ldots, h$, the real numbers α_s and β_s satisfy $|\alpha_s| \leq \frac{1}{2}$. If we square the expression (1.57), then the right hand side becomes

$$\sum_{s'=1}^{h}{}^* \sum_{s''=1}^{h}{}^* (\alpha_{s'} - \psi(2^{-s'}(y - \beta_{s'})))(\alpha_{s''} - \psi(2^{-s''}(y - \beta_{s''}))).$$

Expanding the summand, this gives rise eventually to a constant term

$$\sum_{s'=1}^{h}{}^* \sum_{s''=1}^{h}{}^* \alpha_{s'}\alpha_{s''}$$

which ultimately leads to the term $2^{-6}h^2$ in (1.56).

Note that this constant term arises from our assumption that all the aligned rectangles under consideration are anchored at the origin, and recall that Roth's attempt to overcome this handicap, discussed in Sect. 1.9, involves the introduction of a translation variable t. So let us attempt to describe Roth's incorporation of this idea of a translation variable into the argument here.

To pave the way for a smooth introduction of a probabilistic variable, we shall modify the van der Corput point set somewhat. Let $N \geq 2$ be a given integer, and let the integer h be determined uniquely by

$$2^{h-1} < N \leq 2^h. \tag{1.58}$$

For every $n = 0, 1, 2, \ldots, 2^h - 1$, we define c_n as before by (1.45) and (1.46). We then extend the definition of c_n to all other integers using periodicity by writing

$$c_{n+2^h} = c_n \quad \text{for every } n \in \mathbf{Z},$$

and consider the extended van der Corput point set

$$\mathcal{D}_h = \{(c_n, n) : n \in \mathbf{Z}\}.$$

Furthermore, for every real number $t \in \mathbf{R}$, we consider the translated van der Corput point set

$$\mathcal{D}_h(t) = \{(c_n, n + t) : n \in \mathbf{Z}\}.$$

It is clear that there is an average of one point of the translated van der Corput point set $\mathcal{D}_h(t)$ per unit area in $[0, 1) \times (-\infty, \infty)$. For any measurable set A in $[0, 1) \times (-\infty, \infty)$, we now let

$$E[\mathcal{D}_h(t); A] = \#(\mathcal{D}_h(t) \cap A) - \mu(A)$$

denote the discrepancy of $\mathcal{D}_h(t)$ in A.

Consider a rectangle of the form

$$B(x_1, y) = [0, x_1) \times [0, y) \subseteq [0, 1) \times [0, N).$$

As before, let $x_1^{(0)} = 0$. For every $s = 1, \ldots, h$, let $x_1^{(s)} = 2^{-s}[2^s x_1]$ denote the greatest integer multiple of 2^{-s} not exceeding x_1. Then, analogous to (1.51), we have the trivial bound

$$|E[\mathcal{D}_h(t); [x_1^{(h)}, x_1) \times [0, y)]| \leq 1, \tag{1.59}$$

so we shall henceforth assume that $x_1 = x_1^{(h)}$, so that

$$E[\mathcal{D}_h(t); B(x_1, y)] = \sum_{s=1}^{h}{}^{*} E[\mathcal{D}_h(t); [x_1^{(s-1)}, x_1^{(s)}) \times [0, y)]. \tag{1.60}$$

Corresponding to Lemma 15, we can establish the following result without too much difficulty.

Lemma 17. *For all positive real numbers y and all non-negative integers s and ℓ such that $s \leq h$ and $\ell < 2^s$ hold, there exist real numbers β_0 and γ_0, depending at most on s, ℓ and y, such that*

$$E[\mathcal{Q}_h(t); [\ell \, 2^{-s}, (\ell+1)2^{-s}) \times [0, y)] = \psi(2^{-s}(t - \beta_0)) - \psi(2^{-s}(t - \gamma_0))$$

at all points of continuity of the right hand side.

Combining (1.60) and Lemma 17, we have

$$E[\mathcal{Q}_h(t); B(x_1, y)] = \sum_{s=1}^{h} {}^{*} (\psi(2^{-s}(t - \beta_s)) - \psi(2^{-s}(t - \gamma_s))) \qquad (1.61)$$

for some real numbers β_s and γ_s depending at most on x_1 and y. We shall square this expression and integrate with respect to the translation variable t over the interval $[0, 2^h)$, an interval of length equal to the period of the set $\mathcal{Q}_h(t)$. We therefore need to study integrals of the form

$$\int_0^{2^h} \psi(2^{-s'}(t - \beta_{s'}))\psi(2^{-s''}(t - \beta_{s''})) \, dt,$$

or when either or both of $\beta_{s'}$ and $\beta_{s''}$ are replaced by $\gamma_{s'}$ and $\gamma_{s''}$ respectively.

Lemma 18. *Suppose that the integers s' and s'' satisfy $0 \leq s', s'' \leq h$, and that the real numbers $\beta_{s'}$ and $\beta_{s''}$ are fixed. Then*

$$\int_0^{2^h} \psi(2^{-s'}(t - \beta_{s'}))\psi(2^{-s''}(t - \beta_{s''})) \, dt = O(2^{h-|s'-s''|}).$$

Proof. The result is obvious if $s' = s''$. Without loss of generality, let us assume that $s' > s''$. For every $a = 0, 1, 2, \ldots, 2^{s'-s''} - 1$, in view of periodicity, we have

$$\int_0^{2^h} \psi(2^{-s'}(t - \beta_{s'}))\psi(2^{-s''}(t - \beta_{s''})) \, dt$$

$$= \int_0^{2^h} \psi(2^{-s'}(t + a2^{s''} - \beta_{s'}))\psi(2^{-s''}(t + a2^{s''} - \beta_{s''})) \, dt$$

$$= \int_0^{2^h} \psi(2^{-s'}(t + a2^{s''} - \beta_{s'}))\psi(2^{-s''}(t - \beta_{s''})) \, dt,$$

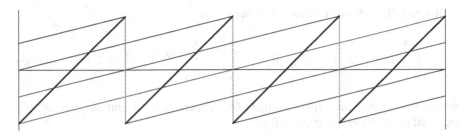

Fig. 1.4 An illustration of the summation (1.62)

with the last equality arising from the observation that

$$\psi(2^{-s''}(t + a2^{s''} - \beta_{s''})) = \psi(a + 2^{-s''}(t - \beta_{s''})) = \psi(2^{-s''}(t - \beta_{s''})).$$

It follows that

$$
2^{s'-s''} \int_0^{2^h} \psi(2^{-s'}(t - \beta_{s'}))\psi(2^{-s''}(t - \beta_{s''}))\, dt
$$

$$
= \sum_{a=0}^{2^{s'-s''}-1} \int_0^{2^h} \psi(2^{-s'}(t + a2^{s''} - \beta_{s'}))\psi(2^{-s''}(t - \beta_{s''}))\, dt
$$

$$
= \int_0^{2^h} \left(\sum_{a=0}^{2^{s'-s''}-1} \psi(2^{-s'}(t + a2^{s''} - \beta_{s'})) \right) \psi(2^{-s''}(t - \beta_{s''}))\, dt.
$$

It is not difficult to see that

$$
\sum_{a=0}^{2^{s'-s''}-1} \psi(2^{-s'}(t + a2^{s''} - \beta_{s'})) = \psi(2^{-s''}(t - \beta_{s'})) \tag{1.62}
$$

at all points of continuity, as shown in Fig. 1.4.

We therefore conclude that

$$
2^{s'-s''} \int_0^{2^h} \psi(2^{-s'}(t - \beta_{s'}))\psi(2^{-s''}(t - \beta_{s''}))\, dt
$$

$$
= \int_0^{2^h} \psi(2^{-s''}(t - \beta_{s'}))\psi(2^{-s''}(t - \beta_{s''}))\, dt = O(2^h),
$$

and the desired result follows immediately.   □

It now follows from (1.61) and Lemma 18 that

$$\int_0^{2^h} |E[\mathcal{Q}_h(t); B(x_1, y)]|^2 \, dt \ll \sum_{s'=1}^{h}{}^* \sum_{s''=1}^{h}{}^* 2^{h-|s'-s''|} \ll 2^h h, \qquad (1.63)$$

noting that the diagonal terms contribute $O(2^h h)$, and the contribution from the off-diagonal terms decays geometrically.

Note that the estimate (1.63) is independent of the choice of x_1 and y. We also recall the trivial estimate (1.59). It follows that integrating (1.63) trivially with respect to x_1 over the interval $[0, 1)$ and with respect to y over the interval $[0, N)$, we conclude that

$$\int_0^N \int_0^1 \int_0^{2^h} |E[\mathcal{Q}_h(t); B(x_1, y)]|^2 \, dt \, dx_1 \, dy$$

$$= \int_0^{2^h} \left(\int_0^N \int_0^1 |E[\mathcal{Q}_h(t); B(x_1, y)]|^2 \, dx_1 \, dy \right) dt \ll 2^h h N.$$

Hence there exists $t^* \in [0, 2^h)$ such that

$$\int_0^N \int_0^1 |E[\mathcal{Q}_h(t^*); B(x_1, y)]|^2 \, dx_1 \, dy \ll h N. \qquad (1.64)$$

Finally, we note that the set $\mathcal{Q}_h(t^*) \cap ([0, 1) \times [0, N))$ contains precisely N points. Rescaling in the vertical direction by a factor N^{-1}, we observe that the set

$$\mathcal{P}^* = \{(z_1, N^{-1} z_2) : (z_1, z_2) \in \mathcal{Q}_h(t^*)\}$$

contains precisely N points in $[0, 1)^2$, and the estimate (1.64) now translates to

$$\int_{[0,1]^2} |D[\mathcal{P}^*; B(\mathbf{x})]|^2 \, d\mathbf{x} \ll h \ll \log N,$$

in view of (1.58). This completes the proof of Theorem 10 for $k = 2$.

We conclude this section by trying to obtain a different interpretation of the effect of the translation variable t. Consider a typical term

$$E[\mathcal{Q}_h(t); [x_1^{(s-1)}, x_1^{(s)}) \times [0, y)]$$

in the sum (1.60). If $x_1^{(s-1)} \neq x_1^{(s)}$, then $x_1^{(s)}$ cannot be an integer multiple of $2^{-(s-1)}$ and therefore must be an odd integer multiple of 2^{-s}, and so

$$[x_1^{(s-1)}, x_1^{(s)}) = [\ell \, 2^{-s}, (\ell + 1) 2^{-s}) \subset \left[\frac{\ell}{2} 2^{-(s-1)}, (\frac{\ell}{2} + 1) 2^{-(s-1)} \right)$$

for some even integer ℓ. One can then show that

$$E[\mathscr{D}_h(2^{s-1}); [\ell\, 2^{-s}, (\ell+1)2^{-s}) \times [0, y)] = E[\mathscr{D}_h; [(\ell+1)2^{-s}, (\ell+2)2^{-s}) \times [0, y)].$$

This means that instead of translating vertically, as on the left hand side above, one may shift horizontally, as on the right hand side above. Another way to see this is to note from Lemma 13 that the interval $[\ell\, 2^{-s}, (\ell+1)2^{-s})$ is associated with a residue class R_s modulo 2^s, whereas the interval $[\ell\, 2^{-s}, (\ell+2)2^{-s})$ is associated with a residue class R_{s-1} modulo 2^{s-1}, so the interval $[(\ell+1)2^{-s}, (\ell+2)2^{-s})$ must be associated with the residue class $R_{s-1} \setminus R_s$ modulo 2^s. But then $R_{s-1} \setminus R_s$ is clearly R_s translated by 2^{s-1}.

1.12 Digit Shifts

In this section, we shall attempt to replace the vertical translation studied in the last section by horizontal shifts, as pioneered by Chen [9].

Let $N \geq 2$ be a given integer, and let the integer h be determined uniquely by

$$2^{h-1} < N \leq 2^h. \tag{1.65}$$

For every $n = 0, 1, 2, \ldots, 2^h - 1$, we define c_n as before by (1.45) and (1.46). As we are not translating vertically, there is no need[4] to extend the definition of c_n to other integers as in the last section, and we consider the set[5]

$$\mathscr{D}_h = \{(c_n, n) : n = 0, 1, 2, \ldots, 2^h - 1\}$$
$$= \{(0.a_1 a_2 a_3 \ldots a_h, a_h \ldots a_3 a_2 a_1) : a_1, \ldots, a_h \in \{0, 1\}\},$$

in terms of binary digits. Furthermore, for every $\mathbf{t} = (t_1, \ldots, t_h) \in \mathbf{Z}_2^h$, where $\mathbf{Z}_2 = \{0, 1\}$, write

$$c_n^{(\mathbf{t})} = 0.(a_1 \oplus t_1)(a_2 \oplus t_2)(a_3 \oplus t_3) \ldots (a_h \oplus t_h) \quad \text{if} \quad c_n = 0.a_1 a_2 a_3 \ldots a_h$$

in binary notation, where $\oplus$ denotes addition modulo 2, and consider the shifted van der Corput point set

---

[4]This is not the case if we wish to study Theorem 10 for $k > 2$.

[5]Note that the set $\mathscr{D}_h$ here is different from that in the last section. However, since we are working with rectangles inside $[0, 1) \times [0, 2^h)$, our statements here concerning $\mathscr{D}_h$ remain valid for the set $\mathscr{D}_h$ defined in the last section.

$$\mathcal{Q}_h^{(\mathbf{t})} = \{(c_n^{(\mathbf{t})}, n) : n = 0, 1, 2, \ldots, 2^h - 1\},$$

obtained from $\mathcal{Q}_h$ by a digit shift $\mathbf{t}$.

It is clear that there is an average of one point of the shifted van der Corput point set $\mathcal{Q}_h^{(\mathbf{t})}$ per unit area in $[0, 1) \times [0, 2^h)$. For any measurable set A in $[0, 1) \times [0, 2^h)$, we study the discrepancy function

$$E[\mathcal{Q}_h^{(\mathbf{t})}; A] = \#(\mathcal{Q}_h^{(\mathbf{t})} \cap A) - \mu(A).$$

Consider a rectangle of the form

$$B(x_1, y) = [0, x_1) \times [0, y) \subseteq [0, 1) \times [0, N).$$

Analogous to (1.59), we have the trivial bound

$$|E[\mathcal{Q}_h^{(\mathbf{t})}; [x_1^{(h)}, x_1) \times [0, y)]| \le 1, \tag{1.66}$$

so we shall henceforth assume that $x_1 = x_1^{(h)}$, so that

$$E[\mathcal{Q}_h^{(\mathbf{t})}; B(x_1, y)] = \sum_{s=1}^{h}{}^* E[\mathcal{Q}_h^{(\mathbf{t})}; [x_1^{(s-1)}, x_1^{(s)}) \times [0, y)]. \tag{1.67}$$

We now square this expression and sum it over all digit shifts $\mathbf{t} \in \mathbf{Z}_2^h$. For simplicity and convenience, let us omit reference to $\mathcal{Q}_h$ and y, and write

$$E[\mathcal{Q}_h^{(\mathbf{t})}; [x_1^{(s-1)}, x_1^{(s)}) \times [0, y)] = E_s[t_1, \ldots, t_h].$$

Then we need to study sums of the form

$$\sum_{\mathbf{t} \in \mathbf{Z}_2^h} E_{s'}[t_1, \ldots, t_h] E_{s''}[t_1, \ldots, t_h].$$

Analogous to Lemma 18, we have the following estimate.

Lemma 19. *Suppose that the real number $y \in [0, N)$ is fixed, and that the integers s' and s'' satisfy $0 \le s', s'' \le h$. Then*

$$\sum_{\mathbf{t} \in \mathbf{Z}_2^h} E_{s'}[t_1, \ldots, t_h] E_{s''}[t_1, \ldots, t_h] = O(2^{h - |s' - s''|}). \tag{1.68}$$

Proof. First of all, for fixed $t_1, \ldots, t_s$, the value of $E_s[t_1, \ldots, t_h]$ remains the same for every choice of $t_{s+1}, \ldots, t_h$, as these latter variables only shift the digits of c_n after the s-th digit, and so

$$c_n^{(t_1,\ldots,t_s,t_s+1,\ldots,t_h)} \in [x_1^{(s-1)}, x_1^{(s)}) \quad \text{if and only if} \quad c_n^{(t_1,\ldots,t_s,0,\ldots,0)} \in [x_1^{(s-1)}, x_1^{(s)}).$$

Next, the case when $x_1^{(s'-1)} = x_1^{(s')}$ or $x_1^{(s''-1)} = x_1^{(s'')}$ is also trivial, as the summand is clearly equal to zero, so we shall assume that $x_1^{(s'-1)} \neq x_1^{(s')}$ and $x_1^{(s''-1)} \neq x_1^{(s'')}$. Now the case when $s' = s''$ is easy, since we have $E[t_1,\ldots,t_h; x_1^{(s-1)}, x_1^{(s)}] = O(1)$ trivially. Without loss of generality, let us assume that $s' > s''$. For fixed $t_1,\ldots,t_{s''}$, in view of the comment at the beginning of the proof, we have

$$\sum_{t_{s''}+1,\ldots,t_h \in \mathbf{Z}_2} E_{s'}[t_1,\ldots,t_h] E_{s''}[t_1,\ldots,t_h]$$

$$= 2^{h-s'} \left(\sum_{t_{s''}+1,\ldots,t_{s'} \in \mathbf{Z}_2} E_{s'}[t_1,\ldots,t_{s'},0,\ldots,0] \right) E_{s''}[t_1,\ldots,t_{s''},0,\ldots,0].$$

We shall show that

$$\sum_{t_{s''}+1,\ldots,t_{s'} \in \mathbf{Z}_2} E_{s'}[t_1,\ldots,t_{s'},0,\ldots,0]$$

$$= \sum_{t_{s''}+1,\ldots,t_{s'} \in \mathbf{Z}_2} E[\mathscr{Q}_h^{(t_1,\ldots,t_{s''},t_{s''}+1,\ldots,t_{s'},0,\ldots,0)}; [x_1^{(s'-1)}, x_1^{(s')}) \times [0, y)]$$

$$= E[\mathscr{Q}_h^{(t_1,\ldots,t_{s''},0,\ldots,0)}; [\ell 2^{-s''}, (\ell+1)2^{-s''}) \times [0, y)], \qquad (1.69)$$

where ℓ is an integer and $[x_1^{(s'-1)}, x_1^{(s')}) \subset [\ell 2^{-s''}, (\ell+1)2^{-s''})$. Then

$$\sum_{t_{s''}+1,\ldots,t_h \in \mathbf{Z}_2} E_{s'}[t_1,\ldots,t_h] E_{s''}[t_1,\ldots,t_h] = O(2^{h-s'}),$$

from which it follows that

$$\sum_{t_1,\ldots,t_h \in \mathbf{Z}_2} E_{s'}[t_1,\ldots,t_h] E_{s''}[t_1,\ldots,t_h] = O(2^{h-s'+s''}),$$

giving the desired result. To establish (1.69), simply note that for fixed $t_1,\ldots,t_{s''}$, if a point

$$c_n^{(t_1,\ldots,t_{s''},0,\ldots,0)} \in [x_1^{(s'-1)}, x_1^{(s')}),$$

then each distinct choice of $t_{s''}+1,\ldots,t_{s'}$ will shift this point into one of the $2^{s'-s''}$ distinct intervals of length $2^{-s'}$ that make up the interval $[\ell 2^{-s''}, (\ell+1)2^{-s''})$. □

It now follows from (1.67) and Lemma 19 that

$$\sum_{\mathbf{t}\in\mathbf{Z}_2^h} |E[\mathscr{Q}_h^{(\mathbf{t})}; B(x_1, y)]|^2 \ll \sideset{}{^*}\sum_{s'=1}^{h} \sideset{}{^*}\sum_{s''=1}^{h} 2^{h-|s'-s''|} \ll 2^h h, \qquad (1.70)$$

noting that the diagonal terms contribute $O(2^h h)$, and the contribution from the off-diagonal terms decays geometrically.

Note that the estimate (1.70) is independent of the choice of x_1 and y. We also recall the trivial estimate (1.66). It follows that integrating (1.70) trivially with respect to x_1 over the interval $[0, 1)$ and with respect to y over the interval $[0, N)$, we conclude that

$$\int_0^N \int_0^1 \sum_{\mathbf{t}\in\mathbf{Z}_2^h} |E[\mathscr{Q}_h^{(\mathbf{t})}; B(x_1, y)]|^2 \, dx_1 \, dy$$

$$= \sum_{\mathbf{t}\in\mathbf{Z}_2^h} \int_0^N \int_0^1 |E[\mathscr{Q}_h^{(\mathbf{t})}; B(x_1, y)]|^2 \, dx_1 \, dy \ll 2^h h N.$$

Hence there exists $\mathbf{t}^* \in \mathbf{Z}_2^h$ such that

$$\int_0^N \int_0^1 |E[\mathscr{Q}_h^{(\mathbf{t}^*)}; B(x_1, y)]|^2 \, dx_1 \, dy \ll h N. \qquad (1.71)$$

Finally, we note that the set $\mathscr{Q}_h^{(\mathbf{t}^*)} \cap ([0, 1) \times [0, N))$ contains precisely N points. Rescaling in the vertical direction by a factor N^{-1}, we observe that the set

$$\mathscr{P}^* = \{(z_1, N^{-1}z_2) : (z_1, z_2) \in \mathscr{Q}_h^{(\mathbf{t}^*)}\}$$

contains precisely N points in $[0, 1)^2$, and the estimate (1.71) now translates to

$$\int_{[0,1]^2} |D[\mathscr{P}^*; B(\mathbf{x})]|^2 \, d\mathbf{x} \ll h \ll \log N,$$

in view of (1.65). This completes the proof of Theorem 10 for $k = 2$.

1.13 A Fourier–Walsh Approach to van der Corput Sets

In this section, we sketch yet another proof of Theorem 10 for $k = 2$ by highlighting the interesting group structure of the van der Corput point set

$$\mathscr{P}(2^h) = \{(0.a_1a_2a_3\ldots a_h, 0.a_h\ldots a_3a_2a_1) : a_1,\ldots,a_h \in \{0,1\}\}.$$

This is a finite abelian group isomorphic to the group $\mathbf{Z}_2^h$. We shall make use of the characters of these groups. These are the Walsh functions.

To define the Walsh functions, we first consider binary representation of any integer $\ell \in \mathbf{N}_0$, written uniquely in the form

$$\ell = \sum_{i=1}^{\infty} \lambda_i(\ell)2^{i-1}, \tag{1.72}$$

where the coefficient $\lambda_i(\ell) \in \{0,1\}$ for every $i \in \mathbf{N}$. On the other hand, every real number $y \in [0,1)$ can be represented in the form

$$y = \sum_{i=1}^{\infty} \eta_i(y)2^{-i}, \tag{1.73}$$

where the coefficient $\eta_i(y) \in \{0,1\}$ for every $i \in \mathbf{N}$. This representation is unique if we agree that the series in (1.73) is finite for every $y = m2^{-s}$ where $s \in \mathbf{N}_0$ and $m \in \{0,1,\ldots,2^s-1\}$.

For every $\ell \in \mathbf{N}_0$ of the form (1.72), we define the Walsh function $w_\ell : [0,1) \to \mathbf{R}$ by writing

$$w_\ell(y) = (-1)^{\sum_{i=1}^{\infty} \lambda_i(\ell)\eta_i(y)}. \tag{1.74}$$

Since (1.72) is essentially a finite sum, the Walsh function is well defined, and takes the values ± 1. It is easy to see that $w_0(y) = 1$ for every $y \in [0,1)$. It is well known that under the inner product

$$\langle w_k, w_\ell \rangle = \int_0^1 w_k(y)w_\ell(y)\,dy,$$

the collection of Walsh functions form an orthonormal basis of $L^2[0,1]$.

For every $\ell, k \in \mathbf{N}_0$, we can define $\ell \oplus k$ by setting

$$\lambda_i(\ell \oplus k) = \lambda_i(\ell) + \lambda_i(k) \bmod 2$$

for every $i \in \mathbf{N}$. Then it is easy to see that for every $y \in [0,1)$, we have

$$w_{\ell \oplus k}(y) = w_\ell(y)w_k(y). \tag{1.75}$$

For every $x, y \in [0, 1)$, we can define $x \oplus y$ be setting

$$\eta_i(x \oplus y) = \eta_i(x) + \eta_i(y) \bmod 2$$

for every $i \in \mathbf{N}$. Then it is easy to see that for every $\ell \in \mathbf{N}_0$, we have

$$w_\ell(x \oplus y) = w_\ell(x)w_\ell(y). \tag{1.76}$$

We shall be concerned with the characteristic function

$$\chi_{B(\mathbf{x})}(\mathbf{y}) = \begin{cases} 1, & \text{if } \mathbf{y} \in B(\mathbf{x}), \\ 0, & \text{otherwise,} \end{cases}$$

of the aligned rectangle $B(\mathbf{x}) = [0, x_1) \times [0, x_2)$, where $\mathbf{x} = (x_1, x_2)$. Then we have the discrepancy function

$$D[\mathscr{P}(2^h); B(\mathbf{x})] = \sum_{\mathbf{p} \in \mathscr{P}(2^h)} \chi_{B(\mathbf{x})}(\mathbf{p}) - 2^h x_1 x_2. \tag{1.77}$$

Clearly the characteristic function in question can be written as a product of one-dimensional characteristic functions in the form

$$\chi_{B(\mathbf{x})}(\mathbf{y}) = \chi_{[0,x_1)}(y_1)\chi_{[0,x_2)}(y_2),$$

where $\mathbf{y} = (y_1, y_2)$. Since the Walsh functions form an orthonormal basis for the space $L^2[0, 1]$, we shall use Fourier–Walsh analysis[6] to study a characteristic function of the form $\chi_{[0,x)}(y)$. We have the Fourier–Walsh series

$$\chi_{[0,x)}(y) \sim \sum_{\ell=0}^{\infty} \widetilde{\chi_\ell}(x)w_\ell(y),$$

where, for every $\ell \in \mathbf{N}_0$, the Fourier–Walsh coefficients are given by

$$\widetilde{\chi_\ell}(x) = \int_0^x w_\ell(y)\,dy.$$

In particular, we have $\widetilde{\chi_0}(x) = x$ for every $x \in [0, 1)$.

Instead of using the full Fourier–Walsh series, we shall truncate it and use the approximation

---

$$\chi_{[0,x)}^{(h)}(y) = \sum_{\ell=0}^{2^h-1} \widetilde{\chi_\ell}(x) w_\ell(y).$$ (1.78)

Note that there exists a unique $m \in \mathbf{N}_0$ such that $m2^{-h} \le x < (m+1)2^{-h}$. Then

$$\chi_{[0,x)}^{(h)}(y) = \begin{cases} 1, & \text{if } 0 \le y < m2^{-h}, \\ 2^h x - m, & \text{if } m2^{-h} \le y < (m+1)2^{-h}, \\ 0, & \text{if } (m+1)2^{-h} \le y < 1, \end{cases}$$

where the quantity

$$2^h x - m = 2^h \int_{m2^{-h}}^{(m+1)2^{-h}} \chi_{[0,x)}(y)\,dy$$

represents the average value of $\chi_{[0,x)}(y)$ in the interval $[m2^{-h}, (m+1)2^{-h})$.

The approximation (1.78) in turn leads to the approximation

$$\chi_{B(\mathbf{x})}^{(h)}(\mathbf{y}) = \chi_{[0,x_1)}^{(h)}(y_1)\chi_{[0,x_2)}^{(h)}(y_2) = \sum_{\ell_1=0}^{2^h-1}\sum_{\ell_2=0}^{2^h-1} \widetilde{\chi_{\mathbf{l}}}(\mathbf{x}) W_{\mathbf{l}}(\mathbf{y})$$

of the characteristic function $\chi_{B(\mathbf{x})}(\mathbf{y})$. Here $\mathbf{l} = (\ell_1, \ell_2)$,

$$\widetilde{\chi_{\mathbf{l}}}(\mathbf{x}) = \widetilde{\chi_{\ell_1}}(x_1)\widetilde{\chi_{\ell_2}}(x_2) \quad \text{and} \quad W_{\mathbf{l}}(\mathbf{y}) = w_{\ell_1}(y_1)w_{\ell_2}(y_2).$$ (1.79)

Corresponding to this, we approximate the discrepancy function (1.77) by

$$D^{(h)}[\mathscr{P}(2^h); B(\mathbf{x})] = \sum_{\mathbf{p} \in \mathscr{P}(2^h)} \chi_{B(\mathbf{x})}^{(h)}(\mathbf{p}) - 2^h x_1 x_2$$

$$= \sum_{\mathbf{p} \in \mathscr{P}(2^h)} \sum_{\ell_1=0}^{2^h-1}\sum_{\ell_2=0}^{2^h-1} \widetilde{\chi_{\mathbf{l}}}(\mathbf{x}) W_{\mathbf{l}}(\mathbf{p}) - 2^h \widetilde{\chi_{\mathbf{0}}}(\mathbf{x})$$

$$= \sum_{\substack{\ell_1=0 \\ (\ell_1,\ell_2)\ne(0,0)}}^{2^h-1}\sum_{\ell_2=0}^{2^h-1} \left(\sum_{\mathbf{p} \in \mathscr{P}(2^h)} W_{\mathbf{l}}(\mathbf{p}) \right) \widetilde{\chi_{\mathbf{l}}}(\mathbf{x}),$$

noting that

$$\sum_{\mathbf{p} \in \mathscr{P}(2^h)} W_{\mathbf{0}}(\mathbf{p}) = \#\mathscr{P}(2^h) = 2^h.$$ (1.80)

It is well known in the theory of abelian groups that the sum

$$\sum_{\mathbf{p}\in\mathscr{P}(2^h)} W_{\mathbf{l}}(\mathbf{p}) \in \{0, 2^h\}; \tag{1.81}$$

see, for instance, [25, Chapters 5 and 9] or [26, Chapter 5]. We therefore need to have some understanding on the set

$$L(h) = \left\{ \mathbf{l} \in [0, 2^h) \times [0, 2^h) : \mathbf{l} \neq \mathbf{0} \text{ and } \sum_{\mathbf{p}\in\mathscr{P}(2^h)} W_{\mathbf{l}}(\mathbf{p}) = 2^h \right\}.$$

Then

$$D^{(h)}[\mathscr{P}(2^h); B(\mathbf{x})] = 2^h \sum_{\mathbf{l}\in L(h)} \widetilde{\chi}_{\mathbf{l}}(\mathbf{x}). \tag{1.82}$$

Recall the discussion at the beginning of Sect. 1.11. The estimate (1.56) shows that the set $\mathscr{P}(2^h)$ is insufficient for us to establish Theorem 10 in the case $k = 2$. To overcome this problem, we use digit shifts in Sect. 1.12. Here, for every $\mathbf{t} \in \mathbf{Z}_2^{2h}$, we consider the set

$$\mathscr{P}(2^h) \oplus \mathbf{t} = \{\mathbf{p} \oplus \mathbf{t} : \mathbf{p} \in \mathscr{P}(2^h)\}$$

where, for every

$$\mathbf{p} = (0.a_1 \ldots a_h, 0.a_h \ldots a_1) \in \mathscr{P}(2^h) \quad \text{and} \quad \mathbf{t} = (t_1', \ldots, t_h', t_h'', \ldots, t_1'') \in \mathbf{Z}_2^{2h},$$

we have the shifted point[7]

$$\mathbf{p} \oplus \mathbf{t} = (0.b_1' \ldots b_h', 0.b_h'' \ldots b_1''),$$

with the digits $b_1', \ldots, b_h', b_1'', \ldots, b_h'' \in \{0, 1\}$ satisfying

$$b_s' \equiv a_s + t_s' \bmod 2 \quad \text{and} \quad b_s'' \equiv a_s + t_s'' \bmod 2$$

for every $s = 1, \ldots, h$. Then

---

[7]Here we somewhat abuse notation, as $\mathbf{t}$ clearly has more coordinates than $\mathbf{p}$. In the sequel, $W_{\mathbf{l}}(\mathbf{t})$ is really $W_{\mathbf{l}}(\mathbf{0} \oplus \mathbf{t})$, notation abused again.

$$D^{(h)}[\mathscr{P}(2^h) \oplus \mathbf{t}; B(\mathbf{x})] = \sum_{\mathbf{p} \in \mathscr{P}(2^h)} \chi^{(h)}_{B(\mathbf{x})}(\mathbf{p} \oplus \mathbf{t}) - 2^h x_1 x_2$$

$$= \sum_{\substack{\ell_1=0 \\ (\ell_1,\ell_2) \neq (0,0)}}^{2^h-1} \sum_{\ell_2=0}^{2^h-1} \left(\sum_{\mathbf{p} \in \mathscr{P}(2^h)} W_\mathbf{l}(\mathbf{p} \oplus \mathbf{t}) \right) \widetilde{\chi}_\mathbf{l}(\mathbf{x})$$

$$= \sum_{\substack{\ell_1=0 \\ (\ell_1,\ell_2) \neq (0,0)}}^{2^h-1} \sum_{\ell_2=0}^{2^h-1} W_\mathbf{l}(\mathbf{t}) \left(\sum_{\mathbf{p} \in \mathscr{P}(2^h)} W_\mathbf{l}(\mathbf{p}) \right) \widetilde{\chi}_\mathbf{l}(\mathbf{x}),$$

in view of (1.76) and the second identity in (1.79). It follows that

$$D^{(h)}[\mathscr{P}(2^h); B(\mathbf{x})] = 2^h \sum_{\mathbf{l} \in L(h)} W_\mathbf{l}(\mathbf{t}) \widetilde{\chi}_\mathbf{l}(\mathbf{x}).$$

Squaring this expression and summing over all $\mathbf{t} \in \mathbf{Z}_2^{2h}$, we obtain

$$\sum_{\mathbf{t} \in \mathbf{Z}_2^{2h}} |D^{(h)}[\mathscr{P}(2^h) \oplus \mathbf{t}; B(\mathbf{x})]|^2 = 4^h \sum_{\mathbf{t} \in \mathbf{Z}_2^{2h}} \left(\sum_{\mathbf{l} \in L(h)} W_\mathbf{l}(\mathbf{t}) \widetilde{\chi}_\mathbf{l}(\mathbf{x}) \right)^2$$

$$= 4^h \sum_{\mathbf{t} \in \mathbf{Z}_2^{2h}} \sum_{\mathbf{l}',\mathbf{l}'' \in L(h)} W_{\mathbf{l}'}(\mathbf{t}) W_{\mathbf{l}''}(\mathbf{t}) \widetilde{\chi}_{\mathbf{l}'}(\mathbf{x}) \widetilde{\chi}_{\mathbf{l}''}(\mathbf{x})$$

$$= 4^h \sum_{\mathbf{l}',\mathbf{l}'' \in L(h)} \left(\sum_{\mathbf{t} \in \mathbf{Z}_2^{2h}} W_{\mathbf{l}'}(\mathbf{t}) W_{\mathbf{l}''}(\mathbf{t}) \right) \widetilde{\chi}_{\mathbf{l}'}(\mathbf{x}) \widetilde{\chi}_{\mathbf{l}''}(\mathbf{x}). \tag{1.83}$$

Lemma 20. *For every* $\mathbf{l}', \mathbf{l}'' \in \mathbf{N}_0^2$, *we have*

$$\sum_{\mathbf{t} \in \mathbf{Z}_2^{2h}} W_{\mathbf{l}'}(\mathbf{t}) W_{\mathbf{l}''}(\mathbf{t}) = \begin{cases} 4^h, & \text{if } \mathbf{l}' = \mathbf{l}'', \\ 0, & \text{otherwise.} \end{cases}$$

Proof. Note first of all that in view of (1.75) and the second identity in (1.79), with $\mathbf{l}' \oplus \mathbf{l}'' = (\ell_1', \ell_2') \oplus (\ell_1'', \ell_2'') = (\ell_1' \oplus \ell_1'', \ell_2' \oplus \ell_2'')$, we have $W_{\mathbf{l}'}(\mathbf{t}) W_{\mathbf{l}''}(\mathbf{t}) = W_{\mathbf{l}' \oplus \mathbf{l}''}(\mathbf{t})$. For simplicity, write

$$S = \sum_{\mathbf{t} \in \mathbf{Z}_2^{2h}} W_{\mathbf{l}'}(\mathbf{t}) W_{\mathbf{l}''}(\mathbf{t}) = \sum_{\mathbf{t} \in \mathbf{Z}_2^{2h}} W_{\mathbf{l}' \oplus \mathbf{l}''}(\mathbf{t}).$$

If $\mathbf{l}' = \mathbf{l}''$, so that $\mathbf{l}' \oplus \mathbf{l}'' = \mathbf{0}$, then $W_{\mathbf{l}' \oplus \mathbf{l}''}(\mathbf{t}) = W_{\mathbf{0}}(\mathbf{t}) = 1$ for every $\mathbf{t} \in \mathbf{Z}_2^{2h}$, and so clearly $S = \#\mathbf{Z}_2^{2h} = 4^h$. If $\mathbf{l}' \neq \mathbf{l}''$, so that $\mathbf{l}' \oplus \mathbf{l}'' \neq \mathbf{0}$, then there exists $\mathbf{t}_0 \in \mathbf{Z}_2^{2h}$ such that $W_{\mathbf{l}' \oplus \mathbf{l}''}(\mathbf{t}_0) \neq 1$. As $\mathbf{t}$ runs through the group $\mathbf{Z}_2^{2h}$, so does $\mathbf{t} \oplus \mathbf{t}_0$, so that

$$S = \sum_{\mathbf{t} \in \mathbf{Z}_2^{2h}} W_{\mathbf{l}' \oplus \mathbf{l}''}(\mathbf{t} \oplus \mathbf{t}_0) = \sum_{\mathbf{t} \in \mathbf{Z}_2^{2h}} W_{\mathbf{l}' \oplus \mathbf{l}''}(\mathbf{t}) W_{\mathbf{l}' \oplus \mathbf{l}''}(\mathbf{t}_0) = S W_{\mathbf{l}' \oplus \mathbf{l}''}(\mathbf{t}_0),$$

in view of (1.76) and the second identity in (1.79). Clearly $S = 0$ in this case. □

Combining (1.83) and Lemma 20, we deduce that

$$\frac{1}{4^h} \sum_{\mathbf{t} \in \mathbf{Z}_2^{2h}} |D^{(h)}[\mathscr{P}(2^h) \oplus \mathbf{t}; B(\mathbf{x})]|^2 = 4^h \sum_{\mathbf{l} \in L(h)} |\widetilde{\chi}_{\mathbf{l}}(\mathbf{x})|^2, \tag{1.84}$$

so that on integrating trivially with respect to $\mathbf{x} \in [0,1]^2$, we have

$$\frac{1}{4^h} \sum_{\mathbf{t} \in \mathbf{Z}_2^{2h}} \int_{[0,1]^2} |D^{(h)}[\mathscr{P}(2^h) \oplus \mathbf{t}; B(\mathbf{x})]|^2 \, d\mathbf{x} = 4^h \sum_{\mathbf{l} \in L(h)} \int_{[0,1]^2} |\widetilde{\chi}_{\mathbf{l}}(\mathbf{x})|^2 \, d\mathbf{x}. \tag{1.85}$$

To estimate the right hand side of (1.85), we need to use a formula of Fine [17] on the Fourier–Walsh coefficients of the characteristic function $\chi_{[0,x)}(y)$.

Let $\rho(0) = 0$. For any integer $\ell \in \mathbf{N}$ with representation (1.72), let

$$\rho(\ell) = \max\{i \in \mathbf{N} : \lambda_i(\ell) \neq 0\}, \quad \text{so that} \quad 2^{\rho(\ell)-1} \leq \ell < 2^{\rho(\ell)}. \tag{1.86}$$

Then the formula of Fine gives

$$\int_0^1 |\widetilde{\chi}_\ell(x)|^2 \, dx = \frac{4^{-\rho(\ell)}}{3}.$$

If we write $\rho(\mathbf{l}) = \rho(\ell_1) + \rho(\ell_2)$ for $\mathbf{l} = (\ell_1, \ell_2)$, then in view of the first identity in (1.79), we have

$$\int_{[0,1]^2} |\widetilde{\chi}_{\mathbf{l}}(\mathbf{x})|^2 \, d\mathbf{x} = \frac{4^{-\rho(\mathbf{l})}}{9},$$

and the identity (1.85) becomes

$$\frac{1}{4^h} \sum_{\mathbf{t} \in \mathbf{Z}_2^{2h}} \int_{[0,1]^2} |D^{(h)}[\mathscr{P}(2^h) \oplus \mathbf{t}; B(\mathbf{x})]|^2 \, d\mathbf{x} = \frac{4^h}{9} \sum_{\mathbf{l} \in L(h)} 4^{-\rho(\mathbf{l})}. \tag{1.87}$$

To estimate the sum on the right hand side of (1.87), we need some reasonably precise information on the set $L(h)$. The following result is rather useful.

Lemma 21. *For every* $y \in [0, 1)$ *and every* $s \in \mathbf{N}_0$, *we have*

$$\sum_{\ell=0}^{2^s-1} w_\ell(y) = 2^s \chi_{[0,2^{-s})}(y).$$

Proof. If $y \in [0, 2^{-s})$, then it follows from (1.73) that $\eta_i(y) = 0$ whenever $1 \leq i \leq s$. On the other hand, for every $\ell = 0, 1, 2, \ldots, 2^s - 1$, it follows from (1.72) that $\lambda_i(\ell) = 0$ for every $i > s$. It follows that for every $\ell = 0, 1, 2, \ldots, 2^s - 1$, we have

$$\sum_{i=1}^{\infty} \lambda_i(\ell)\eta_i(y) = 0,$$

and so $w_\ell(y) = 1$. On the other hand, if $y \in [2^{-s}, 1)$, then it follows from (1.73) that there exists some $j \in \{1, \ldots, s\}$ such that $\eta_j(y) = 1$. We now choose $k \in \{1, 2, \ldots, 2^s - 1\}$ such that $\lambda_j(k) = 1$ and $\lambda_i(k) = 0$ for every $i \neq j$. Then $w_k(y) \neq 1$. It is easy to see that as ℓ runs through the set $0, 1, 2, \ldots, 2^s - 1$, then so does $\ell \oplus k$, so that

$$\sum_{\ell=0}^{2^s-1} w_\ell(y) = \sum_{\ell=0}^{2^s-1} w_{\ell \oplus k}(y) = w_k(y) \sum_{\ell=0}^{2^s-1} w_\ell(y),$$

in view of (1.75). The result follows immediately.   □

Lemma 22. *For every* $s_1, s_2 \in \{0, 1, \ldots, h\}$, *let*

$$\varXi(s_1, s_2) = \sum_{\ell_1=0}^{2^{s_1}-1} \sum_{\ell_2=0}^{2^{s_2}-1} \sum_{\mathbf{p}\in\mathscr{P}(2^h)} W_{\mathbf{l}}(\mathbf{p}).$$

Then

$$\varXi(s_1, s_2) = \begin{cases} 2^{s_1+s_2}, & \text{if } s_1 + s_2 \geq h, \\ 2^h, & \text{if } s_1 + s_2 \leq h. \end{cases}$$

Proof. Writing $\mathbf{p} = (p_1, p_2)$ and $\mathbf{l} = (\ell_1, \ell_2)$ and noting the second identity in (1.79) and Lemma 21, we have

$$\sum_{\ell_1=0}^{2^{s_1}-1} \sum_{\ell_2=0}^{2^{s_2}-1} \sum_{\mathbf{p}\in\mathscr{P}(2^h)} W_{\mathbf{l}}(\mathbf{p}) = \sum_{\mathbf{p}\in\mathscr{P}(2^h)} \left(\sum_{\ell_1=0}^{2^{s_1}-1} w_{\ell_1}(p_1) \right) \left(\sum_{\ell_2=0}^{2^{s_2}-1} w_{\ell_2}(p_2) \right)$$

$$= 2^{s_1+s_2} \sum_{\mathbf{p}\in\mathscr{P}(2^h)} \chi_{[0,2^{-s_1})}(p_1)\chi_{[0,2^{-s_2})}(p_2)$$

$$= 2^{s_1+s_2} \sum_{\mathbf{p}\in\mathscr{P}(2^h)} \chi_{[0,2^{-s_1})\times[0,2^{-s_2})}(\mathbf{p}).$$

It is not difficult to deduce from Lemma 14 that every rectangle of the form

$$[m_1 2^{-s}, (m+1)2^{-s}) \times [m_2 2^{s-h}, (m_2+1)2^{s-h}) \subseteq [0,1)^2$$

where $m_1, m_2 \in \mathbf{N}_0$, and area 2^{-h}, contains precisely one point of $\mathscr{P}(2^h)$. Let us say that such a rectangle is an elementary rectangle. Suppose first of all that $s_1 + s_2 \geq h$. Then the rectangle $[0, 2^{-s_1}) \times [0, 2^{-s_2})$ is contained in one elementary rectangle anchored at the origin, and so contains at most one point of $\mathscr{P}(2^h)$. Clearly it contains the point $\mathbf{0} \in \mathscr{P}(2^h)$, and so

$$\sum_{\mathbf{p} \in \mathscr{P}(2^h)} \chi_{[0,2^{-s_1}) \times [0,2^{-s_2})}(\mathbf{p}) = 1.$$

Suppose then that $s_1 + s_2 \leq h$. Then the rectangle $[0, 2^{-s_1}) \times [0, 2^{-s_2})$ is a union of precisely $2^{h-s_1-s_2}$ elementary rectangles, and so contains precisely $2^{h-s_1-s_2}$ points of $\mathscr{P}(2^h)$, whence

$$\sum_{\mathbf{p} \in \mathscr{P}(2^h)} \chi_{[0,2^{-s_1}) \times [0,2^{-s_2})}(\mathbf{p}) = 2^{h-s_1-s_2}.$$

This completes the proof.   □

Note that with $s_1 = s_2 = h$, Lemma 22 gives

$$\sum_{\ell_1=0}^{2^h-1} \sum_{\ell_2=0}^{2^h-1} \sum_{\mathbf{p} \in \mathscr{P}(2^h)} W_1(\mathbf{p}) = 4^h.$$

In view of (1.80) and (1.81), we conclude that $\#L(h) = 2^h - 1$. We now study the set $L(h)$ in greater detail.

Lemma 23. *For every $s_1, s_2 \in \{1, \ldots, h\}$, let*

$$L(s_1, s_2) = \left\{ \mathbf{l} \in [2^{s_1-1}, 2^{s_1}) \times [2^{s_2-1}, 2^{s_2}) : \sum_{\mathbf{p} \in \mathscr{P}(2^h)} W_1(\mathbf{p}) = 2^h \right\}.$$

Then

a) for every $\mathbf{l} \in L(s_1, s_2)$, we have $\rho(\mathbf{l}) = s_1 + s_2$;
b) we have

$$\#L(s_1, s_2) = \begin{cases} 2^{s_1+s_2-h-2}, & \text{if } s_1 + s_2 \geq h+2, \\ 1, & \text{if } s_1 + s_2 = h+1, \\ 0, & \text{otherwise.} \end{cases}$$

Furthermore, every $\mathbf{l} \in L(h)$ *belongs to* $L(s_1, s_2)$ *for some* $s_1, s_2 \in \{1, \ldots, h\}$ *that satisfy* $s_1 + s_2 \geq h + 1$.

Proof. Note that if $\mathbf{l} \in L(s_1, s_2)$, then $\rho(\mathbf{l}) = \rho(\ell_1) + \rho(\ell_2) = s_1 + s_2$, in view of (1.86). This establishes part (a). To prove part (b), note that in view of (1.81), we have, in the notation of Lemma 22,

$$\#L(s_1, s_2) = 2^{-h} \sum_{\ell_1=2^{s_1-1}}^{2^{s_1}-1} \sum_{\ell_2=2^{s_2-1}}^{2^{s_2}-1} \sum_{\mathbf{p} \in \mathscr{P}(2^h)} W_{\mathbf{l}}(\mathbf{p})$$

$$= 2^{-h}(\varXi(s_1, s_2) - \varXi(s_1 - 1, s_2) - \varXi(s_1, s_2 - 1) + \varXi(s_1 - 1, s_2 - 1)).$$

Part (b) now follows easily from Lemma 22. Finally, it is easily checked that

$$\sum_{\substack{s_1=1 \\ s_1+s_2=h+1}}^{h} \sum_{s_2=1}^{h} 1 + \sum_{\substack{s_1=1 \\ s_1+s_2\geq h+2}}^{h} \sum_{s_2=1}^{h} 2^{s_1+s_2-h-2} = 2^h - 1 = \#L(h).$$

The last assertion follows immediately. $\qquad\square$

Using Lemma 23, we deduce that

$$\sum_{\mathbf{l}\in L(h)} 4^{-\rho(\mathbf{l})} = \sum_{\substack{s_1=1 \\ s_1+s_2=h+1}}^{h} \sum_{s_2=1}^{h} 4^{-h-1} + \sum_{\substack{s_1=1 \\ s_1+s_2\geq h+2}}^{h} \sum_{s_2=1}^{h} 2^{s_1+s_2-h-2} 4^{-s_1-s_2}$$

$$= \sum_{\substack{s_1=1 \\ s_1+s_2=h+1}}^{h} \sum_{s_2=1}^{h} 4^{-h-1} + \sum_{\substack{s_1=1 \\ s_1+s_2\geq h+2}}^{h} \sum_{s_2=1}^{h} 2^{-s_1-s_2-h-2}$$

$$= \sum_{\substack{s_1=1 \\ s_1+s_2=h+1}}^{h} \sum_{s_2=1}^{h} 4^{-h-1} + \sum_{k=2}^{h} \sum_{\substack{s_1=1 \\ s_1+s_2=h+k}}^{h} \sum_{s_2=1}^{h} 2^{-h-k-h-2}$$

$$= 4^{-h-1}h + 4^{-h-1} \sum_{k=2}^{h} \sum_{\substack{s_1=1 \\ s_1+s_2=h+k}}^{h} \sum_{s_2=1}^{h} 2^{-k}$$

$$< 4^{-h-1}h + 4^{-h-1}h \sum_{k=2}^{h} 2^{-k} < 4^{-h}h.$$

Combining this with (1.87), we obtain

$$\frac{1}{4^h} \sum_{\mathbf{t}\in\mathbb{Z}_2^{2h}} \int_{[0,1]^2} |D^{(h)}[\mathscr{P}(2^h) \oplus \mathbf{t}; B(\mathbf{x})]|^2 \, d\mathbf{x} < \frac{h}{9} \ll \log N,$$

noting that $N = 2^h$ in this case. Hence there is a digit shift $\mathbf{t}^* \in \mathbf{Z}_2^{2h}$ such that

$$\int_{[0,1]^2} |D^{(h)}[\mathscr{P}(2^h) \oplus \mathbf{t}^*; B(\mathbf{x})]|^2 \, d\mathbf{x} \ll \log N,$$

essentially establishing Theorem 10 in the case $k = 2$, apart from our not having properly analyzed the effect of the approximation of the certain characteristic functions by their truncated Fourier–Walsh series.

We complete this section by making an important comment for later use. Let us return to (1.82) and make the hypothetical assumption that the functions $\widetilde{\chi}_{\mathbf{l}}(\mathbf{x})$, where $\mathbf{l} \in L(h)$, are orthogonal. Then

$$\int_{[0,1]^2} |D^{(h)}[\mathscr{P}(2^h); B(\mathbf{x})]|^2 \, d\mathbf{x} = 4^h \sum_{\mathbf{l} \in L(h)} \int_{[0,1]^2} |\widetilde{\chi}_{\mathbf{l}}(\mathbf{x})|^2 \, d\mathbf{x}.$$

Note that the right hand side is exactly the same as the right hand side of (1.85), so that we can analyze this as before.

Unfortunately, the functions $\widetilde{\chi}_{\mathbf{l}}(\mathbf{x})$, where $\mathbf{l} \in L(h)$, are not orthogonal in this instance, so we cannot proceed in this way. Our technique in overcoming this handicap is to make use of the digit shifts $\mathbf{t} \in \mathbf{Z}_2^{2h}$, and bring into the argument, one may say through the back door, some orthogonality in the form of Lemma 20. We shall return to this in Sects. 1.15 and 1.16.

1.14 Generalizations of van der Corput Point Sets

In our discussion of the van der Corput sequence and van der Corput point sets in Sects. 1.10 and 1.11, we have restricted our discussion to dimension $k = 2$. Indeed, historically, the van der Corput sequence is constructed dyadically, and offers no generalization to the multi-dimensional case without going beyond dyadic constructions, except for one instance which we shall describe later in this section.

To study the general case in Theorems 10 and 11, one way is to generalize the van der Corput sequence. Here we know two ways of doing so, one by Halton [19] and the other by Faure [16]. The Halton construction enables Halton to establish Theorem 11 in its generality and forms the basis for the proof of Theorem 10 in its generality by Roth [31]. The Faure construction enables Faure to give an alternative proof of Theorem 11 in its generality, enables Chen [9] soon afterwards to give an alternative proof of Theorem 10 in its generality and, more recently, forms the basis for the explicit construction proof of Theorem 10 by Chen and Skriganov [11, 13].

The generalizations by Halton and by Faure both require the very natural p-adic generalization of the van der Corput construction. The difference is that while Halton uses many different primes p, Faure uses only one such prime p but chosen to be sufficiently large.

1.14.1 Halton Point Sets

We first discuss Halton's contribution. Recall the dyadic construction (1.45) and (1.46) of the classical van der Corput sequence. Suppose now that we wish to study Theorem 10 or 11 in arbitrary dimension $k \geq 2$. Let p_i, where $i = 1, \ldots, k - 1$, denote the first $k - 1$ primes, with $p_1 < \ldots < p_{k-1}$. For every non-negative integer $n \in \mathbf{N}_0$ and every $i = 1, \ldots, k - 1$, we write

$$n = \sum_{j=1}^{\infty} a_j^{(i)} p_i^{j-1} \tag{1.88}$$

as a p_i-adic expansion. Then we write

$$c_n^{(i)} = \sum_{j=1}^{\infty} a_j^{(i)} p_i^{-j}. \tag{1.89}$$

Finally we write

$$\mathbf{c}_n = (c_n^{(1)}, \ldots, c_n^{(k-1)}).$$

Note that $\mathbf{c}_n \in [0, 1)^{k-1}$. The infinite sequence $\mathbf{c}_0, \mathbf{c}_1, \mathbf{c}_2, \ldots$ is usually called a Halton sequence, and the infinite set

$$\mathscr{H} = \{(\mathbf{c}_n, n) : n = 0, 1, 2, \ldots\} \tag{1.90}$$

in $[0, 1)^{k-1} \times [0, \infty)$ is usually called a Halton point set.

Corresponding to Lemma 13, we have the following multi-dimensional version.

Lemma 24. *For all non-negative integers $s_1, \ldots, s_{k-1}$ and $\ell_1, \ldots, \ell_{k-1}$ satisfying $\ell_i < p_i^{s_i}$ for every $i = 1, \ldots, k - 1$, the set*

$$\left\{ n \in \mathbf{N}_0 : \mathbf{c}_n \in \prod_{i=1}^{k-1} [\ell_i \, p_i^{-s_i}, (\ell_i + 1) p_i^{-s_i}) \right\}$$

contains precisely all the elements of a residue class modulo $p_1^{s_1} \ldots p_{k-1}^{s_{k-1}}$ in $\mathbf{N}_0$.

Proof. For fixed $i = 1, \ldots, k - 1$, the p_i-adic version of Lemma 13 says that the set

$$\{ n \in \mathbf{N}_0 : c_n^{(i)} \in [\ell_i \, p_i^{-s_i}, (\ell_i + 1) p_i^{-s_i}) \}$$

contains precisely all the elements of a residue class modulo $p_i^{s_i}$ in $\mathbf{N}_0$. The result now follows from the Chinese remainder theorem. $\qquad \square$

We say that a rectangular box of the form

$$\prod_{i=1}^{k-1}[\ell_i\, p_i^{-s_i}, (\ell_i + 1)p_i^{-s_i}) \subseteq [0, 1)^{k-1}$$

for some integers $\ell_1, \ldots, \ell_{k-1}$ is an elementary $(p_1, \ldots, p_{k-1})$-adic box of volume $p_1^{-s_1} \ldots p_{k-1}^{-s_{k-1}}$. Hence Lemma 24 says that the given Halton sequence has very good distribution among such elementary $(p_1, \ldots, p_{k-1})$-adic boxes for all non-negative integer values of $s_1, \ldots, s_{k-1}$.

Lemma 25. *For all non-negative integers $s_1, \ldots, s_{k-1}, \ell_1, \ldots, \ell_{k-1}$ and m satisfying $\ell_i < p_i^{s_i}$ for every $i = 1, \ldots, k - 1$, the rectangular box*

$$\prod_{i=1}^{k-1}[\ell_i\, p_i^{-s_i}, (\ell_i + 1)p_i^{-s_i}) \times \left[m \prod_{i=1}^{k-1} p_i^{s_i}, (m + 1) \prod_{i=1}^{k-1} p_i^{s_i} \right)$$

contains precisely one point of the Halton point set $\mathcal{H}$.

Clearly there is an average of one point of the Halton point set $\mathcal{H}$ per unit volume in $[0, 1)^{k-1} \times [0, \infty)$. For any measurable set A in $[0, 1)^{k-1} \times [0, \infty)$, let

$$E[\mathcal{H}; A] = \#(\mathcal{H} \cap A) - \mu(A)$$

denote the discrepancy of $\mathcal{H}$ in A.

We have the following generalization of Lemma 15.

Lemma 26. *For all non-negative integers $s_1, \ldots, s_{k-1}$ and $\ell_1, \ldots, \ell_{k-1}$ satisfying $\ell_i < p_i^{s_i}$ for every $i = 1, \ldots, k - 1$, there exist real numbers α_0, β_0, depending at most on $s_1, \ldots, s_{k-1}$ and $\ell_1, \ldots, \ell_{k-1}$, such that $|\alpha_0| \le \frac{1}{2}$ and*

$$E\left[\mathcal{H}; \prod_{i=1}^{k-1}[\ell_i\, p_i^{-s_i}, (\ell_i + 1)p_i^{-s_i}) \times [0, y) \right] = \alpha_0 - \psi(p_1^{-s_1} \ldots p_{k-1}^{-s_{k-1}}(y - \beta_0))$$

$$(1.91)$$

at all points of continuity of the right hand side.

We can now prove Theorem 11. Let $N \ge 2$ be a given integer. It follows at once from the definition of $\mathcal{H}$ that the set

$$\mathcal{H}_0 = \mathcal{H} \cap ([0, 1)^{k-1} \times [0, N))$$

contains precisely N points. Let the integer h be determined uniquely by

$$p_1^{h-1} < N \le p_1^h. \qquad (1.92)$$

Consider a rectangular box of the form

$$B(x_1, \ldots, x_{k-1}, y) = [0, x_1) \times \ldots \times [0, x_{k-1}) \times [0, y) \subseteq [0, 1)^{k-1} \times [0, N).$$

Similar to our technique in Sect. 1.10, we shall approximate each interval $[0, x_i)$, where $i = 1, \ldots, k - 1$, by the subinterval $[0, x_i^{(h)})$, where $x_i^{(h)} = p_i^{-h}[p_i^h x_i]$ is the greatest integer multiple of p_i^{-h} not exceeding x_i, and then consider the smaller rectangular box

$$B(x_1^{(h)}, \ldots, x_{k-1}^{(h)}, y) = [0, x_1^{(h)}) \times \ldots \times [0, x_{k-1}^{(h)}) \times [0, y)$$

as an approximation of $B(x_1, \ldots, x_{k-1}, y)$. A slight elaboration of the corresponding argument in Sect. 1.10 will show that the difference

$$B(x_1, \ldots, x_{k-1}, y) \setminus B(x_1^{(h)}, \ldots, x_{k-1}^{(h)}, y)$$

is contained in a union of at most $k - 1$ sets of the type discussed in Lemma 25, and so

$$|E[\mathscr{H}; B(x_1, \ldots, x_{k-1}, y)] - E[\mathscr{H}; B(x_1^{(h)}, \ldots, x_{k-1}^{(h)}, y)]| \leq k - 1; \qquad (1.93)$$

note that since $y \leq N$, it makes no difference whether we write $\mathscr{H}$ or $\mathscr{H}_0$ in our argument.

It remains to estimate $E[\mathscr{H}; B(x_1^{(h)}, \ldots, x_{k-1}^{(h)}, y)]$. To do so, we need to write each interval $[0, x_i^{(h)})$, where $i = 1, \ldots, h - 1$, as a union of elementary p_i-adic intervals, each of length p_i^{-s} for some integer s satisfying $0 \leq s \leq h$.

If $x_i^{(h)} = 1$, then $[0, x_i^{(h)})$ is a union of precisely one elementary p_i-adic interval of unit length, so we now assume that $0 \leq x_i^{(h)} < 1$.

Lemma 27. *Suppose that $0 \leq x_i^{(h)} < 1$, with*

$$x_i^{(h)} = \sum_{s=1}^{h} b_s p_i^{-s}$$

as a p_i-adic expansion. Then $[0, x_i^{(h)})$ can be written as a union of

$$\sum_{s=1}^{h} b_s < h p_i$$

elementary p_i-adic intervals, namely b_1 elementary p_i-adic intervals of length p_i^{-1}, together with b_2 elementary p_i-adic intervals of length p_i^{-2}, and so on.

Hence the set $B(x_1^{(h)}, \ldots, x_{k-1}^{(h)}, y)$ is a disjoint union of fewer than $h^{k-1} p_1 \ldots p_{k-1}$ sets of the type discussed in Lemma 26. Hence

$$|E[\mathscr{H}; B(x_1^{(h)}, \ldots, x_{k-1}^{(h)}, y)]| < h^{k-1} p_1 \ldots p_{k-1} \ll_k (\log N)^{k-1}. \qquad (1.94)$$

Combining (1.93) and (1.94), we conclude that

$$|E[\mathscr{H}; B(x_1, \ldots, x_{k-1}, y)]| \ll_k (\log N)^{k-1}. \qquad (1.95)$$

Finally, rescaling the second coordinate of the points of $\mathscr{H}_0$ by a factor N^{-1}, we obtain a set

$$\mathscr{P} = \{(\mathbf{c}_n, N^{-1}n) : n = 0, 1, 2, \ldots, N - 1\}$$

of precisely N points in $[0, 1)^k$. For every $\mathbf{x} = (x_1, \ldots, x_k) \in [0, 1]^k$, we have

$$D[\mathscr{P}; B(\mathbf{x})] = E[\mathscr{H}_0; [0, x_1) \times \ldots \times [0, x_{k-1}) \times [0, Nx_k)] \ll_k (\log N)^{k-1},$$

in view of (1.95) and noting that $0 \le Nx_k \le N$. This now completes the proof of Theorem 11.

Next we discuss Roth's ideas in shaping this Halton construction to give a proof of Theorem 10. As in the special case $k = 2$, one needs to introduce a probabilistic variable. To pave the way for this, we shall modify the Halton point set somewhat. Let $N \ge 2$ be a given integer, and let the integer h be determined uniquely by

$$p_1^{h-1} < N \le p_1^h, \qquad (1.96)$$

as before. For every $i = 1, \ldots, k - 1$ and every $n = 0, 1, 2, \ldots, p_i^h - 1$, we define $c_n^{(i)}$ as before by (1.88) and (1.89). We then extend the definition of $c_n^{(i)}$ to all other integers using periodicity by writing

$$c_{n+p_i^h} = c_n \quad \text{for every } n \in \mathbf{Z},$$

write $\mathbf{c}_n = (c_n^{(1)}, \ldots, c_n^{(k-1)})$, and consider the extended Halton point set

$$\mathscr{H}_h = \{(c_n, n) : n \in \mathbf{Z}\}.$$

Remark. In Roth [31], as well as Chen [8], the construction of the set $\mathscr{H}_h$ is slightly different, but the difference does not affect the argument in any way. Let $M = p_1 \ldots p_{k-1}$. One then defines $c_n^{(i)}$ for $n = 0, 1, 2, \ldots, M^h - 1$ by (1.88) and (1.89), write $\mathbf{c}_n = (c_n^{(1)}, \ldots, c_n^{(k-1)})$ for these values of n, and define $\mathbf{c}_n$ for all other integer values of n by the periodicity relationship $\mathbf{c}_{n+M^h} = \mathbf{c}_n$ for every $n \in \mathbf{Z}$.

Furthermore, for every real number $t \in \mathbf{R}$, we consider the translated Halton point set

$$\mathcal{H}_h(t) = \{(\mathbf{c}_n, n + t) : n \in \mathbf{Z}\}.$$

It is clear that there is an average of one point of the translated Halton point set $\mathcal{H}_h(t)$ per unit volume in $[0, 1)^{k-1} \times (-\infty, \infty)$. For any measurable set A in $[0, 1)^{k-1} \times (-\infty, \infty)$, we now let

$$E[\mathcal{H}_h(t); A] = \#(\mathcal{H}_h(t) \cap A) - \mu(A)$$

denote the discrepancy of $\mathcal{H}_h(t)$ in A.

Consider a rectangular box of the form

$$B(x_1, \dots, x_{k-1}, y) = [0, x_1) \times \dots \times [0, x_{k-1}) \times [0, y) \subseteq [0, 1)^{k-1} \times [0, N).$$

As in the earlier proof of Theorem 11, we shall consider the smaller rectangular box $B(x_1^{(h)}, \dots, x_{k-1}^{(h)}, y)$ and, corresponding to (1.93), we have

$$|E[\mathcal{H}_h(t); B(x_1, \dots, x_{k-1}, y)] - E[\mathcal{H}; B(x_1^{(h)}, \dots, x_{k-1}^{(h)}, y)]| \le k - 1. \quad (1.97)$$

Next, we study $E[\mathcal{H}_h(t); B(x_1^{(h)}, \dots, x_{k-1}^{(h)}, y)]$ in detail, and require an analogue of the expansion (1.60). It is not difficult to see that

$$E[\mathcal{H}; B(x_1^{(h)}, \dots, x_{k-1}^{(h)}, y)] = \sum_{I_1 \in \mathscr{I}_1} \dots \sum_{I_{k-1} \in \mathscr{I}_{k-1}} E[\mathcal{H}_h(t); \mathbf{I} \times [0, y)],$$

where $\mathbf{I} = I_1 \times \dots \times I_{k-1}$ and where, for every $i = 1, \dots, k - 1$, $\mathscr{I}_i$ denotes the collection of elementary p_i-adic intervals in the union that makes up the interval $[0, x_i^{(h)})$ in Lemma 27.

Corresponding to Lemma 17, one can show that each summand

$$E[\mathcal{H}_h(t); \mathbf{I} \times [0, y)]$$

can be written in the form

$$\psi(p_1^{-s_1} \dots p_{k-1}^{-s_{k-1}} (t - \beta_{\mathbf{I}})) - \psi(p_1^{-s_1} \dots p_{k-1}^{-s_{k-1}} (t - \gamma_{\mathbf{I}})),$$

where the real numbers $\beta_{\mathbf{I}}$ and $\gamma_{\mathbf{I}}$ depend at most on $\mathbf{I}$ and y, and where, for every $i = 1, \dots, k - 1$, the elementary p_i-adic interval I_i has length $p_i^{-s_i}$. Making use of this, one can then proceed to show, corresponding to Lemma 18, that

$$\int_0^{M^h} E[\mathcal{H}_h(t); \mathbf{I}' \times [0, y)] E[\mathcal{H}_h(t); \mathbf{I}'' \times [0, y)] \, dt = O\left(M^h \prod_{i=1}^{k-1} p_i^{-|s_i' - s_i''|} \right)$$

for any $\mathbf{I}' = I'_1 \times \ldots \times I'_{k-1}$ and $\mathbf{I}'' = I''_1 \times \ldots \times I''_{k-1}$ where, for every $i = 1, \ldots, k-1$, the elementary p_i-adic intervals $I'_i, I''_i \in \mathscr{I}_i$ have lengths $p_i^{-s'_i}$ and $p_i^{-s''_i}$ respectively. One then goes on to show that

$$\int_0^{M^h} |E[\mathscr{H}; B(x_1^{(h)}, \ldots, x_{k-1}^{(h)}, y)]|^2 \, dt$$

$$\ll \sum_{I'_1 \in \mathscr{I}_1} \cdots \sum_{I'_{k-1} \in \mathscr{I}_{k-1}} \sum_{I''_1 \in \mathscr{I}_1} \cdots \sum_{I''_{k-1} \in \mathscr{I}_{k-1}} M^h \prod_{i=1}^{k-1} p_i^{-|s'_i - s''_i|}$$

$$\ll_k M^h h^{k-1}.$$

Taking the bound (1.97) into account and then integrating trivially with respect to $x_1, \ldots, x_{k-1}$, each over the interval $[0, 1)$, and with respect to y over the interval $[0, N)$, we conclude that

$$\int_0^N \int_0^1 \cdots \int_0^1 \int_0^{M^h} |E[\mathscr{H}_h(t); B(x_1, \ldots, x_{k-1}, y)]|^2 \, dt \, dx_1 \ldots dx_{k-1} \, dy$$

$$= \int_0^{M^h} \left(\int_0^N \int_0^1 \cdots \int_0^1 |E[\mathscr{H}_h(t); B(x_1, \ldots, x_{k-1}, y)]|^2 \, dx_1 \ldots dx_{k-1} \, dy \right) dt$$

$$\ll_k M^h h^{k-1} N.$$

Hence there exists $t^* \in [0, M^h)$ such that

$$\int_0^N \int_0^1 \cdots \int_0^1 |E[\mathscr{H}_h(t^*); B(x_1, \ldots, x_{k-1}, y)]|^2 \, dx_1 \ldots dx_{k-1} \, dy$$

$$\ll_k h^{k-1} N. \tag{1.98}$$

Finally, we note that the set $\mathscr{H}_h(t^*) \cap ([0, 1)^{k-1} \times [0, N))$ contains precisely N points. Rescaling in the vertical direction by a factor N^{-1}, we observe that the set

$$\mathscr{P}^* = \{(z_1, \ldots, z_{k-1}, N^{-1} z_k) : (z_1, \ldots, z_k) \in \mathscr{H}_h(t^*)\}$$

contains precisely N points in $[0, 1)^k$, and the estimate (1.98) now translates to

$$\int_{[0,1]^k} |D[\mathscr{P}^*; B(\mathbf{x})]|^2 \, d\mathbf{x} \ll_k h^{k-1} \ll_k (\log N)^{k-1},$$

in view of (1.96). This completes our brief sketch of the proof of Theorem 10.

1.14.2 Faure Point Sets

We now discuss Faure's contribution. Suppose again that we wish to study Theorem 10 or 11 in arbitrary dimension $k \geq 2$. Let p denote a prime such that[8] $p \geq k - 1$. For every non-negative integer $n \in \mathbf{N}_0$, we write

$$n = \sum_{j=1}^{\infty} a_j^{(1)} p^{j-1} \tag{1.99}$$

as a p-adic expansion. Then we write

$$c_n^{(1)} = \sum_{j=1}^{\infty} a_j^{(1)} p^{-j}. \tag{1.100}$$

For $i = 2, \ldots, k - 1$, we shall write

$$c_n^{(i)} = \sum_{j=1}^{\infty} a_j^{(i)} p^{-j}, \tag{1.101}$$

where the coefficients $a_j^{(i)}$ are defined inductively using the infinite upper triangular matrix

$$\mathscr{B} = \begin{bmatrix} \binom{0}{0} & \binom{1}{0} & \binom{2}{0} & \binom{3}{0} & \cdots \\ & \binom{1}{1} & \binom{2}{1} & \binom{3}{1} & \cdots \\ & & \binom{2}{2} & \binom{3}{2} & \cdots \\ & & & \binom{3}{3} & \cdots \\ & & & & \ddots \end{bmatrix} \tag{1.102}$$

made up of binomial coefficients.

It is convenient to use matrix multiplication modulo p to define the coefficients $a_j^{(i)}$ when $i > 1$. For every $i = 1, \ldots, k - 1$, consider the infinite column matrix

$$\mathbf{a}^{(i)} = \begin{bmatrix} a_1^{(i)} \\ a_2^{(i)} \\ a_3^{(i)} \\ a_4^{(i)} \\ \vdots \end{bmatrix}.$$

---

[8]The assumption that $p \geq k - 1$ cannot be relaxed, as noted by Chen [9].

Then for every $i = 2, \ldots, k - 1$, we write

$$\mathbf{a}^{(i)} \equiv \mathscr{B}\mathbf{a}^{(i-1)} \bmod p;$$

in other words, we write

$$
\begin{bmatrix}
a_1^{(i)} \\
a_2^{(i)} \\
a_3^{(i)} \\
a_4^{(i)} \\
\vdots
\end{bmatrix}
\equiv
\begin{bmatrix}
\binom{0}{0} & \binom{1}{0} & \binom{2}{0} & \binom{3}{0} & \cdots \\
& \binom{1}{1} & \binom{2}{1} & \binom{3}{1} & \cdots \\
& & \binom{2}{2} & \binom{3}{2} & \cdots \\
& & & \binom{3}{3} & \cdots \\
& & & & \ddots
\end{bmatrix}
\begin{bmatrix}
a_1^{(i-1)} \\
a_2^{(i-1)} \\
a_3^{(i-1)} \\
a_4^{(i-1)} \\
\vdots
\end{bmatrix}
\bmod p.
$$

For every $n \in \mathbf{N}_0$, write

$$\mathbf{c}_n = (c_n^{(1)}, \ldots, c_n^{(k-1)}).$$

The set

$$\mathscr{F} = \{(\mathbf{c}_n, n) : n = 0, 1, 2, \ldots\}$$

in $[0, 1)^{k-1} \times [0, \infty)$ is usually called a Faure point set.

Analogous to Lemma 25, we have the following result.

Lemma 28. *For all non-negative integers $s_1, \ldots, s_{k-1}, \ell_1, \ldots, \ell_{k-1}$ and m such that $\ell_i < p^{s_i}$ holds for every $i = 1, \ldots, k - 1$, the rectangular box*

$$\prod_{i=1}^{k-1} [\ell_i \, p^{-s_i}, (\ell_i + 1) p^{-s_i}) \times [m p^{s_1 + \ldots + s_{k-1}}, (m + 1) p^{s_1 + \ldots + s_{k-1}}) \qquad (1.103)$$

contains precisely one point of the Faure point set $\mathscr{F}$.

To prove Lemma 28, we need a simple result concerning the matrix $\mathscr{B}$.

Lemma 29. *For the matrix $\mathscr{B}$ given by (1.102), we have, for every $i = 1, \ldots, k-1$,*

$$
\mathscr{B}^{i-1} =
\begin{bmatrix}
\binom{0}{0} & \binom{1}{0}(i - 1) & \binom{2}{0}(i - 1)^2 & \binom{3}{0}(i - 1)^3 & \cdots \\
& \binom{1}{1} & \binom{2}{1}(i - 1) & \binom{3}{1}(i - 1)^2 & \cdots \\
& & \binom{2}{2} & \binom{3}{2}(i - 1) & \cdots \\
& & & \binom{3}{3} & \cdots \\
& & & & \ddots
\end{bmatrix}.
$$

Proof (of Lemma 28). Suppose that suitable integers $s_1, \ldots, s_{k-1}, \ell_1, \ldots, \ell_{k-1}$ and m are chosen and fixed. For a point $(\mathbf{c}_n, n)$ to lie in the rectangle (1.103), we must have

$$c_n^{(i)} \in [\ell_i \, p^{-s_i}, (\ell_i + 1) p^{-s_i}) \tag{1.104}$$

for every $i = 1, \ldots, k - 1$, as well as

$$n \in [m p^{s_1 + \ldots + s_{k-1}}, (m + 1) p^{s_1 + \ldots + s_{k-1}}). \tag{1.105}$$

Comparing (1.99) and (1.105), it is clear that the value of the coefficient $a_j^{(1)}$ for every $j > s_1 + \ldots + s_{k-1}$ is uniquely determined. It therefore remains to show that there is one choice of the vector

$$(a_1^{(1)}, \ldots, a_{s_1 + \ldots + s_{k-1}}^{(1)})$$

that satisfies the requirement (1.104) for every $i = 1, \ldots, k - 1$.

Note next that for every $i = 1, \ldots, k - 1$, we have

$$
\begin{bmatrix} a_1^{(i)} \\ a_2^{(i)} \\ a_3^{(i)} \\ a_4^{(i)} \\ \vdots \end{bmatrix}
\equiv
\begin{bmatrix}
\binom{0}{0} & \binom{1}{0}(i-1) & \binom{2}{0}(i-1)^2 & \binom{3}{0}(i-1)^3 & \cdots \\
& \binom{1}{1} & \binom{2}{1}(i-1) & \binom{3}{1}(i-1)^2 & \cdots \\
& & \binom{2}{2} & \binom{3}{2}(i-1) & \cdots \\
& & & \binom{3}{3} & \cdots \\
& & & & \ddots
\end{bmatrix}
\begin{bmatrix} a_1^{(1)} \\ a_2^{(1)} \\ a_3^{(1)} \\ a_4^{(1)} \\ \vdots \end{bmatrix} \bmod p.
$$

Let us consider the p-adic expansion

$$\ell_i \, p^{-s_i} = \beta_1^{(i)} p^{-1} + \ldots + \beta_{s_i}^{(i)} p^{-s_i}.$$

If (1.104) holds, then in view of (1.100) or (1.101), we must have $a_j^{(i)} = \beta_j^{(i)}$ for every $j = 1, \ldots, s_i$. This can be summarized by writing

$$
\mathscr{W}_i
\begin{bmatrix} a_1^{(1)} \\ a_2^{(1)} \\ a_3^{(1)} \\ a_4^{(1)} \\ \vdots \end{bmatrix}
\equiv
\begin{bmatrix} \beta_1^{(i)} \\ \beta_2^{(i)} \\ \beta_3^{(i)} \\ \vdots \\ \beta_{s_i}^{(i)} \end{bmatrix} \bmod p, \tag{1.106}
$$

where the matrix $\mathscr{W}_i$ contains precisely the first s_i rows of the matrix $\mathscr{B}^{i-1}$. Now recall that $a_j^{(1)}$ are already uniquely determined for every $j > S = s_1 + \ldots + s_{k-1}$

by (1.105), and clearly there are at most finitely many non-zero terms among these. The system (1.106) can therefore be simplified to one of the form

$$
\mathcal{V}_i \begin{bmatrix} a_1^{(1)} \\ a_2^{(1)} \\ a_3^{(1)} \\ \vdots \\ a_S^{(1)} \end{bmatrix} \equiv \begin{bmatrix} \gamma_1^{(i)} \\ \gamma_2^{(i)} \\ \gamma_3^{(i)} \\ \vdots \\ \gamma_{s_i}^{(i)} \end{bmatrix} \bmod p, \tag{1.107}
$$

where the matrix $\mathcal{V}_i$ contains precisely the first S columns of the matrix $\mathcal{W}_i$. On combining (1.107) for every $i = 1, \ldots, k-1$, we arrive at a system of S linear congruences in the S variables $a_1^{(1)}, \ldots, a_S^{(1)}$, with the matrix given by

$$
\mathcal{V} = \begin{bmatrix} \mathcal{V}_1 \\ \vdots \\ \mathcal{V}_{k-1} \end{bmatrix}.
$$

It is not difficult to see that for every $i = 1, \ldots, k-1$, we have

$$
\mathcal{V}_i = \begin{bmatrix} \binom{0}{0} & \binom{1}{0}(i-1) & \binom{2}{0}(i-1)^2 & \cdots & \binom{S-1}{0}(i-1)^{S-1} \\ & \binom{1}{1} & \binom{2}{1}(i-1) & \cdots & \binom{S-1}{1}(i-1)^{S-2} \\ & & \ddots & & \vdots \\ & & & \binom{s_i-1}{s_i-1} \cdots & \binom{S-1}{s_i-1}(i-1)^{S-s_i} \end{bmatrix},
$$

a matrix with s_i rows and S columns. It follows that the matrix $\mathcal{V}$ is of generalized Vandermonde type, with determinant

$$
\prod_{1 \le i' < i'' \le k-1} (i'' - i')^{s_{i'} s_{i''}} \not\equiv 0 \bmod p,
$$

in view of the assumption that $p \ge k-1$. Hence the system of S linear congruences in the S variables $a_1^{(1)}, \ldots, a_S^{(1)}$ has unique solution. Recall once again that the coefficients $a_j^{(1)}$ are already uniquely determined for every $j > S$, we conclude that there is precisely one value of n that satisfies all the requirements.   □

The following analogue of Lemma 26 is a simple consequence of Lemma 29.

Lemma 30. *For all non-negative integers $s_1, \ldots, s_{k-1}$ and $\ell_1, \ldots, \ell_{k-1}$ satisfying $\ell_i < p^{s_i}$ for every $i = 1, \ldots, k-1$, and for every real number $y > 0$, we have*

$$
\left| E \left[\mathcal{F}; \prod_{i=1}^{k-1} [\ell_i \, p^{-s_i}, (\ell_i + 1) p^{-s_i}) \times [0, y) \right] \right| \le 1.
$$

To study Theorem 11, let $N \geq 2$ be a given integer. It follows at once from the definition of $\mathscr{F}$ that the set

$$\mathscr{F}_0 = \mathscr{F} \cap ([0, 1)^{k-1} \times [0, N))$$

contains precisely N points. Let the integer h be determined uniquely by

$$p^{h-1} < N \leq p^h.$$

We can now deduce Theorem 11 from Lemmas 28 and 30 in a way similar to our deduction of the same result from Lemmas 25 and 26 in Sect. 1.14.1, noting that Lemma 27 remains valid with p_i replaced by p. Indeed, rescaling the second coordinate of the points of $\mathscr{F}_0$ by a factor N^{-1}, we obtain a set

$$\mathscr{P} = \{(\mathbf{c}_n, N^{-1}n) : n = 0, 1, 2, \ldots, N - 1\},$$

of precisely N points in $[0, 1)^k$ and which satisfies the conclusion of Theorem 11.

1.14.3 A General Point Set and a Digit Shift Argument

In this section, we briefly describe a rather general digit shift argument developed by Chen [9] which enables us to establish Theorem 10 using Halton point sets discussed in Sect. 1.14.1 or Faure point sets discussed in Sect. 1.14.2. Recall that these point sets satisfy Lemmas 25 and 28 respectively.

Let $p_1 \leq \ldots \leq p_{k-1}$ be primes, not necessarily distinct, and let h be a non-negative integer. We shall say that a set of the form

$$\mathscr{L} = \{(\mathbf{c}_n, n) : n = 0, 1, 2, \ldots\} \tag{1.108}$$

in $[0, 1)^{k-1} \times [0, \infty)$ is a 1-set of order h with respect to the primes $p_1, \ldots, p_{k-1}$ if the following condition is satisfied. For all non-negative integers $s_1, \ldots, s_{k-1}$, $\ell_1, \ldots, \ell_{k-1}$ and m satisfying $s_i \leq h$ and $\ell_i < p_i^{s_i}$ for every $i = 1, \ldots, k - 1$, the rectangular box

$$\prod_{i=1}^{k-1} [\ell_i \, p_i^{-s_i}, (\ell_i + 1)p_i^{-s_i}) \times \left[m \prod_{i=1}^{k-1} p_i^{s_i}, (m + 1) \prod_{i=1}^{k-1} p_i^{s_i} \right)$$

contains precisely one point of $\mathscr{L}$.

If the primes $p_1, \ldots, p_{k-1}$ are distinct, then the Halton set $\mathscr{H}$ is a 1-set of every non-negative order with respect to $p_1, \ldots, p_{k-1}$. If the primes $p_1, \ldots, p_{k-1}$ are all identical and equal to p, then the Faure set $\mathscr{F}$ is 1-set of every non-negative order with respect to $p, \ldots, p$, provided that $p \geq k - 1$.

The property below follows almost immediately from the definition.

Lemma 31. *Suppose that* h *be a non-negative integer, and that* $\mathscr{Z}$ *is a* 1-*set of order* h *with respect to the primes* $p_1, \ldots, p_{k-1}$. *Then for all non-negative integers* $s_1, \ldots, s_{k-1}$ *and* $\ell_1, \ldots, \ell_{k-1}$ *satisfying* $s_i \le h$ *and* $\ell_i < p_i^{s_i}$ *for every* $i = 1, \ldots, k-1$, *and for every real number* $y > 0$, *we have*

$$\left| E\left[\mathscr{Z}; \prod_{i=1}^{k-1} [\ell_i \, p_i^{-s_i}, (\ell_i + 1)p_i^{-s_i}) \times [0, y) \right] \right| \le 1.$$

Let $N \ge 2$ be a given integer, and let the integer h be determined uniquely by

$$p_1^{h-1} < N \le p_1^h. \tag{1.109}$$

For any 1-set (1.108) of order h with respect to the primes $p_1, \ldots, p_{k-1}$, the set

$$\mathscr{Z}_0 = \mathscr{Z} \cap ([0, 1)^{k-1} \times [0, N))$$

contains precisely N points. Then it can be shown easily that the set

$$\mathscr{P} = \{(\mathbf{c}_n, N^{-1}n) : n = 0, 1, 2, \ldots, N - 1\},$$

of precisely N points in $[0, 1)^k$ and which satisfies the conclusion of Theorem 11.

To study Theorem 10, we again choose the integer h to satisfy (1.109). However, we need to modify the 1-set $\mathscr{Z}$.

Let $\mathscr{M}$ denote the collection of all $(k-1) \times h$ matrices $\mathbf{T} = (t_{i,j})$ where, for every $i = 1, \ldots, k-1$ and $j = 1, \ldots, h$, the entry $t_{i,j} \in \{0, 1, 2, \ldots, p_i - 1\}$. Clearly the collection $\mathscr{M}$ has $(p_1 \ldots p_{k-1})^h$ elements.

For every $n = 0, 1, 2, \ldots$, let us write

$$\mathbf{c}_n = (c_1(n), \ldots, c_{k-1}(n)).$$

For every $i = 1, \ldots, k-1$, we consider the base p_i expansion

$$c_i(n) = 0.a_{i,1}a_{i,2} \ldots a_{i,h}a_{i,h+1} \ldots.$$

For every $\mathbf{T} \in \mathscr{M}$ and every $n = 0, 1, 2, \ldots,$, we shall write

$$\mathbf{c}_n^{\mathbf{T}} = (c_1^{\mathbf{T}}(n), \ldots, c_{k-1}^{\mathbf{T}}(n)),$$

where, for every $i = 1, \ldots, k-1$, we have

$$c_i^{\mathbf{T}}(n) = 0.(a_{i,1} \oplus t_{i,1})(a_{i,2} \oplus t_{i,2}) \ldots (a_{i,h} \oplus t_{i,h})a_{i,h+1} \ldots,$$

where $\oplus$ denotes addition modulo p_i. It is not difficult to show that the shifted set

$$\mathscr{L}^{\mathbf{T}} = \{(\mathbf{c}_n^{\mathbf{T}}, n) : n = 0, 1, 2, \ldots\}$$

in $[0, 1)^{k-1} \times [0, \infty)$ is also a 1-set of order h with respect to the primes $p_1, \ldots, p_{k-1}$.
  Consider a rectangular box of the form

$$B(x_1, \ldots, x_{k-1}, y) = [0, x_1) \times \ldots \times [0, x_{k-1}) \times [0, y) \subseteq [0, 1)^{k-1} \times [0, N).$$

As in the earlier proof of Theorem 10, we shall again consider the smaller rectangular box $B(x_1^{(h)}, \ldots, x_{k-1}^{(h)}, y)$, where, for every $i = 1, \ldots, k - 1$, we replace the point x_i by $x_i^{(h)} = p_i^{-h}[p_i^h x_i]$, the greatest integer multiple of p_i^{-h} not exceeding x_i. Then for every $\mathbf{T} \in \mathscr{M}$, we have

$$|E[\mathscr{L}^{\mathbf{T}}; B(x_1, \ldots, x_{k-1}, y)] - E[\mathscr{L}^{\mathbf{T}}; B(x_1^{(h)}, \ldots, x_{k-1}^{(h)}, y)]| \le k - 1,$$

so it remains to study $E[\mathscr{L}^{\mathbf{T}}; B(x_1^{(h)}, \ldots, x_{k-1}^{(h)}, y)]$ in detail. It can be shown that

$$\sum_{\mathbf{T} \in \mathscr{M}} |E[\mathscr{L}^{\mathbf{T}}; B(x_1^{(h)}, \ldots, x_{k-1}^{(h)}, y)]|^2 \ll_k (p_1 \ldots p_{k-1})^h h^{k-1},$$

from which it follows that

$$\int_0^N \int_0^1 \cdots \int_0^1 \left(\sum_{\mathbf{T} \in \mathscr{M}} |E[\mathscr{L}^{\mathbf{T}}; B(x_1^{(h)}, \ldots, x_{k-1}^{(h)}, y)]|^2 \right) dx_1 \ldots dx_{k-1}\, dy$$

$$= \sum_{\mathbf{T} \in \mathscr{M}} \left(\int_0^N \int_0^1 \cdots \int_0^1 |E[\mathscr{L}^{\mathbf{T}}; B(x_1^{(h)}, \ldots, x_{k-1}^{(h)}, y)]|^2 dx_1 \ldots dx_{k-1}\, dy \right)$$

$$\ll_k (p_1 \ldots p_{k-1})^h h^{k-1} N.$$

Hence there exists $\mathbf{T}^* \in \mathscr{M}$ such that

$$\int_0^N \int_0^1 \cdots \int_0^1 |E[\mathscr{L}^{\mathbf{T}^*}; B(x_1, \ldots, x_{k-1}, y)]|^2 dx_1 \ldots dx_{k-1}\, dy \ll_k h^{k-1} N.$$

Finally, we note that the set $\mathscr{L}^{\mathbf{T}^*} \cap ([0, 1)^{k-1} \times [0, N))$ contains precisely N points. Rescaling in the vertical direction by a factor N^{-1}, we observe that the set

$$\mathscr{P}^* = \{(z_1, \ldots, z_{k-1}, N^{-1} z_k) : (z_1, \ldots, z_k) \in \mathscr{L}^{\mathbf{T}^*}\}$$

contains precisely N points in $[0, 1)^k$, and satisfies the conclusion of Theorem 10.

1.15   Group Structure and p-adic Fourier–Walsh Analysis

In Sect. 1.13, we exploit the group structure of the van der Corput set $\mathscr{P}(2^h)$ to sketch a proof of Theorem 10 for $k = 2$. The central argument there is to use Fourier–Walsh analysis to show that an approximation $D^{(h)}[\mathscr{P}(2^h); B(\mathbf{x})]$ of the discrepancy function $D[\mathscr{P}(2^h); B(\mathbf{x})]$ satisfies the identity (1.85) which involves digit shifts. Under certain hypothetical orthogonality assumptions, we can further deduce the simpler identity

$$\int_{[0,1]^2} |D^{(h)}[\mathscr{P}(2^h); B(\mathbf{x})]|^2 \, d\mathbf{x} = 4^h \sum_{\mathbf{l} \in L(h)} \int_{[0,1]^2} |\widehat{\chi}_{\mathbf{l}}(\mathbf{x})|^2 \, d\mathbf{x}.$$

Unfortunately, these hypothetical orthogonality assumptions do not hold.

To have a better understanding of the underlying ideas, it is necessary to study p-adic versions of the analysis carried out earlier.

For simplicity, let us again restrict our attention to Theorem 10 for $k = 2$. Let p be a prime, and consider the base p van der Corput point set

$$\mathscr{P}(p^h) = \{(0.a_1a_2a_3 \ldots a_h, 0.a_h \ldots a_3a_2a_1) : a_1, \ldots, a_h \in \{0, 1, \ldots, p-1\}\}.$$

This is a finite abelian group isomorphic to the group $\mathbf{Z}_p^h$. We shall make use of the characters of these groups. These are the base p Walsh functions, usually known as the Chrestenson or Chrestenson–Levy functions. For simplicity, we refer to them all as Walsh functions here.

To define these Walsh functions, we first consider p-ary representation of any integer $\ell \in \mathbf{N}_0$, written uniquely in the form

$$\ell = \sum_{i=1}^{\infty} \lambda_i(\ell) p^{i-1}, \tag{1.110}$$

where the coefficient $\lambda_i(\ell) \in \{0, 1, \ldots, p-1\}$ for every $i \in \mathbf{N}$. On the other hand, every real number $y \in [0, 1)$ can be represented in the form

$$y = \sum_{i=1}^{\infty} \eta_i(y) p^{-i}, \tag{1.111}$$

where the coefficient $\eta_i(y) \in \{0, 1, \ldots, p-1\}$ for every $i \in \mathbf{N}$. This representation is unique if we agree that the series in (1.111) is finite for every $y = mp^{-s}$ where $s \in \mathbf{N}_0$ and $m \in \{0, 1, \ldots, p^s - 1\}$.

For every $\ell \in \mathbf{N}_0$ of the form (1.110), we define the Walsh function $w_\ell : [0, 1) \to \mathbf{R}$ by writing

$$w_\ell(y) = e_p \left(\sum_{i=1}^{\infty} \lambda_i(\ell) \eta_i(y) \right), \tag{1.112}$$

where $e_p(z) = e^{2\pi i z/p}$ for every real number z. Since (1.110) is essentially a finite sum, the Walsh function is well defined, and takes the p-th roots of unity as its values. It is easy to see that $w_0(y) = 1$ for every $y \in [0, 1)$. It is well known that under the inner product

$$\langle w_k, w_\ell \rangle = \int_0^1 w_k(y) \overline{w_\ell(y)} \, dy,$$

the collection of Walsh functions form an orthonormal basis of $L^2[0, 1]$.

The operation $\oplus$ defined modulo 2 previously can easily be suitably modified to an operation modulo p. Then (1.75) and (1.76) remain valid in this new setting.

As before, we shall use Fourier–Walsh analysis to study characteristic functions of the form $\chi_{[0,x)}(y)$. We have the Fourier–Walsh series

$$\chi_{[0,x)}(y) \sim \sum_{\ell=0}^{\infty} \widetilde{\chi}_\ell(x) \overline{w_\ell(y)},$$

where, for every $\ell \in \mathbf{N}_0$, the Fourier–Walsh coefficients are given by

$$\widetilde{\chi}_\ell(x) = \int_0^x w_\ell(y) \, dy.$$

In particular, we have $\widetilde{\chi}_0(x) = x$ for every $x \in [0, 1)$. Again, as before, instead of using the full Fourier–Walsh series, we shall truncate it and use the approximation

$$\chi_{[0,x)}^{(h)}(y) = \sum_{\ell=0}^{p^h-1} \widetilde{\chi}_\ell(x) \overline{w_\ell(y)}.$$

This approximation in turn leads to the approximation

$$\chi_{B(\mathbf{x})}^{(h)}(\mathbf{y}) = \chi_{[0,x_1)}^{(h)}(y_1) \chi_{[0,x_2)}^{(h)}(y_2) = \sum_{\ell_1=0}^{p^h-1} \sum_{\ell_2=0}^{p^h-1} \widetilde{\chi}_\mathbf{l}(\mathbf{x}) \overline{W_\mathbf{l}(\mathbf{y})}$$

of the characteristic function $\chi_{B(\mathbf{x})}(\mathbf{y})$. Here $\mathbf{l} = (\ell_1, \ell_2)$,

$$\widetilde{\chi}_\mathbf{l}(\mathbf{x}) = \widetilde{\chi}_{\ell_1}(x_1) \widetilde{\chi}_{\ell_2}(x_2) \quad \text{and} \quad W_\mathbf{l}(\mathbf{y}) = w_{\ell_1}(y_1) w_{\ell_2}(y_2).$$

Consequently, we approximate the discrepancy function

$$D[\mathscr{P}(p^h); B(\mathbf{x})] = \sum_{\mathbf{p} \in \mathscr{P}(p^h)} \chi_{B(\mathbf{x})}(\mathbf{p}) - p^h x_1 x_2$$

by

$$D^{(h)}[\mathscr{P}(p^h); B(\mathbf{x})] = \sum_{\mathbf{p} \in \mathscr{P}(p^h)} \chi_{B(\mathbf{x})}^{(h)}(\mathbf{p}) - p^h x_1 x_2$$

$$= \sum_{\mathbf{p} \in \mathscr{P}(p^h)} \sum_{\ell_1=0}^{p^h-1} \sum_{\ell_2=0}^{p^h-1} \widetilde{\chi}_{\mathbf{l}}(\mathbf{x}) W_{\mathbf{l}}(\mathbf{p}) - p^h \widetilde{\chi}_{\mathbf{0}}(\mathbf{x})$$

$$= \sum_{\substack{\ell_1=0 \\ (\ell_1,\ell_2) \neq (0,0)}}^{p^h-1} \sum_{\ell_2=0}^{p^h-1} \left(\sum_{\mathbf{p} \in \mathscr{P}(p^h)} W_{\mathbf{l}}(\mathbf{p}) \right) \widetilde{\chi}_{\mathbf{l}}(\mathbf{x}),$$

noting that

$$\sum_{\mathbf{p} \in \mathscr{P}(p^h)} W_{\mathbf{0}}(\mathbf{p}) = \#\mathscr{P}(p^h) = p^h.$$

It is well known in the theory of abelian groups that the sum

$$\sum_{\mathbf{p} \in \mathscr{P}(p^h)} W_{\mathbf{l}}(\mathbf{p}) \in \{0, p^h\}.$$

We therefore need to have some understanding on the set

$$L(h) = \left\{ \mathbf{l} \in [0, p^h) \times [0, p^h) : \mathbf{l} \neq \mathbf{0} \text{ and } \sum_{\mathbf{p} \in \mathscr{P}(p^h)} W_{\mathbf{l}}(\mathbf{p}) = p^h \right\}.$$

Then

$$D^{(h)}[\mathscr{P}(p^h); B(\mathbf{x})] = p^h \sum_{\mathbf{l} \in L(h)} \widetilde{\chi}_{\mathbf{l}}(\mathbf{x}).$$

We have the following special case of a general result of Skriganov [34].

Lemma 32. *Suppose that the prime p satisfies $p \geq 8$. Then the functions $\widetilde{\chi}_{\mathbf{l}}(\mathbf{x})$, where $\mathbf{l} \in L(h)$, are orthogonal, so that*

$$\int_{[0,1]^2} |D^{(h)}[\mathscr{P}(p^h); B(\mathbf{x})]|^2 \, d\mathbf{x} = p^{2h} \sum_{\mathbf{l} \in L(h)} \int_{[0,1]^2} |\widetilde{\chi}_{\mathbf{l}}(\mathbf{x})|^2 \, d\mathbf{x}. \qquad (1.113)$$

To progress further, we need to estimate each of the integrals

$$\int_{[0,1]^2} |\widetilde{\chi}_{\mathbf{i}}(\mathbf{x})|^2 \, d\mathbf{x} = \left(\int_0^1 |\widetilde{\chi}_{\ell_1}(x_1)|^2 \, dx_1 \right) \left(\int_0^1 |\widetilde{\chi}_{\ell_2}(x_2)|^2 \, dx_2 \right) \qquad (1.114)$$

on the right hand side of (1.113).

Lemma 33. *We have*

$$\int_0^1 |\widetilde{\chi}_0(x)|^2 \, dx = \frac{1}{4} + \frac{1}{4(p^2 - 1)} \sum_{j=1}^{p-1} \csc^2 \frac{\pi j}{p}. \qquad (1.115)$$

Furthermore, for every $\ell \in \mathbf{N}$, we have

$$\int_0^1 |\widetilde{\chi}_\ell(x)|^2 \, dx = p^{-2\rho(\ell)} \left(\frac{1}{2} \csc^2 \frac{\pi \lambda(\ell)}{p} - \frac{1}{4} + \frac{1}{4(p^2 - 1)} \sum_{j=1}^{p-1} \csc^2 \frac{\pi j}{p} \right),$$

$$(1.116)$$

where

$$\rho(\ell) = \begin{cases} 0, & \text{if } \ell = 0, \\ \max\{i \in \mathbf{N} : \lambda_i(\ell) \neq 0\}, & \text{if } \ell \in \mathbf{N}, \end{cases}$$

denotes the position of the leading coefficient of ℓ given by (1.110) and $\lambda(\ell) = \lambda_{\rho(\ell)}(\ell)$ denotes its value.

Proof. We have the Fine–Price formula, that for every $\ell \in \mathbf{N}_0$,

$$\widetilde{\chi}_\ell(x) = p^{-\rho(\ell)} u_\ell(x), \qquad (1.117)$$

where

$$u_0(x) = \frac{1}{2} w_0(x) + \sum_{i=1}^{\infty} p^{-i} \sum_{j=1}^{p-1} \zeta^j (1 - \zeta^j)^{-1} w_{jp^{i-1}}(x), \qquad (1.118)$$

and where for every $\ell \in \mathbf{N}$,

$$u_\ell(x) = (1 - \zeta^{\lambda(\ell)})^{-1} w_{\tau(\ell)}(x) + \left(\frac{1}{2} - (1 - \zeta^{\lambda(\ell)})^{-1} \right) w_\ell(x)$$

$$+ \sum_{i=1}^{\infty} p^{-i} \sum_{j=1}^{p-1} \zeta^j (1 - \zeta^j)^{-1} w_{\ell + jp^{\rho(\ell)+i-1}}(x). \qquad (1.119)$$

Here $\tau(\ell) = \ell - \lambda(\ell)p^{\rho(\ell)-1}$, and $\zeta = e^{2\pi i/p}$ is a primitive p-th root of unity. For details, see Fine [17] and Price [28]. The right hand side of (1.119) is a linear combination of distinct Walsh functions. It follows that for every $\ell \in \mathbf{N}$, we have

$$\int_0^1 |u_\ell(x)|^2 \, dx = \frac{1}{(1 - \zeta^{\lambda(\ell)})(1 - \zeta^{-\lambda(\ell)})} + \left(\frac{1}{2} - \frac{1}{1 - \zeta^{\lambda(\ell)}}\right)\left(\frac{1}{2} - \frac{1}{1 - \zeta^{-\lambda(\ell)}}\right)$$

$$+ \sum_{i=1}^\infty p^{-2i} \sum_{j=1}^{p-1} |1 - \zeta^j|^{-2}$$

$$= 2|1 - \zeta^{\lambda(\ell)}|^{-2} - \frac{1}{4} + \frac{1}{p^2 - 1} \sum_{j=1}^{p-1} |1 - \zeta^j|^{-2}. \tag{1.120}$$

The identity (1.116) follows on combining (1.117) and (1.120) with the observation

$$|1 - \zeta^j|^2 = \left(1 - \cos\frac{2\pi j}{p}\right)^2 + \sin^2\frac{2\pi j}{p} = 4\sin^2\frac{\pi j}{p}. \tag{1.121}$$

Similarly, we have

$$\int_0^1 |u_0(x)|^2 \, dx = \frac{1}{4} + \sum_{i=1}^\infty p^{-2i} \sum_{j=1}^{p-1} |1 - \zeta^j|^{-2} = \frac{1}{4} + \frac{1}{p^2 - 1} \sum_{j=1}^{p-1} |1 - \zeta^j|^{-2}. \tag{1.122}$$

The identity (1.115) follows on combining (1.117), (1.121) and (1.122). □

Lemma 34. *For every $\ell \in \mathbf{N}_0$, we have*

$$\int_0^1 |\widetilde{\chi}_\ell(x)|^2 \, dx \le \frac{p^{2-2\rho(\ell)}}{4}.$$

Proof. Suppose first of all that $\ell \ne 0$. Then using the inequality that

$$\csc^2\frac{\pi j}{p} \le \frac{p^2}{4}$$

for every $j = 1, \ldots, p - 1$, we see from (1.116) that

$$\int_0^1 |\widetilde{\chi}_\ell(x)|^2 \, dx \le p^{-2\rho(\ell)}\left(\frac{p^2}{8} + \frac{1}{4} + \frac{p^2(p-1)}{16(p^2-1)}\right) \le \frac{p^{2-2\rho(\ell)}}{4}.$$

On the other hand, it follows similarly from (1.115) that

$$\int_0^1 |\widetilde{\chi}_0(x)|^2 \, dx \le \frac{1}{4} + \frac{p^2(p-1)}{16(p^2-1)} \le \frac{p^2}{4} = \frac{p^{2-2\rho(0)}}{4}$$

as required.   □

Combining (1.114) and Lemma 34, we conclude that

$$\int_{[0,1]^2} |\widetilde{\chi}_l(\mathbf{x})|^2 \, d\mathbf{x} \le \frac{p^{4-2\rho(l)}}{16},$$

where $\rho(\mathbf{l}) = \rho(\ell_1) + \rho(\ell_2)$. Thus we need to estimate the sum

$$\sum_{l \in L(h)} p^{-2\rho(l)}. \tag{1.123}$$

Here $\rho(\mathbf{l})$ is a non-Hamming weight that arises from the Rosenblum–Tsfasman weight in coding theory. The idea here is that if the distribution dual to $\mathscr{P}(p^h)$ has sufficiently large Rosenblum–Tsfasman weight, then we can obtain a good estimate for the sum (1.123).

For a brief discussion on how we may complete our proof, the reader is referred to the paper of Chen and Skriganov [13].

1.16   Explicit Constructions and Orthogonality

The first proof of Theorem 10 for arbitrary $k \ge 2$ by Roth [31] is probabilistic in nature, as are the subsequent proofs by Chen [9] and Skriganov [33]. The disadvantage of such probabilistic arguments is that while we can show that a good point set exists, we cannot describe it explicitly.

On the other hand, the proof by Davenport [15] of Theorem 10 in dimension $k = 2$ is not probabilistic in nature, and one can describe the point set explicitly. However, finding explicit constructions in dimensions $k \ge 3$ turns out to be rather hard. Its eventual solution by Chen and Skriganov [11] is based on the observation that provided that the prime p is sufficiently large, then the functions $\widetilde{\chi}_l(\mathbf{x})$, where $\mathbf{l} \in L(h)$, are *quasi-orthogonal*, so that some weaker version of Lemma 32 in arbitrary dimensions holds.

However, if we are not able to establish any orthogonality or quasi-orthogonality, then our techniques thus far fail to give any explicit constructions in dimensions $k \ge 3$. To establish an appropriate upper bound, we may resort to digit shifts, and our argument is underpinned by the general result below for arbitrary dimensions $k \ge 2$ for some suitably defined Walsh function $W_l(\mathbf{t})$.

Lemma 35. *For every* $\mathbf{l}', \mathbf{l}'' \in \mathbf{N}_0^k$, *we have*

$$\sum_{\mathbf{t} \in \mathbf{Z}_p^{kh}} W_{\mathbf{l}'}(\mathbf{t}) W_{\mathbf{l}''}(\mathbf{t}) = \begin{cases} p^{kh}, & \text{if } \mathbf{l}' = \mathbf{l}'', \\ 0, & \text{otherwise.} \end{cases}$$

This result can be viewed as an orthogonality result. We may therefore conclude that orthogonality or quasi-orthogonality in some form is central to our upper bound arguments here, whether we consider explicit constructions or otherwise.

Acknowledgements The research of the second author has been supported by RFFI Project No. 08-01-00182.

References

1. J. Beck, W.W.L. Chen, Note on irregularities of distribution. II. Proc. Lond. Math. Soc. III. Ser. **61**(2), 251–272 (1990). doi:10.1112/plms/s3-61.2.251
2. J. Beck, W.W.L. Chen, Irregularities of point distribution relative to convex polygons. III. J. Lond. Math. Soc. II. Ser. **56**(2), 222–230 (1997). doi:10.1112/S0024610797005267
3. J. Beck, Roth's estimate of the discrepancy of integer sequences is nearly sharp. Combinatorica **1**, 319–325 (1981). doi:10.1007/BF02579452
4. J. Beck, Irregularities of distribution. I. Acta Math. **159**, 1–49 (1987). doi:10.1007/BF02392553
5. J. Beck, W.W.L. Chen, *Irregularities of Distribution*. Cambridge Tracts in Mathematics, vol. 89 (Cambridge University Press, Cambridge, 1987)
6. D. Bilyk, M.T. Lacey, On the small ball inequality in three dimensions. Duke Math. J. **143**(1), 81–115 (2008). doi:10.1215/00127094-2008-016
7. D. Bilyk, M.T. Lacey, A. Vagharshakyan, On the small ball inequality in all dimensions. J. Funct. Anal. **254**(9), 2470–2502 (2008). doi:10.1016/j.jfa.2007.09.010
8. W.W.L. Chen, On irregularities of distribution. Mathematika **27**, 153–170 (1980). doi:10.1112/S0025579300010044
9. W.W.L. Chen, On irregularities of distribution. II. Q. J. Math. Oxf. II. Ser. **34**, 257–279 (1983). doi:10.1093/qmath/34.3.257
10. W.W.L. Chen, Fourier techniques in the theory of irregularities of point distribution, in *Fourier Analysis and Convexity*, ed. by L. Brandolini, L. Colzani, A. Iosevich, G. Travaglini, Applied and Numerical Harmonic Analysis (Birkhäuser, Boston, MA, 2004), pp. 59–82
11. W.W.L. Chen, M.M. Skriganov, Explicit constructions in the classical mean squares problem in irregularities of point distribution. J. Reine Angew. Math. **545**, 67–95 (2002). doi:10.1515/crll.2002.037
12. W.W.L. Chen, G. Travaglini, Discrepancy with respect to convex polygons. J. Complexity **23**(4–6), 662–672 (2007). doi:10.1016/j.jco.2007.03.006
13. W.W.L. Chen, M.M. Skriganov, Orthogonality and digit shifts in the classical mean squares problem in irregularities of point distribution, in *Diophantine Approximation: Festschrift for Wolfgang Schmidt*, ed. by H.P. Schlickewei, K. Schmidt, R.F. Tichy, Developments in Mathematics, vol. 16 (Springer, Wien, 2008), pp. 141–159. doi:10.1007/978-3-211-74280-8_7
14. W.W.L. Chen, G. Travaglini, Deterministic and probabilistic discrepancies. Ark. Mat. **47**(2), 273–293 (2009). doi:10.1007/s11512-008-0091-z
15. H. Davenport, Note on irregularities of distribution. Mathematika Lond. **3**, 131–135 (1956). doi:10.1112/S0025579300001807

16. H. Faure, Discrépance de suites associées à un système de numération (en dimension s). Acta Arith. **41**, 337–351 (1982)
17. N.J. Fine, On the Walsh functions. Trans. Am. Math. Soc. **65**, 372–414 (1949). doi:10.2307/1990619
18. G. Halász, On Roth's method in the theory of irregularities of point distributions, in *Recent Progress in Analytic Number Theory*, vol. 2, ed. by H. Halberstam, C. Hooley, vol. 2 (Academic Press, London, 1981), pp. 79–94
19. J.H. Halton, On the efficiency of certain quasi-random sequences of points in evaluating multi-dimensional integrals. Numer. Math. **2**, 84–90 (1960)
20. J.H. Halton, S.K. Zaremba, The extreme and L^2 discrepancies of some plane sets. Monatsh. Math. **73**, 316–328 (1969). doi:10.1007/BF01298982
21. G.H. Hardy, J.E. Littlewood, Some problems of diophantine approximation: The lattice-points of a right-angled triangle I. Proc. London Math. Soc. (2) **20**(1), 15–36 (1921). doi:10.1112/plms/s2-20.1.15
22. G.H. Hardy, J.E. Littlewood, Some problems of diophantine approximation: The lattice-points of a right-angled triangle II. Abh. Math. Sem. Hamburg **1**(1), 212–249 (1921)
23. S.V. Konyagin, M.M. Skriganov, A.V. Sobolev, On a lattice point problem arising in the spectral analysis of periodic operators. Mathematika **50**(1–2), 87–98 (2003). doi:10.1112/S0025579300014819
24. M. Lerch, Question 1547. L'Intermediaire Math. **11**, 144–145 (1904)
25. R. Lidl, H. Niederreiter, *Finite Fields. Foreword by P. M. Cohn.* (Addison-Wesley, Reading, 1983)
26. F.J. MacWilliams, N.J.A. Sloane, *The Theory of Error-Correcting Codes. Parts I, II.* (North-Holland, Amsterdam, 1977)
27. D. Pollard, *Convergence of Stochastic Processes.* (Springer, New York, 1984)
28. J.J. Price, Certain groups of orthonormal step functions. Can. J. Math. **9**, 413–425 (1957). doi:10.4153/CJM-1957-049-x
29. K.F. Roth, On irregularities of distribution. Mathematika **1**, 73–79 (1954). doi:10.1112/S0025579300000541
30. K.F. Roth, On irregularities of distribution. III. Acta Arith. **35**, 373–384 (1979)
31. K.F. Roth, On irregularities of distribution. IV. Acta Arith. **37**, 67–75 (1980)
32. W.M. Schmidt, Irregularities of distribution. VII. Acta Arith. **21**, 45–50 (1972)
33. M.M. Skriganov, Constructions of uniform distributions in terms of geometry of numbers. St. Petersbg. Math. J. **6**(3), 635–664 (1995). Translation from Algebra i Analiz 6(3), 200–230 (1994)
34. M.M. Skriganov, Harmonic analysis on totally disconnected groups and irregularities of point distributions. J. Reine Angew. Math. **600**, 25–49 (2006). doi:10.1515/CRELLE.2006.085
35. T. van Aardenne-Ehrenfest, Proof of the impossibility of a just distribution of an infinite sequence of points over an interval. Proc. Kon. Ned. Akad. v. Wetensch. **48**, 266–271 (1945)
36. T. van Aardenne-Ehrenfest, On the impossibility of a just distribution. Proc. Kon. Ned. Akad. v. Wetensch. **52**, 734–739 (1949)
37. J.G. van der Corput, Verteilungsfunktionen. I. Mitt. Proc. Kon. Ned. Akad. v. Wetensch. **38**, 813–821 (1935)
38. J.G. van der Corput, Verteilungsfunktionen. II. Proc. Kon. Ned. Akad. v. Wetensch. **38**, 1058–1066 (1935)

Chapter 2
Roth's Orthogonal Function Method in Discrepancy Theory and Some New Connections

Dmitriy Bilyk

Abstract In this survey we give a comprehensive, but gentle introduction to the circle of questions surrounding the classical problems of discrepancy theory, unified by the same approach originated in the work of Klaus Roth (Mathematika 1:73–79, 1954) and based on multiparameter Haar (or other orthogonal) function expansions. Traditionally, the most important estimates of the discrepancy function were obtained using variations of this method. However, despite a large amount of work in this direction, the most important questions in the subject remain wide open, even at the level of conjectures. The area, as well as the method, has enjoyed an outburst of activity due to the recent breakthrough improvement of the higher-dimensional discrepancy bounds and the revealed important connections between this subject and harmonic analysis, probability (small deviation of the Brownian motion), and approximation theory (metric entropy of spaces with mixed smoothness). Without assuming any prior knowledge of the subject, we present the history and different manifestations of the method, its applications to related problems in various fields, and a detailed and intuitive outline of the latest higher-dimensional discrepancy estimate.

2.1 Introduction

The subject and the structure of the present chapter is slightly unconventional. Instead of building the exposition around the results from one area, united by a common topic, we concentrate on problems from different fields which all share a common method.

D. Bilyk (✉)
School of Mathematics, University of Minnesota, Minneapolis, MN 55408, USA
e-mail: dbilyk@math.umn.edu

W. Chen et al. (eds.), *A Panorama of Discrepancy Theory*, Lecture Notes
in Mathematics 2107, DOI 10.1007/978-3-319-04696-9_2,
© Springer International Publishing Switzerland 2014

The starting point of our discussion is one of the earliest results in discrepancy theory, Roth's 1954 L^2 bound of the discrepancy function in dimensions $d \geq 2$ [81], as well as the tool employed to obtain this bound, which later evolved into a powerful orthogonal function method in discrepancy theory. We provide an extensive overview of numerous results in the subject of irregularities of distribution, whose proofs are based on this method, from the creation of the field to the latest achievements.

In order to highlight the universality of the method, we shall bring out and emphasize analogies and connections of discrepancy theory and Roth's method to problems in a number of different fields, which include numerical integration (errors of cubature formulas), harmonic analysis (the small ball inequality), probability (small deviations of multiparameter Gaussian processes), approximation theory (metric entropy of spaces with dominating mixed smoothness). While some of these problems are related by direct implications, others are linked only by the method of proof, and perhaps new relations are yet to be discovered.

We also present a very detailed and perceptive account of the proof of one of the most recent important developments in the theory, the improved L^∞ bounds of the discrepancy function, and the corresponding improvements in other areas. We focus on the heuristics and the general strategy of the proof, and thoroughly explain the idea of every step of this involved argument, while skipping some of the technicalities, which could have almost doubled the size of this chapter.

We hope that the content of the volume will be of interest to experts and novices alike and will reveal the omnipotence of Roth's method and the fascinating relations between discrepancy theory and other areas of mathematics. We have made every effort to make our exposition clear, intuitive, and essentially self-contained, requiring familiarity only with the most basic concepts of the underlying fields.

2.1.1 The History and Development of the Field

Geometric *discrepancy theory* seeks answers to various forms of the following questions: *How accurately can one approximate a uniform distribution by a finite discrete set of points? And what are the errors and limitations that necessarily arise in such approximations?* The subject naturally grew out of the notion of uniform distribution in number theory. A sequence $\omega = \{\omega_n\}_{n=1}^{\infty} \subset [0, 1]$ is called uniformly distributed if, for any subinterval $I \subset [0, 1]$, the proportion of points ω_n that fall into I approximates its length, i.e.

$$\lim_{N \to \infty} \frac{\#\{\omega_n \in I : 1 \leq n \leq N\}}{N} = |I|. \tag{2.1}$$

This property can be easily quantified using the notion of *discrepancy*:

$$D_N(\omega) = \sup_{I \subset [0,1]} |\#\{\omega_n \in I : 1 \le n \le N\} - N \cdot |I||, \qquad (2.2)$$

where I is an interval. In fact, it is not hard to show that ω is uniformly distributed if and only if $D_N(\omega)/N$ tends to zero as $N \to \infty$ (see e.g. [60]).

In [115, 1935], van der Corput posed a question whether there exists a sequence ω for which the quantity $D_N(\omega)$ stays bounded as N gets large. More precisely, he mildly conjectured that the answer is "No" by stating that he is unaware of such sequences. Indeed, in [113, 1945], [114], van Aardenne-Ehrenfest gave a negative answer to this question, which meant that no sequence can be distributed too well. This result is widely regarded as a predecessor of the theory of *irregularities of distribution*.

This area was turned into a real theory with precise quantitative estimates and conjectures by Roth, who in particular, see [81], greatly improved van Aardenne-Ehrenfest's result by demonstrating that for any sequence ω the inequality

$$D_N(\omega) \ge C \sqrt{\log N} \qquad (2.3)$$

holds for infinitely many values of N. These results signified the birth of a new theory.

Roth in fact worked on the following, more geometrical version of the problem. Let $\mathscr{P}_N \subset [0,1]^d$ be a set of N points and consider the discrepancy function

$$D_N(x_1, \dots, x_d) = \#\{\mathscr{P}_N \cap [0, x_1) \times \cdots \times [0, x_d)\} - N \cdot x_1 \cdots \cdot x_d, \qquad (2.4)$$

i.e. the difference of the actual and expected number of points of $\mathscr{P}_N$ in the box $[0, x_1) \times \cdots \times [0, x_d)$. Notice that, in contrast to some of the standard references, we are working with the unnormalized version of the discrepancy function, i.e. we do not divide this difference by N as it is often done. Obviously, the most natural norm of this function is the L^∞ norm, i.e. the supremum of $|D_N(x)|$ over $x \in [0,1]^d$, often referred to as *the star-discrepancy*. In fact the term *star-discrepancy* is reserved for the sup-norm of the normalized discrepancy function, i.e. $\frac{1}{N} \|D_N\|_\infty$, however since we only use the unnormalized version in this text, we shall abuse the language and apply this term to $\|D_N\|_\infty$.

Instead of directly estimating the L^∞ norm of the discrepancy function $\|D_N\|_\infty = \sup_{x \in [0,1]^d} |D_N(x)|$, Roth considered a smaller quantity, namely its L^2 norm $\|D_N\|_2$. This substitution allowed for an introduction of a variety of Hilbert space techniques, including orthogonal decompositions. In this setting Roth proved.

Theorem 1 (Roth [81]). *In all dimensions $d \ge 2$, for any N-point set $\mathscr{P}_N \subset [0,1]^d$, one has*

$$\|D_N\|_2 \ge C_d \log^{\frac{d-1}{2}} N, \qquad (2.5)$$

where C_d is an absolute constant that depends only on the dimension d. This in particular implies that

$$\sup_{x \in [0,1]^d} |D_N(x)| \geq C_d \log^{\frac{d-1}{2}} N. \tag{2.6}$$

It was also shown that, when $d = 2$, inequality (2.6) is equivalent to (2.3). More generally, uniform lower bounds for the discrepancy function of finite point sets (for all values of N) in dimension d are equivalent to lower estimates for the discrepancy of infinite sequences (2.2) (for infinitely many values of N) in dimension $d - 1$. These two settings are sometimes referred to as 'static' (fixed finite point sets) and 'dynamic' (infinite sequences). In these terms, one can say that the dynamic and static problems are equivalent at the cost of one dimension—the relation becomes intuitively clear if one views the index of the sequence (or time) as an additional dimension. In this text, we adopt the former geometrical, 'static' formulation of the problems.

According to Roth's own words, these results "started a new theory" [32]. The paper [81] in which it was presented, entitled "On irregularities of distribution", has had a tremendous influence on the further development of the field. Even the number of papers with identical or similar titles, that appeared in subsequent years, attests to its importance: 4 papers by Roth himself (*On irregularities of distribution. I-IV*, [81–84]), one by H. Davenport (*Note on irregularities of distribution*, [38]), 10 by W. M. Schmidt (*Irregularities of distribution. I-X*, [86–95]), 2 by J. Beck (*Note on irregularities of distribution. I-II*, [6, 7]), 4 by W. W. L. Chen (*On irregularities of distribution. I-IV*, [27–30]), at least 2 by Beck and Chen (*Note on irregularities of distribution. I-II*, [3, 4] and several others with similar, but more specific names, as well as the fundamental monograph on the subject by Beck and Chen, "*Irregularities of distribution*", [10].

The technique proposed in the aforementioned paper was no less important than the results themselves. Roth was the first to apply the expansion of the discrepancy function D_N in the classical orthogonal Haar basis. Furthermore, he realized that in order to obtain good estimates of $\|D_N\|_2$ it suffices to consider just its projection onto the span of those Haar functions which are supported on dyadic rectangles of volume roughly equal to $\frac{1}{N}$. This is heuristically justified by the fact that, for a well distributed set, each such rectangle contains approximately one point. To be even more precise, the size of the rectangles R was chosen so that $|R| \approx \frac{1}{2N}$, ensuring that about half of all rectangles are free of points of $\mathscr{P}_N$. The Haar coefficients of D_N, corresponding to these empty rectangles, are then easy to compute, which leads directly to the estimate (2.5). This idea is the main theme of Sect. 2.2. Roth's approach strongly resonates with Kolmogorov's method of proving lower error bounds for cubature formulas, see e.g. [105, Chapter IV]. We shall discuss these ideas in more detail in Sect. 2.2.3.

A famous quote attributed to G. Polya [79] says,

What is the difference between method and device? A method is a device which you used twice.

In agreement with this statement, over the past years Roth's clever device has indeed evolved into a powerful and versatile method: it has been applied an enormous number of times to various problems and questions in discrepancy theory and other areas. Our survey is abundant in such applications: discrepancy estimates in other function spaces Sect. 2.3.4, estimates of the star-discrepancy Sects. 2.4.4, 2.5, the small ball inequality Sects. 2.4.3, 2.5, constructions of low-discrepancy distributions Sect. 2.6.

Roth's L^2 result has been extended to other L^p norms, $1 < p < \infty$, only significantly later by W. Schmidt in [95, 1977], who showed that in all dimensions $d \geq 2$, for all $p \in (1, \infty)$ the inequality

$$\|D_N\|_p \geq C_{d,p} \log^{\frac{d-1}{2}} N, \tag{2.7}$$

holds for some constant $C_{d,p}$ independent of the collection of points $\mathscr{P}_N$. Schmidt's approach was a direct extension of Roth's method: rather then working with arbitrary integrability exponents p, he considers only those p's for which the dual exponent q is an even integer. This allows one to iterate the orthogonality arguments. Even though it took more than 20 years to extend Roth's L^2 inequality to other L^p spaces, a contemporary harmonic analyst may realize that such an extension can be derived in just a couple of lines using Littlewood–Paley inequalities. A comprehensive discussion will be provided in Sect. 2.3.

While the case $1 < p < \infty$ is thoroughly understood, the endpoint case $p = \infty$, i.e. the *star-discrepancy*, is much more mysterious, despite the fact that it is most natural and important in the theory as it describes the worst possible discrepancy. It turns out that Roth's inequality (2.6) is not sharp for the sup-norm of the discrepancy function. It is perhaps not surprising: intuitively, the discrepancy function is highly irregular and comes close to its maximal values only on small sets. Hence, its extremal (i.e. L^∞) norm must necessarily be much larger than its average (e.g. L^2) norm. This heuristics also guides the use of some of the methods that have been exploited in the proofs of the star-discrepancy estimates, such as Riesz products.

In 1972, W. M. Schmidt proved that in dimension $d = 2$ one has the following lower bound:

$$\sup_{x \in [0,1]^d} |D_N(x)| \geq C \log N, \tag{2.8}$$

which is known to be sharp. Indeed, two-dimensional constructions, for which $\|D_N\|_\infty \leq C \log N$ holds for all N (or, equivalently, one-dimensional sequences ω for which $D_N(\omega) \leq C \log N$ infinitely often), have been known for a long time and go back to the works of Lerch [65, 1904], van der Corput [115, 1935] and others, see e.g. Sect. 2.6.

Several other proofs of Schmidt's inequality (2.8) have been given later [68, 1979], [11, 1982], [48, 1981]. The latter (due to Halász) presents great interest to us as it has been built upon Roth's Haar function method—we will reproduce and analyze the argument in Sect. 2.4.4. Incidentally, the title of Halász's article [48]

("On Roth's method in the theory of irregularities of point distributions") almost coincides with the title of this chapter.

Higher dimensional analogs of Schmidt's estimate (2.8), however, turned out to be extremely proof-resistant. For a long time inequality (2.6) remained the best known bound in dimensions three and above. In fact, the first gain over the L^2 estimate was obtained only 35 years after Roth's result by Beck [8, 1989], who proved that in dimension $d = 3$, discrepancy function satisfies

$$\|D_N\|_\infty \geq C \log N \cdot (\log \log N)^{\frac{1}{8}-\varepsilon}. \tag{2.9}$$

Almost 20 years later, in 2008, the author jointly with M. Lacey and A. Vagharshakyan [15, $d = 3$]; [17, $d \geq 4$] obtained the first significant improvement of the L^∞ bound in *all dimensions $d \geq 3$*:

Theorem 2 (Bilyk, Lacey, Vagharshakyan). *For all $d \geq 3$, there exists some $\eta = \eta(d) > 0$, such that for all $\mathscr{P}_N \subset [0, 1]^d$ with $\#\mathscr{P}_N = N$ we have the estimate:*

$$\|D_N\|_\infty \geq C_d (\log N)^{\frac{d-1}{2}+\eta}. \tag{2.10}$$

The exact rate of growth of the star-discrepancy in higher dimensions remains an intriguing question; in their book [10], Beck and Chen named it "the great open problem" and called it "excruciatingly difficult".

Even the precise form of the conjecture is a subject of ongoing debate among the experts in the field. The opinions are largely divided between two possible formulations of this conjecture. We start with the form which is directly pertinent to the orthogonal function method.

Conjecture 3. For all $d \geq 3$ and all $\mathscr{P}_N \subset [0, 1]^d$ with $\#\mathscr{P}_N = N$ we have the estimate:

$$\|D_N\|_\infty \geq C_d (\log N)^{\frac{d}{2}}. \tag{2.11}$$

This conjecture is motivated by connections of this field to other areas of mathematics and, in particular, by a related conjecture in analysis, the *small ball conjecture* (2.111), which is known to be sharp, see Sect. 2.4.2. Unfortunately, this relation is not direct—it is not known whether the validity of the small ball conjecture implies the discrepancy estimate (2.11), the similarity lies just in the methods of proof. But, at the very least, this connection suggests that Conjecture 3 is the best result that one can achieve using Roth's Haar function method.

On the other hand, the best known examples [49, 51] of well distributed sets in higher dimensions have star-discrepancy of the order

$$\|D_N\|_\infty \leq C_d (\log N)^{d-1}. \tag{2.12}$$

Numerous constructions of such sets are known and are currently a subject of massive ongoing research, see e.g. the book [39]. These upper bounds together with the estimates for a "smooth" version of discrepancy (see Temlyakov [110]), provide grounds for an alternative form of the conjecture (which is actually older and more established).

Conjecture 4. For all $d \geq 3$ and all $\mathscr{P}_N \subset [0, 1]^d$ with $\#\mathscr{P}_N = N$ we have the estimate:

$$\|D_N\|_\infty \geq C_d (\log N)^{d-1}. \tag{2.13}$$

One can notice that both conjectures coincide with Schmidt's estimate (2.8) when $d = 2$. Skriganov has proposed yet another form of the conjecture [99]:

$$\|D_N\|_\infty \geq C_d (\log N)^{\frac{d-1}{2} + \frac{d-1}{d}}, \tag{2.14}$$

which is exact both in $d = 1$ and $d = 2$.

In contrast to the L^∞ inequalities, it is well known that in the average (L^2 or L^p) sense Roth's bound (2.3), as well as inequality (2.7), is sharp. This was initially proved by Davenport [38] in two dimensions for $p = 2$, who constructed point distributions with $\|D_N\|_2 \leq C \sqrt{\log N}$. Subsequently, different constructions have been obtained by numerous other authors, including Roth [83, 84], Chen [27], Frolov [44]. It should be noted that most of the optimal constructions in higher dimensions $d \geq 3$ are probabilistic in nature and are obtained as randomizations of some classic low-discrepancy sets. In fact, deterministic examples of sets with $\|D_N\|_p \leq C_{d,p} \log^{\frac{d-1}{2}} N$ have been constructed only in the last decade by Chen and Skriganov [33, 34] ($p = 2$) and Skriganov [98] ($p > 1$). It would be interesting to note that their results are also deeply rooted in Roth's orthogonal function method— they use the orthogonal system of Walsh functions to analyze the discrepancy function and certain features of the argument remind one of the ideas that appear in Roth's proof.

The other endpoint of the L^p scale, $p = 1$, is not any less (and perhaps even more) difficult than the star-discrepancy estimates. The only information that is available is the two-dimensional inequality (proved in the aforementioned paper of Halász [48]), which also makes use of Roth's orthogonal function method:

$$\|D_N\|_1 \geq C \sqrt{\log N}, \tag{2.15}$$

This means that the L^1 norm of discrepancy behaves roughly like its L^2 norm. It is conjectured that the same similarity continues to hold in higher dimensions.

Conjecture 5. For all $d \geq 3$ and all sets of N points in $[0, 1]^d$:

$$\|D_N\|_1 \geq C_d (\log N)^{\frac{d-1}{2}}. \tag{2.16}$$

However, almost no results pertaining to this conjecture have been discovered for $d \geq 3$. The only known relevant fact is that $\sqrt{\log N}$ bound still holds in higher dimensions, i.e. it is not even known if the exponent increases with dimension. The reader is referred to Sect. 2.4.5 for Halász's L^1 argument.

2.1.2 Preliminary Discussion

While the main subject of this chapter is Roth's method in discrepancy theory, we are also equally concentrated on its applications and relations to a wide array of problems extending to topics well beyond discrepancy. One of our principal intentions is to stress the connections between different areas of mathematics and accentuate the use of the methods of harmonic analysis in discrepancy and related fields. Having aimed to cover such a broad range of topics, we left ourselves with little chance to make the exposition very detailed and full of technicalities. Instead, we decided to focus on the set of ideas, connections, arguments, and conjectures that permeate discrepancy theory and several other subjects.

We assume only very basic prior knowledge of any of the underlying fields, introducing and explaining the new concepts as they appear in the text, discussing basic properties, and providing ample references. In particular, we believe this chapter to be a very suitable reading for graduate students as well as for mathematicians of various backgrounds interested in discrepancy or any of the discussed areas. In an effort to make our exposition reader-friendly and accessible, we often sacrifice generality, and sometimes even rigor, in favor of making the presentation more intuitive, providing simpler and more transparent arguments, or explaining the heuristics and ideas behind the proof. The reader however should *not* get the impression that this chapter is void of mathematical content. In fact, a great number of results are meticulously proved in the text and numerous computations, which could have been skipped in a technical research paper, are carried out in full detail.

2.1.2.1 A Brief Outline of the Chapter

Even though our exposition consists of several distinct sections which sometimes deal with seemingly unrelated subjects, every section naturally continues and interlaces with the discussion of the previous ones. In the next several paragraphs we give a brief 'sneak preview' of the content of this chapter.

- In Sect. 2.2 we introduce the reader to the main ideas of Roth's L^2 method. We start with the necessary definitions and background information on Haar functions and product orthogonal bases and then proceed to explain a general principle behind Roth's argument. We then give the proof of the L^2 discrepancy bound, Theorem 1. We present Roth's original proof which relies on duality and the Cauchy–Schwarz inequality, as well as a slightly different argument which makes use of orthogonality and Bessel's inequality directly. In the end of Sect. 2.2 we turn to Kolmogorov's method of obtaining lower bounds for errors of cubature formulas on various function classes governed by the behavior of the mixed derivative. The method is based on the same idea as Roth's method in discrepancy theory and provides an important connection between these two intimately related areas.

- Extensions of Theorem 1 even to L^p spaces with $p \neq 2$ turned out to be somewhat delicate and not immediate. However, harmonic analysis provides means to make these extensions almost automatic. This instrument, the Littlewood–Paley inequalities, is the subject of Sect. 2.3. The Littlewood–Paley serves as a natural substitute for orthogonality in non-Hilbert spaces, e.g. L^p. In Sect. 2.3 we discuss the relevant version of this theory—the dyadic Littlewood–Paley inequalities, starting with the one-dimensional case and then moving forward to the multiparameter setting. We also discuss the connections of this topic to objects in probability theory such as the famous Khintchine inequality and the martingale difference square function. Unfortunately, unlike many other methods of harmonic analysis, Littlewood–Paley theory has not yet become a "household name" among experts in various fields outside analysis. It is our sincere hope that our exposition will further publicize and popularize this powerful method.
- Next, we demonstrate how these tools can be used to extend Roth's L^2 discrepancy estimate to L^p essentially in one line. Further, a large portion of Sect. 2.3 is devoted to the discussion of discrepancy estimates analogous to Theorem 1 in various function spaces, such as Hardy, Besov, BMO, weighted L^p, and exponential Orlicz spaces. All of these results, in one way or another, take their roots in Roth's method and the Littlewood–Paley (or similar in spirit) inequalities.
- In Sect. 2.4 we turn to arguably the most important problem of discrepancy theory—sharp estimates of the star-discrepancy (L^∞ norm of the discrepancy function). We introduce the *small ball inequality*—a purely analytic inequality which is concerned with lower bounds of the supremum norm of sums of Haar functions supported by rectangles of fixed size. The very structure of these sums suggests certain connections with Roth's method in discrepancy. And indeed, even though it is not known if one problem directly implies the other, there are numerous similarities in the known methods of proof and the small ball inequality may be viewed as a linear model of the star-discrepancy method. We state the small ball conjecture and discuss known results and its sharpness, which indirectly bears some effect on the sharpness of the relevant discrepancy conjectures.
- In Sect. 2.4.3 we present a beautiful proof of the small ball conjecture in dimension $d = 2$. We then proceed to demonstrate an amazingly similar proof of Schmidt's lower bound (2.8) for the star-discrepancy in $d = 2$ as well as a proof of the L^1 discrepancy bound (2.15). All three proofs are based on an ingenious method known as the Riesz product. To reinforce the connections of these problems with the classical problems of analysis, in Sect. 2.4.6 we briefly discuss the area in which Riesz product historically first appeared—lacunary

Fourier series. We give a proof of Sidon's theorem whose statement, as well as the argument used to prove it, resemble both the small ball inequality and the discrepancy estimates in great detail and perhaps have inspired their respective proofs.

- The small ball inequality turns out to be connected to other areas of mathematics besides discrepancy—in particular, approximation theory and probability. In Sects. 2.4.7–2.4.8 we describe the relevant problems: the small deviation probabilities for the Brownian sheet and the metric entropy of function classes with mixed smoothness. We demonstrate that the small ball inequality directly implies lower bounds in both of these problems and hence indirectly ties them to discrepancy.

- A substantial part of this chapter, Sect. 2.5, focuses on the important recent developments in the subject, namely the first significant improvement in the small ball inequality and the L^∞ discrepancy estimates in all dimensions $d \geq 3$. We thoroughly discuss the main steps and ingredients of the proof, intuitively explain many parts of the argument and pinpoint the arising difficulties without going too deep into the technical details. This approach, in our opinion, will allow one to comprehend the 'big picture' and the strategy of the proof. An interested reader will then be well-equipped and prepared to fill in the complicated technicalities by consulting the provided references.

- Finally, in Sect. 2.6 our attention makes a 180-degree turn from lower bounds to constructions of well-distributed point sets and upper discrepancy estimates. We introduce one of the most famous low-discrepancy distributions in two dimensions—the van der Corput digit reversing set, whose binary structure makes it a perfect fit for the tools of dyadic analysis and Roth's method. We describe certain modifications of this set, which achieve the optimal order of discrepancy in various function spaces, in particular, demonstrating the sharpness of some of the results in Sect. 2.3.4.

The aim of this survey is really two-fold: to acquaint specialists in discrepancy theory with some of the techniques of harmonic analysis which may be used in this subject, as well as to present the circle of problems in the field of irregularities of distribution to the analysts. Numerous books written on discrepancy theory present Roth's method and related arguments, see [10, 25, 39, 60, 72, 105]; the book [73] studies the relations between uniform distribution and harmonic analysis, [111] views the subject through the lens of the function space theory, while [100] specifically investigates the connections between discrepancy and Haar functions. In addition, the survey [35] explores various ideas of Roth in discrepancy theory, including the method discussed here. Finally, [61] and [14] are very similar in spirit to this chapter; however, the survey [14] is much more concise than the present text, and the set of notes [61] focuses primarily on the underlying harmonic analysis. We have tried to make the to presentation accessible to a wide audience, rather than experts in one particular area, yet at the same time inclusive, embracing and accentuating the connections between numerous topics. We sincerely hope that, despite a vast amount of literature on the subject, this chapter will provide some novel ideas and useful insights and will be of interest to both novices and specialists in the field.

2.1.2.2   Some Other Problems Related to Roth's Method

Unfortunately, there are still a number of topics that either grew directly out of Roth's method or are tangentially, but strongly correlated with it, which we will not be touching upon in this survey, since these discussions would have taken us very far afield. They include, in particular, Beck's beautiful lower bound on the growth of polynomials with roots on the unit circle [9]. By an argument, very similar to the Halász's proof of the two-dimensional star-discrepancy estimate, Beck showed that there exists a constant $\delta > 0$ such that for any infinite sequence $\{z_n\}_{n=1}^{\infty}$ of unimodular complex numbers and polynomials $P_N(z) = \prod_{n=1}^{N}(z - z_n)$ the bound

$$\sup_{|z| \le 1} |P_N(z)| > N^{\delta} \tag{2.17}$$

holds for infinitely many values of N, thus giving a negative answer to a question of Erdős.

Another problem considered by Beck and Roth deals with the so-called combinatorial discrepancy, a natural companion of the geometric discrepancy. Let the function $\lambda : \mathscr{P}_N \to \{\pm 1\}$ represent a "red-blue" coloring of an N-point set $\mathscr{P}_N \subset [0, 1]^d$. The *combinatorial discrepancy* of $\mathscr{P}_N$ with respect to a family of sets $\mathscr{B}$ is defined as $T(\mathscr{P}_N) = \inf_{\lambda} \sup_{B \in \mathscr{B}} \left| \sum_{p \in \mathscr{P}_N \cap B} \lambda(p) \right|$, i.e. the minimization of the largest disbalance of colors in sets from $\mathscr{B}$ over all possible colorings. In [5], Beck discovered that, when $\mathscr{B}$ is the family of axis-parallel boxes, the quantity $T(N) := \sup_{\mathscr{P}_N} T(\mathscr{P}_N)$ is tightly related to the discrepancy function estimates. In particular, in $d = 2$ one has $T(N) \gtrsim \log N$. In [85] Roth has extended this to real-valued functions λ (continuous coloring) showing that

$$T(N) \gtrsim \frac{(\log N)}{N} \sum_{p \in \mathscr{P}_N} |\lambda(p)|. \tag{2.18}$$

Roth's argument relied on Haar expansions and Riesz product and almost repeated the proof of the L^{∞} discrepancy bound in dimension two with an addition of some new ideas. Recent progress (2.10) on the discrepancy function directly yields an analogous improvement in the "red-blue" case for $d \ge 3$ and can be adjusted to provide a similar estimate for "continuous" colorings in dimension $d = 3$.

There are numerous other examples. Chazelle [26] has applied a discrete version of Roth's orthogonal function method to a problem in computational geometry, obtaining a lower bound for the complexity of orthogonal range searching. The Riesz product techniques, similar to Halász's, have been used in approximation theory for a long time to obtain Bernstein-type inequalities, estimates for entropy numbers and Kolmogorov widths of certain function classes, see e.g. [106,108,109]. We shall only briefly discuss some of these connections in Sect. 2.4.8.

This diverse set of topics shows the universality and ubiquitousness of the method and ideas under discussion.

2.1.2.3 Notation and Conventions

Before we proceed to the mathematical part of the text, we would like to explain some of the notation and conventions that we shall be actively using. Since many different constants arise in our discussion, we often make use of the symbol "$\lesssim$": $F \lesssim G$ means that there exists a constant $C > 0$ such that $F \leq CG$. The relation $F \approx G$ means that $F \lesssim G$ and $G \lesssim F$. The implicit constants in such inequalities will be allowed to depend on the dimension and, perhaps, some other parameters, but never on the number of points N.

In other words, in this survey we are interested in the asymptotic behavior of the discrepancy when the dimension is fixed and the number of points increases. Therefore, such effects as the *curse of dimensionality* do not come into play. Finding optimal estimates as the dimension goes to infinity is a separate, very interesting and important subject, see e.g. [52]. While one may argue that these questions are sometimes more useful for applications, we firmly insist that the questions discussed here, which go back to van der Corput, van Aardenne-Ehrenfest, and Roth, are at least equally as important, especially considering the fact that in such natural (and low!) dimensions as, say, 3 or 4 the exact rate of growth of discrepancy is far from being understood and the relative gap between the lower and upper estimates is quite unsatisfactory.

Throughout the text several variables will have robustly reserved meanings. The dimension will always be denoted by d. Capital N will always stand for the number of points, while n will represent the scale and will usually satisfy $n \approx \log N$. Unless otherwise specified, all logarithms are taken to be natural, although this is not so important since we are not keeping track of the constants. The discrepancy function of an N-point set $\mathscr{P}_N \subset [0, 1]^d$ will be denoted either by $D_{\mathscr{P}_N}$ or, more often, if this creates no ambiguity, simply by D_N. Recall that, unlike a number of standard references, we are considering the unnormalized version of discrepancy, i.e. we do not divide by N in the definition (2.4). The term *star-discrepancy* refers to the L^∞ norm of D_N.

For a set $A \subset \mathbb{R}^d$, its Lebesgue measure is denoted either by $|A|$ or by $\mu(A)$. For a finite set F, we use $\#F$ to denote its cardinality—the number of elements of F. Whenever we have to resort to probabilistic concepts, $\mathbb{P}$ will stand for probability and $\mathbb{E}$ for expectation.

2.2 Roth's Orthogonal Function Method and the L^2 Discrepancy

Before we begin a detailed discussion of Roth's method, we need to introduce and define its main tool, Haar functions. We shall then explain Roth's main idea and proceed to reproduce his original proof of Theorem 1, although our exposition will slightly differ from the style of the original paper [81] (the argument, however, will be identical to Roth's). We shall make use of somewhat more modern notation which is closer in spirit to functional and harmonic analysis. Hopefully, this will allow us to make the idea of the proof more transparent. Along the way, we shall try to look at the argument at different angles and to find motivation behind some of the steps of the proof.

2.2.1 Haar Functions and Roth's Principle

We start by defining the Haar basis in $L^2[0, 1]$. Let $\mathbf{1}_I(x)$ stand for the characteristic function of the interval I. Consider the collection of all *dyadic* subintervals of $[0,1]$:

$$\mathscr{D} = \left\{ I = \left[\frac{m}{2^n}, \frac{m+1}{2^n} \right) : m, n \in \mathbb{Z}, n \geq 0, 0 \leq m < 2^n \right\}. \tag{2.19}$$

Dyadic intervals form a *grid*, meaning that any two intervals in $\mathscr{D}$ are either disjoint, or one is contained in another. In addition, for every interval $I \in \mathscr{D}$, its left and right halves (we shall denote them by I_l and I_r) are also dyadic. The Haar function corresponding to the interval I is then defined as

$$h_I(x) = -\mathbf{1}_{I_l}(x) + \mathbf{1}_{I_r}(x). \tag{2.20}$$

Notice that in our definition Haar functions are normalized to have unit norm in L^∞ (their L^2 norm is $\|h_I\|_2 = |I|^{1/2}$). This will cause some of the classical formulas to look a little unusual to those readers who are accustomed to the L^2 normalization.

These functions have been introduced by Haar [47, 1910] and have played an extremely important role in analysis, probability, signal processing etc. They are commonly viewed as the first example of *wavelets*. Their orthogonality, i.e. the relation

$$\langle h_{I'}, h_{I''} \rangle = \int_0^1 h_{I'}(x) \cdot h_{I''}(x) \, dx = 0, \qquad I', I'' \in \mathscr{D}, I' \neq I'', \tag{2.21}$$

follows easily from the facts that $\mathscr{D}$ is a grid and that the condition $I' \subsetneq I''$, I', $I'' \in \mathscr{D}$ implies that I' is contained either in the left or right half of I'', hence $h_{I''}$ is constant on the support of $h_{I'}$. It is well known that the system $\mathscr{H} = \mathbf{1}_{[0,1]} \cup$

$\{h_I : I \in \mathscr{D}\}$ forms an orthogonal basis in $L^2[0, 1]$ and an unconditional basis in $L^p[0, 1]$, $1 < p < \infty$.

In order to simplify the notation and make it more uniform, we shall sometimes employ the following trick. Denote by $\mathscr{D}_* = \mathscr{D} \cup \{[-1, 1]\}$ the dyadic grid on $[0, 1]$ with the interval $[-1, 1]$ added to it. Then the family $\mathscr{H} = \{h_I\}_{I \in \mathscr{D}_*}$ forms an orthogonal basis of $L^2([0, 1])$. In other words, the constant function on $[0, 1]$ can be viewed as a Haar function of order -1.

In higher dimensions, we consider the family of *dyadic rectangles* $\mathscr{D}^d = \{R = R_1 \times \cdots \times R_d : R_j \in \mathscr{D}\}$. For a dyadic rectangle R, the Haar function supported by R is defined as a coordinatewise product of the one-dimensional Haar functions:

$$h_R(x_1, \ldots, x_d) = h_{I_1}(x_1) \cdot \ldots \cdot h_{I_d}(x_d). \qquad (2.22)$$

The orthogonality of these functions is easily derived from the one dimensional property. It is also well known that the 'product' Haar system $\mathscr{H}^d = \{f(x) = f_1(x_1) \cdot \ldots \cdot f_d(x_d) : f_k \in \mathscr{H}\}$ is an orthogonal basis of $L^2([0, 1]^d)$—often referred to as the *product Haar basis*. The construction of product bases starting from a one-dimensional orthogonal basis is also valid for more general systems of orthogonal functions. In view of the previous remark, one can write $\mathscr{H}^d = \{h_R\}_{R \in \mathscr{D}_*^d}$, although most of the times we shall restrict our attention to rectangles in $\mathscr{D}^d$. Thus, every function $f \in L^2([0, 1]^d)$ can be written as

$$f = \sum_{R \in \mathscr{D}_*^d} \frac{\langle f, h_R \rangle}{|R|} h_R, \qquad (2.23)$$

where the series converges in L^2. If this expression seems slightly unconventional, this is a result of the L^∞ normalization of h_R. We note that this is not the only way to extend wavelet bases to higher dimensions [37], but this multiparameter approach is the correct tool for the problems at hand, where the dimensions of the underlying rectangles are allowed to vary absolutely independently (e.g. some rectangles may be long and thin, while others may resemble a cube). This is precisely the setting of the product (multiparameter) harmonic analysis—we shall keep returning to this point throughout the text.

One of the numerous important contributions of Klaus Roth to discrepancy theory is the idea of using orthogonal function (in particular, Haar) decompositions in order to obtain discrepancy estimates. This idea was introduced already in his first paper on irregularities of distribution [81]. Even though Haar functions have been introduced almost simultaneously to some questions connected with uniform distribution theory and numerical integration (see Sobol's book [100]), their power for discrepancy estimates only became apparent with Roth's proof of the lower bound for the L^2 bound of the discrepancy function.

In addition to introducing a new tool to the field, Roth has clearly demonstrated a proper way to use it. An orthogonal expansion may be of very little use to us, unless we know how to extract information from it and which coefficients play the most important role. The method of proof of the L^2 bound (2.5) unambiguously suggests where one should look for the most relevant input of the decomposition to the intrinsic features of the discrepancy function. Further success of this approach in various discrepancy setting and connections to other areas and problems, described throughout this chapter, validates the correctness of the idea and turns it into a method. We formulate it here as a general principle.

Roth's principle: The behavior of the discrepancy function is essentially defined by its projection onto the span of Haar functions h_R supported by rectangles of volume $|R| \approx \frac{1}{N}$, i.e.

$$D_N \sim \sum_{R \in \mathscr{D}^d : |R| \approx \frac{1}{N}} \frac{\langle D_N, h_R \rangle}{|R|} h_R. \qquad (2.24)$$

In the formulation of this principle, we interpret the symbols '$\sim$' and '$\approx$' very loosely and broadly. This principle as such should not be viewed as a rigorous mathematical statement. It is rather a circle of ideas and a heuristic approach. In this chapter we shall see many manifestations of this principle in discrepancy theory (both for upper and lower estimates) and will draw parallels with similar methods and ideas in other fields, such as approximation theory, probability, and harmonic analysis.

An intuitive explanation of this principle, perhaps, lies in the fact that, for 'nice' distributions of points $\mathscr{P}_N$, any dyadic rectangle of area $|R| \approx \frac{1}{N}$ would contain roughly one point (or the number of empty rectangles is comparable to the number of points). At fine scales, the boxes are too small and most of the time they contain no points of $\mathscr{P}_N$ and hence do not carry much information about the discrepancy. While rectangles that are too big (coarse scales) incorporate too much cancellation: the discrepancy of $[0, 1]^d$, for example, is always zero. (We should note that large rectangles, however, often give important additional information, see e.g. [54]). Therefore, the intermediate scales are the most important ones. Of course, this justification is too naive and simplistic and does not provide a complete picture. Some details will become more clear after discussing the proof of (2.5) which we turn to now.

2.2.2   The Proof of the L^2 Discrepancy Estimate

As promised we shall now reconstruct Roth's original proof of the L^2 estimate (2.5). Following the general lines of Roth's principle (2.24), we consider dyadic rectangles

$R \in \mathscr{D}^d$ of volume $|R| = 2^{-n} \approx \frac{1}{N}$. To be more exact, let us choose the number $n \in \mathbb{N}$ so that

$$2^{n-2} \leq N < 2^{n-1}, \tag{2.25}$$

i.e. $n \approx \log_2 N$ (although the precise choice of n is important for the argument).

These rectangles come in a variety of shapes, especially in higher dimension. This fact dramatically increases the combinatorial complexity of the related problems. To keep track of these rectangles we introduce a special bookkeeping device—a collection of vectors with non-negative integer coordinates

$$\mathbb{H}_n^d = \{\mathbf{r} = (r_1, \ldots, r_d) \in \mathbb{Z}_+^d : \|\mathbf{r}\|_1 = n\}, \tag{2.26}$$

where the ℓ_1 norm is defined as $\|\mathbf{r}\|_1 = |r_1| + \cdots + |r_d|$. These vectors will specify the shape of the dyadic rectangles in the following sense: for $R \in \mathscr{D}^d$, we say that $R \in \mathscr{D}_{\mathbf{r}}^d$ if $|R_j| = 2^{-r_j}$ for $j = 1, \ldots, d$. Obviously, if $R \in \mathscr{D}_{\mathbf{r}}^d$ and $\mathbf{r} \in \mathbb{H}_n^d$, then $|R| = 2^{-n}$. Besides, it is evident that, for a fixed $\mathbf{r}$, all the rectangles $R \in \mathscr{D}_{\mathbf{r}}^d$ are disjoint. It is also straightforward to see that the cardinality

$$\#\mathbb{H}_n^d = \binom{n+d-1}{d-1} \approx n^{d-1}, \tag{2.27}$$

which agrees with the simple logic that we have $d-1$ "free" parameters: the first $d-1$ coordinates can be chosen essentially freely, while the last one would be fixed due to the condition $\|\mathbf{r}\|_1 = n$ or $|R| = 2^{-n}$.

We shall say that a function f on $[0,1]^d$ is an r-function with parameter $\mathbf{r} \in \mathbb{Z}_+^d$ if f is of the form

$$f(x) = \sum_{R \in \mathscr{D}_{\mathbf{r}}^d} \varepsilon_R h_R(x), \tag{2.28}$$

for some choice of signs $\varepsilon_R = \pm 1$. These functions are generalized Rademacher functions (hence the name)—indeed, setting all the signs $\varepsilon_R = 1$, one obtains the familiar Rademacher functions. It is trivial that if f is an r-function, then $f^2 = 1$ and thus $\|f\|_2 = 1$. Such functions play the role of building blocks in numerous discrepancy arguments, therefore their L^2 normalization justifies the choice of the L^∞ normalization for the Haar functions. In addition, the fact that two r-functions corresponding to different vectors $\mathbf{r}$ are orthogonal readily follows from the orthogonality of the family of Haar functions.

Next, we would like to compute how the discrepancy function D_N interacts with Haar functions in certain cases. Notice that discrepancy can be written in the form

$$D_N(x) = \sum_{p \in \mathscr{P}_N} \mathbf{1}_{[p,1]}(x) - N \cdot x_1 \cdots x_d, \tag{2.29}$$

where $\mathbf{1} = (1, \ldots, 1)$ and $[p, \mathbf{1}] = [p_1, 1] \times \cdots \times [p_d, 1]$. We shall refer to the first term as the *counting* part and the second as the *volume (area)* or the *linear* part.

It is easy to see that, in one dimension, we have

$$\int \mathbf{1}_{[q,1]}(x) \cdot h_I(x)\, dx = \int_q^1 h_I(x)\, dx = 0 \tag{2.30}$$

unless I contains the point q. This implies that for $p \in [0, 1]^d$

$$\int_{[0,1]^d} \mathbf{1}_{[p,1]}(x) \cdot h_R(x)\, dx = \prod_{j=1}^d \int_{p_j}^1 h_{R_j}(x_j)\, dx_j = 0 \tag{2.31}$$

when $p \notin R$. Assume now that a rectangle $R \in \mathscr{D}^d$ is empty, i.e. does not contain points of $\mathscr{P}_N$. It follows from the previous identity that for such a rectangle, the inner product of the corresponding Haar function with the counting part of the discrepancy function is zero:

$$\left\langle \sum_{p \in \mathscr{P}_N} \mathbf{1}_{[p,1]},\, h_R \right\rangle = 0. \tag{2.32}$$

In other words, if R is free of points of $\mathscr{P}_N$, the inner product $\langle D_N, h_R \rangle$ is determined purely by the linear part of D_N.

It is however a simple exercise to compute the inner product of the linear part with any Haar function:

$$\langle N x_1 \ldots x_d, h_R \rangle = N \prod_{j=1}^d \langle x_j, h_{R_j}(x_j) \rangle = N \cdot \frac{|R|^2}{4^d}. \tag{2.33}$$

Hence we have shown that if a rectangle $R \in \mathscr{D}^d$ does not contain points of $\mathscr{P}_N$ in its interior, we have

$$\langle D_N, h_R \rangle = -N |R|^2 4^{-d}. \tag{2.34}$$

These, somewhat mysterious, computations can be explained geometrically (see [95], also [25, Chapter 3]). For simplicity, we shall do it in dimension $d = 2$, but this argument easily extends to higher dimensions. Let $R \subset [0, 1]^2$ be an arbitrary dyadic rectangle of dimensions $2h_1 \times 2h_2$ which does not contain any points of $\mathscr{P}_N$ and let $R' \subset R$ be the lower left quarter of R. Notice that, for any point $x = (x_1, x_2) \in R'$, the expression

$$D_N(x) - D_N\big(x + (h_1, 0)\big) + D_N\big(x + (h_1, h_2)\big) - D_N\big(x + (0, h_2)\big)$$

$$= -N \cdot h_1 h_2 = -N \cdot \frac{|R|}{4}. \tag{2.35}$$

Indeed, since R is empty, the counting parts will cancel out, and the area parts will yield precisely the area of the rectangle with vertices at the four points in the identity above. Hence, it is easy to see that

$$\int_{R'} \Big(D_N(x) - D_N\big(x + (h_1, 0)\big) + D_N\big(x + (h_1, h_2)\big) - D_N\big(x + (0, h_2)\big) \Big)\, dx$$

$$= -N \cdot \frac{|R|}{4} \cdot |R'| = -N \cdot \frac{|R|^2}{4^2},$$
$$(2.36)$$

while, on the other hand,

$$\int_{R'} \Big(D_N(x) - D_N\big(x + (h_1, 0)\big) + D_N\big(x + (h_1, h_2)\big) - D_N\big(x + (0, h_2)\big) \Big)\, dx$$

$$= \int_R D_N(x) \cdot h_R(x)\, dx = \langle D_N, h_R \rangle.$$
$$(2.37)$$

In other words, the inner product of discrepancy with the Haar function supported by an empty rectangle picks up the local discrepancy arising purely from the area of the rectangle.

We are now ready to prove a crucial preliminary lemma.

Lemma 6. *Let $\mathcal{P}_N \subset [0, 1]^d$ be a distribution of N points and let $n \in \mathbb{N}$ be such that $2^{n-2} \le N < 2^{n-1}$. Then, for any $\mathbf{r} \in \mathbb{H}_n^d$, there exists an $\mathbf{r}$-function $f_{\mathbf{r}}$ with parameter $\mathbf{r}$ such that*

$$\langle D_N, f_{\mathbf{r}} \rangle \ge c_d > 0, \tag{2.38}$$

where the constant c_d depends on the dimension only.

Proof. Construct the function $f_{\mathbf{r}}$ in the following way:

$$f_{\mathbf{r}} = \sum_{R \in \mathcal{D}_{\mathbf{r}}^d : R \cap \mathcal{P}_N = \emptyset} (-1) \cdot h_R + \sum_{R \in \mathcal{D}_{\mathbf{r}}^d : R \cap \mathcal{P}_N \neq \emptyset} \mathrm{sgn}(\langle D_N, h_R \rangle) \cdot h_R \tag{2.39}$$

By our choice of n (2.25), at least 2^{n-1} of the 2^n rectangles in $\mathcal{D}_{\mathbf{r}}^d$ must be free of points of $\mathcal{P}_N$. It then follows from (2.32) and (2.33) that

$$\langle D_N, f_{\mathbf{r}} \rangle \ge - \sum_{R \cap \mathcal{P}_N = \emptyset} \langle D_N, h_R \rangle = \sum_{R \cap \mathcal{P}_N = \emptyset} \langle N x_1 \ldots x_d, h_R \rangle \tag{2.40}$$

$$= \sum_{R \cap \mathcal{P}_N = \emptyset} N \cdot \frac{|R|^2}{4^d} \ge 2^{n-1} \cdot 2^{n-2} \cdot \frac{2^{-2n}}{4^d} = c_d.$$

Remark. Roth [81] initially defined the functions $f_{\mathbf{r}}$ slightly differently: he set them equal to zero on those dyadic rectangles which *do* contain points of $\mathscr{P}_N$, i.e. Roth's functions consisted only of the first term of (2.39). While this bears no effect on this argument, it was later realized by Schmidt [95] that in more complex situations it is desirable to have more uniformity in the structure of these building blocks. He simply chose the sign that increases the inner product on non-empty rectangles (the second term in (2.39)). Schmidt's paper, as well as subsequent papers by Halász [48], Beck [8], the author of this chapter and collaborators [15–18], make use of the r-functions as defined here (2.28). As we shall see in Sect. 2.5.5, in certain cases this definition brings substantial simplifications, whereas allowing even a small number of zeros in the definition may significantly complicate matters.

We are now completely equipped to prove Roth's theorem. Lemma 6 produces a rather large collection of *orthogonal*(!) functions such that the projections of D_N onto each of them is big, hence the norm of D_N must be big: this is the punchline of Roth's proof.

Proof of Theorem 1. Roth's original proof made use of duality. Let us construct the following test function:

$$F = \sum_{\mathbf{r} \in \mathbb{H}_n^d} f_{\mathbf{r}}, \qquad (2.41)$$

where $f_{\mathbf{r}}$ are the r-functions provided by Lemma 6. Orthogonality of $f_{\mathbf{r}}$'s yields:

$$\|F\|_2 = \left(\sum_{\mathbf{r} \in \mathbb{H}_n^d} \|f_{\mathbf{r}}\|_2^2 \right)^{1/2} = (\#\mathbb{H}_n^d)^{1/2} \approx n^{\frac{d-1}{2}}, \qquad (2.42)$$

while Lemma 6 guarantees that

$$\langle D_N, F \rangle \geq (\#\mathbb{H}_n^d) \cdot c_d \approx n^{d-1}. \qquad (2.43)$$

Now Cauchy–Schwarz inequality easily implies that:

$$\|D_N\|_2 \geq \frac{\langle D_N, F \rangle}{\|F\|_2} \gtrsim n^{\frac{d-1}{2}} \approx \left(\log N \right)^{\frac{d-1}{2}}, \qquad (2.44)$$

which finishes the proof. □

As one can see, the role of the building blocks is played by the generalized Rademacher functions $f_{\mathbf{r}}$, which we shall observe again in many future arguments. Therefore it is naturally convenient that they are normalized both in L^∞ and in L^2.

One, of course, does not have to use duality to obtain this inequality: we could use orthogonality directly. This proof initially appeared in an unpublished manuscript of A. Pollington and its analogs are often useful when one wants to prove estimates in quasi-Banach spaces and is thus forced to avoid duality arguments, see e.g. (2.95). For the sake of completeness, we also include this variation of the proof.

Second proof of Theorem 1. The proof is based on the same idea. Let n be chosen as in (2.25). We use orthogonality, Bessel's inequality and (2.33) to write

$$\|D_N\|_2^2 \geq \sum_{|R|=2^{-n},\, R\cap \mathscr{P}_N=\emptyset} \frac{|\langle D_N, h_R\rangle|^2}{|R|} = \sum_{r\in \mathbb{H}_n^d} \sum_{R\in \mathscr{D}_r^d:\, R\cap \mathscr{P}_N=\emptyset} N^2 \cdot \frac{2^{-4n}}{2^{-n}\cdot 4^{2d}}$$

(2.45)

$$\gtrsim (\#\mathbb{H}_n^d)\cdot 2^{n-1}\cdot 2^{2n-4}2^{-3n} \approx n^{d-1} \approx \left(\log N\right)^{d-1}.$$

The first line of the above calculation may look a bit odd: this is a consequence of the L^∞ normalization of the Haar functions. □

One can easily extend the first proof to an L^p bound, $1 < p < \infty$, provided that one has the estimate for the L^q norm of the test function F, where q is the dual index to p, i.e. $1/p + 1/q = 1$. Indeed, it will be shown in the next section as a simple consequence of the Littlewood–Paley inequalities that for any $q \in (1,\infty)$ we have the same estimate as for the L^2 norm: $\|F\|_q \approx n^{\frac{d-1}{2}}$, see (2.85). Hence, replacing Cauchy–Schwarz by Hölder's inequality in (2.44), one immediately recovers Schmidt's result:

$$\|D_N\|_p \gtrsim (\log N)^{\frac{d-1}{2}}.$$

(2.46)

Schmidt had originally estimated the L^q norms of the function F in the case when $q = 2m$ is an even integer, by using essentially L^2 techniques: squaring out the integrands and analyzing the orthogonality of the obtained terms. We point out that an analog of the second proof (2.45) can be carried out also in L^p using the device of the product Littlewood–Paley square function instead of orthogonality. The reader is invited to proceed to the next section, Sect. 2.3, for details.

Recently, Hinrichs and Markhasin [54] have slightly modified Roth's method to obtain the best known value of the constant C_d in Theorem 1. Their idea is quite clever and simple. They have noticed that one can extend the summation in (2.45) to also include finer scales, i.e. rectangles with smaller volume $|R| \leq 2^{-n}$. A careful computation then yields $C_2 = 0.0327633\ldots$ and $C_d = \frac{1}{\sqrt{21}\cdot 2^{2d-1}\sqrt{(d-1)!}(\log 2)^{\frac{d-1}{2}}}$ for $d \geq 3$, where all logarithms are taken to be natural.

2.2.3 Lower Bounds for Cubature Formulas on Function Classes: Kolmogorov's Method

Before finishing the discussion of Roth's proof, we would like to highlight its striking similarity to some arguments in the closely related field of numerical integration: namely, Kolmogorov's proof of the lower estimate for the error of cubature formulas in the class $MW_r^p([0,1]^d)$ of functions whose rth mixed derivative has L^p norm at most one. For the purposes of our introductory exposition, we shall define these spaces in a naive fashion. For more general settings and a wider array of numerical integration results and their relations to discrepancy theory, the reader is referred to e.g. [39, 75, 105, 110].

Define the integration operator $(\mathscr{I}_d f)(x_1, \ldots, x_d) := \int_0^{x_1} \ldots \int_0^{x_d} f(y) \, dy_1 \ldots dy_d$. For $p \geq 1$ and an integer $r \geq 1$, define the space $MW_r^p([0, 1]^d) = (\mathscr{I}_d)^r(L^p([0, 1]^d))$, i.e. the image of L^p under the action of an r-fold composition of the integration operators. Let $B(L^p)$ be the unit ball of L^p and $B(MW_r^p) = (\mathscr{I}_d)^r(B(L^p))$ be its image, i.e. the unit ball of MW_r^p or the set of functions whose rth mixed derivative has L^p norm at most one. We shall encounter these classes again in Sect. 2.4.8.

The field of numerical integration is concerned with approximate computations of integrals and evaluations of the arising errors. Let $\mathscr{F}$ be a class of functions on $[0, 1]^d$ and $\mathscr{P}_N \subset [0, 1]^d$ be a set of N points. For an arbitrary function f on $[0, 1]^d$, define the cubature formula associated to $\mathscr{P}_N$ as

$$\Lambda(f, \mathscr{P}_N) = \frac{1}{N} \sum_{p \in \mathscr{P}_N} f(p). \tag{2.47}$$

Denote by $\Lambda_N(\mathscr{F}, \mathscr{P}_N)$ the supremum of the errors of this cubature formula over the class $\mathscr{F}$:

$$\Lambda_N(\mathscr{F}, \mathscr{P}_N) := \sup_{f \in F} \left| \Lambda(f, \mathscr{P}_N) - \int \ldots \int_{[0,1]^d} f(x) dx_1 \ldots dx_d \right|. \tag{2.48}$$

The infimum of this quantity over all choices of the point set $\mathscr{P}_N$ is the optimal error of the N-point cubature formulas on the class $\mathscr{F}$:

$$\delta_N(\mathscr{F}) := \inf_{\mathscr{P}_N : \#\mathscr{P}_N = N} \Lambda_N(\mathscr{F}, \mathscr{P}_N). \tag{2.49}$$

Notice that the star-discrepancy, $\|D_N\|_\infty$, is equal to $N \cdot \Lambda_N(\mathscr{F}, \mathscr{P}_N)$, where $\mathscr{F}$ is the class of characteristic functions of rectangles $[\mathbf{0}, x)$. This is only the most trivial of the vast and numerous connections between numerical integrations and discrepancy theory. We recommend, for example, the book [39] for a very friendly introduction to the relations between these fields. Also, in the present book, the chapter by E. Novak and H. Woźniakowski is devoted to the discussion of discrepancy and integration.

We shall also consider the space of functions whose rth mixed derivative satisfies the product Hölder condition. Recall that a univariate function f is called Hölder if the condition $|\Delta_t f(x)| \lesssim |t|$ for the difference operator $\Delta_t f(x) = f(x+t) - f(x)$ holds for all x. The multiparameter nature of the problems under consideration dictates that rather than using the standard generalization of this concept, we use the product version, where the difference operator is applied iteratively in each coordinate. For a vector $\mathbf{t} = (t_1, \ldots, t_d)$, $t_j > 0$, and a function f on $[0, 1]^d$, define

$\Delta_t f(x) = \Delta_{t_d}^{x_d}(\ldots \Delta_{t_1}^{x_1} f)\ldots)(x)$, where the superscript indicates the variable in which the difference operator is being applied. We denote by $H([0, 1]^d)$ the class of product Hölder functions—those functions for which

$$\|\Delta_t f\|_\infty \leq C |t_1| \cdot \ldots \cdot |t_d|, \tag{2.50}$$

and let $B(H([0, 1]^d))$ be the unit ball of this space, i.e. functions which are product Hölder with constant one: $\|\Delta_t f\|_\infty \leq |t_1| \cdot \ldots \cdot |t_d|$. Furthermore, denote by $B(MH_r([0, 1]^d)) = (\mathscr{T}_d)^r (B(H([0, 1]^d)))$ the class of functions whose rth mixed derivative has Hölder norm one.

It is not hard to check that for a smooth function f we have

$$\Delta_t f(x) = \int_{x_1}^{x_1+t_1} \cdots \int_{x_d}^{x_d+t_d} \frac{\partial^d f(y)}{\partial x_1 \ldots \partial x_d} dy, \tag{2.51}$$

while it is also clear that $|\Delta_t f(x)| \leq 2^d \|f\|_\infty$. Hence

$$|\Delta_t f(x)| \lesssim \min \left\{ \left\| \frac{\partial^d f}{\partial x_1 \ldots \partial x_d} \right\|_\infty \prod_{j=1}^d |t_j|, \ 2^d \|f\|_\infty \right\}. \tag{2.52}$$

We shall now demonstrate a method of proof of the lower bounds for the optimal integration errors $\delta_N(\mathscr{F})$ for some function classes. This method, which was invented by Kolmogorov, resembles Roth's method in discrepancy theory to a great extent. We shall prove the following theorem by means of an argument whose main idea the reader will easily recognize.

Theorem 7. *For any $r \in \mathbb{N}$, the optimal integration errors for the classes $B(MH_r)$ and $B(MW_r^2)$ satisfy the lower estimates*

$$\delta_N(B(MH_r)) \gtrsim N^{-r} (\log N)^{d-1}, \tag{2.53}$$

$$\delta_N(B(MW_r^2)) \gtrsim N^{-r} (\log N)^{\frac{d-1}{2}}. \tag{2.54}$$

Proof. The main idea of the method is to construct a function which is zero at all nodes of the cubature formula, but whose integral is large. Similarly to Roth's original proof of (2.5), this is achieved by appropriately defining the function on the dyadic rectangles which contain no chosen points.

We start by proving (2.53). Fix any positive infinitely-differentiable function $b(x)$ of one variable supported on the interval $[0, 1]$. For a dyadic box $R = R_1 \times \ldots \times R_d \in \mathscr{D}^d$, where $R_j = [k_j 2^{-s_j}, (k_j + 1)2^{-s_j})$, define the functions

$$b_R(x_1, \ldots, x_d) := \prod_{j=1}^d b(2^{s_j} x_j - k_j). \tag{2.55}$$

The function b_R is obviously supported on the rectangle R. As in (2.25) we choose n so that $2N < 2^n \leq 4N$. For each choice of $\mathbf{r} \in \mathbb{H}_n^d$, out of 2^n dyadic boxes $R \in \mathscr{D}_\mathbf{r}^d$, at least a half, 2^{n-1}, do not contain any points of $\mathscr{P}_N$. Set

$$G(x_1, \ldots, x_d) = c2^{-rn} \sum_{s \in \mathbb{H}_n^d} \sum_{R \in \mathscr{D}_s^d : R \cap \mathscr{P}_N = \emptyset} b_R(x_1, \ldots, x_d) \qquad (2.56)$$

for some small constant $c > 0$. It is evident that $\Lambda(G, \mathscr{P}_N) = 0$ because all the terms of G are supported on empty rectangles R, so that $G(p) = 0$ for all $p \in \mathscr{P}_N$. At the same time, denoting $B = \int_0^1 b(x)dx$, we have

$$\int_{[0,1]^d} G(x)dx_1 \ldots dx_d \geq c2^{-rn} \cdot \#\mathbb{H}_n^d \cdot 2^{n-1} \cdot 2^{-n} B^d \gtrsim 2^{-rn} n^{d-1}. \qquad (2.57)$$

Hence we obtain

$$\left| \Lambda(G, \mathscr{P}_N) - \int_{[0,1]^d} G(x)dx \right| \gtrsim 2^{-rn} n^{d-1} \approx N^{-r} (\log N)^{d-1}. \qquad (2.58)$$

It only remains to check that $G \in B(MH_r)$. The Hölder norm of the rth mixed derivative of G can be estimated in the following way

$$\left\| \Delta_\mathbf{t} \left(\left(\frac{\partial^d}{\partial x_1 \ldots \partial x_d} \right)^r G \right) \right\|_\infty$$

$$\leq c \sum_{s \in \mathbb{H}_n^d} \left\| \sum_{R \in \mathscr{D}_s^d : R \cap \mathscr{P}_N = \emptyset} 2^{-rs_j} \Delta_\mathbf{t} \left(\left(\frac{\partial^d}{\partial x_1 \ldots \partial x_d} \right)^r b_R \right) \right\|_\infty$$

$$\leq c \sum_{s \in \mathbb{H}_n^d} \prod_{j=1}^d 2^{-rs_j} \| \Delta_{t_j} (2^{rs_j} b^{(r)} (2^{s_j} x_j)) \|_\infty$$

$$\lesssim \sum_{s \in \mathbb{H}_n^d} \prod_{j=1}^d 2^{-rs_j} \min\{1, 2^{rs_j} |t_j|\}$$

$$\leq \prod_{j=1}^d \sum_{s_j=0}^\infty 2^{-rs_j} \min\{1, 2^{rs_j} |t_j|\} \lesssim \prod_{j=1}^d |t_j|, \qquad (2.59)$$

where we have used the fact that rectangles $R \in \mathscr{D}_\mathbf{r}^d$ are disjoint for fixed $\mathbf{r}$, the product structure of the functions b_R, and the estimate (2.52). Therefore $G \in B(MH_r)$ if the constant c is small enough and hence (2.53) is proved.

We turn to the proof of (2.54). As one can guess from the right-hand side of this inequality, it will resemble Roth's proof of the L^2 discrepancy estimate (2.5) even more. The argument will proceed along the same lines as the proof of (2.53), but the choice of the analog of the function b_R will be more delicate. The rth mixed derivatives of these functions should form an orthogonal family. Unfortunately, we cannot start with the Haar function, because even in one dimension its rth antiderivative $(\mathcal{T}_1)^r h_I$ is not compactly supported anymore if $r \geq 2$. In order to fix this problem, we can define auxiliary functions inductively depending on r. For a dyadic interval I, whose left and right halves are denoted by I_l and I_r, let us set $h_I^0 = h_I$, $h_I^1 = h_{I_l} - h_{I_r}$, and proceeding in a similar fashion $h_I^r = h_{I_l}^{r-1} - h_{I_r}^{r-1}$.

This construction creates the following effect: not only h_I^r itself, but also all of its antiderivatives $(\mathcal{T}_1)^k h_I^r$ of order $k \leq r - 1$ are supported on I and have mean zero, therefore the rth antiderivative $(\mathcal{T}_1)^r h_I^r$ is supported on the interval I. Set $\phi_{[0,1)}^r = (\mathcal{T}_1)^r h_{[0,1)}^r$. For a dyadic interval $I \in \mathcal{D}^d$, $I = [k2^{-j}, (k+1)2^{-j})$, we define $\phi_I^r(x) = \phi_{[0,1)}^r(2^j x - k)$, assuming that $\phi_{[0,1)}^r$ is zero outside $[0,1)$. Then we have

$$\left(\phi_I^r\right)^{(r)}(x) = 2^{jr}\left(\phi_{[0,1)}^r\right)^{(r)}(2^j x - k) = 2^{jr} h_{[0,1)}^r(2^j x - k) = 2^{jr} h_I^r(x) = |I|^{-r} h_I^r(x),$$
(2.60)

i.e. $\phi_I^r = |I|^{-r}(\mathcal{T}_1)^r h_I^r$. As usually, in the multivariate case, for a dyadic box R we define

$$\phi_R^r(x_1, \ldots, x_d) = \prod_{j=1}^d \phi_{R_j}^r(x_j), \qquad h_R^r(x_1, \ldots, x_d) = \prod_{j=1}^d h_{R_j}^r(x_j). \qquad (2.61)$$

The one-dimensional case then implies that $\left(\frac{\partial^d}{\partial x_1 \ldots \partial x_d}\right)^r \phi_R^r(x) = |R|^{-r} h_R^r(x)$. Next, we choose n as before, $2N < 2^n \leq 4N$, and define a function similar to (2.56) and (2.41)

$$W(x_1, \ldots, x_d) = \gamma 2^{-rn} n^{-\frac{d-1}{2}} \sum_{s \in \mathbb{H}_n^d} \sum_{R \in \mathcal{D}_s^d : R \cap \mathcal{P}_N = \emptyset} \phi_R^r(x_1, \ldots, x_d). \qquad (2.62)$$

From the definition of ϕ_R^r, we have $\int_R \phi_R^r(x) = |R| \int_{[0,1]^d} \phi_{[0,1)^d}^r(x) dx$. Repeating the previous reasoning verbatim we find that $\Lambda(W, \mathcal{P}_N) = 0$ and

$$\left| \int_{[0,1]^d} W(x) dx \right| \gtrsim 2^{-rn} n^{-\frac{d-1}{2}} \#\mathbb{H}_n^d \, 2^{n-1} |R| \left| \int_{[0,1]^d} \phi_{[0,1)^d}^r(x) dx \right| \qquad (2.63)$$

$$\approx 2^{-rn} n^{\frac{d-1}{2}} \approx N^{-r}(\log N)^{\frac{d-1}{2}}.$$

To see that $W \in B(MW_r^2)$, we first observe that h_R^r form an orthogonal system. Obviously $W \in MW_r^2$ since each $\phi_R^r = (\mathcal{T}_k)^r h_R^r$. We use orthogonality to estimate the norm of the rth mixed derivative.

$$\left\| \left(\frac{\partial^d}{\partial x_1 \dots \partial x_d} \right)^r W \right\|_2^2 = \gamma^2 2^{-2rn} n^{-(d-1)} \sum_{s \in \mathbb{H}_n^d} \sum_{R \in \mathcal{D}_s^d : R \cap \mathcal{P}_N = \emptyset} 2^{2rn} \|h_R^r\|_2^2 \quad (2.64)$$

$$\approx n^{-(d-1)} \cdot n^{d-1} \cdot 2^n \cdot 2^{-n} \approx 1.$$

Hence $W \in B(MW^2)$ if γ is sufficiently small. This finishes the proof of (2.54). $\square$

We would like to point out that in order for this proof to be extended to the classes $B(MW_r^p)$ for $p \in (1, \infty)$, one should estimate the L^p norm of the mixed derivative of W, which, by the way, has a very similar structure to the test function (2.41) used by Roth. This can be done in a straightforward way using the material of the next section—Littlewood–Paley theory. The computation leading to this estimate is almost identical to (2.88). A more detailed account of various lower bounds for the errors of cubature formulas in classes of functions with mixed smoothness can be found, for example, in [105, 110]. The recent books [39] and [75], as well as the chapters of this book written by the same authors, give very nice accounts of the connections between discrepancy and numerical integration.

2.3 Littlewood–Paley Theory and Applications to Discrepancy

While Roth's method in its original form provides sharp information about the behavior of the L^2 norm of the discrepancy function, additional ideas and tools are required in order to extend the result to other function spaces, such as L^p, $1 < p < \infty$. In particular, the L^2 arguments of the previous section made essential use of orthogonality. Therefore, one needs an appropriate substitute for this notion in the case $p \neq 2$. A hands-on approach to this problem has been discovered by Schmidt in [95], see the discussion after (2.46).

However, harmonic analysis provides a natural tool which allows one to push orthogonality arguments from L^2 to L^p, as well as to more general function spaces. This tool is the so-called Littlewood–Paley theory. In this section, we shall give the necessary definitions, facts, and references relevant to our settings and concentrate on applications of this theory to the irregularities of distribution.

We would like to point out that in general Littlewood–Paley theory is a vast subject in harmonic analysis which arises in various fields and settings, has numerous applications, and is available in many different variations. For the purposes of our exposition we are restricting the discussion just to the dyadic Littlewood–Paley theory, i.e. its version related to the Haar function expansions and other similar dyadic orthogonal decompositions. Other versions of this theory (on Euclidean spaces $\mathbb{R}^n$, on domains, for trigonometric (Fourier) series, in the context of complex analysis) can be found in many modern books on harmonic analysis, e.g. [46, 101]. A more detailed treatment of the dyadic Littlewood–Paley theory can be enjoyed in [77].

2.3.1 One-Dimensional Dyadic Littlewood–Paley Theory

We start by considering the one-dimensional case. Let f be a measurable function on the interval $[0, 1]$. The dyadic (Haar) square function of f is defined as

$$Sf(x) = \left(\left| \int_0^1 f(t)dt \right|^2 + \sum_{I \in \mathscr{D}} \frac{|\langle f, h_I \rangle|^2}{|I|^2} \mathbf{1}_I(x) \right)^{\frac{1}{2}} \tag{2.65}$$

$$= \left(\left| \int_0^1 f(t)dt \right|^2 + \sum_{k=0}^{\infty} \left(\sum_{I \in \mathscr{D}, |I|=2^{-k}} \frac{\langle f, h_I \rangle}{|I|} h_I(x) \right)^2 \right)^{\frac{1}{2}}$$

We stress again that the formula may look unusual to a reader familiar with the subject due to the uncommon (L^∞, not L^2) normalization of the Haar functions. To intuitively justify the correctness of this definition, notice that $Sh_I = \mathbf{1}_I$ for any $I \in \mathscr{D}$. In particular, if the function has the Haar expansion $f = \sum_{I \in \mathscr{D}_*} a_I h_I$, then its square function is

$$Sf = \left(\sum_{I \in \mathscr{D}_*} a_I^2 \mathbf{1}_I \right)^{\frac{1}{2}} = \left(\sum_{k=-1}^{\infty} \left(\sum_{I \in \mathscr{D}: |I|=2^{-k}} a_I h_I \right)^2 \right)^{\frac{1}{2}}. \tag{2.66}$$

Since Haar functions (together with the constant $\mathbf{1}_{[0,1]}$) form an orthogonal basis of $L^2[0, 1]$, Parseval's identity immediately implies that

$$\|Sf\|_2 = \|f\|_2. \tag{2.67}$$

A non-trivial generalization of this fact to an equivalence of L^p norms, $1 < p < \infty$, is referred to as the Littlewood–Paley inequalities.

Theorem 8 (Littlewood–Paley inequalities, [118]). *For $1 < p < \infty$, there exist constants $B_p > A_p > 0$ such that for every function $f \in L^p[0, 1]$ we have*

$$A_p\|Sf\|_p \le \|f\|_p \le B_p\|Sf\|_p. \tag{2.68}$$

The asymptotic behavior of the constants A_p and B_p is known [118] and is very useful in numerous arguments, especially when (2.68) is applied for very high values of p. In particular $B_p \approx \sqrt{p}$ when p is large. Also, a simple duality argument shows that $A_q = B_p^{-1}$, where q is the dual index of p. The reader is invited to consult the following references for more details: [21, 101, 118].

The dyadic square function arises naturally in probability theory. Denote by $\mathscr{D}_k$ the collection of dyadic intervals in $[0, 1]$ of fixed length 2^{-k}. We shall slightly abuse notation and also denote the σ-algebra generated by this family by $\mathscr{D}_k$. Let f be an

L^2 function on $[0, 1]$. We construct the sequence of conditional expectations of f with respect to the families $\mathscr{D}_k$,

$$f_k = \mathbb{E}(f|\mathscr{D}_k) = \sum_{I \in \mathscr{D}_k} \frac{1}{|I|} \int_I f(x)dx \cdot \mathbf{1}_I. \tag{2.69}$$

The sequence $\{f_k\}_{k \geq 0}$ forms a martingale, meaning that $\mathbb{E}(f_{k+1}|\mathscr{D}_k) = f_k$. As usually for a dyadic interval I of length $2^{-(k-1)}$ denote by I_l and I_r its left and right dyadic "children" of length 2^{-k} and let $\langle f \rangle_I$ stand for the average of f over I. Keeping in mind that $2\langle f \rangle_I = \langle f \rangle_{I_l} + \langle f \rangle_{I_r}$, it is then easy to check that the martingale differences for $k \geq 1$ satisfy

$$d_k := f_k - f_{k-1} = \sum_{I \in \mathscr{D}_{k-1}} \left((\langle f \rangle_{I_l} \mathbf{1}_{I_l} + \langle f \rangle_{I_r} \mathbf{1}_{I_r}) - \langle f \rangle_I \mathbf{1}_I \right) \tag{2.70}$$

$$= \sum_{I \in \mathscr{D}_{k-1}} \frac{1}{2}(-\langle f \rangle_{I_l} + \langle f \rangle_{I_r})(-\mathbf{1}_{I_l} + \mathbf{1}_{I_r}) = \sum_{I \in \mathscr{D}_{k-1}} \frac{\langle f, h_I \rangle}{|I|} h_I.$$

Setting $d_0 = f_0$, we define the *martingale difference square function*:

$$Sf = \left(\sum_{k=0}^{\infty} |d_k|^2 \right)^{\frac{1}{2}}. \tag{2.71}$$

One can see from (2.66) that it is exactly the same object as the dyadic Littlewood–Paley square function defined in (2.65).

Littlewood–Paley square function estimates (2.68) can also be viewed as a generalization of the famous Khintchine inequality. Indeed, consider the Rademacher functions $r_k(x) = \sum_{I \in \mathscr{D}_k} h_I(x)$. Then at any point $x \in [0, 1]^d$, since dyadic intervals in $\mathscr{D}_k$ are disjoint, the square function of a linear combination of Rademacher functions is constant:

$$S\left(\sum_k \alpha_k r_k\right)(x) = \left(\sum_{k=0}^{\infty} \sum_{I \in \mathscr{D}_k} |\alpha_k|^2 \mathbf{1}_I \right)^{\frac{1}{2}} = \left(\sum_k |\alpha_k|^2\right)^{\frac{1}{2}}. \tag{2.72}$$

Therefore, Littlewood–Paley inequalities imply

$$\left\| \sum_k \alpha_k r_k \right\|_p \approx \left\| S\left(\sum_k \alpha_k r_k\right) \right\|_p = \left(\sum_k |\alpha_k|^2\right)^{\frac{1}{2}}, \tag{2.73}$$

which is precisely the Khintchine inequality for $p > 1$.

2.3.1.1   The Chang–Wilson–Wolff Inequality

The Littlewood–Paley inequalities are tightly related to the famous Chang–Wilson–
Wolff inequality, which states that if the square function of f is bounded, then f is
exponentially square integrable (subgaussian).

To formulate it rigorously we need to introduce exponential Orlicz function
classes. For a convex function $\psi : \mathbb{R}_+ \to \mathbb{R}_+$ with $\psi(0) = 0$, the Orlicz norm
of a function f on the domain D is defined as

$$\|f\|_\psi := \inf \left\{ K > 0 : \int_D \psi\left(\frac{|f(x)|}{K}\right) dx \leq 1 \right\} \tag{2.74}$$

The corresponding Orlicz space is the space of functions for which the above norm
is finite. For example, if $\psi(t) = t^p$, one recovers the usual L^p spaces. In the case
when $\psi(t) = e^{t^\alpha}$ for large values of t (if $\alpha \geq 1$, one may take $\psi(t) = e^{|t|^\alpha} - 1$,
however for $\alpha < 1$ convexity near zero would be violated) the arising Orlicz spaces
are denoted $\exp(L^\alpha)$. One of the most important members of this scale of function
spaces is $\exp(L^2)$, often referred to as the space of *exponentially square integrable*
or *subgaussian* functions. It is a standard fact that exponential Orlicz norms can be
characterized in the following ways

$$\|F\|_{\exp(L^\alpha)} \approx \sup_{q>1} q^{-1/\alpha} \|F\|_q \approx \sup_{\lambda>0} -\lambda^{-\alpha} \log \left|\{x : |F(x)| > \lambda\}\right| \tag{2.75}$$

The first equivalence here can be easily established using Taylor series for e^x
and Stirling's formula, while the second one is a simple computation involving
distribution functions, see a similar calculation in (2.120). The last expression
explains the term *subgaussian* in the context of functions $f \in \exp(L^2)$: in this
space, $\mathbb{P}(|f| > \lambda) \lesssim e^{-c\lambda^2}$.

We can now state the Chang–Wilson–Wolff inequality:

Theorem 9 (Chang–Wilson–Wolff inequality, [24]). *The following estimate
holds:*

$$\|f\|_{\exp(L^2)} \lesssim \|Sf\|_\infty. \tag{2.76}$$

This fact can be derived extremely easily as a consequence of the Littlewood–
Paley inequality (2.68) with sharp constants and the characterization (2.75) of the
exponential norm.

$$\|f\|_{\exp(L^2)} \approx \sup_{p\geq 1} p^{-\frac{1}{2}} \|f\|_p \lesssim \sup_{p\geq 1} p^{-\frac{1}{2}} \cdot \sqrt{p} \|Sf\|_p = \sup_{p\geq 1} \|Sf\|_p \leq \|Sf\|_\infty,$$

$$\tag{2.77}$$

which proves (2.76). $\square$

Observe that this bound strongly resembles the Khintchine inequality. Indeed, if we use the Littlewood–Paley inequality with sharp constants in (2.73), much in the same fashion as in (2.77), we obtain the exponential form of the Khintchine inequality

$$\left\| \sum_k \alpha_k r_k \right\|_{\exp(L^2)} \lesssim \left(\sum_k |\alpha_k|^2 \right)^{\frac{1}{2}}. \tag{2.78}$$

In other words, a linear combination of independent ± 1 random variables obeys a subgaussian estimate. For a precise quantitative distributional version of this statement see (2.119).

2.3.2 From Vector-Valued Inequalities to the Multiparameter Setting

It is very important for our further discussion that the Littlewood–Paley inequalities continue to hold for the Hilbert space-valued functions (in this case, all the arising integrals are understood as Bochner integrals). This delicate fact, which was proved in [42], allows one to extend the Littlewood–Paley inequalities to the multi-parameter setting in a fairly straightforward way by successively applying (2.68) in each dimension while treating the other dimensions as vector-valued coefficients [78, 101].

We note that in the general case one would apply the one dimensional Littlewood–Paley inequality d times—once in each coordinate, see Sect. 2.3.3. However, in the setting introduced by Roth's method (where the attention is restricted to dyadic boxes R of fixed volume $|R| = 2^{-n}$) one would apply it only $d - 1$ times since this is the number of free parameters—once the lengths of $d - 1$ sides are specified, the last one is determined automatically by the condition $|R| = 2^{-n}$.

Rather then stating the relevant inequalities in full generality (which an interested reader may find in [15, 78]), we postpone this to (2.87) and first illustrate the use of this approach by a simple example, important to the topic of our discussion.

Recall that the test function (2.41) in Roth's proof was constructed as $F = \sum_{\mathbf{r} \in \mathbb{H}_n^d} f_{\mathbf{r}} = \sum_{R:|R|=2^{-n}} \varepsilon_R h_R$, where $\varepsilon_R = \pm 1$. We want to estimate the L^q norm of F. Notice that we can rewrite it as $F = \sum_{I \in \mathscr{D}} \alpha_I h_I(x_1)$, where

$$\alpha_I = \sum_{\substack{R:|R|=2^{-n} \\ R_1 = I}} \varepsilon_R \prod_{j=2}^{\infty} h_{R_j}(x_j), \tag{2.79}$$

which allows one to apply the one-dimensional Littlewood–Paley square function (2.66) in the first coordinate x_1 to obtain

$$\| F \|_q = \left\| \sum_{|R|=2^{-n}} \varepsilon_R h_R \right\|_q \leq B_q \left\| \left[\sum_{r_1=0}^{n} \left| \sum_{\substack{|R|=2^{-n} \\ |R_1|=2^{-r_1}}} \varepsilon_R h_R \right|^2 \right]^{1/2} \right\|_q. \tag{2.80}$$

In the two-dimensional case for any value of r_1 all the rectangles satisfying the conditions of summation are disjoint, and for each point x we have:

$$\sum_{r_1=1}^{n} \left| \sum_{\substack{|R|=2^{-n} \\ |R_1|=2^{-r_1}}} \varepsilon_R h_R(x) \right|^2 = \sum_{r_1=1}^{n} \sum_{\substack{|R|=2^{-n} \\ |R_1|=2^{-r_1}}} |\varepsilon_R|^2 \mathbf{1}_R(x) = \sum_{\substack{R \in \mathscr{D}^2, \\ |R|=2^{-n}}} \mathbf{1}_R(x) = \#\mathbb{H}_n^2 \approx n,$$
$$\tag{2.81}$$

since $\varepsilon_R^2 = 1$ and every point is contained in $\#\mathbb{H}_n^2$ dyadic rectangles (one per each shape).

In the case $d \geq 3$, the expression on the right-hand side of (2.80) can be viewed as a Hilbert space-valued function. Indeed, fix all the coordinates except x_2 and define an ℓ^2-valued function

$$F_2(x_2) = \sum_{I \in \mathscr{D}} \left\{ \sum_{\substack{|R|=2^{-n}, R_2=I \\ |R_1|=2^{-r_1}}} \varepsilon_R \prod_{j \neq 2} h_{R_j}(x_j) \right\}_{r_1=1}^{n} h_I(x_2). \tag{2.82}$$

Then the expression inside the L^q norm on the right hand side of (2.80) is exactly $\| F_2(x_2) \|_{\ell^2}$. Applying the Hilbert space-valued Littlewood–Paley inequality in the second coordinate, we get

$$\| F \|_q = \left\| \sum_{|R|=2^{-n}} \varepsilon_R h_R \right\|_q \leq B_q \left\| \| F_2 \|_{\ell^2} \right\|_q$$

$$\leq B_q^2 \left\| \left[\sum_{r_1=1}^{n} \sum_{r_2=1}^{n} \left| \sum_{\substack{|R|=2^{-n} \\ |R_j|=2^{-r_j}, \ j=1,2}} \varepsilon_R h_R \right|^2 \right]^{1/2} \right\|_q. \tag{2.83}$$

And if $d = 3$, then an analog of (2.81) holds, completing the proof in this case. In the case of general d we continue applying the vector-valued Littlewood–Paley inequalities inductively in a similar fashion a total of $d - 1$ times to obtain

$$\| F \|_q = \left\| \sum_{|R|=2^{-n}} \varepsilon_R h_R \right\|_q \leq \dots$$

$$\leq B_q^{d-1} \left\| \left[\sum_{r_1=1}^{n} \cdots \sum_{r_{d-1}=1}^{n} \left| \sum_{\substack{|R|=2^{-n} \\ |R_j|=2^{-r_j}, j=1,\ldots,d-1}} \varepsilon_R h_R \right|^2 \right]^{1/2} \right\|_q. \tag{2.84}$$

Just as explained in (2.81), in this case all the rectangles in the innermost summation are disjoint and thus

$$\|F\|_q \leq B_q^{d-1} \left\| \left[\sum_{R\in\mathscr{D}^d, |R|=2^{-n}} |\varepsilon_R|^2 \mathbf{1}_R \right]^{\frac{1}{2}} \right\|_q = B_q^{d-1} \left(\#\mathbb{H}_n^d \right)^{\frac{1}{2}} \approx n^{\frac{d-1}{2}}. \tag{2.85}$$

2.3.3   Multiparameter (Product) Littlewood–Paley Theory

For a function of the form $f = \sum_{R\in\mathscr{D}_*^d} a_R h_R$ on $[0,1]^d$, the expression

$$S_d f(x) = \left[\sum_{R\in\mathscr{D}_*^d} |a_R|^2 \mathbf{1}_R(x) \right]^{\frac{1}{2}} = \left(\sum_{\mathbf{r}\in\{\{-1\}\cup\mathbb{Z}_+\}^d} \left| \sum_{R\in\mathscr{D}_{\mathbf{r}}^d} a_R h_R(x) \right|^2 \right)^{\frac{1}{2}} \tag{2.86}$$

is called the *product* dyadic square function of f. We remind that $\mathscr{D}_{\mathbf{r}}^d$ is the collection of dyadic rectangles R whose shape is defined by $|R_j| = 2^{-r_j}$ for $j = 1, \ldots, d$ and the rectangles in this family are disjoint.

The product Littlewood–Paley inequalities (whose proof is essentially identical to the argument presented above) state that

$$A_p^d \|S_d f\|_p \leq \|f\|_p \leq B_p^d \|S_d f\|_p. \tag{2.87}$$

With these inequalities at hand, one can estimate the L^q norm of F in a single line:

$$\|F\|_q = \left\| \sum_{|R|=2^{-n}} \varepsilon_R h_R \right\|_q \approx \|S_d f\|_q = \left\| \left[\sum_{|R|=2^{-n}} |\varepsilon_R|^2 \mathbf{1}_R \right]^{\frac{1}{2}} \right\|_q = \left(\#\mathbb{H}_n^d \right)^{\frac{1}{2}} \approx n^{\frac{d-1}{2}}.$$
$$\tag{2.88}$$

We have chosen to include a separate illustrative proof of this estimate earlier in order to demonstrate the essence of the product Littlewood–Paley theory. In addition, the argument leading to (2.85) gives a better implicit constant than the general inequalities (B_q^{d-1} versus B_q^d, according to the number of free parameters). While we generally are not concerned with the precise values of constants in this note, the behavior of this particular one plays an important role in some further estimates, see (2.99).

The proof of Schmidt's L^p lower bound (2.7) can now be finished immediately. Let q be the dual index of p, i.e. $1/p + 1/q = 1$ and let F be as defined in (2.41). Then, replacing Cauchy–Schwarz with Hölder's inequality in (2.44) and using (2.88), we obtain:

$$\|D_N\|_p \geq \frac{\langle D_N, F \rangle}{\|F\|_q} \gtrsim n^{\frac{d-1}{2}} \approx (\log N)^{\frac{d-1}{2}}. \qquad (2.89)$$

An analog of the second proof (2.45) of Roth's estimate (2.5) can also be carried out easily using the Littlewood–Paley square function. We include it since it provides a foundation for discrepancy estimates in other function spaces. It is particularly useful when one deals with quasi-Banach spaces and is forced to avoid duality arguments. We start with a simple lemma:

Lemma 10. *Let $A_k \subset [0,1]^d$, $k \in \mathbb{N}$, satisfy $\mu(A_k) \geq c$, where μ is the Lebesgue measure, then for any $M \in \mathbb{N}$*

$$\mu\left(\{x \in [0,1]^d : \sum_{k=1}^{M} \mathbf{1}_{A_k}(x) \geq \frac{1}{2}cM\} \right) > \frac{1}{2}c. \qquad (2.90)$$

Proof. Assuming this is not true, we immediately arrive to a contradiction

$$cM \leq \int \sum_{k=1}^{M} \mathbf{1}_{A_k}(x)\, dx < \frac{1}{2}cM \cdot \mu\left(\sum_{k=1}^{M} \mathbf{1}_{A_k} < \frac{1}{2}cM \right) \qquad (2.91)$$

$$+ M \cdot \mu\left(\sum_{k=1}^{M} \mathbf{1}_{A_k} \geq \frac{1}{2}cM \right) \leq \frac{1}{2}cM + M \cdot \frac{1}{2}c = cM. \square$$

We shall apply the lemma as follows: for each $\mathbf{r} \in \mathbb{H}_n^d$, let $A_{\mathbf{r}}$ be the union of rectangles $R \in \mathscr{D}_{\mathbf{r}}^d$ which do not contain points of $\mathscr{P}_N$. Then $\mu(A_{\mathbf{r}}) \geq c = \frac{1}{2}$ and $M = \#\mathbb{H}_n^d \approx n^{d-1}$. Let $E \subset [0,1]^d$ be the set of points where at least $M/4$ empty rectangles intersect. By the lemma above, $\mu(E) > \frac{1}{4}$. On this set, using (2.34):

$$S_d D_N(x) = \left[\sum_{R \in \mathscr{D}^d} \frac{\langle D_N, h_R \rangle^2}{|R|^2} \mathbf{1}_R(x) \right]^{\frac{1}{2}} \gtrsim (M \cdot N^2 2^{-2n})^{\frac{1}{2}} \approx n^{\frac{d-1}{2}}. \qquad (2.92)$$

Integrating this estimate over E and applying the Littlewood–Paley inequality (2.87) finishes the proof of (2.7):

$$\|D_N\|_p \gtrsim \|S_d D_N\|_p \gtrsim n^{\frac{d-1}{2}} \approx (\log N)^{\frac{d-1}{2}}. \square \qquad (2.93)$$

2.3.4   Lower Bounds in Other Function Spaces

The use of the Littlewood–Paley theory opens the door to considering much wider classes of functions than just the L^p spaces. Discrepancy theory has recently witnessed a surge of activity in this direction. We shall give a very brief overview of estimates and conjectures related to various function spaces. All of the results described below are direct descendants of Theorem 1 and Roth's method as every single one of them makes use of the Haar coefficients of the discrepancy function.

2.3.4.1   Hardy Spaces H^p

In particular, a direct extension of the above argument provides a lower bound of the discrepancy function in product Hardy spaces H^p, $0 < p \leq 1$. These spaces are generalizations due to Chang and R. Fefferman of the classical classes introduced by Hardy, see [22, 23]. The discussion of these spaces in the multiparameter dyadic setting, which is relevant to our situation, can be found in [12]. The Hardy space H^p norm of a function $f = \sum_{R \in \mathscr{D}^d} \alpha_R h_R$ is equivalent to the norm of its square function in L^p, i.e.

$$\|f\|_{H^p} \approx \|S_d f\|_p. \tag{2.94}$$

The following result about the Hardy space norm of the discrepancy function was obtained by Lacey [62]. For $0 < p \leq 1$,

$$\|\tilde{D}_N\|_{H^p} \geq C_{d,p} (\log N)^{\frac{d-1}{2}}, \tag{2.95}$$

where $\tilde{D}_N = \displaystyle\sum_{R \in \mathscr{D}^d} \frac{\langle D_N, h_R \rangle}{|R|} h_R$, in other words, $\tilde{D}_N$ is the discrepancy function D_N modified so as to have mean zero over every subset of coordinates. The proof of this result is a verbatim repetition of the previous proof (2.92)—one simply estimates the norm of the square function. Observe that a duality argument in the spirit of (2.44) would not have worked in this case, as H^p is only a *quasi*-Banach space for $p < 1$ and thus no duality arguments are available.

As this example clearly illustrates, in harmonic analysis Hardy spaces H^p serve as a natural substitute for L^p spaces when $p \leq 1$. Indeed, numerous analytic tools, such as square functions, maximal functions, atomic decompositions [101], allow one to extend the L^p estimates to the H^p setting for $0 < p \leq 1$. Similarly, the L^p asymptotics of the discrepancy is continued by the H^p estimates when $p \leq 1$.

The L^p behavior of the discrepancy function for this range of p, however, still remains a mystery. It is conjectured that the L^p norm should obey the same asymptotic bounds in N for *all* values of $p > 0$, which includes Conjecture 5 as a subcase.

Conjecture 11. For all $p \in (0, 1]$ the discrepancy function satisfies the estimate

$$\|D_N\|_p \geq C_{d,p} (\log N)^{\frac{d-1}{2}}. \tag{2.96}$$

2.3.4.2 The Behavior of Discrepancy in and Near L^1

The only currently available information regarding the conjecture above is the result of Halász [48] who proved that (2.96) indeed holds in dimension $d = 2$ for the L^1 norm:

$$\|D_N\|_1 \geq C \sqrt{\log N}. \tag{2.97}$$

We shall discuss his method in Sect. 2.4. Halász was also able to extend this inequality to higher dimensions, but only with the same right-hand side. Thus it is not known whether the L^1 bound even grows with the dimension. As to the case $p < 1$, no information whatsoever is available at this time.

In attempts to get close to L^1, Lacey [62] has proved that if one replaces L^1 with the Orlicz space $L(\log L)^{\frac{d-2}{2}}$, then the conjectured bound holds

$$\|D_N\|_{L(\log L)^{(d-2)/2}} \geq C_d (\log N)^{\frac{d-1}{2}}. \tag{2.98}$$

We remark that an adaptation of the proof of Schmidt's L^p bound given in the previous subsection, specifically estimate (2.85), can easily produce a slightly weaker inequality

$$\|D_N\|_{L(\log L)^{(d-1)/2}} \geq C_d (\log N)^{\frac{d-1}{2}} \tag{2.99}$$

Indeed, let F once again be as defined in (2.41). It is well known that (see e.g. [70]) the dual of $L(\log L)^{(d-1)/2}$ is the exponential Orlicz space $\exp(L^{2/(d-1)})$. Hence we need to estimate the norm of F in this space.

We recall that the constant arising in the Littlewood–Paley inequalities (2.68) is $B_q \approx \sqrt{q}$ for large q and the implicit constant in (2.85) is B_q^{d-1}. Thus using the equivalence between the exponential Orlicz norm and the growth of L^p norms (2.75) we obtain

$$\|F\|_{\exp\left(L^{2/(d-1)}\right)} \approx \sup_{q>1} q^{-\frac{d-1}{2}} \|F\|_q \lesssim \sup_{q>1} q^{-\frac{d-1}{2}} \cdot B_q^{d-1} n^{\frac{d-1}{2}} \tag{2.100}$$

$$\approx \sup_{q>1} q^{-\frac{d-1}{2}} \cdot q^{\frac{d-1}{2}} n^{\frac{d-1}{2}} = n^{\frac{d-1}{2}},$$

and (2.99) immediately follows by duality. Notice that a more straightforward bound (2.88) would not suffice for this estimate, since in the general d-parameter inequality the constant is of the order $q^{d/2}$, not $q^{(d-1)/2}$. These estimates are similar in spirit to the Chang–Wilson–Wolff inequality discussed in Sect. 2.3.1.1.

2.3.4.3 Besov Space Estimates

In a different vein, Triebel has recently studied the behavior of the discrepancy in Besov spaces [111, 112]. He proves, among other things, that

$$\|D_N\|_{S_{p,q}^r B([0,1]^d)} \geq C_{d,p,q,r} \, N^r \, (\log N)^{\frac{d-1}{q}}, \tag{2.101}$$

$$1 < p, q < \infty, \quad \frac{1}{p} - 1 < r < \frac{1}{p}. \tag{2.102}$$

Here the space $S_{p,q}^r B([0,1]^d)$ is the Besov space with dominating mixed smoothness. The exact original definition of this class is technical and would take our discussion far afield. There exists, however, a characterization of the Besov norms in terms of the Haar expansion (which is reminiscent of the Littlewood–Paley square function Sf). For a function $f = \sum_{R \in \mathscr{D}_*^d} \dfrac{\alpha_R}{|R|} h_R$, we have

$$\|f\|_{S_{p,q}^r B([0,1]^d)} \approx \left(\sum_{s \in (\{-1\} \cup \mathbb{Z}_+)^d} 2^{(s_1 + \ldots + s_d)(r - 1/p + 1)q} \left(\sum_{\substack{R \in \mathscr{D}_*^d : \\ |R_j| = 2^{-s_j}}} |\alpha_R|^p \right)^{\frac{q}{p}} \right)^{\frac{1}{q}} \tag{2.103}$$

whenever the right-hand side is finite.

To give a better idea about these spaces, we would mention that the index p represents integrability, r measures smoothness, and q is a certain 'correction' index. In particular, the case $q = 2$ corresponds to the well-known Sobolev spaces which, roughly speaking, consist of functions with rth mixed derivative in L^p and are similar to the previously defined spaces $MW_r^p([0,1]^d)$, see Sect. 2.2.3. Furthermore, when $r = 0$, $S_{p,2}^0 B([0,1]^d)$ is nothing but $L^p([0,1]^d)$. In particular, in the case $p = q = 2$, $r = 0$, the characterization (2.103) simply states that $\{h_R\}_{R \in \mathscr{D}_*^d}$ is an orthogonal basis of L^2.

Thus, if $q = 2$ and $r = 0$, one recovers Roth's L^2 and Schmidt's L^p estimates from (2.101). Inequalities (2.101) are sharp in all dimensions [53, $d = 2$], [71, $d \geq 3$], see Sect. 2.6. For more details, the reader is directed towards Triebel's recent book [111] concentrating on discrepancy and numerical integration in this context as well as to his numerous other famous books for a comprehensive treatise of the theory of function spaces in general.

2.3.4.4 Weighted L^p Estimates

The recent work of Ou [76] deals with the growth of the discrepancy function in weighted L^p spaces. A non-negative measurable function ω on $[0, 1]^d$ is called an A_p *(dyadic product) weight* if the following condition (initially introduced by Muckenhoupt [74]) holds

$$\sup_{R \in \mathscr{D}^d} \left(\int_R \omega(x) dx \right) \left(\int_R \omega^{-\frac{1}{p-1}}(x) dx \right)^{p-1} < \infty. \qquad (2.104)$$

The space $L^p(\omega)$ is then defined as the L^p space with respect to the measure $\omega(x)\,dx$. The class of A_p weights plays a tremendously important role in harmonic analysis: they give the largest reasonable class of measures such that the standard boundedness properties of classical operators (such as maximal functions, singular integrals, square functions) continue to hold in L^p spaces built on these measures. By an adaptation of the square function argument (2.92), Ou was able to show that

$$\|D_N\|_{L^p(\omega)} \geq C_{d,p,\omega} (\log N)^{\frac{d-1}{2}}, \qquad (2.105)$$

i.e. the behavior in weighted L^p spaces is essentially the same as in their Lebesgue-measure prototypes.

2.3.4.5 Approaching L^∞: BMO and Exponential Orlicz Spaces

Moving toward the other end of the L^p scale in attempts to understand the precise nature of the kink that occurs at the passage from the average (L^p) to the maximum (L^∞) norm, Bilyk, Lacey, Parissis, and Vagharshakyan [18] computed the lower bounds of the discrepancy function in spaces which are "close" to L^∞. One such space is the product dyadic BMO (which stands for *bounded mean oscillation*), i.e. the space of functions f for which the following norm is finite:

$$\|f\|_{\mathrm{BMO}} = \sup_{U \subset [0,1]^d} \left(\frac{1}{|U|} \sum_{R \in \mathscr{D}^d} \frac{|\langle f, h_R \rangle|^2}{|R|} \right)^{\frac{1}{2}}, \qquad (2.106)$$

where the supremum is extended over all measurable subsets of $[0, 1]^d$ with positive measure. Notice that in the case $d = 1$, when U is a dyadic interval, the expression inside the parentheses is actually equal to $\frac{1}{|U|} \int_U |f(x) - f_U|^2 dx$, where f_U is the mean of f over U, which yields exactly the standard one-dimensional BMO. The definition above, introduced by Chang and Fefferman [22], is a proper generalization of the classical BMO space to the dyadic multiparameter setting. In particular, the classical $H^1 - \mathrm{BMO}$ duality is preserved.

Just as H^1 often serves as a natural substitute for L^1, in many problems of harmonic analysis BMO naturally replaces L^∞. However, Bilyk, Lacey, Parissis, and Vagharshakyan showed that in this case the BMO norm behaves like L^p norms rather than L^∞:

$$\|D_N\|_{\mathrm{BMO}} \geq C_d (\log N)^{\frac{d-1}{2}}. \tag{2.107}$$

In fact, this estimate is not hard to obtain with the help of the same test function F (2.41) that we have used several times already—all we have to do is estimate its dual (H^1) norm. Just as in (2.88):

$$\|F\|_{H^1} \approx \|SF\|_1 = \left\| \left[\sum_{R \in \mathscr{D}^d, |R|=2^{-n}} |\varepsilon_R|^2 \mathbf{1}_R \right]^{\frac{1}{2}} \right\|_1 = \left(\#\mathbb{H}_n^d \right)^{\frac{1}{2}} \approx n^{\frac{d-1}{2}}, \tag{2.108}$$

which immediately yields the result.

In addition, the authors prove lower bounds in the aforementioned exponential Orlicz spaces, see (2.75). These spaces $\exp(L^\alpha)$ serve as an intermediate scale between the L^p spaces, $p < \infty$, and L^∞. In particular, for all $\alpha > 0$ and for all $1 < p < \infty$, we have $L^\infty \subset \exp(L^\alpha) \subset L^p$. The following estimate is contained in [18]: *in dimension $d = 2$ for all $2 \leq \alpha < \infty$ we have*

$$\|D_N\|_{\exp(L^\alpha)} \geq C (\log N)^{1-\frac{1}{\alpha}}. \tag{2.109}$$

We note that this inequality can be viewed as a smooth interpolation of lower bounds between L^p and L^∞. Indeed, when $\alpha = 2$ (the subgaussian case $\exp(L^2)$), the estimate is $\sqrt{\log N}$—the same as in L^2. On the other hand, as α approaches infinity, the right hand side approaches the L^∞ bound—$\log N$.

The proof of this estimate closely resembles Halász's proof of the L^∞ bound (see (2.128) below), with the obvious modification that the test function has to be estimated in the dual space $\left(\exp(L^\alpha) \right)^* = L(\log L)^{1/\alpha}$. Hence the same problems and obstacles that arise when dealing with the star-discrepancy prevent straightforward extensions of this estimate to higher dimensions. We finish this discussion by mentioning that both of these estimates, (2.107) and (2.109), were shown to be sharp, see Sect. 2.6.

2.4   The Star-Discrepancy (L^∞) Lower Bounds and the Small Ball Inequality

We now turn our attention to the most important case: L^∞ bounds of the discrepancy function. As explained in the introduction, when the set $\mathscr{P}_N$ is distributed rather well, its discrepancy comes close to its maximal values only on a thin set, while staying relatively small on most of $[0, 1]^d$. Therefore the extremal L^∞ norm of this function has to be much larger than the averaging L^2 norm. This heuristic was first confirmed by Schmidt [92] who proved

$$\|D_N\|_\infty \geq C \log N. \tag{2.110}$$

Other proofs of this inequality have been later given by Liardet [68, 1979], Béjian [11, 1982] (who produced the best currently known value of the constant $C = 0.06$), and Halász [48, 1981]. The proof of Halász is the most relevant to the topic of the present survey as it relies on Roth's orthogonal function idea and takes it to a new level. However, before we proceed to Halász's proof of Schmidt's lower bound, we shall discuss another related inequality.

2.4.1   The Small Ball Conjecture: Formulations and Simple Estimates

The *small ball inequality*, which arises naturally in probability and approximation, besides being important and significant in its own right, also serves as a model for the lower bounds of the star-discrepancy (2.11). This inequality is concerned with the lower estimates of the supremum norm of linear combinations of multivariate Haar functions supported by dyadic boxes of fixed volume (we call such sums 'hyperbolic') and can be viewed as a reverse triangle inequality.

Unfortunately, this inequality does not (more precisely, has not been proved to) directly imply the lower bound for the L^∞ norm of the discrepancy function. It is, however, linked to discrepancy through Roth's orthogonal function method. Even though no formal connections are known, most arguments designed for this inequality can be transferred to the discrepancy setting. In a certain sense, it can be viewed as a linear version of the star-discrepancy estimate.

We now state the conjectured inequality:

Conjecture 12 (The small ball conjecture). In dimensions $d \geq 2$, for any choice of the coefficients α_R one has the following inequality:

$$n^{\frac{d-2}{2}} \left\| \sum_{R \in \mathscr{D}^d : |R|=2^{-n}} \alpha_R h_R \right\|_\infty \gtrsim 2^{-n} \sum_{R: |R|=2^{-n}} |\alpha_R|. \tag{2.111}$$

The challenge and the point of interest of the conjecture is the precise value of the exponent of n on the left-hand side. If one replaces $n^{(d-2)/2}$ by $n^{(d-1)/2}$, the inequality becomes almost trivial, and, in fact, holds even for the L^2 norm:

$$n^{\frac{d-1}{2}} \left\| \sum_{R \in \mathscr{D}^d : |R|=2^{-n}} \alpha_R h_R \right\|_2 \gtrsim 2^{-n} \sum_{R: |R|=2^{-n}} |\alpha_R|. \tag{2.112}$$

Proof of (2.112). Indeed, using the orthogonality of Haar functions and keeping in mind that $\|h_R\|_2 = |R|^{1/2}$, we obtain

$$\left\| \sum_{R \in \mathcal{D}^d : |R|=2^{-n}} \alpha_R h_R \right\|_2 = \left(\sum_{|R|=2^{-n}} |\alpha_R|^2 2^{-n} \right)^{\frac{1}{2}} \tag{2.113}$$

$$\gtrsim \frac{\sum_{|R|=2^{-n}} |\alpha_R| 2^{-n/2}}{\left(n^{d-1} 2^n \right)^{\frac{1}{2}}} = n^{-\frac{d-1}{2}} \cdot 2^{-n} \sum_{|R|=2^{-n}} |\alpha_R|,$$

where in the last line we have used the Cauchy–Schwarz inequality and the fact that the number of terms in the sum is of the order $n^{d-1} 2^n$.

Alternatively, this inequality can be proved by duality. Consider the familiar function $F = \sum_{\mathbf{r} \in \mathbb{H}_n^d} f_{\mathbf{r}} = \sum_{|R|=2^{-n}} \varepsilon_R h_R$, where $\varepsilon_R = \text{sgn}(\alpha_R)$. We know very well by now, see (2.42), that $\|F\|_2 \approx n^{\frac{d-1}{2}}$. On the other hand, by orthogonality,

$$\left\langle \sum_{|R|=2^{-n}} \alpha_R h_R, F \right\rangle = \sum_{|R|=2^{-n}} |\alpha_R| \|h_R\|_2^2 = 2^{-n} \sum_{|R|=2^{-n}} |\alpha_R|, \tag{2.114}$$

which immediately implies (2.112). □

As we have already witnessed on several occasions, the presence of the quantity $d - 1$ in this context is absolutely natural, as it is, in fact, the number of free parameters dictated by the condition $|R| = 2^{-n}$. The passage to $d - 2$ for the L^∞ norm requires a much deeper analysis and brings out a number of complications.

The L^2 inequality (2.112) and the conjecture (2.111) should be compared to Roth's L^2 discrepancy estimate (2.5) and Conjecture 3. The computations just presented are very close to the proof (2.45) and (2.44) of (2.5). In fact, the resemblance becomes even more striking if one restricts the attention to the case when all the coefficients $\alpha_R = \pm 1$. In this case $2^{-n} \sum_{|R|=2^{-n}} |\alpha_R| \approx n^{d-1}$ and the L^2 estimate (2.112) becomes

$$\left\| \sum_{R \in \mathcal{D}^d : |R|=2^{-n}} \alpha_R h_R \right\|_2 \gtrsim n^{\frac{d-1}{2}}, \tag{2.115}$$

while the conjectured L^∞ inequality (2.111) for $\alpha_R = \pm 1$ turns into

Conjecture 13 (The signed small ball conjecture). If all the coefficients $\alpha_R = \pm 1$, we have the inequality

$$\left\| \sum_{R \in \mathcal{D}^d : |R|=2^{-n}} \alpha_R h_R \right\|_\infty \gtrsim n^{\frac{d}{2}}. \tag{2.116}$$

Recalling that n in Roth's argument was chosen to be approximately $\log_2 N$, one immediately sees the similarity of these inequalities to (2.5) and (2.11).

We would like to add a few comments about the signed small ball conjecture. There are some indications that this restricted version may turn out to be significantly simpler to prove than the more general Conjecture 12, see Sect. 2.5.5. However, this variation of the conjecture, unlike its full form, does not appear to have any real applications. On the other hand, one can formulate a slightly more generic statement of the conjecture by allowing some coefficients to equal zero, but not allowing the left-hand side to degenerate:

Conjecture 14 (Generic signed small ball conjecture). Assume that the coefficients α_R are either ± 1 or 0, and no more than half of all the coefficients are zero. Then we have the inequality

$$\left\| \sum_{R \in \mathscr{D}^d : |R| = 2^{-n}} \alpha_R h_R \right\|_\infty \gtrsim n^{\frac{d}{2}}. \tag{2.117}$$

This form of the conjecture is strong enough to yield applications, see Sect. 2.4.8. Unfortunately, it seems to be just as hard as the general small ball conjecture (2.111).

2.4.2   Sharpness of the Small Ball Conjecture

Choosing α_R's to be either independent Gaussian random variables or independent random signs $\alpha_R = \pm 1$ verifies that this conjecture is sharp, see e.g. [15] or [108]. We include the proof of the sharpness of inequality (2.111) here for the sake of completeness.

Lemma 15 (Sharpness of the small ball conjecture). *Let $\{\alpha_R\}_{R \in \mathscr{D}^d : |R| = 2^{-n}}$ be independent ± 1 random variables. Then, on the average, the converse of the small ball inequality holds, i.e.*

$$\mathbb{E} \left\| \sum_{|R| = 2^{-n}} \alpha_R h_R(x) \right\|_\infty \lesssim n^{-\frac{d-2}{2}} 2^{-n} \sum_{|R| = 2^{-n}} |\alpha_R| = n^{d/2}. \tag{2.118}$$

Proof. The function $\sum_{|R| = 2^{-n}} \alpha_R h_R(x)$ is constant on dyadic cubes Q_k of sidelength $2^{-(n+1)}$. The total number of such cubes is $M = 2^{(n+1)d}$. Let us define M random variables $X_k = \sum_{|R| = 2^{-n}} \alpha_R h_R |_{Q_k}$. Since X_k is a sum of $\#\mathbb{H}_n^d$ independent ± 1 random variables, by the Khintchine inequality we have $\mathbb{E}|X_k| \approx n^{(d-1)/2}$. Moreover, by a standard inequality (usually attributed to Bernstein, Hoeffding, Chernoff, or Azuma, see e.g. [55]), concerning sums of random variables, we have

$$\mathbb{P}(|X_k| > t) \leq 2 \exp \left(-t^2 / (4 \cdot \#\mathbb{H}_n^d) \right). \tag{2.119}$$

Recalling that $\#\mathbb{H}_n^d \approx n^{d-1}$, it is easy to deduce from this inequality that for some constant $C > 0$, the random variables $Y_k = \frac{1}{Cn^{(d-1)/2}} X_k$ have bounded $\exp(L^2)$ norm, in other words $\|X_k\|_{\exp(L^2)} \lesssim n^{(d-1)/2}$ (this is essentially the exponential form of the Khintchine inequality, see (2.78)). Indeed, denoting $\psi(t) = \exp(t^2)$, we obtain

$$
\mathbb{E}\,\psi(Y_k) = \int_0^\infty \mathbb{P}\big(\psi(Y_k) > t\big) = \int_0^\infty \mathbb{P}\big(|X_k| > Cn^{(d-1)/2}\sqrt{\log t}\,\big)\,dt
$$

$$
\leq \int_0^\infty \min\{1, 2\exp(-C^2 n^{d-1}\log t/(4 \cdot \#\mathbb{H}_n^d))\}\,dt
$$

$$
\leq \int_0^\infty \min\{1, t^{-K}\}\,dt \lesssim 1, \tag{2.120}
$$

where $K > 1$, if C is large enough. Therefore, applying Jensen's inequality with the convex function ψ, we get

$$
\psi\left(\mathbb{E}\sup_{k=1,\ldots,M}|Y_k|\right) \leq \mathbb{E}\psi\left(\sup_{k=1,\ldots,M}|Y_k|\right) \leq \mathbb{E}\sup_{k=1,\ldots,M}\psi(|Y_k|)
$$

$$
\leq \mathbb{E}\sum_{k=1}^M \psi(|Y_k|) \lesssim M = 2^{(n+1)d}. \tag{2.121}
$$

Since $\psi^{-1}(t) = \sqrt{\log t}$, we arrive to

$$
\mathbb{E}\left\|\sum_{|R|=2^{-n}}\alpha_R h_R(x)\right\|_\infty = Cn^{\frac{d-1}{2}}\mathbb{E}\sup|Y_k| \lesssim n^{\frac{d-1}{2}} \cdot \psi^{-1}(2^{(n+1)d}) \approx n^{d/2},
$$

$$
\tag{2.122}
$$

which finishes the proof. $\qquad\qquad\qquad\qquad\qquad\qquad\qquad\qquad\qquad\qquad\square$

The sharpness of the Small Ball Conjecture provides evidence that perhaps the correct estimate for the star-discrepancy should be Conjecture 3: $\|D_N\|_\infty \gtrsim (\log N)^{d/2}$. To validate the evidence we shall now illustrate the connection between this inequality and the discrepancy estimates. As mentioned earlier, the connection is not direct, but rather comes from the method of proof. We have already discussed the similarities between the proofs of the L^2 inequalities. Let us now turn to the case of L^∞.

The small ball conjecture (2.111) has been verified in $d = 2$ by M. Talagrand [102] in 1994. In 1995, V. Temlyakov [107] (see also [106, 108]) has given another, very elegant proof of this inequality in two dimensions, which closely resembled the argument of Halász [48] for (2.8). We shall present Temlyakov's proof first as it is somewhat "cleaner" and avoids some technicalities. Then we shall explain which adjustments need to be made in order to translate this argument into Halász's proof of Schmidt's estimate for $\|D_N\|_\infty$.

2.4.3   Proof of the Small Ball Conjecture in Dimension $d = 2$

The proof is based on Riesz products. An important feature of the two-dimensional case is the following *product rule*.

Lemma 16 (Product rule). *Assume that R, $R' \in \mathscr{D}^2$ are not disjoint, $R \neq R'$, and $|R| = |R'|$, then*

$$h_R \cdot h_{R'} = \pm h_{R \cap R'}, \tag{2.123}$$

i.e. the product of two Haar functions is again a Haar function.

The proof of this fact is straightforward. Unfortunately, this rule does not hold in higher dimensions. Indeed, for $d \geq 3$ one can have two different boxes of the same volume which coincide in one of the coordinates, say $R_1 = R'_1$. Then, $h_{R_1} \cdot h_{R'_1} = h_{R_1}^2 = \mathbf{1}_{R_1}$, so we lose orthogonality in the first coordinate. Since, as the reader will see below, we shall be considering very long products, the orthogonality may be lost completely. The fact that the product rule fails in higher dimensions is a major obstruction on the path to solving the conjecture.

For each $k = 0, \ldots, n$ consider the r-functions $f_k = \sum_{|R|=2^{-n}, |R_1|=2^{-k}} \operatorname{sgn}(\alpha_R) h_R$. Obviously, in two dimensions, the conditions $|R| = 2^{-n}$ and $|R_1| = 2^{-k}$ uniquely define the shape of a dyadic rectangle. Hence these are really r-functions, $f_k = f_{\mathbf{r}}$ with $\mathbf{r} = (k, n-k)$ and $\varepsilon_R = \operatorname{sgn}(\alpha_R)$. We are now ready to construct the test function as a Riesz product:

$$\Psi := \prod_{k=0}^{n} \left(1 + f_k \right). \tag{2.124}$$

First of all, notice that Ψ is non-negative. Indeed, since f_k's only take the values ± 1, each factor above is equal to either 0 or 2. Thus, we can say even more than $\Psi \geq 0$: the only possible values of Ψ are 0 and 2^{n+1}. Next, we observe that $\int \Psi(x) dx = 1$. This can be explained as follows. Expand the product in (2.124). The leading term is equal to 1. All the other terms are products of Haar functions; therefore, by the product rule, they themselves are Haar functions and have integral zero. So, Ψ is a non-negative function with integral 1. In other words, it has L^1 norm 1: $\|\Psi\|_1 = 1$.

A similar argument applies to the inner product of $\sum_{|R|=2^{-n}} \alpha_R h_R$ and Ψ. Multiplying out the product in (2.124) and using the product rule, one can see that

$$\Psi = 1 + \sum_{R \in \mathscr{D}^d : |R|=2^{-n}} \operatorname{sgn}(\alpha_R) h_R + \Psi_{>n}, \tag{2.125}$$

where $\Psi_{>n}$ is a linear combination of Haar functions supported by rectangles of area less than 2^{-n}. The first and the third term are orthogonal to $\sum_{|R|=2^{-n}} \alpha_R h_R$. Hence, using the trivial case of Hölder's inequality, $p = \infty$, $q = 1$,

$$\left\| \sum_{R\in\mathcal{D}^d : |R|=2^{-n}} \alpha_R h_R \right\|_\infty \geq \left\langle \sum_{|R|=2^{-n}} \alpha_R h_R, \Psi \right\rangle \qquad (2.126)$$

$$= \left\langle \sum_{|R|=2^{-n}} \alpha_R h_R, \sum_{|R|=2^{-n}} \mathrm{sgn}(\alpha_R) h_R \right\rangle$$

$$= \sum_{|R|=2^{-n}} \alpha_R \cdot \mathrm{sgn}(\alpha_R) \cdot \|h_R\|_2^2 = 2^{-n} \cdot \sum_{|R|=2^{-n}} |\alpha_R|,$$

$$(2.127)$$

and we are done (notice that for $d = 2$ we have $n^{\frac{d-2}{2}} = 1$). $\qquad\qquad\square$

2.4.4 Halász's Proof of Schmidt's Lower Bound for the Discrepancy

We now explain how the same idea can be used to prove a discrepancy estimate. This argument has, in fact, been created by Halász [48, 1981] even earlier than Temlyakov's proof of the small ball inequality in $d = 2$. In place of the r-functions f_k used above, we shall utilize the r-functions $f_k = \sum_{|R|=2^{-n}} \varepsilon_R h_R$ such that $\langle D_N, f_k \rangle \geq c$, which were used in Roth's proof (2.44) of the L^2 estimate (2.5) and whose existence is guaranteed by Lemma 6. The test function is then constructed in a fashion very similar to (2.124):

$$\Phi := \prod_{k=0}^{n} \left(1 + \gamma f_k\right) - 1 = \gamma \sum_{k=0}^{n} f_k + \Phi_{>n}, \qquad (2.128)$$

where $\gamma > 0$ is a small constant, and $\Phi_{>n}$, by the product rule (2.123), is in the span of Haar functions with support of area less than 2^{-n}. In complete analogy with the previous proof, we find that $\|\Phi\|_1 \leq 2$. Also,

$$\left\langle D_N, \sum_{k=0}^{n} f_k \right\rangle \geq c(n + 1) \geq C' \log N. \qquad (2.129)$$

Up to this point the argument repeated the proof of the two-dimensional small ball conjecture word for word. In this regard, one can view the small ball inequality as the linear part of the star-discrepancy estimate. Notice that subtracting 1 in the definition of Φ eliminated the need to estimate the "constant" term $\int D_N(x)dx$. All that remains is to show that the higher-order terms, $\Phi_{>n}$, yield a smaller input. This can be done by "brute force". We first prove an auxiliary lemma which is a natural extension of Lemma 6.

Lemma 17. *Let $f_\mathbf{s}$ be any* r*-function with parameter* $\mathbf{s}$. *Denote* $s = \|\mathbf{s}\|_1$. *Then, for some constant* $\beta_d > 0$,

$$\langle D_N, f_\mathbf{s} \rangle \leq \beta_d N 2^{-s}. \tag{2.130}$$

Proof. It follows from (2.33), that the area part of D_N satisfies $|\langle N x_1 \cdot \ldots \cdot x_d, f_\mathbf{s} \rangle| \lesssim 2^s \cdot N 2^{-2s} = N 2^{-s}$. As to the counting part, it follows from the proof of Lemma 6 that $\mathbf{1}_{[p,1]}$ is orthogonal to the functions h_R for all $R \in \mathcal{D}_\mathbf{s}^d$ except for the rectangle R which contains the point p. It is then easy to check that

$$\langle \mathbf{1}_{[p,1)}, f_\mathbf{s} \rangle = \langle \mathbf{1}_{[p,1)}, h_R \rangle \lesssim |R| = 2^{-s}. \tag{2.131}$$

The estimate for the counting part of D_N then follows by summing over all the points of $\mathscr{P}_N$. $\square$

We now estimate the higher order terms in $\langle D_N, \Phi \rangle$. Write $\Phi_{>n} = F_2 + F_3 + \ldots + F_n$, where

$$F_k = \gamma^k \sum_{0 \leq j_1 < j_2 < \cdots < j_k \leq n} f_{j_1} \cdot f_{j_2} \cdots f_{j_k}.$$

Notice that, due to the product rule, the product $f_{j_1} \cdot f_{j_2} \ldots f_{j_k}$ is an r-function with parameter $\mathbf{s} = (n - j_1, j_k)$, so $s = n - j_1 + j_k$. We reorganize the sum according to the parameter s, $n + 1 \leq s \leq 2n$. To obtain a term which yields an r-function corresponding to a fixed value of s, we need to have $j_k = j_1 + s - n \leq n$. This can be done in $2n - s + 1$ ways ($j_1 = 0, \ldots, 2n - s$). For each such choice of j_1 and j_k we can choose the "intermediate" $k - 2$ values in $\binom{s-n-1}{k-2}$ ways. Notice that we must have $2 \leq k \leq s - n + 1$. We obtain

$$\langle D_N, \Phi_{>n} \rangle = \sum_{k=2}^{n} \langle D_N, F_k \rangle = \sum_{s=n+1}^{2n} (2n - s + 1) \sum_{k=2}^{s-n+1} \binom{s-n-1}{k-2} \cdot \gamma^k \cdot \beta_2 N 2^{-s}$$

$$\leq \beta_2 n \sum_{s=n+1}^{2n} \gamma^2 (1 + \gamma)^{s-n-1} N 2^{-s} \leq \frac{1}{4} \beta_2 \gamma^2 n \sum_{s=n+1}^{\infty} \left(\frac{1 + \gamma}{2} \right)^{s-n-1}$$

$$= \frac{\gamma^2 \beta_2}{2(1 - \gamma)} n,$$

where we used that $N \leq 2^{n-1}$. Since $n \leq \log_2 N + 2$, by making γ very small we can assure that this quantity is less than $\frac{1}{2} C' \log N$, a half of (2.129). We finally obtain that

$$\|D_N\|_\infty \geq \frac{1}{2} \langle D_N, \Phi \rangle \geq \frac{1}{2} \left(C' \log N - \frac{1}{2} C' \log N \right) \gtrsim \log N, \tag{2.132}$$

which finishes the proof of Schmidt's bound. $\square$

2.4.5   The Proof of the L^1 Discrepancy Bound

To reinforce the potency of the powerful blend of Roth's method and the Riesz product techniques, we describe the proof of the L^1 lower bound (2.15) for the discrepancy function contained in the same fascinating paper by Halász [48] (while the L^∞ bound was already known, this result was completely new at the time). This argument introduces another brilliant idea: using complex numbers. The test function is constructed as follows

$$\Gamma := \prod_{k=0}^{n} \left(1 + \frac{i\gamma}{\sqrt{\log N}} f_k\right) - 1 = \frac{i\gamma}{\sqrt{\log N}} \sum_{k=0}^{n} f_k + \Gamma_{>n}, \qquad (2.133)$$

where a small constant $\gamma > 0$ and the "-1" in the end play the same role as in the previous argument, and $\Gamma_{>n}$ is the sum of the higher-order terms. Then one can see that

$$\|\Gamma\|_\infty \le \left(1 + \frac{\gamma^2}{\log N}\right)^{\frac{n}{2}} + 1 \le e^{\gamma^2/2} + 1 \lesssim 1. \qquad (2.134)$$

Just as before, one can show that the input of $\Gamma_{>n}$ will be small provided that γ is small enough. Hence,

$$\|D_N\|_1 \gtrsim |\langle D_N, \Gamma\rangle| \gtrsim \frac{\gamma}{\sqrt{\log N}} \langle D_N, \sum_{k=0}^{n} f_k\rangle \gtrsim \frac{n+1}{\sqrt{\log N}} \approx \sqrt{\log N}, \quad (2.135)$$

which finishes the proof of (2.15).   □

2.4.6   Riesz Products: Lacunary Fourier Series

It is not surprising that the Riesz product approach is effective in these problems. As discussed earlier, the extremal values of the discrepancy function (as well as of hyperbolic Haar sums) are achieved on very thin sets. Riesz products are known to capture such sets extremely well. In fact, we can see that Temlyakov's test function $\Psi = 2^{n+1} 1_E$, where E is the set on which all the functions f_k are positive, and in particular the L^∞ norm is attained. We shall make a further remark about the structure of this set E in Sect. 2.6.1.3.

But there is an even better explanation of the reason behind the successful application of the Riesz products in these contexts. In order to understand its roots we turn to classical Fourier analysis. Riesz products have initially appeared in connection with lacunary Fourier series [80, 96, 119] and have proved to be an extremely important tool for these objects. It would be interesting to compare the estimates whose proofs we have just discussed with a classical theorem about lacunary Fourier series due to Sidon [96,97]. Its proof can be found in almost every book on Fourier analysis, e.g. [46,56,119]. We shall reproduce it here in order to convince the reader that the proofs of the three previous inequalities (the small ball inequality (2.126) and lower bounds for $\|D_N\|_\infty$ (2.132) and $\|D_N\|_1$ (2.135) in dimension $d = 2$) are natural.

Recall that an increasing sequence $\{\lambda_j\}_{j=1}^\infty \subset \mathbb{N}$ is called *lacunary* if there exists $q > 1$ so that $\lambda_{j+1}/\lambda_j > q$. Let f be a 1-periodic function. We say that f has lacunary Fourier series if there exists a lacunary sequence Λ such that the Fourier coefficients of f,

$$\hat{f}(k) = \int_0^1 f(x)e^{-2\pi i k x}dx, \qquad (2.136)$$

are supported on the sequence Λ. In other words, $\hat{f}(k) = 0$ whenever $k \notin \Lambda$. We have the following theorem.

Theorem 18 (Sidon [96,97]).

1. Let f be a bounded 1-periodic function with lacunary Fourier series. Then we have

$$\|f\|_\infty \gtrsim \sum_{k=1}^\infty |\hat{f}(k)|. \qquad (2.137)$$

2. Assume that a function $f \in L^1[0,1]$ has lacunary Fourier series. Then

$$\|f\|_1 \gtrsim \|f\|_2. \qquad (2.138)$$

In both cases, the implicit constant depends only on the constant of lacunarity $q > 1$.

Proof. The reader will easily recognize the arguments that follow: the previous proofs in this section are their direct offsprings. We shall initially operate under the assumption that $q \geq 3$. This condition guarantees that any integer n can be represented in the form $n = \sum_k \varepsilon_k \lambda_k$, $\varepsilon_k = -1$, 0, 1, in at most one way.

We begin by proving the first part of the theorem. Construct the following Riesz product

$$P_N(x) = \prod_{k=1}^N \left(1 + \cos(2\pi\lambda_k x + \delta_k)\right), \qquad (2.139)$$

where δ_k is chosen so that $e^{i\delta_k} = \hat{f}(k)/|\hat{f}(k)|$. Obviously, $P_N(x)$ is non-negative for all x. It is also easy to see that $\widehat{P_N}(0) = \int_0^1 P(x)\,dx = 1$. Indeed, writing

$\cos t = \frac{1}{2}(e^{it} + e^{-it})$ and multiplying the product out, we see that the leading term is 1 and all others have integral zero. Hence, $\|P_N\| = 1$.

Moreover, for $k \leq N$, we have $\widehat{P_N}(\lambda_k) = \frac{1}{2}e^{i\delta_k}$. This again follows from expanding the Riesz product. We obtain a trigonometric polynomial, in which, due to our assumption that $q \geq 3$, the term $e^{2\pi i \lambda_k x}$ can only arise from the product of the cosine in the kth factor with the 1's coming from all the other factors. Besides, for $k > N$, evidently $\widehat{P_N}(\lambda_k) = 0$. Therefore we can apply the Parseval identity:

$$\|f\|_\infty \geq \left| \int_0^1 f(x)\overline{P_N(x)}dx \right| = \left| \sum_{k \in \mathbb{Z}} \hat{f}(k)\overline{\widehat{P_N}(k)} \right| = \frac{1}{2}\sum_{k=1}^N |\hat{f}(\lambda_k)|. \quad (2.140)$$

Clearly, we can now take the limit as $N \to \infty$. The restriction $q \geq 3$ may be removed in the following fashion. Find the smallest n such that $q^n > 3$, $1 - \frac{1}{q^n-1} > \frac{1}{q}$, $1 + \frac{1}{q^n-1} < q$ and subdivide the sequence $\{\lambda_j\}_{j=1}^\infty$ into n subsequences of the form $\Lambda_m = \{\lambda_{m+jn}\}_{j=1}^\infty$, $m = 0, 1, \ldots, n-1$. Then, repeating the argument above, we can prove an analog of (2.140) for Λ_m, i.e., $\|f\|_\infty \gtrsim \sum_{k \in \Lambda_m} |\hat{f}(k)|$, see [56, Chapter V] for details. Summing these estimates over m finishes the proof.

We now turn to the proof of the second part of the theorem. It will also be achieved using a Riesz product. We first assume that $q \geq 3$. Let $a_N^2 = \sum_{k=1}^N |\hat{f}(\lambda_k)|^2$ and $c_k = |\hat{f}(\lambda_k)|/a_N$. Define the function

$$Q_N(x) = \prod_{k=1}^N \left(1 + ic_k \cos(2\pi i \lambda_k x + \theta_k)\right). \quad (2.141)$$

It is then clear that $|Q_N(x)| \leq \prod_{k=1}^N (1 + c_k^2)^{1/2} \leq e^{\frac{1}{2}\sum c_k^2} = \sqrt{e}$, i.e. $\|Q_N\|_\infty \leq \sqrt{e}$. If $q \geq 3$, we can easily show that $\widehat{Q_N}(\lambda_k) = \frac{1}{2}ic_k e^{i\theta_k} = \frac{1}{2a_N}\overline{\hat{f}(\lambda_k)}$ for a proper choice of θ_k. Parseval's identity then yields

$$\|f\|_1 \gtrsim \int_0^1 f(x)\overline{Q_N(x)}dx = \sum_{k \in \mathbb{Z}} \hat{f}(k)\overline{\widehat{Q_N}(k)}$$

$$= \frac{1}{2a_N}\sum_{k=1}^N |\hat{f}(\lambda_k)|^2 = \frac{1}{2}\left(\sum_{k=1}^N |\hat{f}(\lambda_k)|^2\right)^{\frac{1}{2}}.$$

We finish the proof of (2.138) by letting N approach infinity and recalling that $\|f\|_2^2 = \left(\sum_{k=1}^N |\hat{f}(\lambda_k)|^2\right)^{\frac{1}{2}}$. The assumption $q \geq 3$ is removed in exactly the same way as in the first case. $\square$

One cannot help but notice extremely close similarities between the constructions of Riesz products for the small ball inequality and discrepancy estimates in dimension $d = 2$ and the ones just used in the proof of Sidon's theorem. Indeed, the constructions (2.124) and (2.128) bear strong resemblance to the product (2.139) used to estimate $\|f\|_\infty$, while the idea of the product (2.133) is nearly identical to the Riesz product (2.141) which produces the bound for $\|f\|_1$.

The absolute efficiency of Riesz products in the two-dimensional cases of the small ball inequality and the L^∞ discrepancy bound is justified by the fact that the condition $|R| = 2^{-n}$ effectively leaves only one free parameter (e.g., the value of $|R_1|$ defines the shape of the rectangle) and creates lacunarity ($|R_1| = 2^{-k}$, $k = 0, 1, \ldots, n$, in other words, the consecutive frequencies differ by a factor of 2). As we saw in this subsection, historically Riesz products were specifically designed to work in such settings (lacunary Fourier series, see e.g. [119], [80, 1918]). From the probabilistic point of view, Riesz products work best when the factors behave similarly to independent random variables, which relates perfectly to our problems for $d = 2$, since the functions f_k actually are independent random variables. The failure of the product rule explains the loss of independence in higher dimensions. This approach towards Conjecture 12 is taken in [19].

Before we proceed to the discussion of the recent progress in the multidimensional case, we would like to briefly explain the connections of Conjecture 12 to other areas of mathematics. While the connection of the small ball conjecture to discrepancy function is indirect, it does have important formal implications in probability and approximation theory.

2.4.7 Probability: The Small Ball Problem for the Brownian Sheet

Having read thus far, the reader is perhaps slightly confused by the name *small ball inequality*. It would be worthwhile to explain this nomenclature at this point. It comes from probability theory, namely the small ball problem for the Brownian sheet, which is concerned with finding the exact asymptotic behavior of the small deviation probability $\mathbb{P}(\|\mathbb{B}\|_{L^\infty([0,1]^d)} < \varepsilon)$ as $\varepsilon \to 0$, where $\mathbb{B}$ is the Brownian sheet, i.e. a centered multiparameter Gaussian process characterized by the covariance relation

$$\mathbb{E}\mathbb{B}(s) \cdot \mathbb{B}(t) = \prod_{j=1}^{d} \min(s_j, t_j) \qquad (2.142)$$

for $s, t \in [0, 1]^d$. It is known that the paths of $\mathbb{B}$ are almost surely continuous, so we can safely write $L^\infty([0, 1]^d)$ and $C([0, 1]^d)$ norms interchangeably.

The circle of small deviation (or small ball) problems is an active and rapidly developing area of modern probability theory. The common goal of all of these problems is computing the probability that the values of a random variable or a random process deviate little from the mean in various senses (i.e. stay in a *small ball* for a certain norm). This field is far less understood than the classical area of *large deviation* estimates, and numerous fundamental questions about small deviations are still open. A detailed account of small ball probabilities for Gaussian processes can be found in a nice survey [67]. The Brownian sheet $\mathbb{B}$, being the basic example of a multiparameter process and a natural generalization of the Brownian motion, presents special interest.

For the sake of brevity, let us denote the logarithm of the probability of the small deviation of $\mathbb{B}$ in the sup-norm by $\varphi(\varepsilon) := -\log \mathbb{P}(\|\mathbb{B}\|_{L^\infty([0,1]^d)} < \varepsilon)$. It is well known that in the case when $d = 1$, i.e. $\mathbb{B}$ is the Brownian *motion*, $\varphi(\varepsilon) \approx \varepsilon^{-2}$ for small ε. Moreover, even the precise value of the implicit constant is known in this case: $\lim_{\varepsilon \to 0} \frac{\varphi(\varepsilon)}{\varepsilon^{-2}} = \frac{\pi^2}{8}$, see [43]. In higher dimensions, however, the situation becomes more complicated due to the appearance of logarithmic factors in this asymptotics. In dimension $d = 2$, it was shown by Bass [2, 1988] that $\varphi(\varepsilon) \lesssim \frac{1}{\varepsilon^2} \left(\log \frac{1}{\varepsilon} \right)^3$. This estimate was later extended to all dimensions by Dunker, Kühn, Lifshits, and Linde [40]:

$$\varphi(\varepsilon) \lesssim \frac{1}{\varepsilon^2} \left(\log \frac{1}{\varepsilon} \right)^{2d-1}. \tag{2.143}$$

On the other hand, it was established much earlier [36, 1982] that the probability of the small deviation in the L^2 norm in all dimensions $d \geq 2$ satisfies

$$-\log \mathbb{P}(\|\mathbb{B}\|_{L^2([0,1]^d)} < \varepsilon) \approx \frac{1}{\varepsilon^2} \left(\log \frac{1}{\varepsilon} \right)^{2d-2}, \tag{2.144}$$

and since $\|\mathbb{B}\|_{L^2} \leq \|\mathbb{B}\|_{L^\infty}$, this readily implies $\varphi(\varepsilon) \gtrsim \frac{1}{\varepsilon^2} \left(\log \frac{1}{\varepsilon} \right)^{2d-2}$. Thus, one finds a gap of the order of $\log \frac{1}{\varepsilon}$ between the upper and the lower estimates, and the lower estimate is, in fact, an L^2 bound. This is a situation, which closely mirrors what happens in the case of discrepancy and the small ball inequality. For a while the experts were not sure which of the two bounds, if any, is correct (notice that the upper bound (2.143) is too big when $d = 1$). However, it is now generally believed that the upper bound (2.143) is sharp for $d \geq 2$.

Conjecture 19. In dimensions $d \geq 2$, for the Brownian sheet B we have

$$-\log \mathbb{P}(\|\mathbb{B}\|_{C([0,1]^d)} < \varepsilon) \simeq \varepsilon^{-2} (\log 1/\varepsilon)^{2d-1}, \quad \varepsilon \downarrow 0.$$

The lower bound for $d = 2$ in this conjecture has been obtained by Talagrand [102] using (2.111). The work of Bilyk, Lacey, and Vagharshakyan [15, 17] yields a decrease in the gap between lower and upper bounds in dimensions $d \geq 3$. Namely, there exists $\theta = \theta(d) > 0$ such that for small ε

$$-\mathbb{P}(\|\mathbb{B}\|_{C([0,1]^d)} < \varepsilon) \gtrsim \varepsilon^{-2} (\log 1/\varepsilon)^{2d-2+\theta}. \tag{2.145}$$

This improvement was based on the progress in the higher-dimensional small ball inequality (2.196). We should now explain how the small ball inequality for Haar functions (2.111) enters the picture in this problem. The argument presented here follows Talagrand's ideas.

2.4.7.1 Small Ball Inequality Implies a Lower Bound for the Small Deviation Probability

Consider the integration operator $\mathscr{T}_d$ acting on functions on the unit cube $[0, 1]^d$ and defined as

$$(\mathscr{T}_d f)(x_1, \ldots, x_d) := \int_0^{x_1} \cdots \int_0^{x_d} f(y_1, \ldots, y_d)\, dy_1 \ldots dy_d. \qquad (2.146)$$

Let $\{u_k\}_{k \in \mathbb{N}}$ be any orthonormal basis of $L^2([0, 1]^d)$ and set $\eta_k = \mathscr{T}_d u_k$. Then the Brownian sheet can be represented as

$$\mathbb{B} = \sum_{k \in \mathbb{N}} \gamma_k \eta_k, \qquad (2.147)$$

where γ_k are independent $\mathcal{N}(0, 1)$ (standard Gaussian) random variables. This idea goes back to Levy's construction of the Brownian motion [66]. The Gaussian structure is not hard to check. As to the covariance, writing $\eta_k(s) = \langle \mathbf{1}_{[0,s)}, u_k \rangle$ and taking into account independence of γ_k's, one can easily compute

$$\mathbb{E}\left(\sum_{k \in \mathbb{N}} \gamma_k \eta_k(s) \right)\left(\sum_{k \in \mathbb{N}} \gamma_k \eta_k(t) \right) = \sum_{k \in \mathbb{N}} \mathbb{E}\gamma_k^2 \cdot \eta_k(s)\eta_k(t) \qquad (2.148)$$

$$= \sum_{k \in \mathbb{N}} \langle \mathbf{1}_{[0,s)}, u_k \rangle \langle \mathbf{1}_{[0,t)}, u_k \rangle = \langle \mathbf{1}_{[0,s)}, \mathbf{1}_{[0,t)} \rangle$$

$$= \big|[\mathbf{0}, s) \cap [\mathbf{0}, t)\big| = \prod_{j=1}^{d} \min\{s_j, t_j\},$$

where in the second line we use the fact that u_k's form an orthonormal basis.

We shall use specific functions u_k and η_k. In dimension 1, for a dyadic interval I, consider the function

$$u_I(x) = \frac{1}{|I|^{\frac{1}{2}}}\big(-\mathbf{1}_{I_1}(x) + \mathbf{1}_{I_2}(x) + \mathbf{1}_{I_3}(x) - \mathbf{1}_{I_4}(x) \big), \qquad (2.149)$$

where I_j, $j = 1, \ldots, 4$ are four quarters of I: successives dyadic subintervals of I of length $\frac{1}{4}|I|$. The point of this choice of u is that both u and its antiderivative $\mathscr{T}_1 u$ behave similarly to the Haar function. In particular, the system $\{u_I\}_{I \in \mathscr{D}}$ is also an orthonormal basis of $L^2([0, 1])$. Observe that, up to the normalization, these functions are identical to the functions h_I^r with $r = 1$, defined in Sect. 2.2.3 in the proof of the lower bounds for the errors of cubature formulas in the class $B(MW_r^2)$, Theorem 7. In dimensions $d \geq 2$, one defines the basis functions indexed by dyadic rectangles $R = R_1 \times \cdots \times R_d \in \mathcal{D}^d$ as a tensor product

$$u_R(x_1, \ldots, x_d) = u_{R_1}(x_1) \cdot \ldots \cdot u_{R_d}(x_d). \tag{2.150}$$

The functions $\eta_R = \mathcal{T}_d u_R$ are then continuous; moreover, their mixed derivative $\frac{\partial^d}{\partial x_1 \ldots \partial x_d} \eta_R = u_R$ has L^2 norm equal to 1. We shall now formulate a version of the small ball conjecture for these continuous wavelets.

Conjecture 20. In all dimensions $d \geq 2$, for any choice of coefficients α_R, we have the inequality

$$n^{\frac{d-2}{2}} \left\| \sum_{|R|=2^{-n}} \alpha_R \eta_R \right\|_\infty \gtrsim 2^{-\frac{3n}{2}} \sum_{|R|=2^{-n}} |\alpha_R|. \tag{2.151}$$

Notice that the factor $2^{-\frac{3n}{2}}$ is different from the one in the inequality (2.111). This is a result of normalization: while we have used L^∞-normalized Haar functions, $\|h_R\|_\infty = 1$, the sup-norm of the functions η_R is smaller, $\|\eta_R\|_\infty \approx 2^{-|R|/2} = 2^{-n/2}$.

Even though this conjecture is at the first glance somewhat harder than the small ball conjecture for the Haar functions, the proofs are usually similar. In fact, Talagrand in his paper [102] proves this conjecture for $d = 2$, but first he presents the proof of Conjecture 12 for the Haar functions, (2.111), despite the fact that strictly speaking it was not necessary—it is simply more transparent, less obstructed by the technicalities, and clearly explains the main ideas. The Riesz product arguments can also be adapted to this case. One can even still use Riesz products built with Haar functions, which brings the amount of technical complications to an absolute minimum (see the discussion on the last page of [17]).

For now let us assume that the conjectured inequality (2.151) holds. We shall now show how it implies a lower bound for the small deviation problem. First, we shall need a well-known fact from probability theory, which we state here in a very simple form.

Lemma 21 (Anderson's lemma, [1]). *Let X_t, Y_t, $t \in T$ be independent centered Gaussian random processes. Then for any bounded measurable function $\theta : \mathbb{R} \to \mathbb{R}$*

$$\mathbb{P}(\sup_{t \in T} |X_t + \theta(t)| < c) \leq \mathbb{P}(\sup_{t \in T} |X_t| < c) \qquad \text{and} \tag{2.152}$$

$$\mathbb{P}(\sup_{t \in T} |X_t + Y_t| < c) \leq \mathbb{P}(\sup_{t \in T} |X_t| < c). \tag{2.153}$$

The first inequality of this lemma reflects a general intuition that Gaussian measures are concentrated near zero. The second inequality can be deduced by simply applying the first one conditionally.

We now employ Anderson's lemma to extract just one layer of η_R's from the decomposition (2.147) of $\mathbb{B}$—namely, we shall leave only those functions η_R which are supported on dyadic boxes of volume $|R| = 2^{-n}$ for a carefully chosen value of n. This idea strongly resonates with Roth's principle (2.24): just as in the case of the discrepancy function D_N, the behavior of the small ball probabilities of $\mathbb{B}$ is essentially defined by its projection onto the part of the basis which corresponds to rectangles with fixed volume. We apply (2.153) with $X_t = \sum_{|R|=2^{-n}} \gamma_R \eta_R$ and $Y_t = \sum_{|R|\neq 2^{-n}} \gamma_R \eta_R$. This would enable us to use the small ball inequality (2.151) as our next step.

$$\mathbb{P}(\|\mathbb{B}\|_{L^\infty([0,1]^d)} < \varepsilon) \leq \mathbb{P}\left(\left\|\sum_{|R|=2^{-n}} \gamma_R \eta_R\right\|_\infty < \varepsilon\right) \tag{2.154}$$

$$\leq \mathbb{P}\left(Cn^{-\frac{d-2}{2}}2^{-\frac{3n}{2}}\sum_{|R|=2^{-n}} |\gamma_R| < \varepsilon\right),$$

where C is the implied constant in (2.151). We are left with a standard object in probability theory: the sum of absolute values of independent $\mathcal{N}(0,1)$ random variables. Using the exponential form of Chebyshev's inequality we can write for a sequence of independent standard Gaussians γ_k:

$$\mathbb{P}\left(\sum_{k=1}^M |\gamma_k| \leq A\right) \leq e^A\,\mathbb{E}e^{-\sum_{k=1}^M |\gamma_k|} = e^A\left(\mathbb{E}e^{-|\gamma|}\right)^M. \tag{2.155}$$

We now apply this inequality with $M = \#\{R \in \mathscr{D}^d : |R| = 2^{-n}\} = 2^n \cdot \#\mathbb{H}_n^d \approx 2^n n^{d-1}$ and $A = \frac{\varepsilon}{C}n^{\frac{d-2}{2}}2^{\frac{3n}{2}}$ in order to be able to finish (2.154). We see that the right-hand side of (2.155) is then bounded by $\exp(\frac{1}{C}\varepsilon n^{\frac{d-2}{2}}2^{\frac{3n}{2}} - C_1 2^n n^{d-1})$. Choosing n to be the maximal integer such that

$$\frac{1}{C}\varepsilon n^{\frac{d-2}{2}}2^{\frac{3n}{2}} \leq \frac{1}{2}C_1 2^n n^{d-1}, \quad \text{i.e.} \quad \varepsilon \leq CC_1 2^{-\frac{n}{2}}n^{\frac{d}{2}}, \tag{2.156}$$

we find that, since in this case $\varepsilon \approx 2^{-\frac{n}{2}}n^{\frac{d}{2}}$,

$$\mathbb{P}\left(\sum_{|R|=2^{-n}} |\gamma_R| < \frac{\varepsilon}{C}n^{\frac{d-2}{2}}2^{\frac{3n}{2}}\right) \leq e^{-\frac{1}{2}C_1 2^n n^{d-1}} \leq e^{-\frac{C''}{\varepsilon^2}\left(\log\frac{1}{\varepsilon}\right)^{2d-1}}. \tag{2.157}$$

Therefore,

$$\varphi(\varepsilon) = -\log \mathbb{P}\big(\|\mathbb{B}\|_\infty < \varepsilon\big) \gtrsim \frac{1}{\varepsilon^2}\left(\log\frac{1}{\varepsilon}\right)^{2d-1}. \tag{2.158}$$

This finishes the proof of the lower bound in Conjecture 19 assuming that the smooth (or, rather, continuous) version of the small ball conjecture, Conjecture 20, holds.   □

Notice that, in another close parallel to Roth's method in discrepancy theory, we chose $n \approx \log\frac{1}{\varepsilon}$, although the exact choice of its value here was more delicate.

2.4.8 Approximation Theory: Entropy of Classes with Mixed Smoothness

Consider the integration operator $\mathcal{T}_d$ as described in (2.146). Let us define the function space $MW^p([0,1]^d) = \mathcal{T}_d(L^p([0,1]^d))$ and set $B(MW^p) = \mathcal{T}_d(B(L^p))$ to be the image of the unit ball of $L^p([0,1]^d)$ under the action of $\mathcal{T}_d$. In other words, $MW^p([0,1]^d)$ can be viewed as the space of functions on $[0,1]^d$ with mixed derivative $\frac{\partial^d f}{\partial x_1 \partial x_2 \ldots \partial x_d}$ in L^p, and $B(MW^p)$ is its unit ball. These function classes have already been defined in Sect. 2.2.3. It is not hard to see that $B(MW^p)$ is compact in the L^∞ metric. Its compactness may be quantified using the notion of *covering numbers*. Let B_∞ denote the unit ball of $L^\infty([0,1]^d)$ and define

$$N(\varepsilon, p, d) := \min\left\{N : \exists\{x_k\}_{k=1}^N \subset B(MW^p), \ B(MW^p) \subset \bigcup_{k=1}^N (x_k + \varepsilon B_\infty)\right\} \tag{2.159}$$

to be the least number N of L^∞ balls of radius ε needed to cover the unit ball $B(MW^p([0,1]^d))$, or, equivalently, the size of the smallest ε-net of $B(WM^p)$ in the uniform norm. The task at hand is to determine the correct order of growth of these numbers as $\varepsilon \downarrow 0$. The quantity

$$\psi(\varepsilon) = \log N(\varepsilon, p, d) \tag{2.160}$$

is referred to as the *metric entropy* of $B(MW^p)$ with respect to the L^∞ norm. The inverse of this quantity is known in the literature as *entropy numbers*:

$$\varepsilon_m := \inf\left\{\varepsilon : \exists\{x_k\}_{k=1}^{2^m} \subset B(MW^p), \ B(MW^p) \subset \bigcup_{k=1}^{2^m} (x_k + \varepsilon B_\infty)\right\}, \tag{2.161}$$

in other words, the smallest value of ε for which $\psi(\varepsilon) \leq m$. It is clear that estimates of metric entropy or covering numbers may be reformulated in terms of the entropy numbers, however we shall mostly resort to the former.

Kuelbs and Li [59] have discovered a tight connection between the small ball probabilities and the properties of the corresponding reproducing kernel Hilbert space, which in the case of the Brownian sheet is $WM^2([0, 1]^d)$, see Sect. 2.4.9. We state a partial form of their result tailored to the topic of our presentation.

Theorem 22 (Kuelbs and Li [59]). *The rates of asymptotic growth of the metric entropy $\psi(\varepsilon)$ of the space $MW^2([0, 1]^d)$ and the logarithm of the small ball probability of the d-dimensional Brownian sheet $\varphi(\varepsilon) = -\log \mathbb{P}(\|\mathbb{B}\|_\infty < \varepsilon)$ are related in the following way. For $\alpha > 0$,*

$$\varphi(\varepsilon) \approx \varepsilon^{-2}\left(\log \frac{1}{\varepsilon}\right)^\alpha \quad \text{if and only if} \quad \psi(\varepsilon) \approx \varepsilon^{-1}\left(\log \frac{1}{\varepsilon}\right)^{\alpha/2}. \tag{2.162}$$

We shall explore this connection in a more general setting in Sect. 2.4.9 and, in particular, prove this theorem. For more information and a wider spectrum of inequalities relating the small deviation probabilities and metric entropy, the reader is referred to [59, 67, 69].

Theorem 22 together with Conjecture 19 yields an equivalent conjecture:

Conjecture 23. For $d \geq 2$, we have

$$\log N(\varepsilon, 2, d) \simeq \varepsilon^{-1}(\log 1/\varepsilon)^{d-1/2}, \tag{2.163}$$

as $\varepsilon \downarrow 0$.

Just as in the case of the small ball probabilities for the Brownian sheet, the conjecture is resolved in dimension $d = 2$, which follows from the work of Talagrand [102]. The upper bound is known in all dimensions [40]. The lower bound of the order $\frac{1}{\varepsilon}\left(\log \frac{1}{\varepsilon}\right)^{d-1+\theta/2}$ in dimensions $d \geq 3$ can be 'translated' from the corresponding inequality (2.145) for the Brownian sheet.

We would now like to discuss the relation between this conjecture and the small ball inequality, Conjecture 12. Of course, one can combine the arguments of the previous subsection for the Brownian sheet with the Kuelbs–Li equivalence to demonstrate that the lower bound in Conjecture 23 follows from the small ball conjecture (2.111) or, more precisely, its continuous counterpart (2.151). However, we would like to illustrate how one can use the small ball inequality to directly deduce the lower bound for the metric entropy.

2.4.8.1 Small Ball Conjecture Implies a Lower Bound for Metric Entropy

Estimates akin to the small ball inequality (2.111) or (2.151) have been known for a long time to be useful for obtaining bounds of various approximation theory characteristics, such as metric entropy, entropy numbers, Kolmogorov widths etc., see [106, 108]. We present one possible approach to this connection.

We shall use the basis functions u_R (see (2.150)) and their antiderivatives $\eta_R = \mathcal{T}_d u_R$ defined in the previous subsection. Let $\sigma : \{R \in \mathcal{D}^d, |R| = 2^{-n}\} \to \{\pm 1\}$ be a choice of signs on the set of dyadic rectangles with fixed volume 2^{-n}. Define the functions

$$F_\sigma = \frac{c}{2^{\frac{n}{2}} n^{\frac{d-1}{2}}} \sum_{|R|=2^{-n}} \sigma_R \eta_R, \tag{2.164}$$

where $c > 0$ is a small constant. Then by the orthonormality of the functions u_R we have

$$\left\| \frac{\partial^d F_\sigma}{\partial x_1 \ldots \partial x_d} \right\|_2^2 = \left\| \frac{c}{2^{\frac{n}{2}} n^{\frac{d-1}{2}}} \sum_{|R|=2^{-n}} \sigma_R u_R \right\|_2^2 = \frac{c^2}{2^n n^{d-1}} 2^n \cdot \#\mathbb{H}_n^d \leq 1, \tag{2.165}$$

if c is sufficiently small. Since $\eta_R = \mathcal{T}_d u_R$, this estimate implies that $F_\sigma \in B(MW^2)$. Now assume that the continuous version of the small ball conjecture, Conjecture 20 holds for the functions η_R. Take two different choices of signs σ and σ'. Then (2.151) would imply:

$$\| F_\sigma - F_{\sigma'} \|_\infty = \left\| \frac{c}{2^{\frac{n}{2}} n^{\frac{d-1}{2}}} \sum_{|R|=2^{-n}} (\sigma_R - \sigma'_R) \eta_R \right\|_\infty \tag{2.166}$$

$$\gtrsim 2^{-\frac{n}{2}} n^{-\frac{d-1}{2}} \cdot n^{-\frac{d-2}{2}} 2^{-\frac{3n}{2}} \sum_{|R|=2^{-n}} |\sigma_R - \sigma'_R|$$

$$\gtrsim n^{-\frac{2d-3}{2}} 2^{-2n} \cdot 2^n \#\mathbb{H}_n^d \approx n^{1/2} 2^{-n},$$

where we have additionally assumed that σ and σ' differ on a large portion (e.g., one quarter) of all dyadic rectangles with volume 2^{-n}. We see that, in this case, F_σ and $F_{\sigma'}$ are ε-separated in L^∞ with $\varepsilon = 2^{-n} n^{1/2}$.

In order to construct a large ε net for the set $B(MW^2([0, 1]^d))$, it would be therefore sufficient to produce a large collection $\mathscr{A}$ of choices of sign σ such that any two elements of $\mathscr{A}$ are sufficiently different, i.e. coincide at most on a fixed portion of the rectangles.

Coding theory comes in handy in this setup. In fact, a reader familiar with its basic notions perhaps already recognized the concept of Hamming distance in the previous sentence. Consider a binary code X of length m, i.e. $X \subset \{0, 1\}^m$ is just a collection of strings of m zeros and ones. For any two elements $x, y \in X$, their Hamming distance is defined as

$$d_H(x, y) = \#\{j = 1, \ldots, m : x_j \neq y_j\}, \tag{2.167}$$

in other words, the number of components in which x and y do not coincide. The minimum Hamming distance (weight) of the code X is then defined as the smallest

Hamming distance between its elements, $\min_{x,y\in X,\,x\neq y} d_H(x,y)$. The following classical result in coding theory, which we state in the simplest form adapted to our exposition, provides a lower bound on the size of the maximal code with large minimum Hamming weight.

Lemma 24 (Gilbert–Varshamov bound [45,117]). *Let $A(m,k)$ denote the maximal size of a binary code of length m with the minimum Hamming distance at least k. Then*

$$A(m,k) \geq \frac{2^m}{\sum_{j=0}^{k-1}\binom{m}{j}}. \qquad (2.168)$$

The proof of this estimate is so beautifully simple that we decided to include it here.

Proof. We first observe that given an m-bit string $x \in \{0,1\}^m$, there are precisely $\binom{m}{j}$ strings $y \in \{0,1\}^m$ such that $d_H(x,y) = j$. Indeed, we need to choose j bits out of m that are to be changed. Hence the size of $B_H(x,k)$, the neighborhood of x of radius k in the Hamming metric (all elements y with $d_H(x,y) < k$), is equal to $\sum_{j=0}^{k-1}\binom{m}{j}$.

Let now X be the maximal code of length m with minimum Hamming weight k. Then $\cup_{x\in X} B_H(x,k) = \{0,1\}^m$, for otherwise there would exist another element whose distance to all points of X is at least k, which would violate the maximality of X. Thus,

$$2^m = \#\bigcup_{x\in X} B_H(x,k) \leq \sum_{x\in X} \#B_H(x,k) = \#X \cdot \sum_{j=0}^{k-1}\binom{m}{j}, \qquad (2.169)$$

which proves the lemma. □

We shall apply this lemma to codes X indexed by the family of dyadic rectangles $\{R \in \mathscr{D}^d, |R| = 2^{-n}\}$. Hence, the length of the code is $m = 2^n \#\mathbb{H}_n^d \approx 2^n n^{d-1}$. For any element of such a code $x \in X$, we can define a choice of sign σ^x by setting $\sigma_R^x = (-1)^{x_R}$. We would like the code to have the minimal Hamming weight of the same order of magnitude as the length of the code, i.e. $k \approx m \approx 2^n n^{d-1}$. Take, for example, $k = \frac{m}{4}$. One can easily check using Stirling's formula $m! \approx \frac{1}{\sqrt{2\pi m}}\left(\frac{m}{e}\right)^m$ that $\binom{m}{m/4} \approx \frac{1}{\sqrt{m}}\left(\frac{1}{4}\right)^{-m/4}\left(\frac{3}{4}\right)^{-3m/4}$. The Gilbert–Varshamov bound then guarantees that there exists such a code X with size at least

$$\#X \geq \frac{2^m}{\sum_{j=0}^{k-1}\binom{m}{j}} \geq \frac{2^m}{k\cdot\binom{m}{k}} = \frac{2^m}{m/4\cdot\binom{m}{m/4}} \qquad (2.170)$$

$$\approx \frac{1}{\sqrt{m}}\cdot 2^m \left(\frac{1}{4}\right)^{m/4}\left(\frac{3}{4}\right)^{3m/4} \gtrsim C^m$$

for some constant $C > 1$ when m is large, since $2 \cdot (1/4)^{1/4} \cdot (3/4)^{3/4} > 1$. To summarize, we can find a code such that its Hamming weight is roughly the same as its length m and its size is roughly the same as the size of the largest possible code, $\{0, 1\}^m$ (both are exponential in m).

Having chosen such a code X, we define the collection $\mathscr{A} = \{\sigma^x : x \in X\}$ and consider the set of functions $\mathscr{F} = \{F_\sigma\}_{\sigma \in \mathscr{A}}$. According to (2.166) this family is an ε-net of $B(MW^2)$ in the L^∞ norm with $\varepsilon = 2^{-n}n^{1/2}$. The cardinality of this family satisfies

$$\log \# \mathscr{F} = \log \# X \gtrsim m \approx 2^n n^{d-1} = 2^n n^{-1/2} \cdot n^{d-\frac{1}{2}} \approx \frac{1}{\varepsilon}\left(\log \frac{1}{\varepsilon}\right)^{d-\frac{1}{2}}, \quad (2.171)$$

which yields precisely the lower bound in Conjecture 23.   □

In the end we would like to observe that in the proof of this implication we have employed only a restricted form of the small ball inequality. In the computation (2.166), the coefficients $\alpha_R = \sigma_R - \sigma'_R$ take only three values: ± 2 and 0. Besides, zeros are not allowed to occur too often (at most a fixed proportion of all coefficients). This (up to a factor of 2) is exactly the setting of the *generic* signed small ball conjecture, Conjecture 14. Therefore, this version of the conjecture (but with smooth wavelets η_R in place of the Haar functions) is already sufficient for applications. However, unlike Conjecture 13 (the purely signed variant of the inequality, see Sect. 2.5.5), the generic setting does not seem to produce any real simplifications.

2.4.9 The Equivalence of Small Ball Probabilities and Metric Entropy

The equivalence of Conjecture 19 in probability and Conjecture 23 in approximation theory proved by Kuelbs and Li [59] is a fascinating connection between two problems, which at first glance have little in common. We strongly agree with Michel Talagrand who stated [102]:

> It certainly would be immoral to deprive the reader of a discussion of this beautifully simple fact (that once again demonstrates the power of abstract methods).

Therefore we would like devote a portion of this chapter to the discussion of the proof of this equivalence.

Before we are able to explain the argument however, we need to recall some classical results from the theory of Gaussian measures, which we shall state here without proof. Complete details and background information may be found in such excellent references as [20, 64], or [69]. We, rather than giving the most general definitions and statements, will mostly specialize to the particular problem at hand.

Let $\mathbf{P}$ be a Gaussian measure on the Banach space X. The *small ball problem* for the measure $\mathbf{P}$ is concerned with the asymptotic behavior of the quantity

$$\varphi(\varepsilon) = -\log \mathbf{P}(\varepsilon B_X), \qquad (2.172)$$

where B_X is the unit ball of the space X.

In the case we are interested in, the Brownian sheet, the space X is $C([0, 1]^d)$ and the measure $\mathbf{P}$ is the law of the Brownian sheet $\mathbb{B}$, i.e. for a set $A \in C([0, 1]^d)$, $\mathbf{P}(A) = \mathbb{P}(\mathbb{B} \in A)$. In this notation, the definition of $\varphi(\varepsilon)$ above coincides with the one given in Sect. 2.4.7

$$\varphi(\varepsilon) = -\log \mathbb{P}(\|\mathbb{B}\|_{L^\infty([0,1]^d)} < \varepsilon) = -\log \mathbf{P}(B_\infty(0, \varepsilon)), \qquad (2.173)$$

where $B_\infty(a, r)$ is the L^∞ ball of radius $r > 0$ centered at $a \in C([0, 1]^d)$. Recall that $\mathbb{B}$ has mean zero, so the measure $\mathbf{P}$ is centered.

Assume that X, as in our case, is a space of real valued functions on a domain $D \subset \mathbb{R}^d$ with the property that point evaluations $L_x(f) = f(x)$, $x \in D$, are continuous linear functionals on X. We can then introduce the covariance kernel of $\mathbf{P}$, the function $K_{\mathbf{P}} : D \times D \to \mathbb{R}$ defined by

$$K_{\mathbf{P}}(s, t) = \int_X f(s) f(t) \mathbf{P}(df). \qquad (2.174)$$

By definition, see (2.142), the covariance kernel of the Brownian sheet $\mathbb{B}$ is given by $K_{\mathbf{P}}(s, t) = \mathbb{E}\mathbb{B}(s)\mathbb{B}(t) = \prod_{j=1}^d \min\{s_j, t_j\}$.

The *reproducing kernel Hilbert space* $H_{\mathbf{P}}$ is then defined as the Hilbert space of functions $f \in X$ with the property that the reproducing kernel of $H_{\mathbf{P}}$ is precisely the covariance kernel of $\mathbf{P}$, i.e. for $t \in D$ and any $f \in H_{\mathbf{P}}$, the function evaluation of f at t can be represented as the inner product of f and $K_{\mathbf{P}}(\cdot, t)$,

$$f(t) = \langle f, K_{\mathbf{P}}(\cdot, t) \rangle. \qquad (2.175)$$

In the case when $X = C([0, 1]^d)$ and $\mathbf{P}$ is the law of the Brownian sheet, this space happens to be precisely the Sobolev space of functions with mixed derivative in L^2 as defined in the previous subsection, $H_{\mathbf{P}} = MW^2([0, 1]^d)$. Indeed, MW^2 is a Hilbert space with the inner product given by

$$\langle f, g \rangle_{MW^2} = \langle \phi_f, \phi_g \rangle_{L^2} = \int_{[0,1]^d} \frac{\partial^d f}{\partial x_1 \dots \partial x_d}(x) \cdot \frac{\partial^d g}{\partial x_1 \dots \partial x_d}(x) dx, \qquad (2.176)$$

where $\phi_f \in L^2([0, 1]^d)$ is such that $f = \mathcal{T}_d \phi_f$, in other words ϕ_f is the mixed derivative of f. It is easy to see that in $d = 1$, $\min\{s, t\} =$

$\int_0^s \mathbf{1}_{[0,t)}(\tau)\,d\tau = (\mathscr{T}_1 \mathbf{1}_{[0,t)})(s)$ for $s, t \in [0, 1]$. Therefore, in d dimensions $K_{\mathbf{P}}(s, t) = (\mathscr{T}_d (\prod_{j=1}^d \mathbf{1}_{[0,t_j)}))(s_j)$ and for any $f \in MW^2([0, 1]^d)$ we have

$$\langle f, K_{\mathbf{P}}(\cdot, t)\rangle_{MW^2} = \int_0^{t_1} \cdots \int_0^{t_d} \phi_f(s)ds = f(t), \qquad (2.177)$$

hence $K_{\mathbf{P}}(s, t)$ is in fact the reproducing kernel of MW^2.

In a certain sense, $H_{\mathbf{P}}$ is a subspace of X which carries most of the information about the measure $\mathbf{P}$. We shall need two standard facts which relate the Gaussian measure and its reproducing kernel Hilbert space.

Lemma 25. *Let $\mathbf{P}$ be a centered Gaussian measure on a Banach space X, let $H_{\mathbf{P}}$ be its reproducing kernel Hilbert space and $h \in H_{\mathbb{P}}$. Then, for any symmetric set $A \in X$ we have*

$$\exp\left(-\|h\|_{H_{\mathbf{P}}}^2/2\right) \cdot \mathbf{P}(A) \le \mathbf{P}(A + h) \le \mathbf{P}(A). \qquad (2.178)$$

The right inequality here is simply a restatement of Anderson's lemma, (2.153), which is intuitively natural since a Gaussian measure is concentrated around the mean. The left bound, known as Borell's inequality, shows that the measure of a shifted set decays not too fast, in a fashion suggested by the Gaussian structure of the measure. The assumption $h \in H_{\mathbf{P}}$ is crucial for Borell's inequality as the shifted measure $\mathbf{P}(\cdot + h)$ is not even absolutely continuous with respect to $\mathbf{P}$ unless h lies in the reproducing kernel Hilbert space.

The second fact that we shall rely upon is the isoperimetric inequality.

Theorem 26 (Gaussian isoperimetric inequality). *Let $\mathbf{P}$ be a centered Gaussian measure on the Banach space X and K be the unit ball of $H_{\mathbf{P}}$. For a measurable set $A \subset X$ and $\lambda > 0$, we have*

$$\Phi^{-1}(\mathbf{P}(A + \lambda K)) \ge \Phi^{-1}(\mathbf{P}(A)) + \lambda, \qquad (2.179)$$

where Φ is the distribution function of a $\mathcal{N}(0, 1)$ (standard Gaussian) random variable, i.e. $\Phi(x) = \dfrac{1}{\sqrt{2\pi}} \displaystyle\int_{-\infty}^{x} e^{-t^2/2}dt$. The equality in (2.179) holds whenever A is a half-space.

This inequality is a proper extension of the classical Euclidean isoperimetric inequality to the infinite dimensional setting, where $\mathbb{R}^d$ is replaced by a Banach space X, the volume by the Gaussian measure $\mathbf{P}$, and the surface measure of A by $\lim_{\lambda \downarrow 0} \frac{1}{\lambda}(\mathbf{P}(A + \lambda K) - \mathbf{P}(A))$. Observe that in the Gaussian case the role of Euclidean balls is played by half-spaces.

Such a correspondence allows one to transfer geometric volume arguments to Banach spaces, where volume is not available. Indeed, if one wants to establish the connection between the covering numbers and the size of the small balls, the first impulse is to attempt to compare volumes. We have already given an argument along these lines in the proof of the Gilbert–Varshamov bound (2.168). In the general case, Gaussian measures provide an appropriate substitution for the notion of volume, while the above estimates (2.178) and (2.179) provide the necessary tools.

We are now ready to give the proof of the equivalence between the metric entropy and small ball probability estimates.

Let $N(\varepsilon, K)$ be the covering number of K, the unit ball of $H_\mathbf{P}$, with respect to the norm of X, that is the smallest number N such that for some $\{x_k\}_{k=1}^N \subset K$ we have $K \subset \cup_{k=1}^N B_X(x_k, \varepsilon)$, where $B_X(a, r) = \{x \in X : \|x - a\|_X < r\}$. Consider the quantities $\psi(\varepsilon) = \log N(\varepsilon, K)$ (the *metric entropy*) and $\varphi(\varepsilon) = -\log \mathbf{P}(\varepsilon B_X)$.

Lemma 27. *We have the following two estimates relating the metric entropy and the small ball probability:*

$$\psi(\sqrt{2}\varepsilon / \sqrt{\varphi(\varepsilon)}) \leq 2\varphi(\varepsilon), \tag{2.180}$$

$$\psi(\varepsilon / \sqrt{2\varphi(\varepsilon)}) \geq \varphi(2\varepsilon) - \log 2. \tag{2.181}$$

Proof. Fix a parameter $\lambda > 0$ to be chosen later. Let $M = M(\varepsilon)$ be the largest number of *disjoint* balls of X of radius ε with centers in λK: $B_X(x_k, \varepsilon)$, $x_k \in \lambda K$, $k = 1, \ldots, M$. Then $N(2\varepsilon, \lambda K) = N(2\varepsilon / \lambda, K) \leq M(\varepsilon)$. Indeed, doubling the radii of all $M(\varepsilon)$ disjoint balls we obtain a covering of λK by balls of radius 2ε (if some point x of λK is not covered, then $B_X(x, \varepsilon)$ does not intersect any of the original balls, which contradicts the maximality assumption: we have chosen the *largest* disjoint family). By Borell's inequality, we have $\mathbf{P}(B_X(x_k, \varepsilon)) \geq e^{-\lambda^2/2}\mathbf{P}(\varepsilon B_X)$. Therefore, by disjointness of the balls $B(x_k, \varepsilon)$,

$$1 = \mathbf{P}(X) \geq \sum_{k=1}^M \mathbf{P}(x_k, \varepsilon) \geq N(2\varepsilon / \lambda, K) \cdot e^{-\lambda^2/2} \, \mathbf{P}(\varepsilon B_X). \tag{2.182}$$

Hence, taking logarithms, one obtains

$$\psi(2\varepsilon / \lambda) \leq \frac{\lambda^2}{2} + \varphi(\varepsilon). \tag{2.183}$$

Choosing $\lambda = \sqrt{2\varphi(\varepsilon)}$ results in $\psi(\sqrt{2}\varepsilon / \sqrt{\varphi(\varepsilon)}) \leq 2\varphi(\varepsilon)$, which proves (2.180).

In the opposite direction, let the family of balls $\{B(x_k, \varepsilon)\}_{k=1}^N$, $x_k \in \lambda K$, be a covering of λK. Then $N \geq N(\varepsilon, \lambda K) = N(\varepsilon / \lambda, K)$. Besides, the doubled balls $\{B(x_k, 2\varepsilon)\}_{k=1}^N$ obviously form a covering of a "thickened" set $\lambda K + \varepsilon B_X$. Therefore, using Anderson's lemma (the second inequality in (2.178)), we arrive to

$$\mathbf{P}(\lambda K + \varepsilon B_X) \leq \sum_{k=1}^N \mathbf{P}(B_X(x_k, 2\varepsilon)) \leq N(\varepsilon / \lambda, K) \cdot \mathbf{P}(2\varepsilon B_X). \tag{2.184}$$

We now only need to show that the left-hand side is bounded below by some constant. Notice that the thickening was necessary, since $\mathbf{P}(\lambda K) = 0$. We shall apply the isoperimetric inequality (2.179) with $A = \varepsilon B_X$ and $\lambda = \sqrt{2\varphi(\varepsilon)}$. We have

$$\mathbf{P}(\lambda K + \varepsilon B_X) \geq \Phi(\Phi^{-1}(\mathbf{P}(\varepsilon B_X)) + \lambda) = \Phi\big(\Phi^{-1}\big(e^{-\varphi(\varepsilon)}\big) + \sqrt{2\varphi(\varepsilon)}\big) \quad (2.185)$$

$$\geq \Phi\big(-\sqrt{2\varphi(\varepsilon)} + \sqrt{2\varphi(\varepsilon)}\big) = \Phi(0) = \frac{1}{2},$$

where we have used the fact that $\Phi(-x) \leq e^{-x^2/2}$. Therefore it follows from (2.184) that $\psi(\varepsilon/\sqrt{2\varphi(\varepsilon)}) \geq \varphi(2\varepsilon) - \log 2$, which is precisely (2.181). $\square$

Proof of Theorem 22. We now specialize these estimates to the Brownian sheet $\mathbb{B}$ and its reproducing kernel Hilbert space MW^2. In this situation $\mathbf{P}(\varepsilon B_X) = \mathbb{P}(\|\mathbb{B}\|_{C([0,1]^d)} < \varepsilon)$ and $N(\varepsilon, K) = N(\varepsilon, MW^2([0,1]^d)) = N(\varepsilon, 2, d)$.

Assume that, as suggested by the discussion in Sect. 2.4.7, $\varphi(\varepsilon) \approx \varepsilon^{-2}\big(\log \frac{1}{\varepsilon}\big)^\alpha$. Setting $\delta = \frac{\sqrt{2\varepsilon}}{\sqrt{\varphi(\varepsilon)}} \approx \frac{\varepsilon^2}{(\log \frac{1}{\varepsilon})^{\alpha/2}}$ and using (2.180), we obtain

$$\psi(\delta) \lesssim \varepsilon^{-2}\bigg(\log 1/\varepsilon\bigg)^\alpha \approx \delta^{-1}\bigg(\log \frac{1}{\delta}\bigg)^{\alpha/2}. \quad (2.186)$$

The other parts of the equivalence (2.162) are proved analogously. $\square$

2.4.10   Trigonometric Polynomials with Frequencies in the Hyperbolic Cross

Finally we would like to give a short overview of a different, but closely related analog of the small ball inequality, namely its version for trigonometric polynomials. Consider periodic functions defined on $\mathbb{T}^d$. For an integrable function on $\mathbb{T}^d$, its Fourier coefficients are defined as $\hat{f}_\mathbf{k} = \int_\mathbb{T} f(x)e^{-2\pi i \mathbf{k} \cdot x}\, dx$ where $\mathbf{k} = (k_1, \ldots, k_d) \in \mathbb{Z}^d$. In the case of trigonometric polynomials, unlike the case of Haar functions, frequencies are not readily dyadic. Hence it will be useful to split the frequencies into dyadic blocks. For a vector $\mathbf{s} = (s_1, \ldots, s_d) \in \mathbb{Z}_+^d$ we denote

$$\rho(\mathbf{s}) := \{\mathbf{k} \in \mathbb{Z}^d : [2^{s_j-1}] \leq k_j < 2^{s_j}, \ j = 1, \ldots, d\}, \quad (2.187)$$

where $[x]$ stands for the integer part of x. We then define the dyadic blocks of a function $f \in L^1(\mathbb{T}^d)$ as parts of the Fourier expansion of f which correspond to $\rho(\mathbf{s})$:

$$\delta_\mathbf{s}(f)(x) := \sum_{\mathbf{k}:\, |\mathbf{k}| \in \rho(\mathbf{s})} \hat{f}_\mathbf{k} e^{2\pi i \mathbf{k} \cdot x}, \quad (2.188)$$

where we put $|\mathbf{k}| = (|k_1|, \ldots, |k_d|)$. These blocks play a similar role to the expressions $\sum_{R \in \mathscr{D}_s^d} a_R h_R$, where the summation runs over the family of disjoint dyadic rectangles R with $|R_j| = 2^{-s_j}$ for $j = 1, \ldots, d$. Such linear combinations appeared naturally in the definitions of the r-functions (2.28) and the dyadic Littlewood–Paley square function (2.86).

The Littlewood–Paley inequalities adapted to this trigonometric setting read

$$\|f\|_p \approx \left\| \left(\sum_{\mathbf{s} \in \mathbb{Z}_+^d} |\delta_{\mathbf{s}}(f)|^2 \right)^{\frac{1}{2}} \right\|_p, \tag{2.189}$$

which bears a strong resemblance to (2.86)–(2.87). In particular, when $d = 1$ one recovers the classical Littlewood–Paley inequalities for Fourier series.

For an even number n, denote by $\mathbf{Y}_n = \{\mathbf{s} \in (2\mathbb{Z}_+)^d : s_1 + \ldots + s_d = n\}$ the set of vectors with even coordinates and ℓ_1 norm equal to n. This is essentially the familiar set $\mathbb{H}_n^d$ slightly modified for technical reasons. We shall also define the dyadic *hyperbolic cross* as

$$\mathbb{Q}_n = \bigcup_{\mathbf{s} \in \mathbb{Z}_+^d : s_1 + \ldots + s_d \leq n} \rho(\mathbf{s}). \tag{2.190}$$

In dimension $d = 2$, roughly speaking, it consists of the integer points that lie under the parabola $xy = 2^n$ and satisfy $x, y < 2^n$. Considering integer vectors $\mathbf{k}$ with $|k| \in \mathbb{Q}_n$ produces a symmetrization which visualizes the meaning of the name dyadic *hyperbolic cross*. The pure hyperbolic cross is defined as $\Gamma(N) = \{\mathbf{k} \in \mathbb{Z}^d : \prod_{j=1}^d \max\{1, |k_j|\} \leq N$, which makes the term even more obvious.

The trigonometric analog of the small ball inequality (2.111) in dimension $d = 2$

$$\left\| \sum_{\mathbf{s} \in \mathbb{Y}_n} \delta_{\mathbf{s}}(f) \right\|_\infty \gtrsim \sum_{\mathbf{s} \in \mathbb{Y}_n} \|\delta_{\mathbf{s}}(f)\|_1 \tag{2.191}$$

was obtained by Temlyakov [106] via a Riesz product argument very similar to Sect. 2.4.3. One can notice easily that the small ball inequality for the Haar functions can be rewritten in a very similar form

$$n^{\frac{d-2}{2}} \left\| \sum_{R: |R| = 2^{-n}} \alpha_R h_R \right\|_\infty \gtrsim \sum_{\mathbf{r} \in \mathbb{H}_n^d} \left\| \sum_{R \in \mathscr{D}_\mathbf{r}^d} \alpha_R h_R \right\|_1. \tag{2.192}$$

In fact, (2.191) can be improved to a somewhat stronger version. Define the *best hyperbolic cross approximation* of f as

$$E_{\mathbb{Q}_n}(f)_p = \inf_{t \in T(\mathbb{Q}_n)} \|f - t\|_p, \tag{2.193}$$

where $T(\mathbb{Q}_n) = \{t : t(x) = \sum_{\mathbf{k}: |\mathbf{k}| \in \mathbb{Q}_n} c_{\mathbf{k}} e^{2\pi i \mathbf{k} \cdot x}\}$ is the family of trigonometric polynomials with frequencies in the hyperbolic cross $\mathbb{Q}_n$. Then almost the same argument that proves (2.191) also yields

$$E_{\mathbb{Q}_{n-3}}(f)_\infty \gtrsim \sum_{s \in \mathbb{Y}_n} \|\delta_s(f)\|_1. \tag{2.194}$$

To draw a parallel with the Haar function version, the reader can check that (2.192) holds if the summation on the left-hand side is extended to include rectangles of size $|R| \geq 2^{-n}$—the proof given in Sect. 2.4.3 need not even be changed.

Inequalities (2.191), (2.4.10) have been applied in [106, 108] to obtain estimates of entropy numbers and Kolmogorov widths of certain function classes with mixed smoothness. It was also shown in [108] that inequality (2.191) cannot hold unless $d = 2$. Moreover, it cannot even hold if we replace the L^∞ norm on the left by L^p, $p < \infty$ or the L^1 norm on the right by L^q, $q > 1$. Analogously to Conjecture 12, we can formulate

Conjecture 28 (The trigonometric small ball conjecture). In dimensions $d \geq 2$, the following inequality holds

$$n^{\frac{d-2}{2}} \left\| \sum_{s \in \mathbb{Y}_n} \delta_s(f) \right\|_\infty \gtrsim \sum_{s \in \mathbb{Y}_n} \|\delta_s(f)\|_1 \tag{2.195}$$

The sharpness of (2.195) has been established in [108] by a probabilistic argument of the same flavor as the one presented in Sect. 2.4.2. For more information about these inequalities, their applications, and hyperbolic cross approximations the reader is invited to consult [106, 108] as well as the monographs [104, 105].

2.5 Higher Dimensions

While the failure of the product rule or lack of independence are huge obstacles to the Riesz product method in higher dimensions, they are not intrinsic to our problems. After all, this could be just an artifact of the method.

However, there are direct indications that the small ball inequality is much more difficult and delicate in dimensions $d \geq 3$ than in $d = 2$. Consider the signed ($\alpha_R = \pm 1$) case, see (2.116). In this case, at every point $x \in [0, 1]^d$ the sum on the left-hand side has $\#\mathbb{H}_n^d \approx n^{d-1}$ terms, while the right-hand side of the inequality is $n^{d/2}$. In dimension $d = 2$, these two numbers are equal, which means that the L^∞ norm is achieved at those points where almost all the terms have the same sign (the function Ψ finds precisely those points). In dimensions $d \geq 3$ on the other hand, n^{d-1} is much greater than $n^{d/2}$, while we know that the conjecture is sharp. This means that for certain choices of coefficients, very subtle cancellation will happen at all points of the cube, where even in the worst case one sign will outweigh the other by a very small fraction, $\frac{n^{d/2}}{n^{d-1}}$, of all terms. (Of course, in some specific cases, say $\alpha_R = 1$ for all R, at some points all functions have the same sign and $\left\| \sum_{|R|=2^{-n}} \alpha_R h_R \right\|_\infty \approx n^{d-1}$)

For a long time there have been virtually no improvements over the L^2 bound neither in the small ball conjecture, nor in the star-discrepancy bound. In the seminal 1989 paper on discrepancy [8], J. Beck gains a factor of $(\log \log N)^{\frac{1}{8}-\varepsilon}$ over Roth's L^2 bound. A corresponding logarithmic improvement for the small ball inequality can also be extracted from his argument, although he did not state this result and apparently was not aware of the connections. In turn, the fact that Beck's work implicitly contains progress on Conjectures 19 and 23 in dimension $d = 3$ eluded most of the experts in small deviation probabilities and metric entropy.

In 2008, largely building upon Beck's work and enhancing it with new ideas and methods, the author, M. Lacey, and A. Vagharshakyan [15, 17], obtained the first significant improvement over the 'trivial' estimate in all dimensions greater than two:

Theorem 29. *In all dimensions $d \geq 3$ there exists $\eta(d) > 0$ such that for all choices of coefficients we have the inequality:*

$$n^{\frac{d-1}{2}-\eta(d)} \left\| \sum_{R:\, |R|=2^{-n}} \alpha_R h_R \right\|_\infty \gtrsim 2^{-n} \sum_{R:\, |R|=2^{-n}} |\alpha_R|. \qquad (2.196)$$

A modification of the argument to the discrepancy framework (in a way analogous to the one described in the previous section) was also used to obtain an improvement (2.10) over Roth's estimate (2.5) in all dimensions $d \geq 3$. (This theorem has already been stated in the introduction, see Theorem 2; we simply restate it here in order to show the whole spectrum of theorems obtained by the method.)

Theorem 30. *There exists a constant $\eta = \eta(d)$, such that in all dimensions $d \geq 3$, for any set $\mathscr{P}_N \subset [0, 1]^d$ of N points, the discrepancy function satisfies*

$$\|D_N\|_\infty \gtrsim (\log N)^{\frac{d-1}{2}+\eta}. \qquad (2.197)$$

The inequality (2.196) also directly translates into improved lower bounds of the small deviation probabilities for the Brownian sheet (Conjecture 23) and the metric entropy of the mixed derivative spaces (Conjecture 19).

Theorem 31. *There exists a constant $\theta = \theta(d)$, such that in all dimensions $d \geq 3$, the small ball probability for the Brownian sheet satisfies*

$$-\log \mathbb{P}(\|\mathbb{B}\|_{C([0,1]^d)} < \varepsilon) \gtrsim \frac{1}{\varepsilon^2} \left(\log \frac{1}{\varepsilon} \right)^{2d-2+\theta}. \qquad (2.198)$$

Theorem 32. *In dimensions* $d \geq 3$, *the metric entropy of the unit ball of* $MW^2([0, 1]^d)$ *with respect to the* L^∞ *norm satisfies*

$$\log N(\varepsilon, 2, d) \gtrsim \frac{1}{\varepsilon}\left(\log \frac{1}{\varepsilon}\right)^{d-1+\theta/2}. \tag{2.199}$$

Due to the equivalence between the two problems, the value of $\theta = \theta(d)$ is the same in both theorems above.

Since complete technical details of the proof of (2.196), which can be found in [15–17] as well as Lacey's notes on the subject [61], would take up more space than the rest of this chapter, we shall simply present the main ideas of the argument and the heuristics behind them. An interested reader can then follow the complete proof in the listed references.

2.5.1   A Short Riesz Product

The Riesz product constructed in (2.124) for the proof of the two-dimensional small ball conjecture turns out to be just too long for a higher dimensional problem.

Consider a very simple example when all $\alpha_R > 0$ and the dimension d is even (or $\alpha_R < 0$ for odd d). If we take the same product as in (2.124), $\prod_{\mathbf{r} \in \mathbb{H}_n^d}(1 + f_{\mathbf{r}})$ with $f_{\mathbf{r}} = \sum_{|R|=2^{-n}} \text{sgn}(\alpha_R)h_R$, we can easily see that on the dyadic cube of sidelength $2^{-(n+1)}$ adjacent to the origin all the functions $f_{\mathbf{r}}$ are positive, hence all the factors of the Riesz product are equal to 2. Therefore,

$$\left\| \prod_{\mathbf{r} \in \mathbb{H}_n^d}(1 + f_{\mathbf{r}}) \right\|_1 \gtrsim 2^{\#\mathbb{H}_n^d} \cdot 2^{-d(n+1)}. \tag{2.200}$$

This number becomes huge for large n as $\#\mathbb{H}_n^d \approx n^{d-1}$. Therefore, this construction does not stand a chance in dimensions $d \geq 3$.

Following the idea of Beck, the test function is constructed as a "short" Riesz product. For $\mathbf{r} \in \mathbb{H}_n^d$, we consider the $\mathbf{r}$-functions $f_{\mathbf{r}} = \sum_{R \in \mathscr{D}_{\mathbf{r}}^d} \text{sgn}(\alpha_R)h_R$. Let q be an integer such that $q \approx an^\varepsilon$ for small constants $a, \varepsilon > 0$. Divide the set $\{0, 1, \ldots, n\}$ into q disjoint (almost) equal intervals of length about n/q: $I_1, I_2, \ldots,$ I_q numbered in increasing order. Let $\mathbb{A}_j := \{\mathbf{r} \in \mathbb{H}_n^d \mid r_1 \in I_j\}$. Each group $\mathbb{A}_j$ then contains $\#\mathbb{A}_j \approx n^{d-1}/q$ vectors. Indeed, the first coordinate r_1 can be chosen in n/q ways, the next $d - 2$—roughly in n ways each, and the last one is fixed due to the condition $\|\mathbf{r}\|_1 = n$. We construct the functions

$$F_j = \sum_{\mathbf{r} \in \mathbb{A}_j} f_{\mathbf{r}}. \tag{2.201}$$

Due to orthogonality, $\|F_j\|_2 \approx \sqrt{\#\mathbb{A}_j} \approx n^{(d-1)/2}/\sqrt{q}$. We now introduce the "false" L^2 normalization: $\tilde{\rho} = aq^{1/4}n^{-(d-1)/2}$ ($a > 0$ is a small constant), whereas the "true" normalization would be somewhat larger, $\rho = \sqrt{q}n^{-\frac{d-1}{2}}$. We are now ready to define the Riesz product

$$\Psi := \prod_{j=1}^{q}(1 + \tilde{\rho}F_j). \qquad (2.202)$$

Let us explain the effects that this construction creates and compare it to the two-dimensional Temlyakov's test function (2.124).

First of all, the grouping of r-functions by the values of the first coordinate mildly mirrors the construction of (2.124). Here, rather than specifying the value of $|R_1|$, we indicate the range of values that it may take. This idea allows us to preserve some lacunarity in the Riesz product. In particular, if $i < j$, then, in the first coordinate, the Haar functions involved in F_j are supported on intervals strictly smaller than those that support the Haar functions in F_i. It follows that for any $k \leq q$ and $1 \leq j_1 < j_2 < \ldots < j_k \leq q$

$$\int_{[0,1]^d} F_{j_1}(x) \cdot \ldots \cdot F_{j_k}(x) = 0, \qquad (2.203)$$

since the integral in the first coordinate is already zero (all the Haar functions are distinct). In particular,

$$\int_{[0,1]^d} \Psi(x)dx = 1, \qquad (2.204)$$

as (2.203) implies that all the higher order terms have mean zero. By comparison, Beck's [8] construction of the short Riesz product was probabilistic, which made it much more difficult to collect definitive information about the interaction of different factors in the product.

Secondly, recall that the Riesz product in (2.124) was non-negative allowing one to replace the L^1 norm with the integral which is much easier to compute. While in our case positivity everywhere is too much to hope for, it can be shown that the product is positive with large probability. The "false" L^2 normalization $\tilde{\rho}$ makes the L^2 norm of $\tilde{\rho}F_j$ small: $\|\tilde{\rho}F_j\|_2 \approx q^{-1/4} \approx n^{-\varepsilon/4} \ll 1$. Thus $(1 + \tilde{\rho}F_j)$ is positive on a set of large measure, therefore, so is the product (2.202). This heuristic is quantified in (2.212).

However, we cannot take Ψ to be the test function since we do not know exactly how it interacts with $\sum_{|R|=2^{-n}} \alpha_R h_R$. As explained in the remarks after the product rule (2.123), problems arise when the rectangles supporting the Haar functions coincide in one of the coordinates, in other words, when for two vectors $\mathbf{r}, \mathbf{s} \in \mathbb{H}_n^d$ and for some $k = 1, \ldots, d$, we have $r_k = s_k$. We say that a *coincidence* occurs in this situation. We say that vectors $\{\mathbf{r}_j\}_{j=1}^m \subset \mathbb{H}_n^d$ are *strongly distinct* if no coincidences occur between the elements of the collection, i.e., for all $1 \leq i < j \leq m$, $1 \leq k \leq d$, we have $r_{i,k} \neq r_{j,k}$. We can then write

$$\Psi = 1 + \Psi^{sd} + \Psi^{\neg sd}, \quad \text{where} \tag{2.205}$$

$$\Psi^{sd} = \sum_{k=1}^{q} \tilde{\rho}^{k} \sum_{1 \leq j_1 < j_2 < \ldots < j_k \leq q} \left(\widetilde{\sum f_{\mathbf{r}_{j_1}}} \cdots f_{\mathbf{r}_{j_k}} \right), \tag{2.206}$$

and the tilde above the innermost sum indicates that the sum is extended over all collections of vectors $\{\mathbf{r}_{j_t} \in \mathbb{A}_{j_t} : t = 1, \ldots, k\}$ which are strongly distinct. To put it simpler, $\Psi^{\neg sd}$ consists of the terms that involve coincidences, and Ψ^{sd}—of the ones that don't.

2.5.2 The Beck Gain

The function Ψ^{sd} is then taken to be the test function. Since all the coincidences are eliminated, the product rule (2.123) is applicable and an argument similar to (2.126)–(2.127) can be carried out, provided we can show that $\|\Psi^{sd}\|_1 \lesssim 1$.

2.5.2.1 Simple Coincidences

An enormous part of the proof of Theorem 29 in [15, 17] is devoted to the study of analytic and combinatorial aspects of coincidences, i.e. the behavior of $\Psi^{\neg sd}$. An important starting point is the following non-trivial lemma, which as a tribute to József Beck's ideas [8] we call the *Beck gain*:

Lemma 33 (Beck gain). *For every $p \geq 2$ we have the following inequality*

$$\left\| \sum_{\substack{\mathbf{r} \neq \mathbf{s} \in \mathbb{H}_n^d \\ r_1 = s_1}} f_{\mathbf{r}} \cdot f_{\mathbf{s}} \right\|_p \lesssim p^{\frac{2d-1}{2}} n^{\frac{2d-3}{2}}. \tag{2.207}$$

The main aspect of this lemma is the precise power of n in the estimate. The exponent $\frac{2d-3}{2}$ is in fact very natural. Indeed, d-dimensional vectors $\mathbf{r}$ and $\mathbf{s}$ have d parameters each. The condition $\|\mathbf{r}\|_1 = \|\mathbf{s}\|_1 = n$ eliminates one free parameter in each vector. Additionally, the coincidence $r_1 = s_1$ freezes one more parameter. Hence, the total number of free parameters in the sum is $2d - 3$ and each can take roughly n values. Thus the total number of terms in the sum is of the order of n^{2d-3} and (2.207) essentially says that they behave as if they were orthogonal. The power of p doesn't seem to be sharp (perhaps, $\frac{2d-3}{2}$ should also be the correct exponent of p), but it is important for further estimates that this dependence is polynomial in p, see e.g. computation (2.227) and the discussion thereafter.

Another intuitive explanation may be given from the following point of view. It is not hard to see that

$$\left\| \sum_{\mathbf{r} \neq \mathbf{s} \in \mathbb{H}_n^d} f_{\mathbf{r}} \cdot f_{\mathbf{s}} \right\|_p = \left\| \left(\sum_{\mathbf{r} \in \mathbb{H}_n^d} f_{\mathbf{r}} \right)^2 - \sum_{\mathbf{r} \in \mathbb{H}_n^d} f_{\mathbf{r}}^2 \right\|_p \qquad (2.208)$$

$$\leq \left\| \sum_{\mathbf{r} \in \mathbb{H}_n^d} f_{\mathbf{r}} \right\|_{2p}^2 + \#\mathbb{H}_n^d \approx n^{d-1},$$

since $\#\mathbb{H}_n^d \approx n^{d-1}$ and the L^{2p} norm of $F = \sum_{\mathbf{r} \in \mathbb{H}_n^d} f_{\mathbf{r}}$ is of the order $n^{\frac{d-1}{2}}$ as was shown in (2.88) using the Littlewood–Paley inequalities. Therefore, by imposing the condition $r_1 = s_1$ one *gains* $\sqrt{n}$ in the estimate, which explains the name that the authors have given to this estimate.

This lemma, albeit in a weaker form (just for $p = 2$ and with a larger power of n) appeared in the aforementioned paper of Beck [8]. In his argument, in order to compute the L^2 norm, Beck expands the square of the sum:

$$\left\| \sum_{\substack{\mathbf{r} \neq \mathbf{s} \in \mathbb{H}_n^d \\ r_1 = s_1}} f_{\mathbf{r}} \cdot f_{\mathbf{s}} \right\|_2^2 = \sum_{\substack{\mathbf{r} \neq \mathbf{s}, \, \mathbf{u} \neq \mathbf{v} \\ r_1 = s_1, \, u_1 = v_1}} \int_{[0,1]^d} f_{\mathbf{r}} \cdot f_{\mathbf{s}} \cdot f_{\mathbf{u}} \cdot f_{\mathbf{v}} \, dx. \qquad (2.209)$$

Notice that each integral above is equal to zero unless the four-tuple of vectors $(\mathbf{r}, \mathbf{s}, \mathbf{u}, \mathbf{v}) \in (\mathbb{H}_n^d)^4$ has a coincidence in each coordinate. Careful and lengthy combinatorial analysis of the arising patterns of coincidences then leads to the desired inequality.

The extension and generalization obtained in [15, 17] is achieved by replacing the process of expanding the square by the applications of the Littlewood–Paley square function (2.66), which is a natural substitution in harmonic analysis, when one wants to pass from L^2 to L^p, $p \neq 2$. Every application of the Littlewood–Paley inequality (2.68) yields a constant $B_p \approx \sqrt{p}$. The lemma was initially proved in $d = 3$ [15] and then extended to $d \geq 3$ [17] by a tricky induction argument. The reader is invited to see [15, Lemma 8.2], [17, Lemma 5.2] for complete details.

2.5.2.2   Long Coincidences

As we shall see, Lemma 33 is very powerful and yields important consequences, e.g. (2.212)–(2.213). Yet it is only a starting point in the analysis. One needs to analyze more complicated instances of coincidences which arise in Ψ^{-sd}. Their high combinatorial complexity in large dimensions aggravates the difficulty of the problem. Further success of the Riesz product method requires inequalities of the type

$$\left\| \sum f_{\mathbf{r}_1} \cdot \ldots \cdot f_{\mathbf{r}_k} \right\|_p \lesssim p^{\alpha M} n^{\frac{M}{2}}, \tag{2.210}$$

where the sum is extended over all k-tuples $\mathbf{r}_1, \ldots, \mathbf{r}_k$ with a specified configuration of coincidences and M is the number of free parameters imposed by this configuration; $\alpha > 0$ is a constant which is conjectured to be $\frac{1}{2}$. Estimates of this type suggest that free parameters behave orthogonally even for longer coincidences.

These patterns of coincidences may be described by d-colored graphs $G = (V, E)$, where the set of vertices $V = \{1, \ldots, k\}$ corresponds to vectors $\mathbf{r}_1, \ldots, \mathbf{r}_k$, and two vertices i and j are connected by an edge of color m, $m = 1, \ldots, d$ if the vectors $\mathbf{r}_i$ and $\mathbf{r}_j$ have a coincidence in the mth coordinate: $r_{i,m} = r_{j,m}$.

In the case of a single coincidence, when $k = 2$ and the graph describing the coincidence consists of two vertices and one edge, estimate (2.210) turns precisely into inequality (2.207) of Lemma 33. At present, inequality (2.210) in full generality is only a conjecture. In [15, 17] a partial result with a larger power of n is obtained for $k > 2$. Namely, it is proved that, if the summation is taken over a fixed pattern of coincidences of length k, the following estimate holds for some $\gamma > 0$

$$\left\| \sum f_{\mathbf{r}_1} \cdot \ldots \cdot f_{\mathbf{r}_k} \right\|_p \lesssim p^{Ck} n^{\left(\frac{d-1}{2} - \gamma\right) \cdot k}. \tag{2.211}$$

In other words, we have a gain proportional to the total length of the coincidence. This would later allow one to sum the estimates over all possible patterns of coincidences.

Roughly speaking, this inequality is proved by choosing a large *matching* (disjoint collection of edges) in the associated graph. Each edge in the matching corresponds to a simple coincidence to which an analog of the Beck gain lemma (2.207) may be applied, see [17, Theorem 8.3] for details. This approach, in particular, puts a restriction on the size of the gain. Consider, for example, a star-like graph with d edges of d distinct colors, which connect a single vertex (center) to d other vertices. The largest matching in such a graph consists of one edge. Therefore, in general, one cannot expect a matching of size more than k/d, which immediately yields $\gamma \lesssim 1/d$.

2.5.3 The Proof of Theorem 29

In this subsection we shall outline the main steps and ideas of the proof of Theorem 29 based on the construction of the short Riesz product and the Beck gain.

The ultimate goal of constructing the short Riesz product Ψ (2.202) was to produce an L^1 test function Ψ^{sd}. The fact that Ψ^{sd} has bounded L^1 norm is proved

through a series of estimates which are gathered in the following technical lemma
(see Lemma 4.8 in [17]):

Lemma 34. *We have the following estimates:*

$$\mu(\{\Psi < 0\}) \lesssim \exp(-A\sqrt{q}) ; \qquad (2.212)$$

$$\|\Psi\|_2 \lesssim \exp(a'\sqrt{q}) ; \qquad (2.213)$$

$$\int \Psi(x) dx = 1 ; \qquad (2.214)$$

$$\|\Psi\|_1 \lesssim 1 ; \qquad (2.215)$$

$$\|\Psi^{-sd}\|_1 \lesssim 1 ; \qquad (2.216)$$

$$\|\Psi^{sd}\|_1 \lesssim 1 , \qquad (2.217)$$

*where $0 < a' < 1$ is a small constant, $A > 1$ is a large constant, and μ is the
Lebesgue measure.*

While we shall not give complete proofs of most of these inequalities, some remarks,
explaining their nature and the main ideas, are in order.

We start with the first two inequalities (2.212)–(2.213) which are consequences
of the Beck gain (2.207) for simple coincidences.

2.5.3.1 The Distributional Estimate (2.212)

Inequality (2.212) is a quantification of the fact discussed earlier that, due to the
false L^2 normalization $\tilde{\rho}$, Ψ is negative on a very small set. Indeed, since $\Psi = \prod_{j=1}^{q}(1 + \tilde{\rho}F_j)$, we have

$$\mu(\{\Psi < 0\}) \leq \sum_{j=1}^{q} \mu(\{\tilde{\rho}F_j < -1\}) = \sum_{j=1}^{q} \mu\left(\{\rho F_j < -\frac{1}{a}\sqrt[4]{q}\}\right), \qquad (2.218)$$

where we have replaced the 'false' L^2 normalization $\tilde{\rho} = aq^{1/4}/n^{(d-1)/2}$ by the
'true' one $\rho = \sqrt{q}/n^{\frac{d-1}{2}}$. Let us view the functions F_j as a sum of ± 1 random
variables. If all of them were independent, we would be able to deduce estimate
(2.212) immediately using the large deviation bounds of Chernoff-Hoeffding type,
see e.g. (2.119), much in the same way as in (2.120). However, the presence of
coincidences destroys independence. The Beck gain estimate (2.207) allows one to
surpass this obstacle.

In fact, a weaker version of (2.212) can be proved without referring to the Beck
gain. We have, for all $p > 1$,

$$\|\rho F_j\|_p = \frac{\sqrt{q}}{n^{\frac{d-1}{2}}}\left\|\sum_{\mathbf{r}\in\mathbb{A}_j} f_{\mathbf{r}}\right\|_p \lesssim \frac{\sqrt{q}}{n^{\frac{d-1}{2}}} p^{\frac{d-1}{2}}(\#\mathbb{A}_j)^{\frac{1}{2}} \lesssim p^{\frac{d-1}{2}}. \qquad (2.219)$$

This estimate follows from successive applications of the Littlewood–Paley inequality (2.68) in the first $d-1$ coordinates (the last one is not needed due to the restriction $|R| = 2^{-n}$) and is identical to the calculation leading to (2.85). A constant of the order $\sqrt{p}$ arises each time we apply the square function. This shows, using the equivalent definitions of the exponential Orlicz norms (2.75), that $\|\rho F_j\|_{\exp(L^{2/(d-1)})} \lesssim 1$ and hence $\mu\big(\{\rho F_j < -\frac{1}{a}\sqrt[4]{q}\}\big) \lesssim \exp(-Cq^{1/2(d-1)})$.

To get the desired $\exp(L^2)$ bound, one would have to use Littlewood–Paley just once in order to get the constant of $p^{1/2}$ on the right-hand side. Therefore, the strategy to obtain the sharper inequality (2.212) is the following: we apply the Littlewood–Paley inequality to ρF_j just in the first coordinate. The "diagonal" terms yield a constant, while the rest of the terms are precisely the ones that have a coincidence in the first coordinate and are governed by the Beck gain. To be more precise, recall that $F_j = \sum_{\mathbf{r}:\, r_1\in I_j} f_{\mathbf{r}}$ and apply the Littlewood–Paley square function in the first coordinate

$$\|\rho F_j\|_p \lesssim \sqrt{p}\|S_1(F_j)\|_p = \sqrt{p}\left\|\left(\sum_{t\in I_j} \rho^2\Big(\sum_{\mathbf{r}:\, r_1=t} f_{\mathbf{r}}\Big)^2\right)^{1/2}\right\|_p \qquad (2.220)$$

$$= \sqrt{p}\left\|\rho^2\sum_{\mathbf{r}\in\mathbb{A}_j} f_{\mathbf{r}}^2 + \rho^2\sum_{\substack{\mathbf{r}\neq\mathbf{s}\in\mathbb{A}_j\\ r_1=s_1}} f_{\mathbf{r}}\cdot f_{\mathbf{s}}\right\|_{p/2}^{1/2}$$

$$\lesssim \sqrt{p}\left(1 + \rho^2\left\|\sum_{\substack{\mathbf{r}\neq\mathbf{s}\in\mathbb{A}_j\\ r_1=s_1}} f_{\mathbf{r}}\cdot f_{\mathbf{s}}\right\|_{p/2}^{1/2}\right).$$

The diagonal term above is bounded by a constant since $f_{\mathbf{r}}^2 = 1$ and $\rho^{-2} = n^{d-1}/q$ is roughly equal to the number of elements of $\mathbb{A}_j$. The Beck gain estimate (2.207) can be applied to the second term to obtain

$$\rho^2\left\|\sum_{\substack{\mathbf{r}\neq\mathbf{s}\in\mathbb{A}_j\\ r_1=s_1}} f_{\mathbf{r}}\cdot f_{\mathbf{s}}\right\|_{p/2} \lesssim \frac{q}{n^{d-1}} p^{\frac{2d-1}{2}} n^{\frac{2d-3}{2}} = qp^{d-\frac{1}{2}}n^{-\frac{1}{2}} \lesssim 1 \qquad (2.221)$$

when p is not too big. Hence for relatively small values of p, the Beck gain term will not dominate over the diagonal term. For this range of exponents p we obtain $\|\rho F_j\|_p \lesssim \sqrt{p}$. This inequality for the full range of p by (2.75) would have implied $\|\rho F_j\|_{\exp(L^2)} \lesssim 1$. Even though this estimate cannot be deduced in full generality, repeating the proof of (2.75) we can find that $\mu(\{\rho F_j < -t\}) \lesssim \exp(-Ct^2)$ for moderate values of t, and (2.212) will follow. $\square$

2.5.3.2   The L^2 Bound (2.213)

An explanation for the L^2 bound (2.213) may again be given using the heuristics of probability theory. If F_j's were independent random variables, we would immediately have (2.213):

$$\int \prod_{j=1}^{q} (1 + \tilde{\rho} F_j)^2 \, dx = \prod_{j=1}^{q} \int (1 + \tilde{\rho} F_j)^2 \, dx \tag{2.222}$$

$$\leq \prod_{j=1}^{q} (1 + \tilde{\rho}^2 \|F_j\|_2^2) \leq \left(1 + \frac{a^2}{\sqrt{q}}\right)^q \leq e^{a^2 \sqrt{q}}.$$

While they are not independent, one can apply a conditional expectation argument and Beck gain (2.207), since the lack of independence is the result of coincidences.

We can see from the discussion of the first two conclusions of Lemma 34 that, from the probabilistic point of view, the Beck gain estimate (2.207) compensates for the lack of independence.

2.5.3.3   The Integral and the L^1 Norm of the Riesz Product Ψ (2.214)–(2.215)

Equality (2.214) has already been explained, see (2.204). It follows from the fact that the functions F_j, $j = 1, \ldots, q$ are orthogonal already in the first coordinate, since they consist of Haar functions of different frequencies.

Even though Ψ is not positive unlike in the two-dimensional case, the L^1 estimate (2.215) easily follows from the previous three inequalities (2.212)–(2.214) using Cauchy–Schwarz inequality:

$$\|\Psi\|_1 = \int \Psi(x) dx - 2 \int_{\{\Psi < 0\}} \Psi(x) dx \leq 1 + 2\mu(\{\Psi < 0\})^{1/2} \cdot \|\Psi\|_2 \tag{2.223}$$

$$\lesssim 1 + \exp(-A\sqrt{q}/2 + a'\sqrt{q}) \lesssim 1. \quad \square$$

2.5.3.4   The L^1 Norm of Coincidences (2.216)

Estimate (2.216) is the deepest part of this result and follows from the scrupulous analysis of coincidences which was outlined in Sect. 2.5.2, especially the bounds for long coincidences.

Recall that, as explained in Sect. 2.5.2.2, we describe long coincidences by d-colored graphs. Let the set of vertices be $V = V(G) \subset \{1, \ldots, q\}$ and impose an additional condition that $s \in V(G)$ implies $\mathbf{r}_s \in \mathbb{A}_s$. This assumption reflects the way the vectors in the Riesz product are grouped. Denote by

$$SumProduct(G) := \sum_{s \in V(G)} \prod f_{\mathbf{r_s}}, \qquad (2.224)$$

where the sum is extended over all tuples of vectors $\{r_s\}_{s \in V(G)}$ with $\mathbf{r_s} \in \mathbb{A}_s$ whose pattern of coincidences is described by the graph G. This is precisely the object whose norm is estimated in the Beck gain inequality for longer coincidences (2.211).

We can then represent the non-distinct part of the Riesz product Ψ as a sum over all possible configurations of coincidences as follows

$$\Psi^{-sd} = \sum_G \tilde{\rho}^{|V(G)|} (-1)^{\text{ind}(G)+1} \cdot SumProd(G) \cdot \prod_{s \notin V(G)} (1 + \tilde{\rho} F_j). \qquad (2.225)$$

Here the sum is taken over all 'admissible' graphs—graphs that describe a realizable pattern of coincidences. The parameter $\text{ind}(G)$ is simply a proper parameter needed in order to take care of the overlaps of different patterns of coincidences and to produce a correct version of the inclusion-exclusion formula. It is defined as the total number of equalities which describe the given arrangement of coincidences.

For a given pattern G, the factor $SumProd(G)$ gives all possible products arising from this pattern, while $\prod_{s \notin V(G)}(1 + \tilde{\rho} F_j)$ is the part of the Riesz product which is not involved in the given configuration of coincidences. Observe that in general the function $\prod_{s \notin V(G)}(1 + \tilde{\rho} F_j)$ satisfies more or less the same estimates as the full Riesz product Ψ itself, since it is of nearly identical form.

We shall interpolate between L^1 (2.215) and L^2 (2.213) estimates of the Riesz product $\prod_{s \notin V(G)}(1 + \tilde{\rho} F_j)$ to bound its L^p norm and find that, when p gets sufficiently close to 1, it is bounded by a constant. This is quite natural since its L^1 norm is bounded by a constant and it is a limit of L^p norms as p approaches 1. To be more precise, we take $p = (\sqrt{q})' = \frac{\sqrt{q}}{\sqrt{q}-1}$. In this case, $\frac{1}{p} = \frac{\sqrt{q}-2}{\sqrt{q}} \cdot 1 + \frac{2}{\sqrt{q}} \cdot \frac{1}{2}$. For the sake of brevity we denote $\Psi_{V(G)^c} := \prod_{s \notin V(G)}(1 + \tilde{\rho} F_j)$. We then obtain

$$\|\Psi_{V(G)^c}\|_{(\sqrt{q})'} \leq \|\Psi_{V(G)^c}\|_1^{(\sqrt{q}-2)/\sqrt{q}} \cdot \|\Psi_{V(G)^c}\|_2^{2/\sqrt{q}} \lesssim 1 \cdot e^{a'\sqrt{q} \cdot \frac{2}{\sqrt{q}}} \lesssim 1. \qquad (2.226)$$

We are now ready to estimate the L^1 norm of Ψ^{-sd}, the non-distinct part of Ψ. From (2.225) we have

$$\|\Psi^{-sd}\|_1 \leq \sum_G \tilde{\rho}^{|V(G)|} \|SumProd(G) \cdot \Psi_{V(G)^c}\|_1 \qquad (2.227)$$

$$\leq \sum_G \tilde{\rho}^{|V(G)|} \|SumProd(G)\|_{\sqrt{q}} \cdot \|\Psi_{V(G)^c}\|_{(\sqrt{q})'}$$

$$\lesssim \sum_{v=2}^{q} \sum_{G:|V(G)|=v} q^{\frac{v}{4}} n^{-\frac{d-1}{2} \cdot v} \cdot q^{Cv} n^{\left(\frac{d-1}{2}-\gamma\right) \cdot v} \cdot 1$$

$$\lesssim \sum_{v=2}^{q} \binom{q}{v} q^{vd} n^{(C'\varepsilon-\gamma)\cdot v} \leq \left(n^{-\gamma'}+1\right)^q - 1$$

$$\lesssim q n^{-\gamma'} \lesssim 1$$

provided that ε is small enough. Here we have applied the Beck gain for long coincidences (2.211) to $SumProd(G)$ and the interpolation estimate (2.226) to $\Psi_{V(G)^c}$.

The number of admissible graphs with the given vertex set of v vertices can be controlled in the following way. Let us initially look at coincidences in a single coordinate. The number of ways to have a single coincidence is at most 2^v (every vertex either participates in a coincidence or not), for two coincidences the number of possibilities is at most 3^v (every vertex is in the first, second, or none of the coincidences), etc. Hence the total number of possibilities is no more than $2^v + 3^v + \ldots + (v/2)^v \lesssim v^v$. If we now consider coincidences in all coordinates, the number of patterns is bounded by $(v^v)^d \leq q^{vd}$.

This computation reveals the motivation for some of the previously discussed estimates as well as the arising limitations.

- First of all, we see from the last two lines of the inequality that the amount of gain in (2.196) is forced to be bounded by the amount of the Beck gain (2.211). More precisely, we need $\varepsilon \lesssim \gamma$ in order to have $q^C \ll n^\gamma$. Moreover, the estimate on the total number of graphs $q^{vd} \approx n^{\varepsilon vd}$ suggests that $\varepsilon \lesssim \frac{1}{d}\gamma$. Since $\gamma \lesssim \frac{1}{d}$ as explained in Sect. 2.5.2.2, this tells us that the gain $\eta(d)$ in Theorem 29 coming from thus argument is at most $\varepsilon \lesssim 1/d^2$.

- Besides, we see that in order to bound the norm of $\Psi_{V(G)^c}$, the index $(\sqrt{q})'$ needs to be sufficiently close to 1, hence $\sqrt{q}$ is rather large. Therefore, it is really important to be able not only to estimate the norm of the terms involving coincidences $SumProd(G)$ in L^p spaces for $p \geq 2$, but also to be able to track how the implicit constants depend on the integrability index p.

The computation (2.227) finishes the proof of (2.216) and leaves us just one little step away from the proof of Theorem 29.

2.5.4   The L^1 Norm of Ψ^{sd} (2.217) and the Conclusion of the Proof

Since $\Psi^{sd} = \Psi - 1 - \Psi^{\neg sd}$, the sought bound (2.217) is trivially implied by the previous two. We can now conclude the proof of Theorem 29 following the lines of (2.126)–(2.127):

$$\left\| \sum_{R \in \mathscr{D}^d : |R|=2^{-n}} \alpha_R h_R \right\|_\infty \gtrsim \left\langle \sum_{|R|=2^{-n}} \alpha_R h_R, \Psi^{sd} \right\rangle \tag{2.228}$$

$$= \left\langle \sum_{|R|=2^{-n}} \alpha_R h_R, \tilde{\rho} \sum_{|R|=2^{-n}} \mathrm{sgn}(\alpha_R) h_R \right\rangle$$

$$= \tilde{\rho} \sum_{R \in \mathscr{D}^d : |R|=2^{-n}} \alpha_R \cdot \mathrm{sgn}(\alpha_R) \cdot \|h_R\|_2^2$$

$$\approx n^{-\frac{d-1}{2}+\frac{\varepsilon}{4}} 2^{-n} \cdot \sum_{|R|=2^{-n}} |\alpha_R|,$$

so, (2.196) holds with $\eta = \varepsilon/4$.   □

2.5.5 The Signed Small Ball Inequality

The *signed* small ball inequality, i.e. a version with $\alpha_R = \pm 1$ for each R, see (2.116) may be viewed as a toy model of Conjecture 12. It avoids numerous technicalities, while preserving most of the complications arising from the combinatorial complexity of the higher dimensional dyadic boxes. In [16], the same authors came up with a significant simplification of the arguments in [15, 17] for the signed case—it only required the simplest estimate for coincidences (2.207), and not the more complicated (2.210). It yielded the bound

$$\left\| \sum_{|R|=2^{-n}} \alpha_R h_R \right\|_\infty \gtrsim n^{\frac{d-1}{2}+\eta} \tag{2.229}$$

for $\alpha_R = \pm 1$ in all dimensions and allowed them to obtain an explicit value of the gain $\eta(d) = \frac{1}{8d} - \varepsilon$.

In fact, given Lemma 34, it is quite easy to produce a proof of the "improvement of the L^2 estimate" (2.229), which is just (2.196) restricted to the signed case $\alpha_R = \pm 1$. We provide this proof below.

2.5.5.1 The Proof of the Signed Version of Theorem 29

We shall use the same short Riesz product $\Psi = \prod_{j=1}^q (1 + \tilde{\rho} F_j)$ defined in (2.202). Recall that $F_j = \sum_{\mathbf{r} \in \mathbb{A}_j} f_{\mathbf{r}}$, where $\mathbb{A}_j := \{\mathbf{r} \in \mathbb{H}_n^d : \frac{n(j-1)}{q} \le r_1 < \frac{nj}{q}\}$, i.e. the first component of $\mathbf{r}$ lies in the jth subinterval of $\{1, 2, \dots, n\}$. Notice that in the signed case the expression inside the L^∞ norm in the small ball inequality simply equals the sum of all F_j's:

$$H_n = \sum_{|R|=2^{-n}} \alpha_R h_R = \sum_{j=1}^{q} F_j. \tag{2.230}$$

Unlike the general case, we can now take the product Ψ itself to be the dual test function, rather than extracting its "coincidence-free" part. The coincidences will be taken care of inside the argument. We first look at the inner product of a single F_j with Ψ. Denote by $\Psi_{\neq j} = \prod_{i=1, i \neq j}^{q} (1 + \tilde{\rho} F_i)$ the part of the Riesz product which consists of all factors except the jth one. Note that its structure is virtually indistinguishable from that of the full product, hence $\Psi_{\neq j}$ satisfies essentially the same estimates as Ψ itself, see Lemma 34. Another important observation is that F_j is orthogonal to $\Psi_{\neq j}$: because of the structure of the product there are no coincidences in the first coordinate, thus, in the first component, F_j and Ψ_j consist of Haar functions of different frequencies. We then obtain

$$\begin{aligned}
\langle F_j, \Psi \rangle &= \langle \sum_{r \in \mathbb{A}_j} f_r, \Psi \rangle = \sum_{r \in \mathbb{A}_j} \langle f_r, (1 + \tilde{\rho} F_j) \cdot \Psi_{\neq j} \rangle \\
&= \tilde{\rho} \sum_{r \in \mathbb{A}_j} \langle f_r^2, \Psi_{\neq j} \rangle + \tilde{\rho} \langle \Phi_j, \Psi_{\neq j} \rangle \\
&= \tilde{\rho}(\#\mathbb{A}_j) + \tilde{\rho} \langle \Phi_j, \Psi_{\neq j} \rangle,
\end{aligned} \tag{2.231}$$

where Φ_j is exactly the expression arising in the Beck gain estimate (2.207)

$$\Phi_j = \sum_{r,s \in \mathbb{A}_j, r \neq s, r_1 = s_1} f_r \cdot f_s. \tag{2.232}$$

The second line of the above computation (2.231) reflects the fact that the integral of $f_r \Psi$ is equal to zero unless we get a coincidence in the first coordinate; and this coincidence may arise in two ways—when f_r hits itself (in which case, since $f_r^2 = 1$ and $\int \Psi_{\neq j} = 1$, we simply pick up the cardinality of $\mathbb{A}_j$) or when it is paired with a different vector from $\mathbb{A}_j$ with the same first coordinate (so that the Beck gain (2.207) may be applied). We shall see that the former will be the main term in the estimate, while the latter may be treated as the error term.

Just as in (2.226), interpolation between L^1 (2.215) and L^2 (2.213) estimates of the Riesz product $\Psi_{\neq j}$ shows that the L^p norm $\|\Psi_{\neq j}\|_p$ is at most a constant. Copying (2.226), we need to take $p = (\sqrt{q})' = \frac{\sqrt{q}}{\sqrt{q}-1}$ to obtain

$$\|\Psi_{\neq j}\|_{(\sqrt{q})'} \leq \|\Psi_{\neq j}\|_1^{(\sqrt{q}-2)/\sqrt{q}} \|\Psi_{\neq j}\|_2^{2/\sqrt{q}} \lesssim 1 \cdot e^{a'\sqrt{q} \cdot \frac{2}{\sqrt{q}}} \lesssim 1. \tag{2.233}$$

We can now apply Hölder's inequality, the Beck gain (2.207), and the previous inequality to estimate

$$|\langle \Phi_j, \Psi_{\neq j}\rangle| \leq \|\Phi\|_{\sqrt{q}}\|\Psi_{\neq j}\|_{(\sqrt{q})'} \lesssim (\sqrt{q})^{d-1/2}n^{d-3/2} \cdot q^{-1/2}, \qquad (2.234)$$

where the extra factor of $q^{-1/2}$ comes from the restriction $\mathbf{r}, \mathbf{s} \in \mathbb{A}_j$, which means that the parameter $r_1 = s_1 \in \left[\frac{n(j-1)}{q}, \frac{nj}{q}\right)$ can actually be chosen in n/q ways rather than n. Recalling that $\tilde{\rho} \approx q^{1/4}n^{-(d-1)/2}$, $q \approx n^{\varepsilon}$, and $(\#\mathbb{A}_j) \approx n^{d-1}/q$, together with the fact that $\|\Psi\|_1 \lesssim 1$, we obtain

$$\|H_n\|_\infty \gtrsim |\langle H_n, \Psi\rangle| = \left|\sum_{j=1}^{q}\langle F_j, \Psi\rangle\right| \gtrsim q\tilde{\rho}\left(\#\mathbb{A}_j - q^{\frac{d}{2}-\frac{3}{4}}n^{d-\frac{3}{2}}\right)$$

$$\approx q^{\frac{1}{4}}n^{\frac{d-1}{2}} - q^{\frac{d+1}{2}}n^{\frac{d-2}{2}} \gtrsim n^{\frac{d-1}{2}+\frac{\varepsilon}{4}}, \qquad (2.235)$$

provided that ε is small enough, so that the second term is of smaller order of magnitude than the first one. This happens precisely when $\varepsilon < \frac{2}{d}$ which already yields the aforementioned restriction $\eta(d) \approx \frac{1}{d}$. An even more stringent condition on ε (yet still yielding the same rate of decay in terms of the dimension) arises from the proof of the L^2 estimate of Ψ (2.213).

The reader is reminded that estimate (2.215), $\|\Psi\|_1 \lesssim 1$, which is used in this proof, only relied on the L^2 bound (2.213), which in turn exploited only the simplest case of the Beck gain (2.207) for a single coincidence. In other words, one does not need to consider long coincidences—dealing with Ψ^{-sd} can be avoided altogether. Thus the proof of estimate (2.196) in the signed case circumvents the heavy analytic and combinatorial investigation of coincidences and indeed allows for tremendous simplifications of the argument.

We close the discussion of the signed version of the small ball inequality by outlining two other potential points of view and approaches to the problem.

2.5.5.2 A New Approach: Independence and Conditional Expectation

In dimension $d = 2$, the signed small ball inequality (2.111) can be easily proved as a consequence of the independence of the random variables $f_{\mathbf{r}}$, which is easy to check. Independence implies

$$\mathbb{P}(f_{\mathbf{r}} = 1 : \mathbf{r} = (k, n-k), k = 0, \ldots, n) = \prod_{k=0}^{n}\mathbb{P}(f_{(k,n-k)} = 1) = \frac{1}{2^{n+1}} > 0, \qquad (2.236)$$

i.e. on a set of positive measure the all the functions $f_{\mathbf{r}}$ are positive. On this set therefore

$$\sum_{|R|=2^{-n}} \varepsilon_R h_R(x) = \sum_{k=0}^{n} f_{(k,n-k)}(x) = n + 1, \qquad (2.237)$$

which proves Conjecture 13 in dimension $d = 2$.

In higher dimensions, due to possible coincidences in vectors $\mathbf{r} \in \mathbb{H}_n^d$, independence of the functions $f_{\mathbf{r}}$ no longer holds. This shortcoming can be partially compensated for by delicate conditional expectation arguments. The proof of the three-dimensional version of inequality (2.229) in [19] yields the best currently known gain: $\eta(3) = \frac{1}{8}$. Unfortunately, at this time it is not clear how to transfer this method to the discrepancy setting or extend it to higher dimensions.

2.5.5.3 L^1 Approximation

An alternative viewpoint stems from the close examination of the structure of the two-dimensional Riesz products Ψ. Consider again the signed case $\alpha_R = \pm 1$ and denote $H_n = \sum_{|R|=2^{-n}} \alpha_R h_R$. It can be shown that $\|H_n\|_1 \approx \|H_n\|_2 \approx n^{1/2}$. Indeed, Hölder's inequality implies that $\|H_n\|_2 \leq \|H_n\|_1^{1/3} \cdot \|H_n\|_4^{2/3}$. It is easy to see that $\|H_n\|_2 \approx \|H_n\|_4 \approx n^{(d-1)/2} = n^{1/2}$ (the computation of the L^4 norm is identical to (2.88)). The estimate for the L^1 norm of H_n then follows.

Equality (2.125), on the other hand, implies that the L^1 norm of $H_n - (-\Psi_{>n})$ is at most $1 + \|\Psi\|_1 = 2$, i.e. H_n, the hyperbolic sum of Haar functions of order n, can be well approximated in the L^1 norm by a linear combination of Haar functions of higher order. In fact, the Small Ball Conjecture 12 would follow if we can prove that for any choice of $\alpha_R = \pm 1$ we have

$$\text{dist}_{L^1}\left(\sum_{R:\, |R|=2^{-n}} \alpha_R h_R, \, H_{>n} \right) \lesssim n^{\frac{d-2}{2}}, \qquad (2.238)$$

where $H_{>n}$ is the span of Haar functions supported by rectangles of size $|R| < 2^{-n}$.

These ideas are not new. In fact, in [103, 1980] (see also [104]), more than 10 years prior to the proof of the small ball inequality (2.111) in dimension $d = 2$, Temlyakov has used a very similar Riesz product construction in order to prove an analog of the statement described above, namely, that trigonometric polynomials with frequencies in a hyperbolic cross (see Sect. 2.4.10) can be well approximated in the L^1 norm by a linear combination of harmonics of higher order.

2.6 Low Discrepancy Distributions and Dyadic Analysis

Most of the content of this chapter so far has been concerned with proofs of various lower bounds for the discrepancy. In the last section we would like to illustrate how Roth's idea of incorporating dyadic harmonic analysis into discrepancy theory helps in proving some upper discrepancy estimates.

2.6.1 The Van Der Corput Set

We recall a very standard construction, the so-called "digit-reversing" van der Corput set [116], also known as the Hammersley point set. This distribution of points is constructed in the following simple, yet very clever fashion. For $N = 2^n$ define a set $\mathcal{V}_n$ consisting of 2^n points

$$\mathcal{V}_n = \{(0.x_1x_2 \ldots x_n, 0.x_nx_{n-1} \ldots x_1) : x_k = 0, 1; k = 1, \ldots, n\}, \quad (2.239)$$

where the coordinates are written as binary fractions. That means that the binary digits of the y-coordinate are exactly the digits of the x-coordinate written in the reverse order. Very roughly speaking, the effect of this construction is the following: if the x-coordinate changes a little, the y-coordinate changes significantly (although this is not exactly true), hence this set is well spread over the unit square.

Indeed, its star-discrepancy is optimal in the order of magnitude,

$$\|D_{\mathcal{V}_n}\|_\infty \leq n + 1 \approx \log N. \quad (2.240)$$

This fact has been shown by van der Corput for the corresponding one-dimensional infinite sequence. Halton [49] and Hammersley [51] later transferred the idea to the multidimensional setting to construct the sets with the best currently known order of magnitude of the star-discrepancy, $(\log N)^{d-1}$.

A crucial property of the van der Corput set, which allows one to deduce such a favorable discrepancy bound is the fact that it forms a dyadic (or binary) net of order n: any dyadic rectangle R of area $|R| = 2^{-n}$ contains precisely one point of $\mathcal{V}_n$, and hence the discrepancy of $\mathcal{V}_n$ with respect to such rectangles is zero. For more information on nets, their constructions, and properties, the reader is referred to Sect. 3 of the Chap. 9 by J. Dick and F. Pillichshammer in the current book as well as [39].

2.6.1.1 The L^2 Discrepancy of the Van Der Corput Set

Different norms of the discrepancy function of variations of this set have been studied by many authors: [13,18,34,41,50,53,57,63,82,116] to name just a few. We do not claim to present a complete survey of these results here—a comprehensive survey of numerous interesting properties of this elementary and at the same time wonderful set is yet to be written. Instead, we concentrate only on some estimates which we find most relevant to the theme of this chapter. Naturally, we shall start with the L^2 discrepancy.

It is well known that, while $\mathcal{V}_n$ has optimal star-discrepancy, its L^2 discrepancy is also of the order $\log N$ as opposed to the optimal $\sqrt{\log N}$. The problem actually lies in the fact that

$$\int D_{\mathcal{V}_n}(x)dx = \frac{n}{8} + \mathcal{O}(1) \approx \log N \qquad (2.241)$$

as observed in [13, 34, 50, 53]. Therefore, of course, $\|D_{\mathcal{V}_n}\|_2 \gtrsim \log N$.

One can look at this from a different point of view: (2.241) means that in any reasonable orthogonal (Haar, Fourier, wavelet, Walsh etc) decomposition of $D_{\mathcal{V}_n}$ the zero-order coefficient is already too big, so, by Plancherel's theorem, the L^2 norm is big. However, it turns out that the input of all the other coefficients is exactly of the right order, see e.g. [13, 18, 50], hence (2.241) is the *only* obstacle. Halton and Zaremba [50] showed that

$$\|D_{\mathcal{V}_n}\|_2^2 = \frac{n^2}{2^6} + \mathcal{O}(n), \qquad (2.242)$$

which in conjunction with (2.241) proves this point.

There are several standard remedies which allow one to alter the van der Corput set so as to achieve the optimal order of the L^2 discrepancy. All of them, explicitly or implicitly, deal with reducing the quantity $\int D_{\mathcal{V}_n}(x)dx$. Here is a brief list of these methods.

(i) **Random shifts.**
   Roth [82, 83] has demonstrated that there exists a shift of $\mathcal{V}_n$ modulo 1 which achieves optimal L^2 discrepancy. The proof was probabilistic: it was shown that the expectation over random shifts has the right order of magnitude

$$\mathbb{E}_\alpha \|D_{\mathcal{V}_{n,\alpha}}\|_2 \lesssim \sqrt{\log N}, \qquad (2.243)$$

   where $\mathcal{V}_{n,\alpha} = \{((x+\alpha) \bmod 1, y) : (x,y) \in \mathcal{V}_n\}$. A straightforward calculation shows that $\mathbb{E}_\alpha \int D_{\mathcal{V}_n}(x)dx = \mathcal{O}(1)$. A deterministic example of such a shift was constructed recently in [13].

(ii) **Symmetrization.**
   This idea was introduced by Davenport [38] to construct the first example of a set with optimal order of L^2 discrepancy in dimension $d = 2$. His example was a symmetrized irrational lattice, i.e. $\left\{\left(\pm\frac{k}{N}, \{k\alpha\}\right)\right\}_{k=1}^{N}$, where α is an irrational number with bounded partial quotients of the continued fraction expansion and $\{x\}$ is the fractional part of x. Roughly speaking, the

symmetrization 'cancels out' the zero-order term of the Fourier expansion of D_N. A similar idea was applied to the van der Corput set in [34].

(iii) Digit shifts (digit scrambling).

The method goes back to [50] in dimension $d = 2$ and [28] in higher dimensions. In the case of the van der Corput set it works extremely well and may be easily described. Fix an n-bit sequence of zeros and ones $\sigma = (\sigma_k)_{k=1}^n \in \{0, 1\}^n$. We alter $\mathscr{V}_n$ as follows:

$$\mathscr{V}_n^\sigma = \{(0.x_1x_2\ldots x_n, \ 0.(x_n \oplus \sigma_n)(x_{n-1} \oplus \sigma_{n-1})\ldots(x_1 \oplus \sigma_1)): \quad (2.244)$$

$$x_k = 0, 1; k = 1,\ldots,n\},$$

where $\oplus$ denotes addition modulo 2. To put this definition into simple words, we can say that after flipping the digits, we also change some of them to the opposite (we 'scramble' or 'shift' precisely those digits for which $\sigma_k = 1$). This procedure has been thoroughly studied for the van der Corput set. It is well known that it improves its distributional qualities in many different senses [41, 57]. In particular, when approximately half of the digits are shifted, i.e. $\sum \sigma_k \approx \frac{n}{2}$, this set has optimal order of magnitude of the L^2 discrepancy [18, 50, 58].

There is a natural explanation for this phenomenon which continues the line of reasoning started by (2.241). If one views the digits x_i as independent $0 - 1$ random variables and tries to compute the quantity $\int D_{\mathscr{V}_n}(x)dx$, one inevitably encounters expressions of the type $\mathbb{E}x_i \cdot x_j$. And while for $i \neq j$ we obtain $\mathbb{E}x_i \cdot x_j = \frac{1}{4}$, in the 'diagonal' case this quantity is twice as big, $\mathbb{E}x_i^2 = \frac{1}{2}$. And this occurs $n = \log_2 N$ times which leads to the estimate (2.241). However, if the digit x_i is scrambled, we have $\mathbb{E}x_i \cdot (1 - x_i) = 0$. Therefore, one should scramble approximately one half of all digits in order to compensate for the 'diagonal' effect. The details are left to the reader and can be also found in the aforementioned references.

The nice dyadic structure of this set makes it perfectly amenable to the methods of harmonic analysis. For example, in [13] it is analyzed using Fourier series, in [33, 34, 39, 63] the authors exploit Walsh functions (the Walsh analysis of the van der Corput sets is nicely described in the chapter by W. Chen and M. Skriganov in the current volume), while the estimates in [18, 53] are based on the Haar coefficients of $D_{\mathscr{V}_n}$. We shall focus on the latter results as they directly relate to Roth's method, the main topic of our chapter, and complement previously discussed lower bounds.

2.6.1.2 Discrepancy of the Van Der Corput Set in Other Function Spaces

It has been shown in [18] that the BMO (2.107) and $\exp(L^\alpha)$ (2.109) lower estimates in dimension $d = 2$, which we presented in Sect. 2.3.4 are sharp. In particular, for the digit-shifted van der Corput set $\mathscr{V}_n^\sigma$ with $\sum \sigma_k \approx \frac{n}{2}$ and for $\alpha \geq 2$ we have

$$\|D_{\mathscr{V}_n^\sigma}\|_{\exp(L^\alpha)} \lesssim (\log N)^{1-\frac{1}{\alpha}}. \qquad (2.245)$$

In the case of the BMO norm, the standard van der Corput set satisfies

$$\|D_{\mathscr{V}_n}\|_{\text{BMO}} \lesssim \sqrt{\log N}. \tag{2.246}$$

These inequalities were based on estimates of the Haar coefficients of the discrepancy function, namely

$$|\langle D_{\mathscr{V}_n^\sigma}, h_R \rangle| \lesssim \min\{1/N, |R|\}. \tag{2.247}$$

This estimate for small rectangles is straightforward. The counting and linear part can be bounded separately. The estimate for the counting part relies on the fact that $\mathscr{V}_n^\sigma$ is a dyadic net and thus there cannot be too many points in a small dyadic box, while the contribution of the linear part as computed in (2.33) is of the order $N|R|^2 \lesssim |R|$. In turn, coefficients corresponding to large rectangles involve subtle cancellations suggested by the structure and self-similarities of $\mathscr{V}_n^\sigma$. We point out that, in accordance with Roth's principle (2.24), the cutoff between 'small' and 'large' rectangles occurs at the scale $|R| \approx \frac{1}{N}$. The BMO and $\exp(L^\alpha)$ can then be obtained by applying arguments of Littlewood–Paley type.

Almost simultaneously to these results, the Besov norm of the same digit-shifted van der Corput set has been estimated using a very similar method in [53], see also [71]. In fact, this work went much further: all the Haar coefficients of $D_{\mathscr{V}_n^\sigma}$ have been computed exactly. This led to showing that the lower Besov space estimate (2.101) of Triebel [112] is sharp in $d = 2$, more precisely

$$\|D_{\mathscr{V}_n^\sigma}\|_{S_{pq}^r B([0,1]^d)} \lesssim N^r (\log N)^{\frac{1}{q}} \tag{2.248}$$

for $1 \le p, q \le \infty, 0 \le r < \frac{1}{p}$.

2.6.1.3 The Structure of the Riesz Product and the Van Der Corput Set

We close our discussion of the van der Corput set with an amusing observation which pinpoints yet another connection between the small ball inequality (2.111) and discrepancy.

Consider the two-dimensional case of the small ball inequality and assume that all the coefficients α_R are non-negative. Recall Temlyakov's test function (2.124): $\Psi = \prod_{k=1}^{n} (1 + f_k)$. In this case, since $\text{sgn}(\alpha_R) = +1$, the r-functions $f_k = \sum_{|R|=2^{-n}, |R_1|=2^{-k}} h_R$ are actually Rademacher functions. As explained in the very beginning of Sect. 2.4.6, the Riesz product Ψ captures the set where all the functions f_k are positive. To be more precise, $\Psi = 2^{n+1}\mathbf{1}_E$, where $E = \{x \in [0, 1]^2 : f_k(x) = +1, k = 0, 1, \ldots, n\}$.

We shall describe the geometry of the set E. Evidently, it consists of 2^{n+1} dyadic squares of area $2^{-2(n+1)}$. We characterize the locations of the lower left corners of these squares. If $t \in [0, 1]$ and a dyadic interval I of length 2^{-k} contains t, then $h_I(t) = -1$ if the $(k + 1)$st binary digit of t is 0, and $h_I(t) = 1$ if it is 1. Thus $f_k(x_1, x_2) = +1$ exactly when the $(k + 1)$st digit of x_1 and the $(n - k + 1)$st digit of x_2 are the same, either both 0, or both 1. Therefore, $(x_1, x_2) \in E$ when this holds for all $k = 0, 1, \ldots, n$, i.e. the first $n + 1$ binary digits of x_2 are formed as the reversed sequence of the first $n + 1$ digits of x_1—but this is precisely the definition of the van der Corput set $\mathscr{V}_{n+1}$! Therefore

$$E = \mathscr{V}_{n+1} + [0, 2^{-(n+1)}) \times [0, 2^{-(n+1)}), \qquad (2.249)$$

i.e. the Riesz product, which produces the proof of the small ball Conjecture 12, is essentially supported on the standard van der Corput set. Notice also that replacing f_k by $-f_k$ results in 'scrambling' the kth digit in the van der Corput set.

References

1. T.W. Anderson, The integral of a symmetric unimodal function over a symmetric convex set and some probability inequalities. Proc. Am. Math. Soc. **6**, 170–176 (1955). doi:10.2307/2032333
2. R.F. Bass, Probability estimates for multiparameter Brownian processes. Ann. Probab. **16**(1), 251–264 (1988). doi:10.1214/aop/1176991899
3. J. Beck, W.W.L. Chen, Note on irregularities of distribution. Mathematika **33**, 148–163 (1986). doi:10.1112/S0025579300013966
4. J. Beck, W.W.L. Chen, Note on irregularities of distribution. II. Proc. Lond. Math. Soc. III. Ser. **61**(2), 251–272 (1990). doi:10.1112/plms/s3-61.2.251
5. J. Beck, Balanced two-colorings of finite sets in the square. I. Combinatorica **1**(4), 327–335 (1981). doi:10.1007/BF02579453
6. J. Beck, Irregularities of distribution. I. Acta Math. **159**(1–2), 1–49 (1987). doi:10.1007/BF02392553
7. J. Beck, Irregularities of distribution. II. Proc. Lond. Math. Soc. III. Ser. **56**(1), 1–50 (1988). doi:10.1112/plms/s3-56.1.1
8. J. Beck, A two-dimensional van Aardenne-Ehrenfest theorem in irregularities of distribution. Compos. Math. **72**(3), 269–339 (1989)
9. J. Beck, The modulus of polynomials with zeros on the unit circle: A problem of Erdős. Ann. Math. **134**(2), 609–651 (1991). doi:10.2307/2944358
10. J. Beck, W.W.L. Chen, *Irregularities of Distribution*. Cambridge Tracts in Mathematics, vol. 89 (Cambridge University Press, Cambridge, 1987)
11. R. Béjian, Minoration de la discrepance d'une suite quelconque sur T. Acta Arith. **41**(2), 185–202 (1982)
12. A. Bernard, Espaces H^1 de martingales à deux indices. Dualité avec les martingales de type "BMO". Bull. Sci. Math. II. Ser. **103**(3), 297–303 (1979)
13. D. Bilyk, Cyclic shifts of the van der Corput set. Proc. Am. Math. Soc. **137**(8), 2591–2600 (2009). doi:10.1090/S0002-9939-09-09854-2
14. D. Bilyk, On Roth's orthogonal function method in discrepancy theory. Uniform Distrib. Theory **6**(1), 143–184 (2011)
15. D. Bilyk, M.T. Lacey, On the small ball inequality in three dimensions. Duke Math. J. **143**(1), 81–115 (2008). doi:10.1215/00127094-2008-016

16. D. Bilyk, M.T. Lacey, A. Vagharshakyan, On the signed small ball inequality. Online J. Analytic Combinator. **3** (2008)
17. D. Bilyk, M.T. Lacey, A. Vagharshakyan, On the small ball inequality in all dimensions. J. Funct. Anal. **254**(9), 2470–2502 (2008). doi:10.1016/j.jfa.2007.09.010
18. D. Bilyk, M.T. Lacey, I. Parissis, A. Vagharshakyan, Exponential squared integrability of the discrepancy function in two dimensions. Mathematika **55**(1–2), 1–27 (2009). doi:10.1112/S0025579300000930
19. D. Bilyk, M.T. Lacey, I. Parissis, A. Vagharshakyan, *A Three-Dimensional Signed Small Ball Inequality.* (Kendrick Press, Heber City, UT, 2010)
20. V.I. Bogachev, *Gaussian Measures. Transl. from the Russian by the author.* Mathematical Surveys and Monographs, vol. 62 (American Mathematical Society (AMS), Providence, RI, 1998)
21. D.L. Burkholder, *Sharp Inequalities for Martingales and Stochastic Integrals.* Les processus stochastiques, Coll. Paul Lévy, Palaiseau/Fr. 1987, Astérisque 157–158, 75–94, 1988
22. S.-Y.A. Chang, R. Fefferman, A continuous version of duality of H^1 with BMO on the bidisc. Ann. Math. **112**(1), 179–201 (1980). doi:10.2307/1971324
23. S.-Y.A. Chang, R. Fefferman, Some recent developments in Fourier analysis and H^p-theory on product domains. Bull. Am. Math. Soc. New Ser. **12**(1), 1–43 (1985). doi:10.1090/S0273-0979-1985-15291-7
24. S.-Y.A. Chang, J.M. Wilson, T.H. Wolff, Some weighted norm inequalities concerning the Schrödinger operators. Comment. Math. Helv. **60**(2), 217–246 (1985). doi:10.1007/BF02567411
25. B. Chazelle, *The Discrepancy Method. Randomness and Complexity.* (Cambridge University Press, Cambridge, 2000)
26. B. Chazelle, Complexity bounds via Roth's method of orthogonal functions, in *Analytic Number Theory. Essays in Honour of Klaus Roth on the Occasion of his 80th Birthday*, ed. by W.W.L. Chen (Cambridge University Press, Cambridge, 2009), pp. 144–149
27. W.W.L. Chen, On irregularities of distribution. Mathematika **27**(2), 153–170 (1980). doi:10.1112/S0025579300010044
28. W.W.L. Chen, On irregularities of distribution. II. Q. J. Math. Oxford. II. Ser. **34**(135), 257–279 (1983). doi:10.1093/qmath/34.3.257
29. W.W.L. Chen, On irregularities of distribution. III. J. Aust. Math. Soc. Ser. A **60**(2), 228–244 (1996)
30. W.W.L. Chen, On irregularities of distribution. IV. J. Number Theory **80**(1), 44–59 (2000). doi:10.1006/jnth.1999.2442
31. W.W.L. Chen, Fourier techniques in the theory of irregularities of point distribution, in *Fourier Analysis and Convexity*, ed. by L. Brandolini, Applied and Numerical Harmonic Analysis (Birkhäuser, Boston, MA, 2004), pp. 59–82
32. W.W.L. Chen, private communication, Palo Alto, CA (2008)
33. W.W.L. Chen, M.M. Skriganov, Explicit constructions in the classical mean squares problem in irregularities of point distribution. J. Reine Angew. Math. **545**, 67–95 (2002). doi:10.1515/crll.2002.037
34. W.W.L. Chen, M.M. Skriganov, Davenport's theorem in the theory of irregularities of point distribution. J. Math. Sci. New York **115**(1), 2076–2084 (2003). doi:10.1023/A:1022668317029
35. W.W.L. Chen, G. Travaglini, Some of Roth's ideas in discrepancy theory, in *Analytic Number Theory. Essays in Honour of Klaus Roth on the Occasion of his 80th Birthday*, ed. by W.W.L. Chen (Cambridge University Press, Cambridge, 2009), pp. 150–163
36. E. Csaki, *On Small Values of the Square Integral of a Multiparameter Wiener Process.* (Reidel, Dordrecht, 1984)
37. I. Daubechies, *Ten Lectures on Wavelets.* CBMS-NSF Regional Conference Series in Applied Mathematics, vol. 61. (SIAM, Society for Industrial and Applied Mathematics, Philadelphia, PA, 1992)

38. H. Davenport, Note on irregularities of distribution. Mathematika Lond. **3**, 131–135 (1956). doi:10.1112/S0025579300001807

39. J. Dick, F. Pillichshammer, *Digital Nets and Sequences. Discrepancy Theory and Quasi-Monte Carlo Integration.* (Cambridge University Press, Cambridge, 2010)

40. T. Dunker, T. Kühn, M. Lifshits, W. Linde, Metric entropy of the integration operator and small ball probabilities for the Brownian sheet. C. R. Acad. Sci. Paris Sér. I Math. **326**(3), 347–352 (1998). doi:10.1016/S0764-4442(97)82993-X

41. H. Faure, F. Pillichshammer, L_p discrepancy of generalized two-dimensional Hammersley point sets. Monatsh. Math. **158**(1), 31–61 (2009). doi:10.1007/s00605-008-0039-1

42. R. Fefferman, J. Pipher, Multiparameter operators and sharp weighted inequalities. Am. J. Math. **119**(2), 337–369 (1997). doi:10.1353/ajm.1997.0011

43. W. Feller, *An Introduction to Probability Theory and Its Applications*, vol. II. (Wiley, New York-London-Sydney, 1966)

44. K.K. Frolov, An upper estimate of the discrepancy in the L_p-metric, $2 \leq p < \infty$. Dokl. Akad. Nauk SSSR **252**, 805–807 (1980)

45. E.N. Gilbert, A comparison of signalling alphabets. Bell Syst. Tech. J. **3**, 504–522 (1952)

46. L. Grafakos, *Classical and Modern Fourier Analysis.* (Pearson/Prentice Hall, Upper Saddle River, NJ, 2004)

47. A. Haar, Zur Theorie der orthogonalen Funktionensysteme. (Erste Mitteilung.). Math. Annal. **69**, 331–371 (1910). doi:10.1007/BF01456326

48. G. Halász, On Roth's method in the theory of irregularities of point distributions, in *Recent Progress in Analytic Number Theory*, vol. 2, ed. by H. Halberstam, C. Hooley, vol. 2 (London Academic Press, Durham, 1981), pp. 79–94

49. J.H. Halton, On the efficiency of certain quasi-random sequences of points in evaluating multi-dimensional integrals. Num. Math. **2**, 84–90 (1960)

50. J.H. Halton, S.K. Zaremba, The extreme and L^2 discrepancies of some plane sets. Monatsh. Math. **73**, 316–328 (1969). doi:10.1007/BF01298982

51. J.M. Hammersley, Monte Carlo methods for solving multivariable problems. Ann. N.Y. Acad. Sci. **86**, 844–874 (1960). doi:10.1111/j.1749-6632.1960.tb42846.x

52. S. Heinrich, Some open problems concerning the star-discrepancy. J. Complexity **19**(1), 416–419 (2003). doi:10.1016/S0885-064X(03)00014-1

53. A. Hinrichs, Discrepancy of Hammersley points in Besov spaces of dominating mixed smoothness. Math. Nachr. **283**(3), 478–488 (2010). doi:10.1002/mana.200910265

54. A. Hinrichs, L. Markhasin, On lower bounds for the L_2-discrepancy. J. Complexity **27**(2), 127–132 (2011). doi:10.1016/j.jco.2010.11.002

55. W. Hoeffding, Probability inequalities for sums of bounded random variables. J. Am. Stat. Assoc. **58**, 13–30 (1963). doi:10.2307/2282952

56. Y. Katznelson, *An Introduction to Harmonic Analysis*, 3rd edn. Cambridge Mathematical Library (Cambridge University Press, Cambridge, 2004)

57. P. Kritzer, On some remarkable properties of the two-dimensional Hammersley point set in base 2. J. Thor. Nombres Bordx. **18**(1), 203–221 (2006). doi:10.5802/jtnb.540

58. P. Kritzer, F. Pillichshammer, An exact formula for L_2 discrepancy of the shifted Hammersley point set. Unif. Distrib. Theory **1**(1), 1–13 (2006)

59. J. Kuelbs, W.V. Li, Metric entropy and the small ball problem for Gaussian measures. J. Funct. Anal. **116**(1), 133–157 (1993). doi:10.1006/jfan.1993.1107

60. L. Kuipers, H. Niederreiter, *Uniform Distribution of Sequences.* Pure and Applied Mathematics, a Wiley-Interscience Publication (Wiley, New York-London-Sydney, 1974)

61. M.T. Lacey, Small ball and discrepancy inequalities. arXiv: **abs/math/0609816** (2006)

62. M.T. Lacey, On the discrepancy function in arbitrary dimension, close to L^1. Anal. Math. **34**(2), 119–136 (2008). doi:10.1007/s10476-008-0203-9

63. G. Larcher, F. Pillichshammer, Walsh series analysis of the L_2-discrepancy of symmetrisized point sets. Monatsh. Math. **132**(1), 1–18 (2001). doi:10.1007/s006050170054

64. M. Ledoux, M. Talagrand, *Probability in Banach Spaces. Isoperimetry and Processes.* Ergebnisse der Mathematik und ihrer Grenzgebiete, vol. 3. Folge, 23 (Springer, Berlin, 1991)

65. M. Lerch, Question 1547. L'Intermediaire Math. **11**, 144–145 (1904)
66. P. Lévy, *Problèmes Concrets d'analyse Fonctionelle.* (Gautheir-Villars, Paris, 1951)
67. W.V. Li, Q.-M. Shao, *Gaussian Processes: Inequalities, Small Ball Probabilities and Applications.*, ed. by D.N. Shanbhag, et al., Stochastic Processes: Theory and Methods, vol. 19 (North-Holland/Elsevier, Amsterdam, 2001), pp. 533–597. Handb. Stat.
68. P. Liardet, *A Three-Dimensional Signed Small Ball Inequality.* Discrépance sur le cercle. Primaths. I (Univ. Marseille, 1979)
69. M.A. Lifshits, *Gaussian Random Functions.* Mathematics and Its Applications (Dordrecht), vol. 322 (Kluwer Academic, Dordrecht, 1995)
70. J. Lindenstrauss, L. Tzafriri, *Classical Banach Spaces I. Sequence Spaces.* Ergebnisse der Mathematik und ihrer Grenzgebiete, vol. 92. (Springer, Berlin-Heidelberg-New York, 1977)
71. L. Markhasin, Quasi-monte carlo methods for integration of functions with dominating mixed smoothness in arbitrary dimension. arXiv: **abs/1201.2311v3** (2012)
72. J. Matoušek, *Geometric Discrepancy. An Illustrated Guide.* Algorithms and Combinatorics, vol. 18 (Springer, Berlin, 1999)
73. H.L. Montgomery, *Ten Lectures on the Interface Between Analytic Number Theory and Harmonic Analysis.* Regional Conference Series in Mathematics, vol. 84 (American Mathematical Society (AMS), Providence, RI, 1994)
74. B. Muckenhoupt, Weighted norm inequalities for the Hardy maximal function. Trans. Am. Math. Soc. **165**, 207–226 (1972). doi:10.2307/1995882
75. E. Novak, H. Woźniakowski, *Tractability of Multivariate Problems*, Vol. II: Standard information for functionals. EMS Tracts in Mathematics, vol. 12 (European Mathematical Society (EMS), Zürich, 2010)
76. W. Ou, Irregularity of distributions and multiparameter A_p weights. Uniform Distrib. Theory **5**(2), 131–139 (2010)
77. M.C. Pereyra, *Lecture Notes on Dyadic Harmonic Analysis* (American Mathematical Society (AMS), Providence, RI, 2001). doi:10.1090/conm/289
78. J. Pipher, Bounded double square functions. Ann. Inst. Fourier **36**(2), 69–82 (1986). doi:10.5802/aif.1048
79. G. Pólya, *How to Solve It. A New Aspect of Mathematical Method. Reprint of the 2nd edn.* Princeton Science Library (Princeton University Press, Princeton, NJ, 1988), p. 253
80. F. Riesz, Über die *Fourier*koeffizienten einer stetigen Funktion von beschränkter Schwankung. Math. Zs. **2**(3–4), 312–315 (1918). doi:10.1007/BF01199414
81. K.F. Roth, On irregularities of distribution. Mathematika **1**, 73–79 (1954). doi:10.1112/S0025579300000541
82. K.F. Roth, On irregularities of distribution. II. Commun. Pure Appl. Math. **29**(6), 749–754 (1976). doi:10.1002/cpa.3160290614
83. K.F. Roth, On irregularities of distribution. III. Acta Arith. **35**, 373–384 (1979)
84. K.F. Roth, On irregularities of distribution. IV. Acta Arith. **37**, 67–75 (1980)
85. K.F. Roth, On a theorem of Beck. Glasgow Math. J. **27**, 195–201 (1985). doi:10.1017/S0017089500006182
86. W.M. Schmidt, Irregularities of distribution. Q. J. Math. Oxford II. Ser. **19**, 181–191 (1968). doi:10.1093/qmath/19.1.181
87. W.M. Schmidt, Irregularities of distribution. II. Trans. Am. Math. Soc. **136**, 347–360 (1969). doi:10.2307/1994719
88. W.M. Schmidt, Irregularities of distribution. III. Pacific J. Math. **29**, 225–234 (1969). doi:10.2140/pjm.1969.29.225
89. W.M. Schmidt, Irregularities of distribution. IV. Invent. Math. **7**, 55–82 (1969). doi:10.1007/BF01418774
90. W.M. Schmidt, Irregularities of distribution. V. Proc. Am. Math. Soc. **25**, 608–614 (1970). doi:10.2307/2036653
91. W.M. Schmidt, Irregularities of distribution. VI. Compos. Math. **24**, 63–74 (1972)
92. W.M. Schmidt, Irregularities of distribution. VII. Acta Arith. **21**, 45–50 (1972)

93. W.M. Schmidt, Irregularities of distribution. VIII. Trans. Am. Math. Soc. **198**, 1–22 (1974). doi:10.2307/1996744
94. W.M. Schmidt, Irregularities of distribution. IX. Acta Arith. **27**, 385–396 (1975)
95. W.M. Schmidt, *Irregularities of Distribution. X.* (Academic Press, New York, 1977)
96. S. Sidon, Verallgemeinerung eines Satzes über die absolute Konvergenz von Fourierreihen mit Lücken. Math. Ann. **97**, 675–676 (1927). doi:10.1007/BF01447888
97. S. Sidon, Ein Satz über trigonometrische Polynome mit Lücken und seine Anwendung in der Theorie der Fourier-Reihen. J. Reine Angew. Math. **163**, 251–252 (1930). doi:10.1515/crll. 1930.163.251
98. M.M. Skriganov, Harmonic analysis on totally disconnected groups and irregularities of point distributions. J. Reine Angew. Math. **600**, 25–49 (2006). doi:10.1515/CRELLE.2006.085
99. M.M. Skriganov, private communication, Palo Alto, CA (2008)
100. I.M. Sobol, *Multidimensional Quadrature Formulas and Haar Functions.* (Biblioteka Prikladnogo Analiza i Vychislitel'noĭ Matematiki. Izdat. 'Nauka', FizMatLit., Moscow, 1969), p. 288
101. E.M. Stein, *Harmonic Analysis: Real-Variable Methods, Orthogonality, and Oscillatory Integrals. With the Assistance of Timothy S. Murphy.* Princeton Mathematical Series, vol. 43 (Princeton University Press, Princeton, NJ, 1993), p. 695
102. M. Talagrand, The small ball problem for the Brownian sheet. Ann. Probab. **22**(3), 1331–1354 (1994). doi:10.1214/aop/1176988605
103. V.N. Temlyakov, Approximation of periodic functions of several variables with bounded mixed difference. Math. USSR Sb. **41**, 53–66 (1982). doi:10.1070/SM1982v041n01ABEH002220
104. V.N. Temlyakov, *Approximation of Functions with a Bounded Mixed Derivative. Transl. from the Russian by H. H. McFaden.* Proceedings of the Steklov Institute of Mathematics, vol. 1 (American Mathematical Society (AMS), Providence, RI, 1989), p. 121
105. V.N. Temlyakov, *Approximation of Periodic Functions.* Computational Mathematics and Analysis Series (Nova Science Publishers, Commack, NY, 1993), p. 419
106. V.N. Temlyakov, An inequality for trigonometric polynomials and its application for estimating the entropy numbers. J. Complexity **11**(2), 293–307 (1995). doi:10.1006/jcom.1995.1012
107. V.N. Temlyakov, Some inequalities for multivariate Haar polynomials. East J. Approx. **1**(1), 61–72 (1995)
108. V.N. Temlyakov, An inequality for trigonometric polynomials and its application for estimating the Kolmogorov widths. East J. Approx. **2**(2), 253–262 (1996)
109. V.N. Temlyakov, On two problems in the multivariate approximation. East J. Approx. **4**(4), 505–514 (1998)
110. V.N. Temlyakov, Cubature formulas, discrepancy, and nonlinear approximation. J. Complexity **19**(3), 352–391 (2003). doi:10.1016/S0885-064X(02)00025-0
111. H. Triebel, *Bases in Function Spaces, Sampling, Discrepancy, Numerical Integration.* EMS Tracts in Mathematics, vol. 11 (European Mathematical Society (EMS), Zürich, 2010), p. 296. doi:10.4171/085
112. H. Triebel, Numerical integration and discrepancy, a new approach. Math. Nachr. **283**(1), 139–159 (2010). doi:10.1002/mana.200910842
113. T. van Aardenne-Ehrenfest, Proof of the impossibility of a just distribution of an infinite sequence of points over an interval. Proc. Kon. Ned. Akad. v. Wetensch. **48**, 266–271 (1945)
114. T. van Aardenne-Ehrenfest, On the impossibility of a just distribution. Proc. Nederl. Akad. Wetensch. Amsterdam **52**, 734–739 (1949)
115. J.G. van der Corput, Verteilungsfunktionen. I. Proc. Akad. Wetensch. Amsterdam **38**, 813–821 (1935)
116. J.G. van der Corput, Verteilungsfunktionen. II. Proc. Akad. Wetensch. Amsterdam **38**, 1058–1066 (1935)
117. R.R. Varshamov, The evaluation of signals in codes with correction of errors. Dokl. Akad. Nauk SSSR **117**, 739–741 (1957)

118. G. Wang, Sharp square-function inequalities for conditionally symmetric martingales. Trans. Am. Math. Soc. **328**(1), 393–419 (1991). doi:10.2307/2001887
119. A. Zygmund, *Trigonometric Series. Volumes I and II Combined. With a foreword by Robert Fefferman. 3rd edn.* Cambridge Mathematical Library (Cambridge University Press, Cambridge, 2002), p. 364

Chapter 3
Irregularities of Distribution and Average Decay of Fourier Transforms

Luca Brandolini, Giacomo Gigante, and Giancarlo Travaglini

Abstract In Geometric Discrepancy we usually test a distribution of N points against a suitable family of sets. If this family consists of dilated, translated and rotated copies of a given d-dimensional convex body $D \subset [0, 1)^d$, then a result proved by W. Schmidt, J. Beck and H. Montgomery shows that the corresponding L^2 discrepancy cannot be smaller than $c_d N^{(d-1)/2d}$. Moreover, this estimate is sharp, thanks to results of D. Kendall, J. Beck and W. Chen. Both lower and upper bounds are consequences of estimates of the decay of $\|\hat{\chi}_D (\rho \cdot)\|_{L^2(\Sigma_{d-1})}$ for large ρ, where $\hat{\chi}_D$ is the Fourier transform (expressed in polar coordinates) of the characteristic function of the convex body D, while Σ_{d-1} is the unit sphere in $\mathbb{R}^d$. In this chapter we provide the Fourier analytic background and we carefully investigate the relation between the L^2 discrepancy and the estimates of $\|\hat{\chi}_D (\rho \cdot)\|_{L^2(\Sigma_{d-1})}$.

3.1 Introduction

More than 40 years ago W. Schmidt [51, 52] proved the following theorem on irregularities of point distribution related to discs.

Theorem 1 (Schmidt). *For every distribution $\mathscr{P}$ of N points in the torus $\mathbb{T}^2$ there exists a disc $D \subset \mathbb{T}^2$ of diameter less than 1 such that, for every $\varepsilon > 0$,*

L. Brandolini (✉) • G. Gigante
Dipartimento di Ingegneria, Università di Bergamo, Viale Marconi 5, 24044 Dalmine,
Bergamo, Italia
e-mail: luca.brandolini@unibg.it; giacomo.gigante@unibg.it

G. Travaglini (✉)
Dipartimento di Statistica e Metodi Quantitativi, Edificio U7, Università di Milano-Bicocca,
Via Bicocca degli Arcimboldi 8, 20126 Milano, Italia
e-mail: giancarlo.travaglini@unimib.it

W. Chen et al. (eds.), *A Panorama of Discrepancy Theory*, Lecture Notes
in Mathematics 2107, DOI 10.1007/978-3-319-04696-9_3,
© Springer International Publishing Switzerland 2014

$$\left| \operatorname{card}\left(\mathscr{P} \cap D\right) - N \left| D \right| \right| \geq c_\varepsilon \, N^{(1/4)-\varepsilon} \, ,$$

where $|A|$ denotes the volume.

This result has to be compared with the following earlier results of K. Roth [50] and H. Davenport [26].

Theorem 2 (Roth). *For every distribution $\mathscr{P}$ of N points in $[0,1]^2$ we have the following lower bound*

$$\int_{\mathbb{T}^2} \left| \operatorname{card}\left(\mathscr{P} \cap I_x\right) - N x_1 x_2 \right|^2 dx_1 \, dx_2 \geq c \, \log N \, ,$$

where $I_x = [0, x_1] \times [0, x_2]$ for every $x = (x_1, x_2) \in [0,1]^2$. Hence for every distribution $\mathscr{P}$ of N points in the torus $\mathbb{T}^2$ there exists a rectangle $R \subset \mathbb{T}^2$, having sides parallel to the axes and such that

$$\left| \operatorname{card}\left(\mathscr{P} \cap R\right) - N \left| R \right| \right| \geq c \, \log^{1/2} N \, .$$

Theorem 3 (Davenport). *For every integer $N \geq 2$ there exists a distribution $\mathscr{P}$ of N points in the torus $\mathbb{T}^2$ such that*

$$\int_{\mathbb{T}^2} \left| \operatorname{card}\left(\mathscr{P} \cap I_x\right) - N x_1 x_2 \right|^2 dx_1 \, dx_2 \leq c \, \log N \, .$$

Schmidt's theorem has been improved and extended by J. Beck [3] and H. Montgomery [42], who have independently obtained the following L^2 result (see also [2, 4, 11]).

Theorem 4 (Beck, Montgomery). *Let $B \subset \mathbb{T}^d$ be a convex body. Then for every distribution $\mathscr{P}$ of N points in $\mathbb{T}^d$ we have*

$$\int_0^1 \int_{SO(d)} \int_{\mathbb{T}^d} \left| \operatorname{card}\left(\mathscr{P} \cap (\lambda \sigma \, (B + t))\right) - \lambda^d N \left| B \right| \right|^2 dt \, d\sigma \, d\lambda \geq c_d N^{(d-1)/d} \, .$$

Hence for every distribution $\mathscr{P}$ of N points in $\mathbb{T}^d$ there exists a translated, rotated and dilated copy B' of B such that

$$\left| \operatorname{card}\left(\mathscr{P} \cap B'\right) - N \left| B' \right| \right| \geq c N^{(d-1)/2d} \, .$$

J. Beck and W. Chen have proved that the above L^2 estimate is sharp (see [2], see also [13, 22, 35]).

Theorem 5 (Beck and Chen). *Let $B \subset \mathbb{R}^d$ be a convex body having diameter less than 1. Then for every positive integer N there exists a distribution $\mathscr{P}$ of N points in $\mathbb{T}^d$ such that*

$$\int_0^1 \int_{SO(d)} \int_{\mathbb{T}^d} |\text{card}\,(\mathscr{P} \cap (\sigma\,(B+t))) - |B||^2 \; dt\,d\sigma \le c_d \; N^{(d-1)/d} \; .$$

The large gap between the sharp L^2 estimates which appear in Theorem 2 and in Theorem 4 seems to be related to the different behaviors of the Fourier transforms of the characteristic functions of balls and polyhedra. The case of the ball is enlightening: the main ingredients in the proofs of the results in Theorems 4 and 5 are provided by the sharp estimates of the L^2 average decay of the Fourier transform $\hat{\chi}_B$ of the characteristic function of a convex body B:

$$\int_{\Sigma_{d-1}} |\hat{\chi}_B\,(\rho\sigma)|^2 \; d\sigma \tag{3.1}$$

(here $\Sigma_{d-1} = \{x \in \mathbb{R}^d : |x| = 1\}$ is the unit sphere in $\mathbb{R}^d$) so that the study of the above problem on irregularities of distribution turns out to be strictly related to the study of (3.1) (see e.g. [6, 11, 13, 61, 62]).

The purpose of this chapter is to exploit the above relation in a detailed and self-contained way. In the second section we prove the L^2 results for the average decay of Fourier transforms of characteristic functions of convex bodies. The third section contains L^p results for polyhedra. In the fourth section we deduce lattice point results. The fifth and the sixth section are the main part of this chapter and show how to obtain different proofs of Theorems 4 and 5, depending on the estimates proved before.

During this chapter positive constants are denoted by $c, c', c_1, \ldots$ (they may vary at every occurrence). By $c_d, c_\varepsilon, c_B, \ldots$ we denote constants which depend on d, ε, B, $\ldots$ For positive A and B, we write $A \approx B$ when there exist positive constants c_1 and c_2 such that $c_1 A \le B \le c_2 A$.

3.2 Decay of the Fourier Transform: L^2 Estimates for Characteristic Functions of Convex and More General Bodies

3.2.1 Introduction

Let $B \subset \mathbb{R}^d$ be a convex body, i.e. a convex bounded set with non empty interior, and let $d\mu$ be the surface measure on ∂B. The study of the decay of the Fourier transforms $\hat{\chi}_B\,(\xi)$ and $\hat{\mu}\,(\xi)$ has a long history and provides several applications to different fields in mathematics (see [56, Ch. VIII, 5, B]). Of course we have $\hat{\chi}_B\,(\xi) \to 0$ as $|\xi| \to +\infty$, by the Riemann-Lebesgue lemma. However more is true, since

$$|\hat{\chi}_B\,(\xi)| \le c_B\,|\xi|^{-1}\,, \tag{3.2}$$

for every $\xi \in \mathbb{R}^d$. Indeed, write $\xi = \rho\sigma$ in polar coordinates ($\rho \geq 0$, $\sigma \in \Sigma_{d-1}$) and for every $\sigma \in \Sigma_{d-1}$ define, for $s \in \mathbb{R}$, the *parallel section function* $\Lambda(s) = \Lambda_\sigma(s)$ equal to the $(d-1)$-volume of the set $B \cap \{\sigma^\perp + s\sigma\}$. In order to prove (3.2) it is enough to assume $\sigma = (1, 0, \ldots, 0)$, so that $\xi = (\rho, 0, \ldots, 0)$. Then, if $x = (x_1, x_2, \ldots, x_d)$, we have

$$\hat{\chi}_B(\xi) = \int_B e^{-2\pi i \xi \cdot x}\, dx = \int_{\mathbb{R}} e^{-2\pi i \rho x_1}\, \Lambda(x_1)\, dx_1 = \hat{\Lambda}(\rho) \ . \tag{3.3}$$

Observe that the variation of the function Λ_σ is bounded uniformly in σ, then (see e.g. [64, p.221]) we get (3.2). The case of the cube $Q = [-1/2, 1/2]^d$, shows that (3.2) cannot be improved. Indeed

$$\hat{\chi}_Q(\xi) = \prod_{j=1}^d \frac{\sin(\pi\xi_j)}{\pi\xi_j}\ ,$$

so that, for the directions orthogonal to the facets (i.e the $(d-1)$-faces) of this cube, e.g. for $\xi = (\rho, 0, \ldots, 0)$, we have $\hat{\chi}_Q(\rho, 0, \ldots, 0) = \sin(\pi\rho)/\pi\rho$ and then we have

$$\limsup_{|\xi| \to +\infty} |\xi|\, |\hat{\chi}_Q(\xi)| > 0\ .$$

In the same way it is easy to see that if $\xi = \rho\sigma$ and σ is not orthogonal to any facet of the cube, then $|\hat{\chi}_Q(\xi)| \leq c_\sigma\, \rho^{-2}$. More generally, if σ is not orthogonal to any face (of any dimension), then $|\hat{\chi}_Q(\xi)| \leq c_\sigma\, \rho^{-d}$, hence this last inequality holds for almost all directions.

The case of the (unit) ball $D = \{x \in \mathbb{R}^d : |x| \leq 1\}$ is of course peculiar, $\hat{\chi}_D$ is a radial function and we have

$$\hat{\chi}_D(\xi) = |\xi|^{-d/2}\, J_{d/2}(2\pi |\xi|)\ , \tag{3.4}$$

for every $\xi \in \mathbb{R}^d$. Here $J_{d/2}$ is the Bessel function of order $d/2$. By the asymptotics of Bessel functions (see [57, Ch. IV, Lemma 3.11], see also [63] for the basic reference on Bessel functions) we know that

$$\hat{\chi}_D(\xi) = \pi^{-1}\, |\xi|^{-(d+1)/2}\, \cos(2\pi |\xi| - \pi (d+1)/4) + \mathcal{O}_d\left(|\xi|^{-(d+3)/2}\right) \tag{3.5}$$

as $|\xi| \to +\infty$.

In certain cases $\hat{\chi}_B(\xi)$ admits interesting upper bounds of geometric nature. When $d = 2$ we shall see in Lemma 14 that for every convex body $B \subset \mathbb{R}^2$ we have, for large $\xi = \rho\sigma$,

$$\left|\hat{\chi}_B(\xi)\right| \le c_B \, \rho^{-1} \left\{ \Lambda\left(-\rho^{-1} + \sup_{y \in B} y \cdot \sigma\right) + \Lambda\left(\rho^{-1} + \inf_{y \in B} y \cdot \sigma\right) \right\} , \quad (3.6)$$

where Λ is the parallel section function. It is easy to show that (3.6) is false when $d \ge 3$. Indeed let P be the octahedron in $\mathbb{R}^3$ given by the convex hull of the six points $(\pm 1, \pm 1, 0)$, $(0, 0, \pm 1)$, and let $\sigma = (0, 0, 1)$. Then

$$\hat{\chi}_P(\rho\sigma) = \int_P e^{-2\pi i \rho x_3} \, dx_1 \, dx_2 \, dx_3 = \hat{\Lambda}(\rho) .$$

Since

$$\Lambda(\rho) = (1 - |\rho|)_+^2 ,$$

then the RHS of (3.6) is $\approx \rho^{-3}$. Now observe that the piecewise smooth function $\Lambda(\rho)$ has continuous derivative at ± 1, but it is only continuous at 0. Then an integration by parts shows that

$$\limsup_{\rho \to +\infty} \rho^2 \left|\hat{\Lambda}(\rho)\right| > 0 .$$

Then neither (3.6) nor an average version of it can be true. On the other hand a deeper analysis shows that (3.6) holds true for every d as long as ∂B is smooth and it has everywhere finite order of contact (see [1] and [16]).

In general (3.6) cannot be reverted. Indeed let B be a ball and recall (3.4), then the zeros of the Bessel function (see [63]) show that the inequality (3.6) can be reverted for no d. A. Podkorytov has shown that (3.6) can be inverted "in mean" (Podkorytov, 2001, personal communication).

A very important case is given by the class of convex bodies B such that ∂B is smooth with everywhere positive Gaussian curvature. In this case the decay of $\hat{\chi}_B(\xi)$ resembles the decay for the ball. Indeed we have (see [16, 30, 32, 33] or [56, Ch. VIII, 5, B])

$$\left|\hat{\chi}_B(\xi)\right| \le c_B \, |\xi|^{-(d+1)/2} , \quad (3.7)$$

for every $\xi \in \mathbb{R}^d$.

When ∂B is flat at some points or irregular, the bound in (3.7) may fail and a pointwise estimate for $\hat{\chi}_B(\xi)$ may lead to poor results in the applications. As a way to overcome this difficulty, we observe that in several problems (see e.g. [10, 11, 15, 46, 48, 49, 62]) the Fourier transform has to be integrated over the rotations, so that it may be enough to study suitable spherical averages of $\hat{\chi}_B(\xi)$. In the next subsection we will study the L^2 spherical means

$$\left\{ \int_{\Sigma_{d-1}} \left|\hat{\chi}_B(\rho\sigma)\right|^2 \, d\sigma \right\}^{1/2}$$

for the case of arbitrary convex bodies, while in the following section we will consider L^p spherical means for polyhedra.

3.2.2 L^2 Spherical Estimates for Convex Bodies

The main result in this field shows that if $B \subset \mathbb{R}^d$ is a convex body, then the L^2 spherical average of $\hat{\chi}_B$ decays of order $(d + 1)/2$. Of course this agrees with the case of the ball, where no spherical average is necessary. The following theorem has been proved by A. Podkorytov in the case $d = 2$ [45] and L. Brandolini, S. Hofmann and A. Iosevich for any dimension d [6].

Theorem 6. *Let $B \subset \mathbb{R}^d$ be a convex body. Then there exists a positive constant $c = c_d$ such that*

$$\left\| \hat{\chi}_B \left(\rho \cdot \right) \right\|_{L^2(\Sigma_{d-1})} \le c \left(\mathrm{diam}\left(B \right) \right)^{(d-1)/2} \rho^{-(d+1)/2} . \tag{3.8}$$

Proof. For every $\varepsilon > 0$ consider a convex body $B' \subset B$ such that $\partial B'$ is smooth with positive Gaussian curvature and $|B \backslash B'| < \varepsilon$ (here $|A|$ denotes the Lebesgue measure of the set A). Assume

$$\left\| \hat{\chi}_{B'} \left(\rho \cdot \right) \right\|_{L^2(\Sigma_{d-1})} \le c \rho^{-(d+1)/2}$$

with c depending on B, but not on B'. Then

$$\left\| \hat{\chi}_B \left(\rho \cdot \right) \right\|_{L^2(\Sigma_{d-1})} \le \left\| \hat{\chi}_{B'} \left(\rho \cdot \right) \right\|_{L^2(\Sigma_{d-1})} + \left\| \hat{\chi}_{B \backslash B'} \left(\rho \cdot \right) \right\|_{L^2(\Sigma_{d-1})} \le c \rho^{-(d+1)/2} + \varepsilon$$

and (3.8) follows by choosing suitable B' (and ε) as ρ diverges. Then it is enough to prove (3.8) assuming B smooth, since the constant $c \left(\mathrm{diam}\left(B \right) \right)^{(d-1)/2}$ must be independent of the smoothness of ∂B. For $\xi \ne 0$ let

$$\omega \left(t \right) = \frac{e^{-2\pi i t \cdot \xi}}{-2\pi i |\xi|^2} \xi.$$

Then $\mathrm{div} \, \omega \left(t \right) = e^{-2\pi i t \cdot \xi}$ and the divergence theorem yields

$$\hat{\chi}_B \left(\rho \sigma \right) = \int_B e^{-2\pi i \rho \sigma \cdot t} \, dt = -\frac{1}{2\pi i \rho} \int_{\partial B} e^{-2\pi i \rho \sigma \cdot t} \left(\nu(t) \cdot \sigma \right) \, d\mu \left(t \right) , \tag{3.9}$$

where $\nu(t)$ is the outward unit normal to ∂B at t and $d\mu$ denotes the surface measure on ∂B. Now write the unit sphere Σ_{d-1} as a finite union of spherical caps U_j having small radius and centers at points $\gamma_j \in \Sigma_{d-1}$, in such a way that every spherical cap U_j supports a cutoff function η_j, so that the η_j's provide a smooth partition of unity

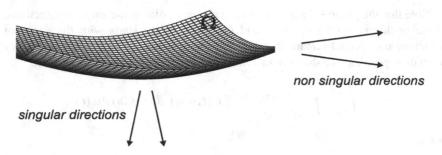

singular directions

non singular directions

Fig. 3.1 The set Ω

of Σ_{d-1}. Although this partition of unity is independent of B, the family $\{\eta_j(\nu(t))\}$ is a partition of unity on ∂B. We then write

$$\hat{\chi}_B(\rho\sigma) = -\frac{1}{2\pi i\rho} \sum_j \int_{\partial B} e^{-2\pi i\rho\sigma\cdot t}(\nu(t)\cdot\sigma)\,\eta_j(\nu(t))\,d\mu(t)$$

and it is enough to prove that for every j we have

$$\int_{\Sigma_{d-1}} \left| \int_{\partial B} e^{-2\pi i\rho\sigma\cdot t}(\nu(t)\cdot\sigma)\,\eta_j(\nu(t))\,d\mu(t) \right|^2 d\sigma \le c\,(\mathrm{diam}\,(B))^{(d-1)/2}\,\rho^{-(d-1)}\,.$$
$$(3.10)$$

Now suppose j is given, write η for η_j, γ for γ_j, and let $\Omega \subset \partial B$ be the support of $\eta(\nu(t))$, so that from now on the inner integral in (3.10) will be on Ω. We may assume η supported in a small spherical cap having center at $\gamma = (0,\ldots,0,-1)$. We need to consider directions which are essentially orthogonal and directions which are essentially non orthogonal to Ω, and tell them apart. In order to do this, let $\psi : \mathbb{R} \to [0,1]$ be a $\mathscr{C}^\infty$ cutoff function such that $\psi(t) = 1$ for $|t| \le c_1$ and $\psi(t) = 0$ for $|t| \ge c_2$, for $0 < c_1 < c_2 < 1$. We write

$$\int_{\Sigma_{d-1}} \left| \int_\Omega e^{-2\pi i\rho\sigma\cdot t}(\nu(t)\cdot\sigma)\,\eta(\nu(t))\,d\mu(t) \right|^2 d\sigma$$

$$= \int_{\Sigma_{d-1}} \left| \int_\Omega e^{-2\pi i\rho\sigma\cdot t}(\nu(t)\cdot\sigma)\,\eta(\nu(t))\,d\mu(t) \right|^2 (1-\psi(-\sigma_d))\,d\sigma$$

$$+ \int_{\Sigma_{d-1}} \left| \int_\Omega e^{-2\pi i\rho\sigma\cdot t}(\nu(t)\cdot\sigma)\,\eta(\nu(t))\,d\mu(t) \right|^2 \psi(-\sigma_d)\,d\sigma$$

$$= S + NS\,.$$

We term "singular" the directions essentially orthogonal to the hyperplanes tangent to Ω and "non singular" the remaining ones (Fig. 3.1).

Note that the phase $-2\pi i \rho \sigma \cdot t$ has a stationary point in the singular directions. However this is not an obstacle, and the proof in [6] starts with the easy but somehow unexpected remark that the L^2 spherical mean "makes this stationary point disappear", as we shall see in a moment. In order to estimate S we write

$$S = \int_\Omega \int_\Omega \int_{\Sigma_{d-1}} e^{-2\pi i \rho \sigma \cdot (t-u)} f(t,u,\sigma) \, d\sigma d\mu(u) \, d\mu(t) \, ,$$

where

$$f(t,u,\sigma) = (\nu(t) \cdot \sigma) \, \eta(\nu(t)) \, (\nu(u) \cdot \sigma) \, \eta(\nu(u)) \, (1 - \psi(\sigma \cdot y))$$

is smooth in σ. Note that $t - u$ in the above integral is essentially parallel to Ω. Then, writing the integral in $d\sigma$ in local coordinates, we can apply [56, Ch. 8, Prop. 4] and obtain

$$\left| \int_{\Sigma_{d-1}} e^{-2\pi i \rho \sigma \cdot (t-u)} f(t,u,\sigma) \, d\sigma \right| \le c \left((1 + \rho |t - u|)^{-N} \right)$$

for a large positive integer N. Then

$$S \le c \int_\Omega \int_\Omega \frac{1}{(1 + \rho |t - u|)^N} \, d\mu(u) \, d\mu(t)$$

$$\le c \int \int_{\{|t-u| \le \rho^{-1}\}} \frac{1}{(1 + \rho |t - u|)^N} \, d\mu(u) \, d\mu(t)$$

$$+ c \int \int_{\{|t-u| \ge \rho^{-1}\}} \frac{1}{(1 + \rho |t - u|)^N} \, d\mu(u) \, d\mu(t)$$

$$\le c\mu(\Omega) \left(\int_{\{x \in \mathbb{R}^{d-1} : |x| \le \rho^{-1}\}} dx + \rho^{-N} \int_{\{x \in \mathbb{R}^{d-1} : |x| > \rho^{-1}\}} |x|^{-N} \, dx \right)$$

$$\le c\mu(\Omega) \rho^{-(d-1)} \, .$$

Now we need to prove the same estimate for NS. If we were free to integrate by parts several times,[1] it should then be easy to handle NS and to end the proof. Since the constants in our estimates need to be independent of the smoothness of ∂B, we need a more refined argument. As a first step, let us see Ω as the graph of a convex smooth function $x \mapsto \Phi(x)$. Then, writing $\Sigma_{d-1} \ni \sigma = (\sigma_1, \ldots, \sigma_d) = (\sigma', \sigma_d)$ we have

---

[1] Actually the convexity hypothesis allows us to integrate by parts at least once without using any regularity assumption on ∂B. In this way we get the bound ρ^{-1} (uniformly in σ), which is enough to prove the theorem in the dimensions $d = 2$ and $d = 3$.

$$
NS = \int_{\Sigma_{d-1}} \left| \int_{\mathbb{R}^{d-1}} e^{-2\pi i \rho(\sigma' \cdot x + \sigma_d \Phi(x))} \left(\frac{(\nabla\Phi(x), -1)}{\sqrt{|\nabla\Phi(x)|^2 + 1}} \cdot \sigma \right) \right.
$$

$$
\left. \times\, \eta\left(\frac{(\nabla\Phi(x), -1)}{\sqrt{|\nabla\Phi(x)|^2 + 1}} \right) dx \right|^2 \psi(\sigma \cdot \gamma)\, d\sigma
$$

$$
= \int_{\Sigma_{d-1}} \left| \int_{\mathscr{A}} e^{-2\pi i \rho(\sigma' \cdot x + \sigma_d \Phi(x))} h(\sigma, \nabla\Phi(x))\, dx \right|^2 \psi(\sigma \cdot \gamma)\, d\sigma ,
$$

where $\mathscr{A}$ is the support of

$$
\eta\left(\frac{(\nabla\Phi(x), -1)}{\sqrt{|\nabla\Phi(x)|^2 + 1}} \right)
$$

and h is a smooth function in the variables σ and $\nabla\Phi(x)$. Note that our choice of Ω implies that $\nabla\Phi$ is uniformly bounded on $\mathscr{A}$ and that $|\sigma'| \geq c > 0$ for a suitable choice of c.

We will work uniformly in $\sigma \cdot \gamma = -\sigma_d$, so that σ_d will not play a role. We will then concentrate on σ' or, better, on $\sigma'/|\sigma'| \in \Sigma_{d-2}$. As we did for Σ_{d-1} we now write Σ_{d-2} as a finite union of spherical caps having small radius and supporting cutoff functions ζ which give a smooth partition of unity on Σ_{d-2}. It is enough to consider the cutoff function ζ supported on a small spherical cap centered at $(1, 0, \ldots, 0) \in \Sigma_{d-2}$. We then have to bound

$$
\int_{\Sigma_{d-1}} \left| \int_{\mathscr{A}} e^{-2\pi i \rho(\sigma' \cdot x + \sigma_d \Phi(x))} h(\sigma, \nabla\Phi(x))\, dx \right|^2 \psi(\sigma_d)\, \zeta\left(\frac{\sigma'}{|\sigma'|} \right) d\sigma . \quad (3.11)
$$

None of the previous steps has said anything on the coordinates $\sigma_2, \ldots, \sigma_{d-1}$ inside σ. We then introduce the change of variables $\sigma = \Xi(\tau, \theta)$, where θ is a real variable, $\tau = (\tau_1, \tau_2, \ldots, \tau_{d-3}, \tau_{d-2})$, with (τ, θ) defined in a neighborhood V of the origin in $\mathbb{R}^{d-1}$ and

$$
\sigma = (\sigma', \sigma_d) = (\sigma_1, \sigma_2, \sigma_3, \ldots, \sigma_{d-2}, \sigma_{d-1}, \sigma_d)
$$

$$
= \left(\sqrt{\frac{1 - |\tau|^2}{1 + \theta^2}}, \tau_1, \tau_2, \ldots, \tau_{d-3}, \tau_{d-2}, \theta\sqrt{\frac{1 - |\tau|^2}{1 + \theta^2}} \right)
$$

$$
= \left(\sqrt{\frac{1 - |\tau|^2}{1 + \theta^2}}, \tau, \theta\sqrt{\frac{1 - |\tau|^2}{1 + \theta^2}} \right) .
$$

Then (3.11) takes the form

$$\int_V \left| \int_{\mathbb{R}^{d-1}} e^{-2\pi i \rho (\sigma' \cdot x + \sigma_d \Phi(x))} h\left(\sigma, \nabla \Phi\left(x\right)\right) dx \right|^2 J\left(\tau, \theta\right) \, d\tau \, d\theta \qquad (3.12)$$

where $J\left(\tau, \theta\right)$ is the Jacobian of the change of variables, times a smooth function. Let $x' = (x_2, \ldots, x_{d-1})$. Since $\sigma_d = \theta \sigma_1$, the inner integral in (3.12) equals

$$\int_{\mathbb{R}^{d-2}} e^{-2\pi i \rho \tau \cdot x'} \int_{\mathbb{R}} e^{-2\pi i \rho \sigma_1 (x_1 + \theta \Phi(x_1, x'))} h\left(\sigma, \nabla \Phi\left(x_1, x'\right)\right) \, dx_1 dx' \ . \qquad (3.13)$$

Now let

$$s = g_{\theta, x'}\left(x_1\right) = x_1 + \theta \Phi\left(x_1, x'\right) \ .$$

Since $\nabla \Phi$ is small we have $g'_{\theta, x'} > c > 0$, so that we may write (3.13) as

$$\int_{\mathbb{R}^{d-2}} e^{-2\pi i \rho \tau \cdot x'} \int_{\mathbb{R}} e^{-2\pi i \rho \sigma_1 s} H\left(\tau, \theta, s, x'\right) \, ds \, dx' \ , \qquad (3.14)$$

where

$$H\left(\tau, \theta, s, x'\right) = \frac{h\left(\Xi\left(\tau, \theta\right), \nabla \Phi\left(g_{\theta, x'}^{-1}\left(s\right), x'\right)\right)}{g'_{\theta, x'}\left(g_{\theta, x'}^{-1}\left(s\right)\right)}$$

is smooth in τ and bounded.

Let us introduce the difference operator Δ_ρ:

$$\Delta_\rho\left[f\left(s\right)\right] = f\left(s + (2\rho)^{-1}\right) - f(s) \ .$$

Since $\Delta_{-\rho}\left[e^{-2\pi i \rho \sigma_1 s}\right] = \left(e^{\pi i \sigma_1} - 1\right) e^{-2\pi i \rho \sigma_1 s}$ and since

$$\int_{\mathbb{R}} \Delta_{-\rho}\left(f\right) g = \int_{\mathbb{R}} f \Delta_{-\rho}\left(g\right) \ ,$$

then (3.14) equals

$$\frac{1}{e^{\pi i \sigma_1} - 1} \int_{\mathbb{R}^{d-2}} e^{-2\pi i \rho \tau \cdot x'} \int_{\mathbb{R}} \Delta_{-\rho}\left[e^{-2\pi i \rho \sigma_1 s}\right] H\left(\tau, \theta, s, x'\right) \, ds \, dx'$$

$$= \frac{1}{e^{\pi i \sigma_1} - 1} \int_{\mathbb{R}^{d-2}} e^{-2\pi i \rho \tau \cdot x'} \int_{\mathbb{R}} e^{-2\pi i \rho \sigma_1 s} \Delta_\rho\left[H\left(\tau, \theta, s, x'\right)\right] \, ds \, dx' \ .$$

Then, by Minkowski integral inequality and by the boundedness of $\left(e^{\pi i \sigma_1} - 1\right)^{-1}$ on V, we have

$$\sqrt{NS} \le c \int_{\mathbb{R}} \left\{ \int_V \left| \int_{\mathbb{R}^{d-2}} e^{-2\pi i \rho \tau \cdot x'} \Delta_\rho \left[H\left(\tau, \theta, s, x'\right)\right] dx' \right|^2 J\left(\tau, \theta\right) \, d\tau \, d\theta \right\}^{1/2} ds \ .$$

Let us rewrite the inner integrals as

$$\int_V \left| \int_{\mathbb{R}^{d-2}} e^{-2\pi i \rho \tau \cdot x'} \Delta_\rho \left[H\left(\tau, \theta, s, x'\right)\right] dx' \right|^2 J\left(\tau, \theta\right) \, d\tau \, d\theta$$

$$= \int_{\mathbb{R}^{d-2}} \int_{\mathbb{R}^{d-2}} \int_V e^{-2\pi i \rho \tau \cdot (x'-y')} \Delta_\rho \left[H\left(\tau, \theta, s, x'\right)\right] \Delta_\rho \left[H\left(\tau, \theta, s, y'\right)\right]$$

$$\times J\left(\tau, \theta\right) \, d\tau \, d\theta \, dx' \, dy' \ .$$

We define

$$D^N f = \sum_{k=0}^{N} \sum_{|\alpha|=k} \sup_\tau \left| \frac{\partial^\alpha}{\partial \tau^\alpha} f(\tau) \right|$$

so that we can integrate by parts several times in τ and obtain, for every positive integer N,

$$\left| \int_V e^{-2\pi i \rho \tau \cdot (x'-y')} \Delta_\rho \left[H\left(\tau, \theta, s, x'\right)\right] \Delta_\rho \left[H\left(\tau, \theta, s, y'\right)\right] J\left(\tau, \theta\right) \, d\tau \, d\theta \right|$$

$$\le c \int_V \frac{1}{\left(1 + \rho |x' - y'|\right)^N}$$

$$\times D^N \left(\Delta_\rho \left[H\left(\tau, \theta, s, x'\right)\right] \Delta_\rho \left[H\left(\tau, \theta, s, y'\right)\right] J\left(\tau, \theta\right)\right) \, d\tau \, d\theta$$

$$\le c \int_V \frac{1}{\left(1 + \rho |x' - y'|\right)^N}$$

$$\times D^N \left(\Delta_\rho \left[H\left(\tau, \theta, s, x'\right)\right]\right) D^N \left(\Delta_\rho \left[H\left(\tau, \theta, s, y'\right)\right] J\left(\tau, \theta\right)\right) \, d\tau \, d\theta \ .$$

Since H and J are smooth in τ, the term

$$D^N \left(\Delta_\rho \left[H\left(\tau, \theta, s, y'\right)\right] J\left(\tau, \theta\right)\right)$$

is bounded. For the remaining term

$$D^N \left(\Delta_\rho \left[H\left(\tau, \theta, s, x'\right)\right]\right)$$

we seek a better estimate. Observe that for every α we have

$$\frac{\partial^\alpha}{\partial \tau^\alpha} \Delta_\rho \left[H \left(\tau, \theta, s, x' \right) \right] = \Delta_\rho \frac{\partial^\alpha H}{\partial \tau^\alpha} \left(\tau, \theta, s, x' \right)$$

$$= \frac{\partial^\alpha H}{\partial \tau^\alpha} \left(\tau, \theta, s + \frac{1}{2\rho}, x' \right) - \frac{\partial^\alpha H}{\partial \tau^\alpha} \left(\tau, \theta, s, x' \right)$$

$$= \frac{1}{2\rho} \int_0^1 \left(\frac{d}{dr} \frac{\partial^\alpha H}{\partial \tau^\alpha} \right) \left(\tau, \theta, s + \frac{r}{2\rho}, x' \right) \, dr \, .$$

Since $\frac{\partial^\alpha H}{\partial \tau^\alpha}$ is smooth in $\nabla \Phi$, we can bound $\frac{d}{dr} \frac{\partial^\alpha H}{\partial \tau^\alpha}$ (uniformly in τ and θ) by a linear combination of $\frac{\partial^2 \Phi}{\partial x_i \partial x_j}$. Being Φ convex, its Hessian matrix is positive definite and we can bound every matrix entry by the trace $\Delta \Phi$, so that we have

$$D^N \Delta_\rho \left[H \left(\tau, \theta, s, x' \right) \right] \le c \, \frac{1}{\rho} \int_0^1 K \left(g_{\theta,x'}^{-1} \left(s + \frac{r}{2\rho} \right), x' \right) \, dr \, ,$$

where

$$K = \chi_{\mathscr{A}} \Delta \Phi \, .$$

Summarizing,

$$\sqrt{NS}$$

$$\le c \int_{\mathbb{R}} \left\{ \int_V \left| \int_{\mathbb{R}^{d-2}} e^{-2\pi i \rho \tau \cdot x'} \Delta_\rho \left[H \left(\tau, \theta, s, x' \right) \right] dx' \right|^2 J \left(\tau, \theta \right) d\tau \, d\theta \right\}^{1/2} ds$$

$$\le c \int_{\mathbb{R}} \left\{ \int_{\mathbb{R}^{d-2}} \int_{\mathbb{R}^{d-2}} \int_V e^{-2\pi i \rho \tau \cdot (x' - y')} \Delta_\rho \left[H \left(\tau, \theta, s, x' \right) \right] \Delta_\rho \left[H \left(\tau, \theta, s, y' \right) \right] \right.$$

$$\times \, J \left(\tau, \theta \right) d\tau \, d\theta \, dx' \, dy' \Big\}^{1/2} \, ds$$

$$\le c \int_{\mathbb{R}} \left\{ \int_{\mathbb{R}^{d-2}} \int_{\mathbb{R}^{d-2}} \frac{1}{\left(1 + \rho |x' - y'| \right)^N} \right.$$

$$\times \int_V D^N \left(\Delta_\rho \left[H \left(\tau, \theta, s, x' \right) \right] \right) d\tau \, d\theta \, dx' \, dy' \Big\}^{1/2} \, ds$$

$$\le c \rho^{-1/2} \sup_\theta \int_{\mathbb{R}} \left\{ \int_{\mathbb{R}^{d-2}} \int_0^1 K \left(g_{\theta,x'}^{-1} \left(s + \frac{r}{2\rho} \right), x' \right) \, dr \right.$$

$$\times \int_{\mathbb{R}^{d-2}} \frac{1}{\left(1 + \rho |x' - y'| \right)^N} \, dy' \, dx' \Big\}^{1/2} \, ds \, .$$

By the Cauchy-Schwarz inequality the last term is smaller than

$$c\rho^{-1/2} \sup_{\theta} \left\{ \int_{\mathbb{R}^{d-2}} \frac{1}{(1+\rho\,|z'|)^N} \, dz' \right\}^{1/2}$$

$$\times \int_{\mathbb{R}} \left\{ \int_{\mathbb{R}^{d-2}} \int_0^1 K\left(g_{\theta,x'}^{-1}\left(s + \frac{r}{2\rho} \right), x' \right) dr \, dx' \right\}^{1/2} ds$$

$$\leq c\rho^{-(d-1)/2} \sup_{\theta} \sqrt{\text{diam}\,(B)}$$

$$\times \left\{ \int_{\mathbb{R}} \int_{\mathbb{R}^{d-2}} \int_0^1 K\left(g_{\theta,x'}^{-1}\left(s + \frac{r}{2\rho} \right), x' \right) dr \, dx' \, ds \right\}^{1/2}$$

$$\leq c\rho^{-(d-1)/2} \sup_{\theta} \sqrt{\text{diam}\,(B)}$$

$$\times \left\{ \int_0^1 \int_{\mathbb{R}} \int_{\mathbb{R}^{d-2}} K\left(g_{\theta,x'}^{-1}\,(s), x' \right) dx' \, ds \, dr \right\}^{1/2}$$

$$\leq c\rho^{-(d-1)/2} \sqrt{\text{diam}\,(B)} \left\{ \int_{\mathbb{R}^{d-2}} K\,(y) \, dy \right\}^{1/2}.$$

Finally,

$$\int_{\mathbb{R}^{d-2}} K\,(y) \, dy = \int_{\mathscr{A}} \Delta\Phi\,(y) \, dy = \sum_{j=1}^d \int_{\mathscr{A}} \frac{\partial^2 \Phi}{\partial y_j^2}\,(y) \, dy$$

$$= \int_{\mathscr{A}_1'} \int_{\mathscr{A}_1(y')} \frac{\partial^2 \Phi}{\partial y_1^2}\,(y_1, y') \, dy_1 \, dy' + \dots$$

where $\mathscr{A}_1'$ is the projection of $\mathscr{A}$ on the hyperplane $y_1 = 0$, and

$$\mathscr{A}_1\,(y') = \{ y_1 : (y_1, y') \in \mathscr{A} \}.$$

Since $\frac{\partial^2 \Phi}{\partial y_1^2} \geq 0$ then

$$\int_{\mathscr{A}_1(y')} \frac{\partial^2 \Phi}{\partial y_1^2}\,(y_1, y') \, dy_1 \leq \frac{\partial\Phi}{\partial y_1}\,(\sup \mathscr{A}_1\,(y'), y') - \frac{\partial\Phi}{\partial y_1}\,(\inf \mathscr{A}_1\,(y'), y')$$

$$\leq 2 \sup_{\mathscr{A}} |\nabla\Phi|$$

and therefore

$$\int_{\mathbb{R}^{d-2}} K\,(y) \, dy \leq c\,|\mathscr{A}'|.$$

Thus

$$\sqrt{NS} \le c\rho^{-(d-1)/2} \sqrt{\operatorname{diam}(B)} \sqrt{|\mathscr{A}'|} \le c\rho^{-(d-1)/2} (\operatorname{diam}(B))^{(d-1)/2} .$$

$\square$

Remark 7. The above proof shows that the term $(\operatorname{diam}(B))^{(d-1)/2}$ in the statement of Theorem 6 can be replaced by the term

$$(\mu(\partial B) + \operatorname{diam}(B) \, p)^{1/2} ,$$

where p is the maximum of $(d-2)$-dimensional surface area of the projections of B on $(d-1)$-dimensional hyperplanes. When B has large eccentricity, this provides a better estimate.

3.2.3 Estimates for Bounded Sets

In certain problems the spherical mean $\|\hat{\chi}_B(\rho\cdot)\|_{L^2(\Sigma_{d-1})}$ can be replaced by "easier" averages such as

$$\int_{A\rho \le |\xi| \le B\rho} |\hat{\chi}_B(\xi)|^2 \, d\xi.$$

In this way we can get non trivial lower bounds (which for spherical averages are impossible e.g. because of the zeros of the Bessel functions) and also deal with sets more general than convex bodies. The following result is taken from [11], see also [27].

Theorem 8. *Let $B \subset \mathbb{R}^d$ and assume the existence of positive constants c_1 and c_2 such that*

$$c_1 |h| \le |(B \setminus (B+h)) \cup ((B+h) \setminus B)| \le c_2 |h| \qquad (3.15)$$

for every $h \in \mathbb{R}^d$. Then there exist four positive constants $\alpha, \beta, \gamma, \delta$ such that

$$\alpha \, \rho^{-1} \le \int_{\{\gamma\rho \le |\xi| \le \delta\rho\}} |\hat{\chi}_B(\xi)|^2 \, d\xi \le \beta \, \rho^{-1} \qquad (3.16)$$

for every $\rho \ge 1$.

Proof. We first show that

$$\int_{\{|\xi| \ge \rho\}} |\hat{\chi}_B(\xi)|^2 \, d\xi \le c \, \rho^{-1} . \qquad (3.17)$$

In order to prove (3.17) it is enough to show that for every integer $k \geq 0$ we have

$$\int_{\{2^k \leq |\xi| \leq 2^{k+1}\}} |\hat{\chi}_B(\xi)|^2 \, d\xi \leq c \, 2^{-k} . \tag{3.18}$$

By (3.15) and the Parseval identity we have

$$c_2 \, |h| \geq \int_{\mathbb{R}^d} |\chi_B(x+h) - \chi_B(x)|^2 \, dx = \int_{\mathbb{R}^d} \left| e^{2\pi i \xi \cdot h} - 1 \right|^2 |\hat{\chi}_B(\xi)|^2 \, d\xi .$$

We split the set $\{2^k \leq |\xi| \leq 2^{k+1}\}$ into a bounded number of subsets such that in each one of them we have (for a suitably chosen h with $|h| \approx 2^{-k}$) the inequality $\left| e^{2\pi i \xi \cdot h} - 1 \right| \geq c$. This proves (3.18), so that the estimate from above in (3.16) follows from (3.17). Again (3.18) implies

$$\int_{\{|\xi| \leq \gamma \rho\}} |\xi|^2 \, |\hat{\chi}_B(\xi)|^2 \, d\xi \leq c_3 + c_4 \sum_{k=1}^{\log_2(\gamma \rho)} \int_{\{2^k \leq |\xi| \leq 2^{k+1}\}} |\xi|^2 \, |\hat{\chi}_B(\xi)|^2 \, d\xi \tag{3.19}$$

$$\leq c_3 + c_4 \sum_{k=1}^{\log_2(\gamma \rho)} 2^{2k} \int_{\{2^k \leq |\xi| \leq 2^{k+1}\}} |\hat{\chi}_B(\xi)|^2 \, d\xi \leq c_3 + c_5 \sum_{k=1}^{\log_2(\gamma \rho)} 2^{2k} 2^{-k} \leq c \gamma \rho .$$

Then, by (3.17) and (3.19),

$$c_1 \, |h| \leq \int_{\mathbb{R}^d} |\chi_B(x+h) - \chi_B(x)|^2 \, dx$$

$$= \int_{\mathbb{R}^d} \left| e^{2\pi i \xi \cdot h} - 1 \right|^2 |\hat{\chi}_B(\xi)|^2 \, d\xi$$

$$\leq 4\pi^2 \, |h|^2 \int_{\{|\xi| \leq \gamma \rho\}} |\xi|^2 \, |\hat{\chi}_B(\xi)|^2 \, d\xi + 4 \int_{\{\gamma \rho \leq |\xi| \leq \delta \rho\}} |\hat{\chi}_B(\xi)|^2 \, d\xi$$

$$+ 4 \int_{\{|\xi| \geq \delta \rho\}} |\hat{\chi}_B(\xi)|^2 \, d\xi$$

$$\leq c \left[\gamma \rho \, |h|^2 + \delta^{-1} \rho^{-1} \right] + 4 \int_{\{\gamma \rho \leq |\xi| \leq \delta \rho\}} |\hat{\chi}_B(\xi)|^2 \, d\xi ,$$

so that, if $|h| = \rho^{-1}$, γ is suitably small and δ suitably large, we have

$$\int_{\{\gamma \rho \leq |\xi| \leq \delta \rho\}} |\hat{\chi}_B(\xi)|^2 \, d\xi \geq \frac{c_1}{4} \, |h| - \frac{c}{4} \left[\gamma \rho \, |h|^2 + \delta^{-1} \rho^{-1} \right] \geq \frac{c_1}{8} \, \rho^{-1} ,$$

and this ends the proof.   □

It is easy to see that a convex body satisfies (3.15).

Remark 9. M. Kolountzakis and T. Wolff [37] have proved that for every set $B \subset \mathbb{R}^d$ having positive finite measure we have

$$\int_{\{|\xi| \geq \rho\}} |\hat{\chi}_B(\xi)|^2 \, d\xi \geq c \, \rho^{-1} .$$

We can use Theorem 8 to prove that (3.8) is best possible up to the constant involved.

Theorem 10. *Let $B \subset \mathbb{R}^d$ be a convex body. Then*

$$\limsup_{\rho \to +\infty} \rho^{(d+1)/2} \|\hat{\chi}_B(\rho \cdot)\|_{L^2(\Sigma_{d-1})} > 0 . \tag{3.20}$$

Proof. If (3.20) fails, then there exists a function $\varepsilon(\rho)$ such that $\varepsilon(\rho) \to 0$ as $\rho \to +\infty$ and

$$\|\hat{\chi}_B(\rho \cdot)\|_{L^2(\Sigma_{d-1})} \leq \varepsilon(\rho) \rho^{-(d+1)/2}$$

for $\rho > 1$. This contradicts the lower bound in (3.16). $\qquad \square$

We have pointed out that when B is a ball we cannot bound the spherical mean $\|\hat{\chi}_B(\rho \cdot)\|_{L^2(\Sigma_{d-1})}$ from below by $\rho^{-(d+1)/2}$ because of the zeroes of the Bessel function. The next result shows that this lower estimate fails also for a cube, so that it fails for the two most popular convex bodies.

Lemma 11. *Let $d \geq 2$ and $Q = Q_d = [-1/2, 1/2]^d$. Then for every positive integer k we have*

$$\|\hat{\chi}_Q(k \cdot)\|_{L^2(\Sigma_{d-1})} \leq c \, k^{-(d+3/2)/2} .$$

Proof. Let

$$\Sigma' = \Sigma_{d-1} \cap \{x \in \mathbb{R}^d : x_1 \geq |x_k|, \ k = 2, \ldots, d\} .$$

By the symmetries of Q and by Theorem 6 applied to the $(d-1)$-dimensional cube Q_{d-1} we have

$$\|\hat{\chi}_Q(k \cdot)\|_{L^2(\Sigma_{d-1})}^2 \leq c \|\hat{\chi}_Q(k \cdot)\|_{L^2(\Sigma')}^2$$

$$\leq c \int_0^{\pi/4} \int_{\Sigma_{d-2}} \left| \frac{\sin(\pi k \cos(\phi))}{\pi k \cos(\phi)} \hat{\chi}_{Q_{d-1}}(k \sin(\phi)\eta) \right|^2 \sin^{d-2}(\phi) \, d\eta d\phi$$

$$= c \, k^{-2} \int_0^{\pi/4} |\sin(\pi k \cos(\phi))|^2 \phi^{d-2} \int_{\Sigma_{d-2}} |\hat{\chi}_{Q_{d-1}}(k \sin(\phi)\eta)|^2 \, d\eta d\phi$$

$$\leq c \, k^{-d-2} \int_0^{\pi/4} \left|\sin(2\pi k \sin^2(\phi/2))\right|^2 \phi^{-2} \, d\phi$$

$$\leq c \, k^{-d-2} \left(\int_0^{k^{-1/2}} k^2 \phi^2 \, d\phi + \int_{k^{-1/2}}^{\pi/4} \phi^{-2} \, d\phi \right) \leq c \, k^{-d-3/2} .$$

$\square$

3.2.4 A Maximal Estimate for the Planar Case

In Theorem 6 we have seen that for every convex body B we have the upper bound $\|\hat{\chi}_B(\rho \cdot)\|_{L^2(\Sigma_{d-1})} \leq c\rho^{-(d+1)/2}$. On the other hand we shall see that in certain lattice point problems it is important to have a bound in the angular variable σ which is uniform with respect to ρ. This means to study the maximal function

$$\mathcal{M}_B(\sigma) = \sup_{\rho > 0} \rho^{(d+1)/2} |\hat{\chi}_B(\rho\sigma)| .$$

We need the following definition (see [57]).

Definition 12. Let X be a measure space and let $0 < p < \infty$. We define the space $L^{p,\infty}(X)$ (also called weak L^p) by the quasi norm

$$\|f\|_{L^{p,\infty}(X)} = \sup_{\lambda > 0} \lambda \left|\{x \in X : |f(x)| > \lambda\}\right|^{1/p} . \tag{3.21}$$

We shall prove that $\mathcal{M}_B \in L^{2,\infty}(\Sigma_1)$, i.e. that

$$\sup_{\lambda > 0} \lambda^2 \left|\theta \in [0, 2\pi] : \mathcal{M}_B(\Theta) > \lambda\right| < \infty ,$$

where $\Theta = (\cos\theta, \sin\theta)$. Observe that $L^2(\Sigma_1) \subset L^{2,\infty}(\Sigma_1)$, but $\mathcal{M}_B$ does not necessarily belong to $L^2(\Sigma_1)$. Indeed, consider the unit square $Q = [-1/2, 1/2]^2$, then

$$\hat{\chi}_Q(\rho\Theta) = \frac{\sin(\pi\rho\cos\theta)}{\pi\rho\cos\theta} \frac{\sin(\pi\rho\sin\theta)}{\pi\rho\sin\theta} . \tag{3.22}$$

By symmetry it is enough to consider $\theta \in \left(0, \frac{\pi}{4}\right)$, and observe that, for any such θ, there exists ρ_θ satisfying the following conditions (for a suitable integer $k \geq 0$):

$$\frac{1}{4} \leq \rho_\theta \sin(\theta) \leq \frac{3}{4} ,$$

$$\frac{1}{4} + k \leq \rho_\theta \cos(\theta) \leq \frac{3}{4} + k$$

(because the line $\rho\Theta$ intersects at least one of the squares $\left[\frac{1}{4}+k,\frac{3}{4}+k\right]\times\left[\frac{1}{4},\frac{3}{4}\right]$).
Hence, by (3.22) we have

$$\mathscr{M}_B\left(\Theta\right)\geq c\,\rho_\theta^{-1/2}\frac{1}{\sin\theta}\geq c\,\theta^{-1/2}\,.$$

so that $\mathscr{M}_B\notin L^2$.

Before proving the weak type estimate we need the following two results, due to A. Podkorytov (see [45], see also [13]).

Lemma 13. *Let* $f:\mathbb{R}\to[0,+\infty)$ *be supported and concave in* $[-1,1]$. *Then, for every* $|\xi|\geq 1$,

$$\left|\hat{f}(\xi)\right|\leq\frac{1}{|\xi|}\left[f\left(1-\frac{1}{2\,|\xi|}\right)+f\left(-1+\frac{1}{2\,|\xi|}\right)\right].\qquad(3.23)$$

Proof. It is enough to prove (3.23) when $\xi>1$. The assumption on the concavity of f allows us to integrate by parts obtaining

$$\left|\hat{f}(\xi)\right|\leq\frac{1}{2\pi\xi}f(1^-)+\frac{1}{2\pi\xi}f(-1^+)+\frac{1}{2\pi\xi}\left|\int_{-1}^1 f'(t)e^{-2\pi i\xi t}\,dt\right|\,.$$

Let α be a point where f attains its maximum. Then f will be non-decreasing in $[-1,\alpha]$ and non-increasing in $[\alpha,1]$. We can assume $0\leq\alpha\leq1$, so that $f(-1^+)\leq f(-1+1/(2\xi))$. To estimate $f(1^-)$ we observe that when $\alpha\leq 1-1/(2\xi)$ one has $f(1^-)\leq f(1-1/(2\xi))$. On the other hand, since f is concave, in case $\alpha>1-1/(2\xi)$ we have $f(1^-)\leq f(\alpha)\leq 2f(0)\leq 2f(1-1/(2\xi))$.

To estimate the integral we observe that, by a change of variable,

$$I=\int_{-1}^1 f'(t)e^{-2\pi i\xi t}\,dt=-\int_{-1+\frac{1}{2\xi}}^{1+\frac{1}{2\xi}}f'\left(t-\frac{1}{2\xi}\right)e^{-2\pi i\xi t}\,dt\,.$$

So that

$$2I=\int_{-1}^1 f'(t)e^{-2\pi i\xi t}\,dt-\int_{-1+\frac{1}{2\xi}}^{1+\frac{1}{2\xi}}f'\left(t-\frac{1}{2\xi}\right)e^{-2\pi i\xi t}\,dt$$

$$=\int_{-1}^{-1+\frac{1}{2\xi}}f'(t)e^{-2\pi i\xi t}\,dt+\int_{-1+\frac{1}{2\xi}}^1\left[f'(t)-f'\left(t-\frac{1}{2\xi}\right)\right]e^{-2\pi i\xi t}\,dt$$

$$+\int_1^{1+\frac{1}{2\xi}}f'\left(t-\frac{1}{2\xi}\right)e^{-2\pi i\xi t}\,dt$$

$$=I_1+I_2+I_3\,.$$

To estimate I_1 from above we note that

$$|I_1| \leq \int_{-1}^{-1+\frac{1}{2\xi}} f'(t)dt = f\left(-1+\frac{1}{2\xi}\right) - f(-1^+) \leq f\left(-1+\frac{1}{2\xi}\right),$$

since $0 \leq \alpha \leq 1$.

The estimate for I_3 is similar in case $\alpha \leq 1 - 1/(2\xi)$. If $\alpha > 1 - 1/(2\xi)$, then

$$|I_3| \leq \int_{1}^{\alpha+\frac{1}{2\xi}} f'\left(t-\frac{1}{2\xi}\right) dt - \int_{\alpha+\frac{1}{2\xi}}^{1+\frac{1}{2\xi}} f'\left(t-\frac{1}{2\xi}\right) dt$$

$$= 2f(\alpha) - f\left(1-\frac{1}{2\xi}\right) - f(1^-) \leq 2f(\alpha) \leq 4f(0) \leq 4f\left(1-\frac{1}{2\xi}\right).$$

As for I_2, since f' is non increasing, we have

$$|I_2| \leq \int_{-1+\frac{1}{2\xi}}^{1} \left[f'\left(t-\frac{1}{2\xi}\right) - f'(t) \right] dt$$

$$= f\left(1-\frac{1}{2\xi}\right) - f(-1^+) - f(1^-) + f\left(-1+\frac{1}{2\xi}\right)$$

$$\leq f\left(1-\frac{1}{2\xi}\right) + f\left(-1+\frac{1}{2\xi}\right),$$

ending the proof. Note that no constant c is missing in (3.23).   □

Lemma 14. *Let B be a convex body in $\mathbb{R}^2$ and $\Theta = (\cos\theta, \sin\theta)$. For a small $\delta > 0$ we consider the chord*

$$\lambda_B(\delta, \theta) = \lambda(\delta, \theta) = \left\{ x \in B : x \cdot \Theta = -\delta + \sup_{x \in B} x \cdot \Theta \right\}. \tag{3.24}$$

Then

$$|\hat{\chi}_B(\rho\Theta)| \leq \frac{1}{\rho}\left(\left|\lambda\left(\frac{1}{2\rho}, \theta\right)\right| + \left|\lambda\left(\frac{1}{2\rho}, \theta + \pi\right)\right| \right),$$

where $|\lambda|$ denotes the length of the chord (Fig. 3.2).

Proof. Without loss of generality we choose $\Theta = (1, 0)$. Then, as in (3.3),

$$\hat{\chi}_B(\xi_1, 0) = \int_{-\infty}^{+\infty} \left(\int_{-\infty}^{+\infty} \chi_B(x_1, x_2)\, dx_2 \right) e^{-2\pi i x_1 \xi_1}\, dx_1 = \hat{h}(\xi_1), \tag{3.25}$$

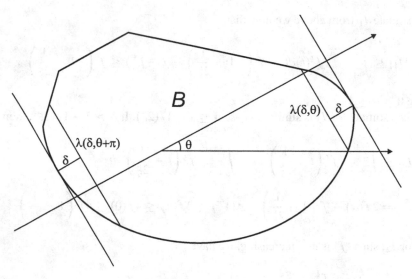

Fig. 3.2 Geometric estimate of $\hat{\chi}_B$

where $h(s)$ is the length of the segment obtained intersecting B with the line $x_1 = s$. Observe that h is concave on its support, say $[a, b]$. We can therefore apply Lemma 13 to obtain, after a change of variable,

$$\left| \hat{h}(\xi_1) \right| \leq \frac{1}{|\xi_1|} \left[h \left(b - \frac{1}{2\,|\xi_1|} \right) + h \left(a + \frac{1}{2\,|\xi_1|} \right) \right]$$

$$\leq \frac{1}{|\xi_1|} \left(\left| \lambda_B \left(\frac{1}{2\,|\xi_1|}, 0 \right) \right| + \left| \lambda_B \left(\frac{1}{2\,|\xi_1|}, \pi \right) \right| \right) .$$

$\square$

We can now prove the following maximal estimate (see [8]).

Theorem 15. *Let $B \subset \mathbb{R}^2$ be a convex body. Then the maximal function*

$$\mathscr{M}_B (\Theta) = \sup_{\rho > 0} \rho^{3/2} \left| \hat{\chi}_B (\rho \Theta) \right|$$

belongs to $L^{2,\infty} (\Sigma_1)$, see (3.21).

Proof. As in the proof of Theorem 6 we assume ∂B smooth with everywhere non-vanishing curvature (and the constants in our inequalities will not depend on the smoothness of ∂B). By Lemma 14 we have, for $\Theta = (\cos \theta, \sin \theta)$,

$$\sup_{\rho > 0} \rho^{3/2} \left| \hat{\chi}_B (\rho \Theta) \right| \leq \sup_{\rho > 0} \rho^{1/2} \left| \lambda_B (\rho^{-1}, \theta) \right| + \sup_{\rho > 0} \rho^{1/2} \left| \lambda_B (\rho^{-1}, \theta + \pi) \right| ,$$

so that we study the maximal function

$$\Omega_B(\theta) = \sup_{\delta > 0} \delta^{-1/2} |\lambda_B(\delta, \theta)| .$$

By the above non-vanishing assumption, the chord $\lambda_B(\delta, \theta)$ reduces to a single point as $\delta \to 0$. Let $z(\theta)$ be this point. Let us choose a direction θ_0 and for every θ close to θ_0 let $\gamma(\theta)$ denote the arc-length on ∂B between $z(\theta_0)$ and $z(\theta)$. Assume that we have proved the inequality

$$\Omega_B^2(\theta) \leq 2 \sup_{\alpha \neq 0} \frac{|\gamma(\theta + \alpha) - \gamma(\theta)|}{\alpha} . \tag{3.26}$$

Then we have

$$\Omega_B^2(\theta) \leq 2 \sup_{\alpha > 0} \frac{1}{\alpha} \int_\theta^{\theta + \alpha} |\gamma'(\varphi)| \, d\varphi ,$$

so that, by the Hardy-Littlewood maximal function theorem (see e.g. [64, 7.9]), we have

$$\sup_{\beta > 0} \beta^2 |\{\theta \in [0, 2\pi) : \mathscr{M}_B(\cos \theta, \sin \theta) > \beta\}|$$

$$\leq c \sup_{\beta > 0} \beta^2 \left| \left\{ \theta \in [0, 2\pi) : \sup_{\alpha > 0} \frac{1}{\alpha} \int_\theta^{\theta + \alpha} |\gamma'(\varphi)| \, d\varphi > \beta^2 \right\} \right|$$

$$\leq c \int_0^{2\pi} |\gamma'(\varphi)| \, d\varphi \leq c .$$

In order to prove (3.26) we observe that if δ is small, then the normal to ∂B at the point $z(\theta)$ cuts the chord $\lambda_B(\delta, \theta)$ into two parts $\lambda_-(\delta, \theta)$ and $\lambda_+(\delta, \theta)$. Let us consider only the segment $\lambda_+(\delta, \theta)$ and let

$$\Omega_+(\theta) = \sup_{\delta > 0} \delta^{-1/2} |\lambda_+(\delta, \theta)| .$$

We may assume that ∂B is locally the graph of a smooth function f defined on an interval $[0, a]$ with $f(0) = f'(0^+) = 0$. Then, by the mean value theorem,

$$\Omega_+^2(\theta) \leq \sup_{0 < x < a} \frac{x^2}{f(x)} \leq \sup_{0 < z < a} \frac{2z}{f'(z)} \leq \sup_{0 < z < a} \frac{2}{f'(z)} \int_0^z \left(1 + [f'(t)]^2\right)^{1/2} dt$$

$$\leq 2 \sup_{0 < \alpha < \frac{\pi}{4}} \frac{|\gamma(\theta + \alpha) - \gamma(\theta)|}{\alpha} .$$

$\square$

3.3 Decay of the Fourier Transform: L^p Estimates for Characteristic Functions of Polyhedra

In Theorem 6 we have seen that $\|\hat{\chi}_B(\rho\cdot)\|_{L^2(\Sigma_{d-1})} \leq c\, \rho^{-(d+1)/2}$ independently of the shape of the convex body B. If we replace B by a ball or, more generally by a convex body with smooth boundary ∂B which has everywhere positive Gaussian curvature, then, by (3.7), the same estimate holds true for every $1 \leq p \leq +\infty$. However, if we replace B by a polyhedron P, then the situation should be different (see Sect. 3.2.1, where we have observed that $\hat{\chi}_P(\xi)$ decays as fast as $|\xi|^{-d}$ along almost all directions, but only as $|\xi|^{-1}$ along the directions perpendicular to the facets). In this section we will prove sharp estimates for the decay of $\|\hat{\chi}_P(\rho\cdot)\|_{L^p(\Sigma_{d-1})}$ and in particular we shall see that this decay is faster than $\rho^{-(d+1)/2}$ when $1 \leq p < 2$ and it is slower than $\rho^{-(d+1)/2}$ when $2 < p \leq +\infty$.

Theorem 16. *Let P be a convex polyhedron in $\mathbb{R}^d$, $d \geq 1$. Write $\xi \in \mathbb{R}^d$ in polar coordinates, $\xi = \rho\sigma$ ($\rho \geq 0$, $\sigma \in \Sigma_{d-1}$). Then, for $\rho \geq 2$, we have*

$$\|\hat{\chi}_P(\rho\cdot)\|_{L^1(\Sigma_{d-1})} \leq c\, \frac{\log^{d-1}(\rho)}{\rho^d}\,, \tag{3.27}$$

$$\|\hat{\chi}_P(\rho\cdot)\|_{L^p(\Sigma_{d-1})} \leq c_p\, \rho^{-1-(d-1)/p}\,, \quad \text{for } 1 < p \leq \infty\,. \tag{3.28}$$

Proof. The proof is by induction on the dimension d. For $d = 1$ the bound is true since in this case the average is trivial and we have

$$\hat{\chi}_{[-1/2,1/2]}(\xi) = \frac{\sin(\pi\xi)}{\pi\xi}\,.$$

We then assume the result true for $d - 1$. Let P have m facets $F_1, \ldots, F_m$ with outward unit normal vectors $\nu_1, \ldots, \nu_m$. As in (3.9) the divergence theorem yields

$$\hat{\chi}_P(\xi) = \int_P e^{-2\pi i \xi \cdot x}\, dx = \sum_{j=1}^m \frac{i\xi \cdot \nu_j}{2\pi\,|\xi|^2} \int_{F_j} e^{-2\pi i \xi \cdot x}\, dx\,. \tag{3.29}$$

Let $x = (x_1, x_2, \ldots, x_d) = (x_1, x')$ and write $\sigma = (\cos(\varphi), \sin(\varphi)\eta)$, with $0 \leq \varphi \leq \pi$ and $\eta \in \Sigma_{d-2}$. We single out one facet F, which we may assume to stay in the hyperplane $x_1 = 0$, with outward normal $\nu = (1, 0, \ldots, 0)$. Then

$$\frac{i\xi \cdot \nu}{2\pi\,|\xi|^2} \int_F e^{-2\pi i \xi \cdot x}\, dx = \frac{i\cos(\varphi)}{2\pi\rho} \int_F e^{-2\pi i \rho \sin(\varphi)\eta \cdot x'}\, dx' \tag{3.30}$$

$$= \frac{i\cos(\varphi)}{2\pi\rho}\, \hat{\chi}_F(\rho\sin(\varphi)\eta)\,,$$

where we see $\hat{\chi}_F$ as a $(d-1)$-dimensional Fourier transform. Then, by the induction hypothesis,

$$\frac{1}{\rho} \int_0^\pi \int_{\Sigma_{d-2}} |\hat{\chi}_F (\rho \sin (\varphi) \, \eta)| \sin^{d-2} (\varphi) \ d\eta d\varphi$$

$$\leq c \, \frac{1}{\rho} \int_0^{2/\rho} \varphi^{d-2} \, d\varphi + c \, \frac{1}{\rho} \int_{2/\rho}^{\pi/2} \frac{\log^{d-2} (\rho \sin (\varphi))}{(\rho \sin (\varphi))^{d-1}} \sin^{d-2} (\varphi) \ d\varphi$$

$$\leq c \, \rho^{-d} + c \, \frac{\log^{d-2} (\rho)}{\rho^d} \int_{2/\rho}^{\pi/2} \frac{1}{\varphi} \, d\varphi \leq c \, \frac{\log^{d-1} (\rho)}{\rho^d} \, ,$$

while for $1 < p < +\infty$ we have

$$\frac{1}{\rho^p} \int_0^\pi \int_{\Sigma_{d-2}} |\hat{\chi}_F (\rho \sin (\varphi) \, \eta)|^p \sin^{d-2} (\varphi) \ d\eta d\varphi$$

$$\leq c_p \, \frac{1}{\rho^p} \int_0^{1/\rho} \varphi^{d-2} \, d\varphi + c \, \frac{1}{\rho^p} \int_{1/\rho}^{\pi/2} (\rho \sin (\varphi))^{-p-(d-2)} \sin^{d-2} (\varphi) \ d\varphi$$

$$\leq c_p \, \rho^{-p-(d-1)} + c_p \, \frac{1}{\rho^{-2p+d-2}} \int_{1/\rho}^{\pi/2} \varphi^{-p} \, d\varphi \leq c_p \, \rho^{-p-(d-1)}$$

so that (3.29) and (3.30) give (3.27) and (3.28). □

The following weak type estimates (see (3.21)) will be useful too.

Theorem 17. *Let P be a polyhedron in $\mathbb{R}^d$, $d \geq 2$. Write $\xi \in \mathbb{R}^d$ in polar coordinates, $\xi = \rho\sigma$ ($\rho \geq 0$, $\sigma \in \Sigma_{d-1}$). Then, for $\rho \geq 2$, we have*

$$\|\hat{\chi}_P (\rho \cdot)\|_{L^{1,\infty}(\Sigma_{d-1})} \leq c \, \frac{\log^{d-2} (\rho)}{\rho^d} \, .$$

Proof. Since here $d \geq 2$, the first step of the induction needs some work. Assume $d = 2$, and let P be a polygon in $\mathbb{R}^2$ with counterclockwise oriented vertices $\{a_j\}_{j=1}^m$. For each side $[a_j, a_{j+1}]$ (assume $a_{m+1} = a_1$) let u_j be a unit vector parallel to this side and with the same orientation, and let v_j be the outside unit normal to this side. Then the divergence theorem gives

$$\hat{\chi}_P (\rho\sigma) = \int_P e^{-2\pi i \rho\sigma \cdot x} \, dx = -\frac{1}{2\pi i \rho} \sum_{j=1}^m \sigma \cdot v_j \int_{[a_j, a_{j+1}]} e^{-2\pi i \rho\sigma \cdot x} \, dx$$

$$= -\frac{1}{4\pi^2 \rho^2} \sum_{j=1}^m \rho\sigma \cdot v_j \, \frac{e^{-2\pi i \rho\sigma \cdot a_j} - e^{-2\pi i \xi \cdot a_{j+1}}}{\rho\sigma \cdot u_j} \, .$$

Hence $\hat{\chi}_P (\rho \cos (\varphi), \rho \sin (\varphi))$ is dominated by a finite sum of terms of the form $\rho^{-2} |\cos (\varphi - \varphi_j)|^{-1}$ and since the functions $\cos^{-1} (\varphi - \varphi_j)$ are in $L^{1,\infty} (\mathbb{T})$, the result for $d = 2$ follows. For $d > 2$ we argue as in Theorem 16 and we reduce to a finite sum of terms of the form

$$\frac{i\cos(\varphi)}{2\pi\rho}\,\hat{\chi}_F\,(\rho\sin(\varphi)\,\eta)\ ,$$

where F is a facet of P. Then, by induction, we have

$$\lambda\left|\left\{(\cos(\varphi),\sin(\varphi)\,\eta)\in\Sigma_{d-1}:\left|\frac{i\cos(\varphi)}{2\pi\rho}\hat{\chi}_F\,(\rho\sin(\varphi)\,\eta)\right|>\lambda\right\}\right|$$

$$=\lambda\int_0^\pi\left|\left\{\eta\in\Sigma_{d-2}:|\hat{\chi}_F\,(\rho\sin(\varphi)\,\eta)|>\frac{2\pi\rho\lambda}{|\cos(\varphi)|}\right\}\right|\sin^{d-2}(\varphi)\ d\varphi$$

$$\leq c\,\rho^{-d}+c\int_{2/\rho}^{\pi/2}\frac{\cos(\varphi)}{\rho}\frac{\log^{d-3}(\rho\sin(\varphi))}{(\rho\sin(\varphi))^{d-1}}\sin^{d-2}(\varphi)\ d\varphi$$

$$\leq c\,\rho^{-d}\int_{2/\rho}^{\pi/2}\cos(\varphi)\,\frac{\log^{d-3}(\rho\varphi)}{\varphi}\ d\varphi\leq c\rho^{-d}\int_2^{\rho\pi/2}\frac{\log^{d-3}(t)}{t}\ dt$$

$$\leq c\,\frac{\log^{d-2}(\rho)}{\rho^d}\ .$$

$\square$

The estimates from above in Theorems 16 and 17 are sharp in many, but not all, cases. We first consider simplices but the proof of the following theorem works with no modifications for polyhedra having a facet not parallel to any other. We need a technical lemma which may be well known.

Lemma 18. *Let Σ be a finite measure space and let $f\in L^\infty(\Sigma)$. Then for any $0<\alpha<|\Sigma|$ we have*

$$\|f\|_1\leq\alpha\,\|f\|_\infty+\log\left(\frac{|\Sigma|}{\alpha}\right)\|f\|_{1,\infty}\ .$$

Proof. Let g be the non-increasing rearrangement of f (see [57]). Then

$$\|g\|_\infty=\|f\|_\infty,$$
$$ug(u)\leq\|g\|_{1,\infty}=\|f\|_{1,\infty},$$

so that

$$\|f\|_1=\int_0^{|\Sigma|}g(u)\,du=\int_0^\alpha g(u)\,du+\int_\alpha^{|\Sigma|}g(u)\,du$$

$$\leq\alpha\,\|f\|_\infty+\|f\|_{1,\infty}\int_\alpha^{|\Sigma|}\frac{1}{u}\,du=\alpha\,\|f\|_\infty+\log\frac{|\Sigma|}{\alpha}\,\|f\|_{1,\infty}\ .$$

$\square$

Theorem 19. *Let P be a simplex in $\mathbb{R}^d$, $d \geq 2$. Again we write $\xi \in \mathbb{R}^d$ in polar coordinates, $\xi = \rho\sigma$ ($\rho \geq 0$, $\sigma \in \Sigma_{d-1}$). Then, for every $\rho \geq 1$,*

i) $\|\hat{\chi}_P(\rho \cdot)\|_{L^{1,\infty}(\Sigma_{d-1})} \geq c \dfrac{\log^{d-2}(\rho)}{\rho^d}$

ii) $\|\hat{\chi}_P(\rho \cdot)\|_{L^1(\Sigma_{d-1})} \geq c \dfrac{\log^{d-1}(\rho)}{\rho^d}$

iii) $\|\hat{\chi}_P(\rho \cdot)\|_{L^p(\Sigma_{d-1})} \geq c_p \, \rho^{-1-(d-1)/p}$,   *if* $1 < p \leq \infty$.

Proof. We prove *ii)* and *iii)* first. The proof is by induction on the dimension d and we first consider the planar case, showing that a triangle $T \subset \mathbb{R}^2$ satisfies

$$\int_0^{2\pi} |\hat{\chi}_T(\rho\Theta)|^p \, d\theta \geq c\rho^{-p-1} , \qquad\qquad (3.31)$$

for $p > 1$, where $\Theta = (\cos\theta, \sin\theta)$ and $\rho \geq 1$. As in the proofs of the previous theorems we use the divergence theorem. Let

$$\omega(t) = \frac{i}{2\pi\rho} e^{-2\pi i \rho\Theta \cdot t} \Theta ,$$

with $t = (t_1, t_2)$. Then

$$\mathrm{div}\,(\omega(t)) = \frac{\partial}{\partial t_1}\left(\frac{i}{2\pi\rho} e^{-2\pi i \rho\Theta \cdot t} \cos\theta\right) + \frac{\partial}{\partial t_2}\left(\frac{i}{2\pi\rho} e^{-2\pi i \rho\Theta \cdot t} \sin\theta\right)$$

$$= e^{-2\pi i \rho\Theta \cdot t} \cos^2\theta + e^{-2\pi i \rho\Theta \cdot t} \sin^2\theta = e^{-2\pi i \rho\Theta \cdot t} ,$$

and by the divergence theorem we obtain

$$\hat{\chi}_T(\rho\Theta) = \int_T e^{-2\pi i \rho\Theta \cdot t} \, dt = \int_{\partial T} \omega(t) \cdot \nu(t) \, dt ,$$

where ν is the outward unit vector, which takes only the three values ν_1, ν_2, ν_3 on the three sides $\lambda_1, \lambda_2, \lambda_3$ respectively. Then, if ds is the measure on ∂T,

$$\hat{\chi}_T(\rho\Theta) = \frac{\Theta \cdot \nu_1}{2\pi\rho} i \int_{\lambda_1} e^{-2\pi i \rho\Theta \cdot s} \, ds + \frac{\Theta \cdot \nu_2}{2\pi\rho} i \int_{\lambda_2} e^{-2\pi i \rho\Theta \cdot s} \, ds$$

$$+ \frac{\Theta \cdot \nu_3}{2\pi\rho} i \int_{\lambda_3} e^{-2\pi i \rho\Theta \cdot s} \, ds$$

$$= A(\rho, \Theta) + B(\rho, \Theta) + C(\rho, \Theta) .$$

We may assume that λ_1 has extremes $(\pm\frac{1}{2}, 0)$. Of course it suffices to show that for a given small $\delta > 0$ we have

$$\int_{-\frac{\pi}{2}-\delta}^{-\frac{\pi}{2}+\delta} |\hat{\chi}_T(\rho\Theta)|^p \, d\theta \geq c_p \, \rho^{-p-1} \, .$$

Indeed $|\Theta \cdot \nu_1| = |\sin\theta|$ and changing variables we obtain

$$\int_{-\frac{\pi}{2}-\delta}^{-\frac{\pi}{2}+\delta} |A(\rho,\Theta)|^p \, d\theta = \frac{1}{(2\pi\rho)^p} \int_{-\frac{\pi}{2}-\delta}^{-\frac{\pi}{2}+\delta} \left| \sin\theta \int_{-1/2}^{1/2} e^{-2\pi i\rho s \cos\theta} \, ds \right|^p d\theta$$

$$= \frac{1}{(2\pi\rho)^p} \int_{-\frac{\pi}{2}-\delta}^{-\frac{\pi}{2}+\delta} \left| \frac{\sin(\pi\rho\cos\theta)}{\pi\rho\cos\theta} \sin\theta \right|^p d\theta$$

$$\geq c_p \frac{1}{\rho^{p+1}} \int_0^{c_1\rho} \left| \frac{\sin(u)}{u} \right|^p du \geq c_p \, \rho^{-p-1} \, .$$

As for $B(\rho,\Theta)$ and $C(\rho,\Theta)$, if $\left|\theta - \frac{\pi}{2}\right| \leq \delta$ we reduce to terms of the form

$$\frac{1}{\rho^p} \int_c^{c'} \left| \frac{\sin(2\pi\rho x)}{\rho x} \right|^p dx$$

with $0 < c < c' < \pi/4$, so that

$$\int_{-\frac{\pi}{2}-\delta}^{-\frac{\pi}{2}+\delta} |B(\rho,\Theta)|^p \, d\theta + \int_{-\frac{\pi}{2}-\delta}^{-\frac{\pi}{2}+\delta} |C(\rho,\Theta)|^p \, d\theta \leq c\rho^{-2p}$$

and (3.31) follows. The proof of the planar case when $p = 1$ is similar.

Now let S be a simplex in $\mathbb{R}^d$ with facets $F_1, \ldots, F_{d+1}$. We may assume F_1 contained in the hyperplane $x_1 = 0$ with outward normal $\nu_1 = (1, 0, \ldots, 0)$. Let U be a small neighborhood of ν_1 in Σ_{d-1}, then by (3.29) we have

$$\|\hat{\chi}_P(\rho\cdot)\|_{L^p(\Sigma_{d-1})}$$

$$\geq c \frac{1}{\rho} \left| \left\{ \int_U \left| \int_{F_1} e^{-2\pi i\rho\sigma \cdot x} dx \right|^p d\sigma \right\}^{1/p} - \sum_{j=2}^m \left\{ \int_U \left| \int_{F_j} e^{-2\pi i\rho\sigma \cdot x} dx \right|^p d\sigma \right\}^{1/p} \right| \, .$$

As in the proof of Theorem 16, the induction assumption implies

$$\frac{1}{\rho} \int_U \left| \int_{F_1} e^{-2\pi i\rho\sigma \cdot x} dx \right| d\sigma \geq c \, \rho^{-d} \log^{d-1}(\rho)$$

and

$$\frac{1}{\rho} \left\{ \int_U \left| \int_{F_1} e^{-2\pi i\rho\sigma \cdot x} dx \right|^p d\sigma \right\}^{1/p} \geq c\rho^{-1-(d-1)/p} \, , \quad \text{for } 1 < p \leq \infty \, .$$

We now have to estimate each term

$$\int_U \left| \int_F e^{-2\pi i \rho \sigma \cdot x} dx \right|^p d\sigma$$

from above, when $F = F_2, \ldots, F_m$. Since we are integrating each facet separately we may rotate and translate F until it belongs to the hyperplane $x_1 = 0$. After this transformation the normal ν_1 to the facet F_1 is no longer parallel to $(1, 0, \ldots, 0)$. Being U a neighborhood of ν_1 in Σ_{d-1} we can choose a small $\delta > 0$ such that

$$U \subset \{(\cos(\varphi), \sin(\varphi)\eta) : \delta \leq \varphi \leq \pi - \delta, \ \eta \in \Sigma_{d-2}\} .$$

Applying Theorem 16 to the $(d-1)$-dimensional Fourier transform of the characteristic function of F we get

$$\frac{1}{\rho} \int_U \left| \int_F e^{-2\pi i \rho \sigma \cdot x} dx \right| d\sigma \leq c \frac{1}{\rho} \int_\delta^{\pi-\delta} \int_{\Sigma_{d-2}} |\hat{\chi}_F(\rho \sin(\varphi)\eta)| \sin^{d-2}(\varphi) \, d\varphi d\eta$$

$$\leq c \frac{1}{\rho} \int_\delta^{\pi-\delta} \frac{\log^{d-2}(\rho \sin(\varphi))}{(\rho \sin(\varphi))^{d-1}} \sin^{d-2}(\varphi) \, d\varphi \leq c \frac{\log^{d-2}(\rho)}{\rho^d} ,$$

while, for $1 < p \leq +\infty$,

$$\frac{1}{\rho^p} \int_U \left| \int_F e^{-2\pi i \rho \sigma \cdot x} dx \right|^p d\sigma$$

$$\leq c \frac{1}{\rho^p} \int_\delta^{\pi-\delta} \int_{\Sigma_{d-2}} |\hat{\chi}_F(\rho \sin(\varphi)\eta)|^p \sin^{d-2}(\varphi) \, d\varphi d\eta$$

$$\leq c \frac{1}{\rho^p} \int_\delta^{\pi-\delta} (\rho \sin(\varphi))^{-p-(d-2)} \sin^{d-2}(\varphi) \, d\varphi \leq c \rho^{-2p-(d-2)} .$$

Hence *ii)* and *iii)* are proved.

To prove *i)* assume, by way of contradiction, that for any arbitrary small ε there exists a suitable large ρ such that

$$\|\hat{\chi}_P(\rho \cdot)\|_{L^{1,\infty}(\Sigma_{d-1})} \leq \varepsilon \rho^{-d} \log^{d-2}(\rho) .$$

By Lemma 18 we have

$$\|\hat{\chi}_P(\rho \cdot)\|_{L^1(\Sigma_{d-1})}$$

$$\leq \rho^{-d} \|\hat{\chi}_P(\rho \cdot)\|_{L^\infty(\Sigma_{d-1})} + \varepsilon \rho^{-d} \log^{d-2}(\rho) \int_{\rho^{-d}}^{|\Sigma_{d-1}|} u^{-1} \, du$$

$$\leq |P| \rho^{-d} + \varepsilon \log(|\Sigma_{d-1}|) \rho^{-d} \log^{d-2}(\rho) + \varepsilon d \rho^{-d} \log^{d-1}(\rho) ,$$

which, for small ε contradicts *ii)*. $\square$

The above theorem is false in the case $d = 1$, and this is simply due to the zeros of $\hat{\chi}_P$ when P is an segment. When $d \geq 2$ the lower bound (*iii*) in Theorem 19 is false for a cube. The following analog of Lemma 11 can be easily proved.

Theorem 20. *Let $Q = Q_d = [-1/2, 1/2]^d$ be the unit cube in $\mathbb{R}^d$, $d \geq 2$. Then for $1 < p \leq +\infty$ and for every positive integer k we have*

$$\left\| \hat{\chi}_Q(k\cdot) \right\|_{L^p(\Sigma_{d-1})} \leq c\, k^{-(3p+2d-3)/2p} .$$

So far we have seen that balls and polyhedra share the same spherical L^p order of decay if and only if $p = 2$. It is natural to look for convex bodies with "intermediate" order of decay. On this problem we have significative results only for $d = 2$ (see, [7, 12, 13, 62]). It can be shown that for every $2 < p \leq +\infty$ and every order of decay $a \in (1 + 1/p, 3/2)$ there exists a convex planar body B having piecewise smooth boundary and satisfying

$$\left\| \hat{\chi}_B(\rho\cdot) \right\|_{L^p(\Sigma_1)} \leq c\, \rho^{-a} , \qquad \limsup_{\rho \to +\infty} \rho^a \left\| \chi_B(\rho\cdot) \right\|_{L^p(\Sigma_1)} > 0 . \qquad (3.32)$$

For $p < 2$ the situation is different: if we keep the piecewise smooth boundary assumption, then there is no intermediate decay between the one of the disc and the one of the polygons (observe that in (3.32) we have a lim sup). The reason is that if ∂B is piecewise smooth, but B is not a polygon, then ∂B contains an arc with positive curvature, and by the argument in [9], this is enough to get the lower estimate $\rho^{-3/2}$ for the lim sup. If we pass to arbitrary convex bodies, then it is possible to construct convex bodies with intermediate L^p order of decay. Moreover, for no convex planar body B we have $\left\| \hat{\chi}_B(\rho\cdot) \right\|_{L^1(\Sigma_1)} = o\left(\rho^{-2} \log \rho\right)$ (see [7, 62]).

3.4 Lattice Points: Estimates from Above

The literature on lattice points in multi-dimensional domains is very impressive and deep, see e.g. [28, 34, 39]. Here we focus on the topics which are necessary for (or close to) the goal of this chapter, i.e. the relation between discrepancy problems and the average decay of Fourier transforms. First we shall see that the upper bounds in Theorems 6 and 16 readily provide estimates from above for L^2 or L^p discrepancy problems related to rotations and translations of convex bodies. The lower bounds for the discrepancy related to the lattice $\mathbb{Z}^d$ are not strictly necessary for our purpose, since the typical results on irregularities of point set distribution involve *arbitrary* choices of points. However we will present some results of this kind, on the one hand because the choice of points related to a lattice are very important, on the other hand because in some cases they compensate the lack of lower bounds for arbitrary choices of points.

For a convex body $B \subset \mathbb{R}^d$ and for a large dilation R, let $\tau(RB) + t$ be the rotated and translated copy of RB. Here $\tau \in SO(d)$, and since we are interested in the cardinality of the set $\mathbb{Z}^d \cap (\tau(RB) + t)$, which is $\mathbb{Z}^d$-periodic in the variable t, we take $t \in \mathbb{T}^d$. Define the discrepancy function D_R on $SO(d) \times \mathbb{T}^d$

$$D_R(\tau, t) = \text{card}\left(\mathbb{Z}^d \cap (\tau(RB) + t)\right) - R^d |B| \tag{3.33}$$

$$= \sum_{k \in \mathbb{Z}^d} \chi_{\tau(RB)}(k - t) - R^d |B| .$$

The Fourier coefficients of the periodic function $D_{\tau,R}(t) = D_R(\tau, t)$ take values

$$\hat{D}_{\tau,R}(m) = \begin{cases} 0 & \text{if } m = 0 \\ R^d \, \hat{\chi}_{\tau(B)}(Rm) & \text{if } m \neq 0 \end{cases} . \tag{3.34}$$

Indeed

$$\hat{D}_{\tau,R}(0) = \int_{[-\frac{1}{2},\frac{1}{2})^d} \left(\text{card}\left(\mathbb{Z}^d \cap \tau(RB) + t\right) - R^d |B|\right) dt$$

$$= -R^d |B| + \sum_{k \in \mathbb{Z}^d} \int_{[-\frac{1}{2},\frac{1}{2})^d} \chi_{\tau(RB)}(k - t) \, dt$$

$$= -R^d |B| + \int_{\mathbb{R}^d} \chi_{\tau(RB)}(t) \, dt = 0 ,$$

while for $m \neq 0$

$$\hat{D}_{\tau,R}(m) = \int_{[-\frac{1}{2},\frac{1}{2})^d} \left(\text{card}\left(\mathbb{Z}^d \cap \tau(RB) + t\right) - R^d |B|\right) e^{-2\pi i m \cdot t} \, dt$$

$$= \sum_{k \in \mathbb{Z}^d} \int_{[-\frac{1}{2},\frac{1}{2})^d} \chi_{\tau(RB)}(k - t) \, e^{-2\pi i m \cdot t} \, dt = \int_{\mathbb{R}^d} \chi_{\tau(RB)}(t) \, e^{-2\pi i m \cdot t} \, dt$$

$$= \hat{\chi}_{\tau(RB)}(m) = R^d \, \hat{\chi}_{\tau(B)}(Rm) .$$

Then $D_{\tau,R}(t)$ has Fourier series

$$R^d \sum_{0 \neq m \in \mathbb{Z}^d} \hat{\chi}_{\tau(B)}(Rm) \, e^{2\pi i m \cdot t} .$$

The following result is due to D. Kendall[2] (see [35], see also [13]).

---

[2]D. Kendall seems to have been the first one to realize that certain lattice points problems can be handled using multi-dimensional Fourier analysis.

Theorem 21. *Let B be a convex body in $\mathbb{R}^d$, $d \geq 1$, and let D_R be as in (3.33).*
Then there exists a positive constant c, depending on d but not on B, such that for
every $R \geq 1$ we have

$$\|D_R\|_{L^2(SO(d) \times \mathbb{T}^d)} \leq c \ (\text{diam}(B))^{(d-1)/2} \ R^{(d-1)/2} .$$

Proof. By Parseval identity we obtain

$$\|D_R\|^2_{L^2(SO(d) \times \mathbb{T}^d)} = \int_{SO(d)} \int_{\mathbb{T}^d} D_R^2(\tau, t) \ dt \, d\tau \qquad (3.35)$$

$$= \int_{SO(d)} \sum_{0 \neq m \in \mathbb{Z}^d} \left| \hat{D}_{\tau, R}(m) \right|^2 \, d\tau = R^{2d} \sum_{0 \neq m \in \mathbb{Z}^d} \int_{SO(d)} \left| \hat{\chi}_{\tau(B)}(Rm) \right|^2 \, d\tau$$

$$= R^{2d} \sum_{0 \neq m \in \mathbb{Z}^d} \int_{SO(d)} \left| \hat{\chi}_B(\tau^{-1}(Rm)) \right|^2 \, d\tau$$

because the Fourier transform commutes with rotations. Then Theorem 6 gives

$$\|D_R\|^2_{L^2(SO(d) \times \mathbb{T}^d)} \leq c \ (\text{diam}(B))^{d-1} R^{2d} \sum_{0 \neq m \in \mathbb{Z}^d} (R |m|)^{-(d+1)} \qquad (3.36)$$

$$\leq c \ (\text{diam}(B))^{d-1} R^{d-1} \int_{x \in \mathbb{R}^d, \ |x| \geq 1} |x|^{-(d+1)} \ dx \leq c' \ (\text{diam}(B))^{d-1} R^{d-1} .$$

$\square$

Remark 22. The above argument can be applied to a more general setting (see [25]).
First consider a body $B \subset \mathbb{R}^d$ and let $0 \leq \alpha \leq 1$ satisfy

$$\left| \{ t \in \mathbb{R}^d : \text{dist}(t, \partial B) \leq \delta \} \right| \leq c_d \ \delta^\alpha$$

for every small $\delta > 0$. Let $D_R(\tau, t)$ be as in (3.33). Then

$$\left\{ \frac{1}{R} \int_0^R \int_{SO(d)} \int_{\mathbb{T}^d} |D_\rho(\tau, t)|^2 \ dt \, d\sigma \, d\rho \right\}^{1/2} \leq c \ R^{(d-\alpha)/2} .$$

Moreover, the characteristic function χ_B can be replaced by an arbitrary integrable
function. In this case a modulus of continuity appears in the upper bound.

3.4.1 The Curious Case of the Ball When $d \equiv 1 \ (\text{mod} \, 4)$

The above upper estimate is best possible, but it is not always sharp. Indeed, let B be
a ball in $\mathbb{R}^d$ and $1 < d \equiv 1 \ (\text{mod} \, 4)$, then there exists a diverging sequence R_j such

that the upper bound R_j^{d-1} in (3.36) can be replaced by $c_\varepsilon\, R_j^{d-1} \log^{-1/(d+\varepsilon)} (R_j)$, where $\varepsilon > 0$ is arbitrarily small. This interesting fact has been proved by L. Parnovski and A. Sobolev in [44] (see also [38] and [43]).

We first need the following approximation result (see [44]).

Lemma 23. *Let* $\alpha_1, \alpha_2, \ldots, \alpha_n$ *be real numbers. Then for every positive integer* j *there exist integers* $p_1, p_2, \ldots, p_n, q$ *such that*

$$j \le q \le j^{n+1}, \qquad |\alpha_k q - p_k| < j^{-1} \text{ for every } k = 1, \ldots, n.$$

Proof. As usual we write $\{x\} = x - [x]$ for the fractional part of a real number x. Split

$$[0,1)^n = \bigcup_{k=1}^{j^n} Q_k,$$

where the Q_k's are cubes of sides parallel to the axes and of length j^{-1}. For every integer $0 \le \ell \le j^{n+1}$ consider

$$(\{\ell\alpha_1\}, \{\ell\alpha_2\}, \ldots, \{\ell\alpha_n\}) = a_\ell \in [0,1)^n.$$

Since the number of the a_ℓ's is $j^{n+1} + 1$, there exists k_0 such that the cube Q_{k_0} contains at least $j + 1$ points $a_{\ell_1}, a_{\ell_2}, \ldots, a_{\ell_{j+1}}$, say with $\ell_1 < \ell_2 < \ldots < \ell_{j+1}$. Then $\ell_{j+1} - \ell_1 \ge j$ and, since the above points stay in Q_{k_0}, we have

$$j^{-1} \ge \left|\{\ell_{j+1}\alpha_k\} - \{\ell_1\alpha_k\}\right| = \left|(\ell_{j+1} - \ell_1)\alpha_k - ([\ell_{j+1}\alpha_k] - [\ell_1\alpha_k])\right|$$

for every $k = 1, \ldots, n$. To end the proof we choose $q = \ell_{j+1} - \ell_1$ and $p_k = [\ell_{j+1}\alpha_k] - [\ell_1\alpha_k]$. $\square$

Theorem 24. *Let* $1 < d \equiv 1 \pmod 4$, *let* $B = \{u \in \mathbb{R}^d : |u| \le 1\}$ *be the unit ball and for every* $t \in \mathbb{T}^d$ *consider the discrepancy*

$$D_R(t) = \operatorname{card}\left(\mathbb{Z}^d \cap (RB + t)\right) - R^d |B|.$$

Then for every $\varepsilon > 0$ *there exists a sequence of integers* $R_j \to +\infty$ *such that*

$$\left\| D_{R_j} \right\|_{L^2(\mathbb{T}^d)} \le c_\varepsilon\, R_j^{(d-1)/2} \log^{\frac{-1}{d+\varepsilon}} (R_j).$$

Proof. For every positive integer j let

$$H_j = \{m \in \mathbb{Z}^d : 0 < |m| \le j^2\}.$$

Then $\operatorname{card} H_j \le 2^d j^{2d}$. Lemma 23 implies the existence of a positive integer R_j such that

$$j \le R_j \le j^{2^d j^{2d}+1}, \qquad \left|\sin\left(2\pi R_j |m|\right)\right| \le j^{-1} \tag{3.37}$$

for every $|m| \le j^2$. Then by (3.35), (3.4), (3.5), (3.34), the assumption $d \equiv 1 \pmod 4$ and (3.37) we obtain

$$\left\|D_{R_j}\right\|_{L^2(\mathbb{T}^d)}^2 = \sum_{m \in \mathbb{Z}^d} \left|\hat{D}_{R_j}(m)\right|^2 = R_j^d \sum_{0 \ne m \in \mathbb{Z}^d} |m|^{-d} J_{d/2}^2\left(2\pi R_j |m|\right) \tag{3.38}$$

$$= R_j^d \sum_{0<|m|\le j^2} |m|^{-d} J_{d/2}^2\left(2\pi R_j |m|\right) + R_j^d \sum_{|m|>j^2} |m|^{-d} J_{d/2}^2\left(2\pi R_j |m|\right)$$

$$\le R_j^{d-1} \sum_{0<|m|\le j^2} \pi^{-2} |m|^{-(d+1)} \sin^2\left(2\pi R_j |m|\right)$$

$$+ R_j^{d-1} \sum_{|m|>j^2} \pi^{-2} |m|^{-(d+1)} + \mathscr{O}\left(R_j^{d-2}\right)$$

$$\le c\, R_j^{d-1} j^{-2} \int_1^{j^2} r^{-2}\, dr + c\, R_j^{d-1} \int_{j^2}^{+\infty} r^{-2}\, dr + \mathscr{O}\left(R_j^{d-2}\right)$$

$$\le c\, j^{-2} R_j^{d-1} + \mathscr{O}\left(R_j^{d-2}\right).$$

Since (3.37) implies

$$\log\left(R_j\right) < \left(\left(2j^2\right)^d + 1\right) \log j < c_\varepsilon \left(2j^2\right)^{d+\varepsilon},$$

$$j^2 > c_\varepsilon' \log^{\frac{1}{d+\varepsilon}}\left(R_j\right)$$

for every $\varepsilon > 0$, we end the proof. $\square$

3.4.2 Lattice Points in Polyhedra

For general p we have some rather sharp results in the case of polyhedra.

Theorem 25. *Let P be a convex polyhedron in $\mathbb{R}^d$ $d \ge 1$, and let $D_R = D_{P,R}$ as in (3.33). Then there exist positive constants c and c_p such that for every $R \ge 2$ we have*

$$\left\|D_R\right\|_{L^1(SO(d)\times\mathbb{T}^d)} \le c\, \log^d(R) \tag{3.39}$$

and for $1 < p \le +\infty$

$$\left\|D_R\right\|_{L^p(SO(d)\times\mathbb{T}^d)} \le c_p\, R^{(d-1)(1-1/p)}. \tag{3.40}$$

Proof. The bound in (3.39) is a particular case of Theorem 30 below. We prove (3.40) first in the case $1 < p \leq 2$. Then by (3.34), Parseval identity, Hölder inequality and the inequality $\|\cdot\|_{\ell^2} \leq \|\cdot\|_{\ell^p}$ we obtain

$$
\left\{ \int_{SO(d)} \int_{\mathbb{T}^d} |D_R(\tau,t)|^p \, dt \, d\tau \right\}^{1/p}
$$

$$
\leq \left\{ \int_{SO(d)} \left\{ \int_{\mathbb{T}^d} |D_R(\tau,t)|^2 \, dt \right\}^{p/2} d\tau \right\}^{1/p}
$$

$$
= \left\{ \int_{SO(d)} \left\{ \sum_{0 \neq m \in \mathbb{Z}^d} |R^d \, \hat{\chi}_{\tau(B)}(Rm)|^2 \right\}^{p/2} d\tau \right\}^{1/p}
$$

$$
\leq \left\{ \int_{SO(d)} \sum_{0 \neq m \in \mathbb{Z}^d} |R^d \, \hat{\chi}_{\tau(B)}(Rm)|^p \, d\tau \right\}^{1/p}
$$

$$
= R^d \left\{ \sum_{0 \neq m \in \mathbb{Z}^d} \int_{SO(d)} |\hat{\chi}_{\tau(B)}(Rm)|^p \, d\tau \right\}^{1/p} .
$$

By Theorem 16 the last term is bounded by

$$
c \, R^d \left\{ \sum_{0 \neq m \in \mathbb{Z}^d} |Rm|^{-p-d+1} \right\}^{1/p} \leq c \, R^{(d-1)(1-1/p)} \int_1^{+\infty} r^{-p} \, dr
$$

$$
= c_p \, R^{(d-1)(1-1/p)} .
$$

For the case $p = +\infty$ a geometric consideration shows the existence of a positive constant c such that for every $\tau \in SO(d)$ and every $t \in \mathbb{T}^d$ we have

$$
|D_R(\tau,t)| \leq c \, R^{d-1} .
$$

We end the proof obtaining the case $2 < p < +\infty$ by interpolation:

$$
\left\{ \int_{SO(d) \times \mathbb{T}^d} |D_R|^p \right\}^{1/p} = \left\{ \int_{SO(d) \times \mathbb{T}^d} |D_R|^2 |D_R|^{p-2} \right\}^{1/p}
$$

$$
\leq \|D_R\|_{L^\infty(SO(d) \times \mathbb{T}^d)}^{(p-2)/p} \, \|D_R\|_{L^2(SO(d) \times \mathbb{T}^d)}^{2/p} \leq c \, R^{(d-1)(p-2)/p} \, R^{(d-1)/p}
$$

$$
= c \, R^{(d-1)(1-1/p)} .
$$

$\square$

A modification of the above argument can be used to study the so-called half-space discrepancy (see [23, 40]).

The proof of the weak-L^1 estimate requires a more delicate argument.

Theorem 26. *Let P be a convex polyhedron in $\mathbb{R}^d$ and let $D_R = D_{P,R}$ as in (3.33). Then there exists a positive constant c such that for every $R \geq 2$ we have*

$$\|D_R\|_{L^{1,\infty}(SO(d) \times \mathbb{T}^d)} \leq c \, \log^{d-1}(R) \, .$$

The proof of this theorem requires two preliminary results.

Lemma 27. *Let X, Y be finite measure spaces, and let*

$$\|F\|_{L^{1,\infty}(X,L^2(Y))} = \sup_{\lambda > 0} \lambda \left| \{ x \in X : \|F(x, \cdot)\|_{L^2(Y)} > \lambda \} \right| < +\infty \, .$$

Then

$$\|F\|_{L^{1,\infty}(X \times Y)} \leq c \, \|F\|_{L^{1,\infty}(X,L^2(Y))} \, .$$

Proof. Without loss of generality we may assume $\|F\|_{L^{1,\infty}(X,L^2(Y))} = 1$. Being the statement rearrangement invariant, we may assume $X = [0, 1]$, $Y = [0, 1]$, endowed with Lebesgue measure and $\|F(x, \cdot)\|_{L^2(Y)} \leq 1/x$. Then, by Chebyshev inequality we obtain

$$|\{(x, y) : 0 \leq x \leq 1, \, 0 \leq y \leq 1, \, |F(x, y)| > \lambda\}|$$

$$\leq \lambda^{-1} + |\{(x, y) : \lambda^{-1} \leq x \leq 1, \, 0 \leq y \leq 1, \, |F(x, y)| > \lambda\}|$$

$$= \lambda^{-1} + \int_{\lambda^{-1}}^{1} |\{y : 0 \leq y \leq 1, \, |F(x, y)| > \lambda\}| \, dx$$

$$\leq \lambda^{-1} + \int_{\lambda^{-1}}^{1} \left(\lambda^{-2} \int_0^1 |F(x, y)|^2 \, dy \right) dx \leq \lambda^{-1} + \lambda^{-2} \int_{\lambda^{-1}}^{1} \frac{1}{x^2} \, dx \leq 2\lambda^{-1} \, .$$

$\square$

The triangle inequality for $\|\cdot\|_{L^{1,\infty}}$ fails when we add infinitely many terms (see [57, p. 215]). The following lemma is a kind of substitute.

Lemma 28. *Let f_m be a sequence of functions in $L^{1,\infty}(X)$. Then*

$$\left\| \left\{ \sum_m |f_m|^2 \right\}^{1/2} \right\|_{L^{1,\infty}(X)} \leq c \sum_m \|f_m\|_{L^{1,\infty}(X)} \, .$$

Proof. We have

$$
\left\| \left\{ \sum_m |f_m|^2 \right\}^{1/2} \right\|_{L^{1,\infty}(X)} = \sup_{\lambda>0} \lambda \left| \left\{ x \in X : \left\{ \sum_m |f_m(x)|^2 \right\}^{1/2} > \lambda \right\} \right| \quad (3.41)
$$

$$
= \sup_{\lambda>0} \lambda \left| \left\{ x \in X : \sum_m |f_m(x)|^2 > \lambda^2 \right\} \right|
$$

$$
= \sup_{\lambda>0} \lambda^{1/2} \left| \left\{ x \in X : \sum_m |f_m(x)|^2 > \lambda \right\} \right| = \left\| \sum_m |f_m|^2 \right\|_{L^{1/2,\infty}(X)}^{1/2} .
$$

Now we recall that the following q-triangular inequality holds true when $0 < q < 1$ (see e.g. [59, Lemma 1.8]):

$$
\left\| \sum_m g_m \right\|_{L^{q,\infty}(X)} \leq c \sum_m \|g_m\|_{L^{q,\infty}(X)} .
$$

Then, as in (3.41),

$$
\left\| \sum_m |f_m|^2 \right\|_{L^{1/2,\infty}(X)}^{1/2} \leq c \sum_m \|f_m^2\|_{L^{1/2,\infty}(X)}^{1/2} = c \sum_m \|f_m\|_{L^{1,\infty}(X)} .
$$

$\square$

Proof (of Theorem 26). By Lemma 27 we have

$$
\|D_R\|_{L^{1,\infty}(SO(d)\times\mathbb{T}^d)} \leq c \, \|D_R\|_{L^{1,\infty}(SO(d),L^2(\mathbb{T}^d))}
$$

$$
= c \left\| \left\{ \sum_{0\neq m\in\mathbb{Z}^d} |\hat{\chi}_{\tau(B)}(Rm)|^2 \right\}^{1/2} \right\|_{L^{1,\infty}(SO(d))}
$$

$$
\leq c \, R^d \left\| \left\{ \sum_{0<|m|\leq R^{d-1}} |\hat{\chi}_{\tau(B)}(Rm)|^2 \right\}^{1/2} \right\|_{L^{1,\infty}(SO(d))}
$$

$$
+ c \, R^d \left\| \left\{ \sum_{|m|>R^{d-1}} |\hat{\chi}_{\tau(B)}(Rm)|^2 \right\}^{1/2} \right\|_{L^{1,\infty}(SO(d))} .
$$

By Lemma 28 we have

$$
R^d \left\| \left\{ \sum_{0<|m|\leq R^{d-1}} \left| \hat{\chi}_{\tau(B)}(Rm) \right|^2 \right\}^{1/2} \right\|_{L^{1,\infty}(SO(d))}
$$

$$
\leq c\, R^d \sum_{0<|m|\leq R^{d-1}} \left\| \hat{\chi}_{\tau(B)}(Rm) \right\|_{L^{1,\infty}(SO(d))} \leq c\, R^d \sum_{0<|m|\leq R^{d-1}} \frac{\log^{d-2}(R\,|m|)}{R^d\,|m|^d}
$$

$$
\leq c\, \log^{d-2}(R) \int_1^{+\infty} \frac{1}{r}\, dr = c\, \log^{d-1}(R) \ .
$$

On the other hand, by Chebyshev inequality and (3.40),

$$
R^d \left\| \left\{ \sum_{|m|>R^{d-1}} \left| \hat{\chi}_{\tau(B)}(Rm) \right|^2 \right\}^{1/2} \right\|_{L^{1,\infty}(SO(d))}
$$

$$
\leq R^d \left\| \left\{ \sum_{|m|>R^{d-1}} \left| \hat{\chi}_{\tau(B)}(Rm) \right|^2 \right\}^{1/2} \right\|_{L^1(SO(d))}
$$

$$
\leq R^d \left\| \left\{ \sum_{|m|>R^{d-1}} \left| \hat{\chi}_{\tau(B)}(Rm) \right|^2 \right\}^{1/2} \right\|_{L^2(SO(d))}
$$

$$
= R^d \left\{ \int_{SO(2)} \sum_{|m|>R^{d-1}} \left| \hat{\chi}_{\tau(B)}(Rm) \right|^2 d\tau \right\}^{1/2}
$$

$$
\leq c\, R^d \left\{ \sum_{|m|>R^{d-1}} |Rm|^{-(d+1)} \right\}^{1/2} \leq c\, R^{(d-1)/2} \left\{ \int_{R^{d-1}}^{+\infty} r^{-2}\, dr \right\}^{1/2} \leq c \ .
$$

$$\square$$

We now prove an upper bound where the discrepancy is averaged only over rotations. The proof follows a known argument which is usually applied to get a short proof of Sierpinski's 1903 estimate for the circle problem (see e.g. [48, 55, 60, 61]). For a convex polyhedron $P \subset \mathbb{R}^d$, for $\tau \in SO(d)$, and for a large dilation R, let $\tau(RP)$ be the rotated copy of RP. Define the discrepancy function $D_R = D_{P,R}$ on $SO(d)$

$$
D_R(\tau) = \text{card}\left(\mathbb{Z}^d \cap \tau(R\,P) \right) - R^d\,|P| \ .
$$

The following result has been pointed out to us by Leonardo Colzani.

Lemma 29. *Let C be a convex body in $\mathbb{R}^d$ such that Interior $(C) \supseteq B(0,1)$, the unit ball centered at the origin. Then for large R and small ε we have*

$$B(q,\varepsilon) \subseteq (R+\varepsilon) C \setminus \text{Interior}(R-\varepsilon) C$$

for every $q \in \partial(RC)$.

Proof. Since C is convex we have

$$\frac{R}{R+\varepsilon} C + \frac{\varepsilon}{R+\varepsilon} C \subseteq C$$

so that

$$(R+\varepsilon) C \supseteq RC + \varepsilon C \supseteq RC + B(0,\varepsilon) \tag{3.42}$$

and therefore $B(q,\varepsilon) \subseteq (R+\varepsilon) C$ for every $q \in \partial(RC)$. Applying (3.42) to Interior (C) with R in place of $R+\varepsilon$ we obtain

$$\text{Interior}(RC) \supseteq \text{Interior}(R-\varepsilon) C + B(0,\varepsilon).$$

Assume there exists $y \in B(q,\varepsilon) \cap \text{Interior}(R-\varepsilon) C$. It follows that

$$q \in \text{Interior}(R-\varepsilon) C + B(0,\varepsilon) \subseteq \text{Interior}(RC)$$

so that $q \notin \partial(RC)$. $\qquad\square$

Theorem 30. *Let $d \geq 2$ and let P be a convex polyhedron in $\mathbb{R}^d$. Then there exists a positive constant c such that, for large R,*

$$\|D_R\|_{L^1(SO(d))} \leq c \, \log^d(R) .$$

Proof. Let $B = \{t \in \mathbb{R}^d : |t| \leq 1\}$ and let $\varphi = c\chi_{\frac{1}{2}B} * \chi_{\frac{1}{2}B}$ where we choose c so that $\int \varphi(x) \, dx = 1$. For every small $\varepsilon > 0$ let $\varphi_\varepsilon(t) = \varepsilon^{-d} \varphi(t/\varepsilon)$, so that for every $\varepsilon > 0$ we have $\int_{\mathbb{R}^d} \varphi_\varepsilon = 1$ and $\hat{\varphi}_\varepsilon(\xi) = \hat{\varphi}(\varepsilon\xi)$. Let $R \geq 2$ and let χ_{RP} be the characteristic function of the dilated polyhedron P. We start the proof introducing the smooth functions

$$\chi_{R,\varepsilon,\tau}^{\pm} = \chi_{(R\pm\varepsilon)\tau^{-1}P} * \varphi_\varepsilon .$$

By (3.4) and (3.5) we know that

$$|\hat{\varphi}(\xi)| = c \left| \hat{\chi}_{\frac{1}{2}B}(\xi) \right|^2 \leq \frac{c'}{1 + |\xi|^{d+1}} .$$

Then, writing in polar coordinates $\xi = \rho\sigma$,

$$\left| \widehat{\chi_{R,\varepsilon,\tau}^{\pm}} (\xi) \right| = \left| \hat{\chi}_{(R\pm\varepsilon)\tau^{-1}P} (\xi)\, \hat{\varphi}_{\varepsilon} (\xi) \right| \tag{3.43}$$

$$\leq c\, R^d\, \left| \hat{\chi}_{\tau^{-1}P} ((R\pm\varepsilon)\rho\sigma) \right| \frac{1}{1+|\varepsilon\rho|^{d+1}}\, .$$

By Lemma 29, the support of $\chi_{R,\varepsilon,\tau}^{-}$ is contained in $R\tau^{-1}P$, while $R\tau^{-1}P$ is contained in the set where $\chi_{R,\varepsilon,\tau}^{+}$ takes the value 1. Therefore, for all $t \in \mathbb{R}^d$ we have

$$\chi_{R,\varepsilon,\tau}^{-} (t) \leq \chi_{R\tau^{-1}P} (t) \leq \chi_{R,\varepsilon,\tau}^{+} (t)\, .$$

By the Poisson summation formula we have

$$D_R (\tau) = -R^d\, |P| + \sum_{m\in\mathbb{Z}^d} \chi_{R\tau^{-1}P} (m) \leq -R^d\, |P| + \sum_{m\in\mathbb{Z}^d} \chi_{R,\varepsilon,\tau}^{+} (m)$$

$$= -R^d\, |P| + \sum_{m\in\mathbb{Z}^d} \widehat{\chi_{R,\varepsilon,\tau}^{+}} (m) = \left((R+\varepsilon)^d - R^d \right) |P| + \sum_{m\neq 0} \widehat{\chi_{R,\varepsilon,\tau}^{+}} (m)\, ,$$

and similarly,

$$D_R (\tau) \geq \left((R-\varepsilon)^d - R^d \right) |P| + \sum_{m\neq 0} \widehat{\chi_{R,\varepsilon,\tau}^{-}} (m)\, .$$

Thus,

$$|D_R (\tau)| \leq c\, R^{d-1}\varepsilon + c \sum_{m\neq 0} \left| \widehat{\chi_{R,\varepsilon,\tau}^{\pm}} (m) \right|\, .$$

Hence, by Theorem 16 and (3.43),

$$\int_{SO(d)} \left| \mathrm{card} \left(\mathbb{Z}^d \cap \tau (RP) \right) - R^d\, |P| \right|\, d\tau$$

$$\leq c\, R^{d-1}\varepsilon + c\, R^d \sum_{0\neq m\in\mathbb{Z}^d} |\hat{\varphi}(\varepsilon m)| \int_{SO(d)} |\hat{\chi}_P ((R\pm\varepsilon)\tau(m))|\, d\tau$$

$$\leq c\, R^{d-1}\varepsilon + c \sum_{0\neq m\in\mathbb{Z}^d} \frac{1}{1+|\varepsilon m|^{d+1}}\, |m|^{-d}\, \log^{d-1} (R\,|m|)\, .$$

Now choose $\varepsilon = R^{1-d}$. Then a repeated integration by parts yields

$$\int_{SO(d)} \left| \text{card}\left(\mathbb{Z}^d \cap \tau\,(RP)\right) - R^d\,|P| \right|\,d\tau$$

$$\leq c + c\,\log^{d-1}(R) \int_1^{R^{d-1}} \frac{1}{r}\,dr + c\,R^{d^2-1} \int_{R^{d-1}}^{+\infty} \frac{\log^{d-1}(r)}{r^{d+2}}\,dr$$

$$\leq c + c\,\log^d(R)$$

$$+ c\,R^{d^2-1} \left(R^{1-d^2}\log^{d-1}(R) + \int_{R^{d-1}}^{+\infty} \frac{\log^{d-2}(r)}{r^{d+2}}\,dr \right)$$

$$\leq \ldots \leq c_d\,\log^d(R) .$$

$\square$

Remark 31. Note that the estimate in the above theorem coincides with the upper L^1 estimate in Theorem 25 where the discrepancy has been averaged also over translations. The case $1 < p < \infty$ seems to be different, since either repeating the steps of the above proof for L^p norms or interpolating between L^1 and L^∞ we get estimates larger than the one in (3.40).

The previous theorem shows that the discrepancy of a convex body with respect to $\mathbb{Z}^d$ can be quite small after we have averaged over the rotations. Let us make some remarks on this point. Let us consider for simplicity a square in $\mathbb{R}^2$ with sides parallel to the axes: the two close dilations (say R and $R + \varepsilon$) of the square in the picture, have almost the same area, but the number of integer points inside differ for $\approx R$ (Fig. 3.3).

The same happens for every rational rotation of the square. On the other hand we know (see Theorem 3) that in certain directions the discrepancy can be as small

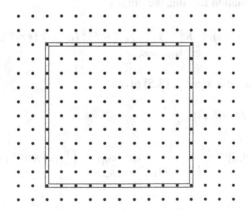

Fig. 3.3 Integer points and squares with sides parallel to the axes

as $\sqrt{\log R}$. Then we may expect that the discrepancy of a convex body C with respect to $\mathbb{Z}^d$ is reasonably small for *almost every rotation* of C. This is a very deep problem, since when C is the unit disc centered at the origin, then the rotation θ disappears and we have the classical Gauss' circle problem (so far the best bound for this problem is due to M. Huxley and it is close to $R^{0.629\cdots}$). We are now ready to state the following result (see [8], see also [24, 28, 47]). Let

$$D_R(\theta) = \text{card}\left(\mathbb{Z}^2 \cap R\theta(C)\right) - R^2 |C| = -R^2 |C| + \sum_{m \in \mathbb{Z}^2} \chi_{R\theta(C)}(m) ,$$

where C is a convex planar body and $\theta \in SO(2)$.

Theorem 32. *Let $C \subset \mathbb{R}^2$ be a convex body, let $\delta > 1/2$ and $R \geq 2$. Then for almost every $\theta \in SO(2)$ there exists a constant $c = c_{\theta,\delta}$ such that*

$$|D_R(\theta)| \leq c\, R^{2/3} \log^\delta R . \tag{3.44}$$

Proof. We use Theorem 15 and a smoothing argument similar to the one we have used in Theorem 30. Let $\psi = \pi^{-1}\chi_{\{t \in \mathbb{R}^2 : |t| \leq 1\}}$ be the normalized characteristic function of the unit disc. For every small $\varepsilon > 0$ let $\psi_\varepsilon(t) = \varepsilon^{-2}\psi(t/\varepsilon)$, so that for every $\varepsilon > 0$ we have $\int_{\mathbb{R}^2} \psi_\varepsilon = 1$ and $\hat{\psi}_\varepsilon(\xi) = \hat{\psi}(\varepsilon\xi)$. Let

$$D_R(\theta, \varepsilon) = -R^2 |C| + \sum_{m \in \mathbb{Z}^2} \left(\chi_{R\theta(C)} * \psi_\varepsilon\right)(m) ,$$

observe that, as in the proof of Theorem 30,

$$D_{R-\varepsilon}(\theta, \varepsilon) + \left(-2R\varepsilon + \varepsilon^2\right)|C| \leq D_R(\theta) \leq D_{R+\varepsilon}(\theta, \varepsilon) + \left(2R\varepsilon + \varepsilon^2\right)|C| . \tag{3.45}$$

By the Poisson summation formula we obtain

$$D_R(\theta, \varepsilon) = R^2 \sum_{0 \neq m \in \mathbb{Z}^2} \hat{\chi}_C\left(R\theta^{-1}(m)\right) \hat{\psi}(\varepsilon m) .$$

Then, for every positive integer j, (3.5) gives

$$\sup_{2^j \leq R \leq 2^{j+1}} R^{-2/3} |D_R(\theta, \varepsilon)| \tag{3.46}$$

$$\leq c\, 2^{-j/6} \sum_{0 \neq m \in \mathbb{Z}^2} |m|^{-3/2} \frac{1}{1 + |\varepsilon m|^{3/2}} \sup_{2^j \leq R \leq 2^{j+1}} \left(\left|\hat{\chi}_C\left(R\theta^{-1}(m)\right)\right| |Rm|^{3/2}\right) .$$

By Theorem 15 the function

$$\theta \longmapsto \sup_{2^j \le R \le 2^{j+1}} \left(\left| \hat{\chi}_C \left(R\theta^{-1} (m) \right) \right| |Rm|^{3/2} \right)$$

belongs to $L^{2,\infty} (SO(2))$, uniformly with respect to j and m. Since $L^{2,\infty}$ is a Banach space, then also the sum in (3.46) belongs to $L^{2,\infty} (SO(2))$, with norm bounded up to a constant by

$$2^{-j/6} \sum_{0 \ne m \in \mathbb{Z}^2} |m|^{-3/2} \frac{1}{1 + |\varepsilon m|^{3/2}}$$

$$= 2^{-j/6} \sum_{0 < |m| \le \varepsilon^{-1}} |m|^{-3/2} + 2^{-j/6} \varepsilon^{-3/2} \sum_{|m| > \varepsilon^{-1}} |m|^{-3} \le c \, 2^{-j/6} \, \varepsilon^{-1/2} .$$

Choosing $\varepsilon = 2^{-j/3}$ and using (3.45) we obtain

$$\left\| \sup_{2^j \le R \le 2^{j+1}} R^{-2/3} |D_R (\theta)| \right\|_{L^{2,\infty}(SO(2))} \le c . \tag{3.47}$$

Then

$$\sup_{R \ge 2} \left(\log^{-\delta} (R) R^{-2/3} |D_R (\theta, \varepsilon)| \right)^2 = \sup_{R \ge 2} \left(\log^{-2\delta} (R) R^{-4/3} |D_R (\theta, \varepsilon)|^2 \right)$$

$$\le \sum_{j=1}^{+\infty} j^{-2\delta} \sup_{2^j \le R \le 2^{j+1}} \left(R^{-4/3} |D_R (\theta, \varepsilon)|^2 \right)$$

belongs to $L^{1,\infty} (SO(2))$ since by (3.47) the function

$$\sup_{2^j \le R \le 2^{j+1}} \left(R^{-4/3} |D_R (\theta, \varepsilon)|^2 \right)$$

is uniformly in $L^{1,\infty}$ and can therefore be summed by the sequence $j^{-2\delta}$ if $\delta > 1/2$ (see [58, Lemma 2.3]). Then the function

$$\sup_{R \ge 2} \left(\log^{-\delta} (R) R^{-2/3} |D_R (\theta, \varepsilon)| \right)$$

belongs to $L^{2,\infty} (SO(2))$ and therefore is a.e. bounded. This proves (3.44). □

Remark 33. M. Skriganov [54] has shown that when C is a polygon we have $|D_R (\theta)| \le c_\varepsilon \log^{1+\varepsilon} R$ for any $\varepsilon > 0$ and almost every θ. Our technique can be applied also in the case of a polygon, but we only get a power of the logarithm larger than the one in [54].

3.5   Lattice Points: Estimates from Below

In this section we prove that the previous upper bounds are essentially best possible.
We will consider the balls (with the intriguing case $d \equiv 1 \pmod 4$ introduced in
Theorem 24) and the simplices.

We need a technical result (see [44]) where, as usual, $\|\beta\|$ denotes the minimal
distance of a real number β from the integers.

Lemma 34. *For every $\varepsilon > 0$ there exist $R_0 \geq 1$ and $0 < \alpha < 1/2$ such that for
every $R \geq R_0$ there exists $m \in \mathbb{Z}^d$ such that*

$$|m| \leq R^\varepsilon , \qquad \|R\,|m|\| \geq \alpha . \tag{3.48}$$

Proof. We introduce positive integers $n = n(R,\varepsilon)$ and $k_0 = k_0(\varepsilon)$ which will be
chosen later. For every integer $k \in [0, k_0]$ we consider the point

$$m_k = (n, k, 0, \ldots, 0) \in \mathbb{Z}^d$$

and write $B(k) = \sqrt{n^2 + k^2} = |m_k|$. We are going to show that for all $\varepsilon > 0$
there exist $R_0 \geq 1$, $\alpha \in (0, 1/2)$ and $k_0 \in \mathbb{N}$ such that for all $R \geq R_0$ there exist
$n \leq R^\varepsilon/2$ and $k \in [0, k_0]$ such that $\|R\,|m_k|\| \geq \alpha$. Assume the contrary, so that
there exists $\varepsilon > 0$ such that for every $R_0 \geq 1$, $\alpha \in (0, 1/2)$ and $k_0 \in \mathbb{N}$ there exist
$R \geq R_0$ such that for all $n \leq R^\varepsilon/2$ and $k \in [0, k_0]$ we have $\|R\,|m_k|\| < \alpha$. Let

$$B^{(1)}(k) = B(k+1) - B(k) , \qquad k = 0, 1, \ldots, k_0 - 1$$

$$B^{(2)}(k) = B^{(1)}(k+1) - B^{(1)}(k) , \qquad k = 0, 1, \ldots, k_0 - 2$$

$$\vdots$$

$$B^{(\ell)}(k) = B^{(\ell-1)}(k+1) - B^{(\ell-1)}(k) , \qquad k = 0, 1, \ldots, k_0 - \ell$$

$$\vdots$$

$$B^{(k_0)}(0) = B^{(k_0-1)}(1) - B^{(k_0-1)}(0)$$

Since $\|R\,B(k)\| < \alpha$ we have $\|R\,B^{(\ell)}(k)\| < 2^\ell \alpha$. Now replace k with a real
variable ny and let, for $|y| < 1$,

$$\tilde{B}(y) = \sqrt{1 + y^2} = \sum_{j=0}^{+\infty} \binom{1/2}{j} y^{2j} .$$

Differentiating, we obtain

$$\frac{d^{2j}\tilde{B}}{dx^{2j}}\left(\frac{x}{n}\right) = (2j)!\binom{1/2}{j} + \mathcal{O}\left(\frac{1}{n^2}\right)$$

uniformly in $x \in [0, k_0]$. Since $B(x) = n\tilde{B}(x/n)$, we have

$$\frac{d^{2j}B}{dx^{2j}}(x) = n^{1-2j}\left((2j)!\binom{1/2}{j} + \mathcal{O}\left(\frac{1}{n^2}\right)\right) .$$

Now observe that

$$B^{(\ell)}(x) = \int_x^{x+1}\int_{x_1}^{x_1+1}\cdots\int_{x_{\ell-1}}^{x_{\ell-1}+1}\frac{d^\ell B}{dx_\ell^\ell}\, dx_\ell\ldots dx_2 dx_1 ,$$

so that

$$B^{(2j)}(x) = n^{1-2j}\left((2j)!\binom{1/2}{j} + \mathcal{O}\left(\frac{1}{n^2}\right)\right)$$

uniformly in $x \in [0, 1/2]$. Now let j^* be the smallest integer j such that $j \geq 1+\varepsilon^{-1}$ and choose

$$k_0 = 2j^*$$

$$n = \left[\left(2(2j^*)!R\left|\binom{1/2}{j^*}\right|\right)^{\frac{1}{2j^*-1}}\right].$$

Then

$$R\,B^{(2j^*)}(x)$$

$$= \left(2(2j^*)!\left|\binom{1/2}{j^*}\right|\right)^{-1}(2j^*)!\binom{1/2}{j^*} + o(1) = \frac{1}{2}\text{sign}\binom{1/2}{j^*} + o(1)$$

as $R \to +\infty$, so that

$$\left\|R\,B^{(2j^*)}(x)\right\| = \frac{1}{2} + o(1) .$$

Observe that

$$n = \left[\left(2 \left(2j^* \right)! \left| \binom{1/2}{j^*} \right| R \right)^{\frac{1}{2j^*-1}} \right] \le \left(2 \left(2j^* \right)! \left| \binom{1/2}{j^*} \right| \right)^{\frac{\varepsilon}{\varepsilon+2}} R_0^{-\frac{\varepsilon^2+\varepsilon}{\varepsilon+2}} R^\varepsilon .$$

Choosing α such that $2^{2j^*}\alpha < 1/2$ and R_0 such that

$$\left(2 \left(2j^* \right)! \left| \binom{1/2}{j^*} \right| \right)^{\frac{\varepsilon}{\varepsilon+2}} R_0^{-\frac{\varepsilon^2+\varepsilon}{\varepsilon+2}} < \frac{1}{2}$$

we obtain a contradiction.   □

Again, let $B = \{t \in \mathbb{R}^d : |t| \le 1\}$ be the unit ball and for every $t \in \mathbb{T}^d$ consider the discrepancy

$$D_R (t) = \text{card} \left(\mathbb{Z}^d \cap (RB + t) \right) - R^d |B| .$$

We have the following result.

Theorem 35. *Let $d > 1$.*

(i) If $d \not\equiv 1 \pmod 4$, then there exists a positive constant c such that for every $R \ge 1$ we have

$$\|D_R\|_{L^2(\mathbb{T}^d)} \ge c \, R^{(d-1)/2} .$$

(ii) If $d \equiv 1 \pmod 4$, then for every small $\varepsilon > 0$ there exists a positive constant c_ε such that for every $R \ge 1$ we have

$$\|D_R\|_{L^2(\mathbb{T}^d)} \ge c_\varepsilon \, R^{(d-1)/2-\varepsilon} .$$

Proof. We prove (i).

Arguing as in (3.38) we obtain

$$\|D_R\|_{L^2(\mathbb{T}^d)}^2 = \sum_{0 \ne m \in \mathbb{Z}^d} \left| \hat{D}_R (m) \right|^2 = R^d \sum_{0 \ne m \in \mathbb{Z}^d} |m|^{-d} J_{d/2}^2 (2\pi R |m|)$$

$$= \pi^{-2} R^{d-1} \sum_{0 \ne m \in \mathbb{Z}^d} |m|^{-d-1} \cos^2 (2\pi R |m| - \pi (d + 1) /4) + \mathcal{O} \left(R^{d-2} \right) .$$

Now let $m' = (1, 0, \dots, 0)$ and assume

$$\left| \cos \left(2\pi R |m'| - \pi (d + 1) /4 \right) \right| = |\cos (2\pi R - \pi (d + 1) /4)| > \frac{1}{100} .$$

Then

$$\|D_R\|^2_{L^2(\mathbb{T}^d)} \geq \pi^{-2} R^{d-1} |m'|^{-d-1} \cos^2\left(2\pi R |m'| - \pi(d+1)/4\right) + \mathcal{O}\left(R^{d-2}\right)$$

$$= \pi^{-2} 10^{-2} R^{d-1} + \mathcal{O}\left(R^{d-2}\right) \geq c R^{d-1} .$$

Now assume

$$\left|\cos\left(2\pi R |m'| - \pi(d+1)/4\right)\right| = \left|\sin\left(2\pi R - \pi(d-1)/4\right)\right| \leq \frac{1}{100} .$$

Then there exists an integer ℓ such that $2R = \ell + (d-1)/4 \pm \delta$, for a suitable $|\delta| \leq 1/50$. Then $4R = 2\ell + (d-1)/2 \pm 2\delta$, and since $(d-1)/4$ is not an integer we have

$$\left|\cos\left(2\pi R |2m'| - \pi(d+1)/4\right)\right|$$

$$= \left|\sin\left(\pi \{2\ell + (d-1)/2 \pm 2\delta - \pi(d-1)/4\}\right)\right|$$

$$\geq \frac{1}{2} \left\|\pm 2\delta + (d-1)/4\right\| \geq \frac{1}{10} .$$

Then choosing $\bar{m}$ equal to m' or to $2m'$, we have

$$\|D_R\|^2_{L^2(\mathbb{T}^d)} \geq c R^{d-1} .$$

We now prove (ii). Let m be as in Lemma 34. Since $d \equiv 1 \pmod 4$ we have

$$\|D_R\|^2_{L^2(\mathbb{T}^d)} \geq \pi^{-2} R^{d-1} |m|^{-d-1} \cos^2\left(2\pi R |m| - \pi(d+1)/4\right) + \mathcal{O}\left(R^{d-2}\right)$$

$$= \pi^{-2} R^{d-1} |m|^{-d-1} \sin^2\left(2\pi R |m|\right) + \mathcal{O}\left(R^{d-2}\right)$$

$$\geq c_\varepsilon R^{d-1} R^{-(d+1)\varepsilon} .$$

$\square$

For the simplices we have results complementary to the ones in Theorems 25 and 26. Let S be a simplex in $\mathbb{R}^d$ and for every $t \in \mathbb{T}^d$ let

$$D_R(t) = \text{card}\left(\mathbb{Z}^d \cap (\tau(R S) + t)\right) - R^d |S| .$$

Theorem 36. *For every simplex S in $\mathbb{R}^d$ ($d \geq 2$) and $R \geq 2$ we have*

i) $\|D_R\|_{L^{1,\infty}(SO(d) \times \mathbb{T}^d)} \geq c \, \log^{d-2}(R)$

ii) $\|D_R\|_{L^1(SO(d) \times \mathbb{T}^d)} \geq c \, \log^{d-1}(R)$

iii) $\|D_R\|_{L^p(SO(d) \times \mathbb{T}^d)} \geq R^{(d-1)(1-1/p)} , \quad$ *if $1 < p \leq +\infty$.*

Proof. For every $m' \neq 0$ we have

$$\|D_R\|_{L^p(SO(d) \times \mathbb{T}^d)} \geq R^d \left\{ \int_{SO(d)} \int_{\mathbb{T}^d} \left| \sum_{m \neq 0} \hat{\chi}_S (R\tau(m)) \, e^{2\pi i m \cdot t} \right|^p dt \, d\tau \right\}^{1/p}$$

$$\geq R^d \left\{ \int_{\mathbb{T}^d} \int_{SO(d)} \left| \hat{\chi}_S (R\tau(m')) \right|^p d\tau \, dt \right\}^{1/p}.$$

Then (*ii*) and (*iii*) follow from Theorem 19. The proof of (*i*) is a consequence of Lemma 18 as in the proof of the corresponding part of Theorem 19.   □

3.6   Irregularities of Distribution: Estimates from Below

It is time to go back to the Introduction, where we have referred to the fundamental results of K. Roth (Theorem 2) and W. Schmidt (Theorem 1). In this section we present two different approaches to Theorem 4, due to J. Beck and H. Montgomery respectively. For convenience we shall apply Beck's argument to prove Theorem 4 and Montgomery's argument to prove a stronger version of the theorem which holds true in the particular case when B is a simplex.

We now repeat the main part of the statement of Theorem 4.

Theorem 37. *Let $d \geq 2$ and let $B \subset \mathbb{R}^d$ be a body of diameter smaller than 1 which satisfies (3.15). Then for every distribution $\mathscr{P}$ of N points in $\mathbb{T}^d$ we have*

$$\int_0^1 \int_{SO(d)} \int_{\mathbb{T}^d} \left| \mathrm{card}\left(\mathscr{P} \cap (\lambda \tau (B + t)) \right) - \lambda^d N |B| \right|^2 dt \, d\tau \, d\lambda \geq c_d N^{(d-1)/d}.$$

$$(3.49)$$

Proof. For every $0 < \lambda \leq 1$, $\tau \in SO(d)$, $t \in \mathbb{T}^d$ the projection of $\lambda \tau (B) - t$ is injective from $\mathbb{R}^d$ to $\mathbb{T}^d$. Given a finite distribution $\mathscr{P} = \{t(j)\}_{j=1}^N$ of N points in $\mathbb{T}^d$ we consider the discrepancies

$$D_N^{B,\mathscr{P}} = D_N = -N |B| + \sum_{j=1}^N \chi_B (t(j))$$

$$D_N^{B,\mathscr{P}} (\lambda, \tau, t) = D_N (\lambda, \tau, t) = -N \lambda^d |B| + \sum_{j=1}^N \chi_{\lambda \tau^{-1}(B) - t} (t(j)).$$

First we show that the function $t \mapsto D_N(\lambda, \tau, t)$ has Fourier series

$$\sum_{m \neq 0} \left(\sum_{j=1}^{N} e^{2\pi i m \cdot t(j)} \right) \lambda^d \, \hat{\chi}_B(\lambda \tau(m)) \, e^{2\pi i m \cdot t} . \tag{3.50}$$

Indeed

$$\int_{\mathbb{T}^d} \left(-N\lambda^d |B| + \sum_{j=1}^{N} \chi_{\lambda\tau^{-1}(B)-t}(t(j)) \right) dt$$

$$= -N\lambda^d |B| + \sum_{j=1}^{N} \int_{\mathbb{T}^d} \chi_{\lambda\tau^{-1}(B)}(t(j)+t) \, dt$$

$$= -N\lambda^d |B| + N \int_{\mathbb{R}^d} \chi_{\lambda\tau^{-1}(B)}(u) \, du = 0 ,$$

while for $m \neq 0$

$$\int_{\mathbb{T}^d} \left(-N\lambda^d |B| + \sum_{j=1}^{N} \chi_{\lambda\tau^{-1}(B)-t}(t(j)) \right) e^{-2\pi i m \cdot t} \, dt$$

$$= \sum_{j=1}^{N} \int_{\mathbb{T}^d} \chi_{\lambda\tau^{-1}(B)}(t(j)+t) \, e^{-2\pi i m \cdot t} \, dt$$

$$= \sum_{j=1}^{N} \int_{\mathbb{T}^d} \chi_{\lambda\tau^{-1}(B)}(u) \, e^{-2\pi i m \cdot (u-t(j))} \, du = \sum_{j=1}^{N} e^{2\pi i m \cdot t(j)} \lambda^d \, \hat{\chi}_B(\lambda \tau(m)) .$$

Let $0 < q < 1$ and $0 < r < 1$. By (3.50) and Theorem 8 we have

$$\frac{1}{r} \int_{qr}^{r} \int_{SO(d)} \int_{\mathbb{T}^d} |D_N(\lambda, \tau, t)|^2 \, dt \, d\tau \, d\lambda \tag{3.51}$$

$$= \sum_{m \neq 0} \left| \sum_{j=1}^{N} e^{2\pi i m \cdot t(j)} \right|^2 \frac{1}{r} \int_{qr}^{r} \int_{SO(d)} \left| \lambda^d \, \hat{\chi}_B(\lambda\tau(m)) \right|^2 \, d\tau \, d\lambda$$

$$= c \sum_{m \neq 0} \left| \sum_{j=1}^{N} e^{2\pi i m \cdot t(j)} \right|^2 \frac{1}{r} \int_{\{qr \leq |\xi| \leq r\}} |\hat{\chi}_B(|m|\xi)|^2 \, |\xi|^{d+1} \, d\xi$$

$$\approx \sum_{m\neq 0} \left| \sum_{j=1}^{N} e^{2\pi i m \cdot t(j)} \right|^2 r^d \int_{\{qr\leq |\xi| \leq r\}} |\hat{\chi}_B(|m|\xi)|^2 \, d\xi$$

$$\approx \sum_{m\neq 0} \left| \sum_{j=1}^{N} e^{2\pi i m \cdot t(j)} \right|^2 r^d |m|^{-d} \int_{\{qr|m|\leq |\xi| \leq r|m|\}} |\hat{\chi}_B(\eta)|^2 \, d\eta$$

$$\approx \sum_{m\neq 0} \left| \sum_{j=1}^{N} e^{2\pi i m \cdot t(j)} \right|^2 r^d |m|^{-d} (1 + r|m|)^{-1} .$$

(again, $A \approx B$ means that there exist two positive constants c_1 and c_2 which do not depend on N and r and satisfy $c_1 A \leq B \leq c_2 A$). Now we apply (3.51) first with $r = 1$ and then with $r = kN^{-1/d}$ (we shall choose the constant k later on). We obtain

$$\int_q^1 \int_{SO(d)} \int_{\mathbb{T}^d} |D_N(\lambda, \tau, t)|^2 \, dt \, d\tau \, d\lambda \tag{3.52}$$

$$\approx \sum_{m\neq 0} \left| \sum_{j=1}^{N} e^{2\pi i m \cdot t(j)} \right|^2 |m|^{-d-1}$$

$$\geq c \left\{ \inf_{m\neq 0} \frac{1 + kN^{-1/d}|m|}{k^d N^{-1}|m|} \right\} \left\{ \sum_{m\neq 0} \left| \sum_{j=1}^{N} e^{2\pi i m \cdot t(j)} \right|^2 \left(\frac{k^d N^{-1}|m|^{-d}}{1 + kN^{-1/d}|m|} \right) \right\}$$

$$\approx \left\{ N^{1-1/d} k^{1-d} \right\} \left\{ k^{-1} N^{1/d} \int_{qkN^{-1/d}}^{kN^{-1/d}} \int_{SO(d)} \int_{\mathbb{T}^d} |D_N(\lambda, \tau, t)|^2 \, dt \, d\tau \, d\lambda \right\} .$$

Since

$$qkN^{-1/d} \leq \lambda \leq kN^{-1/d}$$

there exists a small constant $\delta > 0$ such that, for suitable choices of the constants q and k we have

$$\delta \leq q^d k^d |B| \leq N\lambda^d |B| \leq k^d |B| \leq 1 - \delta .$$

Being

$$\sum_{j=1}^{N} \chi_{\lambda \tau^{-1}(B)-t}(t(j))$$

an integer, we deduce that

$$|D_N(\lambda, \tau, t)| = \left| -N\lambda^d |B| + \sum_{j=1}^{N} \chi_{\lambda\tau^{-1}(B)-t}(t(j)) \right| \geq \delta$$

for every τ, t and $\lambda \in [qkN^{-1/d}, kN^{-1/d}]$. Then (3.52) gives

$$\int_q^1 \int_{SO(d)} \int_{\mathbb{T}^d} |D_N(\lambda, \tau, t)|^2 \, dt \, d\tau \, d\lambda \geq cN^{1-1/d} .$$

□

Because of Theorem 24, the dilation in λ in (3.49) cannot be deleted. In the sequel of this section we shall see how to avoid the dilation in particular cases. The starting point is a lemma due to J. Cassels (see [17, 42]).

Lemma 38. *For every positive integer N let*

$$Q_N = \left\{ x = (x_1, x_2, \ldots, x_d) \in \mathbb{R}^d : |x_j| \leq \sqrt[d]{2N} \quad \text{for every } j = 1, 2, \ldots, d \right\} .$$

Then for every finite set $\{t(j)\}_{j=1}^N \subset \mathbb{T}^d$

$$\sum_{0 \neq m \in Q_N \cap \mathbb{Z}^d} \left| \sum_{j=1}^{N} e^{2\pi i m \cdot t(j)} \right|^2 \geq N^2 . \tag{3.53}$$

Proof. Let $m = (m_1, m_2, \ldots, m_d)$ an element of $\mathbb{Z}^d$. Adding N^2 on both sides of (3.53), we have to prove that

$$\sum_{|m_1| \leq \sqrt[d]{2N}} \cdots \sum_{|m_d| \leq \sqrt[d]{2N}} \left| \sum_{j=1}^{N} e^{2\pi i m \cdot t(j)} \right|^2 \geq 2N^2 ,$$

and this is a consequence of the following inequality:

$$\sum_{|m_1| \leq [\sqrt[d]{2N}]} \cdots \sum_{|m_d| \leq [\sqrt[d]{2N}]} \left| \sum_{j=1}^{N} e^{2\pi i m \cdot t(j)} \right|^2 \geq N \left([\sqrt[d]{2N}] + 1 \right)^d , \tag{3.54}$$

where $[\gamma]$ denotes the integral part of the real number γ. The LHS in (3.54) is larger than

$$\sum_{|m_1|\leq\left[\sqrt[d]{2N}\right]}\left(1-\frac{|m_1|}{\left[\sqrt[d]{2N}\right]+1}\right)\cdots \tag{3.55}$$

$$\times \sum_{|m_d|\leq\left[\sqrt[d]{2N}\right]}\left(1-\frac{|m_d|}{\left[\sqrt[d]{2N}\right]+1}\right)\left|\sum_{j=1}^{N}e^{2\pi i m\cdot t(j)}\right|^2$$

$$= \sum_{|m_1|\leq\left[\sqrt[d]{2N}\right]}\left(1-\frac{|m_1|}{\left[\sqrt[d]{2N}\right]+1}\right)\cdots \sum_{|m_d|\leq\left[\sqrt[d]{2N}\right]}\left(1-\frac{|m_d|}{\left[\sqrt[d]{2N}\right]+1}\right)$$

$$\times \sum_{j=1}^{N}\sum_{\ell=1}^{N}e^{2\pi i m\cdot(t(j)-t(\ell))}$$

$$= \sum_{j=1}^{N}\sum_{\ell=1}^{N}\sum_{|m_1|\leq\left[\sqrt[d]{2N}\right]}\left(1-\frac{|m_1|}{\left[\sqrt[d]{2N}\right]+1}\right)e^{2\pi i m_1(t_1(j)-t_1(\ell))}\cdots$$

$$\times \sum_{|m_d|\leq\left[\sqrt[d]{2N}\right]}\left(1-\frac{|m_d|}{\left[\sqrt[d]{2N}\right]+1}\right)e^{2\pi i m_d(t_d(j)-t_d(\ell))}$$

$$= \sum_{j=1}^{N}\sum_{\ell=1}^{N}K_{\left[\sqrt[d]{2N}\right]}(t_1(j)-t_1(\ell))\cdots K_{\left[\sqrt[d]{2N}\right]}(t_d(j)-t_d(\ell)) ,$$

where

$$K_M(t) = \sum_{j=-M}^{M}\left(1-\frac{|j|}{M+1}\right)e^{2\pi i j t} = \frac{1}{M+1}\left(\frac{\sin(\pi(M+1)t)}{\sin(\pi t)}\right)^2$$

is the Fejér kernel ($M \in \mathbb{N}$, $t \in \mathbb{T}$). Since $K_M(t) \geq 0$ for every t, we may bound the terms in (3.55) from below by the "diagonal"

$$\sum_{j=1}^{N}K_{\left[\sqrt[d]{2N}\right]}(t_1(j)-t_1(j))\cdots K_{\left[\sqrt[d]{2N}\right]}(t_d(j)-t_d(j)) = N\left(K_{\left[\sqrt[d]{2N}\right]}(0)\right)^d$$

$$= N\left(\left[\sqrt[d]{2N}\right]+1\right)^d .$$

$\square$

Theorem 39. *Let S be a simplex in $\mathbb{R}^d$, $d \geq 2$, the sides of which have length smaller than 1. Then there exists a constant $c_d > 0$ such that for every finite set $\{t(j)\}_{j=1}^N \subset \mathbb{T}^d$ we have*

$$\int_{SO(d)} \int_{\mathbb{T}^d} \left| -N|S| + \sum_{j=1}^N \chi_{\tau^{-1}(S)-t}(t(j)) \right|^2 dt\, d\tau \geq c_d\, N^{(d-1)/d}.$$

Proof. As a consequence of Parseval identity, Theorem 19 and Lemma 38 we obtain

$$\int_{SO(d)} \int_{\mathbb{T}^d} \left| -N|S| + \sum_{j=1}^N \chi_{\tau^{-1}(S)-t}(t(j)) \right|^2 dt\, d\tau$$

$$= \sum_{m \neq 0} \left| \sum_{j=1}^N e^{2\pi i m \cdot t(j)} \right|^2 \int_{SO(d)} |\hat{\chi}_S(\tau(m))|^2\, d\tau$$

$$\geq \sum_{0 \neq m \in Q_N} \left| \sum_{j=1}^N e^{2\pi i m \cdot t(j)} \right|^2 \int_{SO(d)} |\hat{\chi}_S(\tau(m))|^2\, d\tau$$

$$\geq c \sum_{0 \neq m \in Q_N} \left| \sum_{j=1}^N e^{2\pi i m \cdot t(j)} \right|^2 |m|^{-d-1}$$

$$\geq c \inf_{m \in Q_N} \left(|m|^{-d-1} \right) \sum_{0 \neq m \in Q_N} \left| \sum_{j=1}^N e^{2\pi i m \cdot t(j)} \right|^2$$

$$\geq c\, N^{-1-1/d} N^2 = c\, N^{1-1/d}.$$

$\square$

Remark 40. Using Theorem 8 the above argument gives a new proof of Theorem 37 (see [42]).

Corollary 41. *Let S be a simplex in $\mathbb{R}^d$ the sides of which have length smaller than 1. Then, for every finite set $\{t(j)\}_{j=1}^N \subset \mathbb{T}^d$ there exists a (translated and rotated) copy S' of S such that*

$$\left| -N|S| + \mathrm{card}\left(S' \cap \{t(j)\}_{j=1}^N \right) \right| \geq c_d\, N^{(d-1)/2d}.$$

3.7   Irregularities of Distribution: Estimates from Above

We are going to check the quality of the estimates from below obtained in the previous section. We will see that for any given body B and for every positive integer N one can find a finite set $\{t\,(j)\}_{j=1}^{N} \subset \mathbb{T}^d$, such that a suitable L^2 mean of the discrepancy is smaller than $cN^{(d-1)/2d}$. Observe that we cannot choose the points at random. Indeed, such a Monte Carlo choice of the N points gives a $\sqrt{N}$ discrepancy (see e.g. (3.61) below), and this is not enough to match the $N^{(d-1)/2d}$ lower estimates in Theorems 37 and 39 (although for large dimension d the exponent $(d-1)/2d$ approaches $1/2$). We shall get the $N^{(d-1)/2d}$ estimate first using an argument related to lattice points problems, and then a probabilistic argument.

For an overview of upper estimates related to irregularities of distribution, see [20].

3.7.1   Applying Lattice Points Results

A very natural way to choose N points in a cube consists in putting them on a grid. Suppose for the time being that we have $N = M^d$ points (Fig. 3.4). Then let

$$\mathscr{P} = \mathscr{P}_M = \frac{1}{M}\mathbb{Z}^d \cap \left[-\frac{1}{2}, \frac{1}{2}\right)^d = \{t\,(j)\}_{j=1}^{N} \qquad (3.56)$$

(the ordering of the $t\,(j)$'s is irrelevant).

This choice of $\mathscr{P}$ immediately relates our point distribution problem to certain lattice point problems similar to the ones that we have considered in the previous sections. Indeed, if B is a body in $\left[-\frac{1}{2}, \frac{1}{2}\right)^d$ and $\mathscr{P}$ is as in (3.56) we have

$$\operatorname{card}\left(B \cap \mathscr{P}_M\right) = \operatorname{card}\left(MB \cap \mathbb{Z}^d\right) .$$

Fig. 3.4 Points on a grid

Before going on, observe that here M is an integer, while in the lattice point problems we have considered so far, the dilation parameter R is real. In other words, the choice of the piece of lattice in (3.56) implicitly contains some (but not all the) dilations.

Now we show that the lower estimate in Theorem 37 cannot be improved. This result has been originally proved by J. Beck and W. Chen [2]. We give two proofs: the first one is based on Theorem 21, while the second one is probabilistic in nature (see [2, 14, 22], see also [36, 41]). Since the second proof works under assumptions more general than convexity, we will state two different theorems (the first one is contained in the second one).

Theorem 42. *Let $B \subset \mathbb{R}^d$ be a convex body of diameter smaller than 1. Then for every positive integer N there exists a finite set $\{t(j)\}_{j=1}^N \subset \mathbb{T}^d$ such that*

$$\int_{SO(d)} \int_{\mathbb{T}^d} \left| -N|B| + \sum_{j=1}^N \chi_{\tau^{-1}(B)-t}(t(j)) \right|^2 dt\, d\tau \le c_d\, N^{(d-1)/d}.$$

Here c_d depends only on the dimension d.

Proof. We apply Theorem 21. Assume first that $N = M^d$ for a positive integer M. For any $a \in \left[-\frac{1}{2}, \frac{1}{2}\right)^d$ let

$$A_N = \{t(j)\}_{j=1}^N = \left(a + M^{-1}\mathbb{Z}^d\right) \cap \left[-\frac{1}{2}, \frac{1}{2}\right)^d.$$

(the role of a will be clear later on). Then

$$\int_{SO(d)} \int_{\mathbb{T}^d} \left| \text{card}\left(A_{M^d} \cap (\tau(B)+t)\right) - M^d|B| \right|^2 dt\, d\tau$$

$$= \int_{SO(d)} \int_{\mathbb{T}^d} \left| \text{card}\left(A_{M^d} \cap (\tau(B)+t+a)\right) - M^d|B| \right|^2 dt\, d\tau$$

$$= M^d \int_{SO(d)} \int_{\left[-\frac{1}{2M}, \frac{1}{2M}\right)^d} \left| \text{card}\left(M^{-1}\mathbb{Z}^d \cap (\tau(B)+t)\right) - M^d|B| \right|^2 dt\, d\tau$$

$$= M^d \int_{SO(d)} \int_{\left[-\frac{1}{2M}, \frac{1}{2M}\right)^d} \left| \text{card}\left(\mathbb{Z}^d \cap (\tau(MB)+Mt)\right) - M^d|B| \right|^2 dt\, d\tau$$

$$= \int_{SO(d)} \int_{\mathbb{T}^d} \left| \text{card}\left(\mathbb{Z}^d \cap (\tau(MB)+v)\right) - M^d|B| \right|^2 dv\, d\tau,$$

since the function

$$t \mapsto \text{card}\left(M^{-1}\mathbb{Z}^d \cap (\tau(B)+t)\right) - M^d|B|$$

is $M^{-1}\mathbb{Z}^d$ periodic and the cube $\left[-\frac{1}{2}, \frac{1}{2}\right)^d$ contains M^d disjoint copies of $\left[-\frac{1}{2M}, \frac{1}{2M}\right)^d$. By Theorem 21 we have

$$\int_{SO(d)} \int_{\mathbb{T}^d} \left|\mathrm{card}\left(A_{M^d} \cap (\tau(B) + t)\right) - M^d |B|\right|^2 \, dt \, d\tau \le c_d M^{d-1} \le c_d N^{1-1/d}.$$

$$(3.57)$$

To end the proof we need to pass from $N = M^d$ to an arbitrary positive integer N. By a theorem of Hilbert (Waring problem, see [29]) there exists a constant $H = H_d$ such that every positive integer N can be written a sum of at most H dth powers:

$$N = \sum_{j=1}^{H} M_j^d$$

with $M_1, M_2, \ldots, M_H$ positive integers. Now choose $a_1, a_2, \ldots, a_H \in \left[-\frac{1}{2}, \frac{1}{2}\right)^d$ such that

$$\left(a_j + M_j^{-1}\mathbb{Z}^d\right) \cap \left(a_k + M_k^{-1}\mathbb{Z}^d\right) = \varnothing \qquad (3.58)$$

whenever $j \ne k$. For $j = 1, 2, \ldots, H$ let

$$A_{M_j^d} = \left(a_j + M_j^{-1}\mathbb{Z}^d\right) \cap \left[-\frac{1}{2}, \frac{1}{2}\right)^d.$$

By (3.58) the union

$$A_N = \bigcup_{j=1}^{H} A_{M_j^d}$$

is disjoint, so that A_N contains exactly N points. Since

$$\mathrm{card}\left(A_N \cap B\right) - N|B| = \sum_{j=1}^{H} \left(\mathrm{card}\left(A_{M_j^d} \cap B\right) - M_j^d |B|\right),$$

the theorem follows from (3.57). □

3.7.2 Deterministic and Probabilistic Discrepancies

In the proof of the next theorem the points will be chosen in a probabilistic way. Since we will start from the piece of lattice $\{t(j)\}_{j=1}^{N}$ introduced in (3.56), it will

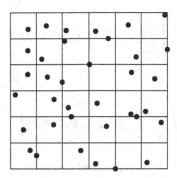

Fig. 3.5 Jittered sampling

be possible, as in the first proof, to assume $N = M^d$. We will choose one point at random inside each one of the N small cubes having sides parallel to the axes and of length $1/M$ (Fig. 3.5).

The above choice is sometimes called jittered sampling.

We have the following generalization of Theorem 42.

Theorem 43. *Let $B \subset \mathbb{T}^d$ be a body of diameter smaller than 1 satisfying (3.15). Then for every positive integer N there exists a finite set $\{t(j)\}_{j=1}^{N} \subset \mathbb{T}^d$ such that*

$$\int_{SO(d)} \int_{\mathbb{T}^d} \left| -N |B| + \sum_{j=1}^{N} \chi_{\tau^{-1}(B)-t}(t(j)) \right|^2 dt\, d\tau \leq c_d\, N^{(d-1)/d} \ .$$

Before starting the proof we introduce the following slightly more general point of view.

Given the finite point set distribution $\mathscr{P} = \{t(j)\}_{j=1}^{N}$ in (3.56), we introduce the following randomization of $\mathscr{P}$, see [2, 14, 22] and also [5, 36, 41]. Let $d\mu$ denote a probability measure on $\mathbb{T}^d$ and for every $j = 1, \dots, N$, let $d\mu_j$ denote the measure obtained after translating $d\mu$ by $t(j)$. More precisely, for any integrable function g on $\mathbb{T}^d$, let

$$\int_{\mathbb{T}^d} g(t)\, d\mu_j = \int_{\mathbb{T}^d} g(t - t(j))\, d\mu \ . \tag{3.59}$$

As before, let dt denote the Lebesgue measure on $\mathbb{T}^d$. For every sequence $V_N = \{v_1, \dots, v_N\}$ in $\mathbb{T}^d$ and every $t \in \mathbb{T}^d$, $\tau \in SO(d)$ let

$$D(t, \tau, V_N) = D(V_N) = -N |B| + \sum_{j=1}^{N} \chi_{\tau^{-1}(B)-t}(v_j) \ .$$

As in (3.50), $D_N(t) = D(t, \tau, V_N)$ has Fourier series

$$\sum_{0 \neq m \in \mathbb{Z}^d} \left(\sum_{j=1}^{N} e^{2\pi i m \cdot v_j} \right) \hat{\chi}_B(\tau(m)) \, e^{2\pi i m \cdot t} \, .$$

We now average

$$\left\{ \int_{SO(d)} \int_{\mathbb{T}^d} D^2(t, \tau, V_N) \, dt \, d\tau \right\}^{1/2}$$

in $L^2(\mathbb{T}^d, d\mu_j)$ for every $j = 1, \ldots, N$, and consider

$$D_{d\mu}(N) = \left\{ \int_{\mathbb{T}^d} \cdots \int_{\mathbb{T}^d} \int_{SO(d)} \int_{\mathbb{T}^d} D^2(t, \tau, V_N) \, dt \, d\tau \, d\mu_1(v_1) \ldots d\mu_N(v_N) \right\}^{1/2} \, .$$

We now use an orthogonality argument to obtain an explicit identity for $D_{d\mu}(N)$. Indeed by Parseval identity, (3.59) and (3.56) we have

$$D_{d\mu}^2(N)$$

$$= \int_{SO(d)} \int_{\mathbb{T}^d} \cdots \int_{\mathbb{T}^d} \sum_{0 \neq m \in \mathbb{Z}^d} \left| \sum_{j=1}^{N} e^{2\pi i m \cdot v_j} \right|^2 |\hat{\chi}_B(\tau(m)|^2 \, d\mu_1(v_1) \ldots d\mu_N(v_N) d\tau$$

$$= \int_{SO(d)} \sum_{0 \neq m \in \mathbb{Z}^d} |\hat{\chi}_B(\tau(m))|^2 \sum_{j,\ell=1}^{N} \int_{\mathbb{T}^d} \int_{\mathbb{T}^d} e^{2\pi i m \cdot v_j} e^{-2\pi i m \cdot v_\ell} \, d\mu_j(v_j) d\mu_\ell(v_\ell) d\tau$$

$$= \int_{SO(d)} \sum_{0 \neq m \in \mathbb{Z}^d} |\hat{\chi}_B(\tau(m))|^2$$

$$\times \left(N + \sum_{\substack{j,\ell=1 \\ j \neq \ell}}^{N} \int_{\mathbb{T}^d} \int_{\mathbb{T}^d} e^{2\pi i m \cdot (v_j - t(j))} e^{-2\pi i m \cdot (v_\ell - t(\ell))} \, d\mu(v_j) d\mu(v_\ell) \right) d\tau \, .$$

Then

$$D_{d\mu}^2(N)$$

$$= \int_{SO(d)} \sum_{0 \neq m \in \mathbb{Z}^d} |\hat{\chi}_B(\tau(m))|^2 \left(N + |\hat{\mu}(m)|^2 \sum_{\substack{j,\ell=1 \\ j \neq \ell}}^{N} e^{2\pi i m \cdot (t(\ell) - t(j))} \right) d\tau$$

$$= \int_{SO(d)} \sum_{0 \neq m \in \mathbb{Z}^d} |\hat{\chi}_B(\tau(m))|^2 \left(N + |\hat{\mu}(m)|^2 \left(\sum_{j,\ell=1}^{N} e^{2\pi i m \cdot (t(j) - t(\ell))} - N \right) \right) d\tau$$

$$= N \sum_{0 \neq m \in \mathbb{Z}^d} \left(1 - |\hat{\mu}(m)|^2 \right) \int_{SO(d)} |\hat{\chi}_B(\tau(m))|^2 \, d\tau$$

$$+ \sum_{0 \neq m \in \mathbb{Z}^d} |\hat{\mu}(m)|^2 \left| \sum_{j=1}^{N} e^{2\pi i m \cdot t(j)} \right|^2 \int_{SO(d)} |\hat{\chi}_B(\tau(m))|^2 \, d\tau \, .$$

So that

$$D_{d\mu}^2(N) = N \left(|B| - \|\chi_{\tau^{-1}B} * \mu\|_{L^2(SO(d) \times \mathbb{T}^d)}^2 \right) + \|D(\cdot, \tau, \mathscr{P}) * \mu\|_{L^2(SO(d) \times \mathbb{T}^d)}^2 \, . \tag{3.60}$$

The following are particular cases.

(a) Let $d\mu = dt$ (the Lebesgue measure on $\mathbb{T}^d$). Then the second term in the RHS of (3.60) vanishes since $\hat{\mu}(m) = 0$ for $m \neq 0$, and we find the *Monte-Carlo discrepancy*:

$$D_{dt}^2(N) = N \int_{SO(d)} \sum_{0 \neq m \in \mathbb{Z}^d} |\hat{\chi}_B(\tau(m))|^2 \, d\tau = N \left(|B| - |B|^2 \right) \, . \tag{3.61}$$

(b) Let $d\mu = \delta_0$ (the Dirac δ at 0). Then we have the piece of grid

$$M^{-1}\mathbb{Z}^d \cap \left[-\frac{1}{2}, \frac{1}{2} \right)^d \, ,$$

and the first term in the RHS of (3.60) vanishes because $\hat{\mu}(m) = 1$ for every m. As for the second term, note that for $t(j)$ in (3.56) we have

$$\sum_{j=1}^{N} e^{2\pi i m \cdot t(j)} = \begin{cases} N & \text{if } m \in M\mathbb{Z}^d \\ 0 & \text{otherwise,} \end{cases} \tag{3.62}$$

We then obtain the *grid discrepancy*:

$$D_{\delta_0}^2(N) = N^2 \int_{SO(d)} \sum_{0 \neq m \in \mathbb{Z}^d} |\hat{\chi}_B(M\tau(m))|^2 \, d\tau \, .$$

In the next case we shall choose one point at random inside each one of the M^d small cubes having sides parallel to the axes and length $1/M$.

(c) Let $d\mu = d\lambda = \lambda(t)dt$, with

$$\lambda(t) = N\chi_{[-1/(2M),1/(2M)]^d}(t).$$

Then, for $m = (m_1, m_2, \ldots, m_d)$ we have

$$\hat{\lambda}(m) = N\, \frac{\sin(\pi m_1/M)}{\pi m_1}\, \frac{\sin(\pi m_2/M)}{\pi m_2}\, \cdots\, \frac{\sin(\pi m_d/M)}{\pi m_d}$$

and, for every $m \neq 0$, (3.62) gives

$$\hat{\mu}(m)\widehat{D(\cdot, \tau, \mathscr{P})}(m)$$

$$= \hat{\lambda}(m)\widehat{D(\cdot, \tau, \mathscr{P})}(m)$$

$$= \left(N\, \frac{\sin(\pi m_1/M)}{\pi m_1}\, \frac{\sin(\pi m_2/M)}{\pi m_2}\, \cdots\, \frac{\sin(\pi m_d/M)}{\pi m_d}\right)\left(\hat{\chi}_B(\tau(m))\sum_{j=1}^{N} e^{2\pi im\cdot t}(j)\right)$$

$$= 0,$$

so that the second term in the RHS of (3.60) vanishes. In this way we have the *jittered sampling discrepancy*:

$$D_{d\lambda}^2(N) = N\left(|B| - \int_{SO(d)} \int_{\mathbb{T}^d} \left|\chi_{\tau^{-1}(B)} * \lambda\right|^2\, dt\, d\tau\right) \tag{3.63}$$

$$= N \sum_{0 \neq m \in \mathbb{Z}^d} \left(1 - \left|\hat{\lambda}(m)\right|^2\right) \int_{SO(d)} |\hat{\chi}_B(\tau(m))|^2\, d\tau.$$

Proof (of Theorem 43). By (3.63) we can select a point u_j from each one of the cubes

$$\left\{t(j) + \left[-\frac{1}{2M}, \frac{1}{2M}\right)^d\right\}_{j=1}^{N}$$

in such a way that

$$\int_{SO(d)} \int_{\mathbb{T}^d} \left|-N|B| + \sum_{j=1}^{N} \chi_{\tau^{-1}(B)-t}(u_j)\right|^2\, dt\, d\tau$$

$$\leq N\left(|B| - \int_{SO(d)} \int_{\mathbb{T}^d} \left(\chi_{\tau^{-1}(B)} * \lambda\right)^2(t)\, dt\, d\tau\right).$$

Since the support of the function $\lambda(t)$ has diameter $\sqrt{d}/M$ we have

$$\left(\chi_{\tau^{-1}(B)} * \lambda\right)^2 (t) = \chi_{\tau^{-1}(B)} (t)$$

for every t not belonging to the set

$$\left\{ x \in \mathbb{R}^d : \min_{y \in \partial(\tau^{-1}(B))} |x - y| \le \sqrt{d}/M \right\}.$$

By our assumptions this set has measure $\le c_d M^{-1}$, uniformly in τ, so that

$$N \left(|B| - \int_{SO(d)} \int_{\mathbb{T}^d} \left(\chi_{\tau^{-1}(B)} * \lambda\right)^2 (t) \, dt \, d\tau \right) \le c_d N M^{-1} = c_d N^{1-1/d}.$$

$$(3.64)$$

$\square$

Remark 44. When B is a ball of radius r and $N = M^d$ the inequality (3.64) can be reversed (see [22]), and there exist positive constants c_1 and c_2, depending at most on d and on r, such that

$$c_1 N^{1-1/d} \le N \left(|B| - \int_{\mathbb{T}^d} (\chi_B * \lambda)^2 (t) \, dt \right) \le c_2 N^{1-1/d}.$$

Remark 45. In the case of the ball it is possible to show that the discrepancy described in Theorem 42 is larger than the one described in Theorem 43 for small d and it is smaller for large d (see [22]). For $d \equiv 1 \pmod 4$ the situation is more intricate because of the results in Theorem 24 (see [22] or [21]).

References

1. J.-G. Bak, D. McMichael, J. Vance, S. Wainger, Fourier transforms of surface area measure on convex surfaces in $\mathbb{R}^3$. Am. J. Math. **111**(4), 633–668 (1989). doi:10.2307/2374816
2. J. Beck, W.W.L. Chen, Note on irregularities of distribution. II. Proc. Lond. Math. Soc. III. Ser. **61**(2), 251–272 (1990). doi:10.1112/plms/s3-61.2.251
3. J. Beck, Irregularities of distribution. I. Acta Math. **159**, 1–49 (1987). doi:10.1007/BF02392553
4. J. Beck, W.W.L. Chen, *Irregularities of Distribution*. Cambridge Tracts in Mathematics vol. 89 (Cambridge University Press, Cambridge, 1987)
5. D.R. Bellhouse, Area estimation by point-counting techniques. Biometrics **37**, 303–312 (1981). doi:10.2307/2530419
6. L. Brandolini, S. Hofmann, A. Iosevich, Sharp rate of average decay of the Fourier transform of a bounded set. Geom. Funct. Anal. **13**(4), 671–680 (2003). doi:10.1007/s00039-003-0426-7

7. L. Brandolini, A. Iosevich, G. Travaglini, Planar convex bodies, Fourier transform, lattice points, and irregularities of distribution. Trans. Am. Math. Soc. **355**(9), 3513–3535 (2003). doi:10.1090/S0002-9947-03-03240-9

8. L. Brandolini, L. Colzani, A. Iosevich, A. Podkorytov, G. Travaglini, Geometry of the Gauss map and lattice points in convex domains. Mathematika **48**(1–2), 107–117 (2001). doi:10.1112/S0025579300014376

9. L. Brandolini, Estimates for Lebesgue constants in dimension two. Ann. Mat. Pura Appl. IV. Ser. **156**, 231–242 (1990). doi:10.1007/BF01766981

10. L. Brandolini, G. Travaglini, Pointwise convergence of Fejér type means. Tohoku Math. J. **49**(3), 323–336 (1997). doi:10.2748/tmj/1178225106

11. L. Brandolini, L. Colzani, G. Travaglini, Average decay of Fourier transforms and integer points in polyhedra. Ark. Mat. **35**(2), 253–275 (1997). doi:10.1007/BF02559969

12. L. Brandolini, A. Greenleaf, G. Travaglini, $L^p - L^{p'}$ estimates for overdetermined Radon transforms. Trans. Am. Math. Soc. **359**(6), 2559–2575 (2007). doi:10.1090/S0002-9947-07-03953-0

13. L. Brandolini, M. Rigoli, G. Travaglini, Average decay of Fourier transforms and geometry of convex sets. Rev. Mat. Iberoam. **14**(3), 519–560 (1998). doi:10.4171/RMI/244

14. L. Brandolini, W.W.L. Chen, G. Gigante, G. Travaglini, Discrepancy for randomized Riemann sums. Proc. Am. Math. Soc. **137**(10), 3187–3196 (2009). doi:10.1090/S0002-9939-09-09975-4

15. L. Brandolini, G. Gigante, S. Thangavelu, G. Travaglini, Convolution operators defined by singular measures on the motion group. Indiana Univ. Math. J. **59**(6), 1935–1946 (2010). doi:10.1512/iumj.2010.59.4100

16. J. Bruna, A. Nagel, S. Wainger, Convex hypersurfaces and Fourier transforms. Ann. Math. **127**(2), 333–365 (1988). doi:10.2307/2007057

17. J.W.S. Cassels, On the sums of powers of complex numbers. Acta Math. Acad. Sci. Hung. **7**, 283–289 (1956). doi:10.1007/BF02020524

18. W.W.L. Chen, *Lectures on Irregularities of Point Distribution* (2000). http://rutherglen.ics.mq.edu.au/wchen/researchfolder/iod00.pdf,2000

19. W.W.L. Chen, Fourier techniques in the theory of irregularities of point distribution, in *Fourier Analysis and Convexity*, ed. by L. Brandolini L. Colzani, A. Iosevich, G. Travaglini, Applied and Numerical Harmonic Analysis (Birkhäuser, Boston, MA, 2004), pp. 59–82

20. W.W.L. Chen, *Upper Bounds in Discrepancy Theory*, ed. by L. Plaskota, et al., MCQMC: Monte Carlo and Quasi-Monte Carlo Methods in Scientific Computing. Proceedings of the 9th International Conference, MCQMC 2010, Warsaw, Poland, August 15–20, 2010 (Springer, Berlin). Springer Proceedings in Mathematics and Statistics Notes, vol. 23 (2012). doi:10.1007/978-3-642-27440-4_2

21. W.W.L. Chen, M.M. Skriganov, *Upper Bounds in Classical Discrepancy Theory*, ed. by W.W.L. Chen, A. Srivastav, G. Travaglini, Panorama of Discrepancy Theory (Springer, New York, 2013)

22. W.W.L. Chen, G. Travaglini, Deterministic and probabilistic discrepancies. Ark. Mat. **47**(2), 273–293 (2009). doi:10.1007/s11512-008-0091-z

23. W.W.L. Chen, G. Travaglini, An L^1 estimate for half-space discrepancy. Acta Arith. **146**(3), 203–214 (2011). doi:10.4064/aa146-3-1

24. Y. Colin de Verdière, Nombre de points entiers dans une famille homothétique de domaines de $\mathbb{R}^n$. Invent. Math. **43**, 15–52 (1977)

25. L. Colzani, Approximation of Lebesgue integrals by Riemann sums and lattice points in domains with fractal boundary. Monatsh. Math. **123**(4), 299–308 (1997). doi:10.1007/BF01326765

26. H. Davenport, Note on irregularities of distribution. Mathematika Lond. **3**, 131–135 (1956). doi:10.1112/S0025579300001807

27. D. Gioev, *Moduli of Continuity and Average Decay of Fourier Transforms: Two-Sided Estimates.*, ed. by J. Baik, et al., Integrable systems and random matrices. In honor of Percy Deift. Conference on integrable systems, random matrices, and applications in honor of Percy

Deift's 60th birthday, New York, NY, USA, May 22–26, 2006 (American Mathematical Society (AMS), Providence, RI, 2008). Contemporary Mathematics 458, 377–391 (2008)

28. P.M. Gruber, C.G. Lekkerkerker, *Geometry of Numbers, 2nd edn*. North-Holland Mathematical Library, vol. 37 (North-Holland, Amsterdam, 1987), p. 732

29. G.H. Hardy, E.M. Wright, *An Introduction to the Theory of Numbers, 5th edn*. (Clarendon Press, Oxford, 1979)

30. C.S. Herz, Fourier transforms related to convex sets. Ann. Math. **75**, 81–92 (1962). doi:10.2307/1970421

31. F.J. Hickernell, H. Woźniakowski, The price of pessimism for multidimensional quadrature. J. Complexity **17**(4), 625–659 (2001). doi:10.1006/jcom.2001.0593

32. E. Hlawka, Über Integrale auf konvexen Körpern. I. Monatsh. Math. **54**, 1–36 (1950). doi:10.1007/BF01304101

33. E. Hlawka, Über Integrale auf konvexen Körpern. II. Monatsh. Math. **54**, 81–99 (1950). doi:10.1007/BF01304101

34. M.N. Huxley, *Area, Lattice Points, and Exponential Sums*. London Mathematical Society Monographs. New Series, vol. 13 (Clarendon Press, Oxford, 1996)

35. D.G. Kendall, On the number of lattice points inside a random oval. Q. J. Math., Oxford Series **19**, 1–26 (1948)

36. T. Kollig, A. Keller, Efficient Multidimensional Sampling. Comput. Graphic Forum **21**(3), 557–563 (2002)

37. M.N. Kolountzakis, T. Wolff, On the Steinhaus tiling problem. Mathematika **46**(2), 253–280 (1999). doi:10.1112/S0025579300007750

38. S.V. Konyagin, M.M. Skriganov, A.V. Sobolev, On a lattice point problem arising in the spectral analysis of periodic operators. Mathematika **50**(1–2), 87–98 (2003). doi:10.1112/S0025579300014819

39. E. Krätzel, *Lattice Points*. Mathematics and Its Applications: East European Series, vol. 33 (Kluwer Academic, Berlin; VEB Deutscher Verlag der Wissenschaften, Dordrecht, 1988)

40. J. Matoušek, *Geometric Discrepancy. An Illustrated Guide. Revised paperback reprint of the 1999 original*. Algorithms and Combinatorics, vol. 18 (Springer, Dordrecht, 2010). doi:10.1007/978-3-642-03942-3

41. D.P. Mitchell, Consequences of Stratified Sampling in Graphics, in *SIGGRAPH '96 Proceedings of the 23rd Annual Conference on Computer Graphics and Interactive Techniques* (Association for Computing Machinery, New York, 1996), pp. 277–280

42. H.L. Montgomery, *Ten Lectures on the Interface Between Analytic Number Theory and Harmonic Analysis*. Regional Conference Series in Mathematics, vol. 84 (American Mathematical Society, Providence, RI, 1994)

43. L. Parnovski, N. Sidorova, Critical dimensions for counting lattice points in Euclidean annuli. Math. Model. Nat. Phenom. **5**(4), 293–316 (2010). doi:10.1051/mmnp/20105413

44. L. Parnovski, A.V. Sobolev, On the Bethe-Sommerfeld conjecture for the polyharmonic operator. Duke Math. J. **107**(2), 209–238 (2001). doi:10.1215/S0012-7094-01-10721-7

45. A.N. Podkorytov, On the asymptotic behaviour of the Fourier transform on a convex curve. Vestn. Leningr. Univ. Math. **24**(2), 57–65 (1991)

46. O. Ramaré, On long κ-tuples with few prime factors. Proc. Lond. Math. Soc. **104**(1), 158–196 (2012). doi:10.1112/plms/pdr026

47. B. Randol, A lattice-point problem. II. Trans. Am. Math. Soc. **125**, 101–113 (1966). doi:10.2307/1994590

48. B. Randol, On the Fourier transform of the indicator function of a planar set. Trans. Am. Math. Soc. **139**, 271–278 (1969). doi:10.2307/1995319

49. F. Ricci, G. Travaglini, Convex curves, Radon transforms and convolution operators defined by singular measures. Proc. Am. Math. Soc. **129**(6), 1739–1744 (2001). doi:10.1090/S0002-9939-00-05751-8

50. K.F. Roth, On irregularities of distribution. Mathematika **1**, 73–79 (1954). doi:10.1112/S0025579300000541

51. W.M. Schmidt, Irregularities of distribution. IV. Invent. Math. **7**, 55–82 (1969). doi:10.1007/BF01418774
52. W.M. Schmidt, *Lectures on Irregularities of Distribution. (Notes by T. N. Shorey)*. Tata Institute of Fundamental Research, Lectures on Mathematics and Physics: Mathematics (Tata Institute of Fundamental Research, Bombay, 1977)
53. R. Schneider, *Convex Bodies: The Brunn-Minkowski Theory*. Encyclopedia of Mathematics and Its Applications, vol. 44 (Cambridge University Press, Cambridge, 1993)
54. M.M. Skriganov, Ergodic theory on $SL(n)$, diophantine approximations and anomalies in the lattice point problem. Invent. Math. **132**(1), 1–72 (1998). doi:10.1007/s002220050217
55. C.D. Sogge, *Fourier Integrals in Classical Analysis*. Cambridge Tracts in Mathematics, vol. 105 (Cambridge University Press, Cambridge, 1993)
56. E.M. Stein, *Harmonic Analysis: Real-Variable Methods, Orthogonality, and Oscillatory Integrals. With the Assistance of Timothy S. Murphy*. Princeton Mathematical Series, vol. 43 (Princeton University Press, Princeton, NJ, 1993)
57. E.M. Stein, G. Weiss, *Introduction to Fourier Analysis on Euclidean Spaces*. Princeton Mathematical Series (Princeton University Press, Princeton, NJ, 1971)
58. E.M. Stein, N.J. Weiss, On the convergence of Poisson integrals. Trans. Am. Math. Soc. **140**, 35–54 (1969). doi:10.2307/1995121
59. E.M. Stein, M.H. Taibleson, G. Weiss, Weak type estimates for maximal operators on certain H^p classes. Rend. Circ. Mat. Palermo **1**, 81–97 (1981)
60. M. Tarnopolska-Weiss, On the number of lattice points in a compact n-dimensional polyhedron. Proc. Am. Math. Soc. **74**, 124–127 (1979). doi:10.2307/2042117
61. G. Travaglini, *Number theory, Fourier analysis and Geometric discrepancy*. London Mathematical Society Student Texts, vol. 81 (Cambridge University Press, Cambridge, 2014)
62. G. Travaglini, Average decay of the Fourier transform, in *Fourier Analysis and Convexity*, ed. by L. Brandolini, L. Colzani, A. Iosevich, G. Travaglini Applied and Numerical Harmonic Analysis (Birkhäuser, Boston, MA, 2004), pp. 245–268
63. G.N. Watson, *A Treatise on the Theory of Bessel Functions* (Cambridge University Press, Cambridge, 1922)
64. R.L. Wheeden, A. Zygmund, *Measure and Integral. An Introduction to Real Analysis*. Monographs and Textbooks in Pure and Applied Mathematics, vol. 43 (Marcel Dekker, New York/Basel, 1977)

Chapter 4
Superirregularity

József Beck

Abstract Finding the integer solutions of a Pell equation is equivalent to finding the integer lattice points in a long and narrow tilted hyperbolic region, where the slope is a quadratic irrational. Motivated by this relationship, we carry out here a systematic study of point counting with respect to translated or congruent families of any given long and narrow hyperbolic region. First we discuss the important special case when the underlying point set is the set of integer lattice points in the plane and the slope of the given hyperbolic region is arbitrary but fixed; see Theorems 3–21. Then we switch to the general case of an arbitrary point set of density one in the plane, and study point counting with respect to congruent copies of a given hyperbolic region; see Theorem 30. The main results are about the extra large discrepancy that we call *superirregularity*. This means that there is always a translated/congruent copy of any given long and narrow hyperbolic region of large area, for which the actual number of points in the copy differs from the area *as much as possible*, i.e. the discrepancy is at least a constant multiple of the area. Our theorems demonstrate, in a quantitative sense, that in point counting with respect to translated/congruent copies of any long and narrow hyperbolic region, superirregularity is inevitable.

4.1   Introduction

Notation. For any real valued function f and positive function g, we write $f = O(g)$ to indicate that there exists a positive constant c such that $|f| < cg$, and also write $f = o(g)$ to indicate that $f/g \to 0$. We write $\|z\|$ to denote the distance of a real number z to the nearest integer. Furthermore, $c_0, c_1, c_2, \ldots$ denote positive constants which may depend on some of the parameters that arise from our discussion.

J. Beck (✉)
Department of Mathematics, Rutgers University, New Brunswick, NJ 08903, USA
e-mail: jbeck@math.rutgers.edu

W. Chen et al. (eds.), *A Panorama of Discrepancy Theory*, Lecture Notes in Mathematics 2107, DOI 10.1007/978-3-319-04696-9_4,
© Springer International Publishing Switzerland 2014

4.1.1 Pell's Equation: Bounded Fluctuations

Our starting point is the well-known Pell's equation, a standard part of any introductory course on number theory. The theory of Pell's equation, while mostly elementary, is nevertheless one of the most beautiful chapters in the whole of mathematics. Also, it is very important, since the concept of *units* plays a key role in algebraic number theory.

We briefly recall the main results. Consider, for simplicity, the concrete equation $x^2 - 2y^2 = \pm 1$. This equation has infinitely many integral solutions; in fact, the set of all integral solutions $(x_k, y_k) \in \mathbf{Z}^2$ forms a cyclic group generated by the least positive solution. More precisely, we have

$$x_k + y_k \sqrt{2} = \pm(1 + \sqrt{2})^k, \quad k \in \mathbf{Z}.$$

All integral solutions of $x^2 - 2y^2 = 1$ are given by $x_k + y_k \sqrt{2} = \pm(1 + \sqrt{2})^{2k}$, while all integral solutions of $x^2 - 2y^2 = -1$ are given by $x_k + y_k \sqrt{2} = \pm(1 + \sqrt{2})^{2k+1}$. In particular, all positive integer solutions of $x^2 - 2y^2 = 1$ are given by

$$x_k + y_k \sqrt{2} = (1 + \sqrt{2})^{2k} = (3 + 2\sqrt{2})^k, \quad k = 1, 2, 3, \ldots.$$

Taking the algebraic conjugate $x_k - y_k \sqrt{2} = (3 - 2\sqrt{2})^k$, and combining these two equations, we obtain the explicit formulas

$$x_k = \frac{(3 + 2\sqrt{2})^k + (3 - 2\sqrt{2})^k}{2} \quad \text{and} \quad y_k = \frac{(3 + 2\sqrt{2})^k - (3 - 2\sqrt{2})^k}{2\sqrt{2}}.$$

Since $0 < 3 - 2\sqrt{2} < \frac{1}{5}$, we have

$$x_k = \text{the nearest integer to } \frac{1}{2}(3 + 2\sqrt{2})^k$$

and

$$y_k = \text{the nearest integer to } \frac{1}{2\sqrt{2}}(3 + 2\sqrt{2})^k.$$

If k is large, the error is very small. For example, the 10-th solution of $x^2 - 2y^2 = 1$ in positive integers is the pair $x_{10} = 22{,}619{,}537$ and $y_{10} = 15{,}994{,}428$. Here we find

$$\frac{1}{2}(3 + 2\sqrt{2})^{10} = 22619536.99999998895\ldots$$

and

$$\frac{1}{2\sqrt{2}}(3 + 2\sqrt{2})^{10} = 15994428.000000007815\ldots.$$

Let $F(N) = F(\sqrt{2}; 1; N)$ denote the number of positive integer solutions of the Pell equation $x^2 - 2y^2 = 1$ up to N, in the sense[1] that $x \geq 1$ and $1 \leq y \leq N$. We have

$$k \leq F(N) \quad \text{if and only if} \quad \frac{(3 + 2\sqrt{2})^k - (3 - 2\sqrt{2})^k}{2\sqrt{2}} \leq N,$$

which implies the asymptotic formula

$$F(N) = F(\sqrt{2}; 1; N) = \frac{\log N}{\log(3 + 2\sqrt{2})} + O(1). \tag{4.1}$$

The formula (4.1) says that the counting function $F(N) = F(\sqrt{2}; 1; N)$ has an extremely predictable, almost deterministic behavior: it is $c_2 \log N$ plus some bounded error term.

Note that (4.1) has some far-reaching generalizations. Let $[\gamma_1, \gamma_2]$ be an arbitrary interval, and let $F(\sqrt{2}; [\gamma_1, \gamma_2]; N)$ denote the number of positive integer solutions of the Pell inequality $\gamma_1 \leq x^2 - 2y^2 \leq \gamma_2$, with $x \geq 1$ and $1 \leq y \leq N$. By using the theory of indefinite binary quadratic forms, it is easy to prove the following analog of (4.1). We have

$$F(\sqrt{2}; [\gamma_1, \gamma_2]; N) = c_0 \log N + O(1), \tag{4.2}$$

where the constant factor $c_0 = c_0(\sqrt{2}; \gamma_1, \gamma_2)$ is independent of N.

Furthermore, we can switch from $\sqrt{2}$ to any other *quadratic irrational* α. This means that α is a root of a quadratic equation $Ax^2 + Bx + C = 0$ with integral coefficients such that the discriminant $B^2 - 4AC \geq 2$ is not a complete square. An equivalent definition is that $\alpha = (a + \sqrt{d})/b$ for some integers a, b, d such that $b \neq 0$ and $d \geq 2$ is not a complete square. Note that the quadratic irrationals are characterized by their continued fractions. The continued fractions of α is finally periodic if and only if α is a quadratic irrational. For example,

$$\frac{24 - \sqrt{15}}{17} = 1 + \frac{1}{5+} \frac{1}{2+} \frac{1}{3+} \frac{1}{2+} \frac{1}{3+} \cdots = [1; 5, 2, 3, 2, 3, 2, 3, \ldots] = [1; 5, \overline{2, 3}].$$

Let us go back to (4.2) and to the special case $\alpha = \sqrt{2}$. If $-2 < \gamma_1 \leq -1$ and $1 \leq \gamma_2 < 2$, then

$$c_0(\sqrt{2}; \gamma_1, \gamma_2) = \frac{1}{\log(1 + \sqrt{2})} = \frac{2}{\log(3 + 2\sqrt{2})}. \tag{4.3}$$

---

[1]For simplicity of notation, it is more convenient to restrict the second variable y.

If $-1 < \gamma_1 \leq 1 \leq \gamma_2 < 2$, then

$$c_0(\sqrt{2}; \gamma_1, \gamma_2) = \frac{1}{\log(3 + 2\sqrt{2})}. \tag{4.4}$$

Finally, if $-1 < \gamma_1 \leq \gamma_2 < 1$, then of course

$$c_0(\sqrt{2}; \gamma_1, \gamma_2) = 0. \tag{4.5}$$

4.1.2 The Naive Area Principle

It is very interesting to compare these well-known asymptotic results about the number of solutions of the Pell equation/inequality to what we like to call the *Naive Area Principle*, a natural guiding intuition in lattice point theory. It goes roughly as follows. If a nice region has a large area, then it should contain a large number of lattice points, and the number of lattice points is close to the area.

Of course, the heart of the matter is how we define a nice region precisely. Consider, for example, the infinite open horizontal strip of height one, given by $0 < y < 1$, $-\infty < x < \infty$. It has infinite area, but it does not contain any lattice point. The reader is likely to agree that the infinite strip is a nice region, so the Naive Area Principle is clearly violated here.

A less trivial example comes from the Pell inequality

$$-\frac{1}{2} \leq x^2 - 2y^2 \leq \frac{1}{2}. \tag{4.6}$$

This is a hyperbolic region of infinite area, and contains no lattice point except the origin. The reader is again likely to agree that the hyperbolic region (4.6) is also nice, so this is again a violation of the Naive Area Principle.

Next we switch from (4.6) to the general Pell inequality

$$\gamma_1 \leq x^2 - 2y^2 \leq \gamma_2, \tag{4.7}$$

where $-\infty < \gamma_1 < \gamma_2 < \infty$ are arbitrary real numbers. Of course, the hyperbolic region (4.7) has infinite area. What we want to compute is the area of a finite segment. Consider the finite region

$$H(\sqrt{2}; [\gamma_1, \gamma_2]; N) = \{(x, y) \in \mathbf{R}^2 : \gamma_1 \leq x^2 - 2y^2 \leq \gamma_2, \ x \geq 1, \ 1 \leq y \leq N\}. \tag{4.8}$$

If N is very large compared to the pair of constants γ_1 and γ_2, then the finite region $H(\sqrt{2}; [\gamma_1, \gamma_2]; N)$ looks like a *hyperbolic needle*. It is easy to give a good estimate for the area of this hyperbolic needle. We have

$$\text{area}(H(\sqrt{2}; [\gamma_1, \gamma_2]; N)) = \frac{\gamma_2 - \gamma_1}{2\sqrt{2}} \log N + O(1), \tag{4.9}$$

where the implicit constant in the term $O(1)$ is independent of N, but may depend on γ_1 and γ_2.

The proof of (4.9) is based on the familiar factorization

$$x^2 - 2y^2 = (x + y\sqrt{2})(x - y\sqrt{2}), \tag{4.10}$$

and on the computation of the Jacobian of the corresponding substitution; this explains the factor $2\sqrt{2}$ in the denominator in (4.9). The details are easy, and go as follows. In view of the factorization (4.10), it is more convenient to compute the area of the following slight variant of the region (4.9). Let

$$H^*(\sqrt{2}; [\gamma_1, \gamma_2]; N)$$
$$= \{(x, y) \in \mathbf{R}^2 : \gamma_1 \le x^2 - 2y^2 \le \gamma_2, \ 1 \le x + y\sqrt{2} \le 2\sqrt{2}N\}. \tag{4.11}$$

Consider the substitution

$$u_1 = x + y\sqrt{2}, \quad u_2 = x - y\sqrt{2}, \tag{4.12}$$

which is equivalent to

$$x = \frac{u_1 + u_2}{2}, \quad y = \frac{u_1 - u_2}{2\sqrt{2}}.$$

The corresponding determinant is

$$\frac{\partial(u, v)}{\partial(x, y)} = \begin{vmatrix} 1 & -\sqrt{2} \\ 1 & \sqrt{2} \end{vmatrix} = 2\sqrt{2}.$$

Applying the substitution (4.12), we have

$$\text{area}(H^*(\sqrt{2}; [\gamma_1, \gamma_2]; N)) = \frac{1}{2\sqrt{2}} \int_1^{2\sqrt{2}N} \left(\int_{\gamma_1/u_1}^{\gamma_2/u_1} du_2 \right) du_1$$

$$= \frac{1}{2\sqrt{2}} \int_1^{2\sqrt{2}N} \frac{\gamma_2 - \gamma_1}{u_1} du_1 = \frac{\gamma_2 - \gamma_1}{2\sqrt{2}} \log N + O(1). \tag{4.13}$$

Simple geometric consideration shows that

$$\text{area}(H(\sqrt{2}; [\gamma_1, \gamma_2]; N)) = \text{area}(H^*(\sqrt{2}; [\gamma_1, \gamma_2]; N)) + O(1),$$

and so (4.13) implies (4.9).

Now let us return to the Naive Area Principle. Comparing (4.2), (4.8) and (4.9), it is reasonable to expect, in view of the Naive Area Principle, that the counting function $F(\sqrt{2}; [\gamma_1, \gamma_2]; N)$ is close to the area of the hyperbolic needle $H(\sqrt{2}; [\gamma_1, \gamma_2]; N)$. In other words, it is reasonable to expect that

$$c_0(\sqrt{2}; \gamma_1, \gamma_2) = \frac{\gamma_2 - \gamma_1}{2\sqrt{2}}. \tag{4.14}$$

Unfortunately, the Naive Area Principle is almost always violated in the quantitative sense that (4.14) fails for the overwhelming majority of the choices $-\infty < \gamma_1 < \gamma_2 < \infty$. In fact, the two sides of (4.14) have completely different behavior. The left-hand side of has discrete jumps and the right-hand side is a continuous function of γ_1 and γ_2. For example, as γ_1 and γ_2 run in the interval $-2 < \gamma_1 < \gamma_2 < 2$, the constant factor $c_0(\sqrt{2}; \gamma_1, \gamma_2)$ has only 3 possible values, namely

$$0, \quad \frac{1}{\log(3 + 2\sqrt{2})}, \quad \frac{2}{\log(3 + 2\sqrt{2})};$$

see (4.3)–(4.5). This shows, in a quantitative way, how the general Pell inequality (4.7) violates the Naive Area Principle.

4.1.3   The Giant Leap in the Inhomogeneous Case: Extra Large Fluctuations

Using the familiar factorization (4.10), we can rewrite the Pell equation $x^2 - 2y^2 = \pm 1$, restricted to positive integers, as

$$|x^2 - 2y^2| \leq 1 \quad \text{or} \quad |y\sqrt{2} - x|(y\sqrt{2} + x) \leq 1 \quad \text{or} \quad \|y\sqrt{2}\|(y\sqrt{2} + x) \leq 1, \tag{4.15}$$

where $\|z\|$ denotes, as usual, the distance of a real number z from the nearest integer. Notice that in (4.15), x is the nearest integer to $y\sqrt{2}$, which is an irrational number. Since $y\sqrt{2} \approx x$, the inequality (4.15) is basically equivalent to the vague inequality

$$\|y\sqrt{2}\| \leq \frac{1 + o(1)}{2\sqrt{2}y}. \tag{4.16}$$

The vagueness of (4.16) comes from the additional term $o(1)$, which tends to 0 as $y \to \infty$. The formula (4.16) is ambiguous, but surely every mathematician understands what we are talking about here.

An expert in number theory would classify (4.16) as a typical problem in diophantine approximation. Next we give a nutshell summary of diophantine approximation.

The classical problem in the theory of diophantine approximation is to find good rational approximations of irrational numbers. More precisely, we want to decide whether an inequality

$$\|n\alpha\| < \frac{1}{n\varphi(n)} \quad \text{or} \quad \left|\alpha - \frac{m}{n}\right| < \frac{1}{n^2\varphi(n)}, \tag{4.17}$$

or in general,

$$\|n\alpha - \beta\| < \frac{1}{n\varphi(n)}, \tag{4.18}$$

where α is a given irrational number and β is a given real number, has infinitely many integral solutions in n, and if this is the case, to determine the solutions, or at least the asymptotic number of integral solutions. Here $\varphi(n)$ is a positive increasing function of n.

The diophantine inequality (4.17) is said to be homogeneous, whereas the diophantine inequality (4.18) is said to be inhomogeneous. For example, in the homogeneous case, the best possible result is Hurwitz's well-known theorem, that for any irrational number α, the inequality

$$\|n\alpha\| < \frac{1}{\sqrt{5}n}$$

has infinitely many positive integer solutions.

In the inhomogeneous case, we can mention an old result of Kronecker, that for any irrational number α and any real number β, the inequality

$$\|n\alpha - \beta\| < \frac{3}{n}$$

has infinitely many positive integer solutions. Perhaps the strongest inhomogeneous result is Minkowski's theorem, that for any irrational number α, the inequality

$$\|n\alpha - \beta\| < \frac{1}{4n}$$

has infinitely many integer but not necessarily positive solutions, unless $0 < \beta < 1$ is an integer multiple of α modulo one.

The homogeneous case (4.17) has a complete theory based on the effectiveness of the tool of continued fractions. These are classical results due mostly to Euler and Lagrange. Unfortunately, we know much less about the inhomogeneous case. Very recently, the author proved some new results in this direction, and basically covered the case when α is an arbitrary quadratic irrational and β is a typical real number. These results form a large part of the forthcoming book [2]; see also the recent papers [8, 9].

Before formulating our main results, we want to first elaborate on the connection between homogeneous/inhomogeneous diophantine inequalities, such as (4.17) and (4.18), and homogeneous/inhomogeneous Pell inequalities.

4.1.3.1 Homogeneous and Inhomogeneous Pell Inequalities

The general form of a quadratic curve on the plane is

$$a_{11}x^2 + a_{12}xy + a_{22}y^2 + a_{13}x + a_{23}y + a_{33} = 0. \tag{4.19}$$

We are interested in the integral solutions $(x, y) \in \mathbf{Z}^2$ of an arbitrary inequality

$$\gamma_1 \leq a_{11}x^2 + a_{12}xy + a_{22}y^2 + a_{13}x + a_{23}y \leq \gamma_2, \tag{4.20}$$

where $\gamma_1 < \gamma_2$ are given real numbers. Note that the inequality (4.20) defines a plane region, and the boundary consists of two curves of the type (4.19). In the case of negative discriminant $D = a_{12}^2 - 4a_{11}a_{22} < 0$, the inequality (4.20) defines a bounded region where the boundary curves are two ellipses. The case of positive discriminant $D = a_{12}^2 - 4a_{11}a_{22} > 0$ is much more interesting, because then the inequality (4.20) defines an unbounded region, where the boundary curves are two hyperbolas, and thus we have a chance for infinitely many integral solutions of (4.20).

For simplicity, assume that the coefficients a_{11}, a_{12}, a_{22} in (4.20) are integers and $D = a_{12}^2 - 4a_{11}a_{22} > 0$. We can factorize the quadratic part in the form

$$a_{11}x^2 + a_{12}xy + a_{22}y^2 = a_{11}(x - \alpha y)(x - \alpha' y), \tag{4.21}$$

where

$$\alpha = \frac{-a_{12} + \sqrt{D}}{2a_{11}} \quad \text{and} \quad \alpha' = \frac{-a_{12} - \sqrt{D}}{2a_{11}}. \tag{4.22}$$

Using (4.21), we can rewrite (4.20) in the form

$$\gamma_1 \leq (x - \alpha y + \rho_1)(x - \alpha' y + \rho_2) \leq \gamma_2, \tag{4.23}$$

where

$$\rho_1 + \rho_2 = \frac{a_{13}}{a_{11}} \quad \text{and} \quad \alpha' \rho_1 + \alpha \rho_2 = -\frac{a_{23}}{a_{11}}.$$

Note that γ_1, γ_2 are generic numbers; the pair γ_1, γ_2 in (4.20) is not necessarily the same as the pair γ_1, γ_2 in (4.23).

Without loss of generality we can assume[2] that $|a_{12}| \leq a_{11} \leq \sqrt{D/3}$, and then we have $\alpha > 0 > \alpha'$.

For simplicity, assume that the interval $[\gamma_1, \gamma_2]$ is symmetric with respect to 0, so that it is of the form $[\gamma_1, \gamma_2] = [-\gamma, \gamma]$. Assume also that we are interested in the positive integral solutions of (4.23). Since $\alpha > 0 > \alpha'$, for large positive x and y, the second factor $(x - \alpha' y + \rho_2)$ in (4.23) is also large and positive, implying that the first factor $(x - \alpha y + \rho_1)$ in (4.23) has to be very small. In other words, x has to be the nearest integer to $(\alpha y - \rho_1)$. It follows that the symmetric version of (4.20), namely

$$-\gamma \leq a_{11}x^2 + a_{12}xy + a_{22}y^2 + a_{13}x + a_{23}y \leq \gamma,$$

where $\gamma > 0$ is a given real number, is equivalent to the diophantine inequality

$$\|y\alpha - \rho_1\| < \frac{c}{y + O(1)}, \quad \text{where} \quad c = \frac{\gamma}{\alpha - \alpha'} = \frac{\gamma a_{11}}{\sqrt{D}}. \tag{4.24}$$

Let us return to the inequality (4.20). If the linear part $a_{13}x + a_{23}y$ in the middle is missing, i.e. $a_{13} = a_{23} = 0$, then we have a complete theory based on Pell's equation. More precisely, write $Q(x, y) = a_{11}x^2 + a_{12}xy + a_{22}y^2$. Then $\gamma_1 \leq Q(x, y) \leq \gamma_2$ if and only if

$$Q(x, y) = m \quad \text{for some } m \in \mathbb{Z} \text{ satisfying } \gamma_1 \leq m \leq \gamma_2.$$

We have a complete characterization of the integral solutions of $Q(x, y) = m$ for any integer m as follows. For any integer m, there is a finite list of primary solutions, say, (x_j, y_j), $j \in J$, where $|J| < \infty$, such that every solution $x = u, y = v$ of $Q(x, y) = m$ can be written in the form

$$u - \alpha v = \pm \left(\frac{u_0 + v_0 \sqrt{D}}{2} \right)^n (x_j - \alpha y_j)$$

for some $j \in J$ and $n \in \mathbb{Z}$, where $x = u_0 > 0$, $y = v_0 > 0$ is the least positive solution of Pell's equation $x^2 - Dy^2 = 4$. As a byproduct, we deduce[3] that the number of positive integral solutions of the inequality

$$\gamma_1 \leq Q(x, y) \leq \gamma_2, \quad 1 \leq x \leq N, \ 1 \leq y \leq N$$

has the simple asymptotic form $c \log N + O(1)$, where $c = c(a_{11}, a_{12}, a_{22}, \gamma_1, \gamma_2)$ is a constant and the error term $O(1)$ is uniformly bounded as $N \to \infty$.

---

[2]This is a well-known fact from the reduction theory of binary quadratic forms. We omit the proof; see, for example, [31].

[3]For a more detailed proof; see [23].

Exactly the same holds if there is a non-zero linear part $a_{13}x + a_{23}y$ in (4.20), but its effect cancels out. Note that ρ_1 in (4.23) is an integer.

Finally, if ρ_1 is not an integer, then we say that (4.23) is an inhomogeneous Pell inequality. In view of (4.24), an inhomogeneous Pell inequality (4.23) is basically equivalent to an inhomogeneous diophantine inequality

$$\|n\alpha - \beta\| < \frac{c}{n} \tag{4.25}$$

with $c = \gamma a_{11}/\sqrt{D}$, where α is a quadratic irrational defined in (4.22). The inequality (4.25) is a special case of (4.18) where $\varphi(n)$ is a constant.

4.1.3.2 Some Results

One of the main results in the forthcoming book [2] describes the asymptotic behavior of the number of positive integral solutions of (4.20) for every non-square integer discriminant $D > 0$ and almost all a_{13}, a_{23}. The number of solutions

- exhibits extra large fluctuations, proportional to the area,
- satisfies an elegant Central Limit Theorem, and
- satisfies a shockingly precise Law of the Iterated Logarithm; see Theorems 3, A and B below.

For notational simplicity, we formulate the results in the special case of discriminant $D = 8$, which corresponds to the most famous quadratic irrational $\alpha = \sqrt{2}$.

Since the class number of the discriminant $D = 8$ is one, the general form of an inhomogeneous Pell inequality of discriminant $D = 8$ is

$$\gamma_1 \le (x + \beta_1)^2 - 2(y + \beta_2)^2 \le \gamma_2, \tag{4.26}$$

where $\gamma_1 < \gamma_2$ and $\beta_1, \beta_2 \in [0, 1)$ are fixed constants. For notational simplicity, we restrict ourselves to symmetric intervals $[-\gamma, \gamma]$ in (4.26); note that everything works similarly for general intervals $[\gamma_1, \gamma_2]$.

The factorization

$$(x + \beta_1)^2 - 2(y + \beta_2)^2 = (x + \beta - y\sqrt{2})(x + \beta' + y\sqrt{2}), \tag{4.27}$$

where $\beta = \beta_1 - \beta_2\sqrt{2}$ and $\beta' = \beta_1 + \beta_2\sqrt{2}$, clearly indicates that the asymptotic number of integral solutions of (4.26) depends heavily on the local behavior of $n\sqrt{2}$ mod 1. In fact, (4.26) is essentially equivalent to the inhomogeneous diophantine inequality

$$\|n\sqrt{2} - \beta\| < \frac{c}{n}, \tag{4.28}$$

with $c = \gamma/2\sqrt{2}$.

To turn the vague term *essentially equivalent* into a precise statement, we proceed as follows. Let $F(\sqrt{2}; \beta_1, \beta_2; \gamma; N)$ be the number of integral solutions $(x, y) \in \mathbf{Z}^2$ of the inequality (4.26) with $\gamma_2 = \gamma$ and $\gamma_1 = -\gamma$ satisfying $1 \leq y \leq N$ and $x \geq 1$. It means counting lattice points in a long and narrow hyperbola segment. Next let $f(\sqrt{2}; \beta; c; N)$ denote the number of integral solutions n of the inequality (4.28) satisfying $1 \leq n \leq N$, where $\beta = \beta_1 - \beta_2\sqrt{2}$. Now *essentially equivalent* means that for almost all pairs β_1, β_2, we have $F(\sqrt{2}; \beta_1, \beta_2; \gamma; N) - f(\sqrt{2}; \beta; c; N) = O(1)$ as $N \to \infty$, where $c = \gamma/2\sqrt{2}$. More precisely, we have

Lemma 1. *Let $\gamma > 0$ and β_2 be arbitrary real numbers. Then for almost all β_1, there exists a finite $0 < C(\beta_1, \beta_2, \gamma) < \infty$ such that*

$$\int_0^1 C(\beta_1, \beta_2, \gamma) \, d\beta < \infty$$

and

$$|F(\sqrt{2}; \beta_1, \beta_2; \gamma; N) - f(\sqrt{2}; \beta; c; N)| < C(\beta_1, \beta_2, \gamma)$$

for all $N \geq 1$, where $c = \gamma/2\sqrt{2}$ and $\beta = \beta_1 - \beta_2\sqrt{2}$.

We postpone the simple proof to Sect. 4.3.

In view of Lemma 1, it suffices to study the special case $\beta_2 = 0$ and $\beta_1 = \beta$. We have

$$-\gamma \leq (x + \beta)^2 - 2y^2 \leq \gamma, \tag{4.29}$$

where $\gamma > 0$ and $\beta \in [0, 1)$ are fixed constants. For simplicity, let $F(\sqrt{2}; \beta; \gamma; N)$ denote the number of integral solutions $(x, y) \in \mathbf{Z}^2$ of (4.29) satisfying $1 \leq y \leq N$ and $x \geq 1$. Note that $F(\sqrt{2}; \beta; \gamma; N)$ counts the number of lattice points in a long and narrow hyperbola segment, or hyperbolic needle, located along a line[4] of slope $1/\sqrt{2}$; see Fig. 4.1.

In the special case $\gamma = 1$ and $\beta = 0$, the inequality (4.29) becomes the simplest Pell equation $x^2 - 2y^2 = \pm 1$. The integral solutions (x_k, y_k) form a cyclic group generated by the smallest positive solution $x = y = 1$ in the well-known way. We have $x_k + y_k\sqrt{2} = (1 + \sqrt{2})^k$, implying the familiar asymptotic formula

$$F(\sqrt{2}; \beta = 0; \gamma = 1; N) = \frac{\log N}{\log(1 + \sqrt{2})} + O(1), \tag{4.30}$$

where $1 + \sqrt{2}$ is the fundamental unit of the real quadratic field $\mathbf{Q}(\sqrt{2})$.

In sharp contrast to the bounded fluctuation in the homogeneous case $\beta = 0$, the inhomogeneous case can exhibit extra large fluctuations proportional to the area;

---

[4]If $\beta = 0$, then the line is $y = x/\sqrt{2}$.

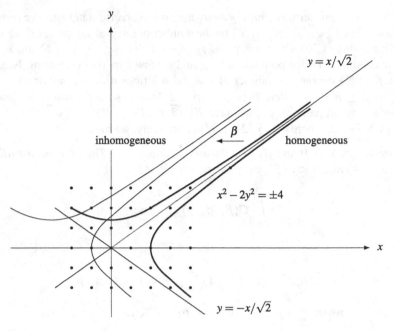

Fig. 4.1 A hyperbolic needle

see Theorem 3 below. To explain this, first we have to compute the mean value of
$F(\sqrt{2}; \beta; \gamma; N)$ as β runs through the unit interval $0 \le \beta < 1$.

Lemma 2. *We have*

$$\int_0^1 F(\sqrt{2}; \beta; \gamma; N)\, d\beta = \frac{\gamma}{\sqrt{2}} \log N + O(1), \qquad (4.31)$$

*where the implicit constant in the term $O(1)$ is independent of N, but may depend
on γ. Moreover, for an arbitrary subinterval $0 \le a < b \le 1$, we have*

$$\lim_{N \to \infty} \frac{\frac{1}{b-a} \int_a^b F(\sqrt{2}; \beta; \gamma; N)\, d\beta}{\log N} = \frac{\gamma}{\sqrt{2}}. \qquad (4.32)$$

The estimates (4.31) and (4.32) express the almost trivial geometric fact that the
average number of lattice points contained in all the translated copies of a given
region, a hyperbola segment in our special case, is precisely the area of the region;
see Lemma 5. We shall give a detailed proof of Lemma 2 in Sect. 4.3.

Now we are ready to formulate our first, and weakest, extra large fluctuation
result, demonstrating that the fluctuations can be proportional to the area. This result
is hardly more than a warmup for, or simplest illustration of, the main results that
will come later.

Theorem 3. *For $\gamma = \frac{1}{2}$, there are continuum many divergence points $\beta^* \in [0, 1)$ in the sense that*

$$\limsup_{n \to \infty} \frac{F(\sqrt{2}; \beta^*; \gamma = 1/2; n)}{\log n} > \liminf_{n \to \infty} \frac{F(\sqrt{2}; \beta^*; \gamma = 1/2; n)}{\log n}. \qquad (4.33)$$

Note that the fluctuation $c_3 \log n$ in $F(\sqrt{2}; \beta^*; \gamma = 1/2; n)$ is as large as possible, apart from a constant factor. This follows from Lemma 4 in the next section. It is fair to say that Theorem 3 represents a sophisticated violation of the Naive Area Principle.

We postpone the proof of Theorem 3 to Sect. 4.3.

Note that Theorem 3 has a far-reaching generalization. It holds for every $\gamma > 0$, and we actually have the stronger inequality

$$\limsup_{n \to \infty} \frac{F(\sqrt{2}; \beta^*; \gamma; n)}{\log n} > \frac{\gamma}{\sqrt{2}} > \liminf_{n \to \infty} \frac{F(\sqrt{2}; \beta^*; \gamma; n)}{\log n}. \qquad (4.34)$$

We shall return to this in Sect. 4.4; see Theorem 12.

Another far-reaching generalization of Theorem 3 will be discussed in Sect. 4.9; see Theorem 21.

Finally, an extra large fluctuation type result for arbitrary point sets, instead of the set $\mathbf{Z}^2$ of lattice points, will be discussed in Sect. 4.10; see Theorem 30.

We refer to these extra large fluctuation type results as *superirregularity*.

4.2 Defending the Naive Area Principle

The estimate (4.30) and inequality (4.33) display the two extreme cases: (1) the negligible bounded fluctuations around the main value which is a constant multiple of $\log N$; and (2) the extra large fluctuations proportional to the area. But what kind of fluctuations do we have for a typical β satisfying $0 < \beta < 1$? We show that for a typical β, the asymptotic number of solutions $F(\sqrt{2}; \beta; \gamma; N)$, as $N \to \infty$, justifies the Naive Area Principle. And beyond that, a more thorough look reveals randomness.

Talking about randomness, note that the two most important parameters of a random variable are the expectation, or mean value, and the variance. For the function $F(\sqrt{2}; \beta; \gamma; N)$, the estimate (4.31) gives the expectation.

Explaining why the natural scaling is exponential. Note that for any $1 < M < N$, the counting function is slowly changing in the sense that

$$F(\sqrt{2}; \beta; \gamma; N) - F(\sqrt{2}; \beta; \gamma; M) = O(\log(N/M)), \qquad (4.35)$$

where $c_4 \log(N/M)$ is the corresponding area. The geometric reason behind this is the exponentially sparse occurrence of lattice points in the corresponding long and narrow tilted hyperbola. The proof of (4.35) is a straightforward application of Lemma 4 below.

We have the following corollary of (4.35). If $M = cN$, i.e. n runs through the interval $cN < n < N$ with some constant $0 < c < 1$, then the fluctuation of $F(\sqrt{2}; \beta; \gamma; N)$ is a trivial $O(1)$. This negligible constant size change $O(1)$ in (4.35), as n runs through $cN < n < N$, explains why it is more natural to switch to the exponential scaling $F(\sqrt{2}; \beta; \gamma; e^N)$. In the rest of this discussion, we shall often prefer the exponential scaling.

The variance comes from the following non-trivial result. For any $\gamma > 0$, there is a positive effective constant $\sigma = \sigma(\gamma) > 0$ such that

$$\lim_{N \to \infty} \frac{1}{N} \int_0^1 \left(F(\sqrt{2}; \beta; \gamma; e^N) - \frac{\gamma}{\sqrt{2}} N \right)^2 \, d\beta = \sigma^2(\gamma).$$

The proof of this limit formula is based on a combination of Fourier analysis (Poisson summation formula, Parseval formula) and the arithmetic of the quadratic number field $\mathbf{Q}(\sqrt{2})$; see [2].

The first probabilistic result, nicely fitting the general scheme of *determinism vs. randomness*, is the following; for the proof, see [2].

Theorem A (Central Limit Theorem). *The renormalized counting function*

$$\frac{F(\sqrt{2}; \beta; \gamma; e^N) - (\gamma/\sqrt{2})N}{\sigma(\gamma)\sqrt{N}}, \quad 0 \leq \beta < 1,$$

has a standard normal limit distribution as $N \to \infty$.

To give at least some vague intuition behind Theorem A, we write

$$G_j(\beta) = F(\sqrt{2}; \beta; \gamma; e^j) - F(\sqrt{2}; \beta; \gamma; e^{j-1}), \quad j = 1, 2, \ldots, N.$$

In other words, $G_j(\beta)$ is the number of integral solutions $n \in \mathbf{N}$ of (4.29) satisfying $e^{j-1} < n \leq e^j$.

Note that $G_j(\beta)$ is a bounded function. This follows from Lemma 4 below, and from the obvious geometric fact that any short hyperbola segment corresponding to G_j is basically a rectangle. More precisely, any short hyperbola segment corresponding to G_j can be approximated by an inscribed rectangle R_1 of slope $1/\sqrt{2}$ and a circumscribed rectangle R_2 of slope $1/\sqrt{2}$ such that the ratio of the two areas is uniformly bounded by an absolute constant.

It is time now to formulate

Lemma 4. *Every tilted rectangle of slope $1/\sqrt{2}$ and area $\frac{1}{5}$ contains at most one lattice point.*

We postpone the proof of this simple but important result to the next section.

Lemma 4 can be easily generalized. The same proof gives that for any quadratic irrational α, there is a positive constant $c_5 = c_5(\alpha) > 0$ such that every tilted rectangle of slope α and area c_5 contains at most one lattice point.

Our key intuition is that the bounded function $G_j(\beta)$ resembles the j-th Rademacher function, so the sum

$$F(\sqrt{2}; \beta; \gamma; e^N) - \frac{\gamma}{\sqrt{2}} N = \sum_{j=1}^{N} \left(G_j(\beta) - \frac{\gamma}{\sqrt{2}} \right),$$

as a function of $\beta \in [0, 1)$, behaves like a sum of N independent Bernoulli variables

$$F(\sqrt{2}; \beta; \gamma; e^N) - \frac{\gamma}{\sqrt{2}} N \approx \underbrace{\pm 1 \pm 1 \pm \ldots \pm 1}_{N},$$

referred to often as an N-step random walk.

Our next result, Theorem B, can be interpreted as a variant of Khintchine's famous Law of the Iterated Logarithm in probability theory; see [21]. We show that the number of solutions $F(\sqrt{2}; \beta; \gamma; e^n)$ of (4.29) oscillates between the sharp bounds

$$\frac{\gamma}{\sqrt{2}} n - \sigma \sqrt{n} \sqrt{(2 + \varepsilon) \log \log n} < F(\sqrt{2}; \beta; \gamma; e^n)$$

$$< \frac{\gamma}{\sqrt{2}} n + \sigma \sqrt{n} \sqrt{(2 + \varepsilon) \log \log n}, \quad (4.36)$$

where $\varepsilon > 0$, as $n \to \infty$ for almost all β. Note that (4.36) fails with $2 - \varepsilon$ in place of $2 + \varepsilon$, where $\varepsilon > 0$. Here the main term $(\gamma/\sqrt{2})n$ means the area, so (4.36) can be considered a highly sophisticated justification of the Naive Area Principle.

The estimate (4.36) is particularly interesting in view of the fact that the classical Circle Problem is unsolved, and seems to be hopeless by current techniques. What (4.36) means is that we can solve a *Hyperbola Problem* instead of the Circle Problem. More precisely, we can prove for long and narrow tilted hyperbola segments what nobody can prove for large concentric circles. Namely, we can show that for almost all centers, i.e. for almost all values of the translation parameter β, the number of lattice points asymptotically equals the area plus an error which, even in the worst case scenario, is about the square root of the area. For circles the corresponding maximum error should be the square root of the circumference.

The Law of the Iterated Logarithm is one of the most famous results in classical probability theory, and describes the maximum fluctuation in the infinite one-dimensional random walk. The term *infinite random walk* refers to an infinite sequence of random Bernoulli trials, where each trial is tossing a fair coin. Of course, coin tossing belongs to the physical world; it is not a mathematical concept. But there is a well-known pure mathematical problem, which is considered equivalent. We can study the digit distribution of a typical real number written in binary form

$$\beta = \frac{b_1}{2} + \frac{b_2}{2^2} + \frac{b_3}{2^3} + \dots,$$

where each $b_i = 0$ or 1; here we have assumed for simplicity that $0 < \beta < 1$. The infinite 0-1 sequence

$$b_1 = b_1(\beta), b_2 = b_2(\beta), b_3 = b_3(\beta), \dots,$$

i.e. the sequence of binary digits of $0 < \beta < 1$, represents an infinite heads-and-tails sequence, say, with 1 as heads and 0 as tails. The sum

$$B_n = B_n(\beta) = b_1 + b_2 + b_3 + \dots + b_n$$

counts the number of 1's, or heads, among the first n binary digits of $0 < \beta < 1$. Borel's classical theorem about normal numbers asserts that

$$\frac{B_n(\beta)}{n} \to \frac{1}{2} \quad \text{for almost all } 0 < \beta < 1.$$

Let $S_n = S_n(\beta)$ denote the corresponding error term

$$S_n = S_n(\beta) = 2B_n(\beta) - n = \text{number of heads} - \text{number of tails},$$

so that $S_n = S_n(\beta)$ represents the number of heads minus the number of tails among the first n random trials, or coin tosses.

A well-known theorem of Khintchine [21] asserts that

$$\limsup_{n \to \infty} \frac{S_n(\beta)}{\sqrt{2n \log \log n}} = 1 \quad \text{for almost all } 0 < \beta < 1.$$

Note that Khintchine's Theorem is a far-reaching quantitative improvement on Borel's famous theorem on normal numbers. The long form of Khintchine's Theorem says that for any $\varepsilon > 0$ and almost all β, we have the following two statements:

- $S_n(\beta) < (1 + \varepsilon)\sqrt{2n \log \log n}$ for all sufficiently large values of n; and
- $S_n(\beta) > (1 - \varepsilon)\sqrt{2n \log \log n}$ for infinitely many values of n.

This strikingly elegant and precise result is the simplest form of the so-called Law of the Iterated Logarithm, usually called Khintchine's form.

Let us return to (4.36). The fact that it is an analog of Khintchine's Law of the Iterated Logarithm suggests the vague intuition that the lattice point counting function $F(\sqrt{2}; \beta; \gamma; e^n)$ behaves like a generalized digit sum as β runs through $0 < \beta < 1$.

What we are going to actually formulate below are two generalizations or refinements of (4.36); see Theorem B. The first generalization is that for almost all β, (4.36) holds for all γ, or in general, for all intervals $[\gamma_1, \gamma_2]$. This is a variant of the so-called Cassels's form of the Law of the Iterated Logarithm; see [12].

The second generalization of (4.36) is the Kolmogorov–Erdős form, an ultimate convergence-divergence criterion, which contains Khintchine's form as a simple corollary; see [14, 15, 22].

Theorem B (Law of the Iterated Logarithm).

(i) *Let $\varepsilon > 0$ be an arbitrarily small but fixed constant. Then for almost all β,*

$$\frac{\gamma}{\sqrt{2}} n - \sigma \sqrt{(2+\varepsilon)n \log\log n} < F(\sqrt{2}; \beta; \gamma; e^n)$$

$$< \frac{\gamma}{\sqrt{2}} n + \sigma \sqrt{(2+\varepsilon)n \log\log n} \qquad (4.37)$$

holds for all $\gamma > 0$ and for all sufficiently large n, i.e. for all $n > n_0(\beta, \gamma)$.

(ii) *Let $\varphi(n)$ be an arbitrary positive increasing function of n. Let $\gamma > 0$ be fixed. Then for almost all β,*

$$F(\sqrt{2}; \beta; \gamma; e^n) > \frac{\gamma}{\sqrt{2}} n + \varphi(n)\sigma\sqrt{n}$$

holds for infinitely many values of n if and only if the series

$$\sum_{n=1}^{\infty} \frac{\varphi(n)}{n} e^{-\varphi^2(n)/2} \qquad (4.38)$$

diverges. The same conclusion holds for the other inequality

$$F(\sqrt{2}; \beta; \gamma; e^n) < \frac{\gamma}{\sqrt{2}} n - \varphi(n)\sigma\sqrt{n}.$$

Note that (4.37) is sharp in the sense that $2 + \varepsilon$ cannot be replaced by $2 - \varepsilon$.

Remarks. (i) By Lemma 1, we have $f(\sqrt{2}; \beta; c; N) = F(\sqrt{2}; \beta; \gamma; N) + O(1)$ as $N \to \infty$, where $c = \gamma/2\sqrt{2}$. So Lemma 1 implies that Theorems A and B remain true if $F(\sqrt{2}; \beta; \gamma; N)$ is replaced by the number of solutions $f(\sqrt{2}; \beta; c; N)$ of the inhomogeneous diophantine inequality (4.28).

(ii) In Theorem B(i), there is a dramatic difference between rational β and almost all β. For every rational β, the counting function has the form

$$F(\sqrt{2}; \beta; \gamma; N) = c(\gamma) \log N + O(1) \quad \text{as } N \to \infty$$

for all $\gamma > 0$, and it remains valid if $\sqrt{2}$ is replaced by any quadratic irrational. This bounded size fluctuation around the main term $c \log N$, which is typically not the area, jumps up considerably. By (4.37), we have square root size fluctuations around the main term, which is the area, so the fluctuations have size the square root of the area, and this holds for almost all β and all $\gamma > 0$.

Let us return to (4.36). It is a special case of Theorem B(ii) with

$$\varphi(n) = ((2 \pm \varepsilon) \log \log n)^{1/2}.$$

Indeed, the series (4.38) is divergent or convergent depending on whether we have $2 + \varepsilon$ or $2 - \varepsilon$ in the definition of $\varphi(n)$.

We can obtain a much more delicate result by choosing a large integer $k \geq 4$ and writing

$$\varphi(n) = (2 \log_2 n + 2 \log_3 n + 2 \log_4 n + \ldots + 2 \log_{k-1} n + (2 \pm \varepsilon) \log_k n)^{1/2}.$$

Beware that here, and here only, we use the space-saving notation $\log_2 n = \log \log n$, i.e. it means the iterated logarithm instead of the usual meaning as base 2 logarithm, and in general, $\log_k n = \log(\log_{k-1} n)$ denotes the k-times iterated logarithm of n. With this choice of $\varphi(n)$, we have

$$\sum_{n=1}^{\infty} \frac{\varphi(n)}{n} e^{-\varphi^2(n)/2} \approx \sum_n \frac{1}{n \log n \log_2 n \log_3 n \ldots \log_{k-1} n (\log_k n)^{1 \pm \varepsilon/2}},$$

which is divergent or convergent depending on whether we have $2 + \varepsilon$ or $2 - \varepsilon$ in the definition of $\varphi(n)$.

This example clearly illustrates the remarkable precision of Theorem B(ii).

Next we focus on a simple consequence of Theorem B. Let $c > 0$ be arbitrarily small but fixed. Then by Theorem B, the inhomogeneous diophantine inequality

$$\|n\sqrt{2} - \beta\| < \frac{c}{n} \tag{4.39}$$

has infinitely many integer solutions $n \geq 1$ for almost all β, in the sense of the Lebesgue measure.

Inequality (4.39) corresponds to the hyperbola segment

$$|y - \beta| < \frac{c}{x}, \quad x \geq 1,$$

where β is fixed, and this has infinite area. But we may go further, and consider smaller regions

$$|y - \beta| < \frac{1}{x \log x}, \quad |y - \beta| < \frac{1}{x \log x \log \log x},$$

and the like. They all have infinite area, since

$$\int_e^N \frac{dx}{x \log x} = \log \log N \quad \text{and} \quad \int_{e^e}^N \frac{dx}{x \log x \log \log x} = \log \log \log N,$$

and the rest all tend to infinity as $N \to \infty$. It is very natural, therefore, to ask the following question.

Question. Consider the inequalities

$$\|n\sqrt{2} - \beta\| < \frac{c}{n \log n}, \quad n \geq n_1, \tag{4.40}$$

$$\|n\sqrt{2} - \beta\| < \frac{c}{n \log n \log \log n}, \quad n \geq n_2, \tag{4.41}$$

and so on, where $0 \leq \beta < 1$ is a fixed constant. Is it true that for almost all β, in the sense of the Lebesgue measure, the inequalities (4.40), (4.41) and the like have infinitely many positive integer solutions n?

Well, the answer is affirmative.

Theorem C (Area Principle for $\sqrt{2}$). *Let $\psi(x)$ be any positive decreasing function of the real variable x satisfying*

$$\sum_{n=1}^{\infty} \psi(n) = \infty. \tag{4.42}$$

Then the inhomogeneous inequality

$$\|n\sqrt{2} - \beta\| < \psi(n)$$

has infinitely many integral solutions for almost all $0 \leq \beta < 1$, in the sense of Lebesgue measure.

Furthermore, there is an interesting generalization of Theorem C where $\sqrt{2}$ is replaced by any real α.

To explain this generalization, Theorem D below, we recall the basic question of diophantine approximation. We want to decide whether an inequality

$$\left| \alpha - \frac{p}{q} \right| < \frac{1}{q^2}, \quad \text{or equivalently,} \quad |q\alpha - p| < \frac{1}{q},$$

with integers p and q, or more generally, an inequality

$$\|q\alpha\| < \psi(q), \tag{4.43}$$

where $\psi(q)$ is a positive decreasing function of q, has infinitely many integral solutions in q, and if this is the case, to determine the solutions, or at least the asymptotic number of integral solutions.

It is perfectly natural to study the inhomogeneous analog of (4.43), the inequality

$$\|q\alpha - \beta\| < \psi(q), \tag{4.44}$$

where β is an arbitrary fixed real number. Of course, we may assume that $0 \le \beta < 1$.

Is there any connection between the solvability of the homogeneous inequality (4.43) and the inhomogeneous inequality (4.44)? Theorem C is about the special case $\alpha = \sqrt{2}$, and it justifies the Naive Area Principle. Recall that the Naive Area Principle is a vague intuition claiming that a nice region of infinite area must contain infinitely many lattice points. We know that the Naive Area Principle is false for the hyperbolic region $-\frac{1}{2} \le x^2 - 2y^2 \le \frac{1}{2}$, which has infinite area and contains only one lattice point, namely the origin. This Pell inequality is basically equivalent to the diophantine inequality

$$\|q\sqrt{2}\| < \frac{c}{q}, \tag{4.45}$$

with $c \le 2^{-5/2}$, and (4.45) does not have infinitely many integral solutions in q if the constant $c < 2^{-5/2}$.

The failure of the Naive Area Principle for (4.45) is compensated by the success of the Naive Area Principle for the inhomogeneous inequality

$$\|q\sqrt{2} - \beta\| < \psi(q),$$

which has infinitely many integral solution q for almost all β, provided that $\psi(x)$ is any positive decreasing function of the real variable x satisfying (4.42). This is the statement of Theorem C. The next result generalizes the special case $\alpha = \sqrt{2}$ to arbitrary real α.

Theorem D (General Area Principle). *Let $\psi(x)$ be any positive decreasing function of the real variable x satisfying (4.42). For any real number α, at least one of the following two cases always holds:*

 (i) The homogeneous inequality (4.43) has infinitely many integral solutions.
 (ii) The inhomogeneous inequality (4.44) has infinitely many integral solutions for almost all $0 \le \beta < 1$, in the sense of Lebesgue measure.

Remark. Note that divergence condition (4.42) is necessary. Indeed, if

$$\sum_{n=1}^{\infty} \psi(n) < \infty, \tag{4.46}$$

then the set of pairs (α, β), for which the inequality (4.44) has infinitely many integral solutions q, has two-dimensional Lebesgue measure zero. This statement immediately follows from the statement that for every fixed β, the set of α which satisfy (4.44) for infinitely many q has Lebesgue measure zero. The second statement has an easy proof as follows. Every such α in $0 < \alpha < 1$ is contained in infinitely many intervals of the form

$$\left[\frac{p+\beta}{q} - \frac{\psi(q)}{q}, \frac{p+\beta}{q} + \frac{\psi(q)}{q} \right]$$

with integers $q \geq N$ and $1 \leq p \leq q$, and the total length of these intervals is less than

$$2 \sum_{q \geq N} \psi(q),$$

which by (4.46) tends to zero as $N \to \infty$. This means that Theorem D is a precise convergence-divergence type result, or we may call it a zero-one law, to borrow a well-known concept from probability theory.

Let us return to the inhomogeneous inequality (4.44). If α is rational and β is irrational, then (4.44) has only finitely many integral solutions for any $\psi(q) \to 0$ as $q \to \infty$. Well, this is trivial. It is less trivial to find an irrational α and a decreasing function $\psi(x)$ satisfying (4.42) such that for almost all β, (4.44) has only finitely many integral solutions. We can take any irrational $0 < \alpha < 1$ with sufficiently large partial quotients in the sense that

$$\alpha = \frac{1}{a_1+} \frac{1}{a_2+} \ldots = [a_1, a_2, a_3, \ldots],$$

where

$$a_k \approx k^{(\log k)^2}, \tag{4.47}$$

and take

$$\psi(q) = \frac{1}{q \log q}. \tag{4.48}$$

Then the denominator q_k of the k-th convergent of α is roughly

$$q_k \approx a_1 a_2 \ldots a_k \approx k^{k(\log k)^2}, \tag{4.49}$$

and so

$$\sum_k \frac{1}{\log q_k} = O\left(\sum_k \frac{1}{k(\log k)^3} \right) < \infty.$$

We recall the well-known fact

$$\left| \alpha - \frac{p_k}{q_k} \right| < \frac{1}{q_k q_{k+1}}$$

which implies

$$\left| n\alpha - \frac{np_k}{q_k} \right| < \frac{n}{q_k q_{k+1}}. \tag{4.50}$$

If $q_k \le n < q_{k+1}k^{-2}$ and

$$\|n\alpha - \beta\| < \frac{1}{n \log n},$$

then by (4.49) and (4.50), we have

$$\left\| \beta - \frac{np_k}{q_k} \right\| < \frac{1}{k^2 q_k} + \frac{1}{n \log n} < \frac{2}{k(\log k)^3 q_k}. \tag{4.51}$$

If $q_{k+1}k^{-2} \le n < q_{k+1}$, then define the set

$$A_k = \bigcup_n \left[n\alpha - \frac{1}{n \log n}, n\alpha + \frac{1}{n \log n} \right] \quad \text{mod } 1, \tag{4.52}$$

where the summation in (4.52) is extended over all n with $q_{k+1}k^{-2} \le n < q_{k+1}$. Motivated by (4.51), define the set

$$B_k = \bigcup_{0 \le j < q_k} \left[\frac{j}{q_k} - \frac{2}{k(\log k)^3 q_k}, \frac{j}{q_k} + \frac{2}{k(\log k)^3 q_k} \right] \quad \text{mod } 1. \tag{4.53}$$

Clearly

$$\sum_k \text{meas}(B_k) \le \sum_k \frac{4}{k(\log k)^3} < \infty, \tag{4.54}$$

where meas denotes the usual Lebesgue measure, and

$$\sum_k \text{meas}(A_k) = O\left(\sum_k \frac{\log(k^2)}{k(\log k)^3} \right) = O\left(\sum_k \frac{1}{k(\log k)^2} \right) < \infty. \tag{4.55}$$

It follows from (4.54) and (4.55) that almost all β are contained in only a finite number of A_k and in a finite number of B_k. In view of (4.51)–(4.53), this implies that for almost all β, the inequality (4.44) has only finitely many integral solutions, where α and ψ are defined by (4.47) and (4.48).

For the proofs of Theorems A and B, we refer the reader to the forthcoming book [2]. For the proofs of Theorems C and D, see the recent paper [8]. This section was a detour, or rather a counterpart; the rest of the chapter is about extra large fluctuations, i.e. sophisticated violations of the Naive Area Principle.

The next section is technical, and contains the proofs of Theorem 3 and Lemmas 1–4. The truly interesting new results come later, starting in Sect. 4.4.

4.3   Proving Theorem 3 and the Lemmas

Proof of Lemma 2. First we establish the estimate (4.31). Consider the hyperbolic needle $H_N(\gamma) = H_N(\sqrt{2}; \gamma)$, defined by

$$H_N(\gamma) = \{(x, y) \in \mathbf{R}^2 : -\gamma \le x^2 - 2y^2 \le \gamma,\ 1 \le x + y\sqrt{2} \le 2\sqrt{2}N\}.\quad (4.56)$$

Comparing (4.11) with (4.56), we see that

$$H_N(\gamma) = H^*(\sqrt{2}; [-\gamma, \gamma]; N),$$

so by (4.13), we deduce that

$$\text{area}(H_N(\gamma)) = \frac{\gamma}{\sqrt{2}} \log N + O(1).\quad (4.57)$$

Next we need the following almost trivial result.

Lemma 5. *Let $A \subset \mathbf{R}^2$ be a Lebesgue measurable set in the plane with finite measure denoted by* $\text{area}(A)$*. Then*

$$\int_0^1 \int_0^1 |(A + \mathbf{x}) \cap \mathbf{Z}^2|\, d\mathbf{x} = \text{area}(A),$$

where $A + \mathbf{x}$ denotes the translation of the set A by the vector $\mathbf{x} \in \mathbf{R}^2$.

Now by Lemma 5, we have

$$\int_0^1 \int_0^1 |(H_N(\gamma) + \mathbf{v}) \cap \mathbf{Z}^2|\, d\mathbf{v} = \text{area}(H_N(\gamma)).\quad (4.58)$$

If $\mathbf{v} = (v_1, v_2) \in [0, 1)^2$ is chosen in such a way that $v_1 - v_2\sqrt{2} \equiv \beta \bmod 1$ is fixed, then clearly

$$|F(\sqrt{2}; \beta; \gamma; N) - |(H_N(\gamma) + \mathbf{v}) \cap \mathbf{Z}^2|| < c_6(\gamma),\quad (4.59)$$

where $c_6(\gamma)$ is a constant independent of β and N. The estimate (4.31) follows on combining (4.57)–(4.59).

Next we prove (4.32). Let $0 \le a < b \le 1$ be fixed. For any $M \ge 1$, consider the parallelogram

$$\mathscr{P}_M = \{\mathbf{v} = (v_1, v_2) \in \mathbf{R}^2 : a \le v_1 - v_2\sqrt{2} \le b, \; 0 \le v_1 + v_2\sqrt{2} \le M\}.$$

If M is large, then $\mathscr{P}_M$ is a long and narrow parallelogram, but we can then turn it into a *round* shape by applying an appropriate automorphism of the quadratic form $x^2 - 2y^2$. The substitution $x_1 = x + 2y$, $y_1 = x + y$ is a fundamental automorphism,[5] and writing

$$A = \begin{pmatrix} 1 & 2 \\ 1 & 1 \end{pmatrix},$$

we note that A^k, $k \in \mathbf{Z}$, give rise to infinitely many automorphisms preserving the lattice points and the area. The eigenvectors of the matrix A are parallel to the sides of parallelogram $\mathscr{P}_M$, so on applying an appropriate power A^k on the long and narrow parallelogram $\mathscr{P}_M$, we obtain a *round* parallelogram $A^k \mathscr{P}_M$ with sides parallel to that of $\mathscr{P}_M$, and

$$\mathrm{area}(A^k \mathscr{P}_M) = \mathrm{area}(\mathscr{P}_M) = c_7 M.$$

Here *round* means that the diameter of parallelogram $A^k \mathscr{P}_M$ is $O(\sqrt{M})$, so the number of unit squares $[0, 1)^2 + \mathbf{n}$, $\mathbf{n} \in \mathbf{Z}^2$, intersecting the boundary of $A^k \mathscr{P}_M$ is $O(\sqrt{M})$.

Combining this geometric fact with (4.58), we have

$$\frac{1}{\mathrm{area}(\mathscr{P}_M)} \int_{\mathscr{P}_M} |(H_N(\gamma) + \mathbf{v}) \cap \mathbf{Z}^2| \, d\mathbf{v} = \mathrm{area}(H_N(\gamma))(1 + O(M^{-1/2})). \quad (4.60)$$

If $\mathbf{v} = (v_1, v_2) \in [0, 1)^2$ is chosen in such a way that $v_1 - v_2\sqrt{2} \equiv \beta \bmod 1$ is fixed, then clearly

$$|F(\sqrt{2}; \beta; \gamma; N) - |(H_N(\gamma) + \mathbf{v}) \cap \mathbf{Z}^2|| < c_8(\gamma, M), \quad (4.61)$$

where $c_8(\gamma, M)$ is a constant independent of β and N. Combining (4.57), (4.60) and (4.61), we have

$$\frac{\frac{1}{b-a} \int_a^b F(\sqrt{2}; \beta; \gamma; N) \, d\beta}{\log N}$$

---

[5]Indeed, we have $x_1^2 - 2y_1^2 = (x + 2y)^2 - 2(x + y)^2 = -(x^2 - 2y^2)$.

$$= \left(\frac{\gamma}{\sqrt{2}} + O\left(\frac{1}{\log N} \right) \right) (1 + O(M^{-1/2})) + \frac{c_8(\gamma, M)}{\log N}. \qquad (4.62)$$

Since M can be arbitrarily large, (4.62) implies (4.32). The proof of Lemma 2 is now complete. $\qquad \square$

Proof of Lemma 5. First assume that A is bounded. Let N be a large integer. In view of the periodicity of $\mathbf{Z}^2$, we have

$$\int_0^N \int_0^N |(A + \mathbf{x}) \cap \mathbf{Z}^2| \, dx = N^2 \int_0^1 \int_0^1 |(A + \mathbf{x}) \cap \mathbf{Z}^2| \, dx.$$

On the other hand,

$$\int_0^N \int_0^N |(A + \mathbf{x}) \cap \mathbf{Z}^2| \, dx = \sum_{\mathbf{n} \in \mathbf{Z}^2} \text{area}\{\mathbf{x} \in [0, N]^2 : \mathbf{n} \in A + \mathbf{x}\}$$

$$= \sum_{\mathbf{n} \in \mathbf{Z}^2} \text{area}\{(\mathbf{n} - A) \cap [0, N]^2\}.$$

Without loss of generality, we can assume that the origin is inside A. Let $d(A)$ denote the diameter of A. Then $(\mathbf{n} - A) \subset [0, N]^2$ if $\mathbf{n} \in [d(A), N - d(A)]^2$. On the other hand, $(\mathbf{n} - A) \cap [0, N]^2 = \emptyset$ if $\mathbf{n} \notin [-d(A), N + d(A)]^2$. Thus we have

$$(N + 2d(A))^2 \cdot \text{area}(A) \geq \sum_{\mathbf{n} \in \mathbf{Z}^2} \text{area}\{(\mathbf{n} - A) \cap [0, N]^2\} \geq (N - 2d(A))^2 \cdot \text{area}(A).$$

Dividing the last inequalities by N^2, and combining with the equations above, we see that Lemma 5 follows as N tends to infinity. If A is unbounded, then we approximate A by an increasing sequence $A_1 \subset A_2 \subset A_3 \subset \ldots$ of subsets of A such that each A_k is bounded and $\text{area}(A \setminus A_k) \to 0$. The last step is then to use the continuity of the Lebesgue measure. $\qquad \square$

Proof of Lemma 1. For notational simplicity, we restrict our proof to the special case $\beta_2 = 0$; the general case is the same. Again the key step is to apply Lemma 5. For $1 \leq K < L \leq \infty$, consider the four regions

$$H_{K,L}(\beta; \gamma) = \{(x, y) \in \mathbf{R}^2 : -\gamma \leq (x + \beta)^2 - 2y^2 \leq \gamma, \ K \leq y \leq L, \ x > 0\},$$

$$\tilde{H}_{K,L}(\beta; \gamma) = \{(x, y) \in \mathbf{R}^2 : 2\sqrt{2}y|x + \beta - y\sqrt{2}| < \gamma, \ K \leq y \leq L, \ x > 0\},$$

$$\tilde{H}_{K,L}^+(\beta; \gamma) = \{(x, y) \in \mathbf{R}^2 : (2\sqrt{2}y+1)|x+\beta-y\sqrt{2}| < \gamma, \ K \leq y \leq L, \ x > 0\},$$

$$\tilde{H}_{K,L}^-(\beta; \gamma) = \{(x, y) \in \mathbf{R}^2 : (2\sqrt{2}y - 1)|x+\beta - y\sqrt{2}| < \gamma, \ K \leq y \leq L, \ x>0\}.$$

In view of the factorization (4.27), the condition $(x, y) \in H_{K,L}(\beta; \gamma)$ gives the estimate $x + \beta = y\sqrt{2} + o(1)$. In fact, we have the stronger form $x + \beta = y\sqrt{2} + O(1/y)$. Thus there is a threshold $c_9 = c_9(\gamma)$ such that

$$\tilde{H}^+_{K,L}(\beta; \gamma) \subset H_{K,L}(\beta; \gamma) \subset \tilde{H}^-_{K,L}(\beta; \gamma)$$

for all $L > K > c_9(\gamma)$. On the other hand, it is trivial that

$$\tilde{H}^+_{K,L}(\beta; \gamma) \subset \tilde{H}_{K,L}(\beta; \gamma) \subset \tilde{H}^-_{K,L}(\beta; \gamma).$$

Consider now the special case $K = 1, L = \infty, \beta = 0$, and study the difference set

$$D(\gamma) = \tilde{H}^-_{1,\infty}(0; \gamma) \setminus \tilde{H}^+_{1,\infty}(0; \gamma).$$

The area of this difference set can be estimated by

$$\text{area}(D(\gamma)) = O\left(\int_1^\infty \left(\frac{1}{2\sqrt{2}y - 1} - \frac{1}{2\sqrt{2}y + 1}\right) dy\right)$$

$$= O\left(\int_1^\infty \frac{dy}{8y^2 - 1}\right) = O(1).$$

Combining this with Lemma 5, we have

$$\int_0^1 \int_0^1 |(D(\gamma) + \mathbf{v}) \cap \mathbf{Z}^2| \, d\mathbf{v} = \text{area}(D(\gamma)) < \infty. \tag{4.63}$$

If $\mathbf{v} = (v_1, v_2) \in [0, 1)^2$ is chosen in such a way that $v_1 - v_2\sqrt{2} \equiv \beta \bmod 1$ is fixed, then

$$D(\gamma) + \mathbf{v} \supset H_{K,L}(\beta; \gamma) \Delta \tilde{H}^+_{K,L}(\beta; \gamma), \tag{4.64}$$

where $A \Delta B = (A \setminus B) \cup (B \setminus A)$ denotes the symmetric difference of the sets A and B. Combining (4.63) and (4.64), Lemma 1 follows easily. □

Proof of Lemma 4. Consider a rectangle of slope $1/\sqrt{2}$ which contains two lattice points $P = (k, \ell)$ and $Q = (m, n)$; in fact, assume that P, Q are two vertices of the rectangle. We denote the vector from P to Q by $\mathbf{v} = (m - k, n - \ell)$, and consider the two perpendicular unit vectors

$$\mathbf{e}_1 = \left(\frac{\sqrt{2}}{\sqrt{3}}, \frac{1}{\sqrt{3}}\right) \quad \text{and} \quad \mathbf{e}_2 = \left(\frac{1}{\sqrt{3}}, -\frac{\sqrt{2}}{\sqrt{3}}\right).$$

Then the two side lengths a and b of the rectangle can be expressed in terms of the inner products

$$a = |\mathbf{e}_1 \cdot \mathbf{v}| = \frac{|p\sqrt{2} + q|}{\sqrt{3}} \quad \text{and} \quad b = |\mathbf{e}_2 \cdot \mathbf{v}| = \frac{|p - q\sqrt{2}|}{\sqrt{3}},$$

where $p = m - k$ and $q = n - \ell$. Thus we have

$$\text{area} = ab = \frac{|(p\sqrt{2} + q)(p - q\sqrt{2})|}{3}.$$

Without loss of generality, we can assume that $p \geq 0$ and $q \geq 0$. Since $(p, q) \neq (0, 0)$, we have $|p - q\sqrt{2}| = 1/(p + q\sqrt{2})$, and so

$$\text{area} = \frac{|(p\sqrt{2}+q)(p - q\sqrt{2})|}{3} = \frac{p\sqrt{2} + q}{3(p+q\sqrt{2})} \geq \frac{p + q}{3(p\sqrt{2}+q\sqrt{2})} = \frac{1}{3\sqrt{2}} > \frac{1}{5},$$

proving Lemma 4. $\qquad\square$

Proof of Theorem 3. We shall show that the set of numbers β in question, the set of divergence points, contains a Cantor set. This guarantees that the cardinality of the set is continuum.

We make a standard Cantor set construction, i.e. we apply the method of nested intervals. For notational convenience, we write $F(\sqrt{2}; \beta; \gamma; N) = F(\beta; \gamma; N)$. By (4.31), we have

$$\int_0^1 F(\beta; \gamma; N) \, d\beta = \frac{\gamma}{\sqrt{2}} \log N + O(1).$$

Applying this with $\gamma = \frac{1}{4}$, we obtain the existence of $0 < \beta_1 < 1$ and an arbitrarily large integer N_1 such that

$$F(\beta_1; \gamma = 1/4; N_1) > \frac{1}{8} \log N_1.$$

Since $\frac{1}{4} < \frac{1}{2}$, there exists an interval $I_1 = [a, b]$ with $0 < a < b < 1$ such that $\beta_1 \in I_1$ and

$$F(\beta; \gamma = 1/2; N_1) > \frac{1}{8} \log N_1 \quad \text{for all } \beta \in I_1. \tag{4.65}$$

Next let $\mathbf{n} = (n_1, n_2) \in \mathbf{Z}^2$ be a lattice point such that $\beta_2 = n_1 - n_2\sqrt{2} \in I_1$. Since the equation $|x^2 - 2y^2| \leq \frac{3}{4}$ does not have a non-zero integral solution, trivially

$$F(\beta_2; \gamma = 3/4; N) < \frac{1}{100} \log N \quad \text{for all } N \geq N_2,$$

where N_2 is a sufficiently large threshold. We can clearly assume that $N_2 > N_1$. Since $\frac{3}{4} > \frac{1}{2}$, there exists[6] an interval $I_2 = [a,b]$ with some $0 < a < b < 1$ such that $\beta_2 \in I_2$ and

$$F(\beta; \gamma = 1/2; N_2) < \frac{1}{100} \log N_2 \quad \text{for all } \beta \in I_2. \tag{4.66}$$

We can clearly assume that I_2 is a proper subinterval of I_1. Let $I(0) = I_2$. Repeating the second argument, we deduce that there exists another closed subinterval $I(1)$ such that $I(0)$ and $I(1)$ are disjoint, $I(0) \cup I(1) \subset I_1$ and

$$F(\beta; \gamma = 1/2; N_2^{(1)}) < \frac{1}{100} \log N_2^{(1)} \quad \text{for all } \beta \in I(1). \tag{4.67}$$

We can clearly assume that $N_2^{(1)} > N_1$.

By (4.32), we have

$$\frac{1}{|I(0)|} \int_{I(0)} F(\beta; \gamma; N) \, d\beta = (1 + o(1)) \frac{\gamma}{\sqrt{2}} \log N,$$

and applying this with $\gamma = \frac{1}{4}$, we obtain the existence of $0 < \beta_3 < 1$ and a large integer N_3 such that

$$F(\beta_3; \gamma = 1/4; N_3) > \frac{1}{8} \log N_3.$$

Since $\frac{1}{4} < \frac{1}{2}$, there exists an interval $I_3 = [a,b]$ with $0 < a < b < 1$ such that $\beta_3 \in I_3$ and

$$F(\beta; \gamma = 1/2; N_3) > \frac{1}{8} \log N_3 \quad \text{for all } \beta \in I_3. \tag{4.68}$$

We can clearly assume that I_3 is a proper subinterval of $I(0)$. Write $I(0,0) = I_3$. Similarly, there exists another subinterval $I(0,1)$ such that $I(0,0)$ and $I(0,1)$ are disjoint, $I(0,0) \cup I(0,1) \subset I(0)$ and

$$F(\beta; \gamma = 1/2; N_3^{(1)}) > \frac{1}{8} \log N_3^{(1)} \quad \text{for all } \beta \in I(0,1). \tag{4.69}$$

There are similar disjoint subintervals $I(1,0)$ and $I(1,1)$ of $I(1)$.

Next, let $\mathbf{n} = (n_1, n_2) \in \mathbf{Z}^2$ be a lattice point such that $\beta_4 = n_1 - n_2\sqrt{2} \in I(0,0)$. Since the inequality $|x^2 - 2y^2| \le \frac{3}{4}$ does not have a non-trivial integral solution,

$$F(\beta_4; \gamma = 3/4; N) < \frac{1}{100} \log N \quad \text{for all } N \ge N_4,$$

---

[6]Here a and b are generic numbers.

where $N_4 < \infty$ is a sufficiently large threshold. We can clearly assume that $N_4 > N_3$. Since $\frac{3}{4} > \frac{1}{2}$, there exists an interval $I_4 = [a,b]$ with $0 < a < b < 1$ such that $\beta_4 \in I_4$ and

$$F(\beta; \gamma = 1/2; N_4) < \frac{1}{100} \log N_4 \quad \text{for all } \beta \in I_4. \tag{4.70}$$

We can clearly assume that I_4 is a proper subinterval of $I(0,0)$. Let $I(0,0,0)) = I_4$. Repeating the last argument, there exists another closed subinterval $I(0,0,1)$ such that $I(0,0,0)$ and $I(0,0,1)$ are disjoint, $I(0,0,0) \cup I(0,0,1) \subset I(0,0)$ and

$$F(\beta; \gamma = 1/2; N_4^{(1)}) < \frac{1}{100} \log N_4^{(1)} \quad \text{for all } \beta \in I(0,0,1), \tag{4.71}$$

and so on. Repeating this argument, we build an infinite binary tree

$$I_1 \supset I_{\varepsilon_1} \supset I_{\varepsilon_1, \varepsilon_2} \supset I_{\varepsilon_1, \varepsilon_2, \varepsilon_3} \supset \dots,$$

where $\varepsilon_1, \varepsilon_2, \varepsilon_3, \dots \in \{0, 1\}$.

For an arbitrary infinite 0-1 sequence $\varepsilon_1, \varepsilon_2, \varepsilon_3, \dots$, let

$$\beta \in I_1 \cap I_{\varepsilon_1} \cap I_{\varepsilon_1, \varepsilon_2} \cap I_{\varepsilon_1, \varepsilon_2, \varepsilon_3} \cap \dots.$$

Then by (4.65)–(4.71), there exists an infinite sequence $1 < M_1 < M_2 < M_3 < M_4 < \dots$ of integers such that

$$F(\beta; \gamma = 1/2; M_{2k-1}) > \frac{1}{8} \log M_{2k-1} \quad \text{and} \quad F(\beta; \gamma = 1/2; M_{2k}) < \frac{1}{100} \log M_{2k},$$

where $k = 1, 2, 3, \dots$. This proves Theorem 3. $\qquad\qquad\square$

4.4 The Riesz Product and Theorem 12

4.4.1 The Method of Nested Intervals vs. the Riesz Product

At the end of Sect. 4.1, we formulated a far-reaching generalization of Theorem 3; see (4.34). It states that Theorem 3 actually holds for every $\gamma > 0$, and we have the stronger inequality

$$\limsup_{n \to \infty} \frac{F(\sqrt{2}; \beta^*; \gamma; n)}{\log n} > \frac{\gamma}{\sqrt{2}} > \liminf_{n \to \infty} \frac{F(\sqrt{2}; \beta^*; \gamma; n)}{\log n}, \tag{4.72}$$

where $(\gamma/\sqrt{2})\log n + O(1)$ is the area of the corresponding hyperbolic region. Indeed, (4.72) holds for continuum many divergence points $\beta^* = \beta^*(\gamma) \in [0, 1)$.

The proof of Theorem 3 was based on an elementary argument that we may call the method of nested intervals. To prove (4.72), we need a new idea, and apply a more sophisticated Riesz product argument. The Riesz product is a powerful tool in Fourier analysis. A typical application is to prove large fluctuations for lacunary trigonometric series. To compare the method of nested intervals to the method of Riesz products, we give a simple illustration; see Facts 1 and 2 below.

Consider a finite cosine sum

$$F(x) = \sum_{j=1}^{N} a_j \cos(2\pi n_j x), \quad \text{where } a_j = \pm 1 \text{ for all } 1 \le j \le N, \quad (4.73)$$

and $1 \le n_1 < n_2 < \ldots < n_N$ are integers. We study the following question. What can we say about $\max_{0 \le x \le 1} F(x)$? Well, under different extra conditions, we have different results. We begin with

Fact 6. *If the strong gap condition $n_{j+1}/n_j \ge 8$ holds for every $1 \le j \le N - 1$, then*

$$\max_{0 \le x \le 1} F(x) \ge \frac{N}{2}.$$

Proof. The proof is almost trivial. Let

$$J_1 = \left\{ x \in [0, 1] : \cos(2\pi n_1 x) \text{ lies between } \frac{a_1}{2} \text{ and } a_1 \right\}.$$

Since $a_1 = \pm 1$, the set J_1 contains a closed subinterval I_1 of length $|I_1| \ge 1/4n_1$. Next let

$$J_2 = \left\{ x \in I_1 : \cos(2\pi n_2 x) \text{ lies between } \frac{a_2}{2} \text{ and } a_2 \right\}.$$

Since $a_2 = \pm 1$, the set J_2 contains a closed subinterval I_2 of length $|I_2| \ge 1/4n_2$. Next let

$$J_3 = \left\{ x \in I_2 : \cos(2\pi n_3 x) \text{ lies between } \frac{a_3}{2} \text{ and } a_3 \right\},$$

and so on. Repeating this process N times, we obtain a nested sequence of closed intervals

$$[0, 1] \supset I_1 \supset I_2 \supset \ldots \supset I_N$$

such that $a_k \cos(2\pi n_k x) \ge \frac{1}{2}$ for all $x \in I_k, k = 1, 2, \ldots, N$. Then clearly $F(x) \ge N/2$ for every $x \in I_N$. $\square$

This is a typical application of the method of nested intervals. Next comes the Riesz product argument. The problem that we study is the following. What will happen if the strong gap condition $n_{j+1}/n_j \geq 8$ is replaced by the weaker condition $n_{j+1}/n_j \geq 1 + \varepsilon > 1$, where $\varepsilon > 0$ is an arbitrarily small but fixed constant? Can we still prove a linear lower bound like $\max_{0 \leq x \leq 1} F(x) \geq cN$ with some constant $c = c(\varepsilon) > 0$ depending only on the value of ε? Unfortunately, the method of nested intervals hopelessly collapses. Our new approach is the Riesz product argument. The following result, a well-known theorem of Sidon in Fourier analysis, is much deeper than Fact 6.

Fact 7 (Sidon's Theorem). *If the weak gap condition*

$$\frac{n_{j+1}}{n_j} \geq 1 + \varepsilon > 1 \tag{4.74}$$

holds for every $1 \leq j \leq N - 1$, where $0 < \varepsilon < \frac{1}{2}$ is a fixed constant, then for $F(x)$ defined in (4.73), we have

$$\max_{0 \leq x \leq 1} F(x) \geq cN \quad \text{with} \quad c = \frac{1}{4\varepsilon^{-1} \log(2\varepsilon^{-1})}.$$

Proof. Let $1 = i(1) < i(2) < \ldots < i(M)$ be a subsequence of $1, 2, 3, \ldots, N$ such that

$$\frac{n_{i(j+1)}}{n_{i(j)}} \geq \frac{2}{\varepsilon}, \quad j = 1, 2, \ldots, M - 1, \tag{4.75}$$

and consider the Riesz product

$$R(x) = \prod_{j=1}^{M} (1 + a_{i(j)} \cos(2\pi n_{i(j)} x)).$$

Since $a_{i(j)} = \pm 1$, we have $R(x) \geq 0$. We shall use this Riesz product $R(x)$ as a test function. First we evaluate the integral

$$\int_0^1 F(x) R(x) \, dx = \sum_{j=1}^{M} a_{i(j)}^2 \int_0^1 \cos^2(2\pi n_{i(j)} x) \, dx = \frac{M}{2}. \tag{4.76}$$

Indeed, multiplying out the Riesz product $R(x)$, and then using Euler's formula $2e^y = e^{iy} + e^{-iy}$, we obtain terms like

$$a_{i(j_1)} a_{i(j_2)} a_{i(j_3)} \cdots a_{i(j_k)} e^{2\pi i(\pm n_{i(j_1)} \pm n_{i(j_2)} \pm n_{i(j_3)} \pm \ldots \pm n_{i(j_k)})}, \tag{4.77}$$

where we shall call (4.77) a product of length $k \geq 1$. We distinguish two cases.

Case 8 (short products). $k = 1$. Multiplying the corresponding terms with $F(x)$ and integrating from 0 to 1, we obtain

$$\sum_{j=1}^{M} a_{i(j)}^2 \int_0^1 \cos^2(2\pi n_{i(j)} x) \, dx = \frac{M}{2},$$

which is precisely (4.76).

Case 9 (long products). $k \geq 2$. We can clearly write $1 \leq j_1 < j_2 < \ldots < j_k$. Then using the elementary inequalities

$$1 + \frac{\varepsilon}{2} + \left(\frac{\varepsilon}{2}\right)^2 + \left(\frac{\varepsilon}{2}\right)^3 + \ldots < 1 + \varepsilon \quad \text{and} \quad 1 - \frac{\varepsilon}{2} - \left(\frac{\varepsilon}{2}\right)^2 - \left(\frac{\varepsilon}{2}\right)^3 - \ldots > \frac{1}{1 + \varepsilon}$$

if $0 < \varepsilon < \frac{1}{2}$, we deduce that

$$|\pm n_{i(j_1)} \pm n_{i(j_2)} \pm n_{i(j_3)} \pm \ldots \pm n_{i(j_k)}| \text{ lies between } (1 + \varepsilon) n_{i(j_k)} \text{ and } \frac{1}{1 + \varepsilon} n_{i(j_k)}.$$

Comparing this to the gap condition (4.74), we see that $F(x)$ and the long products of $R(x)$ represent disjoint sets of exponential functions

$$e^{2\pi i \ell x}, \quad \ell \in \mathbf{Z}.$$

Using the orthogonality of these functions, the contribution of Case 9 to the integral $\int_0^1 F(x) R(x) \, dx$ is zero. This proves (4.76).

The same argument shows that

$$\int_0^1 R(x) \, dx = 1. \tag{4.78}$$

Since $R(x) \geq 0$, the condition (4.78) means that the integral $\int_0^1 F(x) R(x) \, dx$ is a weighted average of $F(x)$, with non-negative weights. It follows from (4.76) that

$$\max_{0 \leq x \leq 1} F(x) \geq \int_0^1 F(x) R(x) \, dx = \frac{M}{2}. \tag{4.79}$$

The inequality $(1 + \varepsilon)^r > 2/\varepsilon$ clearly holds with $r = 2\varepsilon^{-1} \log(2\varepsilon^{-1})$. Thus by (4.74) and (4.75), we can choose

$$M \geq \frac{N}{r} = \frac{N}{2\varepsilon^{-1} \log(2\varepsilon^{-1})}. \tag{4.80}$$

Sidon's theorem then follows from (4.79) and (4.80).   □

4.4.2 The Rectangle Property and Theorem 12

Let us return now to Theorem 3 and (4.72). We restate Theorem 3 in a slightly different form. Recall the notation in (4.56). We have

$$H_N(\sqrt{2}; \gamma) = \{(x, y) \in \mathbf{R}^2 : -\gamma \le x^2 - 2y^2 \le \gamma, \ 1 \le x + y\sqrt{2} \le 2\sqrt{2}N\}, \tag{4.81}$$

that is, $H_N(\sqrt{2}; \gamma)$ is a long, narrow, tilted hyperbolic needle of slope $1/\sqrt{2}$. Its area is $(\gamma/\sqrt{2}) \log N + O(1)$; see (4.57). Theorem 3 states, roughly speaking, that in the special case $\gamma = \frac{1}{2}$, there are two translated copies of the same tilted hyperbolic needle $H_N(\sqrt{2}; \gamma = 1/2)$ such that one is substantially richer in lattice points than the other. The discrepancy is proportional to the area, and we have extra large deviation. More precisely, there is a positive absolute constant $c_{10} > 0$ such that for infinitely many integers N_i, where $N_i \to \infty$, there are translated copies $\mathbf{x}_1^{(i)} + H_{N_i}(\sqrt{2}; \gamma)$ and $\mathbf{x}_2^{(i)} + H_{N_i}(\sqrt{2}; \gamma)$ of the tilted hyperbolic needle $H_{N_i}(\sqrt{2}; \gamma = 1/2)$ such that

$$|\mathbf{Z}^2 \cap (\mathbf{x}_1^{(i)} + H_{N_i}(\sqrt{2}; \gamma = 1/2))| - |\mathbf{Z}^2 \cap (\mathbf{x}_2^{(i)} + H_{N_i}(\sqrt{2}; \gamma = 1/2))|$$

$$> c_{10} \log N_i. \tag{4.82}$$

In view of the periodicity of the lattice points, we can clearly assume that the pairs of vectors $\mathbf{x}_1^{(i)}$ and $\mathbf{x}_2^{(i)}$ are all in the unit square $[0, 1)^2$, with $i \to \infty$.

The extra large deviation result (4.82), which is equivalent to Theorem 3, can be generalized in several stages. The first generalization is (4.72), or at least an equivalent form as follows.

Proposition 10. *Let $\gamma > 0$ be an arbitrary but fixed real number, and let $N \ge 2$ be an integer. Then there exists a positive constant $\delta' = \delta'(\gamma) > 0$, independent of N, such that for the tilted hyperbolic needle $H_N(\sqrt{2}; \gamma)$ of area $(\gamma/\sqrt{2}) \log N + O(1)$, there exist translated copies $\mathbf{x}_1 + H_N(\sqrt{2}; \gamma)$ and $\mathbf{x}_2 + H_N(\sqrt{2}; \gamma)$ such that*

$$|\mathbf{Z}^2 \cap (\mathbf{x}_1 + H_N(\sqrt{2}; \gamma))| > \frac{\gamma}{\sqrt{2}} \log N + \delta' \log N$$

and

$$|\mathbf{Z}^2 \cap (\mathbf{x}_2 + H_N(\sqrt{2}; \gamma))| < \frac{\gamma}{\sqrt{2}} \log N - \delta' \log N.$$

Note that Proposition 10 immediately leads to the existence of a single divergence point $\beta^* = \beta^*(\gamma) \in [0, 1)$ in (4.72). To exhibit continuum many divergence points $\beta^* = \beta^*(\gamma) \in [0, 1)$, we simply have to combine Proposition 10 with the routine Cantor set argument in the proof of Theorem 3.

For the second stage of generalization, we replace the set $\mathbf{Z}^2$ of lattice points in the plane with an arbitrary subset $\mathscr{A} \subset \mathbf{Z}^2$ of positive density. Here is an illustration of such a set $\mathscr{A}$. We say that a lattice point $\mathbf{n} = (n_1, n_2) \in \mathbf{Z}^2$ is *coprime*[7] if the coordinates n_1 and n_2 are relatively prime. Let $\mathbf{Z}^2_{\text{coprime}}$ denote the set of coprime lattice points in the plane. It is well known from number theory that $\mathbf{Z}^2_{\text{coprime}}$ is a subset of $\mathbf{Z}^2$ with positive density $6/\pi^2$.

Now let $\mathscr{A}$ be an arbitrary subset of $\mathbf{Z}^2$ of positive density $\delta = \delta(\mathscr{A}) > 0$. There is a natural generalization of Proposition 10 where we replace $\mathbf{Z}^2$ with $\mathscr{A}$. The price that we have to pay is that, due to the lack of periodicity of a general subset $\mathscr{A}$, the translations are not necessarily in the unit square anymore.

Proposition 11. *Let $\mathscr{A} \subset \mathbf{Z}^2$ be an arbitrary subset of positive density $\delta = \delta(\mathscr{A}) > 0$. Let $\gamma > 0$ be an arbitrary but fixed real number, and let $N \geq 2$ be an integer. Assume further that M/N is sufficiently large, depending only on γ and δ. Then there exists a positive constant $\delta' = \delta'(\gamma, \delta) > 0$, independent of N and M, such that for the tilted hyperbolic needle $H_N(\sqrt{2}; \gamma)$ of area $(\gamma/\sqrt{2}) \log N + O(1)$, there exist translated copies $\mathbf{x}_1 + H_N(\sqrt{2}; \gamma) \subset [0, M]^2$ and $\mathbf{x}_2 + H_N(\sqrt{2}; \gamma) \subset [0, M]^2$ such that*

$$|\mathscr{A} \cap (\mathbf{x}_1 + H_N(\sqrt{2}; \gamma))| > \frac{\delta\gamma}{\sqrt{2}} \log N + \delta' \log N$$

and

$$|\mathscr{A} \cap (\mathbf{x}_2 + H_N(\sqrt{2}; \gamma))| < \frac{\delta\gamma}{\sqrt{2}} \log N - \delta' \log N.$$

It turns out that the only relevant property of a lattice point set $\mathscr{A} \subset \mathbf{Z}^2$ that we really use in the proof of Proposition 11 is the rectangle property in Lemma 4, that every tilted rectangle of slope $1/\sqrt{2}$ and area $\frac{1}{5}$ contains at most one lattice point. Of course, the concrete value $\frac{1}{5}$ of the constant is secondary.

The third stage of generalization goes far beyond the family of lattice point sets $\mathscr{A} \subset \mathbf{Z}^2$. The only requirement is that the point set satisfies the rectangle property.

Theorem 12. *Let $\mathscr{P}$ be a finite set of points in the square $[0, M]^2$ with density δ, so that the number of elements of $\mathscr{P}$ is $|\mathscr{P}| = \delta M^2$. Assume further that $\mathscr{P}$ satisfies the following rectangle property, that there is a positive constant $c_1 = c_1(\mathscr{P}) > 0$ such that every tilted rectangle of slope $1/\sqrt{2}$ and area c_1 contains at most one element of the set $\mathscr{P}$. Let*

---

[7]We also say that such a point is *visible*, explained by the geometric fact that the line segment with $\mathbf{n}$ and the origin as endpoints does not contain another lattice point. If $\mathbf{n} = (n_1, n_2) \in \mathbf{Z}^2$ were not coprime, then the point $(n_1/d, n_2/d) \in \mathbf{Z}^2$, where $d \geq 2$ is the greatest common divisor of n_1 and n_2, would lie on this line segment.

$$\delta' = \delta'(c_1, \gamma, \delta) = 10^{-12}\delta\kappa, \tag{4.83}$$

where

$$\kappa = \min\left\{ \frac{\gamma}{20}, \sqrt{c_1\gamma}, \frac{10^{-7}c_1}{2}, \frac{10^{-7}c_1^2}{2\gamma} \right\}. \tag{4.84}$$

Furthermore, assume that both N and M/N are sufficiently large and satisfy

$$N \geq 2^{10(\gamma+\gamma^{-1})} \quad and \quad M > \frac{10^{11}(\gamma + \gamma^{-1})(N + 2\gamma)}{c_1\delta\kappa}. \tag{4.85}$$

Then for the tilted hyperbolic needle $H_N(\sqrt{2}; \gamma)$ of area $(\gamma/\sqrt{2})\log N + O(1)$, there exist translated copies $x_1 + H_N(\sqrt{2}; \gamma) \subset [0, M]^2$ and $x_2 + H_N(\sqrt{2}; \gamma) \subset [0, M]^2$ such that

$$|\mathscr{P} \cap (x_1 + H_N(\sqrt{2}; \gamma))| > \frac{\delta\gamma}{\sqrt{2}}\log N + \delta'\log N$$

and

$$|\mathscr{P} \cap (x_2 + H_N(\sqrt{2}; \gamma))| < \frac{\delta\gamma}{\sqrt{2}}\log N - \delta'\log N.$$

Note that Propositions 10 and 11 are special cases of Theorem 12, with $\mathscr{P} = \mathbf{Z}^2$ and $\mathscr{P} = \mathscr{A}$ respectively.

Unfortunately, the proof of Theorem 12 is rather difficult and long, and the very complicated details cover the next four sections. But the main idea is quite simple. It is basically a sophisticated application of the Riesz product.

4.5 Proof of Theorem 12 (I): Proving Extra Large Deviations via Riesz Product

Since the proof is long and complicated, a convenient notation here makes a big difference. It is much simpler for us to work with hyperbolic regions in the usual horizontal-vertical position instead of the tilted position. It means that, instead of working with the set $\mathbf{Z}^2$ of lattice points in the plane and the family of tilted hyperbolic needles of a fixed quadratic irrational slope, as in the setting of Theorem 12, we rotate back. In other words, we rotate $\mathbf{Z}^2$ by a quadratic irrational slope, and consider the family of hyperbolic needles in the usual horizontal-vertical position.

Let $\gamma > 0$ be an arbitrary real number, and let $N \geq 2$ be a large integer. Consider the hyperbolic region

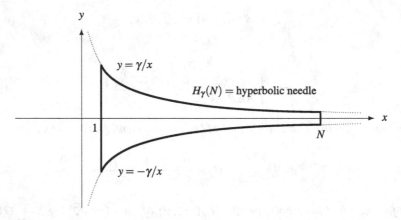

Fig. 4.2 A hyperbolic needle in usual horizontal-vertical position

$$H_\gamma(N) = \{(x, y) \in \mathbf{R}^2 : -\gamma \leq xy \leq \gamma, \; 1 \leq x \leq N\}; \qquad (4.86)$$

see Fig. 4.2. Again we refer to $H_\gamma(N)$ as a hyperbolic needle.

Notice that $H_\gamma(N)$ is basically the horizontal-vertical version of the tilted hyperbolic needle $H_N(\sqrt{2}; \gamma)$; see (4.56) or (4.81). To emphasize the difference between the tilted and the horizontal-vertical versions, we have made a major change in the notation, and switched the location of the parameters γ and N.

The area of $H_\gamma(N)$ equals the integral

$$\mathrm{area}(H_\gamma(N)) = 2 \int_1^N \frac{\gamma}{x} \, \mathrm{d}x = 2\gamma \log N.$$

Let $\mathrm{rot}_\alpha \mathbf{Z}^2$ denote the rotated copy of $\mathbf{Z}^2$ by the angle θ, where $\tan \theta = \alpha$ is the slope and using the origin as the fixed point of the rotation. If $\alpha \neq 0$ is a quadratic irrational, then the continued fractions for α is finally periodic. This is a well known number-theoretic fact; for example, if $\alpha = 1/\sqrt{2}$, then

$$\frac{1}{\sqrt{2}} = \frac{1}{1+} \frac{1}{2+} \frac{1}{2+} \frac{1}{2+} \dots = [1, 2, 2, 2, \dots] = [1, \bar{2}].$$

Periodicity implies that the continued fraction digits, formally known as the partial quotients, form a bounded sequence. It is well known that boundedness yields

$$k \|k\alpha\| \geq c_{11} = c_{11}(\alpha) > 0 \quad \text{for all integers } k \geq 1, \qquad (4.87)$$

where $c_{11} = c_{11}(\alpha) > 0$ is some positive constant depending only on α, and $\|z\|$ denotes the distance of a real number z to the nearest integer. If $\alpha = 1/\sqrt{2}$, then (4.87) follows from the factorization $x^2 - 2y^2 = (x - y\sqrt{2})(x + y\sqrt{2})$. If x and y are integers, then

$$1 \leq |x^2 - 2y^2| = |(x - y\sqrt{2})(x + y\sqrt{2})| = |x\alpha - y|\sqrt{2}|x + y\sqrt{2}|,$$

and we choose $x = k$ and y to be the nearest integer to $k\alpha$. This explains why in the special case $\alpha = 1/\sqrt{2}$ that the choice $c_{11} = \frac{1}{4}$ in (4.87) works.

Inequality (4.87) has an important geometric interpretation, namely that there is another constant $c_{12} = c_{12}(\alpha) > 0$, depending only on α, such that for every axes-parallel rectangle R,

$$|\mathrm{rot}_\alpha \mathbf{Z}^2 \cap R| \leq 1 \quad \text{whenever} \quad \mathrm{area}(R) = c_{12}(\alpha). \tag{4.88}$$

If $\alpha = 1/\sqrt{2}$, then $c_{12} = \frac{1}{5}$ is a good choice in (4.88), in view of Lemma 4.

The following statement is just a slight generalization of Theorem 12.

Proposition 13. *Let $\mathscr{P}$ be a finite set of points in the square $[0, M]^2$ with density δ, so that the number of elements of $\mathscr{P}$ is $|\mathscr{P}| = \delta M^2$. Assume further that $\mathscr{P}$ satisfies the following rectangle property, that there is a positive constant $c_1 = c_1(\mathscr{P}) > 0$ such that every axes-parallel rectangle of area c_1 contains at most one element of the set $\mathscr{P}$. Let $\delta' = \delta'(c_1, \gamma, \delta)$ be defined by (4.83) and (4.84), and assume that both N and M/N are sufficiently large and satisfy (4.85). Then for the hyperbolic needle $H_\gamma(N)$ given by (4.86), there exist translated copies $\mathbf{x}_1 + H_\gamma(N) \subset [0, M]^2$ and $\mathbf{x}_2 + H_\gamma(N) \subset [0, M]^2$ such that*

$$|\mathscr{P} \cap (\mathbf{x}_1 + H_\gamma(N))| > 2\delta\gamma \log N + \delta' \log N \tag{4.89}$$

and

$$|\mathscr{P} \cap (\mathbf{x}_2 + H_\gamma(N))| < 2\delta\gamma \log N - \delta' \log N. \tag{4.90}$$

Remarks. (i) The term $2\delta\gamma \log N$ in (4.89) and (4.90) represents the expectation, since the set $\mathscr{P}$ has density δ and the hyperbolic needle $H_\gamma(N)$ has area $2\gamma \log N$. The extra terms $\pm\delta' \log N$ show that the deviation from the expectation is proportional to the expectation, justifying the terminology *extra large deviation*.

(ii) The constant factors such as 10^{-12} and 10^{11} are certainly very far from best possible. Since the proof is complicated, our primary goal is to present the basic ideas in the simplest form, and we do not care too much about optimizing these constant factors.

We begin our long proof of Proposition 13.

Consider the point-counting function

$$f(\mathbf{x}) = |\mathscr{P} \cap (\mathbf{x} + H_\gamma(N))|. \tag{4.91}$$

If $\mathbf{x} \in [0, M - N] \times [\gamma, M - \gamma]$, then clearly

$$\mathbf{x} + H_\gamma(N) \subset [0, M]^2. \tag{4.92}$$

This explains why we choose the rectangle $[0, M - N] \times [\gamma, M - \gamma]$ to be our underlying domain in the proof.

Let

$$\Delta(\mathbf{x}) = f(\mathbf{x}) - \delta \cdot \text{area}(H_\gamma(N)) = f(\mathbf{x}) - 2\delta\gamma \log N \qquad (4.93)$$

denote the discrepancy function; $\Delta(\mathbf{x})$ deserves its name if (4.92) holds.

In order to show that $\Delta(\mathbf{x}) > \delta' \log N > 0$ holds for some $\mathbf{x} = \mathbf{x}_1$, we apply the test function method initiated by Roth [26]. The basic idea of this method is to construct a positive test function $T(\mathbf{x}) > 0$ such that

$$\frac{1}{(M - N)(M - 2\gamma)} \int_0^{M-N} \int_\gamma^{M-\gamma} \Delta(\mathbf{x}) T(\mathbf{x}) \, d\mathbf{x} > c_{13} \log N > 0, \qquad (4.94)$$

and

$$\frac{1}{(M - N)(M - 2\gamma)} \int_0^{M-N} \int_\gamma^{M-\gamma} T(\mathbf{x}) \, d\mathbf{x} < c_{14}. \qquad (4.95)$$

Combining (4.94) and (4.95) with the general trivial inequality

$$\int \Delta(\mathbf{x}) T(\mathbf{x}) \, d\mathbf{x} \le \max_{\mathbf{x}} \Delta(\mathbf{x}) \int T(\mathbf{x}) \, d\mathbf{x}, \qquad (4.96)$$

which holds for any positive function $T(\mathbf{x}) > 0$, we conclude that

$$\max_{\mathbf{x}} \Delta(\mathbf{x}) > c_{15} \log N$$

with some positive constant $c_{15} > 0$.

Similarly, to show that $\Delta(\mathbf{x}) < -\delta' \log N < 0$ for some $\mathbf{x} = \mathbf{x}_2$, we construct a positive test function $T^*(\mathbf{x}) > 0$ such that

$$\frac{1}{(M - N)(M - 2\gamma)} \int_0^{M-N} \int_\gamma^{M-\gamma} \Delta(\mathbf{x}) T^*(\mathbf{x}) \, d\mathbf{x} < -c_{16} \log N < 0, \qquad (4.97)$$

and again

$$\frac{1}{(M - N)(M - 2\gamma)} \int_0^{M-N} \int_\gamma^{M-\gamma} T^*(\mathbf{x}) \, d\mathbf{x} < c_{17}. \qquad (4.98)$$

Clearly (4.97) and (4.98) lead to the inequality

$$\min_{\mathbf{x}} \Delta(\mathbf{x}) < -c_{18} \log N < 0$$

with some positive constant $c_{18} > 0$.

Let us return to (4.94) and (4.95). We shall express the test function $T(\mathbf{x})$ in terms of modified Rademacher functions, sometimes called Haar wavelet, and this is another idea that we borrow from Roth's pioneering paper [26]. The benefit of working with modified Rademacher functions is that we have orthogonality and, what is more, we have *super-orthogonality*; see the key property below.

Note that Roth simply took the sum of certain modified Rademacher functions, and applied the Cauchy–Schwarz inequality instead of (4.96). For his argument, orthogonality was sufficient. It was Halász's innovation[8] to express $T(\mathbf{x})$ as a Riesz product of modified Rademacher functions; see Halász [19]. The main point is that the Riesz product takes advantage of the super-orthogonality. Here we develop an adaptation of the Roth–Halász method for hyperbolic regions.

Following the Roth–Halász approach, we shall express the test function $T(\mathbf{x})$ as a Riesz product of modified Rademacher functions, in the form

$$T(\mathbf{x}) = \prod_{j \in \mathscr{J}} (1 + \rho R_j(\mathbf{x})), \tag{4.99}$$

where $0 < \rho < 1$ is an appropriate constant to be specified later, $\mathscr{J}$ is some appropriate index-set and $R_j(\mathbf{x})$, $j \in \mathscr{J}$, are certain modified Rademacher functions to be defined below. We assume that the test function $T(\mathbf{x})$ is zero outside the rectangle $[0, M - N] \times [\gamma, M - \gamma]$.

Suppose that $10^{-2} > \eta_1 > 0$ and $10^{-2} > \eta_2 > 0$ are small positive real numbers, to be specified later, such that

$$\frac{M - N}{\eta_1} = \frac{M - 2\gamma}{\eta_2} = 2^m, \tag{4.100}$$

where $m \geq 1$ is an integer. Let j be an arbitrary integer in the interval $0 \leq j \leq n$ where $2^n \approx N$, that is, $n = \log_2 N + O(1)$ in binary logarithm. We decompose the rectangle $[0, M - N] \times [\gamma, M - \gamma]$ into $2^m \times 2^m = 4^m$ disjoint translated copies of the small rectangle

$$[0, 2^j \eta_1] \times [0, 2^{-j} \eta_2], \tag{4.101}$$

and call these congruent copies of the small rectangle (4.101) j-cells. For each of the 4^m j-cells, we independently choose one of the three patterns $+-$, $-+$ and 0; see Fig. 4.3.

As Fig. 4.3 shows, the pattern $+-$ actually means a two-dimensional pattern as follows. We divide the j-cell into four congruent subrectangles, and define a step-function on the j-cell, with value $+1$ on the upper-right and lower-left subrectangles, and value -1 on the upper-left and lower-right subrectangles.

---

[8]Halász used this method, among many other things, to give an elegant new proof of Schmidt's well-known discrepancy theorem; see [27].

Fig. 4.3 The patterns $+-$, $-+$ and 0

Similarly, the pattern $-+$ means the step-function with value -1 on the upper-right and lower-left subrectangles, and value $+1$ on the upper-left and lower-right subrectangles.

Finally, the pattern 0 means that the step-function is zero on the whole j-cell.

In the sequel, we shall simply refer to these two-dimensional patterns as $+-$, $-+$ and 0, representing the bottom rows in Fig. 4.3.

By making an independent choice of $+-$, $-+$ and 0 for each j-cell, we obtain a particular modified Rademacher function $R_j(\mathbf{x})$ of order j, defined over the whole rectangle $[0, M - N] \times [\gamma, M - \gamma]$. We define $R_j(\mathbf{x})$ to be 0 outside the rectangle $[0, M - N] \times [\gamma, M - \gamma]$.

Since for each of the 4^m j-cells there are 3 options, namely $+-$, $-+$ and 0, the total number of modified Rademacher functions $R_j(\mathbf{x})$ of order j is 3^{4^m}. Let $\mathscr{R}(j)$ denote the family of all 3^{4^m} modified Rademacher functions of order j. Note that the notation $R_j(\mathbf{x})$ is somewhat ambiguous in the sense that it represents any element of this huge family $\mathscr{R}(j)$.

Super-Orthogonality: Key Property of the Modified Rademacher Functions.
If $k \geq 1$ and $0 \leq j_1 < \ldots < j_k \leq n$, then in every elementary cell of size $2^{j_1}\eta_1 \times 2^{-j_k}\eta_2$, the product $R_{j_1}(\mathbf{x}) \ldots R_{j_k}(\mathbf{x})$ of k modified Rademacher functions satisfies one of the three familiar patterns in Fig. 4.3.

Note that an elementary cell of size $2^{j_1}\eta_1 \times 2^{-j_k}\eta_2$ arises as a non-empty intersection of a j_1-cell and a j_k-cell, where $j_1 < j_k$. The proof of the above key property is almost trivial. It is based on the fact that for any $k \geq 2$, the intersection of any k cells of different orders $j_1 < \ldots < j_k$ is either empty or equal to the intersection of the j_1-cell and the j_k-cell, i.e. the intersection of the first and the last. We emphasize that in each of the 3 patterns the integral of the corresponding step-function is zero.

Since every modified Rademacher function $R_j(\mathbf{x})$ has values ± 1 or 0, and since $0 < \rho < 1$, it is clear that the Riesz product (4.99) defines a positive test function $T(\mathbf{x})$. The index-set $\mathscr{J}$, a subset of $\{0, 1, 2, \ldots, n\}$, will be specified later. Note in advance that $\mathscr{J}$ is a large subset of $\{0, 1, 2, \ldots, n\}$, in the sense that $|\mathscr{J}| \geq c_{19}(n + 1)$.

Next we check the second requirement (4.95) of the test function. Multiplying out the Riesz product (4.99), we have

$$T(\mathbf{x}) = \prod_{j \in \mathscr{J}} (1 + \rho R_j(\mathbf{x}))$$

$$= 1 + \rho \sum_{j \in \mathscr{J}} R_j(\mathbf{x}) + \rho^2 \sum_{\substack{j_1 < j_2 \\ j_i \in \mathscr{J}}} R_{j_1}(\mathbf{x}) R_{j_2}(\mathbf{x})$$

$$+ \rho^3 \sum_{\substack{j_1 < j_2 < j_3 \\ j_i \in \mathscr{J}}} R_{j_1}(\mathbf{x}) R_{j_2}(\mathbf{x}) R_{j_3}(\mathbf{x}) + \dots, \tag{4.102}$$

in the form 1 plus the linear part plus the quadratic part plus the cubic part and so on. Substituting (4.102) into the left hand side of (4.95), we have

$$\frac{1}{(M-N)(M-2\gamma)} \int_0^{M-N} \int_\gamma^{M-\gamma} T(\mathbf{x}) \, d\mathbf{x}$$

$$= 1 + \sum_{k \geq 1} \frac{\rho^k}{(M-N)(M-2\gamma)} \sum_{\substack{j_1 < \dots < j_k \\ j_i \in \mathscr{J}}} \int_0^{M-N} \int_\gamma^{M-\gamma} R_{j_1}(\mathbf{x}) \dots R_{j_k}(\mathbf{x}) \, d\mathbf{x}$$

$$= 1. \tag{4.103}$$

The vanishing integrals in the last step occurs as a consequence of the super-orthogonality of the modified Rademacher functions. For each of 3 patterns that the integrand takes, the integral is zero. Clearly (4.103) gives (4.95) with $c_{14} = 1$.

Finally, we turn to requirement (4.94). The verification of this is by far the most difficult part of the proof. This is where we make the critical decision on how we choose an appropriate modified Rademacher function $R_j(\mathbf{x})$ from amongst the huge family $\mathscr{R}(j)$ of size 3^{4^m}. We choose the best $R_j(\mathbf{x}) \in \mathscr{R}(j)$ in order to *synchronize the trivial errors*. The synchronization argument is at the very heart of the proof. Note that if we did not synchronize the trivial errors, then they might cancel out, and we would then not be able to guarantee extra large deviation.

The Trivial Errors and Synchronization. By (4.91) and (4.93), the discrepancy function equals

$$\Delta(\mathbf{x}) = |\mathscr{P} \cap (\mathbf{x} + H_\gamma(N))| - \delta \cdot \text{area}(H_\gamma(N)),$$

and so we can write

$$\int_0^{M-N} \int_\gamma^{M-\gamma} \Delta(\mathbf{x}) T(\mathbf{x}) \, d\mathbf{x}$$

$$= \int_0^{M-N} \int_\gamma^{M-\gamma} \left(\sum_{P_i \in \mathscr{P} \cap (\mathbf{x} + H_\gamma(N))} 1 - \delta \cdot \text{area}(H_\gamma(N)) \right) T(\mathbf{x}) \, d\mathbf{x}$$

$$
= \int_0^{M-N} \int_\gamma^{M-\gamma} \left(\sum_{P_i \in \mathscr{P} \cap (\mathbf{x} + H_\gamma(N))} 1 \right) T(\mathbf{x}) \, d\mathbf{x}
$$

$$
-(M-N)(M-2\gamma)\delta \cdot \text{area}(H_\gamma(N)), \tag{4.104}
$$

where in the last step we have used (4.103), and where $P_1, P_2, P_3, \ldots$ denote the elements of the given point set $\mathscr{P}$.

Changing the order of summation and integration, we obtain

$$
\int_0^{M-N} \int_\gamma^{M-\gamma} \left(\sum_{P_i \in \mathscr{P} \cap (\mathbf{x} + H_\gamma(N))} 1 \right) T(\mathbf{x}) \, d\mathbf{x} = \sum_{P_i \in \mathscr{P}} \int_{P_i - H_\gamma(N)} T(\mathbf{x}) \, d\mathbf{x}, \tag{4.105}
$$

where

$$
P_i - H_\gamma(N) = \{ P_i - \mathbf{w} : \mathbf{w} \in H_\gamma(N) \}
$$

denotes a reflected and translated copy of the hyperbolic needle $H_\gamma(N)$. Combining (4.104) and (4.105), we have

$$
\frac{1}{(M-N)(M-2\gamma)} \int_0^{M-N} \int_\gamma^{M-\gamma} \Delta(\mathbf{x}) T(\mathbf{x}) \, d\mathbf{x}
$$

$$
= \sum_{P_i \in \mathscr{P}} \frac{1}{(M-N)(M-2\gamma)} \int_{P_i - H_\gamma(N)} T(\mathbf{x}) \, d\mathbf{x} - \delta \cdot \text{area}(H_\gamma(N)). \tag{4.106}
$$

To evaluate (4.106), we return to the Riesz product (4.102). Note that the term 1 in fact denotes the characteristic function χ_B of the rectangle $B = [0, M-N] \times [\gamma, M-\gamma]$, since by definition the modified Rademacher functions are all zero outside B.

We begin with the contribution of $1 = \chi_B$ in (4.102), and note simply that

$$
\int_{P_i - H_\gamma(N)} \chi_B(\mathbf{x}) \, d\mathbf{x} = \int_{B \cap (P_i - H_\gamma(N))} d\mathbf{x} = \text{area}(B \cap (P_i - H_\gamma(N))). \tag{4.107}
$$

Geometric Ideas. Next we study the contribution of the linear part of (4.102) in (4.106). Synchronization means that we want to make the sum

$$
\sum_{P_i \in \mathscr{P}} \int_{P_i - H_\gamma(N)} R_j(\mathbf{x}) \, d\mathbf{x} \tag{4.108}
$$

large and positive for every $j \in \mathscr{J}$, where the index-set $\mathscr{J} \subset \{0, 1, 2, \ldots, n\}$ will be specified later. We decompose the underlying rectangle $B = [0, M-N] \times [\gamma, M-\gamma]$ into j-cells. Let $\mathscr{C}$ be an arbitrary j-cell; it has size $\eta_1 \eta_2$. Consider a

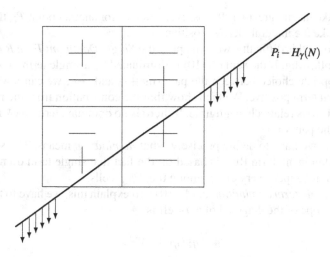

Fig. 4.4 Intersection of a j-cell with a hyperbolic arc $P_i - H_\gamma(N)$

single term in (4.108), and restrict it to the j-cell $\mathscr{C}$. The geometric meaning of the integral

$$\int_{\mathscr{C} \cap (P_i - H_\gamma(N))} R_j(\mathbf{x}) \, d\mathbf{x} \tag{4.109}$$

plays a crucial role in the argument below; see Fig. 4.4.

Since the j-cell is very small, the hyperbola arc $P_i - H_\gamma(N)$ can be approximated by its tangent line locally. This explains the tilted straight line segment in Fig. 4.4. The arrows indicate the inside of the hyperbolic needle, i.e. the arc in the picture is the upper arc of the needle.

The value of integral (4.109) depends heavily on which of the 3 patterns happens to show up in the restriction of $R_j(\mathbf{x})$ to the j-cell $\mathscr{C}$. The patterns $+-$ and $-+$ give two integrals whose sum is 0, whereas the pattern 0 clearly gives an integral with value 0.

How do we choose the right pattern $+-$, $-+$ or 0 in an arbitrary j-cell $\mathscr{C}$? Well, for a fixed point the choice is trivial. For every fixed point $P_i \in \mathscr{P}$, exactly one of the two patterns $+-$ and $-+$ will make the integral (4.109) positive, unless both integrals are equal to 0. The problem is that we are dealing with a large sum

$$\sum_{P_i \in \mathscr{P}} \int_{\mathscr{C} \cap (P_i - H_\gamma(N))} R_j(\mathbf{x}) \, d\mathbf{x} \tag{4.110}$$

instead of just a single term (4.109), and we have to make (4.110) positive. The difficulty is that different points may prefer different patterns; say, for P_{i_1} the pattern

$+-$ may make the integral (4.109) positive, whereas for another point P_{i_2} the pattern $-+$ may make the integral (4.109) positive.

To overcome this difficulty, we will apply the *Single Dominant Term Rule*, which means the following. If the sum (4.110) is dominated by a single term (4.109), then by an appropriate choice between the patterns $+-$ and $-+$, we can always make this dominant term positive. We then show that the contribution from the remaining terms to (4.110) is relatively negligible. If there is no dominant term in (4.110), then we choose the pattern 0.

Of course, we have to define precisely what *domination* means. The success of the Single Dominant Term Rule is based on the fact that single term domination is quite typical: it happens very often among the 4^m j-cells.

What is *single term domination* in (4.110)? To explain this, we have to talk about slopes. The slope of the diagonal of a j-cell is

$$4^{-j} \eta_2 / \eta_1 \approx 4^{-j},$$

since η_1 and η_2 are almost equal.[9] Since the hyperbola is a smooth curve, the intersection of a translated and reflected hyperbolic needle $P_i - H_\gamma(N)$ with the j-cell $\mathscr{C}$ is almost like the intersection of $\mathscr{C}$ with a half-plane, or the intersection of $\mathscr{C}$ with two nearly parallel half-planes. Since half-planes have well-defined constant slopes, as an intuitive oversimplification, we shall use the terms *half-plane* and *slope* for the intersections $\mathscr{C} \cap (P_i - H_\gamma(N))$. Single term domination occurs if

- there is precisely one half-plane $\mathscr{C} \cap (P_i - H_\gamma(N))$ with slope close to 4^{-j} that intersects $\mathscr{C}$; and
- this intersection is a *large triangle* in only one of the four subrectangles of $\mathscr{C}$, namely the lower right subrectangle, where the pattern is constant.

Here the intersection requirement *large triangle from the lower right subrectangle* guarantees that the integral (4.109) is far from zero, and the integral (4.109) of this dominant term is called the *trivial error*.

An Important Consequence of the Rectangle Property. As indicated above, single term domination means that there is exactly one half-plane $\mathscr{C} \cap (P_i - H_\gamma(N))$ with slope close to 4^{-j}. It is important to point out that we cannot have two half-planes with slopes very close to 4^{-j} such that both are upper arcs. As shown Fig. 4.5, if $\mathscr{C} \cap (P_{i_1} - H_\gamma(N))$ and $\mathscr{C} \cap (P_{i_2} - H_\gamma(N))$ are both upper arcs with slopes very close to 4^{-j}, then the two points P_{i_1} and P_{i_2} have to be in the same axes-parallel rectangle of area c_1, namely, in an axes-parallel rectangle where the slope of the

---

[9]We do not distinguish between positive and negative slopes. Note that the reflected hyperbolic needle $-H_\gamma(N)$ has two long arcs: the upper arc, which is increasing, and the lower arc, which is decreasing; here the lower arc is below the upper arc. When we say that $P_i - H_\gamma(N)$ intersects $\mathscr{C}$, then it always means that at least one of the two long arcs of $P_i - H_\gamma(N)$ intersects $\mathscr{C}$. For example, in the *trivial error* discussed at the end of this paragraph, the intersection comes from the upper arc.

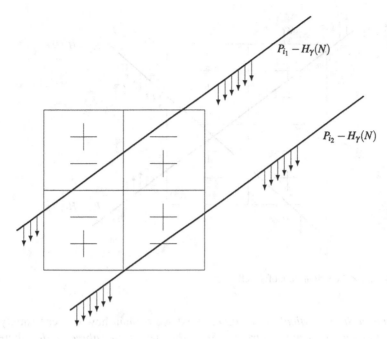

Fig. 4.5 Forbidden configuration

diagonal is close to 4^{-j}. But two points in the same axes-parallel rectangle of area c_1 is impossible: it contradicts the hypothesis of Proposition 13.

What can happen, however, is that we have two half-planes with slopes very close to 4^{-j} such that one is an upper arc and the other one is a lower arc. For example, it can happen that $\mathscr{C} \cap (P_{i_1} - H_\gamma(N))$ is an upper arc and $\mathscr{C} \cap (P_{i_2} - H_\gamma(N))$ is a lower arc with both slopes[10] close to 4^{-j}. To overcome this difficulty, we switch to a 2×2 configuration of j-cells. More precisely, instead of working with a single j-cell $\mathscr{C}$, we switch to a 2×2 configuration of four neighboring j-cells $\mathscr{C}_1$, $\mathscr{C}_2$, $\mathscr{C}_3$ and $\mathscr{C}_4$, where $\mathscr{C}_1$ is the upper left, $\mathscr{C}_2$ is the upper right, $\mathscr{C}_3$ is the lower left and $\mathscr{C}_4$ is the lower right member of the 2×2 configuration. The simple geometric idea is the following. Assume that the upper arc of $P_{i_1} - H_\gamma(N)$ intersects both $\mathscr{C}_2$ and $\mathscr{C}_3$ satisfying the requirement *large triangle from the lower right subrectangle*, where the pattern is constant. Then obviously the lower arc of $P_{i_2} - H_\gamma(N)$ cannot intersect both of $\mathscr{C}_2$ and $\mathscr{C}_3$, since the slopes are close to 4^{-j}. Therefore, either $\mathscr{C}_2$ or $\mathscr{C}_3$ will be a j-cell with single term domination. That is, we can always save at least one of the four neighboring j-cells $\mathscr{C}_1$, $\mathscr{C}_2$, $\mathscr{C}_3$ and $\mathscr{C}_4$. See Fig. 4.6, where $\mathscr{C}_3$ has single term domination.

---

[10]Again, we do not distinguish between positive and negative slopes.

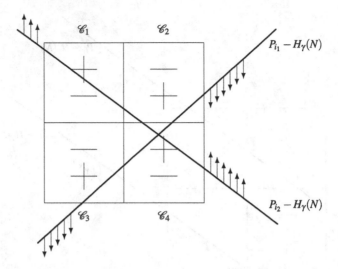

Fig. 4.6 A 2×2 configuration of j-cells

Choosing a Short Vertical Translation. Next we explain how one can satisfy the intersection requirement *large triangle from the lower right subrectangle*, where the pattern is constant. This is very important, since this requirement guarantees that the dominant integral (4.109) is far from zero. First we pick an arbitrary point $P_i \in \mathscr{P}$. Then of course the hyperbolic needle $P_i - H_\gamma(N)$ has a long arc such that the slope is close to 4^{-j}; *long* in fact means length of roughly 2^j. Therefore, for each point $P_i \in \mathscr{P}$, there is a j-cell $\mathscr{C}$ such that the intersection $\mathscr{C} \cap (P_i - H_\gamma(N))$ has slope close to 4^{-j}. Unfortunately, nothing guarantees that $P_i - H_\gamma(N)$ intersects only one of the four subrectangles, where the pattern is constant. The solution is very simple. We apply a short vertical translation of the point set $\mathscr{P}$, but of course the modified Rademacher functions and the test function $T(\mathbf{x})$ remain fixed in the rectangle $B = [0, M - N] \times [\gamma, M - \gamma]$. Here a *short* vertical translation means that the length of the vertical translation runs from 0 to 1. For a j-cell, a translation of length from 0 to $2^{-j} \eta_2$ already suffices: as the point P_i moves up vertically, the intersection $\mathscr{C} \cap (P_i - H_\gamma(N))$ changes, and has good positions where $P_i - H_\gamma(N)$ intersects only the lower right subrectangle, where the pattern is constant, and at the same time, this intersection is a large triangle. Since the slope is close to 4^{-j}, a positive constant percentage of the translations is good. If we apply translations from 0 to 1, then it will work for all j.

It follows from a standard averaging argument that there is[11] a vertical translation $0 < t_0 < 1$ which is good for many pairs (P_i, j) at the same time, where $P_i \in \mathscr{P}$ is a given point and $j \in \{0, 1, 2, \ldots, n\}$ is an order of the modified Rademacher function. Here *many* means a positive constant percentage of all pairs.

---

[11]In fact, the majority will do.

Of course, a vertical translation has a bad side effect. It causes some points to leave the underlying square $[0, M]^2$. However, luckily for us, it suffices to use short translations of length at most 1, so that we lose relatively few points, and only those that are close to the boundary. Note that the rectangle property in the hypothesis of Proposition 13 guarantees that there are at most $O(M)$ points close to the boundary, which clearly is negligible compared to the number δM^2 of points in $\mathcal{P}$.

Summarizing the Vague Geometric Intuition. A typical vertical translation of length $0 < t_0 < 1$ has the property that for a positive constant percentage of the pairs $(j, \mathcal{C})$, where $j \in \{0, 1, 2, \ldots, n\}$ and $\mathcal{C}$ is a j-cell, we have single term domination, so that[12]

$$\sum_{P_i \in \mathcal{P}} \int_{\mathcal{C} \cap (P_i - H_\gamma(N))} R_j(\mathbf{x}) \, d\mathbf{x} \geq \frac{1}{2} \int_{\mathcal{C} \cap (P_{i_0} - H_\gamma(N))} R_j(\mathbf{x}) \, d\mathbf{x} \geq c_{20} > 0, \quad (4.111)$$

where P_{i_0} is the dominating point, i.e. the intersection $\mathcal{C} \cap (P_{i_0} - H_\gamma(N))$ has slope close to 4^{-j}, and this intersection is a large triangle from the lower right subrectangle of $\mathcal{C}$, where the pattern is constant. We shall explain the missing details of (4.111) later, and give an explicit value for c_{20}.

The Single Term Domination Rule and (4.111) give

$$\sum_{j \in \mathcal{J}} \sum_{P_i \in \mathcal{P}} \frac{1}{(M - N)(M - 2\gamma)} \int_{P_i - H_\gamma(N)} R_j(\mathbf{x}) \, d\mathbf{x}$$

$$\geq c_{21} |\mathcal{J}| \geq c_{22}(n + 1) > 0. \quad (4.112)$$

The geometric intuition requires that $j \in \mathcal{J}$ satisfies an inequality like

$$\max\left\{1, \frac{1}{\gamma}\right\} \leq 2^j \leq \min\left\{N, \frac{N}{\gamma}\right\}. \quad (4.113)$$

To guarantee (4.113), we choose $\mathcal{J}$ to be the interval of integers $j \in \{0, 1, 2, \ldots, n\}$ satisfying

$$\log_2\left(\max\left\{1, \frac{1}{\gamma}\right\}\right) \leq j \leq \log_2 N - \log_2(\max\{1, \gamma\}). \quad (4.114)$$

We emphasize that this was just an intuitive proof of (4.112). We shall return to (4.111) and (4.112) later, and show how we can make the whole argument perfectly precise and explicit.

We shall complete the proof of Proposition 13 in the next three sections. Note that (4.112) is the most difficult part.

---

[12]Here we skip a lot of technical details!

4.6 Proof of Theorem 12 (II): More on the Riesz Product

Applying Super-Orthogonality. We next turn to the contribution of the quadratic, cubic and higher order terms of the Riesz product (4.102) to (4.106). Let $k \geq 2$, and let $0 \leq j_1 < \ldots < j_k \leq n$. Suppose that $\mathscr{C}^*$ is the non-empty intersection of k cells of orders $j_1 < \ldots < j_k$. Then $\mathscr{C}^*$ is an elementary cell of size $2^{j_1} \eta_1 \times 2^{-j_k} \eta_2 = 2^{j_1 - j_k} \eta_1 \eta_2$. Super-orthogonality yields that the product $R_{j_1}(\mathbf{x}) \ldots R_{j_k}(\mathbf{x})$ of k modified Rademacher functions of the given orders, restricted to $\mathscr{C}^*$, equals one of the 3 patterns $+-$, $-+$ or 0.

Assume that the translated and reflected hyperbolic needle $P_i - H_\gamma(N)$ intersects $\mathscr{C}^*$, and let slope = slope($\mathscr{C}^* \cap (P_i - H_\gamma(N))$) denote the slope[13] of the intersection $\mathscr{C}^* \cap (P_i - H_\gamma(N))$. Simple geometric consideration shows that, roughly speaking, the integral

$$\frac{1}{\text{area}(\mathscr{C}^*)} \int_{\mathscr{C}^* \cap (P_i - H_\gamma(N))} R_{j_1}(\mathbf{x}) \ldots R_{j_k}(\mathbf{x}) \, d\mathbf{x}$$

is negligible unless the slope of the intersection $\mathscr{C}^* \cap (P_i - H_\gamma(N))$ is close to $2^{-(j_1 + j_k)}$, the slope of the diagonal of $\mathscr{C}^*$. More precisely, we have

$$\frac{1}{\text{area}(\mathscr{C}^*)} \left| \int_{\mathscr{C}^* \cap (P_i - H_\gamma(N))} R_{j_1}(\mathbf{x}) \ldots R_{j_k}(\mathbf{x}) \, d\mathbf{x} \right|$$

$$\leq \min \left\{ \frac{1}{\text{slope} \cdot 2^{j_1 + j_k}}, \text{slope} \cdot 2^{j_1 + j_k} \right\}. \tag{4.115}$$

Note that (4.115) is a straightforward corollary of the geometry of the 3 possible patterns of $R_{j_1}(\mathbf{x}) \ldots R_{j_k}(\mathbf{x})$ in $\mathscr{C}^*$.

The hyperbolic needle $H_\gamma(N)$ is bounded by the long curves $y = \gamma/x$ and its reflection $y = -\gamma/x$, with $1 \leq x \leq N$. The slope is the derivative $(-\gamma/x)' = \gamma x^{-2}$. The number of elementary cells $\mathscr{C}^*$ of size $2^{j_1 - j_k} \eta_1 \eta_2$ intersecting a fixed hyperbolic needle $P_i - H_\gamma(N)$ is estimated from above by the simple expression

$$2 \left(\frac{2N}{2^{j_1} \eta_1} + \frac{2\gamma}{2^{-j_k} \eta_2} \right). \tag{4.116}$$

Here the factor 2 comes from the two long boundary hyperbolic curves, the first term comes from the pointed end of the hyperbolic needle, and the second term comes from the wide part of the hyperbolic needle. A more detailed explanation of (4.116) goes as follows.

---

[13] We do not distinguish between positive and negative slopes.

Let us start with the pointed end of the hyperbolic needle $H_\gamma(N)$.

Case A. As x runs through the interval $N \geq x \geq \sqrt{\gamma}2^{(j_1+j_k)/2}$, the slope of the intersection $\mathscr{C}^* \cap (P_i - H_\gamma(N))$ is γx^{-2}, which is less than $2^{-(j_1+j_k)}$, the slope of the diagonal of $\mathscr{C}^*$. It follows that in this range, $P_i - H_\gamma(N)$ intersects fewer than

$$2 \cdot \frac{2N}{2^{j_1}\eta_1}$$

elementary cells $\mathscr{C}^*$ of size $2^{j_1-j_k}\eta_1\eta_2$, with total area not exceeding $4\eta_2 N 2^{-j_k}$.

Case B. As x runs through the interval $\sqrt{\gamma}2^{(j_1+j_k)/2} \geq x \geq 1$, the slope of the intersection $\mathscr{C}^* \cap (P_i - H_\gamma(N))$ is greater than $2^{-(j_1+j_k)}$, the slope of the diagonal of $\mathscr{C}^*$. It follows that in this range, $P_i - H_\gamma(N)$ intersects fewer than

$$2 \cdot \frac{2\gamma}{2^{-j_k}\eta_2}$$

elementary cells $\mathscr{C}^*$ of size $2^{j_1-j_k}\eta_1\eta_2$, with total area not exceeding $4\eta_1\gamma 2^{j_1}$.

In Case A, we view the hyperbola $xy = \gamma$ as $y = \gamma/x$. In Case B, we switch the role of the coordinate axes and view the same hyperbola as $x = \gamma/y$. Thus by (4.115) and (4.116), we have

$$\left| \int_{P_i - H_\gamma(N)} R_{j_1}(\mathbf{x}) \dots R_{j_k}(\mathbf{x}) \, d\mathbf{x} \right|$$

$$\leq 4\eta_2 N 2^{-j_k} \cdot \frac{2}{n} \int_{\sqrt{\gamma}2^{(j_1+j_k)/2}}^{N} \frac{\gamma 2^{j_1+j_k}}{x^2} \, dx + 4\eta_1\gamma 2^{j_1} \cdot \frac{2}{\gamma} \int_{\sqrt{\gamma}2^{-(j_1+j_k)/2}}^{\gamma} \frac{\gamma 2^{-(j_1+j_k)}}{y^2} \, dy$$

$$= 8\eta_2 2^{-j_k}\left(\sqrt{\gamma}2^{(j_1+j_k)/2} - \frac{\gamma 2^{j_1+j_k}}{N} \right) + 8\eta_1 2^{j_1}\left(\sqrt{\gamma}2^{-(j_1+j_k)/2} - \gamma 2^{-(j_1+j_k)} \right)$$

$$\leq 8\sqrt{\gamma}(\eta_1 + \eta_2)2^{(j_1-j_k)/2}. \tag{4.117}$$

Recall that the contribution $1 = \chi_B$ in (4.102), where $B = [0, M - N] \times [\gamma, M - \gamma]$. Combining (4.102), (4.106) and (4.107), we have

$$\frac{1}{(M-N)(M-2\gamma)} \int_0^{M-N} \int_\gamma^{M-\gamma} \Delta(\mathbf{x})T(\mathbf{x}) \, d\mathbf{x}$$

$$= \sum_{P_i \in \mathscr{P}} \frac{\text{area}(B \cap (P_i - H_\gamma(N)))}{(M-N)(M-2\gamma)} - \delta \cdot \text{area}(H_\gamma(N))$$

$$+\rho \sum_{j \in \mathscr{J}} \sum_{P_i \in \mathscr{P}} \frac{1}{(M-N)(M-2\gamma)} \int_{P_i - H_\gamma(N)} R_j(\mathbf{x}) \, d\mathbf{x}$$

$$+\sum_{k \geq 2} \rho^k \sum_{\substack{j_1 < \dots < j_k \\ j_i \in \mathscr{J}}} \sum_{P_i \in \mathscr{P}} \frac{1}{(M-N)(M-2\gamma)} \int_{P_i - H_\gamma(N)} R_{j_1}(\mathbf{x}) \dots R_{j_k}(\mathbf{x}) \, d\mathbf{x}.$$

$$(4.118)$$

Using (4.117), it is easy to estimate the last term in (4.118). We have

$$\sum_{k \geq 2} \rho^k \sum_{\substack{j_1 < \dots < j_k \\ j_i \in \mathscr{J}}} \sum_{P_i \in \mathscr{P}} \frac{1}{(M-N)(M-2\gamma)} \left| \int_{P_i - H_\gamma(N)} R_{j_1}(\mathbf{x}) \dots R_{j_k}(\mathbf{x}) \, d\mathbf{x} \right|$$

$$\leq \sum_{k \geq 2} \rho^k \sum_{0 \leq j_1 < \dots < j_k \leq n} \sum_{P_i \in \mathscr{P}} \frac{8\sqrt{\gamma}(\eta_1 + \eta_2) 2^{(j_1 - j_k)/2}}{(M-N)(M-2\gamma)}. \qquad (4.119)$$

For convenience, let us write $q = j_k - j_1$. We estimate the sum

$$\sum_{k \geq 2} \rho^k \sum_{j_1 = 0}^{n-k+1} \sum_{q = k-1}^{n-j_1} \sum_{j_1 < j_2 < \dots < j_{k-1} < j_1 + q} 2^{-q/2}. \qquad (4.120)$$

In the innermost sum in (4.120), the indices $j_2, \dots, j_{k-1}$ can be chosen from among the $q-1$ numbers lying between j_1 and $j_1 + q$ in $\binom{q-1}{k-2}$ ways. To simplify (4.120), we can let the indices j_1 and q run up to n. Then we change the order of summation. Thus we have

$$\sum_{k \geq 2} \rho^k \sum_{j_1 = 0}^{n-k+1} \sum_{q = k-1}^{n-j_1} \sum_{j_1 < j_2 < \dots < j_{k-1} < j_1 + q} 2^{-q/2}$$

$$\leq \sum_{k \geq 2} \rho^k \sum_{j_1 = 0}^{n} \sum_{q = k-1}^{n} \binom{q-1}{k-2} 2^{-q/2} = \sum_{j_1 = 0}^{n} \sum_{q=1}^{n} 2^{-q/2} \sum_{k=2}^{q+1} \rho^k \binom{q-1}{k-2}. $$

$$(4.121)$$

Note that the innermost sum

$$\sum_{k=2}^{q+1} \rho^k \binom{q-1}{k-2} = \rho^2 \sum_{k=2}^{q+1} \rho^{k-2} \binom{q-1}{k-2} = \rho^2 (1+\rho)^{q-1}.$$

It follows that if $0 < \rho < \sqrt{2} - 1$, then

$$\sum_{j_1=0}^{n}\sum_{q=1}^{n}2^{-q/2}\sum_{k=2}^{q+1}\rho^k\binom{q-1}{k-2}=\sum_{j_1=0}^{n}\sum_{q=1}^{n}2^{-q/2}\rho^2(1+\rho)^{q-1}$$

$$=\frac{(n+1)\rho^2}{\sqrt{2}}\sum_{q=1}^{n}\left(\frac{1+\rho}{\sqrt{2}}\right)^{q-1}\le\frac{(n+1)\rho^2}{\sqrt{2}}\sum_{q=1}^{\infty}\left(\frac{1+\rho}{\sqrt{2}}\right)^{q-1}$$

$$=\frac{(n+1)\rho^2}{\sqrt{2}}\left(1-\frac{1+\rho}{\sqrt{2}}\right)^{-1}=\frac{(n+1)\rho^2}{\sqrt{2}-1-\rho}. \tag{4.122}$$

Combining (4.119)–(4.122), we obtain

Lemma 14. *If* $0<\rho<\sqrt{2}-1$, *then*

$$\sum_{k\ge2}\rho^k\sum_{\substack{j_1<...<j_k\\ j_i\in\mathscr{J}}}\sum_{P_i\in\mathscr{P}}\frac{1}{(M-N)(M-2\gamma)}\left|\int_{P_i-H_\gamma(N)}R_{j_1}(\mathbf{x})\ldots R_{j_k}(\mathbf{x})\,d\mathbf{x}\right|$$

$$\le\frac{|\mathscr{P}|}{(M-N)(M-2\gamma)}\cdot8\sqrt{\gamma}(\eta_1+\eta_2)\cdot\frac{(n+1)\rho^2}{\sqrt{2}-1-\rho}. \tag{4.123}$$

We return to (4.118). The contribution from the first term on the right hand side is $o(1)$, so that it is negligible. To see this, we recall that $|\mathscr{P}|=\delta M^2$, and also that $P_i-H_\gamma(N)\subset B=[0,M-N]\times[\gamma,M-\gamma]$ for all but $O(M)$ points $P_i\in\mathscr{P}$. Thus

$$\sum_{P_i\in\mathscr{P}}\frac{\text{area}(B\cap(P_i-H_\gamma(N)))}{(M-N)(M-2\gamma)}-\delta\cdot\text{area}(H_\gamma(N))$$

$$=\frac{\delta M^2+O(M)}{(M-N)(M-2\gamma)}\cdot\text{area}(H_\gamma(N))-\delta\cdot\text{area}(H_\gamma(N))$$

$$=O\left(\frac{N\log N}{M}\right)=o(1). \tag{4.124}$$

For the second term on the right hand side of (4.118), we have the estimate (4.112). Thus combining (4.112), (4.118), (4.123) and (4.124), we obtain

$$\frac{1}{(M-N)(M-2\gamma)}\int_0^{M-N}\int_\gamma^{M-\gamma}\Delta(\mathbf{x})T(\mathbf{x})\,d\mathbf{x}$$

$$\ge c_{23}\rho(n+1)-c_{24}\frac{(n+1)\rho^2}{\sqrt{2}-1-\rho}-o(1),$$

where the constants, the first one yet unspecified, are positive and $0 < \rho < \sqrt{2} - 1$. By choosing a sufficiently small ρ in the range $0 < \rho < \sqrt{2} - 1$, we clearly have

$$\frac{1}{(M-N)(M-2\gamma)} \int_0^{M-N} \int_\gamma^{M-\gamma} \Delta(\mathbf{x}) T(\mathbf{x}) \, d\mathbf{x}$$

$$\geq c_{25}\rho(n+1) > c_{26} \log N > 0,$$

proving (4.94), and thus proving Proposition 13 in the positive direction; see (4.89). It remains to clarify the missing details in (4.111) and (4.112); see also the paragraph *Summarizing the Vague Geometric Intuition* at the end of Sect. 4.5.

Single Term Domination: Clarifying the Technical Details. The geometric ideas introduced in Sect. 4.5 lead to the following conclusion. At least half of the short vertical translations $\mathscr{P} + (0, t_0)$, where $0 < t_0 < 1$, of the given point set $\mathscr{P}$ have the property that for at least 1 % of the pairs $(j, \mathscr{C})$, where $j \in \{0, 1, 2, \ldots, n\}$ and $\mathscr{C}$ is a j-cell of the underlying rectangle $B = [0, M-N] \times [\gamma, M-\gamma]$, there is single term domination. This includes, among other requirements to be specified later, that there is a dominating point $P_{i_0} = P_{i_0}(j, \mathscr{C}) \in \mathscr{P}$ such that

- $\mathscr{C} \cap (P_{i_0} - H_\gamma(N))$ has slope between $\frac{5}{6}4^{-j}$ and $\frac{7}{6}4^{-j}$;
- $P_{i_0} - H_\gamma(N)$ intersects only the lower right subrectangle of $\mathscr{C}$, and the intersection is a large triangle, meaning that the area is at least $\frac{1}{32}$ of the area of $\mathscr{C}$, that is, the area is at least $\eta_1\eta_2/32$.

Then, by choosing the pattern $+-$ in the j-cell $\mathscr{C}$, we have

$$\int_{\mathscr{C} \cap (P_{i_0} - H_\gamma(N))} R_j(\mathbf{x}) \, d\mathbf{x} \geq \frac{\eta_1\eta_2}{32}. \tag{4.125}$$

To justify the notion *single term domination*, we shall show that for a typical pair $(j, \mathscr{C})$, the contribution of the remaining points $P_i \in \mathscr{P}$, with $i \neq i_0$, in the j-cell $\mathscr{C}$ is negligible, in the sense that

$$\left| \sum_{\substack{P_i \in \mathscr{P} \\ i \neq i_0}} \int_{\mathscr{C} \cap (P_i - H_\gamma(N))} R_j(\mathbf{x}) \, d\mathbf{x} \right| \leq \frac{\eta_1\eta_2}{40}. \tag{4.126}$$

To prove (4.126), let $P_i \neq P_{i_0}$ be another point in $\mathscr{P}$ such that $P_i - H_\gamma(N)$ intersects $\mathscr{C}$, i.e. the upper or lower arc of the boundary of the hyperbolic needle $P_i - H_\gamma(N)$ intersects the j-cell $\mathscr{C}$. We are going to distinguish four cases, depending on the type of the intersection of $P_i - H_\gamma(N)$ with $\mathscr{C}$, corresponding to upper or lower arc, and close to horizontal or close to vertical, relative to the diagonals of $\mathscr{C}$.

Case 15. The upper arc of $P_i - H_\gamma(N)$ intersects $\mathscr{C}$, and the slope is less than the slope of the dominant needle $P_{i_0} - H_\gamma(N)$; see Fig. 4.7.

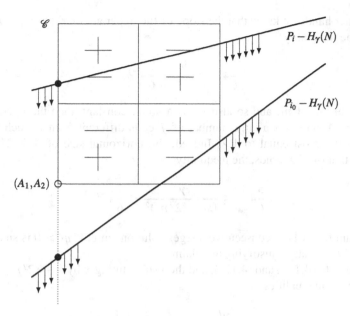

Fig. 4.7 Upper arc of $P_i - H_\gamma(N)$ intersects $\mathscr{C}$, with slope less than slope of $P_{i_0} - H_\gamma(N)$

Let $P_{i_0} = (a_{i_0}, b_{i_0})$ and $P_i = (a_i, b_i)$ denote the coordinates of the two points in question. By the hypothesis of Case 1, we have $a_i > a_{i_0}$. Write

$$h = h_i = a_i - a_{i_0} > 0 \quad \text{and} \quad v = v_i = b_i - b_{i_0},$$

where of course h denotes horizontal and v denotes vertical. The rectangle property guarantees that $h|v| \geq c_1 > 0$.

Let (A_1, A_2) denote the coordinates of the lower left vertex of the j-cell $\mathscr{C}$. The intersection of the line $x = A_1$ with the upper arcs of $P_{i_0} - H_\gamma(N)$ and $P_i - H_\gamma(N)$ give two points, and the hypothesis of Case 1 implies that these intersection points are close to each other. More precisely, with $x = 1 + a_{i_0} - A_1$, where $a_{i_0} - A_1 > 0$ and the additional term 1 comes from the fact that the hyperbolic needle $H_\gamma(N)$ begins at $x = 1$, we have the upper bound

$$\left| \left(b_{i_0} + \frac{\gamma}{x} \right) - \left(b_i + \frac{\gamma}{x+h} \right) \right| < 2 \cdot 2^{-j} \eta_2. \tag{4.127}$$

Since $b_i - b_{i_0} = v$, we can rewrite (4.127) in the form

$$\left| \left(\frac{\gamma}{x} - \frac{\gamma}{x+h} \right) - v \right| = \left| \frac{\gamma h}{x(x+h)} - v \right| < 2^{-j+1} \eta_2. \tag{4.128}$$

On the other hand, we know that the slope of the upper arc of $\mathscr{C} \cap (P_{i_0} - H_\gamma(N))$ satisfies the inequality

$$\frac{5}{6}4^{-j} \le \frac{\gamma}{x^2} \le \frac{7}{6}4^{-j}. \tag{4.129}$$

We claim that if η_1, and so also η_2, is a small constant, then the upper arc of $P_{i_0} - H_\gamma(N)$ intersects a large number of j-cells different from $\mathscr{C}$ such that the slope is still almost equal to 4^{-j}. Indeed, the horizontal size of $\mathscr{C}$ is $2^j \eta_1$ and, assuming that (4.129) holds, the inequality

$$\frac{5}{6}4^{-j} \le \frac{\gamma}{(x + \ell 2^j \eta_1)^2} \le \frac{7}{6}4^{-j} \tag{4.130}$$

has constant times $1/\eta_1$ consecutive integer solutions in ℓ. If $\eta_1 > 0$ is small, then of course $1/\eta_1$ is large, justifying our claim.

Returning to (4.128) and (4.129), and then substituting x by $x + \ell 2^j \eta_1$, we have the respective inequalities

$$\left| \frac{\gamma h}{(x + \ell 2^j \eta_1)(x + \ell 2^j \eta_1 + h)} - v \right| < 2^{-j+1} \eta_2 \tag{4.131}$$

and (4.130). If (4.129) holds, then there are at least $\sqrt{\gamma}/10\eta_1$ consecutive integer solutions ℓ of (4.130).

The basic idea is the following. If ℓ runs through these integer solutions of (4.130) while γ, x, h and v remain fixed, then the function

$$\frac{\gamma h}{(x + \ell 2^j \eta_1)(x + \ell 2^j \eta_1 + h)}, \tag{4.132}$$

as a function of ℓ, has substantially different values, and we expect only very few of them to be very close to a fixed v in the quantitative sense of (4.131). Of course, here we assume that η_2 is small.

Next we work out the details of this intuition. We begin by noting that (4.130) implies

$$\sqrt{\frac{6\gamma}{5}}2^j \ge x + \ell 2^j \eta_1 \ge \sqrt{\frac{6\gamma}{7}}2^j. \tag{4.133}$$

Using this in (4.132), we have the good approximation

$$\frac{\gamma h}{(x + \ell 2^j \eta_1)(x + \ell 2^j \eta_1 + h)} \approx \frac{\gamma h}{\sqrt{\gamma}2^j(\sqrt{\gamma}2^j + h)} = \frac{h}{2^j(2^j + h/\sqrt{\gamma})}. \tag{4.134}$$

We now distinguish two cases. First assume that $0 < h \leq \sqrt{c_1}2^{j-1}$, where $c_1 > 0$ is the positive constant in the rectangle property. Then the rectangle property yields

$$|v| \geq \frac{c_1}{h} \geq \frac{c_1}{\sqrt{c_1}2^{j-1}} = 2\sqrt{c_1}2^{-j} \tag{4.135}$$

and

$$\frac{h}{2^j(2^j + h/\sqrt{\gamma})} < \frac{h}{2^j 2^j} \leq \frac{\sqrt{c_1}}{2}2^{-j}. \tag{4.136}$$

The assumption

$$\eta_2 < \frac{\sqrt{c_1}}{2}, \tag{4.137}$$

together with (4.134)–(4.136), implies that (4.131) has no solution.

We can assume, therefore, that the lower bound

$$h > \sqrt{c_1}2^{j-1} \tag{4.138}$$

holds. Now we go back to the basic idea. We claim that if we switch ℓ to $\ell + 1$ in the function (4.132), then its value changes by at least as much as

$$\frac{\eta_1 2^{-j-2}}{1 + \sqrt{\gamma/c_1}}. \tag{4.139}$$

Indeed, by (4.133), we have

$$\frac{\gamma h}{(x + \ell 2^j \eta_1)(x + \ell 2^j \eta_1 + h)} \approx \frac{1}{\sqrt{\gamma}2^j + 2^j \eta_1} \cdot \frac{\gamma h}{\sqrt{\gamma}2^j + h}. \tag{4.140}$$

We also have the routine estimate

$$\frac{1}{\sqrt{\gamma}2^j} - \frac{1}{\sqrt{\gamma}2^j + 2^j \eta_1} = \frac{1}{\sqrt{\gamma}2^j}\left(1 - \frac{1}{1 + \eta_1/\sqrt{\gamma}}\right)$$

$$= \frac{1}{\sqrt{\gamma}2^j}\left(\frac{\eta_1}{\sqrt{\gamma}} - \left(\frac{\eta_1}{\sqrt{\gamma}}\right)^2 + \left(\frac{\eta_1}{\sqrt{\gamma}}\right)^3 \mp \dots\right) \approx \frac{\eta_1}{\gamma 2^j}. \tag{4.141}$$

Furthermore, by (4.138), we have

$$\frac{\gamma h}{\sqrt{\gamma}2^j + h} > \frac{\gamma}{2\sqrt{\gamma/c_1} + 1}. \tag{4.142}$$

Then the error estimate (4.139) follows on combining (4.140)–(4.142).

Let us return to (4.132) and (4.139), and apply them in (4.131). We deduce that among the constant times $1/\eta_1$ consecutive integer values of ℓ satisfying (4.130), there are only constant times $(1 + \sqrt{\gamma/c_1})$ that will satisfy (4.131). More explicitly, it is safe to say that

$$\text{at most } 10\left(1 + \sqrt{\frac{\gamma}{c_1}}\right) \text{ values of } \ell \text{ will satisfy both (4.130) and (4.131). } \quad (4.143)$$

The next step is

A Combination of the Rectangle Property and the Pigeonhole Principle. We recall (4.138), that $h > \sqrt{c_1}2^{j-1}$. Consider the power-of-two type decomposition

$$2^{r-1}\sqrt{c_1}2^j < h \le 2^r\sqrt{c_1}2^j, \quad r = 0,1,2,\ldots. \quad (4.144)$$

We claim that for a fixed point $P_{i_0} = (a_{i_0}, b_{i_0}) \in \mathscr{P}$ and for a fixed integer $r \ge 0$, there are at most

$$10\sqrt{\frac{\gamma}{c_1}}2^r \quad (4.145)$$

other points $P_i = (a_i, b_i) \in \mathscr{P}$, with $P_i \ne P_{i_0}$, such that $h = h_i = a_i - a_{i_0} > 0$ and $v = v_i = b_i - b_{i_0}$ satisfy (4.131), thus implicitly (4.130) also, and (4.144).

To establish the bound (4.145), first note that if $h = h_i$ satisfies (4.144), then by (4.134) and (4.144), we have

$$\frac{\gamma h}{(x + \ell 2^j \eta_1)(x + \ell 2^j \eta_1 + h)} \approx \frac{h}{2^j(2^j + h/\sqrt{\gamma})}$$

$$\approx \frac{2^r\sqrt{c_1}2^j}{2^j(2^j + 2^r\sqrt{c_1}2^j/\sqrt{\gamma})} = \frac{2^{-j}}{1/\sqrt{\gamma} + 2^{-r}/\sqrt{c_1}},$$

so that a solution of (4.131) gives the approximation

$$v = v_i \approx 2^{-j}\left(\frac{1}{1/\sqrt{\gamma} + 2^{-r}/\sqrt{c_1}} \pm 2\eta_2\right). \quad (4.146)$$

Assuming

$$\eta_2 < \frac{1}{8(1/\sqrt{\gamma} + 1/\sqrt{c_1})}, \quad (4.147)$$

then (4.146) yields the good approximation

$$v = v_i \approx \frac{2^{-j}}{1/\sqrt{\gamma} + 2^{-r}/\sqrt{c_1}}. \quad (4.148)$$

Suppose on the contrary that there are more than (4.145) other points $P_i = (a_i, b_i) \in \mathscr{P}$, with $P_i \neq P_{i_0}$, such that $h = h_i = a_i - a_{i_0} > 0$ and $v = v_i = b_i - b_{i_0}$ satisfy (4.131), thus implicitly (4.130) also, and (4.144). Then by the Pigeonhole Principle and (4.148), there must exist two points $P_{i_1}, P_{i_2} \in \mathscr{P}$, with $i_1 \neq i_2$, such that

$$v_{i_1} \approx \frac{2^{-j}}{1/\sqrt{\gamma} + 2^{-r}/\sqrt{c_1}} \approx v_{i_2} \quad \text{and} \quad |h_{i_1} - h_{i_2}| \leq \frac{2^r \sqrt{c_1} 2^j}{10\sqrt{\gamma/c_1} 2^r} = \frac{c_1 2^j}{10\sqrt{\gamma}}.$$

Since the product

$$\frac{2^{-j}}{1/\sqrt{\gamma} + 2^{-r}/\sqrt{c_1}} \cdot \frac{c_1 2^j}{\sqrt{\gamma}} = \frac{c_1}{1 + 2^{-r}\sqrt{\gamma/c_1}} < c_1,$$

we conclude that there exists an axes-parallel rectangle of area less than c_1 and which contains at least two points of $\mathscr{P}$, namely P_{i_1} and P_{i_2}. This contradicts the rectangle property, and establishes the bound (4.145).

If $h = h_i$ falls into the interval (4.144), then

$$\text{slope}(\mathscr{C} \cap (P_i - H_\gamma(N))) = \frac{\gamma}{(x+h)^2} \leq \frac{\gamma}{h^2} \approx \frac{\gamma}{c_1 4^r} \cdot 4^{-j}, \qquad (4.149)$$

where 4^{-j} almost equals the slope of the diagonals of the j-cell $\mathscr{C}$. By (4.149), we have

$$\frac{1}{\text{area}(\mathscr{C})} \left| \int_{\mathscr{C} \cap (P_i - H_\gamma(N))} R_j(\mathbf{x}) \, d\mathbf{x} \right| \leq \frac{10\gamma}{c_1 4^r}. \qquad (4.150)$$

Furthermore, (4.150) holds for all j-cells $\mathscr{C}$ satisfying

$$\frac{5}{6} 4^{-j} \leq \text{slope}(\mathscr{C} \cap (P_{i_0} - H_\gamma(N))) \leq \frac{7}{6} 4^{-j}. \qquad (4.151)$$

Let us return now to (4.126). Combining (4.143)–(4.145) and (4.150), we have

$$\sum_{\substack{P_i \in \mathscr{P} \\ i \neq i_0 \\ \text{Case 1}}} \sum_{\mathscr{C}} \frac{1}{\text{area}(\mathscr{C})} \left| \int_{\mathscr{C} \cap (P_i - H_\gamma(N))} R_j(\mathbf{x}) \, d\mathbf{x} \right| \leq \sum_{r \geq 0} 10 \left(1 + \sqrt{\frac{\gamma}{c_1}}\right) 10 \sqrt{\frac{\gamma}{c_1}} 2^r \frac{10\gamma}{c_1 4^r}$$

$$= 1000 \left(\left(\frac{\gamma}{c_1}\right)^{3/2} + \left(\frac{\gamma}{c_1}\right)^2\right) \sum_{r \geq 0} 2^{-r} = 2000 \left(\left(\frac{\gamma}{c_1}\right)^{3/2} + \left(\frac{\gamma}{c_1}\right)^2\right). \qquad (4.152)$$

Since there are at least $\gamma/10\eta_1$ consecutive integer solutions ℓ of (4.130), assuming that (4.129) holds, we have

$$\sum_{\mathscr{C}} 1 \geq \frac{\sqrt{\gamma}}{10\eta_1}. \qquad (4.153)$$
$$\text{(4.151)}$$

Recall that in order to prove (4.126), we distinguish four cases. Inequalities (4.152) and (4.153) complete Case 1. The remaining three cases will be discussed in the next section. Note that these cases are quite similar to Case 1, but there are some annoying differences in the minor details. We shall complete the proof of Proposition 13 in Sect. 4.8.

4.7   Proof of Theorem 12 (III): Completing the Case Study

Let us return to (4.125) and (4.126). Again we assume that there is a dominating point $P_{i_0} = P_{i_0}(j, \mathscr{C}) \in \mathscr{P}$ such that

- $\mathscr{C} \cap (P_{i_0} - H_\gamma(N))$ has slope between $\frac{5}{6}4^{-j}$ and $\frac{7}{6}4^{-j}$;
- $P_{i_0} - H_\gamma(N)$ intersects only the lower right subrectangle of $\mathscr{C}$, and the intersection is a large triangle, meaning that the area is at least $\frac{1}{32}$ of the area of $\mathscr{C}$, that is, the area is at least $\eta_1\eta_2/32$.

Again let $P_i \neq P_{i_0}$ be another point in $\mathscr{P}$ such that $P_i - H_\gamma(N)$ intersects $\mathscr{C}$, i.e. the upper or lower arc of the boundary of the hyperbolic needle $P_i - H_\gamma(N)$ intersects the j-cell $\mathscr{C}$. We now discuss the second case, which is quite similar to the first case. Roughly speaking, we switch the roles of the horizontal and the vertical.

Case 16. The upper arc of $P_i - H_\gamma(N)$ intersects $\mathscr{C}$, and the slope is greater than the slope of the dominant needle $P_{i_0} - H_\gamma(N)$; see Fig. 4.8.

Let $P_{i_0} = (a_{i_0}, b_{i_0})$ and $P_i = (a_i, b_i)$ denote the coordinates of the two points in question. By the hypothesis of Case 2, we have $a_{i_0} > a_i$. Write

$$h = h_i = a_{i_0} - a_i > 0 \quad \text{and} \quad v = v_i = b_{i_0} - b_i,$$

where again h denotes horizontal and v denotes vertical. The rectangle property guarantees that $h|v| \geq c_1 > 0$.

Let (A_1, A_2) denote the coordinates of the upper left vertex of the j-cell $\mathscr{C}$. The intersection of the line $y = A_2$ with the upper arcs of $P_{i_0} - H_\gamma(N)$ and $P_i - H_\gamma(N)$ give two points, and the hypothesis of Case 16 implies that these intersection points are close to each other. More precisely, with $y = A_2 - b_{i_0}$, we have the upper bound

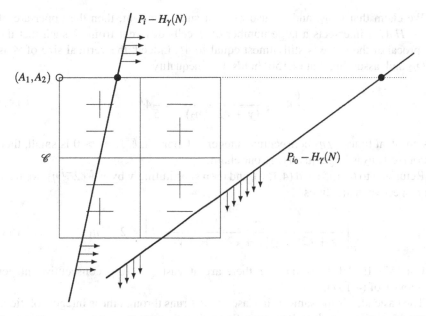

Fig. 4.8 Upper arc of $P_i - H_\gamma(N)$ intersects $\mathscr{C}$, with slope greater than slope of $P_{i_0} - H_\gamma(N)$

$$\left| \left(a_i - \frac{\gamma}{y+v} \right) - \left(a_{i_0} - \frac{\gamma}{y} \right) \right| < 2 \cdot 2^j \eta_1. \qquad (4.154)$$

Since $a_{i_0} - a_i = h > 0$, we can rewrite (4.154) in the form

$$\left| \left(\frac{\gamma}{y} - \frac{\gamma}{y+v} \right) - h \right| = \left| \frac{\gamma v}{y(y+v)} - h \right| < 2^{j+1} \eta_1. \qquad (4.155)$$

We emphasize that $y + v > 0$, otherwise

$$0 \geq y + v = (A_2 - b_{i_0}) + (b_{i_0} - b_i) = A_2 - b_i,$$

so that $b_i \geq A_2$, which means that the whole upper arc of $P_i - H_\gamma(N)$ is above the j-cell $\mathscr{C}$. But this is impossible, since in Case 2 we assume that the upper arc of $P_i - H_\gamma(N)$ intersects $\mathscr{C}$.

Since we switch the roles of the horizontal and the vertical, we focus on the reciprocal of the slope. We know that the reciprocal of the slope of the upper arc of $\mathscr{C} \cap (P_{i_0} - H_\gamma(N))$ satisfies the inequality

$$\frac{6}{7} 4^j \leq \frac{\gamma}{y^2} \leq \frac{6}{5} 4^j. \qquad (4.156)$$

We claim that if η_2, and so also η_1, is a small constant, then the upper arc of $P_{i_0} - H_\gamma(N)$ intersects a large number of j-cells different from $\mathscr{C}$ such that the reciprocal of the slope is still almost equal to 4^j. Indeed, the vertical size of $\mathscr{C}$ is $2^{-j}\eta_2$ and, assuming that (4.156) holds, the inequality

$$\frac{6}{7}4^j \le \frac{\gamma}{(y + \ell 2^{-j}\eta_2)^2} \le \frac{6}{5}4^j \qquad (4.157)$$

has constant times $1/\eta_2$ consecutive integer solutions in ℓ. If $\eta_2 > 0$ is small, then of course $1/\eta_2$ is large, justifying our claim.

Returning to (4.155) and (4.156), and then substituting y by $y + \ell 2^{-j}\eta_2$, we have the respective inequalities

$$\left| \frac{\gamma v}{(y + \ell 2^{-j}\eta_2)(y + \ell 2^{-j}\eta_2 + v)} - h \right| < 2^{j+1}\eta_1 \qquad (4.158)$$

and (4.157). If (4.156) holds, then there are at least $\sqrt{\gamma}/10\eta_2$ consecutive integer solutions ℓ of (4.157).

The basic idea is the same as in Case 15. If ℓ runs through these integer solutions of (4.157) while γ, y, h and v remain fixed, then the function

$$\frac{\gamma v}{(y + \ell 2^{-j}\eta_2)(y + \ell 2^{-j}\eta_2 + v)}, \qquad (4.159)$$

as a function of ℓ, has substantially different values, and we expect only very few of them to be very close to a fixed h in the quantitative sense of (4.158). Of course, here we assume that η_1 is small.

Next we work out the details of this intuition. We begin by noting that (4.157) implies

$$\sqrt{\frac{6\gamma}{7}}2^{-j} \le y + \ell 2^{-j}\eta_2 \le \sqrt{\frac{6\gamma}{5}}2^{-j}. \qquad (4.160)$$

Using this in (4.159), we have the good approximation

$$\frac{\gamma v}{(y + \ell 2^{-j}\eta_2)(y + \ell 2^{-j}\eta_2 + v)} \approx \frac{\gamma v}{\sqrt{\gamma}2^{-j}(\sqrt{\gamma}2^{-j} + v)} = \frac{v}{2^{-j}(2^{-j} + v\sqrt{\gamma})}. \qquad (4.161)$$

We now distinguish three cases. First assume that $v < 0$. Since $y + v > 0$, we have $y^{-1} < (y + v)^{-1}$, and so by (4.158), we have

$$2^{j+1}\eta_1 > |h| = h. \qquad (4.162)$$

Combining (4.162) with the rectangle property, we deduce that

$$|v| \geq \frac{c_1}{h} > \frac{c_1}{2\eta_1} 2^{-j}. \tag{4.163}$$

Substituting (4.163) into (4.161), and assuming that

$$\eta_1 < \frac{c_1}{2\sqrt{\gamma}}, \tag{4.164}$$

we have

$$\frac{v}{2^{-j}(2^{-j} + v/\sqrt{\gamma})} = \frac{|v|}{2^{-j}(v/\sqrt{\gamma} - 2^{-j})} = \frac{2^j}{1/\sqrt{\gamma} - 2^{-j}/|v|} > \sqrt{\gamma} 2^j. \tag{4.165}$$

Combining (4.158), (4.161)–(4.163) and (4.165), we conclude that

$$2^{j+1}\eta_1 > h > \frac{1}{2}\sqrt{\gamma} 2^j - 2^{j+1}\eta_1,$$

which is an obvious contradiction if

$$\eta_1 < \frac{\sqrt{\gamma}}{8}. \tag{4.166}$$

This proves that $v > 0$.

Next assume that $0 < v \leq \sqrt{c_1} 2^{-j-1}$, where $c_1 > 0$ is the positive constant in the rectangle property. Then the rectangle property yields

$$h \geq \frac{c_1}{v} \geq \frac{c_1}{\sqrt{c_1} 2^{-j-1}} = 2\sqrt{c_1} 2^j \tag{4.167}$$

and

$$\frac{v}{2^{-j}(2^{-j} + v/\sqrt{\gamma})} < \frac{v}{2^{-j} 2^{-j}} \leq \frac{\sqrt{c_1}}{2} 2^j. \tag{4.168}$$

The assumption

$$\eta_1 < \frac{\sqrt{c_1}}{2}, \tag{4.169}$$

together with (4.161), (4.167) and (4.168), implies that (4.158) has no solution.

We can assume, therefore, that the lower bound

$$v > \sqrt{c_1} 2^{-j-1} \tag{4.170}$$

holds. Now we go back to the basic idea. We claim that if we switch ℓ to $\ell + 1$ in the function (4.159), then its value changes by at least as much as

$$\frac{\eta_1 2^{j-2}}{1 + \sqrt{\gamma/c_1}}. \tag{4.171}$$

Indeed, by (4.160), we have

$$\frac{\gamma v}{(y + \ell 2^{-j} \eta_2)(y + \ell 2^{-j} \eta_2 + v)} \approx \frac{1}{\sqrt{\gamma} 2^{-j}} \cdot \frac{\gamma v}{\sqrt{\gamma} 2^{-j} + v}. \tag{4.172}$$

We also have the routine estimate

$$\frac{1}{\sqrt{\gamma} 2^{-j}} - \frac{1}{\sqrt{\gamma} 2^{-j} + 2^{-j} \eta_2} = \frac{1}{\sqrt{\gamma} 2^{-j}} \left(1 - \frac{1}{1 + \eta_2/\sqrt{\gamma}} \right)$$

$$= \frac{1}{\sqrt{\gamma} 2^{-j}} \left(\frac{\eta_2}{\sqrt{\gamma}} - \left(\frac{\eta_2}{\sqrt{\gamma}} \right)^2 + \left(\frac{\eta_2}{\sqrt{\gamma}} \right)^3 \mp \ldots \right) \approx \frac{\eta_2 2^j}{\gamma}. \tag{4.173}$$

Furthermore, by (4.170), we have

$$\frac{\gamma v}{\sqrt{\gamma} 2^{-j} + v} > \frac{\gamma}{2\sqrt{\gamma/c_1} + 1}. \tag{4.174}$$

The error estimate (4.171) follows on combining (4.172)–(4.174).

Let us return to (4.159) and (4.171), and apply them in (4.158). We deduce that among the constant times $1/\eta_2$ consecutive integer values of ℓ satisfying (4.157), there are only constant times $(1 + \sqrt{\gamma/c_1})$ that will satisfy (4.158). More explicitly, it is safe to say that

$$\text{at most } 10 \left(1 + \sqrt{\frac{\gamma}{c_1}} \right) \text{ values of } \ell \text{ will satisfy both (4.157) and (4.158).} \tag{4.175}$$

As in Case 15, the next step is

A Combination of the Rectangle Property and the Pigeonhole Principle. We recall (4.170), that $v > \sqrt{c_1} 2^{-j-1}$. Consider the power-of-two type decomposition

$$2^{r-1} \sqrt{c_1} 2^{-j} < v \leq 2^r \sqrt{c_1} 2^{-j}, \quad r = 0, 1, 2, \ldots. \tag{4.176}$$

We claim that for a fixed point $P_{i_0} = (a_{i_0}, b_{i_0}) \in \mathscr{P}$ and for a fixed integer $r \geq 0$, there are at most

$$10 \sqrt{\frac{\gamma}{c_1}} 2^r \tag{4.177}$$

other points $P_i = (a_i, b_i) \in \mathscr{P}$, with $P_i \neq P_{i_0}$, such that $h = h_i = a_{i_0} - a_i > 0$ and $v = v_i = b_{i_0} - b_i > 0$ satisfy (4.158), thus implicitly (4.157) also, and (4.176).

To establish the bound (4.177), first note that if $v = v_i$ satisfies (4.176), then by (4.161) and (4.176), we have

$$\frac{\gamma v}{(y + \ell 2^{-j} \eta_2)(y + \ell 2^{-j} \eta_2 + v)} \approx \frac{v}{2^{-j}(2^{-j} + v/\sqrt{\gamma})}$$

$$\approx \frac{2^r \sqrt{c_1} 2^{-j}}{2^{-j}(2^{-j} + 2^r \sqrt{c_1} 2^{-j}/\sqrt{\gamma})} = \frac{2^j}{1/\sqrt{\gamma} + 2^{-r}/\sqrt{c_1}},$$

so that a solution of (4.158) gives the approximation

$$h = h_i \approx 2^j \left(\frac{1}{1/\sqrt{\gamma} + 2^{-r}/\sqrt{c_1}} \pm 2\eta_1 \right). \tag{4.178}$$

Assuming

$$\eta_1 < \frac{1}{8(1/\sqrt{\gamma} + 1/\sqrt{c_1})}, \tag{4.179}$$

then (4.178) yields the good approximation

$$h = h_i \approx \frac{2^j}{1/\sqrt{\gamma} + 2^{-r}/\sqrt{c_1}}. \tag{4.180}$$

Suppose on the contrary that there are more than (4.177) other points $P_i = (a_i, b_i) \in \mathscr{P}$, with $P_i \neq P_{i_0}$, such that $h = h_i = a_{i_0} - a_i > 0$ and $v = v_i = b_{i_0} - b_i > 0$ satisfy (4.158), thus implicitly (4.157) also, and (4.176). Then by the Pigeonhole Principle and (4.180), there must exist two points $P_{i_1}, P_{i_2} \in \mathscr{P}$, with $i_1 \neq i_2$, such that

$$h_{i_1} \approx \frac{2^j}{1/\sqrt{\gamma} + 2^{-r}/\sqrt{c_1}} \approx h_{i_2} \quad \text{and} \quad |v_{i_1} - v_{i_2}| \leq \frac{2^r \sqrt{c_1} 2^{-j}}{10\sqrt{\gamma/c_1} 2^r} = \frac{c_1 2^{-j}}{10\sqrt{\gamma}}.$$

Since the product

$$\frac{2^j}{1/\sqrt{\gamma} + 2^{-r}/\sqrt{c_1}} \cdot \frac{c_1 2^{-j}}{\sqrt{\gamma}} = \frac{c_1}{1 + 2^{-r}\sqrt{\gamma/c_1}} < c_1,$$

we conclude that there exists an axes-parallel rectangle of area less than c_1 and which contains at least two points of $\mathscr{P}$, namely P_{i_1} and P_{i_2}. This contradicts the rectangle property, and establishes the bound (4.177).

If $v = v_i$ falls into the interval (4.176), then

$$\frac{1}{\text{slope}(\mathscr{C} \cap (P_i - H_\gamma(N)))} = \frac{\gamma}{(y+v)^2} \leq \frac{\gamma}{v^2} \approx \frac{\gamma}{c_1 4^r} \cdot 4^j, \qquad (4.181)$$

where 4^j almost equals the reciprocal of the slope of the diagonals of the j-cell $\mathscr{C}$. By (4.181), we have

$$\frac{1}{\text{area}(\mathscr{C})} \left| \int_{\mathscr{C} \cap (P_i - H_\gamma(N))} R_j(\mathbf{x}) \, d\mathbf{x} \right| \leq \frac{10\gamma}{c_1 4^r}. \qquad (4.182)$$

Furthermore, (4.182) holds for all j-cells $\mathscr{C}$ satisfying (4.151). Let us return now to (4.126). Combining (4.175)–(4.177) and (4.182), we have

$$\sum_{\substack{P_i \in \mathscr{P} \\ i \neq i_0 \\ \text{Case 2}}} \sum_{\mathscr{C}} \frac{1}{\text{area}(\mathscr{C})} \left| \int_{\mathscr{C} \cap (P_i - H_\gamma(N))} R_j(\mathbf{x}) \, d\mathbf{x} \right| \leq \sum_{r \geq 0} 10 \left(1 + \sqrt{\frac{\gamma}{c_1}} \right) 10 \sqrt{\frac{\gamma}{c_1}} 2^r \frac{10\gamma}{c_1 4^r}$$

$$= 1000 \left(\left(\frac{\gamma}{c_1} \right)^{3/2} + \left(\frac{\gamma}{c_1} \right)^2 \right) \sum_{r \geq 0} 2^{-r} = 2000 \left(\left(\frac{\gamma}{c_1} \right)^{3/2} + \left(\frac{\gamma}{c_1} \right)^2 \right), (4.183)$$

a perfect analog of (4.152). This completes Case 16.

Case 17. The lower arc of $P_i - H_\gamma(N)$ intersects $\mathscr{C}$, and the slope is less than the slope of the dominant needle $P_{i_0} - H_\gamma(N)$; see Fig. 4.9.

Let $P_{i_0} = (a_{i_0}, b_{i_0})$ and $P_i = (a_i, b_i)$ denote the coordinates of the two points in question. By the hypothesis of Case 17, we have $a_i > a_{i_0}$. Write

$$h = h_i = a_i - a_{i_0} > 0 \quad \text{and} \quad v = v_i = b_i - b_{i_0},$$

where again h denotes horizontal and v denotes vertical. It is obvious from the geometry of Case 17 that $v > 0$. The rectangle property guarantees that $hv \geq c_1 > 0$.

Let (A_1, A_2) denote the coordinates of the lower left vertex of the j-cell $\mathscr{C}$. The intersection of the line $x = A_1$ with the upper arc of $P_{i_0} - H_\gamma(N)$ and the lower arc of $P_i - H_\gamma(N)$ give two points, and the hypothesis of Case 3 implies that these intersection points are close to each other. More precisely, similar to Case 1, with $x = 1 + a_{i_0} - A_1$, we have the upper bound

$$\left| \left(b_{i_0} + \frac{\gamma}{x} \right) - \left(b_i - \frac{\gamma}{x+h} \right) \right| < 2 \cdot 2^{-j} \eta_2. \qquad (4.184)$$

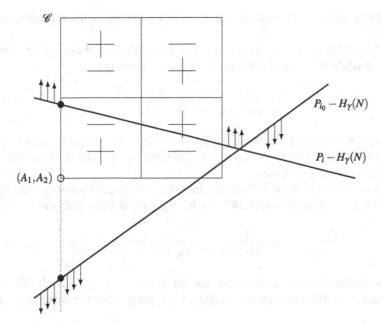

Fig. 4.9 Lower arc of $P_i - H_\gamma(N)$ intersects $\mathscr{C}$, with slope less than slope of $P_{i_0} - H_\gamma(N)$

Since $b_i - b_{i_0} = v$, we can rewrite (4.184) in the form

$$\left| \left(\frac{\gamma}{x} + \frac{\gamma}{x+h} \right) - v \right| < 2^{-j+1} \eta_2. \tag{4.185}$$

Note that (4.185) is an analog of (4.128) in Case 15, the only difference being that a minus sign is replaced by plus sign. This means that we can basically repeat the argument in Case 15. In fact, the plus sign helps and makes Case 17 simpler than Case 15. On the other hand, we know that the slope of the upper arc of $\mathscr{C} \cap (P_{i_0} - H_\gamma(N))$ satisfies the inequality (4.129).

Again, if η_1, and so also η_2, is a small constant, then the upper arc of $P_{i_0} - H_\gamma(N)$ intersects a large number of j-cells different from $\mathscr{C}$ such that the slope is still almost equal to 4^{-j}. Indeed, the horizontal size of $\mathscr{C}$ is $2^j \eta_1$ and, assuming that (4.129) holds, the inequality (4.130) has constant times $1/\eta_1$ consecutive integer solutions in ℓ.

Returning to (4.129) and (4.185), and then substituting x by $x + \ell 2^j \eta_1$, we have the respective inequalities (4.130) and

$$\left| \frac{\gamma}{x + \ell 2^j \eta_1} + \frac{\gamma}{x + \ell 2^j \eta_1 + h} - v \right| < 2^{-j+1} \eta_2. \tag{4.186}$$

If (4.129) holds, then there are at least $\sqrt{\gamma}/10\eta_1$ consecutive integer solutions ℓ of (4.130).

The basic idea is the same as in Case 15. If ℓ runs through these integer solutions of (4.130) while γ, x, h and v remain fixed, then the function

$$\frac{\gamma}{x + \ell 2^j \eta_1} + \frac{\gamma}{x + \ell 2^j \eta_1 + h}, \tag{4.187}$$

as a function of ℓ, has substantially different values, and we expect only very few of them to be very close to a fixed v in the quantitative sense of (4.186). Of course, here we assume that η_2 is small.

Next we work out the details of this intuition. We begin by noting that (4.130) implies (4.133). Using this in (4.187), we have the good approximation

$$\frac{\gamma}{x + \ell 2^j \eta_1} + \frac{\gamma}{x + \ell 2^j \eta_1 + h} \approx \frac{\gamma}{\sqrt{\gamma} 2^j} + \frac{\gamma}{\sqrt{\gamma} 2^j + h}. \tag{4.188}$$

We now distinguish two cases. First assume that $0 < h \leq c_1 2^{j-2}/\sqrt{\gamma}$, where $c_1 > 0$ is the positive constant in the rectangle property. Then the rectangle property yields

$$|v| \geq \frac{c_1}{h} \geq \frac{c_1}{\sqrt{c_1} 2^{j-1}} = 2\sqrt{c_1} 2^{-j}. \tag{4.189}$$

On the other hand, assuming that

$$\eta_2 < \frac{\sqrt{\gamma}}{2}, \tag{4.190}$$

it then follows from (4.186) and (4.188) that

$$v \leq \frac{2\gamma}{\sqrt{\gamma} 2^j} + 2^{-j+1}\eta_2 < 4\sqrt{\gamma} 2^{-j}. \tag{4.191}$$

Since (4.189) and (4.191) contradict each other, we can therefore assume that

$$h > \frac{c_1 2^{j-2}}{\sqrt{\gamma}}, \tag{4.192}$$

which is an analog of (4.138) in Case 1. Now we go back to the basic idea. We claim that if we switch ℓ to $\ell + 1$ in the function (4.187), then its value changes by at least as much as

$$\eta_1 2^{-j-2}, \tag{4.193}$$

an analog of (4.139). Indeed, (4.193) follows immediately from the routine estimate

$$\frac{1}{\sqrt{\gamma}2^j} - \frac{1}{\sqrt{\gamma}2^j + 2^j \eta_1} = \frac{1}{\sqrt{\gamma}2^j}\left(1 - \frac{1}{1 + \eta_1/\sqrt{\gamma}}\right) \approx \frac{\eta_1}{\gamma 2^j}.$$

Let us return to (4.187) and (4.193), and apply them in (4.186). We deduce that

$$\text{at most 10 values of } \ell \text{ will satisfy both (4.130) and (4.186).} \tag{4.194}$$

As in Cases 15–16, the next step is

A Combination of the Rectangle Property and the Pigeonhole Principle. We recall
(4.192), that $h > c_1 2^{j-2}/\sqrt{\gamma}$. Consider the power-of-two type decomposition

$$2^{r-1}\frac{c_1 2^{j-1}}{\sqrt{\gamma}} < h \le 2^r \frac{c_1 2^{j-1}}{\sqrt{\gamma}}, \qquad r = 0, 1, 2, \ldots. \tag{4.195}$$

We claim that for a fixed point $P_{i_0} = (a_{i_0}, b_{i_0}) \in \mathscr{P}$ and for a fixed integer $r \ge 0$,
there are at most

$$10 \cdot 2^r \tag{4.196}$$

other points $P_i = (a_i, b_i) \in \mathscr{P}$, with $P_i \ne P_{i_0}$, such that $h = h_i = a_i - a_{i_0} > 0$
and $v = v_i = b_i - b_{i_0}$ satisfy (4.186), thus implicitly (4.130) also, and (4.195).

To establish the bound (4.196), first note that if $h = h_i$ satisfies (4.195), then by
(4.188) and (4.195), we have

$$\frac{\gamma}{x + \ell 2^j \eta_1} + \frac{\gamma}{x + \ell 2^j \eta_1 + h} \approx \frac{\gamma}{\sqrt{\gamma}2^j} + \frac{\gamma}{\sqrt{\gamma}2^j + h} \approx \sqrt{\gamma}2^{-j}\left(1 + \frac{1}{1 + c_1 2^{r-1}/\gamma}\right),$$

so that a solution of (4.186) gives the approximation

$$v = v_i \approx \sqrt{\gamma}2^{-j}\left(1 + \frac{1}{1 + c_1 2^{r-1}/\gamma}\right) \pm 2^{-j+1}\eta_2. \tag{4.197}$$

Assuming

$$\eta_2 < \frac{\sqrt{\gamma}}{100}, \tag{4.198}$$

then (4.197) yields the good approximation

$$v = v_i \approx \sqrt{\gamma}2^{-j}\left(1 + \frac{1}{1 + c_1 2^{r-1}/\gamma}\right). \tag{4.199}$$

Suppose, contrary to the bound (4.196), that there are more than $10 \cdot 2^r$ other points $P_i = (a_i, b_i) \in \mathscr{P}$, with $P_i \neq P_{i_0}$, such that $h = h_i = a_i - a_{i_0} > 0$ and $v = v_i = b_i - b_{i_0}$ satisfy (4.186), thus implicitly (4.130) also, and (4.195). Then by the Pigeonhole Principle and (4.199), there must exist two points $P_{i_1}, P_{i_2} \in \mathscr{P}$, with $i_1 \neq i_2$, such that

$$v_{i_1} \approx \sqrt{\gamma} 2^{-j} \left(1 + \frac{1}{1 + c_1 2^{r-1}/\gamma} \right) \approx v_{i_2} \quad \text{and} \quad |h_{i_1} - h_{i_2}| \leq \frac{2^r c_1 2^{j-1}/\sqrt{\gamma}}{10 \cdot 2^r} = \frac{c_1 2^j}{20 \sqrt{\gamma}}.$$

Since the product

$$\sqrt{\gamma} 2^{-j} \left(1 + \frac{1}{1 + c_1 2^{r-1}/\gamma} \right) \cdot \frac{c_1 2^j}{2 \sqrt{\gamma}} < c_1,$$

we conclude that there exists an axes-parallel rectangle of area less than c_1 and which contains at least two points of $\mathscr{P}$, namely P_{i_1} and P_{i_2}. This contradicts the rectangle property, and establishes the bound (4.196).

If $h = h_i$ falls into the interval (4.195), then

$$\text{slope}(\mathscr{C} \cap (P_i - H_\gamma(N))) = \frac{\gamma}{(x+h)^2} \leq \frac{\gamma}{h^2} \leq \frac{(\gamma/c_1)^2}{4^{r-2}} \cdot 4^{-j}, \tag{4.200}$$

where 4^{-j} almost equals the slope of the diagonals of the j-cell $\mathscr{C}$. By (4.200), we have

$$\frac{1}{\text{area}(\mathscr{C})} \left| \int_{\mathscr{C} \cap (P_i - H_\gamma(N))} R_j(\mathbf{x}) \, d\mathbf{x} \right| \leq \frac{10(\gamma/c_1)^2}{4^{r-2}}. \tag{4.201}$$

Furthermore, (4.201) holds for all j-cells $\mathscr{C}$ satisfying (4.151). Let us return now to (4.126). Combining (4.194)–(4.196) and (4.201), we have

$$\sum_{\substack{P_i \in \mathscr{P} \\ i \neq i_0 \ (4.151) \\ \text{Case 3}}} \sum_{\mathscr{C}} \frac{1}{\text{area}(\mathscr{C})} \left| \int_{\mathscr{C} \cap (P_i - H_\gamma(N))} R_j(\mathbf{x}) \, d\mathbf{x} \right| \leq \sum_{r \geq 0} 10 \cdot 10 \cdot 2^r \cdot \frac{10(\gamma/c_1)^2}{4^{r-2}}$$

$$= 16000 \left(\frac{\gamma}{c_1} \right)^2 \sum_{r \geq 0} 2^{-r} = 32000 \left(\frac{\gamma}{c_1} \right)^2. \tag{4.202}$$

This completes Case 17.

Case 18. The lower arc of $P_i - H_\gamma(N)$ intersects $\mathscr{C}$, and the slope is greater than the slope of the dominant needle $P_{i_0} - H_\gamma(N)$; see Fig. 4.10.

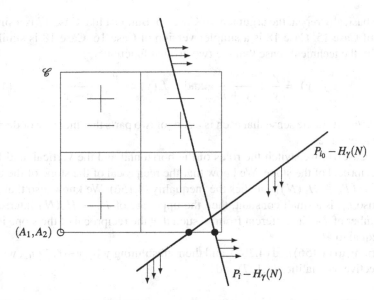

Fig. 4.10 Lower arc of $P_i - H_\gamma(N)$ intersects $\mathscr{C}$, with slope greater than slope of $P_{i_0} - H_\gamma(N)$

Let $P_{i_0} = (a_{i_0}, b_{i_0})$ and $P_i = (a_i, b_i)$ denote the coordinates of the two points in question. By the hypothesis of Case 4, we have $a_{i_0} > a_i$. We want positive real numbers, and write

$$h = h_i = a_{i_0} - a_i > 0 \quad \text{and} \quad v = v_i = b_i - b_{i_0} > 0,$$

where again h denotes horizontal and v denotes vertical. The rectangle property guarantees that $hv \geq c_1 > 0$.

Let (A_1, A_2) denote the coordinates of the lower left vertex of the j-cell $\mathscr{C}$. We have $b_i > A_2 > b_{i_0}$ and $b_i - A_2 > A_2 - b_{i_0}$. The intersection of the line $y = A_2$ with the upper arc of $P_{i_0} - H_\gamma(N)$ and the lower arc of $P_i - H_\gamma(N)$ give two points, and the hypothesis of Case 4 implies that these intersection points are relatively close to each other in the following quantitative sense. Write $y = A_2 - b_{i_0} > 0$. Then $b_i - A_2 = (b_i - b_{i_0}) - y = v - y > y$, and we have the upper bound

$$\left| \left(a_i - \frac{\gamma}{v - y} \right) - \left(a_{i_0} - \frac{\gamma}{y} \right) \right| < 2 \cdot 2^j \eta_1. \tag{4.203}$$

Since $a_{i_0} - a_i = h > 0$, we can rewrite (4.203) in the form

$$\left| \left(\frac{\gamma}{y} - \frac{\gamma}{v - y} \right) - h \right| < 2^{j+1} \eta_1. \tag{4.204}$$

Now we basically repeat the argument of Case 16. But, just like Case 17 is a simpler version of Case 15, Case 18 is a simpler version of Case 16. Case 18 is similar to Case 17 in the technical sense that the two critical functions

$$f_3(y) = \frac{\gamma}{y} + \frac{\gamma}{y+h} \quad \text{and} \quad f_4(y) = \frac{\gamma}{y} - \frac{\gamma}{v-y} \tag{4.205}$$

are in *synchrony*, in the sense that each is a sum of two parts that increase or decrease together as y varies.

As in Case 16, we switch the roles of the horizontal and the vertical, and focus on the reciprocal of the slope. We know that the reciprocal of the slope of the upper arc of $\mathscr{C} \cap (P_{i_0} - H_\gamma(N))$ satisfies the inequality (4.156). We know also that if η_2, and so also η_1, is a small constant, then the upper arc of $P_{i_0} - H_\gamma(N)$ intersects a large number of j-cells different from $\mathscr{C}$ such that the reciprocal of the slope is still almost equal to 4^j.

Returning to (4.156) and (4.204), and then substituting y by $y + \ell 2^{-j} \eta_2$, we have the respective inequalities (4.157) and

$$\left| \frac{\gamma}{y + \ell 2^{-j} \eta_2} - \frac{\gamma}{v - (y + \ell 2^{-j} \eta_2)} - h \right| < 2^{j+1} \eta_1. \tag{4.206}$$

If (4.156) holds, then there are at least $\sqrt{\gamma}/10\eta_2$ consecutive integer solutions ℓ of (4.157).

The basic idea is the same as in Case 16. If ℓ runs through these integer solutions of (4.157) while γ, x, h and v remain fixed, then the function

$$\frac{\gamma}{y + \ell 2^{-j} \eta_2} - \frac{\gamma}{v - (y + \ell 2^{-j} \eta_2)}, \tag{4.207}$$

as a function of ℓ, has substantially different values, and we expect only very few of them to be very close to a fixed h in the quantitative sense of (4.157). Of course, here we assume that η_1 is small.

Next we work out the details of this intuition. We begin by noting that (4.157) implies (4.160). Since the functions $f_3(y)$ and $f_4(y)$ given by (4.205) are in synchrony, we can basically repeat the argument of (4.187), (4.193) and (4.194) in Case 3, and conclude that if we switch ℓ to $\ell + 1$ in the function (4.207), then its value changes by at least as much as

$$\eta_2 2^{j-2},$$

an analog of (4.171) and (4.193). Thus we deduce that

at most 10 values of ℓ will satisfy both (4.157) and (4.206). \hfill (4.208)

As in Cases 15–17, the next step is

A Combination of the Rectangle Property and the Pigeonhole Principle. In this case, since $v - (y + \ell 2^{-j} \eta_2) > y + \ell 2^{-j} \eta_2$, we have

$$v > 2(y + \ell 2^{-j} \eta_2). \tag{4.209}$$

In view of (4.160), we can assume that

$$v > \sqrt{\frac{6\gamma}{7}} 2^{-j+1}.$$

Consider the power-of-two type decomposition

$$2^{r-1} \sqrt{\frac{6\gamma}{7}} 2^{-j+2} < v \le 2^r \sqrt{\frac{6\gamma}{7}} 2^{-j+2}, \quad r = 0, 1, 2, \dots. \tag{4.210}$$

We claim that for a fixed point $P_{i_0} = (a_{i_0}, b_{i_0}) \in \mathscr{P}$ and for a fixed integer $r \ge 0$, there are at most

$$\frac{100\gamma 2^r}{c_1} \tag{4.211}$$

other points $P_i = (a_i, b_i) \in \mathscr{P}$, with $P_i \ne P_{i_0}$, such that $h = h_i = a_{i_0} - a_i > 0$ and $v = v_i = b_i - b_{i_0} > 0$ satisfy (4.206), thus implicitly (4.157) also, and (4.210).

To establish the bound (4.211), first note that if $v = v_i$ satisfies (4.210), then by (4.160), (4.206) and (4.209), and assuming that

$$\eta_1 < \frac{\sqrt{\gamma}}{4}, \tag{4.212}$$

we have

$$h = h_i < \frac{\gamma}{y + \ell 2^{-j} \eta_2} + 2^{j+1} \eta_1 \le \frac{\gamma}{2^{-j} \sqrt{6\gamma/7}} + 2^{j+1} \eta_1 \le 2\sqrt{\gamma} 2^j. \tag{4.213}$$

Suppose, contrary to the bound (4.211), that there are more than $100\gamma 2^r / c_1$ other points $P_i = (a_i, b_i) \in \mathscr{P}$, with $P_i \ne P_{i_0}$, such that $h = h_i = a_{i_0} - a_i > 0$ and $v = v_i = b_i - b_{i_0} > 0$ satisfy (4.206), thus implicitly (4.157) also, and (4.210). Then by the Pigeonhole Principle and (4.213), there must exist two points $P_{i_1}, P_{i_2} \in \mathscr{P}$, with $i_1 \ne i_2$, such that

$$\max\{h_{i_1}, h_{i_2}\} \le 2\sqrt{\gamma} 2^j \quad \text{and} \quad |v_{i_1} - v_{i_2}| \le \frac{2^r \sqrt{6\gamma/7} 2^{-j+2}}{100\gamma 2^r / c_1} = \frac{c_1 \sqrt{6/7}}{25\sqrt{\gamma}} 2^{-j}.$$

Since the product

$$\sqrt{\gamma}2^j \cdot \frac{c_1\sqrt{6/7}}{\sqrt{\gamma}}2^{-j} = \sqrt{\frac{6}{7}}c_1 < c_1,$$

we conclude that there exists an axes-parallel rectangle of area less than c_1 and which contains at least two points of $\mathscr{P}$, namely P_{i_1} and P_{i_2}. This contradicts the rectangle property, and establishes the bound (4.211).

If $v = v_i$ falls into the interval (4.210), then

$$\frac{1}{\text{slope}(\mathscr{C} \cap (P_i - H_\gamma(N)))} = \frac{\gamma}{(y+v)^2} \le \frac{\gamma}{v^2} \approx \frac{1}{4^r} \cdot 4^j, \tag{4.214}$$

where 4^j almost equals the reciprocal of the slope of the diagonals of the j-cell $\mathscr{C}$. By (4.214), we have

$$\frac{1}{\text{area}(\mathscr{C})} \left| \int_{\mathscr{C} \cap (P_i - H_\gamma(N))} R_j(\mathbf{x}) \, d\mathbf{x} \right| \le \frac{10}{4^r}. \tag{4.215}$$

Furthermore, (4.215) holds for all j-cells $\mathscr{C}$ satisfying (4.151). Let us return now to (4.126). Combining (4.208), (4.210), (4.211) and (4.215) we have

$$\sum_{\substack{P_i \in \mathscr{P} \\ i \ne i_0}} \sum_{\substack{\mathscr{C} \\ (4.151) \\ \text{Case 4}}} \frac{1}{\text{area}(\mathscr{C})} \left| \int_{\mathscr{C} \cap (P_i - H_\gamma(N))} R_j(\mathbf{x}) \, d\mathbf{x} \right| \le \sum_{r \ge 0} 10 \cdot 100 \frac{\gamma}{c_1} 2^r \cdot \frac{10}{4^r}$$

$$= 10000 \frac{\gamma}{c_1} \sum_{r \ge 0} 2^{-r} = 20000 \frac{\gamma}{c_1}. \tag{4.216}$$

This completes Case 18.

4.8   Completing the Proof of Theorem 12

In this section, we shall finally complete the proof of Proposition 13. Let us return to (4.125) and (4.126). We are now ready to clarify the technical details of the single term domination.

Let $P_{i_0} \in \mathscr{P}$ and $j \in \mathscr{J}$ be arbitrary.

Property 19. The slope γ/x^2 of the hyperbolic needle $P_{i_0} - H_\gamma(N)$ satisfies

$$\frac{5}{6}4^{-j} \le \frac{\gamma}{x^2} \le \frac{7}{6}4^{-j}. \tag{4.217}$$

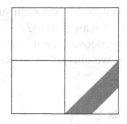 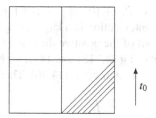

Fig. 4.11 Short vertical translations

Note that (4.217) holds if and only if

$$\sqrt{\frac{6\gamma}{7}} 2^j \le x \le \sqrt{\frac{6\gamma}{5}} 2^j,$$

and this is an interval of length greater than $\sqrt{\gamma} 2^j / 6$. Since a j-cell $\mathscr{C}$ has horizontal side $\eta_1 2^j$, there are more than

$$\frac{\sqrt{\gamma} 2^j / 6}{\eta_1 2^j} = \frac{\sqrt{\gamma}}{6\eta_1}$$

j-cells $\mathscr{C}$ with the slope of the intersection $\mathscr{C} \cap (P_{i_0} - H_\gamma(N))$ satisfying Property 19.

It would be not too difficult to prove directly, by using some familiar arguments from uniform distribution, that among these more than $\sqrt{\gamma}/6\eta_1$ j-cells $\mathscr{C}$, at least 1 % has the following additional property.

Property 20. The hyperbolic needle $P_{i_0} - H_\gamma(N)$ intersects only the lower right subrectangle of $\mathscr{C}$, and the intersection is a large triangle, meaning that the area is at least $\frac{1}{32}$ the area of $\mathscr{C}$, i.e. the area is at least $\eta_1 \eta_2 / 32$.

It is technically simpler, however, to force Property 20 in an indirect way, by using the trick of short vertical translations; see Fig. 4.11. This geometric trick was already mentioned at the end of Sect. 4.5.

More precisely, for every real number t_0 satisfying $0 < t_0 < 1$, consider all j-cells $\mathscr{C}$ such that, with $B = [0, M - N] \times [\gamma, M - \gamma]$, we have

$$\mathscr{C} \cap (P_{i_0} + (0, t_0) - H_\gamma(N)) \subset B \tag{4.218}$$

and

$$\frac{5}{6} 4^{-j} \le \text{slope}(\mathscr{C} \cap (P_{i_0} + (0, t_0) - H_\gamma(N))) \le \frac{7}{6} 4^{-j}. \tag{4.219}$$

Simple geometric consideration shows that for, say, at least 5 % of the pairs $(t_0, \mathscr{C})$, where $\mathscr{C}$ satisfies (4.218) and (4.219), $\mathscr{C} \cap (P_{i_0} + (0, t_0) - H_\gamma(N))$ also satisfies

Property 20, i.e. $P_{i_0} + (0, t_0) - H_\gamma(N)$ intersects only the lower right subrectangle of $\mathscr{C}$, and the intersection is a large triangle of area at least $\eta_1\eta_2/32$.

For the proof of the positive direction (4.89), we choose the pattern $+-$ in every j-cell $\mathscr{C}$ satisfying (4.218) and (4.219). Naturally, we choose the opposite pattern $-+$ for the negative direction (4.90). Then

$$\int_{\mathscr{C} \cap (P_{i_0}+(0,t_0)-H_\gamma(N))} R_j(\mathbf{x})\,d\mathbf{x} \geq \frac{\eta_1\eta_2}{32}. \tag{4.220}$$

Finally, if the j-cell $\mathscr{C}$ does not satisfy both (4.218) and (4.219), then we choose the pattern 0. Therefore, by (4.220) and summarizing Cases 1–4, we have

$$\int_0^1 \left(\sum_{j \in \mathscr{J}} \sum_{P_{i_0} \in \mathscr{P}} \int_{P_{i_0}+(0,t_0)-H_\gamma(N)} R_j(\mathbf{x})\,d\mathbf{x} \right) dt_0$$

$$\geq \sum_{\substack{j \in \mathscr{J} \\ (4.222)}} \sum_{P_{i_0} \in \mathscr{P}} \left(\frac{1}{20} \cdot \frac{\sqrt{\gamma}}{6\eta_1} \cdot \frac{\eta_1\eta_2}{32} \right.$$

$$- \sum_{\substack{P_i \in \mathscr{P} \\ i \neq i_0}} \sum_{\substack{\mathscr{C} \\ (4.219)}} \int_0^1 \left| \int_{\mathscr{C} \cap (P_i+(0,t_0)-H_\gamma(N))} R_j(\mathbf{x})\,d\mathbf{x} \right| dt_0 \right)$$

$$\geq \sum_{\substack{j \in \mathscr{J} \\ (4.222)}} \sum_{P_{i_0} \in \mathscr{P}} \left(\frac{\sqrt{\gamma}\eta_2}{3840} - \eta_1\eta_2 \left(4000 \left(\left(\frac{\gamma}{c_1}\right)^{3/2} + \left(\frac{\gamma}{c_1}\right)^2 \right) \right.\right.$$

$$\left.\left. +32000 \left(\frac{\gamma}{c_1}\right)^2 + 20000 \frac{\gamma}{c_1} \right) \right), \tag{4.221}$$

where the summation over $P_{i_0} \in \mathscr{P}$ is under the restriction

$$P_{i_0} + (0, t_0) - H_\gamma(N) \subset B \quad \text{for all } t_0 \text{ satisfying } 0 < t_0 < 1, \tag{4.222}$$

the summation over $\mathscr{C}$ is under the restriction (4.219), the summation over $P_i \in \mathscr{P}$ with $i \neq i_0$ encompass Cases 15–18, and finally the factor $\frac{1}{20}$ comes from the 5 % mentioned earlier. Furthermore, we have used in the last step the inequalities (4.152), (4.183), (4.202) and (4.216) for every t_0 satisfying $0 < t_0 < 1$.

In our discussion in Sects. 4.6 and 4.7, we have made some assumptions on η_1 and η_2. Corresponding to Cases 15–18, we have assumed respectively that

$$\eta_2 < \min\left\{\frac{\sqrt{c_1}}{2}, \frac{1}{8(1/\sqrt{\gamma}+1/\sqrt{c_1})}\right\},$$

$$\eta_1 < \min\left\{\frac{c_1}{2\sqrt{\gamma}}, \frac{\sqrt{\gamma}}{8}, \frac{\sqrt{c_1}}{2}, \frac{1}{8(1/\sqrt{\gamma}+1/\sqrt{c_1})}\right\},$$

$$\eta_2 < \min\left\{\frac{\sqrt{\gamma}}{2}, \frac{\sqrt{\gamma}}{100}\right\},$$

$$\eta_1 < \frac{\sqrt{\gamma}}{4};$$

see (4.137), (4.147), (4.164), (4.166), (4.169), (4.179), (4.190), (4.198) and (4.212). Since

$$\frac{1}{1/\sqrt{\gamma}+1/\sqrt{c_1}} \geq \frac{\sqrt{\gamma}+\sqrt{c_1}}{2},$$

we can guarantee all of the above requirements on η_1 and η_2 by imposing the single inequality

$$\max\{\eta_1, \eta_2\} < \min\left\{\frac{\sqrt{\gamma}}{100}, \frac{\sqrt{c_1}}{8}, \frac{c_1}{2\sqrt{\gamma}}\right\}. \tag{4.223}$$

Let us return to (4.221). We have

$$\frac{\sqrt{\gamma}\eta_2}{3840} - \eta_1\eta_2\left(4000\left(\left(\frac{\gamma}{c_1}\right)^{3/2} + \left(\frac{\gamma}{c_1}\right)^2\right) + 32000\left(\frac{\gamma}{c_1}\right)^2 + 20000\frac{\gamma}{c_1}\right)$$

$$\geq \frac{\sqrt{\gamma}\eta_2}{7680}, \tag{4.224}$$

assuming that (4.223) holds and η_1 satisfies the additional inequality

$$\frac{1}{\eta_1} \geq \frac{10^8}{\sqrt{\gamma}}\left(\left(\frac{\gamma}{c_1}\right) + \left(\frac{\gamma}{c_1}\right)^2\right). \tag{4.225}$$

Since η_1 and η_2 are almost equal, in view of (4.100), we can satisfy both (4.223) and (4.225) by the choice

$$\eta_1 \approx \eta_2 = \min\left\{\frac{\sqrt{\gamma}}{200}, \frac{\sqrt{c_1}}{10}, \frac{10^{-8}c_1}{2\sqrt{\gamma}}, \frac{10^{-8}c_1^2}{2\gamma^{3/2}}\right\}. \tag{4.226}$$

Substituting (4.226) in (4.224) and then returning to (4.221), we have

$$
\int_0^1 \left(\sum_{j \in \mathscr{J}} \sum_{P_{i_0} \in \mathscr{P}} \int_{P_{i_0} + (0,t_0) - H_\gamma(N)} R_j(\mathbf{x})\, d\mathbf{x} \right) dt_0 \geq \sum_{j \in \mathscr{J}} \sum_{P_{i_0} \in \mathscr{P}} \frac{\sqrt{\gamma}\,\eta_2}{7680},
$$
(4.222)

where i_0 is now a dummy variable. Clearly there exists t_0, satisfying $0 < t_0 < 1$, such that

$$
\sum_{j \in \mathscr{J}} \sum_{P_i \in \mathscr{P}} \int_{P_i + (0,t_0) - H_\gamma(N)} R_j(\mathbf{x})\, d\mathbf{x} \geq \sum_{j \in \mathscr{J}} \sum_{P_i \in \mathscr{P}} \frac{\sqrt{\gamma}\,\eta_2}{7680}.
$$
(4.227)

(4.228)

Note that in (4.227), we have substituted the dummy variable i_0 by i, together with a corresponding summation restriction

$$
P_i + (0, t_0) - H_\gamma(N) \subset B \quad \text{for all } t_0 \text{ satisfying } 0 < t_0 < 1.
$$
(4.228)

Next we return to (4.118), and replace the point set $\mathscr{P}$ by the translated point set $\mathscr{P} + (0, t_0)$. Then Lemma 14 gives

$$
\frac{1}{(M - N)(M - 2\gamma)} \int_0^{M-N} \int_\gamma^{M-\gamma} \Delta(\mathbf{x}) T(\mathbf{x})\, d\mathbf{x}
$$

$$
= \sum_{P_i \in \mathscr{P}} \frac{\text{area}(B \cap (P_i + (0,t_0) - H_\gamma(N)))}{(M - N)(M - 2\gamma)} - \delta \cdot \text{area}(H_\gamma(N))
$$

$$
+ \rho \sum_{j \in \mathscr{J}} \sum_{P_i \in \mathscr{P}} \frac{1}{(M - N)(M - 2\gamma)} \int_{P_i + (0,t_0) - H_\gamma(N)} R_j(\mathbf{x})\, d\mathbf{x} + E_1, \quad (4.229)
$$

where the error E_1 satisfies

$$
|E_1| \leq \frac{|\mathscr{P}|}{(M - N)(M - 2\gamma)} \cdot 8\sqrt{\gamma}(\eta_1 + \eta_2) \cdot \frac{(n + 1)\rho^2}{\sqrt{2} - 1 - \rho}.
$$
(4.230)

Recall that $\mathscr{P}$ is a finite subset of the square $[0, M]^2$ with cardinality $|\mathscr{P}| = \delta M^2$. Since $0 < t_0 < 1$, the rectangle property implies, via elementary calculations, that the condition

$$
P_i + (0, t_0) - H_\gamma(N) \subset B = [0, M - N] \times [\gamma, M - \gamma]
$$
(4.231)

holds for all but at most

$$
\frac{(2N + 4\gamma + 1)M}{c_1}
$$
(4.232)

points $P_i \in \mathscr{P}$. Thus

$$\sum_{P_i \in \mathscr{P}} \frac{\text{area}(B \cap (P_i + (0, t_0) - H_\gamma(N)))}{(M-N)(M-2\gamma)} - \delta \cdot \text{area}(H_\gamma(N))$$

$$= \frac{\delta M^2 + \theta c_1^{-1}(2N + 4\gamma + 1)M}{(M-N)(M-2\gamma)} \cdot \text{area}(H_\gamma(N)) - \delta \cdot \text{area}(H_\gamma(N))$$

$$= \left(\frac{M^2}{(M-N)(M-2\gamma)} - 1 \right) \delta \cdot \text{area}(H_\gamma(N))$$

$$+ \theta \frac{c_1^{-1}(2N + 4\gamma + 1)M}{(M-N)(M-2\gamma)} \cdot \text{area}(H_\gamma(N)),$$

with some constant θ satisfying $-1 \leq \theta \leq 1$. Since $\text{area}(H_\gamma(N)) = 2\gamma \log N$, it then follows that

$$\left| \sum_{P_i \in \mathscr{P}} \frac{\text{area}(B \cap (P_i + (0, t_0) - H_\gamma(N)))}{(M-N)(M-2\gamma)} - \delta \cdot \text{area}(H_\gamma(N)) \right|$$

$$\leq \frac{3N + 6\gamma + 1}{(M-N)(M-2\gamma)} \cdot 2\gamma \log N. \tag{4.233}$$

Combining (4.229), (4.230) and (4.233), we deduce that

$$\frac{1}{(M-N)(M-2\gamma)} \int_0^{M-N} \int_\gamma^{M-\gamma} \Delta(\mathbf{x}) T(\mathbf{x}) \, d\mathbf{x}$$

$$= \rho \sum_{j \in \mathscr{J}} \sum_{P_i \in \mathscr{P}} \frac{1}{(M-N)(M-2\gamma)} \int_{P_i + (0, t_0) - H_\gamma(N)} R_j(\mathbf{x}) \, d\mathbf{x} + E_2, \tag{4.234}$$

where the error E_2 satisfies

$$|E_2| \leq \frac{|\mathscr{P}|}{(M-N)(M-2\gamma)} \cdot 8\sqrt{\gamma}(\eta_1 + \eta_2) \cdot \frac{(n+1)\rho^2}{\sqrt{2} - 1 - \rho}$$

$$+ \frac{3N + 6\gamma + 1}{(M-N)(M-2\gamma)} \cdot 2\gamma \log N. \tag{4.235}$$

Combining (4.227), (4.234) and (4.235), we then conclude that

$$\frac{1}{(M-N)(M-2\gamma)} \int_0^{M-N} \int_\gamma^{M-\gamma} \Delta(\mathbf{x}) T(\mathbf{x}) \, d\mathbf{x}$$

$$\geq \rho \sum_{j \in \mathscr{J}} \frac{1}{(M-N)(M-2\gamma)} \sum_{P_i \in \mathscr{P}} \frac{\sqrt{\gamma} \eta_2}{7680} \tag{4.228}$$

$$-\frac{|\mathscr{P}|}{(M-N)(M-2\gamma)} \cdot 8\sqrt{\gamma}(\eta_1 + \eta_2) \cdot \frac{(n+1)\rho^2}{\sqrt{2}-1-\rho}$$

$$-\frac{3N+6\gamma+1}{(M-N)(M-2\gamma)} \cdot 2\gamma \log N. \qquad (4.236)$$

Recall that $\mathscr{I}$ is an interval of integers satisfying (4.114), so that

$$|\mathscr{I}| \geq (n+1) - \log_2\left(\gamma + \frac{1}{\gamma}\right).$$

On the other hand, it follows from (4.231) and (4.232) that

$$\sum_{\substack{P_i \in \mathscr{P} \\ (4.228)}} 1 \geq \delta M^2 - \frac{(2N+4\gamma+1)M}{c_1}.$$

Thus

$$\sum_{j \in \mathscr{I}} \frac{1}{(M-N)(M-2\gamma)} \sum_{\substack{P_i \in \mathscr{P} \\ (4.228)}} 1$$

$$\geq \left((n+1) - \log_2\left(\gamma + \frac{1}{\gamma}\right)\right)\left(\delta - \frac{2N+4\gamma+1}{c_1 M}\right). \qquad (4.237)$$

Let us now return to (4.236). If ρ is small, then ρ^2 is negligible compared to ρ. Let $\rho = 10^{-6}$, say. Substituting this and the estimate (4.237) into (4.236), and assuming that N and M/N are both large, we deduce that

$$\frac{1}{\text{area}(B)} \int_B \Delta(\mathbf{x})T(\mathbf{x})\,d\mathbf{x}$$

$$\geq \rho\left((n+1) - \log_2\left(\gamma + \frac{1}{\gamma}\right)\right)\left(\delta - \frac{2N+4\gamma+1}{c_1 M}\right)\frac{\sqrt{\gamma}\eta_2}{10^4}.$$

More precisely, the assumptions on N and M are given by (4.84) and (4.85), and the choice for n is made precise by

$$\frac{N}{2} < 2^n \leq N.$$

These choices, together with the definition (4.226) for η_2, ensure that

$$\frac{1}{\text{area}(B)} \int_B \Delta(\mathbf{x})T(\mathbf{x})\,d\mathbf{x} \geq \delta' \log N, \qquad (4.238)$$

where $\delta' = \delta'(c_1, \gamma, \delta) > 0$ is a positive constant independent of N and M, and defined by (4.83) and (4.84).

It now follows from (4.238) that there exists a translated copy $\mathbf{x}_1 + H_\gamma(N)$ of the hyperbolic needle $H_\gamma(N)$ such that $\mathbf{x}_1 + H_\gamma(N) \subset [0, M]^2$ and

$$|\mathscr{P} \cap (\mathbf{x}_1 + H_\gamma(N))| \geq 2\delta\gamma \log N + \delta' \log N.$$

This establishes the inequality (4.89). The proof of the other inequality (4.90) is the same, except that we replace the pattern $+-$ by the opposite pattern $-+$.

Thus the long proof of Proposition 13 is complete. This also completes the proof of Theorem 12.

4.9 Yet Another Generalization of Theorem 3

Let $\alpha > 0, 0 \leq \beta < 1$ and $\gamma > 0$ be arbitrary but fixed real numbers, and let $f(\alpha; \beta; \gamma; N)$ denote the number of integral solutions of the diophantine inequality[14]

$$\|n\alpha - \beta\| < \frac{\gamma}{n}, \quad 1 \leq n \leq N.$$

This inequality motivates the hyperbolic region

$$|y - \beta| < \frac{\gamma}{x}, \quad 1 \leq x \leq N,$$

which has area $2\gamma \log N$.

Let us return to the special case $\alpha = \sqrt{2}$. Combining Lemmas 1 and 2, we have

$$\int_0^1 f(\sqrt{2}; \beta; \gamma; N)\, d\beta = 2\gamma \log N + O(1), \qquad (4.239)$$

and for an arbitrary subinterval $[a, b]$ with $0 \leq a < b \leq 1$, we have the limit formula

$$\lim_{N \to \infty} \frac{\frac{1}{b-a} \int_a^b f(\sqrt{2}; \beta; \gamma; N)\, d\beta}{\log N} = 2\gamma. \qquad (4.240)$$

There is a straightforward generalization of (4.239) and (4.240) for arbitrary $\alpha > 0$, and the proof is the same. We have

$$\int_0^1 f(\alpha; \beta; \gamma; N)\, d\beta = 2\gamma \log N + O(1), \qquad (4.241)$$

---

[14]Note that the special case $\alpha = \sqrt{2}$ was introduced in Sect. 4.1; see (4.28).

and for an arbitrary subinterval $[a, b]$ with $0 \leq a < b \leq 1$, we have the limit formula

$$\lim_{N \to \infty} \frac{\frac{1}{b-a} \int_a^b f(\alpha; \beta; \gamma; N) \, d\beta}{\log N} = 2\gamma. \tag{4.242}$$

The formulas (4.239)–(4.242) express the almost trivial geometric fact that the average number of lattice points contained in all the translated copies of a given region equals the area of the region; see Lemma 5. It is natural, therefore, to study the limit

$$\lim_{N \to \infty} \frac{f(\alpha; \beta; \gamma; N)}{2\gamma \log N}. \tag{4.243}$$

The case of rational α in (4.243) is trivial. Indeed, if $N \to \infty$, then the function $f(\alpha; \beta; \gamma; N)$ remains bounded for all but a finite number of values of $\beta = \beta(\alpha)$ in the unit interval. When $f(\alpha; \beta; \gamma; N)$ tends to infinity, it behaves like a linear function $c_{27} N$, which is much faster than the logarithmic function $\log N$.

If α is irrational, then we have the following non-trivial result, which can be considered a far-reaching generalization of Theorem 3.

Theorem 21. *Let $\alpha > 0$ be an arbitrary irrational, and let $\gamma > 0$ be an arbitrary real number. There are continuum many divergence points $\beta^* = \beta^*(\alpha, \gamma) \in [0, 1)$ such that*

$$\limsup_{n \to \infty} \frac{f(\alpha; \beta^*; \gamma; n)}{\log n} > \liminf_{n \to \infty} \frac{f(\alpha; \beta^*; \gamma; n)}{\log n}.$$

To prove Theorem 21, we can clearly assume that $0 < \alpha < 1$. We need the continued fractions

$$\alpha = \frac{1}{a_1+} \frac{1}{a_2+} \frac{1}{a_3+} \ldots = [a_1, a_2, a_3, \ldots].$$

For irrational α, the digits $a_1, a_2, a_3, \ldots$ form an infinite sequence, with $a_i \geq 1$ for all $i \geq 1$. For $k \geq 2$, the fractions

$$\frac{p_k}{q_k} = [a_1, \ldots, a_k]$$

are known as the convergents to α. It is well known that p_k, q_k are generated by the recurrence relations

$$p_k = a_k p_{k-1} + p_{k-2}, \quad q_k = a_k q_{k-1} + q_{k-2}, \tag{4.244}$$

with the convention that $p_0 = 0, q_0 = 1, p_1 = 1$ and $q_1 = a_1$.

Another well-known fact about the convergents is the inequality

$$\left| \alpha - \frac{p_{k-1}}{q_{k-1}} \right| \le \frac{1}{q_{k-1}q_k},$$

which clearly implies

$$\|q_{k-1}\alpha\| < \frac{1}{a_k q_{k-1}}. \tag{4.245}$$

Write $n = \ell q_{k-1}$. Then by (4.245), we have

$$\|n\alpha\| = \|\ell q_{k-1}\alpha\| < \frac{\ell}{a_k q_{k-1}} = \frac{\ell^2}{a_k \ell q_{k-1}} = \frac{\ell^2}{a_k n},$$

and so $\|n\alpha\| < \gamma/n$ holds whenever $\ell^2/a_k \le \gamma$, i.e. whenever

$$1 \le \ell \le \sqrt{\gamma a_k}. \tag{4.246}$$

Now let

$$N_k = \lfloor \sqrt{\gamma a_k} \rfloor q_{k-1}, \tag{4.247}$$

where $\lfloor z \rfloor$ denotes the lower integral part of a real number z. It then follows from (4.246) that the homogeneous diophantine inequality $\|n\alpha\| < \gamma/n$ has at least

$$\sum_{i=1}^{k} \lfloor \sqrt{\gamma a_i} \rfloor$$

integer solutions n satisfying $1 \le n \le N_k$. Formally, we therefore have

$$f(\alpha; \beta = 0; \gamma; N_k) \ge \sum_{i=1}^{k} \lfloor \sqrt{\gamma a_i} \rfloor. \tag{4.248}$$

We distinguish two cases, and start with the much harder one.

Case 22. For all sufficiently large values of k, we have

$$\sum_{i=1}^{k} \lfloor \sqrt{\gamma a_i} \rfloor \le 100 \cdot 2\gamma \log N_k. \tag{4.249}$$

We proceed in four steps.

Step 1. The crucial first step in the argument is to show that the condition (4.249) implies the exponential upper bound

$$\prod_{i=1}^{k}(a_i + 1) \leq e^{c'k} \tag{4.250}$$

for all sufficiently large values of k, where $c' = c'(\gamma)$ is a finite constant independent of k.

To derive (4.250), we use the well-known principle that the exponential functions grow faster than polynomials, in the form of an elementary inequality as follows.

Lemma 23. *For any fixed positive $c > 0$, the inequality*

$$(x + 1)^c \leq (8c^2 e^{-2})^c e^{\sqrt{x}}$$

holds for every $x \geq 1$.

Proof. We start with the trivial observation that $x + 1 \leq 2x$ for all $x \geq 1$, which leads us to the function $g(x) = (2x)^c e^{-\sqrt{x}}$, which we wish to maximize. It is easy to compute the derivative of $g(x)$ and show that its value is maximized when $x = 4c^2$. The desired inequality follows from $(x + 1)^c e^{-\sqrt{x}} \leq g(x) \leq g(4c^2)$. $\square$

By repeated application of (4.244), we have

$$q_{k-1} = a_{k-1}q_{k-2} + q_{k-3} \leq (a_{k-1} + 1)q_{k-2}$$

$$\leq (a_{k-1} + 1)(a_{k-2} + 1)q_{k-3} \leq \ldots \leq \prod_{i=1}^{k-1}(a_i + 1). \tag{4.251}$$

Combining this with (4.247) and (4.249), we have

$$\sum_{i=1}^{k}(\sqrt{\gamma a_i} - 1) \leq 100 \cdot 2\gamma \left(\log \sqrt{\gamma} + \log \sqrt{a_k} + \log \prod_{i=1}^{k-1}(a_i + 1) \right)$$

$$\leq 200\gamma \left(\log \sqrt{\gamma} + \log \prod_{i=1}^{k}(a_i + 1) \right). \tag{4.252}$$

Applying the exponential function, the inequality (4.252) becomes

$$\prod_{i=1}^{k} e^{\sqrt{\gamma a_i} - 1} \leq \gamma^{100\gamma} \prod_{i=1}^{k}(a_i + 1)^{200\gamma}, \tag{4.253}$$

and this inequality holds for all sufficiently large k, i.e. for all $k \geq k_0$.

Applying Lemma 23 with $c = 400\sqrt{\gamma}$ and $x = a_i$ for each $i = 1, 2, \ldots, k+1$, and then multiplying these inequalities together, we obtain

$$\prod_{i=1}^{k}(a_i + 1)^{400\sqrt{\gamma}} \leq (800\sqrt{\gamma})^{800\sqrt{\gamma}k} \prod_{i=1}^{k} e^{\sqrt{a_i}}.$$

Raising this to the $\sqrt{\gamma}$-th power, we have

$$\prod_{i=1}^{k}(a_i + 1)^{400\gamma} \leq (800\gamma)^{800\gamma k} \prod_{i=1}^{k} e^{\sqrt{\gamma a_i}}. \tag{4.254}$$

We next combine (4.253) and (4.254) to obtain

$$\prod_{i=1}^{k}(a_i + 1)^{400\gamma} \leq (800\gamma)^{800\gamma k} e^k \gamma^{100\gamma} \prod_{i=1}^{k}(a_i + 1)^{200\gamma},$$

which, on removing a common factor and then taking 200γ-th root, becomes

$$\prod_{i=1}^{k}(a_i + 1) \leq (800\gamma)^{4k} e^{k/200\gamma} \sqrt{\gamma} = \sqrt{\gamma}((800\gamma)^4 e^{1/200\gamma})^k. \tag{4.255}$$

Since this holds for all $k \geq k_0$, the inequality (4.250) follows.

Step 2. We shall next show that small digit a_i implies a local rectangle property. It follows from (4.255) that, for all sufficiently large k,

$$a_i + 1 \leq (1000\gamma)^8 e^{\frac{1}{100\gamma}} \tag{4.256}$$

holds for at least $k/2$ values of i in $1 \leq i \leq k$. In other words, at least half of the continued fraction digits a_i of α are small, less than a constant depending only on γ, in the precise quantitative sense of (4.256).

Next we show that, for every small digit a_i, the rectangle property must hold locally, in some power-of-two range around q_i. To prove this, we basically repeat the proof of Lemma 4, and use some facts from the theory of continued fractions; see Lemma 24 below. The details go as follows.

As in the proof of Lemma 4, we consider a rectangle of slope $1/\alpha$ and which contains two lattice points $P = (k, \ell)$ and $Q = (m, n)$; in fact, assume that P and Q are two vertices of the rectangle. We denote the vector from P to Q by $\mathbf{v} = (m - k, n - \ell)$, and consider the two perpendicular unit vectors

$$\mathbf{e}_1 = \left(\frac{\alpha}{\sqrt{1+\alpha^2}}, \frac{1}{\sqrt{1+\alpha^2}}\right) \quad \text{and} \quad \mathbf{e}_2 = \left(\frac{1}{\sqrt{1+\alpha^2}}, -\frac{\alpha}{\sqrt{1+\alpha^2}}\right).$$

Then the two sides a and b of the rectangle can be expressed in terms of the inner products $\mathbf{e}_1 \cdot \mathbf{v}$ and $\mathbf{e}_2 \cdot \mathbf{v}$. We have

$$a = |\mathbf{e}_1 \cdot \mathbf{v}| = \frac{|p\alpha + q|}{\sqrt{1 + \alpha^2}} \quad \text{and} \quad b = |\mathbf{e}_2 \cdot \mathbf{v}| = \frac{|p - q\alpha|}{\sqrt{1 + \alpha^2}},$$

where $p = m - k$ and $q = n - \ell$. Thus the area of the rectangle is equal to

$$\text{area} = ab = \frac{|p\alpha + q||p - q\alpha|}{1 + \alpha^2}. \tag{4.257}$$

Without loss of generality we can assume that $p \geq 0$ and $q \geq 0$, and that p is the nearest integer to $q\alpha$. Then $|p - q\alpha| = \|q\alpha\|$. Next we need the following fact from the theory of continued fractions.

Lemma 24. *If $1 \leq q < q_i$, then*

$$\|q\alpha\| \geq \|q_{i-1}\alpha\| > \frac{1}{(a_i + 2)q_{i-1}}.$$

We postpone the proof of Lemma 24.
Now assume that

$$\frac{q_{i-1}}{4} \leq q < q_i. \tag{4.258}$$

Applying Lemma 24 and (4.258), we have

$$|p - q\alpha| = \|q\alpha\| \geq \|q_{i-1}\alpha\| > \frac{1}{(a_i + 2)q_{i-1}} \geq \frac{1}{4(a_i + 2)q}.$$

Substituting this in (4.257) and assuming (4.258), we have

$$\text{area} = ab = \frac{(p\alpha + q)|p - q\alpha|}{1 + \alpha^2} \geq \frac{q|p - q\alpha|}{1 + \alpha^2} \geq \frac{1}{4(a_i + 2)(1 + \alpha^2)}. \tag{4.259}$$

Let us elaborate on the meaning of (4.259). It is about a rectangle of slope $1/\alpha$ which contains two lattice points $P = (k, \ell)$ and $Q = (m, n)$; in fact, P and Q are two vertices of the rectangle. We write the vector from P to Q as $\mathbf{v} = (p, q)$ and, without loss of generality, we can assume that $p \geq 0$ and $q \geq 0$, and that p is the nearest integer to $q\alpha$. If q is large, then $\sqrt{1 + \alpha^2}q$ is very close to the diameter of this long and narrow rectangle. It means that q is basically a size parameter of the rectangle. Assume that the restriction (4.258) holds. Then the inequality (4.259) tells us that the area of this long and narrow rectangle is at least $1/4(a_i + 2)(1 + \alpha^2)$, that is, the area is not too small if a_i is not too large.

We can therefore rephrase (4.258) and (4.259) together in a nutshell as follows. A small digit a_i yields the rectangle property locally. This means that we have a good chance to adapt the Riesz product technique.

For the convenience of the reader, we interrupt the argument, and include a proof of Lemma 24 which is surprisingly tricky.

Proof of Lemma 24. Recall (4.244), that

$$p_k = a_k p_{k-1} + p_{k-2}, \quad q_k = a_k q_{k-1} + q_{k-2}.$$

These recurrences hold for any a_k, including arbitrary real values. Writing

$$\alpha = [a_1, \ldots, a_{k-1}, \alpha_k],$$

with

$$\alpha_k = a_k + \frac{1}{a_{k+1}+} \frac{1}{a_{k+2}+} \cdots = [a_k; a_{k+1}, a_{k+2}, \ldots],$$

we obtain the useful formula

$$\alpha = \frac{\alpha_k p_{k-1} + p_{k-2}}{\alpha_k q_{k-1} + q_{k-2}},$$

and it follows that

$$q_{k-1}\alpha - p_{k-1} = \frac{q_{k-1}p_{k-2} - p_{k-1}q_{k-2}}{\alpha_k q_{k-1} + q_{k-2}}. \tag{4.260}$$

It is not difficult to show that

$$q_{k-1}p_{k-2} - p_{k-1}q_{k-2} = -(q_{k-2}p_{k-3} - p_{k-2}q_{k-3}). \tag{4.261}$$

Since $p_0 = 0$, $q_0 = 1$, $p_1 = 1$ and $q_1 = a_1$, we have $q_1 p_0 - p_1 q_0 = -1$. It follows by induction, using (4.261), that

$$q_{k-1}p_{k-2} - p_{k-1}q_{k-2} = (-1)^{k-1}. \tag{4.262}$$

Combining this with (4.260), we have

$$q_{k-1}\alpha - p_{k-1} = \frac{(-1)^{k-1}}{\alpha_k q_{k-1} + q_{k-2}}, \tag{4.263}$$

which implies

$$\|q_{k-1}\alpha\| = |q_{k-1}\alpha - p_{k-1}| = \frac{1}{\alpha_k q_{k-1} + q_{k-2}} > \frac{1}{(a_k + 2)q_{k-1}}.$$

It remains to prove that, if p and q are integers with $0 < q < q_k$, then

$$|q\alpha - p| \geq |q_{k-1}\alpha - p_{k-1}|. \qquad (4.264)$$

To prove this, we define integers u and v by the equations

$$p = up_{k-1} + vp_k, \quad q = uq_{k-1} + vq_k. \qquad (4.265)$$

Note that (4.265) is solvable in integers u and v, since the determinant of the system is ± 1, in view of (4.262). Since $0 < q < q_k$, we must have $u \neq 0$. Moreover, if $v \neq 0$, then u and v must have opposite signs. Since $q_{k-1}\alpha - p_{k-1}$ and $q_k\alpha - p_k$ also have opposite signs, in view of (4.263), we conclude that

$$\begin{aligned}
|q\alpha - p| &= |u(q_{k-1}\alpha - p_{k-1}) + v(q_k\alpha - p_k)| \\
&= |u(q_{k-1}\alpha - p_{k-1})| + |v(q_k\alpha - p_k)| \\
&\geq |u(q_{k-1}\alpha - p_{k-1})| \geq |q_{k-1}\alpha - p_{k-1}|,
\end{aligned}$$

proving (4.264).   □

Step 3. We next employ the Riesz product technique. Let us return to Theorem 12, and the basically equivalent Proposition 13. A trivial novelty is that in this section, the slope is $1/\alpha$, whereas in Theorem 12 and Proposition 13, the slopes are respectively $1/\sqrt{2}$ and 0. The Riesz product (4.99) is defined by using some appropriate modified Rademacher functions $R_j(\mathbf{x}) \in \mathcal{R}(j)$ for j with $1 \leq 2^j \leq N$, i.e. for $\log_2 N + O(1)$ values of j. In the hypothesis of Theorem 12 and Proposition 13, we have the unrestricted rectangle property; here we have a restricted rectangle property instead, meaning that the rectangle property holds only for $O(\log N)$ values of the power-of-two parameter j, where $0 \leq j \leq \log_2 N + O(1)$. Indeed, by (4.250) and (4.251), we have

$$\log N_k = \log q_{k-1} + O(1) \leq \log \prod_{i=1}^{k}(a_i + 1) + O(1) = O(\log N),$$

and by (4.256), the continued fraction digit a_i of α is small for at least $k/2$ values of i in $1 \leq i \leq k$, if k is sufficiently large. For these small values of the continued fraction a_i, the rectangle property holds in the power-of-two range around q_{i-1}, i.e. when $2^j \approx q_{i-1}$; see (4.258) and (4.259). This means that we can easily save the Riesz product technique developed earlier in Sects. 4.5–4.8. The minor price that we pay is a constant factor loss, due to the fact that $\log_2 N$ is replaced by $c_{28} \log N$, where $c_{28} = c_{28}(\gamma)$ is a small positive constant depending only on $\gamma > 0$. Thus we obtain the following result.

Lemma 25. *Let* $I = [a, b]$, *where* $0 \le a < b < 1$, *be an arbitrary subinterval of the unit interval. Assume that (4.249) holds. Then there exists a constant* $\delta' = \delta'(\gamma) > 0$, *depending only on* $\gamma > 0$, *such that the following hold:*

(i) *For all sufficiently large integers* N, *there is a subinterval* $I_1 = [a_1, b_1]$ *of* I, *possibly depending on* N *and with* $a < a_1 < b_1 < b$, *such that for all* $\beta_1 \in I_1$,

$$f(\alpha; \beta_1; \gamma; N) > 2\gamma \log N + \delta' \log N.$$

(ii) *For all sufficiently large integers* N, *there is a subinterval* $I_2 = [a_2, b_2]$ *of* I, *possibly depending on* N *and with* $a < a_2 < b_2 < b$, *such that for all* $\beta_2 \in I_2$,

$$f(\alpha; \beta_2; \gamma; N) < 2\gamma \log N - \delta' \log N.$$

Step 4. The last step, the construction of a Cantor set, is routine. Combining the method of nested intervals with Lemma 25, we can easily build an infinite binary tree of nested intervals the same way as in the proof of Theorem 3. The divergence points β^* arise as the intersection of infinitely many decreasing intervals, which correspond to an infinite branch of the binary tree. Since a binary tree of countably infinite height has continuum many infinite branches, we obtain continuum many divergence points, proving Theorem 21 in Case 22.

Case 26. The inequality

$$\sum_{i=1}^{k} \lfloor \sqrt{\gamma a_i} \rfloor > 100 \cdot 2\gamma \log N_k \tag{4.266}$$

holds for infinitely many integers $k \ge 1$, where N_k is defined by (4.247).

The estimate (4.241) tells us that $2\gamma \log N_k$ is the average value of $f(\alpha; \beta; \gamma; N_k)$ as β runs through the unit interval. On the other hand, combining (4.248) and (4.266), we deduce that

$$f(\alpha; \beta = 0; \gamma; N_k) > 100 \cdot 2\gamma \log N_k$$

for infinitely many integers $k \ge 1$. In other words, for infinitely many values $N = N_k$, the homogeneous case $\beta = 0$ gives at least 100 times more integer solutions than the average value $2\gamma \log N_k$. This represents an extreme bias; in fact, an extreme surplus. The proof of Theorem 3 is based on a somewhat similar extreme bias, a violation of the Naive Area Principle, in the sense that the Pell inequality $-1 < x^2 - 2y^2 < 1$ has no integer solution except $x = y = 0$, while the corresponding hyperbolic region has infinite area. The only difference is that whereas in Theorem 3, we have an extreme shortage of solutions for the homogeneous case $\beta = 0$, we have here an extreme surplus. But this difference is irrelevant for the method of nested intervals, as it works in both cases. This means that in Case 2, we can simply repeat the Cantor set construction in the proof of Theorem 3. This completes the proof of Theorem 21.

Theorem 21 is a qualitative result. In contrast, we complete this section with a quantitative result.

Proposition 27. *Let $\alpha > 0$ and $\gamma > 0$ be arbitrary real numbers. Then there is an effectively computable positive constant $\delta' = \delta'(\gamma) > 0$, depending only on $\gamma > 0$, such that for every sufficiently large integer N, there exist two real numbers $\beta_1(N)$ and $\beta_2(N)$ in the unit interval, with $0 \le \beta_1(N) < \beta_2(N) < 1$, such that*

$$|f(\alpha; \beta_1(N); \gamma; N) - f(\alpha; \beta_2(N); \gamma; N)| > \delta' \log N.$$

We just outline the proof in a couple of sentences, since it is basically the same as that of Theorem 21, without the Cantor set construction. Indeed, let $q_{\ell-1} \le N < q_\ell$. Since $q_\ell = a_\ell q_{\ell-1} + q_{\ell-2} \le (a_\ell + 1)q_{\ell-1}$, we have

$$1 \le \frac{N}{q_{\ell-1}} \le a_\ell + 1.$$

Again we distinguish two cases.

Case 28. We have

$$\sum_{i=1}^{\ell-1} \lfloor \sqrt{\gamma a_i} \rfloor + \left\lfloor \sqrt{\frac{\gamma N}{q_{\ell-1}}} \right\rfloor \le 100 \cdot 2\gamma \log N.$$

Then by repeating the argument of Case 1 in the proof of Theorem 21 above, we obtain Proposition 27; see Lemma 25.

Case 29. We have

$$\sum_{i=1}^{\ell-1} \lfloor \sqrt{\gamma a_i} \rfloor + \left\lfloor \sqrt{\frac{\gamma N}{q_{\ell-1}}} \right\rfloor > 100 \cdot 2\gamma \log N.$$

Then

$$f(\alpha; \beta = 0; \gamma; N) > 100 \cdot 2\gamma \log N,$$

and so we can choose $\beta_1(N) = 0$. Finally, for $\beta_2(N)$, we can choose any below average point; in other words, we can choose $\beta_2(N)$ to be any β that satisfies the inequality $f(\alpha; \beta; \gamma; N) \le (2 + o(1))\gamma \log N$; see (4.241).

4.10 General Point Sets: Theorem 30

What will happen if we drop the rectangle property in Theorem 12 or Proposition 13? Can we still exhibit extra large deviations for hyperbolic needles? This is the subject of this last section.

Suppose that $\mathscr{P}$ is a finite point set of density $\delta > 0$ in a large square $[0, M]^2$, so that $|\mathscr{P}| = \delta M^2$. We shall make a very mild technical assumption, that $\mathscr{P}$ is not clustered. More precisely, we introduce a new concept called the *separation constant* and denoted by $\sigma = \sigma(\mathscr{P})$, and say that $\mathscr{P}$ is σ-*separated* if the usual Euclidean distance between any two points of $\mathscr{P}$ is at least σ. For example, the set of integer lattice points in the plane is clearly 1-separated, so that $\sigma(\mathbf{Z}^2) = 1$.

Our basic idea is the following. We show that if $\mathscr{P}$ is σ-separated with some not too small constant $\sigma > 0$, then the rectangle property holds, at least in a weak statistical sense, for the majority of the directions which we shall call the good directions. For example, in Theorem 12, the slope $1/\sqrt{2}$ is a concrete good direction. This is how we will be able to save the Riesz product argument in the proof of Theorem 12 or Proposition 13, and still prove extra large deviations, proportional to the area, for hyperbolic needles, at least for the majority of the directions.

In the rest of the section, we work out the details of the vague intuition, and this will give us Theorem 30. The obvious handicap of this majority approach is that for an arbitrary point set $\mathscr{P}$ which is not clustered, we cannot predict whether a given concrete direction is good or not.

Another, and purely technical, shortcoming is that in Theorem 30, we cannot get rid of the assumption that $\mathscr{P}$ is not clustered. This technical difficulty is rather counterintuitive, since at least at first sight, clusters actually seem to help us create extra large deviations. However, some technical difficulties prevent us from adapting the Riesz product technique for clustered point sets $\mathscr{P}$. It remains an interesting open problem to decide whether or not the separation constant $\sigma = \sigma(\mathscr{P})$ in Theorem 30 plays any role.

In Theorem 30, we change[15] the underlying set, and switch from the large square $[0, M]^2$ to the large disk

$$\text{disk}(\mathbf{0}; M) = \{\mathbf{x} \in \mathbf{R}^2 : |\mathbf{x}| \le M\}$$

of radius M and centered at the origin.

Let $\mathscr{P}$ be a finite point set of density $\delta > 0$ in the large disk $\text{disk}(\mathbf{0}; M)$, so that $|\mathscr{P}| = \delta \pi M^2$; here we assume that the radius M is large. We also assume that $\mathscr{P}$ is not clustered. More precisely, we assume that $\mathscr{P}$ is σ-separated for some positive constant $\sigma = \sigma(\mathscr{P}) > 0$. The goal is to count the number of elements of $\mathscr{P}$ in rotated and translated copies of our usual hyperbolic needle $H_\gamma(N)$.

---

[15]The reason behind this change is rotation-invariance. Theorems 3 and 12 are about translated copies, whereas Theorem 30 is about rotated and translated copies of the hyperbolic needle.

Let $10^{-2} > \eta > 0$ be a small positive real numbers, to be specified later. Let j be an arbitrary integer in the interval $0 \le j \le n$, where $2^n \approx N$, that is, $n = \log_2 N + O(1)$ in binary logarithm. We decompose the large disk $\text{disk}(0; M)$ into disjoint translated copies of the small rectangle

$$[0, 2^j \eta] \times [0, 2^{-j} \eta]; \tag{4.267}$$

in other words, we form a rectangle lattice starting from the origin. We shall focus on the copies of (4.267) which are inside the large disk $\text{disk}(0; M)$, and ignore the copies of (4.267) that intersect the boundary circle or are outside the disk. Note that there are $O(2^j \eta M)$ copies of (4.267) that intersect the boundary circle of the large disk. If $2^j \eta = o(M)$, then there are $(1 + o(1)) \pi M^2 \eta^{-2}$ copies of (4.267) that are inside the large disk $\text{disk}(0; M)$. We call these translated copies of the small rectangle (4.267) j-cells. More precisely, we call them j-cells of angle 0.

In general, let θ be an arbitrary angle, with $0 \le \theta < \pi$. Let Rot_θ denote the rotation of the plane by the angle θ, assuming that the fixed point of the rotation Rot_θ is the origin. We decompose the large disk $\text{disk}(0; M)$ into disjoint translates of the rotated copy

$$\text{Rot}_\theta([0, 2^j \eta] \times [0, 2^{-j} \eta]) \tag{4.268}$$

of the small rectangle (4.267). We shall focus on the translated copies of (4.268) which are inside the large disk $\text{disk}(0; M)$. Again, if $2^j \eta = o(M)$, then there are $(1 + o(1)) \pi M^2 \eta^{-2}$ translated copies of (4.268) that are inside the large disk $\text{disk}(0; M)$. We call these translated copies of the small rectangle (4.268) j-cells of angle θ.

We want to prove, in a quantitative form, that if $\mathscr{P}$ is not clustered, then for a typical angle $\theta \in [0, \pi)$, the overwhelming majority of the j-cells of angle θ that contain at least one point of $\mathscr{P}$ actually contain exactly one point of $\mathscr{P}$. A quantitative result like this, a statistical version of the rectangle property, will serve as a substitute for the rectangle property, and it will suffice to save the Riesz product technique developed in Sects. 4.5–4.8.

Statistical Version of the Rectangle Property: An Average Argument. Suppose that $P_{i_1}, P_{i_2} \in \mathscr{P}$, where $i_1 \ne i_2$, are two arbitrary points. We define the *angle-set* by

$$\text{angle}(P_{i_1}, P_{i_2}; j) = \{\theta \in [0, \pi) : \text{there is a } j\text{-cell of angle } \theta \text{ containing } P_{i_1} \text{ and } P_{i_2}\}.$$

The angle-set $\text{angle}(P_{i_1}, P_{i_2}; j)$ is clearly measurable. Let $|\text{angle}(P_{i_1}, P_{i_2}; j)|$ denote the usual one-dimensional Lebesgue measure, i.e. length.

The basic idea is to estimate the double sum

$$\sum_{\substack{P_{i_1}, P_{i_2} \in \mathscr{P} \\ i_1 \ne i_2}} |\text{angle}(P_{i_1}, P_{i_2}; j)|.$$

Simple geometric consideration shows that

$$|\text{angle}(P_{i_1}, P_{i_2}; j)| < 2 \cdot \frac{2^{-j}\eta}{|P_{i_1} P_{i_2}|},$$

where $2^{-j}\eta$ is the length of the short side of a j-cell and $|P_{i_1} P_{i_2}|$ denotes the usual Euclidean distance between P_{i_1} and P_{i_2}, and so

$$\sum_{\substack{P_{i_1}, P_{i_2} \in \mathscr{P} \\ i_1 \neq i_2}} |\text{angle}(P_{i_1}, P_{i_2}; j)| < 2^{-j}\eta \sum_{P_{i_1} \in \mathscr{P}} \left(\sum_{\substack{P_{i_2} \in \mathscr{P} \\ i_1 \neq i_2}} \frac{1}{|P_{i_1} P_{i_2}|} \right). \qquad (4.269)$$

Since $\mathscr{P}$ is σ-separated, it is easy to give an upper bound to the inner sum in (4.269). Using a standard power-of-two decomposition, we have

$$\sum_{\substack{P_{i_2} \in \mathscr{P} \\ i_1 \neq i_2}} \frac{1}{|P_{i_1} P_{i_2}|} \leq \sum_{1 \leq \ell \leq L} \sum_{\substack{P_{i_2} \in \mathscr{P} \\ i_1 \neq i_2 \\ 2^{\ell-1}\sigma < |P_{i_1} P_{i_2}| \leq 2^{\ell}\sigma}} \frac{1}{|P_{i_1} P_{i_2}|}$$

$$\leq \sum_{1 \leq \ell \leq L} \frac{1}{2^{\ell-1}\sigma} \cdot \pi(2^{\ell+1})^2 = \sum_{1 \leq \ell \leq L} \frac{8\pi}{\sigma} \cdot 2^{\ell} < \frac{16\pi}{\sigma} \cdot 2^L, \qquad (4.270)$$

where L denotes the largest integer such that[16] $2^L\sigma < 2^{j+1}\eta$, and where the estimate $\pi(2^{\ell+1})^2$ arises from the fact that a square of side $\sigma/2$ cannot contain two points from $\mathscr{P}$, since $\mathscr{P}$ is σ-separated. From (4.270), and using the fact that $2^L\sigma < 2^{j+1}\eta$, we conclude that

$$\sum_{\substack{P_{i_2} \in \mathscr{P} \\ i_1 \neq i_2}} \frac{1}{|P_{i_1} P_{i_2}|} < \frac{16\pi}{\sigma} \cdot 2^L < \frac{16\pi}{\sigma} \cdot \frac{2^{j+1}\eta}{\sigma} = \frac{2^5 \pi \eta 2^j}{\sigma^2}. \qquad (4.271)$$

Combining (4.269) and (4.271), and using the fact that $|\mathscr{P}| = 8\pi M^2$, we then obtain

$$\sum_{\substack{P_{i_1}, P_{i_2} \in \mathscr{P} \\ i_1 \neq i_2}} |\text{angle}(P_{i_1}, P_{i_2}; j)| < 2^{-j}\eta |\mathscr{P}| \frac{2^5 \pi \eta 2^j}{\sigma^2} = \frac{2^5 \pi^2 \eta^2 8 M^2}{\sigma^2}. \qquad (4.272)$$

---

[16]Note that $2^j \eta$ is the length of the long side of a j-cell.

Recall that the disk disk$(0; M)$ of radius M contains $(1 + o(1))\pi M^2\eta^{-2}$ j-cells of a given angle θ, and that θ runs through the interval $0 \le \theta < \pi$. It is natural, therefore, to normalize the sum (4.272) and consider the average

$$\frac{1}{\pi^2 M^2\eta^{-2}} \sum_{\substack{P_{i_1}, P_{i_2} \in \mathscr{P} \\ i_1 \ne i_2}} |\text{angle}(P_{i_1}, P_{i_2}; j)| < \eta^4 \cdot \frac{2^5\delta}{\sigma^2}. \tag{4.273}$$

Consequences of Inequality (4.273). Let us return to Sect. 4.8. Recall that the last step in the proof of Proposition 13, and indirectly the proof of Theorem 12, is to choose the parameters η_1 and η_2 as sufficiently small positive constants independent of M and N; see (4.226). In fact, in view of (4.100), η_1 and η_2 are almost equal.

In similar fashion, we assume here that the parameter γ of the hyperbolic needle, the density δ of $\mathscr{P}$ and the separation constant σ of $\mathscr{P}$ are fixed positive constants, and consider η, which of course plays the role of η_1 and η_2, as a parameter that we shall eventually choose as a sufficiently small positive constant independent of M and N.

Since the area of a j-cell is η^2, we can say roughly that the probability that a j-cell of any angle contains a point of $\mathscr{P}$ is

$$\text{density} \times \text{area} = \delta\eta^2. \tag{4.274}$$

On the other hand, in view of (4.273), the probability that a j-cell of any angle contains exactly two points of $\mathscr{P}$ does not exceed $c_{29}\eta^4$, which is negligible compared to $\delta\eta^2$ in (4.274) if η is small enough.

In general, the probability that a j-cell of any angle contains exactly p points of $\mathscr{P}$, where $2^\ell < p \le 2^{\ell+1}$ with $\ell = 1, 2, 3, \ldots$, does not exceed $c_{30}\eta^4 4^{-\ell}$, where the constant factor c_{30} is independent of ℓ. Indeed, p points from $\mathscr{P}$ means that we can choose $\binom{p}{2}$ pairs P_{i_1}, P_{i_2}, implying that those rich j-cells show up with multiplicity

$$\binom{p}{2} > 2^\ell 2^{\ell-1} = \frac{1}{2}4^\ell$$

in (4.273), explaining the factor $4^{-\ell}$ in $c_{30}\eta^4 4^{-\ell}$. The point here is that even the sum of the products

$$\sum_{\ell \ge 1} 2^{\ell+1}\eta^4 4^{-\ell}$$

is negligible compared to the $\delta\eta^2$ in (4.274) if η is small enough.

Summarizing, we can say that (4.273) implies the following general picture about the distribution of the elements of $\mathscr{P}$ in the j-cells of any angle. Let $\theta \in [0, \pi)$ be a typical angle, and consider the j-cells of angle θ. The overwhelming majority of the points $P \in \mathscr{P}$ turn out to be singles, meaning that if the point P is contained in some j-cell $\mathscr{C}$ of angle θ, then $\mathscr{C}$ does not contain any other point of $\mathscr{P}$. Here the vague term *overwhelming majority* in fact has the quantitative meaning of $1 - O(\eta^2)$ part of $\mathscr{P}$. Note that $1 - O(\eta^2)$ is almost 1 if η is small.

Furthermore, rich j-cells turn out to be very rare in the following sense. Let $\ell \geq 0$ be a fixed integer. The proportion of the j-cells $\mathscr{C}$ of angle θ containing p points of $\mathscr{P}$, where $2^\ell < p \leq 2^{\ell+1}$, compared to those j-cells which contain at least one point of $\mathscr{P}$, does not exceed $c_{31}\eta^2 4^{-\ell}$, where the constant factor c_{31} is independent of ℓ. Since 2^ℓ is negligible compared to 4^ℓ if ℓ is large, the term *very rare* is well justified.

We can say, therefore, that a weaker statistical version of the rectangle property holds for the majority of the angles $\theta \in [0, \pi)$, assuming that $\eta > 0$ is a sufficiently small constant depending only on the parameter γ of the hyperbolic needle, the density δ of $\mathscr{P}$ and the separation constant σ of $\mathscr{P}$.

A simple analysis of the Riesz product argument in Sects. 4.5–4.8 shows that this weaker statistical version of the rectangle property is a good substitute for the strict rectangle property, and thus we can prove the following result.

Theorem 30. *Let $\mathscr{P}$ be a finite set of points in the disk disk$(0; M)$ with density δ, so that the number of elements of $\mathscr{P}$ is $|\mathscr{P}| = 8\pi M^2$. Assume that $\mathscr{P}$ is σ-separated for some $\sigma > 0$. Assume further that both N and M/N are sufficiently large, depending only on γ, δ and σ. Then there exist a positive constant $\delta' = \delta'(\sigma, \gamma, \delta) > 0$, independent of N and M, and a measurable subset $\mathscr{A} \subset [0, 2\pi)$, of Lebesgue measure greater than $\frac{99}{100} \cdot 2\pi$, such that for every angle $\theta \in \mathscr{A}$, there exist translated copies $\mathbf{x}_1 + \text{Rot}_\theta H_\gamma(N) \subset$ disk$(0; M)$ and $\mathbf{x}_2 + \text{Rot}_\theta H_\gamma(N) \subset$ disk$(0; M)$ of the rotated hyperbolic needle $\text{Rot}_\theta H_\gamma(N)$ such that*

$$|\mathscr{P} \cap (\mathbf{x}_1 + \text{Rot}_\theta H_\gamma(N))| \geq 2\delta\gamma \log N + \delta' \log N$$

and

$$|\mathscr{P} \cap (\mathbf{x}_2 + \text{Rot}_\theta H_\gamma(N))| \leq 2\delta\gamma \log N - \delta' \log N.$$

As indicated at the beginning of this section, it is reasonable to guess that clusters just help to create extra large fluctuations. This intuition motivates the following

Open Problem. *Can one prove a version of Theorem 30 which makes no reference to the separation constant $\sigma = \sigma(\mathscr{P})$? In other words, can we simply drop $\sigma = \sigma(\mathscr{P})$ from the hypotheses of Theorem 30?*

The author guesses that the answer is affirmative but, unfortunately, cannot prove it.

Finally, we briefly mention a closely related problem, where we cannot drop the separation constant $\sigma = \sigma(\mathscr{P})$ from the hypotheses. Note that Theorems 3–21 all concern the extra large fluctuations of the measure-theoretic discrepancy, meaning the difference between the number of points of $\mathscr{P}$ and its expectation of density times area. What we study last here is the large fluctuations of the ± 1-*discrepancy*, or 2-coloring discrepancy.

This means that we have an arbitrary 2-*coloring* $\varphi : \mathscr{P} \to \{\pm 1\}$ of the given point set $\mathscr{P}$, with $+1$ representing *red* and -1 representing *blue*, say. Extra large fluctuations of the ± 1-discrepancy means that there is a translated, or rotated and translated, copies H' and H'' of the hyperbolic needle $H_\gamma(N)$ such that

$$\sum_{P \in \mathscr{P} \cap H'} \varphi(P) > c_{32} \cdot \text{area}(H') = c_{33} \log N > 0$$

with some positive constants c_{32} and c_{33} and

$$\sum_{P \in \mathscr{P} \cap H''} \varphi(P) < -c_{34} \cdot \text{area}(H'') = -c_{35} \log N < 0$$

with some positive constants c_{34} and c_{35}.

The Riesz product technique can be easily adapted to prove extra large fluctuations of the ± 1-discrepancy. For example, we have the following ± 1-discrepancy analog of Proposition 13.

Proposition 31 (2-Coloring Discrepancy for Translated Copies). *Let $\mathscr{P}$ be a finite set of points in the square $[0, M]^2$ with density δ, so that the number of elements of $\mathscr{P}$ is $|\mathscr{P}| = \delta M^2$. Let $\varphi : \mathscr{P} \to \{\pm 1\}$ be an arbitrary 2-coloring of $\mathscr{P}$. Assume that $\mathscr{P}$ satisfies the following rectangle property, that there is a positive constant $c_1 = c_1(\mathscr{P}) > 0$ such that every axes-parallel rectangle of area c_1 contains at most one element of the set $\mathscr{P}$. As in Proposition 13, let $\delta' = \delta'(c_1, \gamma, \delta)$ be defined by (4.83) and (4.84), and assume that both N and M/N are sufficiently large and satisfy (4.85). Then for the hyperbolic needle $H_\gamma(N)$ given by (4.86), there exist translated copies $\mathbf{x}_1 + H_\gamma(N) \subset [0, M]^2$ and $\mathbf{x}_2 + H_\gamma(N) \subset [0, M]^2$ such that*

$$\sum_{P \in \mathscr{P} \cap (\mathbf{x}_1 + H_\gamma(N))} \varphi(P) \geq \delta' \log N$$

and

$$\sum_{P \in \mathscr{P} \cap (\mathbf{x}_2 + H_\gamma(N))} \varphi(P) \leq -\delta' \log N.$$

Similarly, one can easily prove the following analog of Theorem 30.

Proposition 32 (2-Coloring Discrepancy for Rotated and Translated Copies). *Let $\mathscr{P}$ be a finite set of points in the disk $\text{disk}(\mathbf{0}; M)$ with density δ, so that the number of elements of $\mathscr{P}$ is $|\mathscr{P}| = \delta \pi M^2$. Let $\varphi : \mathscr{P} \to \{\pm 1\}$ be an arbitrary 2-coloring of $\mathscr{P}$. Assume that $\mathscr{P}$ is σ-separated with some $\sigma > 0$. Assume further that both N and M/N are sufficiently large, depending only on γ, δ and σ. Then* *there exist a positive constant* $\delta' = \delta'(\sigma, \gamma, \delta) > 0$, *independent of* N *and* M, *and a measurable subset* $\mathscr{A} \subset [0, 2\pi)$, *of Lebesgue measure greater than* $\frac{99}{100} \cdot 2\pi$, *such that for every angle* $\theta \in \mathscr{A}$, *there exist translated copies* $\mathbf{x}_1 + \text{Rot}_\theta H_\gamma(N) \subset$ disk$(\mathbf{0}; M)$ *and* $\mathbf{x}_2 + \text{Rot}_\theta H_\gamma(N) \subset$ disk$(\mathbf{0}; M)$ *of the rotated hyperbolic needle* $\text{Rot}_\theta H_\gamma(N)$ *such that*

$$\sum_{P \in \mathscr{P} \cap (\mathbf{x}_1 + \text{Rot}_\theta H_\gamma(N))} \varphi(P) \geq \delta' \log N$$

and

$$\sum_{P \in \mathscr{P} \cap (\mathbf{x}_2 + \text{Rot}_\theta H_\gamma(N))} \varphi(P) \leq -\delta' \log N.$$

We want to point out that in Proposition 32 on the ± 1-discrepancy of hyperbolic needles, we definitely need some extra condition implying that $\mathscr{P}$ is not too clustered. Indeed, it is easy to construct an extremely clustered point set $\mathscr{P}$ for which the ± 1-discrepancy of the hyperbolic needles is negligible. For example, we can start with a typical point set in general position, and split up every point into a pair of points being extremely close to each other. The two points in these extremely close pairs are joined with a straight line segment each, and we refer to these line segments as the very short line segments. Consider the particular 2-coloring of the point set where the two points in the extremely close pairs all have different colors, with one $+1$ and the other -1. We can easily guarantee that this particular 2-coloring has negligible ± 1-discrepancy for the family of all hyperbolic needles congruent to $H_\gamma(N)$. If the original point set is in general position and the point pairs are close enough, than the arcs of any congruent copy of $H_\gamma(N)$ intersect at most two very short line segments. Since the boundary of $H_\gamma(N)$ consists of 4 arcs, the ± 1-discrepancy is at most $4 \cdot 2 = 8$, which is indeed negligible.

References

1. J. Beck, Randomness of $n\sqrt{2}$ mod 1 and a Ramsey property of the hyperbola, in *Sets, Graphs and Numbers*, ed. by G. Halász, L. Lovász, D. Miklós, T. Szőnyi. Colloquia Math. Soc. János Bolyai, vol. 60 (North-Holland Publishing, Amsterdam, 1992), pp. 23–66
2. J. Beck, *Randomness in diophantine approximation* (Springer, to appear)
3. J. Beck, Diophantine approximation and quadratic fields, in *Number Theory*, ed. by K. Győry, A. Pethő, V.T. Sós. (Walter de Gruyter, Berlin, 1998), pp. 53–93
4. J. Beck, From probabilistic diophantine approximation to quadratic fields, in *Random and Quasi-Random Point Sets*, ed. by Hellekalek, P., Larcher G. Lecture Notes in Statistics, vol. 138 (Springer, New York, NY, 1998), pp. 1–48
5. J. Beck, Randomness in lattice point problems. Discrete Math. **229**(1–3), 29–55 (2001). doi:10.1016/S0012-365X(00)00200-4

6. J. Beck, Lattice point problems: crossroads of number theory, probability theory and Fourier analysis, in *Fourier Analysis and Convexity*, ed. by L. Brandolini, L. Colzani, A. Iosevich, G. Travaglini (Birkhäuser, Boston, MA, 2004), pp. 1–35
7. J. Beck, *Inevitable randomness in discrete mathematics*. University Lecture Series, vol. 49 (American Mathematical Society (AMS), Providence, RI, 2009)
8. J. Beck, Lattice point counting and the probabilistic method. J. Combinator. **1**(2), 171–232 (2010)
9. J. Beck, Randomness of the square root of 2 and the giant leap. I. Period. Math. Hung. **60**(2), 137–242 (2010). doi:10.1007/s10998-010-2137-9
10. J. Beck, Randomness of the square root of 2 and the giant leap. II. Period. Math. Hung. **62**(2), 127–246 (2011). doi:10.1007/s10998-011-6127-3
11. J. Beck, W.W.L. Chen, *Irregularities of distribution*. Cambridge Tracts in Mathematics vol. 89 (Cambridge University Press, Cambridge, 1987)
12. J.W.S. Cassels, An extension of the law of the iterated logarithm. Math. Proc. Cambridge Philos. Soc. **47**, 55–64 (1951)
13. B. Chazelle, *The discrepancy method. Randomness and complexity* (Cambridge University Press, Cambridge, 2000)
14. P. Erdős, On the law of the iterated logarithm. Ann. Math. (2) **43**, 419–436 (1942). doi:10.2307/1968801
15. W. Feller, The general form of the so-called law of the iterated logarithm. Trans. Am. Math. Soc. **54**, 373–402 (1943). doi:10.2307/1990253
16. W. Feller, *An introduction to probability theory and its applications. I.*, 3rd edn. (Wiley, New York, 1968)
17. W. Feller, *An introduction to probability theory and its applications. II.*, 2nd edn. (Wiley, New York, 1971)
18. R.L. Graham, B.L. Rothschild, J.H. Spencer, *Ramsey theory* (Wiley, New York, 1980)
19. G. Halász, On Roth's method in the theory of irregularities of point distributions, in *Recent Progress in Analytic Number Theory*, vol. 2, ed. by H. Halberstam, C. Hooley, vol. 2 (Academic Press, London, 1981), pp. 79–94
20. M. Kac, Probability methods in some problems of analysis and number theory. Bull. Am. Math. Soc. **55**, 641–665 (1949). doi:10.1090/S0002-9904-1949-09242-X
21. A. Khintchine, Über einen Satz der Wahrscheinlichkeitsrechnung. Fund. math. **6**, 9–20 (1924)
22. A. Kolmogorov, Über das Gesetz des iterierten Logarithmus. Math. Ann. **101**, 126–135 (1929). doi:10.1007/BF01454828
23. S. Lang, *Introduction to diophantine approximations* (Addison-Wesley, Reading, 1966)
24. J. Matoušek, *Geometric discrepancy. An illustrated guide. Revised paperback reprint of the 1999 original*. Algorithms and Combinatorics, vol. 18 (Springer, Berlin, 2010). doi:10.1007/978-3-642-03942-3
25. A. Ostrowski, Bemerkungen zur Theorie der diophantischen Approximationen I. Abh. Math. Sem. Univ. Hamburg **1**, 77–98 (1921). doi:10.1007/BF02940581
26. K.F. Roth, On irregularities of distribution. Mathematika **1**, 73–79 (1954). doi:10.1112/S0025579300000541
27. W.M. Schmidt, Irregularities of distribution. VII. Acta Arith. **21**, 45–50 (1972)
28. J.G. van der Corput, Verteilungsfunktionen. I. Proc. Kon. Ned. Akad. v. Wetensch. Amsterdam **38**, 813–821 (1935)
29. J.G. van der Corput, Verteilungsfunktionen. II. Proc. Kon. Ned. Akad. v. Wetensch. Amsterdam **38**, 1058–1066 (1935)
30. H. Weyl, Über die Gleichverteilung von Zahlen mod. Eins. Math. Ann. **77**, 313–352 (1916). doi:10.1007/BF01475864
31. D.B. Zagier, *Zetafunktionen und quadratische Körper. Eine Einführung in die höhere Zahlentheorie* (Springer, Berlin, 1981)

Part II
Combinatorial Discrepancy

Chapter 5
Multicolor Discrepancy of Arithmetic Structures

Nils Hebbinghaus and Anand Srivastav

Abstract In this chapter we present developments over the last 20 years in the discrepancy theory for hypergraphs with arithmetic structures, e.g. arithmetic progressions in the first N integers and their various generalizations, like Cartesian products, sums of arithmetic progressions, central arithmetic progressions in $\mathbb{Z}_p$ and linear hyperplanes in finite vector spaces. We adopt the notion of multicolor discrepancy and show how the 2-color theory generalizes to multicolors exhibiting new phenomena at several places not visible in the 2-color theory, for example in the coloring of products of hypergraphs. The focus of the chapter is on proofs of lower bounds for the multicolor discrepancy for hypergraphs with arithmetic structures. Here, the application of Fourier analysis or linear algebra techniques is often not sufficient and has to be combined with combinatorial arguments, in the form of an interplay between the examination of suitable color classes and the Fourier analysis.

5.1 Introduction

Several books treat combinatorial discrepancy theory: the "Ten Lectures on the Probabilistic Method" of Joel Spencer [71], the book "The Probabilistic Method" of Noga Alon, Joel Spencer and Paul Erdős [1] and the first and second edition of "Geometric Discrepancy (An Illustrated Guide)" of Jiří Matoušek [51, 53]. The book of Bernard Chazelle [19] places discrepancy theory in a broad context of mathematics and computer science. A concise introduction to Ramsey Theory is the book of Ronald Graham, Bruce Rothschild and Joel Spencer [36]. An excellent survey on geometric and combinatorial discrepancy theory is the article of Beck and Sós [13]. A thorough treatment discussing combinatorial as well as geometric

N. Hebbinghaus • A. Srivastav (✉)
Department of Computer Science, Kiel University, Kiel, Germany
e-mail: nhe@informatik.uni-kiel.de; asr@informatik.uni-kiel.de

W. Chen et al. (eds.), *A Panorama of Discrepancy Theory*, Lecture Notes in Mathematics 2107, DOI 10.1007/978-3-319-04696-9_5,
© Springer International Publishing Switzerland 2014

discrepancy theory is the book of Beck and Chen [11]. The theme discrepancy of arithmetic progressions can be found in many of these references. In this chapter we will discuss the discrepancy of several arithmetic structures generalizing arithmetic progressions in a comprehensive and self-contained way.

We start with the basic notion of combinatorial discrepancy and an overview of fundamental results (Sect. 5.2). Emphasis is given to the introduction of the notion of combinatorial multicolor discrepancy and the generalization of several classical results from 2-color discrepancy theory to c colors. For example, the famous six-standard-deviation upper bound of Spencer for the discrepancy of a hypergraph with n nodes and m hyperedges in 2 colors is $O\left(\sqrt{n \ln(m/n)}\right)$ and for c-colors it becomes $O\left(\sqrt{\frac{n}{c} \ln(cm/n - c)}\right)$. We show at several places, concerning classical upper and lower bounds, how the partitioning factor of $1/c$ has to be invoked into these discrepancy bounds. This forms the basis for further discussions, but the reader more interested in arithmetic structures may skip the technical details.

Section 5.3 introduces the d-fold Cartesian products of hypergraphs, in particularly the d-fold Cartesian product of the hypergraph of arithmetic progressions in the first N integers. This gives a kind of d-dimensional generalization. The starting point of this section is the famous theorem of K. Roth [67] and J. Matoušek and J. Spencer [56] which states that the discrepancy of the hypergraph of arithmetic progressions is $N^{1/4}$, up to a constant. We show that the discrepancy function for Cartesian products is submultiplicative, and almost submultiplicative in c colors, leading to upper bounds for the discrepancy which scale in the dimension d in a natural way. On the other hand we adapt Fourier analysis to prove matching lower bounds. It is also shown that the situation for the c-color discrepancy of Cartesian products is much more involved and Ramsey's theorem is required to describe the complex behavior depending on the very form of the number of colors. Interesting new phenomena appear in mixed situations, where low-dimensional objects are considered in a high-dimensional environment, for example arithmetic progressions that have the same common difference, one-dimensional arithmetic progressions in the $[N]^d$ and d-dimensional symmetric arithmetic progressions in $[N]^d$.

Section 5.4 is devoted to the hypergraph of all sets formed by sums of any k arithmetic progressions in $\{1, \ldots, N\}$, k a fixed integer . Some authors call them high-dimensional arithmetic progressions [59]. We will show that their discrepancy is typically of the order $O(\sqrt{n})$, for $k \geq 2$, thus significantly higher than for $k = 1$. The emphasis is on the proof of lower bounds, with Fourier analysis and combinatorial arguments.

In Sect. 5.5 we consider the problem of finding the c-color discrepancy of arithmetic progressions and centered arithmetic progressions resp. in $\mathbb{Z}_p$, p a prime. We will show that its discrepancy is essentially about $O(\sqrt{p/c})$. Major effort is required to circumvent the fact that Fourier analysis cannot be directly applied to the centered arithmetic progressions because they are not translation invariant. This is done with combinatorial decompositions depending on arithmetic arguments, so that Fourier analysis can be invoked in a suitable way.

A new aspect of combinatorial c-color discrepancy theory is the notion of the one-sided c-color discrepancy. The one-sided discrepancy function cannot be bounded with discrepancy bounds, and has its own structure, demanding for a specific theory. We study in Sect. 5.6 the one-sided c-color discrepancy of the hypergraphs of linear hyperplanes in the finite vector space $\mathbb{F}_q^r$ using Fourier analysis on $\mathbb{F}_q^r$ in an interplay with combinatorial arguments to locate the right color classes leading to large discrepancy hyperedges. We show that the one-sided discrepancy is bounded from below by $\Omega_q\left(\sqrt{nz(1-z)/c}\right)$, where $z = \frac{(q-1)\bmod c}{c}$, and the upper bound is tight up to a logarithmic factor. So again, the one-sided discrepancy obeys the square root behavior.

We conclude the chapter stating some open problems (Sect. 5.7).

5.2 Multicolor Discrepancy

In this section we introduce the notion of combinatorial multicolor discrepancy and generalize several classical results from 2-color discrepancy theory to c colors ($c \geq 2$). We give a method that constructs c-colorings by iteratively computing 2-color discrepancies covering a large class of theorems like the 'six standard deviations' theorem of Spencer [70], the Beck–Fiala theorem [12], the results of Matoušek, Welzl and Wernisch [57] and Matoušek [50]. In particular, the c-color discrepancy of a hypergraph with n vertices and m hyperedges is $O(\sqrt{\frac{n}{c}\log(\frac{cm}{n-c})})$. For $m = O(n)$ and $c \leq \alpha n$ for a constant $\alpha \in (0,1)$ this bound becomes $O(\sqrt{\frac{n}{c}}\log c)$.

In situations where the discrepancy in c colors cannot be bounded in terms of two-color discrepancies in general, this approach fails, as well as for the linear discrepancy version of the Beck–Fiala theorem. Here an extension of the floating color technique via tensor products of matrices is appropriate leading to multicolor versions of the Beck–Fiala theorem and the Bárány–Grinberg theorem. Interestingly, the tensor product technique gives also a lower bound for the c-color discrepancy of general hypergraphs.

5.2.1 From Graphs to Hypergraphs

Combinatorial discrepancy theory has its origin in the coloring of graphs. A (simple, finite) graph is a pair $G = (V, E)$ where V is a finite set and E is a set of 2-element subsets of V, i.e. $E \subseteq \binom{V}{2}$, where $\binom{V}{2}$ denotes the set of all 2-element subsets of V. The elements of V are called *nodes* or *vertices* and the elements of E are called *edges*. A common notation for an edge $e \in E$ is $e = \{u, v\}$, with $u, v \in V$. Thus, the edge e is not directed and here we are not concerned with directed graphs.

For a graph $G = (V, E)$ and $k \in \mathbb{N}$, a function $f : \{1, \ldots k\} \mapsto V$ is called a k-coloring, if for every edge $e = \{u, v\} \in E$ we have $f(u) \neq f(v)$. So the vertices of an edge have different colors. The numbers $1, \ldots, k$ are called colors. With $k = n$ we can trivially color all vertices of G. The challenge is to determine the *smallest* k such that a k-coloring exists. To shorten the notation we set for a non-negative integer l, $[l] := \{1, \ldots, l\}$.

Definition 1. Let $G = (V, E)$ be a graph. The chromatic number $\chi(G)$ is the smallest integer k such that G can be colored with k colors.

A whole branch of combinatorics is devoted to the coloring of graphs. Among the milestones in this area is the famous 4-color theorem that for a planar graph the chromatic number is at most 4. The theorem has been proved in 1976 by Kenneth Appel and Wolfgang Haken [2,3] with a computer-assisted proof, the first of its kind, checking about 1,936 case distinctions. Thereafter, shorter proofs were given. In 1996, Neil Robertson, Daniel Sanders, Paul Seymour and Robin Thomas [66] found a computer-assisted proof with 633 case distinctions. In 2004, Benjamin Werner and Georges Gonthier, see Gonthier (2008) [35], constructed a formal proof using the Coq proof assistant system.

Existence proofs of finite combinatorial objects in combinatorics challenge their construction in an efficient way. In fact, the algorithmic point of view of modern computer science has become a central matter in cross-cutting, interdisciplinary research, both for mathematical insight as well as for applications. For a graph $G = (V, E)$ with $|V| = n$ we may ask if we can construct a coloring with an algorithm using only a polynomial number of steps, e.g. $O(n^2)$, $O(n^3)$ etc. The theory of NP-completeness by Cook (1971) and Karp (1975) is a fundamental discovery to characterize the combinatorial complexity of problems. The coloring problem is hard in the sense that its polynomial-time solution would imply that all other problems in the class NP can be solved in polynomial time as well. Such problems are called NP-hard. It is conjectured that NP-hard problems are not solvable in polynomial time and it comes down to the question if P $=$ NP. It is widely believed and conjectured, that P $\neq$ NP. This conjecture is regarded perhaps as the biggest open problem in theoretical computer science.

On the other hand, if G is bipartite, so $V = A \cup B$, $A \cap B = \emptyset$, and the edges are all of the form $\{a, b\}$, $a \in A$, $b \in B$, then $\chi(G) = 2$. Any graph G can be tested for bipartiteness using breath/depth first search in $O(|E|)$ steps, simultaneously providing a 2-coloring of G, if the test is positive. Since bipartiteness of G is equivalent to $\chi(G) = 2$, the 2-coloring problem of graphs is solvable in polynomial time, in this sense it is an easy problem. The situation changes when passing to hypergraphs.

Definition 2. A (finite) hypergraph $\mathcal{H} = (X, \mathcal{E})$ is a pair where X is a finite set and $\mathcal{E}$ is a subset of the power set $\mathcal{P}(X)$. The elements of X are called nodes or vertices and the elements of $\mathcal{E}$ are called hyperedges (or simply edges). For $Y \subset X$, we define $\mathcal{E}_Y := \{E \cap Y; E \in \mathcal{E}\}$ and $\mathcal{H}_Y = (Y, \mathcal{E}_Y)$ is the hypergraph induced on Y by $\mathcal{H}$.

In combinatorics, $\mathcal{H}$ is known as a set system, and in computational geometry $\mathcal{H}$ is called a range space (here $X \in \mathcal{E}$ is required in addition). Of course, if $|E| = 2$ for all $E \in \mathcal{E}$, $\mathcal{H}$ is just a graph. How can we generalize the coloring concept to hypergraphs? In graphs, a feasible coloring of the vertices of an edge $e = \{v, u\}$ demands that u and v have different colors. So the number of colors in an edge is perfectly balanced. But we can also say that in a feasible graph coloring edges are *not* monochromatic. These two different interpretations lead to the following two alternatives for the generalizations of the coloring concept to hypergraphs. We color the vertices of $\mathcal{H}$ with 2 colors such that

a) (first alternative) the number of different colors for every hyperedge is nearly the same, so the 2-coloring is nearly balanced in every hyperedge,
b) (second alternative) no hyperedge is monochromatic.

The first concept of hypergraph coloring is the roof for combinatorial discrepancy theory, while the second one is fundamental for Ramsey theory. We may ask whether the 2-coloring problem in hypergraphs can be solved in polynomial time. We will show that the complexity jumps by passing from graphs to hypergraphs. In fact the 2-color discrepancy problem is NP-hard (Theorem 26) calling for a mathematical and algorithmic foundation. Now let us proceed to the definition of the discrepancy of a 2-coloring of hypergraphs.

5.2.2 2-Color Discrepancy

The objects of 2-color combinatorial discrepancy theory is the study of 2-colorings of hypergraphs with the property that in all hyperedges the number of vertices in the two colors is roughly the same. In combinatorial discrepancy theory we are interested in tight absolute bounds for the discrepancy function, which depend on the structure of the hypergraphs under consideration. Lower bounds show that for any 2-coloring the discrepancy is at least that lower bound, while upper bounds are derived by the proof of the existence of a particular coloring or even by a construction of the same.

Two Colors: Discrepancy, Linear and Hereditary Discrepancy. Let $\mathcal{H} = (X, \mathcal{E})$ be a *hypergraph*. A partition of X into two color classes can be represented by a function $\chi : X \to \{-1, +1\}$. We identify -1 and $+1$ with colors, say red and blue, and thus call χ a coloring. The color-classes $\chi^{-1}(-1)$ and $\chi^{-1}(+1)$ form a partition of X. For a hyperedge $E \in \mathcal{E}$ let us define $\chi(E) := \sum_{x \in E} \chi(x)$. The imbalance of $E \in \mathcal{E}$ with respect to χ is $|\chi(E)|$. The *discrepancy* of $\mathcal{H}$ with respect to χ is

$$\mathrm{disc}(\mathcal{H}, \chi) := \max_{E \in \mathcal{E}} |\chi(E)| \tag{5.1}$$

Fig. 5.1 A 2-coloring with discrepancy 1, which is optimal since there are hyperedges with odd cardinality

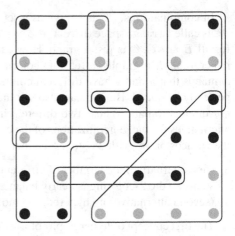

The discrepancy of $\mathcal{H}$ is the hypergraph invariant

$$\mathrm{disc}(\mathcal{H}) := \min_{\chi:X\to\{-1,+1\}} \mathrm{disc}(\mathcal{H},\chi) \qquad (5.2)$$

A two-coloring example is shown in Fig. 5.1.

The concept of hypergraph discrepancy has been generalized to matrices in a natural way. Let $A = (a_{ij})$ be a real $m \times n$-matrix, $\|\cdot\|_\infty$ be the usual L^∞-norm on $\mathbb{R}$ and set

$$\mathrm{disc}(A) := \min_{x\in\{-1,1\}^n} \|Ax\|_\infty \qquad (5.3)$$

Let $\mathcal{H} = (X, \mathcal{E})$ be a hypergraph with $X = \{x_1,\dots,x_n\}$ and $\mathcal{E} = \{E_1,\dots,E_m\}$. The matrix $A = (a_{ij})$, where $a_{ij} = 1$ if $x_j \in E_i$ and $a_{ij} = 0$ otherwise, is called the *incidence matrix* of $\mathcal{H}$. Clearly, we have $\mathrm{disc}(A) = \mathrm{disc}(\mathcal{H})$. Thus the discrepancy of matrices generalizes the discrepancy notion for hypergraphs, and for zero–one-matrices both concepts are equivalent.

There are two related notions: The *linear discrepancy* of an arbitrary matrix A is defined by

$$\mathrm{lindisc}(A) := \max_{p\in[-1,1]^n} \min_{x\in\{-1,1\}^n} \|A(p-x)\|_\infty \qquad (5.4)$$

Linear discrepancy can be regarded as a measure of how well a fractional solution can be rounded to an integer solution. Some authors define linear discrepancy by

$$\max_{p\in[0,1]^n} \min_{x\in\{0,1\}^n} \|A(p-x)\|_\infty \qquad (5.5)$$

Both versions differ only by a constant factor 2. A special case of the second version is the weighted discrepancy, which refers to the problem of splitting the edges in an arbitrary ratio $p : 1 - p$ as in the discrepancy problem. We define the weighted discrepancy in 2 colors: let $\mathbf{1}_n := (1, \ldots, 1)^\top \in \mathbb{R}^n$. For $x \in \{0, 1\}^n$ and $p \in [0, 1]$, the weighted discrepancy with respect to x and p is

$$\text{wd}(A, x, p) := \|A(\mathbf{1}_n p - x)\|_\infty, \tag{5.6}$$

the weighted discrepancy with respect to p is

$$\text{wd}(A, 2, p) := \min_{x \in \{0,1\}^n} \|A(\mathbf{1}_n p - x)\|_\infty, \tag{5.7}$$

and the weighted discrepancy finally is

$$\text{wd}(A, 2) := \max_{p \in [0,1]} \text{wd}(A, 2, p). \tag{5.8}$$

The *hereditary discrepancy* is

$$\text{herdisc}(A) := \max_{J \subseteq [n]} \text{disc}\left((a_{ij})_{i \in [m], j \in J}\right), \tag{5.9}$$

where $[n] := \{1, \ldots, n\}$.

All notions can be formulated for hypergraphs in a natural way. For example, $\text{herdisc}(\mathcal{H})$ is the maximum discrepancy of all induced subgraphs $\mathcal{H}_0$ of $\mathcal{H}$:

$$\text{herdisc}(\mathcal{H}) := \max_{\mathcal{H}_0 \subseteq \mathcal{H}} \text{disc}(\mathcal{H}_0) \tag{5.10}$$

We will write $A_0 \leq A$ to indicate that the matrix A_0 consists of some columns of the matrix A. Similarly for hypergraphs we will write $\mathcal{H}_0 \leq \mathcal{H}$ or $\mathcal{H}_0 \subseteq \mathcal{H}$ if $\mathcal{H}_0$ is an induced subgraph of $\mathcal{H}$.

Proposition 3. *The following relations between the different discrepancy notions hold:*

  (i) $\text{disc}(A) \leq \text{herdisc}(A)$
 (ii) $\text{disc}(A) \leq \text{lindisc}(A)$
(iii) $\text{wd}(A, 2) \leq \frac{1}{2} \text{lindisc}(A)$
 (iv) $\text{lindisc}(A) \leq 2 \text{herdisc}(A)$
  (v) $\text{wd}(A, 2) \leq \text{herdisc}(A)$

Relations (i), (ii) and (iii) are more or less obvious, while the relation between the linear and the hereditary discrepancy is not. Relation (iv) was discovered by Beck and Spencer [14] and Lovász, Spencer and Vesztergombi [48]. In 1986 Spencer conjectured that $\text{lindisc}(A) \leq (1 - \frac{1}{n+1})$. This conjecture is still unsolved, but significant contribution were made meanwhile. Doerr [21] showed that the inequality (iv) can be improved to $\text{lindisc}(A) \leq (1 - \frac{1}{2m}) \text{herdisc}(A)$, see also Bohman and Holzman [16]. And for totally unimodular matrices Spencer's conjecture was proved

in the affirmative by Doerr [21, 24]. The proof uses the fact that the extreme points
of the polyhedron of a totally unimodular matrix are integral.

Theorem 4 (Doerr [21]). *Let A be a totally unimodular $m \times n$ matrix. Then*
$\mathrm{lindisc}(A) \leq (1 - \frac{1}{n+1})$.

Further results on the linear discrepancy for the L^p-norm can be found in Doerr [25].
Matoušek could bound the hereditary discrepancy in terms of the linear discrepancy
in [52].

5.2.3   Some Fundamental Discrepancy Results

5.2.3.1   Upper Bounds for the Combinatorial Discrepancy.

A milestone in combinatorial discrepancy theory is the celebrated "six-standard-
deviation" theorem of Joel Spencer [70].

Theorem 5 (Spencer [70]). *Let $\mathcal{H} = (X, \mathcal{E})$ be a hypergraph with n vertices and
$m \geq n$ hyperedges. Then*

$$\mathrm{disc}(\mathcal{H}) = O\left(\sqrt{n \log(\frac{m}{n})}\right).$$

In particular, if $m = O(n)$, then $\mathrm{disc}(\mathcal{H}) = O(\sqrt{n})$. For $m = n$, the bound is
$6\sqrt{n}$.

The bound of $O(\sqrt{n})$ is tight for an infinite family of hypergraphs, namely those
with a Hadamard matrix as its incidence matrix. Alternative proofs were given by
Gluskin [34] with the Minkowski lattice point theorem and by Giannopoulos [33]
using Gaussian measure and geometry.

It has been a long-standing open problem whether there is a polynomial-time
algorithm which constructs a 2-coloring with discrepancy $O(\sqrt{n})$. In 2010, N.
Bansal solved this problem [6] with a polynomial-time randomized algorithm and
little later, Bansal and Spencer [7] were able to de-randomize the algorithm [7]. We
refer the interested reader to the chapter of Nikhil Bansal.

Theorem 6 (Bansal [6]). *Let $\mathcal{H} = (X, \mathcal{E})$ be a hypergraph with n vertices and
$m = n$ hyperedges. There is a randomized polynomial time algorithm which
computes a 2-coloring with discrepancy $O(\sqrt{n})$.*

Lovett and Meka [49] gave a constructive algorithm which achieves Spencer's
bound for all hypergraphs, i.e. $m \geq n$. The algorithm is based on linear algebra
techniques and random walks. We refer to the chapter of N. Bansal.

When the maximum *vertex degree*[1] of the hypergraph is bounded, another type
of discrepancy result can be proved:

---

[1] The degree of a vertex is the number of hyperedges that contain the vertex.

Theorem 7 (Beck–Fiala [12]). *Let* $\mathcal{H} = (X, \mathcal{E})$ *be a hypergraph and* $\Delta(\mathcal{H}) \in \mathbb{N}$ *be the maximum vertex degree of* $\mathcal{H}$. *Then*

$$\mathrm{disc}(\mathcal{H}) \leq \mathrm{lindisc}(\mathcal{H}) \leq 2\Delta(\mathcal{H}) - 1.$$

Note that the hypergraphs associated to *Hadamard matrices*[2] have discrepancy at least $\Omega(\sqrt{\Delta(\mathcal{H})})$. Beck and Fiala conjectured

Conjecture 8 (Beck–Fiala [12]). $\mathrm{disc}(\mathcal{H}) = O(\sqrt{\Delta(\mathcal{H})})$.

The conjecture is still open. With some effort Bednarchak and Helm [42] improved the bound to $2\Delta(\mathcal{H}) - 3$. A major step was taken by Srinivasan [72]:

Theorem 9 (Srinivasan [72]). *Let* $\mathcal{H} = (X, \mathcal{E})$ *be a hypergraph and* $\Delta(\mathcal{H}) \in \mathbb{N}$ *be the maximum vertex degree of* $\mathcal{H}$. *Then* $\mathrm{disc}(\mathcal{H}) \leq O(\sqrt{\Delta(\mathcal{H})\log n})$.

Bansal [6] gave a randomized polynomial time algorithm which achieves the bound of Srinivasan. The best bound is due to Banaszczyk [5].

Theorem 10 (Banaszczyk [5]). *Let* $\mathcal{H} = (X, \mathcal{E})$ *be a hypergraph and* $\Delta(\mathcal{H}) \in \mathbb{N}$ *be the maximum vertex degree of* $\mathcal{H}$. *Then* $\mathrm{disc}(\mathcal{H}) \leq O(\sqrt{\Delta(\mathcal{H})\log n})$.

Still, no polynomial-time algorithm is known to compute colorings within the discrepancy bound of Banaszczyk.

A discrepancy result widely used in combinatorial discrepancy theory is the following theorem of Beck [10]. We will frequently make use of it in this chapter. Let k be a positive integer. Let us define the hypergraph $\mathcal{H}_k$ by $\mathcal{H}_k := (X, \{E \in \mathcal{E} : |E| \geq k\})$. So only large hyperedges are considered.

Theorem 11 (Beck [10]). *Let* $\mathcal{H} = (X, \mathcal{E})$ *be a finite hypergraph,* $n := |X|$ *and* $m := |\mathcal{E}|$. *Let* t *and* K *be natural numbers such that* $\deg(\mathcal{H}_t) \leq K$. *Then*

$$\mathrm{disc}(\mathcal{H}) \leq c(t + K\log K)^{1/2}\log^{1/2} m \log n,$$

where $c > 0$ *is an absolute constant.*

Theorem 11 implies the following corollary.

Corollary 12. *If there exists a constant* $t > 0$ *with the property* $\deg(\mathcal{H}_t) \leq t$, *then there is an absolute constant* $c > 0$ *such that* $\mathrm{disc}(\mathcal{H}) \leq c\sqrt{t}\log m \log n$.

Let us move to the notion of the discrepancy of a sequence of vectors. This kind of problems were already posed by Dworetzky in 1963, see [13, page 1420]. The theorem of Bárány–Grinberg [9] gives a very general result.

---

[2]A Hadamard matrix of dimension n is a matrix in $\{-1, 1\}^{n \times n}$ whose rows are mutually orthogonal.

Theorem 13 (Bárány–Grinberg [9]). *Let $\| \cdot \|$ be an arbitrary norm on $\mathbb{R}^n$ and let $v_1, v_2, \ldots, v_k$ be vectors in $\mathbb{R}^n$ such that $\|v_i\| \leq 1$ for all $i = 1, \ldots, k$. There are signs $\varepsilon_i \in \{-1, +1\}, i = 1, \ldots, k$, such that for all $l \in [k]$,*

$$\left\| \sum_{i=1}^{l} \varepsilon_i v_i \right\| \leq 2n.$$

The hypergraph setting allows a natural reformulation in terms of vectors by considering the incidence matrix of the hypergraph, resp. its row and column vectors. The Beck–Fiala theorem formulated in the vector setting states that for any vectors $v_1, \ldots, v_k \in \{0, 1\}^n$ with $\|v_i\|_1 \leq 1$ for all i, there are signs $\varepsilon_i \in \{-1, +1\}$ for all i, such that $\left\| \sum_{i=1}^{k} \varepsilon_i v_i \right\|_\infty < 2$. Thus the Beck–Fiala theorem is not a special case of the Bárány–Grinberg theorem, nor vice versa.

Spencer's theorem in the vector setting shows that for any sequence of vectors $v_1, \ldots, v_k \in \{0, 1\}^n$ with $\|v_i\|_\infty \leq 1$ for all i, there are signs $\varepsilon_i \in \{-1, +1\}$ for all i such that $\left\| \sum_{i=1}^{k} \varepsilon_i v_i \right\|_\infty \leq 6\sqrt{n}$. So, informally speaking the Bárány–Grinberg theorem holds true for any norm while the Beck–Fiala theorem combines the L^1-norm and the L^∞-norm. Spencer's theorem combines the L^∞-norm and Grinberg further showed that the combination of the L^2-norm with the L^2-norm gives a bound of $\sqrt{n}$. The only combination for which nothing can be said, and which is a major open problem in discrepancy theory, is the combination of the L^∞-norm and the L^2-norm. The following conjecture is due to János Komlós (see, e.g., the book of Joel Spencer [71]).

Conjecture 14. Let $v_1, v_2, \ldots, v_k$ be vectors in $\mathbb{R}^n$ with $\|v_i\|_2 \leq 1$ for all $i = 1, \ldots, k$. There are signs $\varepsilon_i \in \{-1, +1\}, i = 1, \ldots, k$, and a constant K such that

$$\left\| \sum_{i=1}^{k} \varepsilon_i v_i \right\|_\infty < K$$

Note that the Komlós conjecture implies the Beck–Fiala conjecture. The best result towards the Komlós conjecture is due to Banaszczyk.

Theorem 15 (Banaszczyk [5]). *Let $\| \cdot \|$ be the Euclidean norm on $\mathbb{R}^n$ and γ_n the standard Gaussian measure on $\mathbb{R}^n$ with density $(2\pi)^{-n/2} \exp(-\|x\|^2/2)$. There is a constant $c > 0$ with the following property: if K is an arbitrary convex body in $\mathbb{R}^n$ with $\gamma_n(K) \geq 1/2$, then to each sequence $u_1, \ldots, u_m \in \mathbb{R}^n$ with $\|u_i\| \leq c$ for all i there correspond signs $\epsilon_1, \ldots, \epsilon_m \in \{-1, 1\}$ such that $\sum_i \epsilon_i u_i \in K$.*

Note that this result implies Theorem 10. An interesting further progress was recently made by A. Nikolov [60], who proved that if one assigns unit real vectors rather than signs $+1, -1$, then the Komlós conjecture holds with $K = 1$.

5.2.3.2 Lower Bounds

Throughout this chapter we will make use of a general lower bound of Lovász and T. Sós , see [13, Theorem 2.8 and Corollary 2.9]. Let us define the L^2-discrepancy. For a hypergraph $\mathcal{H} = (X, \mathcal{E})$ with $n = |X|$, $m = |\mathcal{E}|$, the L^2-discrepancy is

$$\mathrm{disc}_2(\mathcal{H}) := \min_{\chi} \left(\frac{1}{m} \sum_{E \in \mathcal{E}} |\chi(E)|^2 \right)^{1/2}. \tag{5.11}$$

We immediately have, $\mathrm{disc}(\mathcal{H}) \geq \mathrm{disc}_2(\mathcal{H})$, so the L^2-discrepancy gives a lower bound for the discrepancy. The advantage of the L^2-discrepancy is that we can invoke linear algebra to compute lower bounds for $\mathrm{disc}_2(\mathcal{H})$.

Theorem 16. *Let $\mathcal{H} = (X, \mathcal{E})$ be a hypergraph with $n = |X|$, $m = |\mathcal{E}|$ and let A be the incidence matrix of $\mathcal{H}$. With Tr we denote the trace of a matrix.*

(i) $\mathrm{disc}(\mathcal{H}) \geq \left(n\lambda_{min}(A^T A) \right)^{1/2}$,

(ii) *If for some diagonal matrix D, the matrix $A^T A - D$ is positive semidefinite, then $\mathrm{disc}_2(\mathcal{H}) \geq (\mathrm{Tr}\, D)^{1/2}$.*

An other fundamental lower bound is due to Lovász, Spencer, and Vesztergombi [48].

For a real matrix A define $\mathrm{detlb}(A) := \max_k \max_B |\det B|^{1/k}$, where the maximum ranges over all $k \times k$ submatrices B of A. For a hypergraph $\mathcal{H}$ let us define $\mathrm{detlb}(\mathcal{H}) := \mathrm{detlb}(A)$ where A is the incidence matrix of $\mathcal{H}$.

Theorem 17 (Lovász, Spencer, and Vesztergombi [48]). *For every hypergraph $\mathcal{H}$, $\mathrm{herdisc}(\mathcal{H}) \geq \frac{1}{2}\mathrm{detlb}(\mathcal{H})$.*

Matoušek [54] showed that this bound is almost tight.

Theorem 18 (Matoušek [54]). *For every hypergraph $\mathcal{H}$ with n nodes and m hyperedges,*

$$\mathrm{herdisc}(\mathcal{H}) \leq O\left(\log n \sqrt{\log m} \right) \mathrm{detlb}(\mathcal{H}).$$

A major breakthrough is the very recent algorithm approximating the hereditary discrepancy within poly-logarithmic factors by Nikolov, Talwar and Zhang [61] (see also the chapter of N. Bansal). Most recently Matoušek and Nikolov [55] introduced the so called ellipsoid-infinity norm and showed that these bounds are asymptotically tight in the worst case. Among the striking applications of this new concept is the new lower bound $\Omega(\log^{d-1} n)$ for the d-dimensional Tusnady problem, asking for the combinatorial discrepancy of an n-point set in $\mathbb{R}^d$ with respect to axis-parallel boxes. For $d \geq 3$ this improves over the previous lower bound of $\Omega(\log^{(d-1)/2} n)$.

5.2.3.3 Arithmetic Progressions

Let us define the hypergraph of arithmetic progressions.

Definition 19. For $a, d, l \in \mathbb{N}$, denote by $A_{a,d,l} := \{a + id : 0 \leq i \leq l - 1\}$ the arithmetic progression with starting point a, difference d and length l. Let $X := [N] = \{0, \ldots, N - 1\}$. Denote by $\mathscr{E}$ the set of all arithmetic progressions in X, that is $\mathscr{E} = \{A_{a,d,l} \cap X : a \in X, d, l \in X \setminus \{0\}\}$. The hypergraph $\mathscr{H}_N := (X, \mathscr{E})$ is called the hypergraph of arithmetic progressions in the first N positive integers.

The investigation of two colorings of arithmetic progressions has been a fundamental problem in the history of combinatorics. In 1927 van der Waerden [75] proved his famous theorem that if the non-negative integers are colored with finitely many colors, then there are arbitrarily long monochromatic arithmetic progressions.

Theorem 20 (Van der Waerden [75]). *For any integers k and c there exist $W(k, c) > 0$, $N > W(k, c)$, N integer, such that for every c-coloring of $\{1, \ldots, N\}$ there exists a monochromatic arithmetic progression of length at least k.*

K. Roth [67] exhibited another aspect of the same phenomenon. Arithmetic progressions have large discrepancy:

Theorem 21 (Roth [67]). *It holds that* $\mathrm{disc}(\mathscr{H}_N) = \Omega\left(N^{\frac{1}{4}}\right)$.

Roth himself did not believe that his lower bound is optimal, most probably due to the fact that the probabilistic method immediately gives the upper bound $O(\sqrt{N \log N})$. A. Sárközy [31] was the first who improved the exponent of N and showed an upper bound of $O(N^{\frac{1}{3}+o(1)})$. A breakthrough was made by J. Beck in [10], who showed that the lower bound is best possible up to a polylogarithmic factor by improving the upper bound to $O(N^{\frac{1}{4}} \log^{\frac{5}{2}} N)$. It lasted 30 years until J. Matoušek and J. Spencer finally solved the problem ("the end of the hunt" [56]).

Theorem 22 (Matoušek, Spencer [56]). *It holds that* $\mathrm{disc}(\mathscr{H}_N) = O\left(N^{\frac{1}{4}}\right)$.

We will give proofs for the lower bound in a more general context. The Roth-Matoušek-Spencer theorem is in fact the motivation and the kernel of the theme of this chapter. Let us conclude this introduction with the most challenging open problem in this context, the old conjecture of P. Erdős, (worth of more than \$ 500), with almost no progress towards an answer so far.

Conjecture (Erdős [30]). For any $\chi : \mathbb{N} \rightarrow \{-1, 1\}$ and every constant $C > 0$ there are d and n such that $\left| \sum_{k=1}^{n} \chi(kd) \right| \geq C$.

5.2.4 Multicolor Discrepancy of Hypergraphs and Matrices

Doerr and Srivastav [26] extended combinatorial discrepancy theory from 2-colors to c-colors. We present some of the basic techniques and results, also as a basis for later sections.

5.2.4.1 Multicolor Discrepancy

A c-coloring of a hypergraph $\mathscr{H} = (X, \mathscr{E})$ is a mapping $\chi : X \to M$, where M is any set of cardinality c. For convenience we set $M = [c]$. Sometimes a different set M can be of advantage. For example, in applications to communication complexity M is a finite Abelian group [4]. We define the *discrepancy of an edge $E \in \mathscr{E}$ in color $i \in M$ with respect to χ* by

$$\operatorname{disc}_{\chi,i}(E) := \left| |\chi^{-1}(i) \cap E| - \frac{|E|}{c} \right|, \tag{5.12}$$

the *discrepancy of $\mathscr{H}$ with respect to χ* by

$$\operatorname{disc}(\mathscr{H}, \chi, c) := \max_{i \in M, E \in \mathscr{E}} \operatorname{disc}_{\chi,i}(E) \tag{5.13}$$

and the *discrepancy of $\mathscr{H}$ in c colors* by

$$\operatorname{disc}(\mathscr{H}, c) := \min_{\chi : X \to [c]} \operatorname{disc}(\mathscr{H}, \chi, c). \tag{5.14}$$

The definition for $c = 2$ implies:

Proposition 23. $\operatorname{disc}(\mathscr{H}, 2) = \frac{1}{2} \operatorname{disc}(\mathscr{H})$.

One might wonder whether multicolor discrepancy can be controlled by the 2-color discrepancy, for example by recursively splitting the color classes induced by an optimal 2-coloring. This is not the case:

Example. Let $k \in \mathbb{N}$ and $n = 4k$. Set

$$\mathscr{H}_n = \left([n], \left\{ X \subseteq [n] : \left| X \cap \left[\tfrac{n}{2} \right] \right| = \left| X \setminus \left[\tfrac{n}{2} \right] \right| \right\} \right).$$

Obviously, $\mathscr{H}_n$ has 2-color discrepancy zero, but $\operatorname{disc}(\mathscr{H}_n, 4) = \frac{1}{8}n$.

Here is a proof: let $\chi : [n] \to [4]$ be any 4-coloring. Let $i \in [4]$ be a color such that $|\chi^{-1}(i)| \leq \frac{1}{4}n$. Then there are sets $E_1 \subseteq \left[\tfrac{n}{2} \right]$, $E_2 \subseteq [n] \setminus \left[\tfrac{n}{2} \right]$ such that $|E_j| = \frac{1}{4}n$ and $\chi^{-1}(i) \cap E_j = \emptyset$. Thus $E_1 \cup E_2$ is an edge in $\mathscr{H}$ and has discrepancy $\frac{1}{8}n$ in color i. On the other hand $\chi : x \mapsto \left\lceil \tfrac{4x}{n} \right\rceil$ is a 4-coloring with discrepancy $\frac{1}{8}n$.

Such examples exist for nearly any two numbers of colors. Unless c_1 divides c_2, there are hypergraphs $\mathscr{H}_n$ on n vertices having discrepancy $\Theta(n)$ in c_1 colors and zero discrepancy in c_2 colors. This has been investigated in [23].

An alternative representation of multicolors are vectors. When concerned with the discrepancy of matrices, this will be more convenient. We describe the color $i \in [c]$ by a special vector $m^{(i)} \in \mathbb{R}^c$ defined by

$$m_j^{(i)} := \begin{cases} \frac{c-1}{c} & : \quad i = j \\ -\frac{1}{c} & : \quad \text{otherwise} \end{cases}$$

Then

$$\text{disc}(\mathcal{H}, \chi, c) = \max_{E \in \mathcal{E}} \left\| \sum_{x \in E} m^{(\chi(x))} \right\|_{\infty} \tag{5.15}$$

Set $M_c := \{m^{(i)} | i \in [c]\}$. Apparently, we have

$$\text{disc}(\mathcal{H}, c) = \min_{\chi : X \to M_c} \max_{E \in \mathcal{E}} \left\| \sum_{x \in E} \chi(x) \right\|_{\infty} \tag{5.16}$$

As for 2 colors, the notion of multicolor discrepancy has a natural extension to matrices. Let $A \in \mathbb{R}^{m \times n}$ be any matrix. Let $\overline{A}$ be the matrix which results from replacing every a_{ij} in A by $a_{ij}I_c$, where I_c shall denote the identity matrix of dimension c. Identifying a $\chi : [n] \to M_c$ by a cn-dimensional vector in the natural way, we get

$$\text{disc}(A, c) := \min_{\chi : [n] \to M_c} \left\| \overline{A} \chi \right\|_{\infty} \tag{5.17}$$

Using tensor products, an elegant reformulation of (5.17) is the following. For any two matrices $A_k \in \mathbb{C}^{m_k \times n_k}$, $k = 1, 2$, the tensor (or Kronecker) product $A_1 \otimes A_2$ is the matrix $B = (b_{ij}) \in \mathbb{C}^{m_1 m_2 \times n_1 n_2}$ such that

$$b_{(i_1-1)m_1+i_2,(j_1-1)n_1+j_2} = a_{i_1 j_1} a_{i_2 j_2}$$

for all $i_k \in [m_k]$, $j_k \in [n_k]$, $k = 1, 2$. So B is formed by replacing every entry a_{ij} of A_1 by $a_{ij}A_2$. With a straightforward calculation we have:

$$\text{disc}(A, c) = \min_{\chi : X \to M_c} \left\| (A \otimes I_c) \chi \right\|_{\infty} \tag{5.18}$$

The Weighted Multicolor Discrepancy. The other notions of discrepancy transform to the multi-color case in a similar way: set $\overline{M}_c = \left\{ \sum_{i \in [c]} \lambda_i m^{(i)} : \lambda \in [0, 1]^c, \sum_{i \in [c]} \lambda_i = 1 \right\}$, which is the convex hull of M_c.

When trying to get a c-coloring by starting with a 2 coloring and partitioning the color classes until c partitions are reached, a new concept appears, the *weighted discrepancy*. For $p \in \overline{M}_c$ set $\overline{p} : [n] \to \overline{M}_c; i \mapsto p$. We define the *weighted discrepancy of A with respect to the weight p and the coloring χ* by

$$\text{wd}(A, \chi, p) := \left\| \overline{A}(\overline{p} - \chi) \right\|_{\infty}, \tag{5.19}$$

the *weighted discrepancy of A with respect to the weight p* by

$$\text{wd}(A, c, p) := \min_{\chi : [n] \to M_c} \text{wd}(A, \chi, p) \tag{5.20}$$

and the *weighted discrepancy of A* by

$$\mathrm{wd}(A, c) := \max_{p \in \overline{M}_c} \mathrm{wd}(A, c, p) \tag{5.21}$$

There is an equivalent way to define weighted discrepancy. Denote by E_c the standard basis of $\mathbb{R}^c$ and by $\overline{E}_c$ its convex hull, so $\overline{E}_c = \{ p \in [0, 1]^c : \|p\|_1 = 1 \}$. We have

$$\mathrm{wd}(A, c) = \max_{p \in \overline{E}_c} \min_{\chi:[n] \to E_c} \|\overline{A}(p - \chi)\|_\infty \tag{5.22}$$

Note that this is an extension of the definition of $\mathrm{wd}(\mathscr{H}, 2)$ in Eq. (5.8). For hypergraphs it reads as follows. We define the weighted discrepancy with respect to a weight p and a coloring χ by

$$\mathrm{wd}(\mathscr{H}, \chi, p) := \max_{j \in [c], E \in \mathscr{E}} \left| |E \cap \chi^{-1}(j)| - p_j |E| \right| \tag{5.23}$$

The weighted discrepancy of the hypergraphs with respect to the weight p is

$$\mathrm{wd}(\mathscr{H}, c, p) := \min_{\chi:X \to M_c} \mathrm{wd}(\mathscr{H}, \chi, p). \tag{5.24}$$

Multicolor Linear and Hereditary Discrepancy. The linear discrepancy in c colors can be defined by

$$\mathrm{lindisc}(A, c) := \max_{p:[n] \to \overline{M}_c} \min_{\chi:[n] \to M_c} \|\overline{A}(p - \chi)\|_\infty. \tag{5.25}$$

Finally, the hereditary discrepancy in c colors is

$$\mathrm{herdisc}(A, c) := \max_{A_0 \leq A} \mathrm{disc}(A_0, c). \tag{5.26}$$

All these notions transfer to hypergraphs by taking the incidence matrix of the hypergraph. For instance, for a hypergraph $\mathscr{H}$ with incidence matrix A we have $\mathrm{lindisc}(\mathscr{H}) := \mathrm{lindisc}(A)$. Like in Proposition 23, these other discrepancy notions are identical with the usual notions up to the constant factor of 2. When citing 2-color results, we will use the conventional notation which has no parameter c in it, e. g. $\mathrm{herdisc}(\mathscr{H})$. In many applications, we will use the hereditary discrepancy and in fact, it bounds the 2-color weighted discrepancy, according to Proposition 3:

Proposition 24. *For all induced subhypergraphs $\mathscr{H}_0$ of $\mathscr{H}$ we have* $\mathrm{wd}(\mathscr{H}_0, 2) \leq \mathrm{herdisc}(\mathscr{H}_0)$.

Doerr [22] showed a kind of converse result: if the hereditary discrepancy in c colors is bounded, one can construct c_2 colorings (and in particular 2-colorings) with low discrepancy. Together with Proposition 24 and Theorem 31 this shows that the hereditary discrepancy in two numbers of colors deviates at most by a constant factor (depending on the numbers of colors, but not on the hypergraph). Furthermore, he studied the c-color linear discrepancy. The following observation suggests that it might behave differently in more than 2 colors. Consider a totally unimodular $m \times n$ matrix A. Various proofs show that $\mathrm{lindisc}(A, 2) \leq 1$ holds. As already mentioned, Doerr [21] proved the sharp bound of $1 - \frac{1}{n+1}$.

NP-Hardness of the Discrepancy Problem. Let us denote by DISCREPANCY the decision problem with a hypergraph along with a non-negative rational number K as the input and the question, whether the discrepancy of this hypergraph is at most K. We show that this decision problem is NP-complete even for a very special class of hypergraphs, namely those with VC-dimension only 2.

Definition 25. Let $\mathcal{H} = (X, \mathcal{E})$ be a hypergraph. Let $Y \subseteq X$.

(i) Y is called shattered, if $\mathcal{H}|_Y = \mathcal{P}(Y)$.
(ii) the VC-dimension (Vapnik-Chervonenkis dimension) of $\mathcal{H}$ is the maximum cardinality of a shattered subset of X.

Hypergraphs with bounded VC-dimension play a central role in computational geometry [58]. They have the nice property that if the VC-dimension is d, then for any $Y \subseteq X$, the number of hyperedges of $\mathcal{H}$ restricted to Y is at most $|Y|^d$ [1]. We have:

Theorem 26. DISCREPANCY *is NP-complete even if the VC-Dimension of the hypergraph is* 2.

For the proof of Theorem 26 we introduce the property B problem. We say a hypergraph has property B, if there is a 2-coloring of the nodes so that no edge is monochromatic [1, page 56]. This problem is NP-hard as shown by Lovász [47]. The proof works by reduction from 3-colorability of a graph: for a graph $G = (V, E)$ a 3-uniform hypergraph $\mathcal{H}(G)$ is introduced, with node set $(\{1, 2, 3\} \times V) \cup \{\infty\}$ and edge set

$$\{\{(i, u), (i, v), \infty\} : i \in \{1, 2, 3\}, (u, v) \in E\} \cup \{\{(1, v), (2, v), (3, v)\} : v \in V\}.$$

It is easy to see that this hypergraph has VC dimension 2, as follows. The VC dimension is at least 2, since $\{(i, u), (i, v)\}$ forms a shattered set for each i and $(u, v) \in E$. Since the largest edge is of cardinality 3, the VC dimension is at most 3. It remains to show that none of the edges forms a shattered set. First consider an edge of the form $Y := \{(i, u), (i, v), \infty\}$ with $(u, v) \in E$. Then $\{(i, u), (i, v)\} \notin \mathcal{H}(G)|_Y$ since there is no other edge than Y where (i, u) and (i, v) occur together. Next consider an edge of the form $Y := \{(1, v), (2, v), (3, v)\}$. Then this is the only edge in which $(1, v)$ and $(2, v)$ occur together, so $\{(1, v), (2, v)\} \notin \mathcal{H}(G)|_Y$.

Hence the property B problem is NP-hard even if restricted to 3-uniform hypergraphs of VC dimension 2. It is now easy to prove Theorem 26:

Proof (of Theorem 26). Obviously, a 3-uniform hypergraph is a yes-instance for the property B problem if and only if it is a yes-instance for the discrepancy problem with parameter $K = 2$. Hence with a polynomial-time algorithm for the discrepancy problem restricted to 3-uniform hypergraphs and VC dimension 2, we can solve the property B problem with the same restrictions in polynomial time.   □

More recently, Charikar et al. [18] proved.

Theorem 27. *It is NP-hard to distinguish between hypergraphs on N nodes with discrepancy zero and those with discrepancy $\Omega(\sqrt{N})$.*

The Basic Probabilistic Bound. This is in fact the starting point of combinatorial discrepancy theory, reflecting the true spirit of the *probabilistic method* invented by Paul Erdős. We consider a random coloring, and color each vertex independently with a random color. Using the Chernoff bound one can prove that with positive probability the random coloring is balanced to a certain extent, more precisely, for a hypergraph $\mathcal{H} = (X, \mathcal{E})$ with $m := |\mathcal{E}|$ and $s = \max_{E \in \mathcal{E}} |E|$, $\mathrm{disc}(\mathcal{H}) \leq \sqrt{2s \ln(2m)}$ [1, Theorem 12.1.1]. For c colors we have:

Proposition 28. $\mathrm{disc}(\mathcal{H}, c) \leq \sqrt{\frac{1}{2}s \ln(2mc)}$.

Proof. Let us define a random c-coloring χ by assigning a color from $[c]$ to every vertex $v \in X$ uniformly at random, independently for all v. We define random variables $X_{i,v}$ by

$$X_{i,v} := \begin{cases} \frac{c-1}{c} & : \quad \chi(v) = i \\ -\frac{1}{c} & : \quad \text{otherwise} \end{cases}$$

for all $v \in X$, $i \in [c]$. Set $X_{i,E} := \sum_{v \in E} X_{i,v}$ for all $E \in \mathcal{E}$, $i \in [c]$. We invoke a large deviation bound due to H. Chernoff [1, Theorem A.1.4]. It states

$$P(|X_{i,E}| > \alpha) < 2e^{-2\alpha^2/|E|}$$

for any real $\alpha > 0$. For $\alpha := \sqrt{\frac{1}{2}s \ln(2mc)}$ we have

$$P(\forall i \in [c], E \in \mathcal{E} : |X_{i,E}| \leq \alpha) = 1 - P(\exists i \in [c], E \in \mathcal{E} : |X_{i,E}| > \alpha)$$

$$\geq 1 - \sum_{i \in [c], E \in \mathcal{E}} P(|X_{i,E}| > \alpha)$$

$$> 1 - \sum_{i \in [c], E \in \mathcal{E}} 2e^{-2\alpha^2/|E|}$$

$$\geq 1 - 2cm \cdot e^{-2\alpha^2/s} = 0.$$

Hence with strictly positive probability χ has discrepancy at most $\sqrt{\frac{1}{2}s \ln(2mc)}$, thus such a coloring exists.   □

The bound of $O(\sqrt{\frac{s}{2}\ln(2mc)})$ is actually not what we would expect, because intuitively c-coloring means partitioning the set X into c color classes, so $\frac{|E|}{c}$ is more likely the expected number of nodes in E in each color class, thus consequently a bound of $O(\sqrt{\frac{s}{c}\ln(mc)})$ would be more natural. In the following Sects. 5.2.5 and 5.2.6 we will show that such a discrepancy bound can be proved.

5.2.5   Recursive Coloring and Its Limitations

A simple way to get a c-coloring is to start with an (optimal) 2-coloring and to partition the two color classes in a suitable way until a c-coloring is reached. What is the success of such an approach, if any? In this subsection we will show that it leads to c-color versions of some classical combinatorial discrepancy bounds, but it is not strong enough to incorporate the number of colors c into the bounds. Nevertheless the recursive approach is the basis for further refinements. But the technical part of Sect. 5.2.5 will not be needed in the sections on arithmetic progressions and their generalizations. For these sections it should be sufficient to note the c-color analogue of the six-standard-deviation theorem (Corollary 44) in Sect. 5.2.6.

Consider an (optimal) 2-coloring and the subhypergraph on the two coloring classes. Now, find a 2-coloring for each of these subhypergraphs. These 2-colorings form a 4-coloring for the original hypergraph. Continuing this partitioning process we arrive at a $c = 2^k$-coloring, k integer. Obviously, such a partitioning works only if c is a power of 2. The hard work is to show that any c can be treated. In Sect. 5.2.6 we will see how to bring c into the discrepancy bound.

We represent the iterated partitioning of the set of colors C by a binary tree. A binary rooted tree $T = (X_T, \mathscr{E}_T)$ is called a *partition tree* for C, if the following conditions are satisfied: the root of T is C, all nodes are subsets of C, all leaves are singletons of C and each two son nodes form a partition of their common father node. For every color $i \in C$ there is a unique path $C = C_0^{(i)} \supset C_1^{(i)} \supset \ldots \supset C_{k(i)}^{(i)} = \{i\}$ in the partition tree. We write $h(T)$ for the height of T, that is the length of a longest path connecting a leaf and the root. We call a function $p : C \to [0, 1]$, a *weight* of the set C of colors if $\sum_{i \in C} p_i = 1$. For a color $i \in C$ set $v(T, p, i) := \sum_{l=1}^{k(i)} \frac{p_i}{p(C_l^{(i)})}$ and $v(T, p) = \max_{i \in C} v(T, p, i)$. The next theorem shows the influence of the partition tree chosen for the recursive coloring process.

Theorem 29. *Let* $\mathrm{wd}(\mathscr{H}_0, 2) \leq K$ *for all induced subgraphs* $\mathscr{H}_0$ *of* $\mathscr{H}$. *Let* C *be a set of colors with* $c = |C|$ *and let* p *be a weight of* C. *Let* $T = (X_T, \mathscr{E}_T)$ *be a partition tree of* C. *Then there is a coloring* $\chi : X \to C$ *such that for all colors* $i \in C$ *and all* $E \in \mathscr{E}$ *we have*

$$\left| |E \cap \chi^{-1}(i)| - p_i |E| \right| \leq K v(T, p, i).$$

In particular, $\mathrm{wd}(\mathscr{H}, c, p) \leq K v(T, p)$. With $K := \max_{\mathscr{H}_0} \mathrm{wd}(\mathscr{H}_0, 2)$, where $\mathscr{H}_0$ is an induced subhypergraph, we have $\mathrm{wd}(\mathscr{H}, c, p) \leq v(T, p) \mathrm{herdisc}(\mathscr{H})$.

Proof. We use induction on the height $h(T)$ of T. For $h(T) = 0$ we have just one color and both sides of the inequality become zero. Let T be of height $h(T) > 0$ and assume that the theorem is true for all partition trees of height strictly less than $h(T)$. Let C_1 and C_2 be the sons of C in T. Set $q_j := p(C_j) = \sum_{k \in C_j} p_k$, $j = 1, 2$. By assumption there is a 2-coloring $\chi_0 : X \to [2]$ such that

$$\left| |E \cap \chi_0^{-1}(j)| - q_j |E| \right| \leq \mathrm{wd}(\mathscr{H}, 2, (q_1, q_2)) \leq K \qquad (5.27)$$

holds for all $j \in [2]$ and $E \in \mathscr{E}$. Put $X_j := \chi_0^{-1}(j)$, $j = 1, 2$. Denote by T_j the subtree having C_j as its root. Then the hypergraph $\mathscr{H}_{|X_j}$ together with the set of colors C_j, the weight $\frac{1}{q_j} p_{|C_j}$ and the partition tree T_j fulfils the assumption of this theorem. By induction there are colorings $\chi_j : X_j \to C_j$, $j \in 1, 2$ such that

$$\left| |E \cap X_j \cap \chi_j^{-1}(i)| - \frac{1}{q_j} p_i |E \cap X_j| \right| \leq K v\left(T_j, \frac{1}{q_j} p_{|C_j}, i\right) \leq K \sum_{l=2}^{k(i)} \frac{\frac{p_i}{q_j}}{\frac{1}{q_j} p\left(C_l^{(i)}\right)} \qquad (5.28)$$

for all $i \in C_j$. Set

$$\chi = \chi_1 \cup \chi_2 : x \mapsto \begin{cases} \chi_1(x) & : \quad x \in X_1 \\ \chi_2(x) & : \quad \text{otherwise.} \end{cases}$$

Let $j \in [2]$ and $i \in C_j$. Then $C_1^{(i)} = C_j$ and $q_j = p(C_1^{(i)})$. Let $E \in \mathscr{E}$. From (5.27), (5.28) and with a straightforward calculation we get

$$\left| |E \cap \chi^{-1}(i)| - p_i |E| \right| \leq \left| |E \cap X_j \cap \chi_j^{-1}(i)| - \frac{p_i}{q_j} |E \cap X_j| \right|$$

$$+ \frac{p_i}{q_j} \left| |E \cap X_j| - q_j |E| \right|$$

$$\overset{(5.27),(5.28)}{\leq} \sum_{l=2}^{k(i)} \frac{\frac{p_i}{q_j}}{\frac{1}{q_j} p(C_l^{(i)})} K + \frac{p_i}{q_j} K$$

$$= K \sum_{l=1}^{k(i)} \frac{p_i}{p(C_l^{(i)})} = K v(T, p, i).$$

Hence χ satisfies the claim. For the specific K, the assertion follows from Proposition 24. $\qquad \square$

The following partition tree gives a bound of $v(T, p, i) < 4$ for all $i \in [c]$: we start with the tree consisting of the unique node C. For a leaf C_0 of cardinality $|C_0|$ greater than 1 let us define sons by the following rule: if there is a color $i \in C_0$ with weight $p_i \geq \frac{1}{2}p(C_0)$, then the sons of C_0 shall be $\{i\}$ and $C_0 \setminus \{i\}$. Otherwise partition C_0 in any way (C_1, C_2) such that $p(C_j) \in \left[\frac{1}{3}p(C_0), \frac{2}{3}p(C_0)\right]$. Repeat this process until all leaves are singletons. Further improvements can be achieved if we consider the equi-weighted discrepancy, where $p = \frac{1}{c}\mathbf{1}_c$.

Definition 30. A partition tree for a positive integer n is a binary tree $T = (X_T, \mathscr{E}_T)$ together with a labelling $l : X_T \to [n]$ such that the following conditions are satisfied:

(i) The root r is labelled $l(r) = n$.
(ii) For every non-leaf v with sons s_1 and s_2 we have

$$l(v) = l(s_1) + l(s_2)$$

(iii) The leaves are labelled 1.

For a path $P : r = v_0^{(i)}, v_1^{(i)}, \ldots, v_{k(i)}^{(i)}$ connecting the root r and a leaf $v_{k(i)}^{(i)}$ labelled i we call

$$v(T, P) = \sum_{l=1}^{k(i)} \frac{1}{l\left(v_l^{(i)}\right)}$$

the value of P and define $v(T)$ as the maximum $v(T, P)$ over all these paths P. Finally $v(n)$ shall be the minimum $v(T)$ over all partition trees T of n. There is a natural correspondence between partition trees for sets of colors and for positive integers. Let $T = (X_T, E_T)$ denote a partition tree for the set of colors C. Define a labelling $l_T : X_T \to [|C|] ; v \mapsto |v|$. Then, T together with l_T is a partition tree for $|C|$.

Now let T together with l denote a partition tree for a positive integer c. Let C be any set of colors such that $|C| = c$. We construct a partition tree T^* for C such that $l_{T^*} = l$. Define $f : X_T \to 2^{[c]}$ recursively: set $f(r) = C$ for the root r of T. For every node v with sons s_1 and s_2 such that $f(v)$ is already defined choose $f(s_1)$ to be any subset of $f(v)$ of size $l(s_1)$ and $f(s_2) = f(v) \setminus f(s_1)$. Note that f is injective, and by replacing every $v \in X_T$ by $f(v)$ we get a partition tree T^* for C. Clearly, $l_{T^*} = l$.

Furthermore, we have

$$v\left(T^*, \frac{1}{c}\mathbf{1}_c\right) = \max_{i \in C} \sum_{l=1}^{k(i)} \frac{\frac{1}{c}}{\left|C_l^{(i)}\right|} = \max_{i \in C} \sum_{l=1}^{k(i)} \frac{1}{l\left(v_l^{(i)}\right)} = v(T).$$

Theorem 31. *Let* $K := \max_{\mathcal{H}_0} \mathrm{wd}(\mathcal{H}_0, 2)$, *where* $\mathcal{H}_0$ *is an induced subhypergraph of* $\mathcal{H}$. *Then*

(i) $\mathrm{disc}(\mathcal{H}, c) \le v(c)K \le 2.0005K$.
(ii) $\mathrm{disc}(\mathcal{H}, c) = \mathrm{wd}(\mathcal{H}, c, p) \le 2.0005\,\mathrm{herdisc}(\mathcal{H})$.

Proof. (i) Let $T = (X_T, \mathcal{E}_T)$ together with l be a partition tree for c such that $v(T) = v(c)$. We build T^* as above and apply Theorem 29 to T^* and $p = \frac{1}{c}\mathbf{1}_c$:

$$\mathrm{disc}(\mathcal{H}, c) = \mathrm{wd}(\mathcal{H}, c, p) \le Kv(T^*, p) = Kv(T) = Kv(c).$$

The upper bound on $v(c)$ can be proved by a specific construction of a partition tree for c (see [26]).
(ii) follows from the argument in (i) using T^* and Theorem 29. □

A first c-color version of the Beck–Fiala theorem is:

Theorem 32. *For any hypergraph* $\mathcal{H}$ *we have*

$$\mathrm{disc}(\mathcal{H}, c) < v(c)\,\Delta(\mathcal{H}) \le 2.0005\,\Delta(\mathcal{H}).$$

Proof. The Beck–Fiala theorem states that $\mathrm{lindisc}(\mathcal{H}) < 2\Delta(\mathcal{H})$ holds for any hypergraph $\mathcal{H}$. In particular, we have $\mathrm{wd}(\mathcal{H}_0, 2) \le \frac{1}{2}\mathrm{lindisc}(\mathcal{H}_0) < \Delta(\mathcal{H}_0) \le \Delta(\mathcal{H})$ for all induced subhypergraphs $\mathcal{H}_0$ of $\mathcal{H}$. The stated inequality then follows from Theorem 31. □

The limitation is clear: at the moment we are not able to bring c into the discrepancy bound. This crucial step is done in the next section.

5.2.6 Refined Recursive Coloring

In this subsection we extend the recursive approach. Roughly speaking we show that if the 2-color discrepancy of the subhypergraphs on any $n_0 \le n$ vertices is bounded by $O\left(n_0^\alpha\right)$, then the c-color discrepancy is bounded by $O\left((\frac{n}{c})^\alpha\right)$.

A useful concept is that of so called fair colorings with respect to a weight p.

Definition 33. Let $p \in [0, 1]^c$ be a c-color weight and $\mathcal{H} = (X, \mathcal{E})$ a hypergraph. A c-coloring χ is a fair p-coloring of $\mathcal{H}$, if $||\chi^{-1}(i)| - p_i|X|| \le 1$ holds for all $i \in [c]$.

Now, if n is not a multiple of c, we cannot split n into c parts of equal size, but on the other hand we wish to get a partition so that each partition class is roughly n/c large. This is the motivation to introduce integral weights.

Definition 34. Let $\mathcal{H} = (X, \mathcal{E})$ be a hypergraph. We call a weight $p \in [0, 1]^c$ *integral* with respect to $\mathcal{H}$ (or $\mathcal{H}$-integral for short) if all p_i, $i \in [c]$, are multiples of $1/|X|$.

It is easy to see that for any weight there is a corresponding integral weight:

Proposition 35. *Let $\mathcal{H} = (X, \mathcal{E})$ be a hypergraph with $|X| = n$. For any weight $p \in [0, 1]^c$, there is a $\mathcal{H}$-integral weight q such that for all $i \in [c]$, $|p_i - q_i| \leq 1/n$.*

The relation between the discrepancy and the weighted discrepancy with respect to a weight and an integral weight is stated by the next proposition.

Proposition 36. *Let $\mathcal{H} = (X, \mathcal{E})$ be a hypergraph. Suppose that for any $\mathcal{H}$-integral weight q there is a fair c-coloring χ with $\mathrm{wd}(\mathcal{H}, \chi, q) \leq k(q)$ for some $k(q) \in \mathbb{R}$. Then for any c-color weight $p \in [0, 1]^c$ there is a $\mathcal{H}$-integral weight p' and a $k = k(p') \in \mathbb{R}$ so that $\mathrm{wd}(\mathcal{H}, \chi, p') \leq k$ and $\mathrm{wd}(\mathcal{H}, \chi, p) \leq k + 1$. In particular, for the weight $p = \frac{1}{c}\mathbf{1}_c$ we get $\mathrm{disc}(\mathcal{H}, \chi_0, c) \leq k + 1$ for some fair c-coloring χ_0.*

Proof. Let $p \in [0, 1]^c$ be a weight and let p' be an associated $\mathcal{H}$-integral weight (Proposition 35). By assumption there is a fair c-coloring χ and $k = k(p') \in \mathbb{R}$ so that $\mathrm{wd}(\mathcal{H}, \chi, p') \leq k$. Let E be a hyperedge and A_i be the i-th color class of χ. By the triangle inequality we have

$$\big||E \cap A_i| - p_i|E|\big| \leq \big||E \cap A_i| - p_i'|E|\big| + |p_i - p_i'||E|$$
$$\leq \mathrm{wd}(\mathcal{H}, \chi, p') + |p_i - p_i'||E| \leq k + 1,$$

since $|p_i - p_i'| \leq 1/n$, according to Proposition 35. Taking the maximum over all i and all E, the claim follows. The statement about the c-color discrepancy follows from the above argument, because for $p = \frac{1}{c}\mathbf{1}_c$ the notion of c-color discrepancy and weighted discrepancy coincide.   □

Using a recoloring argument we can transform arbitrary colorings into fair colorings.

Lemma 37. *Let $\mathcal{H} = (X, \mathcal{E})$ be a hypergraph such that $X \in \mathcal{E}$. Let $p = (q, 1-q)$ be a 2-color weight. Then any 2-coloring χ of $\mathcal{H}$ can be modified in $O(|X|)$ time to a fair p-coloring $\bar{\chi}$ such that*

$$\mathrm{wd}(\mathcal{H}, \bar{\chi}, p) \leq 2\,\mathrm{wd}(\mathcal{H}, \chi, p).$$

Proof. Let χ be a 2-coloring. Set $x := q|X| - |\chi^{-1}(1)|$. Since X is an edge in $\mathcal{H}$, we have $|x| \leq \mathrm{wd}(\mathcal{H}, \chi, p)$. Let $\bar{\chi}$ denote a coloring arising from χ by changing the color of $\lfloor |x| \rfloor$ points in such a way that $\big|q|X| - |\bar{\chi}^{-1}(1)|\big| < 1$. Now $\bar{\chi}$ is a fair coloring with respect to the weight $(q, 1 - q)$. For an edge $E \in \mathcal{E}$ we compute

$$|q|E| - |\overline{\chi}^{-1}(1) \cap E||$$

$$\leq |q|E| - |\chi^{-1}(1) \cap E|| + ||\chi^{-1}(1) \cap E| - |\overline{\chi}^{-1}(1) \cap E||$$

$$\leq |q|E| - |\chi^{-1}(1) \cap E|| + \lfloor |x| \rfloor$$

$$\leq 2 \operatorname{wd}(\mathcal{H}, \chi, p).$$

This completes the proof. □

In the following we present a recursive-coloring algorithm. Its analysis requires the following constants. Let $\alpha \in]0, 1[$. For each $p \in]0, 1[$ define $v_\alpha(p)$ to be the value

$$\max \left\{ \sum_{i=1}^{k} \prod_{j=1}^{i} q_j^\alpha \prod_{j=i+1}^{k} q_j : k \in \mathbb{N}, q_1, \ldots, q_{k-1} \in \left[0, \tfrac{2}{3}\right], q_k \in [0, 1], \prod_{j=1}^{k} q_j = p \right\}.$$

Set $c_\alpha := \frac{2}{2^{1-\alpha}-1} \left(1 + \frac{1}{1-\left(\frac{2}{3}\right)^{(1-\alpha)}}\right)$. Note that $c_{1/2} \leq 31.15$.

Straightforward calculations give [26]:

Lemma 38. *Let* $\alpha \in]0, 1[$.

(i) *For* $0 < p < q \leq \frac{2}{3}$, *we have* $q^\alpha v_\alpha(\frac{p}{q}) + q^\alpha \frac{p}{q} \leq v_\alpha(p)$.
(ii) *For all* $p \in [0, 1]$, $\frac{2}{2^{1-\alpha}-1} v_\alpha(p) \leq c_\alpha p^\alpha$.

We make a technical assumption. It requires that the discrepancy decreases when passing to smaller subhypergraphs. This assumption will later allow us to use hereditary discrepancy bounds.

Assumption 39 (Decreasing-Discrepancies-Assumption). Let $\mathcal{H} = (X, \mathcal{E})$ be a hypergraph. Set $n := |X|$. Let $p_0, \alpha \in]0, 1[$ and $D > 0$. For all $X_0 \subseteq X$ with $|X_0| \geq p_0|X|$ and all $q \in [0, 1]$ such that $(q, 1 - q)$ is $\mathcal{H}_{|X_0}$-integral there is a fair $(q, 1 - q)$-coloring χ of $\mathcal{H}_{|X_0}$ with discrepancy at most $D|X_0|^\alpha$.

The next lemma shows how to get a fair coloring under the decreasing-discrepancy assumption.

Lemma 40. *Suppose that the Decreasing-Discrepancies-Assumption holds with* $p_0, \alpha \in]0, 1[$ *and* $D > 0$. *For each* $\mathcal{H}$-*integral weight* $(q, 1 - q)$, $p_0 \leq q \leq \frac{1}{2}$, *there is a fair* $(q, 1 - q)$-*coloring* χ *with discrepancy at most* $\operatorname{wd}(\mathcal{H}, \chi, p) \leq \frac{2}{2^{1-\alpha}-1} D (qn)^\alpha$.

Algorithm 1 recursively computes a fair coloring with respect to an integral weight.

The main technical theorem is:

Theorem 41. *Suppose that the Decreasing-Discrepancies-Assumption holds. Then for each* $\mathcal{H}$-*integral weight* $p \in [0, 1]^c$ *there is a fair* p-*coloring* χ *of* $\mathcal{H}$ *such*

Algorithm 1: RECURSIVE COLORING

---

   input : A hypergraph $\mathcal{H} = (X, \mathcal{E})$ fulfilling the Decreasing-Discrepancies-Assumption, a
            set C of at least 2 colors and an $\mathcal{H}$-integral weight function $p : C \rightarrow [0, 1]$
   output: A coloring $\chi : X \rightarrow C$ as in Theorem 41
1 Choose a partition $\{C_1, C_2\}$ of the set of colors C such that $\|p_{|C_1}\|_1, \|p_{|C_2}\|_1 \leq \frac{2}{3}$ or C_1
    contains a single color with weight at least $\frac{1}{3}$. Set $(q_1, q_2) := (\|p_{|C_1}\|_1, \|p_{|C_2}\|_1)$;
2 According to Lemma 40, compute a fair (q_1, q_2)-coloring $\chi_0 : X \rightarrow [2]$ that has discrepancy
    at most $\frac{2}{2^{1-\alpha}-1}D(q_i n)^\alpha$ in color $i = 1, 2$ if $q_i \geq p_0$. Set $X_i := \chi^{-1}(i)$ for $i = 1, 2$;
3 for $i = 1, 2$ **do**
4   **if** $|C_i| > 1$ **then**
5   By recursion compute a fair $\frac{1}{q_i}p_{|C_i}$-coloring $\chi_i : X_i \rightarrow C_i$ for $\mathcal{H}_{|X_i}$ with
          discrepancy at most $\frac{2}{2^{1-\alpha}-1}Dv_\alpha(\frac{p_j}{q_i})(q_i n)^\alpha$ in each color $j \in C_i$ such that
          $p_j \geq p_0$;
6   **else if** $C_i = \{j\}$ *for some* $j \in C$ **then** Choose $\chi_i : X_i \rightarrow \{j\}$ as the constant mapping
7 for $x \in X$ **do**
8   **if** $x \in X_1$ **then** $\chi(x) := \chi_1(x)$ **else** $\chi(x) := \chi_2(x)$
9 Return χ;

---

*that the discrepancy is at most $\frac{2}{2^{1-\alpha}-1}Dv_\alpha(p_i)n^\alpha \leq Dc_\alpha(p_i n)^\alpha$ in all those colors
$i \in [c]$ such that $p_i \geq p_0$.*

Proof. Let $p \in [0, 1]^c$ be a $\mathcal{H}$-integral weight. To avoid trivial cases we shall
always assume that for all $i \in C$, $p_i > 0$.

We prove that the algorithm RECURSIVE COLORING computes a coloring as
claimed. Suppose by induction that this holds for sets of less than c colors. We
analyze the algorithm started on a color set C with $c := |C| \geq 2$. We show
correctness. In Step 1 of the algorithm, both C_1 and C_2 are non-empty and $q_2 \leq \frac{2}{3}$
holds. Therefore by Lemma 40 and induction the colorings $\chi_i, i = 0, 1, 2$ can be
computed as desired in Step 2 and 3. Let $E \in \mathcal{E}, i \in [2]$ and $j \in C_i$ such that
$p_j \geq p_0$. If $|C_i| > 1$, then

$$\left| |E \cap \chi^{-1}(j)| - p_j|E| \right|$$

$$= \left| |E \cap \chi_0^{-1}(i) \cap \chi_i^{-1}(j)| - p_j|E| \right|$$

$$\leq \left| |E \cap \chi_0^{-1}(i) \cap \chi_i^{-1}(j)| - \frac{p_j}{q_i}|E \cap \chi_0^{-1}(i)| \right| + \left| \frac{p_j}{q_i}|E \cap \chi_0^{-1}(i)| - p_j|E| \right|$$

$$\leq \left| |(E \cap X_i) \cap \chi_i^{-1}(j)| - \frac{p_j}{q_i}|E \cap X_i| \right| + \frac{p_j}{q_i}\left| |E \cap \chi_0^{-1}(i)| - q_i|E| \right|$$

$$\leq \frac{2}{2^{1-\alpha}-1}Dv_\alpha(\frac{p_j}{q_i})(q_i n)^\alpha + \frac{p_j}{q_i}\frac{2}{2^{1-\alpha}-1}D(q_i n)^\alpha$$

$$\leq \frac{2}{2^{1-\alpha}-1}Dv_\alpha(p_j)n^\alpha$$

by Lemma 38 (i). On the other hand, if C_i contains a single color j, then $p_j = q_i$ and

$$\left| |E \cap \chi^{-1}(j)| - p_j |E| \right| = \left| |E \cap \chi_0^{-1}(i)| - q_i |E| \right|$$

$$\leq \tfrac{2}{2^{1-\alpha}-1} D(q_i n)^\alpha$$

$$\leq \tfrac{2}{2^{1-\alpha}-1} D v_\alpha(p_j) n^\alpha.$$

The last estimate in the theorem follows from Lemma 38 (ii). $\qquad\Box$

We are ready to prove the multicolor analogue to Spencer's six-standard-deviation theorem.

5.2.7 c-Color Six Standard Deviations

The celebrated six standard deviations result of Joel Spencer [70] states that for all hypergraphs $\mathcal{H} = (X, \mathcal{E})$ with $n = |X|$ and $m = |\mathcal{E}|$, $m \geq n$, disc$(\mathcal{H}) \leq K\sqrt{n \ln(\frac{2m}{n})}$ for some constant $K > 0$. An interesting case is $m = O(n)$, and there disc$(\mathcal{H}) = O(\sqrt{n})$. For m significantly larger than n this result is outnumbered (due to the implicit constants) by a simple random coloring. Using the relation between discrepancies respecting a particular weight, the connection to hereditary discrepancy (Proposition 24) and the recoloring argument (Lemma 37), we derive from Spencer's result

Lemma 42. *For any $X_0 \subseteq X$ and $\mathcal{H}_{|X_0}$-integral weight $p = (q, 1 - q)$ there is a fair p-coloring χ of $\mathcal{H}_{|X_0}$ satisfying* wd$(\mathcal{H}_{|X_0}, \chi, p) \leq 2K\sqrt{|X_0| \ln\left(\frac{2m+2}{|X_0|}\right)}$.

Proof. Let $X_0 \subseteq X$. Then any induced subhypergraph of $\mathcal{H}_{|X_0}$ has discrepancy at most $K\sqrt{|X_0| \ln(\frac{2m}{|X_0|})}$, because Spencer's bound is monotone in the number of vertices. According to Proposition 24, we have

$$\text{wd}\left(\mathcal{H}_{|X_0}, 2, p\right) \leq \text{wd}\left(\mathcal{H}_{|X_0}, 2\right) \leq \text{herdisc}\left(\mathcal{H}_{|X_0}\right) \leq K\sqrt{|X_0| \ln\left(\tfrac{2m}{|X_0|}\right)}. \quad (5.29)$$

It remains to show the existence of a fair coloring. Let $\overline{\mathcal{H}}$ denote the hypergraph arising from $\mathcal{H}$ by adding the set X as an additional edge (unless $X \in \mathcal{E}$ already holds). Then $\overline{\mathcal{H}}_{|X_0}$ has at most $m + 1$ edges. Let η be a 2-coloring such that

$$\text{wd}\left(\overline{\mathcal{H}}_{|X_0}, \eta, p\right) = \text{wd}\left(\overline{\mathcal{H}}_{|X_0}, 2, p\right).$$

By (5.29), wd$\left(\overline{\mathcal{H}}_{|X_0}, 2, p\right) \leq K\sqrt{|X_0| \ln\left(\frac{2m+2}{|X_0|}\right)}$. The desired fair p-coloring χ can be constructed from η by Lemma 37. $\qquad\Box$

Theorem 43. *Let $\mathscr{H} = (X, \mathscr{E})$ denote a hypergraph with n vertices, $m \geq n$ and let $p \in [0, 1]^c$ be a $\mathscr{H}|_{X_0}$-integral weight. Set $p_0 := \min_{i \in [c]} p_i$. Then there is a fair p-coloring with discrepancy at most $63K \sqrt{p_i n \ln \left(\frac{2m+2}{p_0 n} \right)}$ for all i.*

Proof. By Lemma 42 we may apply Theorem 41 with $\alpha = \frac{1}{2}$, $D = 2K \sqrt{\ln(\frac{2m+2}{p_0 n})}$ and p_0. This yields a fair p-coloring with discrepancy at most $Dc_\alpha \sqrt{p_i n}$ in color $i \in [c]$, for all i. The claim follows from $c_\alpha \leq 31.15$. $\square$

Corollary 44. *Let $\mathscr{H} = (X, \mathscr{E})$ denote a hypergraph with $|X| = n$ and $|\mathscr{E}| = m \geq n$. Then, $\mathrm{disc}(\mathscr{H}, c) \leq O\left(\sqrt{\frac{n}{c} \ln(cm/n - c)} \right)$. For $m = n$ and $c \leq \gamma n$ for some constant $\gamma < 1$, the bound becomes $O\left(\sqrt{\frac{n}{c} \ln(c)} \right)$.*

Proof. Consider the weight $r = \frac{1}{c} \mathbf{1}_c$. By Proposition 35 there is a $\mathscr{H}$-integral weight p such that for all i

$$|r_i n - p_i| \leq 1/n. \tag{5.30}$$

By Theorem 43, there is a fair coloring χ with $\mathrm{wd}(\mathscr{H}, \chi, p) \leq k$, where

$$k := \max_i 63K \sqrt{p_i n \ln \left(\frac{2m+2}{p_0 n} \right)} = O\left(\sqrt{\frac{n}{c} \ln(cm/n - c)} \right),$$

using (5.30). The statement for $m = n$ and $c \leq \gamma n$ for some constant $\gamma < 1$ now is an immediate consequence. $\square$

In Corollary 44 the log-factor for $m = n^k$, $k > 1$ a constant, is $\ln(cn)$, while for $k = 1$ or $m = n$, it is only $\ln(c)$. A better approximation of only $\ln(n)$, which is independent of c in case of m being large compared to n can be achieved by the next theorem.

Theorem 45. *Let p denote an $\mathscr{H}$-integral c-color weight. Set $p_0 := \min \{ p_i : i \in [c] \}$. Then a c-coloring χ having discrepancy at most $45 \sqrt{p_i n \ln(4m)}$ in color $i \in [c]$ can be computed in expected time $O\left(nm \log \left(\frac{1}{p_0} \right) \right)$. In particular, a c-coloring χ such that*

$$\mathrm{disc}(\mathscr{H}, \chi, c) \leq 45 \sqrt{\frac{n}{c} \ln(4m)} + 1$$

can be computed in expected time $O(nm \log c)$.

Proof. There is little to do for $m = 1$, so let us assume that $m \geq 2$. We show that the colorings required by the Decreasing-Discrepancies-Assumption can be computed in expected time $O(|X_0|m)$. Denote by $\overline{\mathscr{H}}$ the hypergraph obtained from $\mathscr{H}$ by adding the whole vertex set as an additional hyperedge. Let $X_0 \subseteq X$ and $(q, 1 - q)$ be a 2-color weight. Let $\chi : X_0 \to [2]$ be a random coloring, independently coloring the vertices with probabilities $P(\chi(x) = 1) = q$ and $P(\chi(x) = 2) = 1 - q$ for all

$x \in X_0$. A standard application of the Chernoff inequality (cf. [1]) shows that

$$(*) \qquad \mathrm{wd}\left(\overline{\mathcal{H}}_{|X_0}, \chi, (q, 1-q)\right) \le \sqrt{\tfrac{1}{2}|X_0| \ln(4m)}$$

holds with probability at least $\frac{m-1}{2m}$. Hence by repeatedly generating and testing these random colorings until $(*)$ holds we obtain a randomized algorithm computing such a coloring with expected running time $O(nm)$. By Lemma 37 we get a fair $(q, 1-q)$-coloring for $\mathcal{H}_{|X_0}$ having discrepancy at most $\sqrt{2|X_0| \ln(4m)}$. Hence for $\alpha = \frac{1}{2}$, $D = \sqrt{2 \ln(4m)}$ and arbitrary p_0 the colorings required in the Decreasing-Discrepancies-Assumption can be computed in expected time $O(|X_0|m)$.

Therefore we may apply Theorem 41 with $p_0 = \min\{p_i : i \in [c]\}$. The discrepancy bounds follow from $c_\alpha \le 31.15$. Computing such a coloring involves $O\left(\log(\frac{1}{p_0})n\right)$ times computing a color for a vertex. As this can be done in expected time $O(m)$, we have the claimed bound of $O\left(nm \log\left(\frac{1}{p_0}\right)\right)$. $\square$

Note that the construction of the 2-colorings can be derandomized with standard derandomization techniques like an algorithmic version of the Chernoff-Hoeffding inequality (cf. [74] or [73]). Thus the colorings in Theorem 45 can be computed by a deterministic polynomial-time algorithm as well.

5.2.8 Bounded Shatter Functions

We have already defined the notion of the VC-dimension of a hypergraph (Definition 25). There are many natural examples of hypergraphs in geometry with bounded VC-dimension (see [37]). We need the concept of the shatter functions which are closely related to the VC-dimension.

Definition 46. Let $\mathcal{H} = (X, \mathcal{E})$ be a hypergraph.

(i) For a subset $A \subseteq X$ set $\Pi_{\mathcal{H}}(A) := \{A \cap E : E \in \mathcal{E}\}$. The primal shatter function $\pi_{\mathcal{H}}$ is defined by

$$\pi_{\mathcal{H}}(p) := \max\{|\Pi_{\mathcal{H}}(A)| : A \subseteq X, |A| = p\},$$

where $p \le n$.

(ii) The hypergraph $\mathcal{H}^* = (X^*, \mathcal{E}^*)$, where $X^* = \mathcal{E}$, $\mathcal{E}^* = \{E_x : x \in X\}$, and $E_x = \{E \in \mathcal{E} : x \in E\}$, is called the dual hypergraph of $\mathcal{H}$. The primal shatter function of $\mathcal{H}^*$ is called the dual shatter function of $\mathcal{H}$ and is denoted by $\pi_{\mathcal{H}}^*$.

Using our recursive approach we can generalize results due to Matoušek, Welzl and Wernisch [57] and Matoušek [50] connecting discrepancy with the primal shatter function $\pi_{\mathcal{H}}$ and dual shatter function $\pi_{\mathcal{H}}^*$ of a hypergraph.

Theorem 47. *Let $\mathscr{H} = (X, \mathscr{E})$ be a hypergraph and let $d \geq 1$.*

(i) If $\pi_{\mathscr{H}} = O\left(m^d\right)$, then $\mathrm{disc}(\mathscr{H}, c) = O\left((\frac{n}{c})^{\frac{1}{2}-\frac{1}{2d}}\right)$.

(ii) If $\pi^*_{\mathscr{H}} = O(m^d)$, then $\mathrm{disc}(\mathscr{H}, c) = O\left((\frac{n}{c})^{\frac{1}{2}-\frac{1}{2d}} \log n\right)$.

In both cases the implicit constants are independent of c.

Proof. The primal shatter function of an induced subhypergraph is bounded from above by the primal shatter function of the whole hypergraph. Adding the vertex set as additional edge changes the primal shatter function by at most 1, and does not change the dual shatter function. Without loss of generality we may therefore assume $X \in \mathscr{E}$. The remainder of the proof is standard: bound the weighted discrepancies of the induced subhypergraphs using Proposition 24, use (Lemma 37) and apply Theorem 41.   □

5.2.9 The Beck–Fiala Theorem

In this section we extend the Beck–Fiala theorem to multicolors. In the 2-color case the theorem is proved using the technique of 'floating colors' where fractional "colors" in $[-1, 1]$ are successively changed to colors in $\{-1, 1\}$. Linear algebra is the key tool there. For the c-color case we need in addition vector colors and tensor products as well.

Denote by $\Delta(\mathscr{H}) := \max_{x \in X} |\{E \in \mathscr{E} : x \in E\}|$ the maximum degree of the hypergraph $\mathscr{H}$. The Beck–Fiala theorem states that $\mathrm{disc}(\mathscr{H}) < 2\Delta(\mathscr{H})$ for any hypergraph $\mathscr{H}$ (cf. [12]).

Beck and Fiala actually proved a more general result. For any matrix $A = (a_{ij})$ from $\mathbb{R}^{m \times n}$ denote by $\|A\|_1 := \max_{j \in [n]} \sum_{i \in [m]} |a_{ij}|$ the operator norm induced by the 1-norm.

Theorem 48 ([12]). *For any matrix $A \in \mathbb{R}^{m \times n}$, $\mathrm{lindisc}(A) < 2\|A\|_1$.*

The goal of this paragraph is to show:

Theorem 49 (Doerr, Srivastav [26]). *For any matrix $A \in \mathbb{R}^{m \times n}$ it holds that*

$$\mathrm{lindisc}(A, c) < 2\|A\|_1.$$

The following elementary remark plays a crucial role in the proof of the c-color Beck–Fiala theorem.

Lemma 50. *Let $x \in \overline{M}_c$. Assume that there is a $j' \in [c]$ such that $x_{j'} \notin \{-\frac{1}{c}, \frac{c-1}{c}\}$. Then there is at least a second index j'' (different from j') such that $x_{j''} \notin \{-\frac{1}{c}, \frac{c-1}{c}\}$.*

Proof. By assumption we have $cx_{j'} \notin \mathbb{Z}$. As $c \sum_{j \in [c]} x_j = 0 \in \mathbb{Z}$ by definition of $\overline{M}_c$, there exists a $j'' \in [c]$, $j' \neq j''$ such that $cx_{j''} \notin \mathbb{Z}$. In particular, $x_{j''} \notin \{-\frac{1}{c}, \frac{c-1}{c}\}$. □

Proof (of Theorem 49). Set $\Delta := \|A\|_1$ and $\overline{A} = (\bar{a}_{ij}) := A \otimes I_c$. Note that $\Delta = \|\overline{A}\|_1$. Let $p : [n] \to \overline{M}_c$. Set $\chi = p$. Successively we change χ to a mapping $[n] \to \overline{M}_c$. We regard p and χ as cn-dimensional vectors.

Set $J := \{j \in [cn] : \chi_j \notin \{-\frac{1}{c}, \frac{c-1}{c}\}\}$. We call the columns with indices from J floating (the others fixed). Set $I := \{i \in [cm] : \sum_{j \in J} |\bar{a}_{ij}| > 2\Delta\}$, and call the rows from I active (the others ignored). During the rounding process we will ensure that the following conditions are fulfilled (this is clear at the beginning, because $\chi = p$):

1. $(\overline{A}(p - \chi))_{|I} = 0$, i. e. all active rows have discrepancy zero, and
2. all colors are in $\overline{M}_c$, in particular we have $\sum_{k=0}^{c-1} \chi_{cj-k} = 0$ for all $j \in [n]$.

Note that (ii) is the crucial difference to the 2-color case, where we only need a condition of type (i). Condition (ii) increases the number of equations.

We assume that the rounding process is at a stage where J and I are as above and (i) and (ii) hold. If there is no floating color, i. e. $J = \emptyset$, then all χ_j, $j \in [cn]$, are in $\{-\frac{1}{c}, \frac{c-1}{c}\}$ and χ has the desired form.

Assume that there are still floating colors and consider the system

$$\sum_{k=0}^{c-1} \chi_{cj-k} = 0, \ j \in [n] \text{ such that } c(j - 1) + k \in J \text{ for some } k \in [c]. \quad (5.31)$$

By Lemma 50, in every equation of (5.31) there are at least two floating variables $\chi_{j'}, \chi_{j''}$, i. e. $j', j'' \in J$. Thus (5.31) is a system of at most $\frac{1}{2}|J|$ equations. Now,

$$|J| \Delta \geq \sum_{j \in J} \sum_{i \in I} |\bar{a}_{ij}| = \sum_{i \in I} \sum_{j \in J} |\bar{a}_{ij}| > |I| 2\Delta,$$

so $|J| > 2|I|$. Hence

$$\overline{A}_{|I \times J} \ \chi_{|J} = 0 \quad (5.32)$$

$$\sum_{k=0}^{c-1} \chi_{cj-k} = 0, \ j \in [n] \text{ such that } c(j - 1) + k \in J \text{ for some } k \in [c].$$

We have at most $|I| + \frac{1}{2}|J| < |J|$ equations, thus the system is under-determined and there is a non-trivial solution $x \in \mathbb{R}^{|J|}$ for (5.32). We extend x to $x_E \in \mathbb{R}^{cn}$ by

$$(x_E)_j := \begin{cases} x_j & : \ j \in J \\ 0 & : \ \text{otherwise} \end{cases}.$$

By (ii) and the definition of J, all variables χ_j, $j \in J$ are in $]-\frac{1}{c}, \frac{c-1}{c}[$. Now, there must be a $\lambda > 0$ such that at least one component of $\chi + \lambda x_E$ becomes fixed and all colors are in $\overline{M}_c$, i. e. $\chi + \lambda x_E \in \overline{M}_c^n$. $\chi + \lambda x_E$ fulfils (i) since $(\overline{A} x_E)_{|I} = 0$. Define $\chi := \chi + \lambda x_E$. Since (i), (ii) are fulfilled for this new χ, we can continue the rounding process until all χ_j, $j \in [cn]$ are in $\{-\frac{1}{c}, \frac{c-1}{c}\}$.

We can now conclude the proof by showing $\|\overline{A}(p - \chi)\|_\infty < 2\Delta$. Let $i \in [cm]$ and let $\chi^{(0)}$ and $J^{(0)}$ the values of χ and J when the row i first became ignored. We have $\chi_j^{(0)} = \chi_j$ for all $j \notin J^{(0)}$ and $|\chi_j^{(0)} - \chi_j| < 1$ for all $j \in J^{(0)}$. Note that $\sum_{j \in J^{(0)}} |\bar{a}_{ij}| \leq 2\Delta$, since i is ignored. Thus

$$\left|(\overline{A}(p - \chi))_i\right| = \left|(\overline{A}(p - \chi^{(0)}))_i + (\overline{A}(\chi^{(0)} - \chi))_i\right|$$

$$= \left|0 + \sum_{j \in J^{(0)}} \bar{a}_{ij}\left(\chi_j^{(0)} - \chi_j\right)\right| < 2\Delta.$$

This completes the proof.   □

The immediate consequence for the c-color discrepancy is:

Corollary 51. $\mathrm{disc}(\mathcal{H}, c) < 2\Delta(\mathcal{H})$.

5.2.10 The Theorem of Bárány–Grinberg

Let $\|\cdot\|$ denote a norm on $\mathbb{R}^d$. From a partition point of view the theorem of Bárány–Grinberg states that under the given assumptions there is a 2-partition (I_1, I_2) of the set $X = \{v_1, \ldots, v_k\}$ such that for any subset $X_0 = \{v_1, \ldots, v_l\}$

$$\left\| \sum_{v \in I_j \cap X_0} v - \frac{1}{2} \sum_{v \in X_0} v \right\| < n$$

for $j = 1, 2$. This motivates:

Definition 52 (Discrepancy of Vectors). Let X be a finite set of vectors in $\mathbb{R}^n$ and $\mathcal{P} = \{I_1, \ldots, I_c\}$ a c-partition of X. Let $\|\cdot\|$ be an arbitrary norm on $\mathbb{R}^n$. We define the *discrepancy of the set X* w. r. t. $\mathcal{P}$ and $\|\cdot\|$ by

$$\mathrm{disc}(\mathcal{P}, \|\cdot\|) := \max_{j \in [c]} \left\| \sum_{v \in I_j} v - \frac{1}{c} \sum_{v \in X} v \right\|.$$

A further reformulation is useful. For a subset $X_0 \subseteq X$ set

$$\mathscr{P}_{|X_0} := \{I_1 \cap X_0, \ldots, I_c \cap X_0\}.$$

Let $v_1, v_2, \ldots, v_k$ be vectors and $\mathscr{P} = \{I_1, \ldots, I_c\}$ be a c-partition of $\{v_1, v_2, \ldots, v_k\}$. We define the *discrepancy of the sequence* $v_1, v_2, \ldots, v_k$ w. r. t. $\mathscr{P}$ and $\| \cdot \|$ by

$$\text{disc}\left((v_l)_{l \in [k]}, \mathscr{P}, \| \cdot \|\right) := \max_{l \in [k]} \text{disc}\left(\mathscr{P}_{|\{v_1, \ldots, v_l\}}, \| \cdot \|\right).$$

In this notation the Bárány–Grinberg theorem is as follows: there is a 2-partition $\mathscr{P} = \{I_1, I_2\}$ such that $\text{disc}\left((v_l)_{l \in [k]}, \mathscr{P}, \| \cdot \|\right) < n$. We define a norm $\| \cdot \|_c$ on $\mathbb{R}^{cn}$:

Definition 53. Let $w \in \mathbb{R}^{cn}$ and define the n-dimensional vector

$$w_{|\{j, j+c, \ldots, j+(n-1)c\}} := (w_j, w_{j+c}, \ldots, w_{(n-1)c})^\top \in \mathbb{R}^n.$$

Define

$$\|w\|_c := \max_{j \in [c]} \left\| w_{|\{j, j+c, \ldots, j+(n-1)c\}} \right\|.$$

Tensor products allow an elegant reformulation of the discrepancy of a sequence of vectors. A straightforward calculation shows [26]:

Lemma 54. *Let $X \subseteq \mathbb{R}^n$ be a finite set of vectors and $\mathscr{P} = \{I_1, \ldots, I_c\}$ be any c-partition of X. Let $\chi : X \to [c]$ be the corresponding coloring (i. e. for all $v \in X, l \in [c]$ we have $\chi(v) = l$ if and only if $v \in I_l$). Then the discrepancy of X w. r. t. $\mathscr{P}$ and $\| \cdot \|$ is*

$$\text{disc}(\mathscr{P}, \| \cdot \|) = \left\| \sum_{v \in X} v \otimes m^{(\chi(v))} \right\|_c.$$

The c-color version of the Bárány-Grinberg theorem is:

Theorem 55. *Let $\| \cdot \|$ be an arbitrary norm on $\mathbb{R}^n$ and let $v_1, v_2, \ldots, v_k$ be vectors in $\mathbb{R}^n$ such that $\|v_i\| \leq 1$ for all $i = 1, \ldots, k$. Then there is a c-partition $\mathscr{P} = \{I_1, \ldots, I_c\}$ of $\{v_1, v_2, \ldots, v_k\}$ such that*

$$\text{disc}\left(\mathscr{P}, \| \cdot \|\right) < (c-1)n.$$

Proof. According to Lemma 54 we must show the existence of a coloring $\chi : [k] \to M_c$ such that $\left\| \sum_{i \in [l]} v_i \otimes \chi^{(i)} \right\|_c < (c-1)n$ for all $l \in [k]$.

We follow the pattern of the proof of the Bárány-Grinberg theorem. Let $A := [n]$ and $\chi_j^{(i)} := 0$ for all $i \in [k]$, $j \in [c]$. As in the literature we call those $\chi_j^{(i)}$ where $i \in A$ and $\chi_j^{(i)} \notin \{\frac{c-1}{c}, -\frac{1}{c}\}$ *variables* and the corresponding color vector $\chi^{(i)}$

active. At the beginning we have cn variables and n active color vectors and all color vectors $\chi^{(i)}, i \in [k]$ are in $\overline{M_c}$, so we have $\sum_{i \in A} v_i \otimes \chi^{(i)} = 0$.

For the basic rounding procedure, which is called repeatedly, we define

$$A_0 := \left\{ i \in [k] : \exists j \in [c] : \chi_j^{(i)} \notin \left\{ \frac{c-1}{c}, -\frac{1}{c} \right\} \right\}.$$

This is the set of indices of active color vectors. We consider the system of equations:

$$\sum_{i \in A} v_i \otimes \chi^{(i)} = 0 \qquad\qquad (5.33)$$

$$\sum_{j \in [c]} \chi_j^{(i)} = 0 \quad \text{for all } i \in A_0.$$

Let n' be the number of variables and m' the rank of the system (5.33). By Lemma 50, each active vector contains at least two variables, so $n' \geq 2|A_0|$. Furthermore, we have $m' \leq (c-1)n + |A_0|$, since $\sum_{j \in [c]} \chi_j^{(i)} = 0$ for all $i \in [k]$ holds at any phase of the rounding process.

Case 56. There is no nontrivial solution to (5.33). Then there are at most m' variables. The inequality $2|A_0| \leq n' \leq m' \leq (c-1)n + |A_0|$ implies $|A_0| \leq (c-1)n$.

Suppose that there are vectors that have not been active, i. e. $A \neq [k]$. In this case we set $A := A \cup \{\max(A) + 1\}$, update system (5.33) and continue the rounding process. At $A = [k]$ the rounding process is terminated by changing the remaining variables to $\frac{c-1}{c}$ or $-\frac{1}{c}$ in any way such that all $\chi^{(i)}$ are in M_c.

Case 57. There is a nontrivial solution for (5.33). Then—as in the proof of the Beck–Fiala theorem—we can change χ so that some variables become $\frac{c-1}{c}$ or $-\frac{1}{c}$ and all variables stay in $[-\frac{1}{c}, \frac{c-1}{c}]$. Now $\chi^{(i)} \in \overline{M_c}$ for all $i \in [k]$ and $\sum_{i \in A} v_i \otimes \chi^{(i)} = 0$. And we may continue the rounding process.

Consider $l \in [k]$. Let $\tilde{\chi}^{(1)}, \ldots, \tilde{\chi}^{(k)}$ be the value of the color vectors at the phase of the rounding process when $A = [l]$ and no nontrivial solution to (5.33) can be found. Let $\tilde{A}_0$ be the value of A_0 at this stage. Let $\chi_f^{(1)}, \ldots, \chi_f^{(k)}$ be the final values of the color vectors. We know $|\tilde{A}_0| \leq (c-1)n$. As $\chi^{(i)} \in \overline{M_c}$, we have $\left\| \tilde{\chi}^{(i)} - \chi_f^{(i)} \right\|_\infty < 1$ for all $i \in [l]$. Furthermore $\tilde{\chi}^{(i)} = \chi_f^{(i)}$ is true if $i \notin \tilde{A}_0$, because an inactive vector would never become active. According to (5.33) also the equation $\sum_{i \in [l]} v_i \otimes \tilde{\chi}^{(i)} = 0$ holds. The calculations

$$\left\|\sum_{i\in[l]} v_i \otimes \chi^{(i)}\right\|_c \leq \underbrace{\left\|\sum_{i\in[l]} v_i \otimes \tilde{\chi}^{(i)}\right\|_c}_{=\,0\text{ by (5.33)}} + \left\|\sum_{i\in[l]} v_i \otimes (\chi^{(i)} - \tilde{\chi}^{(i)})\right\|_c$$

$$= \left\|\sum_{i\in\tilde{A}_0} v_i \otimes (\chi^{(i)} - \tilde{\chi}^{(i)})\right\|_c$$

$$= \max_{j\in[c]}\left\|\sum_{i\in\tilde{A}_0} (v_i \otimes (\chi^{(i)} - \tilde{\chi}^{(i)}))|_{\{j,\,j+c,\ldots,\,j+(n-1)c\}}\right\|$$

$$= \max_{j\in[c]}\left\|\sum_{i\in\tilde{A}_0} v_i (\chi^{(i)} - \tilde{\chi}^{(i)})_j\right\|$$

$$< \sum_{i\in\tilde{A}_0} \|v_i\|$$

$$\leq (c-1)n$$

finish the proof.   □

Bárány and Doerr [8] showed another multicolor version of the Bárány-Grinberg theorem.

Theorem 58 (Bárány, Doerr [8]). *Let $d, c \in \mathbb{N}$. Let $\|\cdot\|$ be any norm on $\mathbb{R}^d$. Let B denote the unit ball with respect to this norm. Any sequence $v_1, v_2, \ldots$ of vectors in B can be partitioned into c subsequences $V_1, V_2, \ldots, V_c$ in a balanced manner with respect to the partial sums: for all $n \in \mathbb{N}$ and $l \leq c$ we have*

$$\left\|\sum_{i\leq k,\,v_i\in V_l} v_i - \frac{1}{c}\sum_{i\leq k} v_i\right\| \leq 2.005d.$$

5.2.11 Lower Bounds

Let us complement the discussion so far by lower bound considerations. In fact, we will see that almost matching c-color lower bounds can be proved. We need a c-color version of a fundamental result of Lovász and Sós [13], through which the discrepancy function is connected to the minimal eigenvalue of the matrix $A^{\top}A$, where A is the given matrix. We require the following standard results from linear algebra, which we state without proofs.

Lemma 59. *The following laws hold for the tensor product:*

1. Associativity: All matrices A, B, C fulfil $(A \otimes B) \otimes C = A \otimes (B \otimes C)$.

2. *Distributivity with +: For all matrices A, B, C such that $A + B$ is defined we have $(A + B) \otimes C = A \otimes C + B \otimes C$ and $C \otimes (A + B) = C \otimes A + C \otimes B$.*
3. *'Mixed Product Rule': $(AB) \otimes (CD) = (A \otimes C)(B \otimes D)$ for all matrices A, B, C, D such that AB and CD are defined.*
4. *$\otimes$ is compatible with inversion: $(A \otimes B)^{-1} = A^{-1} \otimes B^{-1}$ for all non-singular matrices A and B.*
5. *The (complex) eigenvalues of $A \otimes B$ are exactly the products of an eigenvalue of A and one of B.*
6. $\operatorname{rank}(A \otimes B) = \operatorname{rank}(A)\operatorname{rank}(B)$.
7. $\det(A \otimes B) = (\det A)^{n_B}(\det B)^{n_A}$ *for all matrices $A \in \mathbb{C}^{n_A \times n_A}$ and $B \in \mathbb{C}^{n_B \times n_B}$.*

Eigenvalues associated to the incidence matrix of a hypergraph are connected to the discrepancy function. The c-color version is as follows.

Theorem 60. *For a matrix $A \in \mathbb{R}^{m \times n}$, let $\lambda_{\min} = \lambda_{\min}(A^\top A)$ be the minimal eigenvalue of $A^\top A$. Then $\operatorname{disc}(A, c) \geq \sqrt{\frac{n(c-1)}{mc^2}}\lambda_{\min}$.*

Proof. We fix a c-coloring $\chi : [n] \to M_c$ with minimum c-color discrepancy. Then

$$
\begin{aligned}
\operatorname{disc}(A, c) \quad &= \quad \|(A \otimes I_c)\chi\|_\infty \\[2mm]
&\geq \quad \frac{1}{\sqrt{cm}}\|(A \otimes I_c)\chi\|_2 \\[2mm]
&\geq \quad \frac{1}{\sqrt{cm}}\|\chi\|_2\sqrt{\lambda_{\min}((A \otimes I_c)^\top(A \otimes I_c))} \\[2mm]
&\overset{\text{Lemma 59(iii)}}{=} \quad \frac{1}{\sqrt{cm}}\sqrt{\frac{n(c-1)}{c}}\sqrt{\lambda_{\min}((A^\top A) \otimes I_c)} \\[2mm]
&\overset{\text{Lemma 59(v)}}{=} \quad \sqrt{\frac{n(c-1)}{mc^2}}\sqrt{\lambda_{\min}(A^\top A)}.
\end{aligned}
$$

$\square$

It is a known fact that hypergraphs corresponding to Hadamard matrices show that Spencer's 'six standard deviations' result is tight, up to constant factors. Here is a c-color extension.

Theorem 61. *There is a universal constant $K > 0$ such that for an infinite sequence of $n \in \mathbb{N}$ there is a hypergraph with n vertices, n edges and with discrepancy at least $K\sqrt{\frac{n}{c}}$.*

Proof. We consider $n \in \mathbb{N}$ such that there exists a Hadamard matrix H of dimension n, i. e. $H \in \{+1, -1\}^{n \times n}$ and all rows of H are pairwise orthogonal. By multiplying some rows by -1, we may assume that all entries of the first column v_1 are 1. Let $v_2, \ldots, v_n$ denote the remaining columns. Set $A = \frac{1}{2}(H + J)$, where J is the $n \times n$ matrix consisting of $1'$s only. A is the incidence matrix of a hypergraph $\mathcal{H}$ of n edges on n vertices. We prove that $\mathcal{H}$ has the desired discrepancy.

Fix an arbitrary coloring $\chi : [n] \to M_c$ and let $i \in [c]$ be such that

$$\left| \chi^{-1}(m^{(i)}) \setminus \{1\} \right| \geq \tfrac{n-1}{c}. \tag{5.34}$$

For all $j \in [c]$, define $\chi_j : [n] \to \{-\tfrac{1}{c}, \tfrac{c-1}{c}\}$, $k \mapsto \chi(k)_j$. Let $a_1, \ldots, a_n$ be the row vectors of A and for $x, y \in \mathbb{R}^n$, let $x \cdot y$ be the usual inner product in $\mathbb{R}^n$. Then

$$
\begin{aligned}
\mathrm{disc}(\mathscr{H}, \chi, c) &= \| (A \otimes I_c) \chi \|_\infty \\
&= \left\| (a_1 \cdot \chi_1, \ldots, a_1 \cdot \chi_c, \ldots, a_n \cdot \chi_1, \ldots, a_n \cdot \chi_c)^\top \right\|_\infty \\
&\geq \left\| (a_1 \cdot \chi_i, \ldots, a_n \cdot \chi_i)^\top \right\|_\infty \\
&= \| A \chi_i \|_\infty \\
&\geq \tfrac{1}{\sqrt{n}} \| A \chi_i \|_2.
\end{aligned}
$$

By (5.34) we have

$$\left| \{ k \in [n] \setminus \{1\} : \chi_i(k) = \tfrac{c-1}{c} \} \right| \geq \tfrac{n-1}{c}. \tag{5.35}$$

By definition of A, there exists a $\lambda \in \mathbb{R}$ such that $A\chi_i = \sum_{k=2}^n \tfrac{1}{2} \chi_i(k) v_k + \lambda v_1$. As the $v_1, \ldots, v_n$ are pairwise orthogonal, we may conclude the proof by the following chain of estimations:

$$
\begin{aligned}
\| A \chi_i \|_2 &= \sqrt{\sum_{k=2}^n \chi_i(k)^2 \left\| \tfrac{1}{2} v_k \right\|_2^2 + \lambda^2 \| v_1 \|_2^2} \\
&\geq \sqrt{\sum_{k=2}^n \chi_i(k)^2 \left\| \tfrac{1}{2} v_k \right\|_2^2} \\
&= \tfrac{1}{2} \sqrt{n} \sqrt{\sum_{k=2}^n \chi_i(k)^2} \\
&\geq \tfrac{1}{2} \sqrt{n} \sqrt{\tfrac{n-1}{c} \left(\tfrac{c-1}{c} \right)^2 + \tfrac{(n-1)(c-1)}{c} \left(-\tfrac{1}{c} \right)^2} \quad \text{(by (5.35))} \\
&= \tfrac{1}{2} \sqrt{n} \sqrt{\tfrac{(n-1)(c-1)}{c^2}}.
\end{aligned}
$$

And finally, with the lower bound estimation for the discrepancy above, we obtain

$$\mathrm{disc}(\mathscr{H}, c) \geq \tfrac{1}{2} \sqrt{\tfrac{(n-1)(c-1)}{c^2}}. \qquad \square$$

5.3   Arithmetic Progressions in the Integers

In this section we investigate the hypergraph of arithmetic progressions and their higher dimensional generalizations. Let us consider the hypergraph $\mathcal{H}_N$ of arithmetic progressions in the first N positive integers (Definition 19). While the upper bound of Matoušek and Spencer (Theorem 22) is non constructive, the bound of Sárközy is constructive. In Chap. 7, L. Kliemann presents an elegant probabilistic version of Sárközy's algorithm and its analysis due to V. Sauerland [69], which also has a very efficient implementation (see [44]).

We proceed to the proof of an upper and lower bound for the c-color discrepancy of the hypergraph of arithmetic progressions due to Doerr and Srivastav [26].

Theorem 62. $\mathrm{disc}(\mathcal{H}_N, c) = O(c^{-0.16} N^{0.25})$ *for* $c \leq N^{0.25}$.

Proof. For an absolute constant C' the following holds: Let $p \in [0, 1]^c$ be a weight. Then there is a fair coloring of $\mathcal{H}_N$ with respect to p having discrepancy at most $C' p_i^{0.16} N^{0.25}$ in each color i such that $p_i \geq N^{0.25}$. By Lemma 5.3 of [56] an induced subgraph $\mathcal{H}_0 = (\mathcal{H}_N)_{|X_0}$ of $\mathcal{H}_N$ on $|X_0| = \rho N \geq N^{0.25}$ vertices has discrepancy at most $C_1 \rho^{0.16} N^{0.25}$ for some absolute constant C_1. We show that $\mathrm{herdisc}(\mathcal{H}_0) \leq 2C_1 \rho^{0.16} N^{0.25}$:

Let $\mathcal{H}_1 = (X_1, \mathcal{E}_1)$ be an induced subhypergraph of $\mathcal{H}_0$. If $|X_1| \geq N^{0.25}$ we are done by Lemma 5.3 of [56]. Let us therefore assume $|X_1| < N^{0.25}$. We show that $(\mathcal{H}_1)_{|\left[\frac{N}{2}\right]}$ and $(\mathcal{H}_1)_{|[N]\setminus\left[\frac{N}{2}\right]}$ have discrepancy at most $C_1 \rho^{0.16} N^{0.25}$ and conclude $\mathrm{disc}(\mathcal{H}_1) \leq 2C_1 \rho^{0.16} N^{0.25}$. Consider the hypergraph

$$\mathcal{H}_2 := \mathcal{H}_1\Big|_{\left(X_1 \cap \left[\frac{N}{2}\right]\right) \cup \left\{N - N^{0.25} + \left|X_1 \cap \left[\frac{N}{2}\right]\right| + 1, \ldots, N\right\}}$$

This hypergraph has exactly $N^{0.25} \leq \rho N$ vertices and thus discrepancy at most $C_1 \rho^{0.16} N^{0.25}$. As every edge of $(\mathcal{H}_1)_{|\left[\frac{N}{2}\right]}$ is also an edge of $\mathcal{H}_2$, we conclude $\mathrm{disc}\left((\mathcal{H}_1)_{|\left[\frac{n}{2}\right]}\right) \leq C_1 \rho^{0.16} N^{0.25}$. A similar argument shows $\mathrm{disc}\left((\mathcal{H}_1)_{|[N]\setminus\left[\frac{N}{2}\right]}\right) \leq C_1 \rho^{0.16} N^{0.25}$. Hence $\mathrm{herdisc}(\mathcal{H}_0) \leq 2C_1 \rho^{0.16} N^{0.25}$. The relation between the linear and hereditary discrepancy yields that all weighted discrepancies of $\mathcal{H}_0$ are bounded by $2C_1 \rho^{0.16} N^{0.25}$. As $[N]$ is an arithmetic progression, we may apply Lemma 37 and conclude that twice this discrepancy may be achieved by a fair coloring respecting the underlying weight. We may apply Theorem 41 with $D = 4C_1 N^{0.09}$, $\alpha = 0.16$ and $p_0 = N^{0.25}$, which proves our claim. $\square$

And the lower bound is:

Theorem 63. *The hypergraph of arithmetic progressions fulfils*

$$\text{disc}(\mathcal{H}_N, c) \geq 0.04 \frac{1}{\sqrt{c}} \sqrt[4]{N}.$$

Proof. For the lower bound we will follow the approach of [13]. Set $k = \left\lfloor \sqrt{\frac{1}{6}N} \right\rfloor$.
Let $\mathcal{E}$ be the set of arithmetic progressions of length k and difference less than $6k$ computed modulo N (hence our arithmetic progressions may be over-wrapped from N to 1 at most once). Every arithmetic progression of $\mathcal{E}$ is a union of at most two arithmetic progressions from $\mathcal{E}_N$, so the discrepancy of $\mathcal{H}_N$ is at least half the discrepancy of $\mathcal{E}$.

Recall that a matrix is called circulant if the i-th row can be obtained from the first by shifting it $i - 1$ times to the right. Let us enumerate the arithmetic progressions in $\mathcal{E}$ in a way that if i is not divisible by N, then $E_{i+1} = E_i + 1$ (always computed modulo N), i. e. E_{i+1} is E_i shifted right by one. Thus the incidence matrix $A = (a_{ij}) \in \{0,1\}^{6kN \times N}$ defined by $a_{ij} = 1$ if and only if $j \in E_i$ consists of $6k$ circulant sub-matrices. As sum and product of two circulant matrices are circulant again, $A^\top A$ is circulant. The eigenvectors of circulant matrices are known to be of the form $(1, \varepsilon, \varepsilon^2, \ldots, \varepsilon^{N-1})^\top$, where ε is an Nth root of unity. Thus the minimum eigenvalue $\lambda_{\min}(A^\top A)$ of $A^\top A$ is greater than $\frac{1}{4}k^2$.

Using Theorem 60 we have disc $(([N], \mathcal{E}_N), c)^2 \geq \frac{N(c-1)}{6kNc^2} \frac{1}{4}k^2 = \frac{(c-1)k}{24c^2}$. Hence

$$\text{disc}(\mathcal{H}_N, c) \geq 0.5 \, \text{disc}(([N], \mathcal{E}), c)$$

$$\geq \sqrt{\frac{c-1}{96\sqrt{6}c^2}} \sqrt[4]{N}$$

$$\geq 0.0652 \sqrt{\frac{c-1}{c^2}} \sqrt[4]{N}$$

$$\geq 0.04 \sqrt{\frac{1}{c}} \sqrt[4]{N}.$$

This completes the proof. □

5.3.1 Cartesian Products

This subsection is based on the work of Doerr, Srivastav and Wehr [29]. Let us introduce Cartesian products of arithmetic progressions, a kind of high-dimensional generalization of arithmetic progressions in the integers.

Definition 64. Let $\mathcal{H}_N$ be the hypergraph of arithmetic progressions in the first N integers. A d-dimensional arithmetic progression A in $[N]^d$ is the Cartesian product of d arithmetic progressions in $[N]$, i. e. $A = \prod_{i=1}^{d} A_i$ where A_i is an arithmetic progression in $[N]$ for all $i = 1, \ldots, d$. Let $\mathcal{H}_N^d$ be the hypergraph with node

Fig. 5.2 Some two-dimensional Cartesian products of arithmetic progressions

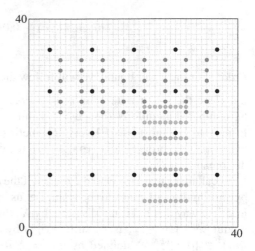

set $[N]^d$ and with the set of hyperedges consisting of Cartesian products of any d arithmetic progressions in $[N]$.

Note that $\mathcal{H}_N^1 = \mathcal{H}_N$. The study of $\mathcal{H}_N^d$ was suggested by Hans Jürgen Prömel in the context of Petra Wehr's (née Knieper) dissertation at the Humboldt University Berlin in 1996. The motivation came from the multi-dimensional version of van der Waerden's theorem, the theorem of Gallai [64] (see also [36]). By Gallai's theorem the following is true: Let t be a positive integer and $X = [t]^d$. If $\mathbb{N}^d$ is finitely colored, then there exist integers $x_1, x_2, \ldots, x_d$ and $\delta \in \mathbb{N}$ such that

$$W = (x_1, \ldots, x_d) + \delta X$$
$$= \left\{ (x_1 + i_1\delta, x_2 + i_2\delta, \ldots, x_d + i_d\delta) : 0 \leq i_j \leq t, j = 1, \ldots, d \right\}$$

is monochromatic. Note that W is a d-dimensional arithmetic progression in the sense of Definition 64.

Examples for two-dimensional Cartesian products of arithmetic progressions are shown in Fig. 5.2.

In the following, we present the proof of Doerr, Srivastav and Wehr [29] for the tight bound $\mathrm{disc}(\mathcal{H}_N^d) = \Theta(N^{\frac{d}{4}})$. This extends the lower bound of $\Omega(N^{1/4})$ of Roth [67] and the matching upper bound $O(N^{1/4})$ of Matoušek and Spencer [56] from $d = 1$ to arbitrary, fixed d. To establish the lower bound harmonic analysis on locally compact Abelian groups is used. For the upper bound a product coloring arising from the theorem of Matoušek and Spencer is sufficient. The main theorem is:

Theorem 65 (Doerr, Srivastav, Wehr [29]). *Let $d \geq 1$ be an integer. We have*

$$\pi^{-d} N^{\frac{d}{4}} \leq \mathrm{disc}(\mathcal{H}_N^d) \leq \alpha^d N^{\frac{d}{4}},$$

where $\alpha > 0$ is an absolute constant. Thus $\mathrm{disc}(\mathcal{H}_N^d) = \Theta(N^{\frac{d}{4}})$ for every fixed d.

We will also show that a kind of submultiplicativity of the c-color discrepancy function allows also to settle a c-color bound using Theorem 65:

Theorem 66. *Let $d \geq 1$ be an integer. Then* $\mathrm{disc}(\mathcal{H}_N^d, c) \leq O(c^{0.84d-1} N^{\frac{d}{4}})$.

Note that a non-trivial lower bound for $\mathrm{disc}(\mathcal{H}_N^d, c)$, $c \geq 3$, is not known. For the proof of the lower bound in Theorem 65 we need some Fourier analysis, useful at many places in this chapter.

5.3.2 Basics of Fourier Analysis on LCA Groups

We give a short overview on the basics of Fourier analysis on locally compact Abelian groups (LCA groups). For an extensive introduction to this field we refer to the book of W. Rudin [68].

Definition 67. Let $(G, +, \mathcal{T})$ be a locally compact Abelian group. A character on G is a function $\gamma : G \to \mathbb{C}$ with

(i) $|\gamma(x)| = 1$ for all $x \in G$,
(ii) $\gamma(x + y) = \gamma(x)\gamma(y)$ for all $x, y \in G$.

Let $\hat{G}$ denote the set of all characters on G.

Remark 68. $\hat{G}$ is a subset of the set $\mathbb{C}^G$ of all complex-valued functions on G. Define an addition of functions by $f + g : G \to \mathbb{C}$, $x \mapsto (f + g)(x) := f(x)g(x)$.

Theorem 69. *Let G be a locally compact Abelian group. Then*

(i) $\hat{G}$ is closed under the addition $+$ in $\mathbb{C}^G$.
(ii) There exists a topology $\mathcal{T}_{\hat{G}}$ on $\hat{G}$ such that $\hat{G}$ is a locally compact Abelian group.

$\hat{G}$ is called the dual group of G.

For a locally compact Abelian group G and all $1 \leq p < \infty$ we denote by $L^p(G)$ the subset of all Borel functions f with $\|f\|_p := (\int_G |f(x)|^p dx)^{1/p} < \infty$, where the used measure is the up to a positive constant unique Haar measure. Now we are able to define the Fourier transform for functions in $L^1(G)$ and the convolution of two functions in $L^1(G)$.

Definition 70 (Fourier Transform). Let G be a locally compact Abelian group and $f \in L^1(G)$. The Fourier transform $\hat{f} : \hat{G} \to \mathbb{C}$ is defined by $\hat{f}(\gamma) := \int_G f(x)\gamma(-x)\, dx$.

Definition 71 (Convolution). Let G be a locally compact Abelian group and $f, g \in L^1(G)$. The convolution $f * g : G \to \mathbb{C}$ is defined by

$$(f * g)(y) := \int_G f(x)g(y - x)\, dx.$$

Remark 72. The Fourier transform and the convolution are well defined. For $f, g \in L^1(G)$ the convolution $f * g$ is also in $L^1(G)$ and it holds $\|f * g\|_1 \leq \|f\|_1 \|g\|_1$ [68].

The following two theorems are the key for the use of Fourier analysis in discrepancy theory. The first shows that the Fourier transform is multiplicative on the Banach algebra $L^1(G)$, where the multiplication on $L^1(G)$ is the convolution. The second is the well-known Plancherel Theorem for locally compact Abelian groups [68].

Theorem 73. *Let G be a locally compact Abelian group. For all $f, g \in L^1(G)$ we have $\widehat{f * g} = \hat{f}\hat{g}$.*

Theorem 74. *The Haar measure on $\hat{G}$ can be normalized such that the Fourier transform is an isometry, i.e. $\|\hat{f}\|_2 = \|f\|_2$, for all $f \in L^1(G) \cap L^2(G)$.*

We have to mention here that, if G is discrete, the integral $\int_G \cdot dx$ is nothing but the sum $\sum_{x \in G}$. In our discrepancy theoretical problems we are only concerned with discrete or even finite Abelian groups. Note that with the discrete topology they are automatically locally compact. The following corollary to Theorem 74 is useful in this setting.

Corollary 75. *Let G be a finite Abelian group of order r. Then for any complex-valued function f on G we have $\|\hat{f}\|_2^2 = r\|f\|_2^2$.*

We also give a basic consequence of the Plancherel theorem. Let $x + E := \{x + y : y \in E\}$ and for an $A \subseteq G$ we define $f(A) := \sum_{a \in A} f(a)$.

Proposition 76. *Let $\mathcal{E}_0$ be a set of subsets of G and let $\gamma > 0$ be chosen such that $\sum_{E \in \mathcal{E}_0} |\widehat{\mathbf{1}_{-E}}(r)|^2 \geq \gamma$ for all $r \in \hat{G}$. Let $f : G \to \mathbb{C}$. Then*

$$\sum_{E \in \mathcal{E}_0} \sum_{x \in G} |f(x + E)|^2 \geq \gamma \|f\|_2^2.$$

Proof. We have

$$f(x + E) = \sum_{y \in x + E} f(y) = \sum_{y \in G} f(y) \mathbf{1}_{x+E}(y)$$

$$= \sum_{y \in G} f(y) \mathbf{1}_{-E}(x - y) = (f * \mathbf{1}_{-E})(x),$$

and

$$
\sum_{E \in \mathscr{E}_0} \sum_{x \in G} |f(x + E)|^2 \;=\; \sum_{E \in \mathscr{E}_0} \sum_{x \in G} |(f * \mathbf{1}_{-E})(x)|^2
$$

$$
\overset{\text{Plancherel}}{=}\; \sum_{E \in \mathscr{E}_0} \sum_{r \in \hat{G}} |\widehat{(f * \mathbf{1}_{-E})}(r)|^2
$$

$$
\overset{\text{multiplic. of conv.}}{=}\; \sum_{r \in \hat{G}} |\hat{f}(r)|^2 \sum_{E \in \mathscr{E}_0} |\widehat{\mathbf{1}}_{-E}(r)|^2
$$

$$
\geq\; \gamma \sum_{r \in \hat{G}} |\hat{f}(r)|^2
$$

$$
\overset{\text{Plancherel}}{=}\; \gamma \|f\|_2^2
$$

proves the claim.   □

5.3.3 Cartesian Products: Lower Bound

The Fourier-analytical proof of the lower bound can also be found in Petra Wehr's dissertation [45]. Roth's proof of the lower bound in the one-dimensional case [67] does not invoke the discrepancy function directly. This might be one reason why it was not possible to generalize Roth's proof to Cartesian products earlier. We use the abstract framework of locally compact Abelian groups. For the remainder of this section let d denote a positive integer. $G := \mathbb{Z}^d$ equipped with the discrete topology is a locally compact Abelian group. Let $T \subset \mathbb{C}$ be the unit circle. We denote the set of characters of G by $\hat{G}$. In this discrete setting the convolution of two functions $f, g \in L^1(G)$ is $(f * g)(y) := \sum_{x \in G} f(x)g(y - x)$. The Fourier transform of f is

$$
\hat{f} : \hat{G} \to \mathbb{C}; \gamma \mapsto \sum_{x \in G} f(x)\gamma(-x).
$$

Note that we have $\widehat{f * g} = \hat{f}\,\hat{g}$. Let $< \cdot, \cdot >$ denote the inner product on $\mathbb{R}^n$. Using the duality $\hat{\mathbb{Z}} \simeq T$, it is straightforward to show the following proposition.

Proposition 77. *For $\alpha \in [0, 1[^d$ let $\gamma_\alpha : \mathbb{Z}^d \to T; z \mapsto e^{2\pi i <\alpha, z>}$ denote the character associated to α and $T^d := \{\gamma_\alpha | \alpha \in [0, 1[^d\}$.*

(i) The dual group $\widehat{\mathbb{Z}^d}$ of $\mathbb{Z}^d$ is T^d.
(ii) The Fourier transform $\hat{f}$ of a function $f \in L^1(\mathbb{Z}^d)$ can be written as

$$
\hat{f}(\gamma_\alpha) = \sum_{z \in \mathbb{Z}^d} e^{-2\pi i <\alpha, z>} f(z).
$$

Proof (of Theorem 65, lower bound). We express the discrepancy of a given d-dimensional arithmetic progression and a given 2-coloring as the convolution of the coloring function and a characteristic function of the arithmetic progression. Then we compute the L^2-norm of this function applying the Plancherel theorem. An average argument (taking the sum over a special set of d-dimensional arithmetic progressions) and an estimate for the sum of unit roots completes the proof. Some notations are needed.

- $L := \frac{1}{2}\sqrt{N}$, $\Delta := \{1, \ldots, \sqrt{N}\}^d$, $J := [N]^d$,
- $A_{j_0, \delta_0} := \{j_0 + \delta_0 i : i \in [L]\} \cap [N]$,
- For $\delta \in \Delta$, $j \in J$ define $A_{j,\delta} := \prod_{i=1}^{d} A_{j_i, \delta_i}$.

Define the extension χ_F of a 2-coloring χ of $[N]^d$ to $\mathbb{Z}^d$ by

$$\chi_F(j) = \begin{cases} \chi(j) & : & j \in J \\ 0 & : & \text{otherwise.} \end{cases}$$

We define a (quasi-)characteristic function of $A_{0,\delta}$ by

$$\eta_\delta(k) = \begin{cases} 1 & : & -k \in A_{0,\delta} \\ 0 & : & \text{otherwise} \end{cases}$$

An easy calculation yields

$$(\chi_F * \eta_\delta)(j) = \chi(A_{j,\delta}) \tag{5.36}$$

for all $\delta \in \Delta$, $j \in J$.

As χ_F and η_δ have finite support, we have $\chi_F * \eta_\delta \in L^1(\mathbb{Z}^d) \cap L^2(\mathbb{Z}^d)$. The Plancherel theorem for locally compact Abelian groups [68] gives:

$$\sum_{j \in J} \chi^2(A_{j,\delta}) \overset{(5.36)}{=} \|\chi_F * \eta_\delta\|_2^2 = \|\widehat{\chi_F * \eta_\delta}\|_2^2$$

$$= \|\hat{\chi}_F \cdot \hat{\eta}_\delta\|_2^2$$

$$= \int_{[0,1]^d} |\hat{\chi}_F(\gamma_\alpha)\, \hat{\eta}_\delta(\gamma_\alpha)|^2 \, d\alpha. \tag{5.37}$$

Roth [67] showed the following estimate for sums of unit roots.

$$\sum_{\delta=1}^{\sqrt{N}} \left| \sum_{j=0}^{L-1} e^{2\pi i \delta j \alpha} \right|^2 \geq \pi^{-2} N = \left(\frac{2}{\pi} L\right)^2 \qquad \text{for arbitrary } \alpha \in \mathbb{R}. \tag{5.38}$$

Thus we have

$$
\sum_{\delta \in \Delta} \left| \hat{\eta}_\delta(\gamma_\alpha) \right|^2 = \sum_{\delta \in \Delta} \left| \sum_{j \in \mathbb{Z}^d} \eta_\delta(j) e^{-2\pi i <j,\alpha>} \right|^2
$$

$$
= \sum_{\delta \in \Delta} \left| \sum_{j_1,\ldots,j_d \in [L]} e^{2\pi i (j_1 \delta_1 \alpha_1 + \cdots + j_d \delta_d \alpha_d)} \right|^2
$$

$$
= \sum_{\delta \in \Delta} \left| \prod_{k=1}^{d} \left(\sum_{j_k=0}^{L-1} e^{2\pi i j_k \delta_k \alpha_k} \right) \right|^2
$$

$$
= \prod_{k=1}^{d} \left(\sum_{\delta_k=1}^{\sqrt{N}} \left| \sum_{j_k=0}^{L-1} e^{2\pi i j_k \delta_k \alpha_k} \right|^2 \right)
$$

$$
\geq (\pi^{-2} N)^d = \pi^{-2d} N^d. \tag{5.39}
$$

The Plancherel theorem says

$$
\|\hat{\chi}_F\|_2^2 = \|\chi_F\|_2^2 = \sum_{j \in J} \chi^2(j) = N^d. \tag{5.40}
$$

Finally

$$
\sum_{\delta \in \Delta} \sum_{j \in J} \chi^2(A_{j,\delta}) \overset{(5.37)}{=} \sum_{\delta \in \Delta} \int_{[0,1]^d} |\hat{\chi}_F(\gamma_\alpha) \hat{\eta}_\delta(\gamma_\alpha)|^2 d\alpha
$$

$$
= \int_{[0,1]^d} |\hat{\chi}_F(\gamma_\alpha)|^2 \left(\sum_{\delta \in \Delta} |\hat{\eta}_\delta(\gamma_\alpha)|^2 \right) d\alpha
$$

$$
\overset{(5.39)}{\geq} (\pi^{-2} N)^d \int_{[0,1]^d} |\hat{\chi}_F(\gamma_\alpha)|^2 d\alpha
$$

$$
\overset{(5.40)}{=} (\pi^{-2} N)^d N^d = \pi^{-2d} N^{2d}.
$$

The sum $\sum_{\delta \in \Delta} \sum_{j \in J} \chi^2(A_{j,\delta})$ consists of $N^{\frac{3d}{2}}$ terms. The pigeonhole principle implies the existence of $\bar{\delta} \in \Delta$ and $\bar{j} \in J$ such that

$$
\chi^2(A_{\bar{j},\bar{\delta}}) \geq \frac{\pi^{-2d} N^{2d}}{N^{\frac{3d}{2}}} = \pi^{-2d} N^{\frac{d}{2}}.
$$

So $|\chi(A_{\bar{j},\bar{\delta}})| \geq \pi^{-d} N^{\frac{d}{4}}$, and thus the discrepancy of $\mathcal{H}_N^d$ is at least $\pi^{-d} N^{\frac{d}{4}}$. $\square$

5.3.4   Cartesian Products: Upper Bound

B. Doerr gave a very general argument which solves the problem also for an arbitrary number of colors. The proof idea came up in a discussion with Nati Linial at the workshop on discrepancy theory in 1998 at Kiel University.

Let $\mathscr{G} = (X, \mathscr{E})$ and $\mathscr{H} = (Y, \mathscr{F})$ be hypergraphs. Define the Cartesian product of $\mathscr{G}$ and $\mathscr{H}$ by

$$\mathscr{G} \times \mathscr{H} := (X \times Y, \{A \times B : A \in \mathscr{E}, B \in \mathscr{F}\}).$$

By this definition, the hypergraph of d-dimensional arithmetic progressions is the d-fold Cartesian product of the hypergraph of (one-dimensional) arithmetic progressions on $[N]$.

Theorem 78. *For any $c \in \mathbb{N}$ and any two hypergraphs $\mathscr{G}$ and $\mathscr{H}$ we have*

$$\mathrm{disc}(\mathscr{G} \times \mathscr{H}, c) \leq c \, \mathrm{disc}(\mathscr{G}, c) \, \mathrm{disc}(\mathscr{H}, c).$$

Proof. Pick a Latin square $Q = (q_{ij})$ of dimension c, i. e. $Q \in [c]^{[c] \times [c]}$ such that every row and column contains every number of $[c]$ exactly once. Note that for every $c \in \mathbb{N}$ there is a Latin square of dimension c: Let $*$ be any group multiplication on $[c]$. Then $q_{ij} := i * j$ defines a Latin square. As Q is a Latin square we may define a permutation π_i of $[c]$ for every $i \in [c]$ by the following rule: $\pi_i(j)$ is the unique $k \in [c]$ such that $q_{jk} = i$.

Choose optimal colorings $\chi_{\mathscr{G}}$ and $\chi_{\mathscr{H}}$ of $\mathscr{G}$ and $\mathscr{H}$ respectively, i. e., $\mathrm{disc}(\mathscr{G}, \chi_{\mathscr{G}}) = \mathrm{disc}(\mathscr{G}, c)$ and $\mathrm{disc}(\mathscr{H}, \chi_{\mathscr{H}}) = \mathrm{disc}(\mathscr{H}, c)$. Define $\chi : X \times Y \to [c]$ by

$$\chi(x, y) := q_{\chi_{\mathscr{G}}(x) \chi_{\mathscr{H}}(y)}$$

for all $x \in X, y \in Y$.

Let $A \in \mathscr{E}, B \in \mathscr{F}$. Set

$$a_i = \left| \chi_{\mathscr{G}}^{-1}(i) \cap A \right| - \frac{|A|}{c},$$

$$b_i = \left| \chi_{\mathscr{H}}^{-1}(i) \cap B \right| - \frac{|B|}{c}$$

for all $i \in [c]$. Then we have

$$\sum_{i=0}^{c-1} a_i = 0 = \sum_{i=0}^{c-1} b_i. \tag{5.41}$$

This yields

$$|\chi^{-1}(i) \cap (A \times B)| = \sum_{j=0}^{c-1} |\chi_G^{-1}(j) \cap A| \cdot |\chi_H^{-1}(\pi_i(j)) \cap B|$$

$$= \sum_{j=0}^{c-1} \left(a_j + \frac{|B|}{c} \right) \left(b_{\pi_i(j)} + \frac{|A|}{c} \right)$$

$$= \sum_{j=0}^{c-1} a_j b_{\pi_i(j)} + \frac{|A|}{c} \sum_{j=0}^{c-1} a_j + \frac{|B|}{c} \sum_{j=0}^{c-1} b_j + c \frac{|A||B|}{c^2}$$

$$= \sum_{j=0}^{c-1} a_j b_{\pi_i(j)} + \frac{|A \times B|}{c} \qquad \text{by (5.41).}$$

As $|a_i| \leq \text{disc}(\mathcal{G}, c)$ and $|b_i| \leq \text{disc}(\mathcal{H}, c)$, we have

$$\left| |\chi^{-1}(i) \cap (A \times B)| - \frac{|A \times B|}{c} \right| = \sum_{j=0}^{c-1} a_j b_{\pi_i(j)} \leq c \, \text{disc}(\mathcal{G}, c) \, \text{disc}(\mathcal{H}, c).$$

This proves the theorem.   □

For two colors we have $\text{disc}(\mathcal{G}) = 2 \, \text{disc}(\mathcal{G}, 2)$, so by Theorem 78:

Corollary 79. *The 2-color discrepancy is sub-multiplicative, i. e.,*

$$\text{disc}(\mathcal{G} \times \mathcal{H}) \leq \text{disc}(\mathcal{G}) \, \text{disc}(\mathcal{H}).$$

We can now finish the proof of Theorem 65.

Proof (of Theorem 65, upper bound). It follows from Definition 64 that the hypergraph of d-dimensional arithmetic progressions is the d-fold Cartesian product of the hypergraph of one-dimensional arithmetic progressions. Using optimal colorings for any of the factors of the hypergraph of d-dimensional arithmetic progressions arising from the theorem of Matoušek and Spencer [56], Corollary 79 implies Theorem 65.   □

We may ask whether there is any further relation between the discrepancy of a general hypergraph $\mathcal{G} \times \mathcal{H}$ and the product $\text{disc}(\mathcal{G}) \, \text{disc}(\mathcal{H})$. Unfortunately, this is not the case:

Example 80. The hypergraph of two-element subsets of a three-element set $\mathcal{G} = ([3], \binom{[3]}{2})$ has discrepancy two (one color class has at least two elements, i.e., it contains a monochromatic two-set). The Cartesian product $\mathcal{G} \times \mathcal{G}$ can be colored in a way that there is no monochromatic rectangle: $\chi(i, i) := 1$ and $\chi(i, j) := -1$ for $i, j \in [3], i \neq j$. So $\text{disc}(\mathcal{G} \times \mathcal{G}) \leq 2 < 4 = \text{disc}(\mathcal{G})^2$. This is easy to see looking at a suitable drawing of $\mathcal{G} \times \mathcal{G}$: the vertices form a 3×3-grid, the hyperedges consist of the corners of the rectangles having axis-parallel edges. All these rectangles have

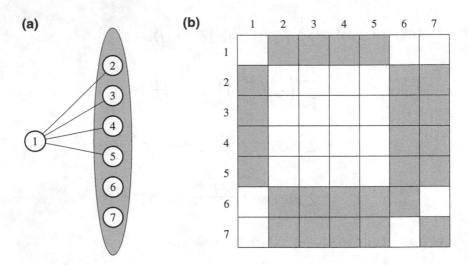

Fig. 5.3 (a) Example hypergraph $\mathscr{G}$ with discrepancy 1; (b) Coloring of the vertices of $\mathscr{H} = \mathscr{G} \times \mathscr{G}$ with discrepancy 0

one or two points on the diagonal of the grid, and thus have discrepancy two or zero with respect to χ.

Another example shows that the discrepancy of a product can become 0, even if the factors have non-zero discrepancy.

Example 81. Let $\mathscr{G}$ be the hypergraph

$$(\{1,\ldots,7\}, \{\{1,2\},\{1,3\},\{1,4\},\{1,5\},\{2,3,4,5,6,7\}\})$$

as depicted in Fig. 5.3a.

One can check that $\mathscr{G}$ does not have discrepancy 0, because otherwise the points $2, 3, 4$ and 5 would be in the same color class, but the edge $E = \{2, 3, 4, 5, 6, 7\}$ is imbalanced. On the other hand, $\mathscr{G} \times \mathscr{G}$ has discrepancy 0 (see the coloring in Fig. 5.3b).

5.3.5 Cartesian Products of Symmetric Progressions

We further follow Doerr, Srivastav and Wehr [29].

A d-dimensional *symmetric* arithmetic progression is defined as follows:

Definition 82. The hypergraph

$$\mathscr{H}^d_{N,\,\mathrm{sym}} := ([N]^d, \{\prod_{i=1}^d A : A \text{ arithmetic progression}\}).$$

is called the hypergraph of d-dimensional symmetric arithmetic progressions in $[N]^d$.

Here a hyperedge is the Cartesian product of the *same* arithmetic progression in $\{1, \ldots, N\}$. At the workshop of the DFG graduate school "Algorithmische Diskrete Mathematik" in Berlin in April 1997, Walter Deuber asked about the discrepancy of this hypergraph. Note that this is a special case of our general problem, because $\mathcal{H}^d_{N,\text{sym}} \subset \mathcal{H}^d_N$. This shows $\text{disc}(\mathcal{H}^d_{N,\text{sym}}) \leq cN^{\frac{d}{4}}$. But a much better bound can be proved.

Theorem 83. *There is a constant $C > 0$ independent of d and N such that*

$$\text{disc}\left(\mathcal{H}^d_{N,\text{sym}}\right) \leq CN^{\frac{1}{4}}.$$

The above theorem is a consequence of a more general result. Let $\mathcal{H} = (X, \mathcal{E})$ be a hypergraph. Set $\mathcal{E}^d_{\text{sym}} := \{E^d \mid E \in \mathcal{E}\}$. We call $\mathcal{H}^d_{\text{sym}} := (X^d, \mathcal{E}^d_{\text{sym}})$ the d-fold symmetric Cartesian product of $\mathcal{H}$.

Theorem 84 (Doerr, Srivastav, Wehr [29]).

$$\text{disc}(\mathcal{H}^d_{\text{sym}}) \leq \text{disc}(\mathcal{H}).$$

Proof. Let $D := \{(x, \ldots, x) : x \in X\}$ be the diagonal of X^d. For $x \in X^d \setminus D$ set

$$a(x) := \min\{i : x_i \neq x_{i+1}\}.$$

Define $f : X^d \to X^d$ by

$$f(x)_i := \begin{cases} x_{a(x)+1} & : & x \notin D, i \leq a(x) \\ x_1 & : & x \notin D, i = a(x) + 1 \\ x_i & : & \text{otherwise} \end{cases}$$

Note that $f(f(x)) = x$ for all $x \in X^d$, so f is a bijection. For all $x \in X^d \setminus D$ the f-orbit $O_f(x)$ of x has order 2 and consists of x and $f(x)$. Further we have

$$\{x_i : i \in [d]\} = \{f(x)_i : i \in [d]\},$$

and thus f leaves the hyperedges of $\mathcal{H}^d_{\text{sym}}$ invariant.

Pick an optimal coloring χ_0 of $\mathcal{H}$. Choose a system R of representatives of the f-orbits in $X^d \setminus D$, i. e., for all $x \in X^d \setminus D$ either x or $f(x)$ lies in R. Define $\chi : X^d \to \{-1, 1\}$ by

$$\chi(x) := \begin{cases} -1 & : & x \in R \\ \chi_0(v) & : & x = (v, \ldots, v) \in D \\ 1 & : & \text{otherwise} \end{cases}$$

Let $E \in \mathcal{E}$. From the properties of f and R we get

$$|E^d \cap R| = |f(E^d \cap R)| = |f(E^d) \cap f(R)| = |E^d \cap (X^d \setminus D \setminus R)| = |E^d \setminus D \setminus R|.$$

Thus

$$\sum_{x \in E^d} \chi(x) = \sum_{x \in E^d \cap R} -1 + \sum_{x \in E^d \cap D} \chi(x) + \sum_{x \in E^d \cap f(R)} 1 = \sum_{v \in E} \chi_0(v),$$

and this concludes the proof. □

Lower bounds are not known. For the general case, not much can be said due to Example 81 in the previous subsection. In the special case of arithmetic progressions, it seems to be difficult to use Fourier analysis, because the convolution and Fourier-transform take place on different groups, namely $\mathbb{Z}^d$ and the diagonal of $\mathbb{Z}^d$. Thus no factoring like $\|\widehat{\chi_F * \eta_\delta}\|_2^2 = \|\hat{\chi}_F \cdot \hat{\eta}_\delta\|_2^2$, i. e., is available, and therefore the coloring function cannot be separated from the characteristic function, which was a crucial step in the proof of the lower bound of Theorem 65.

5.3.6 More on Products: Simplices

So far, we have proved for the 2-color discrepancy of products of hypergraphs and symmetric products

$$\mathrm{disc}(\mathcal{H}^d) \leq \mathrm{disc}(\mathcal{H})^d, \text{ and}$$
$$\mathrm{disc}(\mathcal{H}^d_{\mathrm{sym}}) \leq \mathrm{disc}(\mathcal{H}).$$

There are no such results known for multicolors, and in fact the situation is much more complex. Doerr, Gnewuch and Hebbinghaus [27] showed that $\mathrm{disc}(\mathcal{H}^d_{\mathrm{sym}}, c) = O(\mathrm{disc}(\mathcal{H}, c))$ does not hold in general, but if c divides $k! \, S(d, k)$ for all $k \in \{2, \ldots, d\}$, where $S(d, k)$ is the Stirling number of the second kind, the c-color discrepancy of $\mathcal{H}^d_{\mathrm{sym}}$ can be bounded by the c-color discrepancy of $\mathcal{H}$.

Coloring Simplices. The key idea in [27] stems from the proof of the inequality $\mathrm{disc}(\mathcal{H}^d_{N,\mathrm{sym}}) \leq \mathrm{disc}(\mathcal{H}_N)$ in the case of $d = 2$. The required 2-coloring is constructed as follows. To color the diagonal in $X \times X$, we use an optimal coloring with discrepancy $\mathrm{disc}(\mathcal{H})$. Now color the points above the diagonal blue and the points below the diagonal red. Obviously, it is not clear how to extend this approach to $d \geq 3$ and $c = 3$, the notion "above" resp. "below" the diagonal has no meaning here. The idea of Doerr, Gnewuch and Hebbinghaus is to view the diagonal as a one-dimensional simplex, and the parts above/below the diagonal as two two-dimensional simplices. So the coloring is a suitable coloring of a simplicial decomposition of X^2. For $c = 3$ and $d = 3$, the set X^3 is partitioned into six 3-dimensional simplices obtained from the set $\{x \in X^3 : x_1 < x_2 < x_3\}$ by permuting the coordinates, accordingly the six 2-dimensional simplices permuting the coordinates of the set $\{x \in X^3 : x_1 = x_2 < x_3\}$ and the one-dimensional simplex $\{x \in X^3 : x_1 = x_2 = x_3\}$. With each color one colors exactly two 3-dimensional and two 2-dimensional simplices. The diagonal is colored with an optimal coloring coming from the one-dimensional situation.

We proceed to elaborate on this idea. The detailed discussion will show new and mathematically interesting phenomena regarding the coloring of simplices, in particular, Ramsey's theorem comes into the picture.

A set $\{x_1, \ldots, x_k\}$ of integers with $x_1 < \ldots < x_k$ is denoted by $\{x_1, \ldots, x_k\}_<$. For a set S we put

$$\binom{S}{k} = \{T \subseteq S : |T| = k\}.$$

Furthermore, let S_k be the symmetric group on $[k]$. For $l, d \in \mathbb{N}$ with $l \le d$, $P_l(d)$ denotes the set of all partitions of $[d]$ into l non-empty subsets. Let $e_1 = (1, 0, \ldots, 0), \ldots, e_d = (0, \ldots, 0, 1)$ be the standard basis of $\mathbb{R}^d$. For $c \in \mathbb{N}$ and $\lambda \in \mathbb{N}_0$ we write $c \mid \lambda$ if there exists an $m \in \mathbb{N}_0$ with $mc = \lambda$.

Definition 85. Let $d \in \mathbb{N}$, $l \in [d]$ and $T \subseteq \mathbb{N}$ finite. For $J = \{J_1, \ldots, J_l\} \in P_l(d)$ with $\min J_1 < \ldots < \min J_l$ put $f_i = f_i(J) = \sum_{j \in J_i} e_j$, $i = 1, \ldots, l$. Let $\sigma \in S_l$. We call

$$S_J^\sigma(T) := \left\{ \sum_{i=1}^l \alpha_{\sigma(i)} f_i : \{\alpha_1, \ldots, \alpha_l\}_< \subseteq T \right\}$$

an l-dimensional simplex in T^d. If $l = d$, write $S^\sigma(T)$ instead of $S_J^\sigma(T)$ (as $|P_d(d)| = 1$).

Remark 86. Let $S(d, l)$, $d, l \in \mathbb{N}$ be the Stirling numbers of the second kind, then

$$S(d, l) = \sum_{j=0}^l \frac{(-1)^j (l - j)^d}{j! (l - j)!} \tag{5.42}$$

and $|P_l(d)| = S(d, l)$ (see, e.g. [65]).

Let $T \subseteq \mathbb{N}$ be finite. If $|T| \ge l$, we have $S_I^\sigma(T) \ne S_J^\tau(T)$ as long as $I \ne J$ or $\sigma \ne \tau$. So the number of l-dimensional simplices in T^d is $l! S(d, l)$. If $|T| < l$, then there exists obviously no non-empty l-dimensional simplex in T^d.

The main results in [27] is:

Theorem 87 (Doerr, Gnewuch, Hebbinghaus [27]). *Let $c, d \in \mathbb{N}$.*
If $c \mid k! \, S(d, k)$ for all $k \in \{2, \ldots, d\}$, then every hypergraph $\mathscr{H}$ satisfies

$$\text{disc}(\mathcal{H}^d_{\text{sym}}, c) \leq \text{disc}(\mathcal{H}, c). \tag{5.43}$$

If $c \nmid k! S(d,k)$ for some $k \in \{2,\ldots,d\}$, then there exists a hypergraph $\mathcal{H}$ such that

$$\text{disc}(\mathcal{H}^d_{\text{sym}}, c) \geq \frac{1}{3\,k!} \text{disc}(\mathcal{H}, c)^k, \tag{5.44}$$

and $\mathcal{H}$ can be chosen to have arbitrary large discrepancy $\text{disc}(\mathcal{H}, c)$.

Let us draw some implications of this theorem. Note that (5.43) does not hold for $c = 4$ and for $c = 3$, it holds if d is odd.

Corollary 88. *The following two facts hold.*

 (i) *Let $d \geq 3$ be an odd number. Then for any hypergraph $\mathcal{H}$ it holds that*

$$\text{disc}(\mathcal{H}^d_{\text{sym}}, 3) \leq \text{disc}(\mathcal{H}, 3).$$

(ii) *Let $d \geq 2$ be an even number and $c = 3l$, $l \in \mathbb{N}$. There exists a hypergraph $\mathcal{H}$ with arbitrary large discrepancy that fulfils*

$$\text{disc}(\mathcal{H}^d_{\text{sym}}, c) \geq \frac{1}{6} \text{disc}(\mathcal{H}, c)^2.$$

Proof. Observe that $3 \mid k!$ for all $k \geq 3$. Due to $S(d,2) = 2^{d-1} - 1$, we have $3 \mid S(d,2)$ if and only if d is odd: $2^{3-1} - 1 = 3$, $2^{4-1} - 1 = 7$ and if $d = k + 2$, then $2^{d-1} - 1 = 4(2^{k-1} - 1) + 3$, hence $3 \mid (2^{d-1} - 1)$ iff $3 \mid (2^{k-1} - 1)$. Theorem 87 applies and we are done. □

Corollary 89. *Let $l \in \mathbb{N}$ and $c = 4l$. For all $d \geq 2$ there is a hypergraph $\mathcal{H}$ with $\text{disc}(\mathcal{H}^d_{\text{sym}}, c) \geq \frac{1}{6} \text{disc}(\mathcal{H}, c)^2$.*

Proof. $S(d,2) = 2^{d-1} - 1$ is an odd number, so $4 \nmid 2! S(d,2)$ and Theorem 87 concludes the proof. □

Proof (of Theorem 87 (5.43)). This is the easy part. Let us first consider the case that $c \mid k! S(d,k)$ for all $k \in \{2,\ldots,d\}$. Let $\mathcal{H} = (X, \mathcal{E})$ be a hypergraph and let $\psi : X \to [c]$ be a c-coloring such that $\text{disc}(\mathcal{H}, \psi) = \text{disc}(\mathcal{H}, c)$. For $Y \subseteq X$, put $D(Y) = \{(y,\ldots,y) : y \in Y\}$. We define a c-coloring $\chi : X^d \to [c]$: for $(x,\ldots,x) \in D(X)$, set $\chi(x,\ldots,x) = \psi(x)$. For the remaining vertices, let χ be such that all simplices are monochromatic, and for each k there are exactly $\frac{1}{c} k! S(d,k)$ monochromatic k-dimensional simplices in each color.

Let $E \in \mathcal{E}$ and put $R(E) := E^d \setminus D(E)$. For any $k \in \{2,\ldots,d\}$ and any two k-dimensional simplices S, S' we have $|S \cap R(E)| = |S' \cap R(E)|$. Therefore, our choice of χ implies $|\chi^{-1}(i) \cap R(E)| = \frac{1}{c}|R(E)|$ for all $i \in [c]$. Hence

$$\max_{i \in [c]} \left| |\chi^{-1}(i) \cap E^d| - \frac{|E^d|}{c} \right|$$

$$= \max_{i \in [c]} \left| |\chi^{-1}(i) \cap R(E)| - \frac{|R(E)|}{c} + |\chi^{-1}(i) \cap D(E)| - \frac{|D(E)|}{c} \right|$$

$$= \max_{i \in [c]} \left| |\chi^{-1}(i) \cap D(E)| - \frac{|D(E)|}{c} \right| = \max_{i \in [c]} \left| |\psi^{-1}(i) \cap E| - \frac{|E|}{c} \right|.$$

This shows $\mathrm{disc}(\mathcal{H}^d_{\mathrm{sym}}, c) \leq \mathrm{disc}(\mathcal{H}, c)$. $\square$

For the proof of Theorem 87 (5.44) a Ramsey-type result is required.

Lemma 90. *For all* $m \in \mathbb{N}$, *all* $l \in [d]$, *all* $\sigma \in S_l$, *and all* $J \in P_l(d)$, *there is an* $n \in \mathbb{N}$ *such that for all* $N \subseteq \mathbb{N}$ *with* $|N| = n$ *and each* c-*coloring* $\chi : N^d \to [c]$ *there is a subset* $T \subseteq N$ *with* $|T| = m$ *and* $S^\sigma_J(T)$ *is monochromatic with respect to* χ.

Proof. By Ramsey's theorem (see, e.g. [36], Section 1.2), for every $l \in [d]$ there exists an n such that for each c-coloring $\psi : \binom{[n]}{l} \to [c]$ there is a subset T of $[n]$ with $|T| = m$ and $\binom{T}{l}$ is monochromatic with respect to ψ. Let $N \subseteq \mathbb{N}$ with $|N| = n$. We can assume $N = [n]$ by renaming the elements of N and preserving their order. Let $\chi : [n]^d \to [c]$ be an arbitrary c-coloring. We define $\chi_{l,\sigma,J} : \binom{[n]}{l} \to [c]$ by $\chi_{l,\sigma,J}(\{x_1, \ldots, x_l\}_<) = \chi \left(\sum_{i=1}^{l} x_{\sigma(i)} f_i \right)$, where the $f_i = f_i(J)$ are the vectors corresponding to the partition J introduced in Definition 85. By Ramsey's theorem there is a $T \subseteq N$ with $|T| = m$ and $\chi_{l,\sigma,J}$ is constant on $\binom{T}{l}$. Hence, $S^\sigma_J(T)$ is monochromatic with respect to χ. This proves the claim. $\square$

With induction we get:

Lemma 91. *Let* $c, d \in \mathbb{N}$. *For all* $m \in \mathbb{N}$ *there exists an* $n \in \mathbb{N}$ *having the following property: For each* c-*coloring* $\chi : [n]^d \to [c]$ *there is a subset* $T \subseteq [n]$ *with* $|T| = m$ *such that for all* $l \in [d]$ *each* l-*dimensional simplex in* T^d *is monochromatic with respect to* χ.

Proof. Each simplex is uniquely determined by a pair

$$(\sigma, J) \in \bigcup_{l=1}^{d} (S_l \times P_l(d)).$$

Let $(\sigma_i, J_i)_{i \in [s]}$ be an enumeration of all these pairs. Put $n_0 := m$. We proceed by induction. Let $i \in [s]$ be such that n_{i-1} is already defined and has the property that for any $N \subseteq \mathbb{N}$, $|N| = n_{i-1}$ and any coloring $\chi : N^d \to [c]$ there is a $T \subseteq N$, $|T| = m$ such that for all $j \in [i-1]$, $S^{\sigma_j}_{J_j}(T)$ is monochromatic. According to Lemma 90 choose n_i large enough such that for each $N \subseteq \mathbb{N}$ with $|N| = n_i$ and for each c-coloring $\varphi : N^d \to [c]$ there exists a subset T of N with $|T| = n_{i-1}$ and

$S_{J_i}^{\sigma_i}(T)$ is monochromatic with respect to φ. Note that there is a $T' \subseteq T$, $|T| = m$ such that $S_{J_j}^{\sigma_j}(T')$ is monochromatic for all $j \in [i]$. $n := n_s$ concludes the proof.

□

Proof (of Theorem 87 (5.44)). By assumption $c \nmid k! S(d,k)$ for some $k \in \{2, \ldots, d\}$. Let m be large enough so that

$$\frac{1}{2}\binom{m}{\kappa} - \sum_{l=0}^{\kappa-1} l! \, S(d,l) \binom{m}{l} \geq \frac{1}{3k!} m^k$$

for all $\kappa \in \{k, \ldots, d\}$. This can obviously be done, since the left hand side of the last inequality is of the form $m^\kappa/2\kappa! + O(m^{\kappa-1})$ for $m \to \infty$. Using Lemma 91, we can choose $n \in \mathbb{N}$ such that for any c-coloring $\chi : [n]^d \to [c]$ there is an m-point set $T \subseteq [n]$ with all simplices in T^d being monochromatic with respect to χ.

The special hypergraph $\mathscr{K} := \left([n], \binom{[n]}{m}\right)$ satisfies our claim. Let χ be any c-coloring of $\mathscr{K}$, choose T as in Lemma 91. Let $\kappa \in \{k, \ldots, d\}$ be such that for each $l \in \{\kappa + 1, \ldots, d\}$ there is the same number of l-dimensional simplices in T in each color but not so for the κ-dimensional simplices. With

$$\mathscr{S} := \bigcup_{l=\kappa}^{d} \bigcup_{J \in P_l(d)} \bigcup_{\sigma \in S_l} S_j^\sigma(T)$$

we obtain

$\mathrm{disc}(\mathscr{K}_{\mathrm{sym}}^d, \chi)$

$$\geq \max_{i \in [c]} \left| |\chi^{-1}(i) \cap T^d| - \frac{|T^d|}{c} \right|$$

$$\geq \max_{i \in [c]} \left\{ \left| |\chi^{-1}(i) \cap \mathscr{S}| - \frac{|\mathscr{S}|}{c} \right| - \left| |\chi^{-1}(i) \cap (T^d \setminus \mathscr{S})| - \frac{|T^d \setminus \mathscr{S}|}{c} \right| \right\}$$

$$\geq \max_{i \in [c]} \left| \sum_{J \in P_\kappa(d), \sigma \in S_\kappa} |\chi^{-1}(i) \cap S_J^\sigma(T)| - \frac{\kappa! \, S(d,\kappa)}{c} \binom{m}{\kappa} \right|$$

$$- \frac{c-1}{c} \left(m^d - \sum_{l=\kappa}^{d} l! \, S(d,l) \binom{m}{l} \right)$$

$$\geq \frac{1}{2}\binom{m}{\kappa} - \sum_{l=0}^{\kappa-1} l! \, S(d,l) \binom{m}{l} \geq \frac{1}{3k!} m^k.$$

This proves $\mathrm{disc}(\mathscr{K}_{\mathrm{sym}}^d, c) \geq \frac{1}{3k!} m^k$. The choice of n implies $\mathrm{disc}(\mathscr{K}, c) = \left(1 - \frac{1}{c}\right) m$.

□

5.3.7 Arithmetic Progressions with Common Difference

A special subhypergraph of the hypergraph of Cartesian products of d arithmetic progressions is the hypergraph of arithmetic progressions that have the same common difference.

Definition 92. The hypergraph $\mathcal{H}'_{N,d} := ([N]^d, \mathcal{E}'_{N,d})$ with

$$\mathcal{E}'_{N,d} := \left\{ \prod_{i=1}^{d} A_{a_i,\delta,L_i} : A_{a_i,\delta,L_i} \in \mathcal{E}_N, \delta \in [N] \right\}$$

is called the d-dimensional hypergraph of arithmetic progressions with common differences.

We have $\mathcal{H}^d_{\mathrm{sym}} \subset \mathcal{H}'_{N,d} \subset \mathcal{H}^d_N$. Unfortunately, no lower bound is known for the discrepancy of $\mathcal{H}^d_{\mathrm{sym}}$ which would also be a lower bound for the discrepancy of $\mathcal{H}'_{N,d}$. Interestingly, the c-color discrepancy of $\mathcal{H}'_{N,d}$ can be determined by the following theorem of Hebbinghaus [38]:

Theorem 93 (Hebbinghaus [38]). *Let* $d, N \in \mathbb{N}$. *It holds*

(i) $\mathrm{disc}(\mathcal{H}'_{N,d}, c) = \Omega_d\left(\frac{1}{\sqrt{c}} N^{d/(2d+2)}\right)$.

(ii) $\mathrm{disc}(\mathcal{H}'_{N,d}, c) = O_d\left(N^{d/(2d+2)}(\log N)^{3(d+2)/2}\right)$.

For the proof of the lower bound a special set of hyperedges is considered. Set

$$A_{a,\delta} := \{a + \delta b : b \in \{0, 1, \ldots, L-1\}^d\} \cap [N]^d$$

for $L := \frac{1}{2} N^{1/(d+1)}$, all $\delta \in \Delta := [N^{d/(d+1)}]$ and all $a \in \mathbb{Z}^d$. For convenience let us assume that $\frac{1}{2} N^{1/(d+1)} \in \mathbb{N}$. It is easy to see that there are at most $\left(\frac{3}{2} N\right)^d$ elements $a \in \mathbb{Z}^d$ such that $A_{a,\delta} \neq \emptyset$ for some $\delta \in \Delta$. Thus, we look for the discrepancy of a subhypergraph of $\mathcal{H}'_{N,d}$, which consists of at most $\left(\frac{3}{2}\right)^d N^{d+d/(d+1)}$ hyperedges. A lower discrepancy bound for this subhypergraph is trivially also a lower bound for $\mathrm{disc}(\mathcal{H}'_{N,d}, c)$.

We need two technical lemmas. By the next lemma one can assume for every c-coloring $\chi : [N]^d \to [c]$ of $\mathcal{H}'_{N,d}$ that there is a color $i \in [c]$ such that at least but not much more than $\frac{N^d}{c}$ elements of $[N]^d$ are colored with color i.

Lemma 94. *Let* $\chi : [N]^d \to [c]$ *be a c-coloring of* $\mathcal{H}'_{N,d}$ *and* $\alpha > 0$. *Then it holds* $\mathrm{disc}(\mathcal{H}'_{N,d}, \chi) > \alpha \frac{N^{d/(2d+2)}}{\sqrt{c}}$ *or there exists a color* $i \in [c]$ *such that*

$$0 \leq \delta_A - \frac{1}{c} \leq \alpha c^{-1/2} N^{-d+d/(2d+2)}$$

for $A := \chi^{-1}(i)$ *and* $\delta_A := \frac{1}{N^d}|A|$.

Proof. $\frac{N^d}{c}$ is the average size of a color-class of χ. Therefore, there exists at least one color $i \in [c]$ with $|\chi^{-1}(i)| \geq \frac{N^d}{c}$. If there is a color $i \in [c]$ such that $|\chi^{-1}(i)| - \frac{N^d}{c} > \alpha \frac{N^{d/(2d+2)}}{\sqrt{c}}$, then we get with $[N]^d$ itself as a d-dimensional arithmetic progression that has common difference 1

$$\operatorname{disc}(\mathscr{H}'_{N,d}, \chi) \geq \left| |\chi^{-1}(i) \cap [N]^d| - \frac{|[N]^d|}{c} \right|$$

$$= \left| |\chi^{-1}(i)| - \frac{N^d}{c} \right|$$

$$> \alpha \frac{N^{d/(2d+2)}}{\sqrt{c}}.$$

Thus, we can assume that there is no such color. This yields the existence of a color $i \in [c]$ with $0 \leq |\chi^{-1}(i)| - \frac{N^d}{c} \leq \alpha \frac{N^{d/(2d+2)}}{\sqrt{c}}$. We set $A := \chi^{-1}(i)$ and $\delta_A := \frac{1}{N^d}|A|$ and get

$$0 \leq \delta_A - \frac{1}{c} \leq \alpha c^{-1/2} N^{-d+d/(2d+2)}$$

as desired.   □

In order to estimate Fourier coefficients of the indicator functions of the special hyperedges mentioned above, we need the following lemma which can be proved as (5.38).

Lemma 95. *Let* $\alpha \in [0, 1]^d$ *and* $J := \{0, 1, \ldots, L - 1\}^d$. *There exists* $\delta \in \Delta$ *with*

$$\left| \sum_{j \in J} e^{2\pi i \delta \langle j, \alpha \rangle} \right|^2 \geq \left(\frac{2}{\pi} L \right)^{2d} = \left(\frac{1}{\pi} \right)^{2d} N^{2d/(d+1)}.$$

We are ready to prove Theorem 93 (i).

Proof (of Theorem 93 (i)). Let $\chi : [N]^d \to [c]$ be an arbitrary c-coloring of $\mathscr{H}'_{N,d}$. By Lemma 94 there exists a color $i \in [c]$ such that for $A := \chi^{-1}(i)$ and $\delta_A := \frac{1}{N^d}|A|$

$$0 \leq \delta_A - \frac{1}{c} \leq \alpha c^{-1/2} N^{-d+d/(2d+2)} \tag{5.45}$$

holds for a constant $0 < \alpha \leq \frac{1}{2}$ (that we will fix later). Otherwise, Lemma 94 yields $\operatorname{disc}(\mathscr{H}'_{N,d}, \chi) > \alpha \frac{N^{d/(2d+2)}}{\sqrt{c}}$. Define the function $f_A : \mathbb{Z}^d \to \mathbb{C}$ by

$$f_A(x) := \begin{cases} 1 - \delta_A & : & x \in A, \\ -\delta_A & : & x \in [N]^d \setminus A, \\ 0 & : & x \in \mathbb{Z}^d \setminus [N]^d, \end{cases}$$

for all $x \in \mathbb{Z}^d$. For every subset $X \subseteq [N]^d$ we have

$$f_A(X) := \sum_{x \in X} f_A(x) = \sum_{x \in X \cap A} (1 - \delta_A) + \sum_{x \in X \setminus A} (-\delta_A) = |X \cap A| - \delta_A |X|.$$

Since δ_A is about $\frac{1}{c}$, for every hyperedge $E \in \mathcal{E}'_{N,d}$ the discrepancy of E in the color i is approximately $|f_A(E)|$.

For every $\delta \in \Delta$, we set $E_\delta := A_{0,\delta}$ and get the following chain of equations for the Fourier coefficients of the indicator function $\mathbf{1}_{-E_\delta}$. Let $\alpha \in [0, 1]^d$.

$$\widehat{\mathbf{1}}_{-E_\delta}(\alpha) = \sum_{z \in \mathbb{Z}^d} \mathbf{1}_{-E_\delta}(z) e^{-2\pi i \langle z, \alpha \rangle}$$

$$= \sum_{z \in E_\delta} e^{2\pi i \langle z, \alpha \rangle}$$

$$= \sum_{j \in J} e^{2\pi i \delta \langle j, \alpha \rangle},$$

with $J = \{0, 1, \ldots, L - 1\}^d$. Thus, Lemma 95 yields the existence of a $\delta \in \Delta$ with

$$\left| \widehat{\mathbf{1}}_{-E_\delta}(\alpha) \right|^2 = \left| \sum_{j \in J} e^{2\pi i \delta \langle j, \alpha \rangle} \right|^2 \geq \left(\frac{1}{\pi} \right)^{2d} N^{2d/(d+1)}.$$

Hence, it holds for every $\alpha \in [0, 1]^d$:

$$\sum_{\delta \in \Delta} \left| \widehat{\mathbf{1}}_{-E_\delta}(\alpha) \right|^2 \geq \left(\frac{1}{\pi} \right)^{2d} N^{2d/(d+1)}. \tag{5.46}$$

Using this estimation, with Proposition 76 we get

$$\sum_{\delta \in \Delta} \sum_{j \in \mathbb{Z}^d} \left| f_A(A_{j,\delta}) \right|^2 = \sum_{\delta \in \Delta} \sum_{j \in \mathbb{Z}^d} |f_A(j + E_\delta)|^2$$

$$\geq \left(\frac{1}{\pi} \right)^{2d} N^{2d/(d+1)} \| f_A \|_2^2$$

$$= \left(\frac{1}{\pi}\right)^{2d} N^{2d/(d+1)} \left(\delta_A N^d (1-\delta_A)^2 + (1-\delta_A)N^d(-\delta_A)^2\right)$$

$$= \left(\frac{1}{\pi}\right)^{2d} \delta_A(1-\delta_A)N^{2d/(d+1)+d}.$$

Observe that $|\Delta| = N^{d/(d+1)}$. Hence there exists a $\delta_0 \in \Delta$ such that

$$\sum_{j \in \mathbb{Z}} |f_A(A_{j,\delta_0})|^2 \geq \left(\frac{1}{\pi}\right)^{2d} \delta_A(1-\delta_A)N^{d/(d+1)+d}.$$

As mentioned before, there are at most $\left(\frac{3}{2}N\right)^d$ elements $j \in \mathbb{Z}^d$ such that $f_A(A_{j,\delta_0}) \neq 0$. Therefore we can find a $j_0 \in \mathbb{Z}^d$ satisfying

$$|f_A(A_{j_0,\delta_0})| \geq \frac{1}{\pi^d} \left(\frac{2}{3}\right)^{d/2} \sqrt{\delta_A(1-\delta_A)}N^{d/(2d+2)} \geq \frac{1}{4^d} \sqrt{\delta_A(1-\delta_A)}N^{d/(2d+2)}.$$

Set $x := \delta_A - \frac{1}{c}$. It holds $0 \leq x \leq \alpha c^{-1/2} N^{-d+d/(2d+2)} \leq \frac{1}{2\sqrt{c}}$. Thus, we get

$$|f_A(A_{j_0,\delta_0})| \geq \frac{1}{4^d} \sqrt{\delta_A(1-\delta_A)} \, N^{d/(2d+2)}$$

$$= \frac{1}{4^d} \sqrt{\left(\frac{1}{c}+x\right)\left(\frac{c-1}{c}-x\right)} \, N^{d/(2d+2)}$$

$$= \frac{1}{4^d} \sqrt{\frac{c-1}{c^2} + \frac{c-2}{c}x - x^2} \, N^{d/(2d+2)}$$

$$\geq \frac{1}{4^d} \sqrt{\frac{1}{2c} - \frac{1}{4c}} \, N^{d/(2d+2)}$$

$$= \frac{1}{2^{2d+1}} \frac{N^{d/(2d+2)}}{\sqrt{c}}.$$

We fix the constant α in (5.45). Set $\alpha := \frac{1}{2^{2d+2}}$. Then

$$\mathrm{disc}(\mathcal{H}'_{N,d}, \chi) \geq \left| |A_{j_0,\delta_0}| - \frac{1}{c}|A_{j_0,\delta_0}| \right|$$

$$= \left| |A_{j_0,\delta_0}| - \delta_A |A_{j_0,\delta_0}| + \left(\delta_A - \frac{1}{c}\right)|A_{j_0,\delta_0}| \right|$$

$$\geq \left| |A_{j_0,\delta_0}| - \delta_A |A_{j_0,\delta_0}| \right| - \left|\delta_A - \frac{1}{c}\right| |A_{j_0,\delta_0}|$$

$$\geq \frac{1}{2^{2d+1}} \frac{N^{d/(2d+2)}}{\sqrt{c}} - \frac{1}{2^{2d+2}} c^{-1/2} N^{-d+d/(2d+2)} N^d$$

$$= \frac{1}{2^{2d+2}} \frac{N^{d/(2d+2)}}{\sqrt{c}}.$$

Since χ is an arbitrary c-coloring, we have shown $\text{disc}(\mathcal{H}'_{N,d}, c) \geq \frac{1}{2^{2d+2}} \frac{N^{d/(2d+2)}}{\sqrt{c}}$ and are done.

Proof (of Theorem 93 (ii)). We wish to apply Beck's [10] upper bound for the hereditary discrepancy of the hypergraph $\mathcal{H}'_{N,d}$. We need the subhypergraph of $\mathcal{H}'_{N,d}$ of all *elementary* d-dimensional arithmetic progressions with common difference defined as follows. For every $\delta \in \mathbb{N}$, every $a \in [\delta]^d$, every $s \in \mathbb{N}_0^d$ and every $f \in \mathbb{N}_0^d$ we set

$$AP(a, \delta, s, f) := \underset{i=1}{\overset{d}{\times}} \{a_i + j\delta : f_i 2^{s_i} \leq j \leq (f_i + 1) 2^{s_i}\}.$$

Let $\mathcal{H}_{el} := ([N]^d, \mathcal{E}_{el})$ be the hypergraph, where

$$\mathcal{E}_{el} := \{AP(a, \delta, s, f) \subseteq [N]^d : \delta \in \mathbb{N}, a \in [\delta]^d, s \in \mathbb{N}_0^d, f \in \mathbb{N}_0^d\}.$$

Lemma 96. *There exists a constant $c > 0$ such that*

$$\text{herdisc}(\mathcal{H}'_{N,d}) \leq c \log^d N \, \text{herdisc}(\mathcal{H}_{el}).$$

Proof. Every hyperedge of $\mathcal{H}'_{N,d}$ can be decomposed into at most $c \log^d N$ hyperedges of $\mathcal{H}_{el}$ for an appropriate constant $c > 0$. This decomposition can be found in [45]. Also in every induced subhypergraph of $\mathcal{H}'_{N,d}$ this decomposition can be applied. This proves the assertion of the lemma. $\square$

We continue the proof of Theorem 93 (ii). We apply Beck's theorem, Theorem 11. Accordingly, we have to determine a $t > 0$ such that the maximal degree of the hypergraph $\mathcal{H}_{el,t} := ([N]^d, \mathcal{E}_{el,t})$, where $\mathcal{E}_{el,t} := \{E \in \mathcal{E}_{el} : |E| \geq t\}$, is bounded by t. Let

$$S(a, \delta, t) := \left\{s \in S : 2^{\sum_{i=1}^d s_i} \geq t, a_i + (2^{s_i} - 1)\delta \leq N (i \in [d])\right\}$$

for all $a \in [N]^d$, $\delta \in \mathbb{N}$ and all $t > 0$. Using that for every $m \in [N]^d$ there is only one vector $a \in [\delta]^d$ with $a_i \equiv m_i \pmod{\delta}$ and we get

$$\deg(\mathcal{H}_{el,t}) = \max_{m \in [N]^d} |\{AP(a, \delta, s, f) \in \mathcal{E}_{el,t} : m \in AP(a, \delta, s, f)\}|$$

$$\leq \sum_{\delta=1}^{\lfloor N/\sqrt[d]{t} \rfloor} |S(a, \delta, t)|.$$

Here δ cannot be larger than $\lfloor N/\sqrt[d]{t} \rfloor$, since for every $E = \overset{d}{\underset{i=1}{\mathsf{X}}} A_i \in \mathcal{E}_{el}$ with $|E| \geq t$ there is at least one $i \in [d]$ with $|A_i| \geq \sqrt[d]{t}$. We have $|S(a, \delta, t)| \leq \log^d N$ for all $\delta \geq 2$. Thus, there is a constant $c_1 > 0$ such that

$$\deg(\mathcal{H}_{el,t}) \leq c_1 \frac{N}{\sqrt[d]{t}} (\log N)^d.$$

We set $t := c_1^{d/(d+1)} N^{d/(d+1)} (\log N)^{d^2/(d+1)}$ and get

$$\deg(\mathcal{H}_{el,t}) \leq t.$$

This estimation holds obviously also for all induced subhypergraphs of $\mathcal{H}_{el}$. Thus, Theorem 11 implies

$$\text{herdisc}(\mathcal{H}_{el}) \leq c_2 N^{d/(2d+2)} (\log N)^{d^2/(2d+2)} (\log N)^2$$

$$= c_2 N^{d/(2d+2)} (\log N)^{(d^2+4d+4)/(2d+2)}$$

for a constant $c_2 > 0$ only depending on the dimension d. With Lemma 96 we have

$$\text{herdisc}(\mathcal{H}'_{N,d}) \leq c_3 N^{d/(2d+2)} (\log N)^{(3d^2+6d+4)/(2d+2)}$$

$$\leq c_3 N^{d/(2d+2)} (\log N)^{3(d+2)/2}$$

for a constant $c_3 > 0$ only depending on d. Finally, we apply Theorem 31 (ii) and get

$$\text{disc}(\mathcal{H}'_{N,d}, c) \leq c_0 N^{d/(2d+2)} (\log N)^{3(d+2)/2}$$

for a constant $c_0 > 0$ depending only on the dimension d. $\square$

5.3.8 Arithmetic Progressions in $\mathbb{N}$

Most natural are also arithmetic progressions in the set $\mathbb{N}$. A direct consequence of Roth's lower bound proof for such progressions can be found in [13].

Corollary 97. *Given any 2-coloring* $\chi : \mathbb{N} \to \{-1, +1\}$ *of the non-negative integers, then for infinitely many values of δ there is a (finite) arithmetic progression A of difference δ such that*

$$|\chi(A)| > c\sqrt{\delta}$$

for an absolute constant c.

The best upper bound for the discrepancy in terms of the difference was shown by J. Beck and J. Spencer [15]:

Theorem 98. *There is a constant $c_0 > 0$ such that the following holds: Let n be a positive integer. Then there exists a 2-coloring $\chi : \mathbb{N} \to \{-1, +1\}$ such that for any arithmetic progression A of difference $\delta \leq n$ and of arbitrary length*

$$|\chi(A)| < c_0\sqrt{\delta}(\log n)^{3.5}.$$

An analogous result in the d-dimensional case has been proved by Doerr, Srivastav and Wehr [29]. Let $A_{l,a,\delta} := a + \delta[l]$ denote the arithmetic progression with starting point a, difference δ and length l. For $a, \delta \in \mathbb{N}^d$ and $l \in [N] \setminus 0$, set $A_{l,a,\delta} = A_{l_i,a_i,\delta_i}$. Write $\delta > k$ (resp. $\delta \leq k$) to express that all components δ_i of δ are greater than k (resp. less or equal than k).

Theorem 99. *For any 2-coloring $\chi : \mathbb{N}^d \to \{-1, +1\}$ and every vector $k \in \mathbb{N}^d$ there exists a d-dimensional arithmetic progression $A_{l,a,\delta}$ such that $\delta > k$ and*

$$|\chi(A_{l,a,\delta})| > \pi^{-d}\sqrt{\delta_1 \ldots \delta_d}.$$

Conversely, for any positive integer n there exists a 2-coloring $\chi : \mathbb{N}^d \to \{-1, +1\}$ such that for any arithmetic progression $A_{l,a,\delta}$ of difference $\delta \leq n$ and of arbitrary length and starting point

$$|\chi(A_{l,a,\delta})| < c_0^d\sqrt{\delta_1 \ldots \delta_d}(\log n)^{3.5d}.$$

Proof. The upper bound can be easily proved by the product coloring argument (Theorem 78). For the lower bound fix a 2-coloring χ of $\mathbb{N}^d$ and k. Define a 2-coloring χ_k of $\mathbb{N}^d$ by

$$\chi_k(x) := \chi(kx)$$

for all $x \in \mathbb{N}^d$ (where $kx := (k_i x_i)_{i=1}^d$). Choose an integer $N > \|k\|_\infty^2$. From the proof of the lower bound in Theorem 65 one can conclude the existence of vectors $l, \delta' \in \Delta = \{1, \ldots, \sqrt{N}\}^d$ and $a' \in [N]^d$ such that

$$|\chi_k(A_{l,a',\delta'})| > \pi^{-d} N^{\frac{d}{4}} \geq c \sqrt{\delta'_1 \ldots \delta'_d}.$$

Since $\chi_k(A_{l,a',\delta'}) = \chi(A_{l,ka',k\delta'})$, $A_{l,ka',k\delta'}$ is an arithmetic progression as desired. $\square$

5.4   Sums of Arithmetic Progressions

Another interesting notion of high-dimensional arithmetic progressions are sums of d arithmetic progressions in $\{1, \ldots, N\}$. The famous theorem of G. Freiman in additive number theory states that a sumset with small cardinality contains a d-dimensional arithmetic progression [32, 59]. In 2002, A. Srivastav suggested the study of the discrepancy of the hypergraph of all sets formed by k arithmetic progressions in $\{1, \ldots, N\}$. Subsequently, a theoretical foundation has been achieved in a number of papers by Hebbinghaus [38], Cilleruelo, Hebbinghaus [20] and Přívětivý [62, 63]. We present these developments in this section.

Let us define the sum of arithmetic progressions formally.

Definition 100. Let $\mathcal{A}$ be the set of all arithmetic progressions in $\mathbb{Z}$. Define the hypergraph $\mathcal{H}_{N,k} = ([N], \mathcal{E}_{N,k})$ of sums of k arithmetic progressions, where

$$\mathcal{E}_{N,k} := \{(A_1 + A_2 + \ldots A_k) \cap [N] : A_i \in \mathcal{A}\}.$$

Some examples for sums of 2 arithmetic progressions are shown in Fig. 5.4.

5.4.1   Multicolor Discrepancy of $\mathcal{H}_{N,k}$

Hebbinghaus [38] proved:

Theorem 101 (Hebbinghaus [38]). *For all positive integers k we have*

$$\mathrm{disc}(\mathcal{H}_{N,k}, c) = \Omega_k \left(\frac{N^{k/(2k+2)}}{\sqrt{c}} \right)$$

Let us first describe the proof idea of Theorem 101. We assume that $2^{k-1} | N^{1/(k+1)}$. Bertrand's postulate (also called Chebyshev's theorem) states the existence of prime numbers p_i for all $i \in \{1, 2, \ldots k - 1\}$ with $2^{i-k+1} N^{1/(k+1)} < p_i < 2^{i-k+2} N^{1/(k+1)}$. Every sum of k arithmetic progressions is characterized by a starting point, a k-tuple $\delta = (\delta_1, \delta_2, \ldots, \delta_k)$ of differences and a k-tuple $L = (L_1, L_2, \ldots, L_k)$ which fixes the length of the k arithmetic progressions. Let us introduce here the special set of hyperedges. These hyperedges have the same k-tuple $L = (L_1, L_2, \ldots, L_k)$ fixing the length of the k arithmetic progressions that are summed up. Define the length of the i-th arithmetic progression ($i \in \{1, 2, \ldots k\}$) by

Fig. 5.4 Some arithmetic progressions (*black dots*) and their sums (*gray dots*)

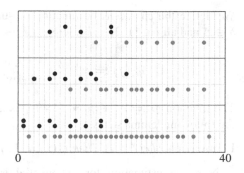

$$L_i := 2^{i-k-1} N^{\frac{1}{k+1}}.$$

Let $\tilde{\Delta} := \prod_{i=1}^{k} \{1, 2, \ldots, 2L_i\}$. We define a set Δ of k-tuples of differences by

$$\Delta := \left\{ (\delta_1, \delta_2, \ldots, \delta_k) : \delta_i = \prod_{j=1}^{i} \tilde{\delta}_j \prod_{j=i}^{k-1} p_j, 1 \leq i \leq k, (\tilde{\delta}_1, \tilde{\delta}_2, \ldots, \tilde{\delta}_k) \in \tilde{\Delta} \right\}.$$

For all $j \in \mathbb{Z}$ and all $\delta = (\delta_1, \delta_2, \ldots, \delta_k) \in \Delta$ we set

$$A_{j,\delta} := \left\{ j + \sum_{i=1}^{k} a_i \delta_i : a_i \in \{0, 1, 2, \ldots, L_i - 1\}, 1 \leq i \leq k \right\} \cap [N].$$

The non-trivial hyperedges $A_{j,\delta}$ with $j \in \mathbb{Z}$ and $\delta \in \Delta$ are building the subhypergraph mentioned above. We set $E_\delta := A_{0,\delta}$ for all $\delta \in \Delta$.

We need a series of lemmas (for proofs see [38]). The first lemma locates j in $A_{j,\delta} \neq \emptyset$.

Lemma 102. *Let $j \in \mathbb{Z}$ and $\delta \in \Delta$ with $A_{j,\delta} \neq \emptyset$. Then $j \in \{-\frac{N}{2} + 1, -\frac{N}{2} + 2, \ldots, N\}$.*

A typical argument in proofs of c-color discrepancy are alternatives of the type that either the discrepancy is of the required order or there is a large color-class. Such arguments will be frequently used in multicolor proofs via Fourier analysis.

Lemma 103. *Let $\chi : X \to [c]$ be a c-coloring of $\mathscr{H}_{N,k}$ and $\alpha > 0$, then it holds* $\mathrm{disc}(\mathscr{H}_{N,k}, c, \chi) > \alpha \frac{N^{k/(2k+2)}}{\sqrt{c}}$ *or there is a color $i \in [c]$ such that it holds for* $A := \chi^{-1}(i)$ *and* $\delta_A := \frac{1}{N}|A|$

$$0 \leq \delta_A - \frac{1}{c} \leq \alpha c^{-1/2} N^{-(k+2)/(2k+2)}.$$

We define the characteristic function of sums of arithmetic progressions. For all $\delta \in \Delta$ and all $i \in \{1, 2, \ldots k\}$, let $\eta_{\delta_i} : \mathbb{Z} \to \{0, 1\}$ be defined by

$$\eta_{\delta_i}(j) := \begin{cases} 1 & : \quad -j \in \delta_i\{0, 1, 2, \ldots, L_i - 1\}, \\ 0 & : \quad \text{otherwise.} \end{cases}$$

The function $\eta_\delta := \eta_{\delta_1} * \eta_{\delta_2} * \ldots * \eta_{\delta_k}$ is an indicator function for the set $-E_\delta$ as the following lemma states.

Lemma 104. *Let $\delta \in \Delta$. Then $\eta_\delta(x) = \mathbf{1}_{-E_\delta}(x)$ for all $x \in \mathbb{Z}$.*

Next, an estimation for exponential sums is required. It is an immediate consequence of Roth's estimate (5.38).

Lemma 105. *Let $\alpha \in \mathbb{R}$, $L \in \mathbb{N}$. There exists an integer $\delta \in \{1, 2, \ldots, 2L\}$ such that*

$$\left| \sum_{j=0}^{L-1} e^{2\pi i \delta j \alpha} \right|^2 \geq \left(\frac{2}{\pi} L \right)^2.$$

The following lemma is the key for the proof of Theorem 101.

Lemma 106. *Let $\alpha \in [0, 1]$. There exists a constant $c_1 > 0$, only depending on k, such that*

$$\sum_{\delta \in \Delta} |\hat{\mathbf{1}}_{-E_\delta}(\alpha)|^2 \geq c_1 N^{2k/(k+1)}.$$

Proof. Observe that for all $\delta \in \Delta$ and all $t \in \{1, 2, \ldots, k\}$

$$\hat{\eta}_{\delta_t}(\alpha) = \sum_{x \in \mathbb{Z}} \eta_{\delta_t}(x) e^{-2\pi i x \alpha}$$

$$= \sum_{j=0}^{L_t - 1} e^{2\pi i j \delta_t \alpha}.$$

We have

$$\sum_{\delta \in \Delta} |\hat{\mathbf{1}}_{-E_\delta}(\alpha)|^2 = \sum_{\delta \in \Delta} |\hat{\eta}_\delta(\alpha)|^2 \tag{5.47}$$

$$= \sum_{\delta \in \Delta} \prod_{t=1}^{k} |\hat{\eta}_{\delta_t}(\alpha)|^2 \qquad \text{by Theorem 73}$$

$$= \sum_{\delta_1 = 1}^{2L_1} \cdots \sum_{\delta_k = 1}^{2L_k} \prod_{t=1}^{k} \left| \sum_{j=0}^{L_t - 1} e^{2\pi i j \alpha \prod_{s=1}^{t} \widetilde{\delta}_s \prod_{s=t}^{k-1} p_s} \right|^2 \tag{5.48}$$

Using Lemma 105 we can find a $\bar{\delta}_1 \in \{1, 2, \ldots, 2L_1\}$ with

$$\left| \sum_{j=0}^{L_1-1} e^{2\pi i j \bar{\delta}_1 \left(\alpha \prod_{s=1}^{k-1} p_s \right)} \right|^2 \geq \left(\frac{2}{\pi} L_1 \right)^2.$$

In the same way we get $\bar{\delta}_2 \in \{1, 2, \ldots, 2L_2\}$, $\bar{\delta}_3 \in \{1, 2, \ldots, 2L_3\}$ up to $\bar{\delta}_k \in \{1, 2, \ldots, 2L_k\}$ such that for all $t \in \{1, 2, \ldots, k\}$

$$\left| \sum_{j=0}^{L_t-1} e^{2\pi i j \bar{\delta}_t \left(\alpha \prod_{s=1}^{t-1} \bar{\delta}_s \prod_{s=t}^{k-1} p_s \right)} \right|^2 \geq \left(\frac{2}{\pi} L_t \right)^2. \tag{5.49}$$

Using (5.47) and (5.49) we get for an appropriate constant $c_1 > 0$ only depending on k:

$$\sum_{\delta \in \Delta} |\hat{\mathbf{1}}_{-E_\delta}(\alpha)|^2 \overset{(5.47)}{\geq} \prod_{t=1}^{k} \left| \sum_{j=0}^{L_t-1} e^{2\pi i j \alpha \prod_{s=1}^{t} \bar{\delta}_s \prod_{s=t}^{k-1} p_s} \right|^2$$

$$\overset{(5.49)}{\geq} \prod_{t=1}^{k} \left(\frac{2}{\pi} L_t \right)^2$$

$$\geq c_1 N^{2k/(k+1)}$$

$\square$

Proof (of Theorem 101). Let $\chi : X \to [c]$ be a c-coloring of $\mathcal{H}_{N,k}$. According to Lemma 103 we can assume that there is a color $i \in [c]$ such that for $A := \chi^{-1}(i)$ and $\delta_A := \frac{1}{N} |A|$

$$0 \leq \delta_A - \frac{1}{c} \leq \alpha c^{-1/2} N^{-(k+2)/(2k+2)} \tag{5.50}$$

for a constant $0 < \alpha \leq \frac{1}{2}$ (that we fix later). Otherwise Lemma 103 yields

$$\operatorname{disc}(\mathcal{H}_{N,k}, c, \chi) > \alpha \frac{N^{k/(2k+2)}}{\sqrt{c}}.$$

Let $f_A : \mathbb{Z} \to \mathbb{C}$ be defined by

$$f_A(x) := \begin{cases} 1 - \delta_A & : \quad x \in A, \\ -\delta_A & : \quad x \in [N] \setminus A, \\ 0 & : \quad x \in \mathbb{Z} \setminus [N], \end{cases}$$

for every $x \in \mathbb{Z}$. For every subset $X \subseteq [N]$ we have

$$f_A(X) := \sum_{x \in X} f_A(x) = \sum_{x \in X \cap A} (1 - \delta_A) + \sum_{x \in X \setminus A} (-\delta_A) = |X \cap A| - \delta_A |X|.$$

The estimation $\sum_{\delta \in \Delta} |\hat{\mathbf{1}}_{-E_\delta}(\alpha)|^2 \geq c_1 N^{2k/(k+1)}$ from Lemma 106 allows us to apply Proposition 76. We get

$$\sum_{\delta \in \Delta} \sum_{j \in \mathbb{Z}} |f_A(A_{j,\delta})|^2 = \sum_{\delta \in \Delta} \sum_{j \in \mathbb{Z}} |f_A(j + E_\delta)|^2$$

$$\geq c_1 N^{2k/(k+1)} \|f_A\|_2^2$$

$$= c_1 N^{2k/(k+1)} \left(\delta_A N (1 - \delta_A)^2 + (1 - \delta_A) N (-\delta_A)^2 \right)$$

$$= c_1 \delta_A (1 - \delta_A) N^{(3k+1)/(k+1)}$$

It holds $|\Delta| = |\tilde{\Delta}| = \prod_{i=1}^{k} (2L_i) = O(N^{k/((k+1))})$. Hence there exists a constant $c_2 > 0$ and a $\delta_0 \in \Delta$ such that

$$\sum_{j \in \mathbb{Z}} |f_A(A_{j,\delta_0})|^2 \geq \frac{c_1}{|\Delta|} \delta_A (1 - \delta_A) N^{(3k+1)/(k+1)} \geq c_2 \delta_A (1 - \delta_A) N^{(2k+1)/(k+1)}.$$

According to Lemma 102, $A_{j,\delta_0} = \emptyset$ for all $j \in \mathbb{Z} \setminus \{-\frac{N}{2} + 1, -\frac{N}{2} + 2, \ldots, N - 1, N\}$. Therefore we can find a $j_0 \in \{-\frac{N}{2} + 1, -\frac{N}{2} + 2, \ldots, N - 1, N\}$ such that

$$|f_A(A_{j_0,\delta_0})| \geq \sqrt{\frac{c_2}{2}} \sqrt{\delta_A (1 - \delta_A)} N^{k/(2k+2)}.$$

Set $x := \delta_A - \frac{1}{c}$. It holds $0 \leq x \leq \alpha c^{-1/2} N^{-(k+2)/(2k+2)} \leq \frac{1}{2\sqrt{c}}$. For $c_3 := \sqrt{\frac{c_2}{8}}$ we get

$$|f_A(A_{j_0,\delta_0})| \geq \sqrt{\frac{c_2}{2}} \sqrt{\delta_A (1 - \delta_A)} N^{k/(2k+2)}$$

$$= \sqrt{\frac{c_2}{2}} \sqrt{\left(\frac{1}{c} + x \right) \left(\frac{c-1}{c} - x \right)} N^{k/(2k+2)}$$

$$= \sqrt{\frac{c_2}{2}} \sqrt{\frac{c-1}{c^2} + \frac{c-2}{c} x - x^2} N^{k/(2k+2)}$$

$$\geq \sqrt{\frac{c_2}{2}} \sqrt{\frac{1}{2c} - \frac{1}{4c}} N^{k/(2k+2)}$$

$$= c_3 \frac{N^{k/(2k+2)}}{\sqrt{c}}.$$

Now we fix the constant α in (5.50). W.l.o.g. we can assume $c_3 \leq 1$ and set $\alpha := \frac{c_3}{2}$. Then

$$
\begin{aligned}
\operatorname{disc}(\mathscr{H}_{N,k}, c, \chi) &\geq \left| |A_{j_0,\delta_0}| - \frac{1}{c}|A_{j_0,\delta_0}| \right| \\
&= \left| |A_{j_0,\delta_0}| - \delta_A|A_{j_0,\delta_0}| + \left(\delta_A - \frac{1}{c}\right)|A_{j_0,\delta_0}| \right| \\
&\geq \left| |A_{j_0,\delta_0}| - \delta_A|A_{j_0,\delta_0}| \right| - \left|\delta_A - \frac{1}{c}\right| |A_{j_0,\delta_0}| \\
&\geq c_3 \frac{N^{k/(2k+2)}}{\sqrt{c}} - \frac{c_3}{2}c^{-1/2}N^{-(k+2)/(2k+2)}N \\
&= \frac{c_3}{2}\frac{N^{k/(2k+2)}}{\sqrt{c}}.
\end{aligned}
$$

Thus, we have shown $\operatorname{disc}(\mathscr{H}_{N,k}, c) \geq \alpha\frac{N^{k/(2k+2)}}{\sqrt{c}}$, where the constant $\alpha > 0$ depends only on k.   □

5.4.2 Přívětivý's 2-Color Improvement for $k \geq 3$

For $k = 2$ the lower bound of Hebbinghaus is $\Omega(N^{1/3})$. For $k \geq 3$, Aleš Přívětivý [62, 63] improved the bound to $\Omega(N^{1/2})$:

Theorem 107 (Přívětivý [62]). *Let N be a prime. For all positive integers $k \geq 3$ we have $\operatorname{disc}(\mathscr{H}_{N,k}) = \Omega\left(N^{1/2}\right)$.*

It is sufficient to show $\operatorname{disc}(\mathscr{H}_{N,3}) = \Omega\left(N^{1/2}\right)$ because $\mathscr{H}_{N,3} \subseteq \mathscr{H}_{N,k}$, $k \geq 3$. The basis of the proof is the eigenvalue method [13]. We briefly review the main ingredients needed here. Let us consider the hypergraph $\mathscr{H} = ([N], \mathscr{E})$ with N vertices and m hyperedges. Let A be the $m \times N$ incidence matrix of $\mathscr{H}$. Recall the notion of the L^2-discrepancy

$$
\operatorname{disc}_2(\mathscr{H})^2 := \min\left\{ \frac{\| Ax \|_2^2}{m} : x \in \{-1, 1\}^N \right\}
$$

Since the matrix $A^T A$ is positive semidefinite, its eigenvalues $\lambda_1 \geq \lambda_2 \geq \cdots \geq \lambda_N$ are non-negative real numbers. Suppose that $x = x_1v_1 + x_2v_2 + \cdots + x_Nv_N$ where $\{v_i\}$ is an orthonormal eigenbasis, the eigenvector v_i being associated to the eigenvalue λ_i. Then

$$\| Ax \|_2^2 = x^T A^T A x = \Big(\sum_i x_i v_i \Big) \Big(\sum_i \lambda_i x_i v_i \Big) = \sum_i^N \lambda_i x_i^2 \geq \lambda_N N$$

Hence

$$\operatorname{disc}(\mathcal{H}) \geq \operatorname{disc}_2(\mathcal{H}) \geq \sqrt{\frac{\lambda_N \cdot N}{m}}. \tag{5.51}$$

We need some facts about *circulant matrices*. A *circulant* matrix is an $N \times N$ matrix whose rows are composed of cyclically shifted copies of the first row. Namely, for an N-dimensional vector $(a_0, a_1, \ldots, a_{N-1})$ we define the circulant matrix B by $b_{ij} = a_{(j-i) \bmod N}$.

Let the N-th roots of unity be $1 = \zeta_0, \zeta_1, \zeta_2, \ldots, \zeta_{N-1}$. It can be shown that $z_i = (1 = \zeta_i^0, \zeta_i, \ldots, \zeta_i^{N-1})^T$ is an eigenvector of B. The eigenvalues $\lambda_0, \lambda_1, \ldots, \lambda_{N-1}$ of a circulant matrices can be expressed as

$$\lambda_i = a_0 + a_1 \zeta_i + a_2 \zeta_i^2 + \cdots + a_{N-1} \zeta_i^{N-1}$$

for all $i \in 0, 1, \ldots, N - 1$.

Let $B = A^T A$ (where A is a incidence matrix of a hypergraph). Assume in addition that $B = A^T A$ is a circulant matrix. Then b_{ij} counts the number of sets $E \in \mathscr{E}$ containing both elements i and j. Now, the following congruences modulo N hold

$$N \lambda_k \equiv N \sum_{i=0}^{N-1} a_i \zeta_k^i \equiv \sum_{i=0}^{N-1} \sum_{j=0}^{N-1} a_{(i-j)} \zeta_k^{(i-j)} \equiv \sum_{i=0}^{N-1} \sum_{j=0}^{N-1} b_{ij} \zeta_k^{(i-j)}$$

$$\equiv \sum_{E \in \mathscr{E}} \sum_{i \in E} \sum_{j \in E} \zeta_k^{(i-j)} \equiv \sum_{E \in \mathscr{E}} \Big| \sum_{i \in E} \zeta_k^i \Big|^2,$$

hence

$$\lambda_k = \frac{1}{N} \sum_{E \in \mathscr{E}} \Big| \sum_{i \in E} \zeta_k^i \Big|^2. \tag{5.52}$$

Let us define a wrapped hypergraph. To shorten the notation set $\mathcal{H} = \mathcal{H}_{N,3} = ([N], \mathscr{E})$ where $\mathscr{E} = \{\mathscr{E}_1, \ldots, \mathscr{E}_m\}$ is a numbering of the hyperedges. Consider for each $\mathscr{E}_i$, its $N - 1$ translated hyperedges: for every $i \in [m]$, $j \in [N]$ and $\mathscr{E}_i \in \mathscr{E}$, the set $\mathscr{E}_{iN+j}$ is defined by

$$\mathscr{E}_{iN+j} = \{(k + j) \bmod N : k \in \mathscr{E}_i\}.$$

The new hypergraph $\mathcal{H}_W = ([N], \mathcal{E}_W)$, where $\mathcal{E}_W = \{\mathcal{E}_0, \mathcal{E}_1, \ldots, \mathcal{E}_{mN-1}\}$, is called the *wrapped hypergraph* of $\mathcal{H}$. The incidence matrix A of $\mathcal{H}_W$ can be expressed as composed of $N \times N$ circulant matrices $A_0, A_1, A_2, \ldots, A_{m-1}$ stacked up vertically

$$A = \begin{pmatrix} A_0 \\ A_1 \\ \vdots \\ A_{m-1} \end{pmatrix}$$

One can show that A is not a circulant matrix, but $B = A^T A$ is a circulant matrix, because B has the form $B = \sum_{i=0}^{m-1} A_i^T A_i$ and the inner product of the two column vectors A_i^k and A_i^l is equal to the inner product of the column vectors A_i^{k+1} and A_i^{l+1}. The next lemma can be proved in a straightforward way, see also [13] or [19].

Lemma 108. *Let* $\mathcal{H}_W = ([N], \mathcal{E}_W)$ *be the wrapped hypergraph, and A be its incidence matrix. Then the $N \times N$ matrix $B = A^T A$ is a circulant matrix and its eigenvalues are* $\lambda_k = \sum_{i=0}^{m-1} \left| \sum_{j \in \mathcal{E}_{iN}} \zeta_k^j \right|^2$ *for all $k \in [N]$.*

Using Lemma 108 and (5.51), it can be established that if for each k there exists a set $\mathcal{E}_q$ such that $\left| \sum_{j \in \mathcal{E}_q} \zeta_k^j \right| > cN^\alpha$ for some $c, \alpha \geq 0$, then we have $\mathrm{disc}(\mathcal{H}) \geq \frac{cN^\alpha}{\sqrt{m}}$.

So we need to find a hyperedge with large value of $\left| \sum_{j \in \mathcal{E}_q} \zeta_k^j \right|$. Such a $\mathcal{E}_q$ has the property that ζ_k^j lies in one part of the unit circle for all $j \in \mathcal{E}_q$. Namely, if all $j \in \mathcal{E}_q$ satisfy

$$-\frac{\pi}{3} \leq \arg \zeta_k^j \leq \frac{\pi}{3} \tag{5.53}$$

then $\mathrm{Re}(\zeta_k^j) \geq \frac{1}{2}$ for all $j \in \mathcal{E}_q$ and $\left| \sum_{j \in \mathcal{E}_q} \zeta_k^j \right|$ will be at least $\frac{|\mathcal{E}_q|}{2}$.

We define for every $\mathcal{E}_q \subseteq [N]$ a new set $\mathcal{E}_q'$:

$$\mathcal{E}_q' := \{jk \mod N : j \in \mathcal{E}_q\}. \tag{5.54}$$

$\mathcal{E}_q'$ is actually the set of indices i of $\zeta_i = \zeta_k^j$ that participate in the sum $\sum_{j \in \mathcal{E}_q} \zeta_k^j$. The condition $\left| \arg \zeta_k^j \right| \leq \frac{\pi}{3}$ for all $j \in \mathcal{E}_q$ is thus equivalent to

$$\mathcal{E}_q' \subseteq \left\{0, 1, \ldots, \left\lfloor \frac{1}{6}N \right\rfloor\right\} \cup \left\{\left\lceil \frac{5}{6}N \right\rceil, \ldots, N-1\right\}$$

Now let us pass to sums of arithmetic progressions. Přívětivý constructed a specific set which is the sum of three arithmetic progressions.

Construction of C_k: For a fixed k, let $c_1, c_2, c_3, d_1, d_2, d_3, n_1, n_2, n_3$ be the integers characterized as follows:

1. Let c_1 be the $j \in \{1, \ldots, \lfloor \sqrt{N} \rfloor\}$ for which the value of $\mathrm{Re}(\zeta_k^j)$ is maximum. Put $d_1 = \min\{-kc_1 \bmod N, kc_1 \bmod N\}$ and $n_1 = \left\lfloor \frac{N}{12 \max\{c_1, d_1\}} \right\rfloor$.

2. If $c_1 \leq 12 d_1$, put $c_2 = d_2 = n_2 = 1$, otherwise set $c_2 = N \bmod c_1$, $d_2 = d_1 \lceil \frac{N}{c_1} \rceil$ and $n_2 = \lfloor \frac{N}{30 d_1} \rfloor$.

3. If $d_1 < 6$, then put $c_3 = d_3 = n_3 = 1$, otherwise let c_3 be the $j \in \{1, \ldots, \lfloor \frac{2N}{d_1} \rfloor\}$ for which the value of $\mathrm{Re}(\zeta_k^j)$ is maximum. Set $d_3 = \min\{-jc_3 \bmod N, jc_3 \bmod N\}$ and $n_3 = \left\lfloor \frac{d_1}{12} \right\rfloor$.

4. Finally, set $C_k := \{i_1 c_1 + i_2 c_2 + i_3 c_3 : i_k \in [n_k], k = 1, 2, 3\}$.

Lemma 109. *Let N be a prime. For each $k \in \{1, 2, \ldots, N-1\}$ we have*

(i) *C_k is a sum of three arithmetic progressions in $[N]$.*
(ii) *$\mathrm{Re}(\zeta_k^j) \geq 1/2$ for every $j \in C_k$.*
(iii) *$|C_k| \geq \frac{1}{5000} N$.*

Proof. We first give some relations among these constants. The constant c_1 is the number $j \in \{1, \ldots, \lfloor \sqrt{N} \rfloor\}$ for which the value of $\mathrm{Re}(\zeta_k^j)$ is maximum. That means we need to find a value of j for which $|\arg(\zeta_k^j)|$ is minimum. By the pigeonhole principle, for at least two distinct $1 \leq j_1 < j_2 \leq \lceil \sqrt{N} \rceil$, the angles $\arg(\zeta_k^{j_1})$ and $\arg(\zeta_k^{j_2})$ fall in the interval of length of $\frac{2\pi}{\lceil \sqrt{N} \rceil}$. Thus we can find c_1 such that

$$-\frac{2\pi}{\lceil \sqrt{N} \rceil} \leq \arg(\zeta_k^{c_1}) \leq \frac{2\pi}{\lceil \sqrt{N} \rceil}$$

By definition of d_1, $|\arg(\zeta_k^{c_1})| = \arg(\zeta_{d_1})$. Hence $d_1 \leq \sqrt{N}$. With the same argument we get for c_3 and d_3, $c_3 \leq \frac{2N}{d_1}$ (by construction) and $d_3 \leq \frac{d_1}{2}$ (by the pigeonhole principle).

Proof of (i). By construction, the set C_k is a sum of three arithmetic progressions. The largest element of C_j is

$$\max C_k = c_1(n_1 - 1) + c_2(n_2 - 1) + c_3(n_3 - 1) \leq \frac{N}{12} + \frac{N}{30} + \frac{N}{6} \leq \frac{N}{2}$$

Proof of (ii). As discussed above, this is the condition we have ensured while picking values for $c_1, c_2, c_3, d_1, d_2, d_3$. Now

$$\max_{j \in C_k} \left| \arg(\zeta_k^j) \right| = \max_{(i_1, i_2, i_3) \in [n_1] \times [n_2] \times [n_3]} \left| \arg(\zeta_k^{c_1 i_1}) + \arg(\zeta_k^{c_2 i_2}) + \arg(\zeta_k^{c_3 i_3}) \right|$$

$$\leq \max_{i_1 \in [n_1]} \left| \arg(\zeta_k^{c_1 i_1}) \right| + \max_{i_2 \in [n_2]} \left| \arg(\zeta_k^{c_2 i_2}) \right| + \max_{i_3 \in [n_3]} \left| \arg(\zeta_k^{c_3 i_3}) \right|$$

$$= \frac{2\pi}{n} \left(d_1(n_1 - 1) + d_2(n_2 - 1) + d_3(n_3 - 1) \right)$$

In the case that $c_1 \leq d_1$

$$\max_{j \in C_k} \left| \arg(\zeta_k^j) \right| = \frac{2\pi}{N} \left(d_1(n_1 - 1) + d_2(n_2 - 1) + d_3(n_3 - 1) \right)$$

$$\leq \frac{2\pi}{N} \left(d_1 \frac{N}{12 d_1} + d_2(1 - 1) + \frac{d_1}{2} \frac{d_1}{12} \right)$$

$$\leq \frac{\pi}{3}$$

otherwise

$$\max_{j \in C_k} \left| \arg(\zeta_k^j) \right| = \frac{2\pi}{N} \left(d_1(n_1 - 1) + d_2(n_2 - 1) + d_3(n_3 - 1) \right)$$

$$\leq \frac{2\pi}{N} \left(d_1 \frac{N}{144 d_1} + d_1 \left\lceil \frac{N}{c_1} \right\rceil \frac{c_1}{30 d_1} + \frac{d_1}{2} \frac{d_1}{12} \right)$$

$$\leq \frac{\pi}{3}.$$

Proof of (iii). Put $D := \{i_1 d_1 + i_2 d_2 : i_1 \in [n_1], i_2 \in [n_2]\}$. From the fact that $d_2 = d_1 \lceil \frac{N}{c_1} \rceil$ and $d_2 > n_1 d_1$ (this ensures disjoint copies of A'_k over $[\lfloor \frac{1}{6} n \rfloor]$), we deduce that D is a subset of arithmetic progression with common difference d_1 and furthermore $|D| = n_1 n_2$.

Now we introduce the set

$$E := \{i_1 d_1 + i_2 d_2 + i_3 d_3 : i_1 \in [n_1], i_2 \in [n_2], i_3 \in [n_3]\}.$$

E is union of n_3 shifted copies of D. We will see that these n_3 copies are mutually disjoint. Assume for a moment that this is not the case and two copies of D have a common element. Then for $p, p' \in n_1$ and $k, k' \in n_3$, we have $(p - p')d_1 = (k' - k)d_3$. This implies $l d_1 = r d_3$, where l, r are appropriate values, since $d_1 > d_3$, $r > l$. On the other hand, let us consider the function $f_k(j) = jk \mod N$. This function is bijective as N is prime. Hence the pre-images of $l d_1, r d_3$ are equal and

$lc_1 = rc_3$. But $c_1 < c_3$, and we have a contradiction. Thus all shifted copies of D are mutually disjoint and hence $|E| = n_1 n_2 n_3 \geq \frac{N}{5000}$. $\qquad\square$

We can finish the proof of Theorem 107

Proof (of Theorem 107). For a fixed N, we construct $\mathscr{F}_N = ([N], S)$, where $S = \{S_0, S_1, \ldots, S_{N^2-1}\}$ as follows: for S_0, take the set $\{0, 1, \ldots, \lfloor \frac{N}{5000} \rfloor\}$ and for $0 < i < N$, we put $S_{iN} = C_i$. By construction we have for all k,

$$\left| \sum_{j \in S_{iN}} \zeta_k^j \right| \geq \sum_{j \in S_{iN}} \mathrm{Re}(\zeta_k^j) \geq \frac{N}{10000},$$

so $\mathrm{disc}(\mathscr{F}_N) > \frac{N^{1/2}}{10000}$. Since any set in $\mathscr{F}_N$ is a union of two sets consisting of sums of three arithmetic progressions, we have (pigeonhole principle) $\mathrm{disc}(\mathscr{H}_{N,3}) \geq \frac{1}{2}\mathrm{disc}(\mathscr{F}_N) = \Omega(N^{1/2})$. $\qquad\square$

5.4.3 The Improvement of Cilleruelo and Hebbinghaus for $k = 2$

Stimulated by Přívětivý's result, Hebbinghaus [39] closed the gap:

Theorem 110 (Hebbinghaus [39]). $\mathrm{disc}(\mathscr{H}_{N,2}) = \Omega\left(N^{1/2}\right)$.

For any $c \geq 2$ Cilleruelo and Hebbinghaus [20] proved

Theorem 111 (Cilleruelo, Hebbinghaus [20]). $\mathrm{disc}(\mathscr{H}_{N,2}, c) = \Omega\left(N^{1/2}\right)$.

We will give the proof for $c = 2$. Recall that the c-color bound of Theorem 101 for $c = 2$ and $k = 2$ is $\Omega(N^{1/3})$. The improvement of Cilleruelo and Hebbinghaus to $\Omega(N^{1/2})$ is based on the following idea. There exist members of a specific set, which we call $\mathscr{E}_0$, consisting of sums of two arithmetic progressions starting at 0, with a large Fourier transform, more exactly, for every $\alpha \in [0, 1)$ there is an $E \in \mathscr{E}_0$ such that

$$|\widehat{\mathbf{1}}_{-E}(\alpha)| \geq \frac{1}{300} N.$$

This leads to the estimate

$$\sum_{E \in \mathscr{E}_0} |\widehat{\mathbf{1}}_{-E}(\alpha)|^2 \geq O(N^2),$$

while in the previous analysis for this sum only a lower bound of $O(N^{4/3})$ was achieved (Lemma 106). The finer analysis benefits from looking inside the set $\mathscr{E}_0$ instead of showing a lower bound for the whole sum $\sum_{E \in \mathscr{E}_0} |\widehat{\mathbf{1}}_{-E}(\alpha)|^2$. The starting point for the refinement is an observation about the structure of such progressions. Let us define $\mathscr{E}_0$ as the set of sums of two arithmetic progressions, each with starting point 0. We can characterize them by the difference and length of the two arithmetic progressions. Further define for all $\delta_1, \delta_2, L_1, L_2 \in \mathbb{N}$:

$$E_{\delta_1, L_1, \delta_2, L_2} := \{j_1 \delta_1 + j_2 \delta_2 : j_1 \in [0, L_1 - 1], j_2 \in [0, L_2 - 1]\}.$$

We represent the set $\mathscr{E}_0$ as the union of three subsets $\mathscr{E}_1$, $\mathscr{E}_2$ and $\mathscr{E}_3$.
The first two sets $\mathscr{E}_1$ and $\mathscr{E}_2$ are as follows.

$$\mathscr{E}_1 := \left\{ E_{\delta_1, L_1, \delta_2, L_2} : \delta_1 \in [24], L_1 = \left\lceil \frac{N}{6\delta_1} \right\rceil, \delta_2 = 1, L_2 = 1 \right\},$$

and

$$\mathscr{E}_2 := \left\{ E_{\delta_1, L_1, \delta_2, L_2} : \delta_1 \in [25, N^{1/2}]_{\mathbb{Z}}, L_1 = \left\lceil \frac{N}{12\delta_1} \right\rceil, \delta_2 \in [\delta_1 - 1], L_2 = \left\lceil \frac{\delta_1 - 1}{12} \right\rceil \right\}.$$

The definition of the last set $\mathscr{E}_3$ is a bit more involved. Consider the sum of two arithmetic progressions. For every difference δ_1 of the first arithmetic progression we determine a set of differences δ_2 for the second arithmetic progression. Let $\delta_1 \in [N^{\frac{1}{2}}]$ and let

$$B(\delta_1) := \{b \in [\delta_1] : (b, \delta_1) = 1\}$$

be the set of all elements of $[\delta_1]$ that are relatively prime to δ_1. Let $b \in B(\delta_1)$. Set $\bar{k} := \lfloor \log(N^{1/2} \delta_1^{-1}) \rfloor$. We define for all $0 \le k \le \bar{k}$ sets, $M(b, k)$, of differences for the second arithmetic progression. The set $M(b, k)$ should cover the range of possible differences for the second arithmetic progression for the interval $(2^k N^{1/2}, 2^{k+1} N^{1/2}]$. Let

$$M(b, k) := \left(b + 2^{2k} \delta_1 \mathbb{Z} \right) \cap \left(2^k N^{1/2}, 2^{k+1} N^{1/2} + 2^{2k} \delta_1 \right].$$

For all $0 \le k \le \bar{k}$ we set $M_{\delta_1}(k) := \bigcup_{b \in B(\delta_1)} M(b, k)$. Now we are able to define the third set $\mathscr{E}_3$:

$$\mathscr{E}_3 := \bigcup_{\delta_1 \in [N^{1/2}]} \bigcup_{k=0}^{\bar{k}} \left\{ E_{\delta_1, L_1, \delta_2, L_2} : L_1 = \left\lceil \frac{2^k N^{1/2}}{12} \right\rceil, \delta_2 \in M_{\delta_1}(k), L_2 = \left\lceil \frac{2^{-k} N^{1/2}}{12} \right\rceil \right\}.$$

We have

Proposition 112. $\mathscr{E}_0 = \mathscr{E}_1 \cup \mathscr{E}_2 \cup \mathscr{E}_3$.

In the next lemma we prove that the cardinality of the sets $\mathscr{E}_i$ is of order $O(N)$.

Lemma 113. $|\mathscr{E}_3| \leq 6N$, *thus* $|\mathscr{E}_0| \leq 7N$.

Proof. By definition of $\mathscr{E}_3$, we have $|\mathscr{E}_3| = \sum_{\delta_1=1}^{N^{1/2}} \sum_{k=0}^{\bar{k}} |M_{\delta_1}(k)|$. Consider $|M(b,k)|$ for all $b \in B(\delta_1)$ and all $0 \leq k \leq \bar{k}$. We first show that the difference $2^{2k}\delta_1$ of two consecutive elements of $M(b,k)$ is at most $2^k N^{1/2}$. Since

$$2^{2k}\delta_1 \leq 2^k 2^{\log(N^{1/2}\delta_1^{-1})}\delta_1 = 2^k N^{1/2}.$$

$$|M(b,k)| \leq \frac{3 \cdot 2^k N^{1/2}}{2^{2k}\delta_1} = 3 \cdot 2^{-k} N^{1/2}\delta_1^{-1}.$$

Further $M_{\delta_1}(k) = \bigcup_{b \in B(\delta_1)} M(b,k)$. This yields

$$|M_{\delta_1}(k)| \leq \delta_1 |M(b,k)| \leq 3 \cdot 2^{-k} N^{1/2}.$$

Thus

$$|\mathscr{E}_3| = \sum_{\delta_1 \in [N^{1/2}]} \sum_{k=0}^{\bar{k}} |M_{\delta_1}(k)|$$

$$\leq \sum_{\delta_1 \in [N^{1/2}]} \sum_{k=0}^{\bar{k}} 3 \cdot 2^{-k} N^{1/2}$$

$$< 3N \sum_{k=0}^{\infty} 2^{-k} \leq 6N.$$

It is easy to see that $|\mathscr{E}_1 \cup \mathscr{E}_2| \leq N$. This proves the lemma. $\qquad\square$

The proof of Theorem 110 is based on the the following lemma.

Lemma 114 (Main Lemma). *For every* $\alpha \in [0,1)$ *there exists an* $E \in \mathscr{E}_0$ *such that*

$$|\widehat{\mathbf{1}}_{-E}(\alpha)| \geq \frac{1}{300}N.$$

With this lemma we can conclude the proof of the claimed lower bound.

Proof (of Theorem 110). For $a \in \mathbb{Z}$ and $E \in \mathscr{E}_{N,2}$ let $E_a := a + E$. Note that $\sum_{a \in \mathbb{Z}} |\chi(E_a)|^2 = \|\chi * \mathbf{1}_{-E}\|_2^2$. Using the Plancherel theorem and the convolution property of the Fourier transform we get

$$\sum_{E \in \mathscr{E}_0} \sum_{a \in \mathbb{Z}} |\chi(E_a)|^2 = \sum_{E \in \mathscr{E}_0} \|\chi * \mathbf{1}_{-E}\|_2^2$$

$$= \sum_{E \in \mathscr{E}_0} \|\widehat{\chi * \mathbf{1}}_{-E}\|_2^2$$

$$= \sum_{E \in \mathcal{E}_0} \|\hat{\chi}\widehat{\mathbf{1}}_{-E}\|_2^2$$

$$= \sum_{E \in \mathcal{E}_0} \int_0^1 |\hat{\chi}(\alpha)|^2 |\widehat{\mathbf{1}}_{-E}(\alpha)|^2 d\alpha$$

$$= \int_0^1 |\hat{\chi}(\alpha)|^2 \left(\sum_{E \in \mathcal{E}_0} |\widehat{\mathbf{1}}_{-E}(\alpha)|^2 \right) d\alpha.$$

The Main Lemma yields for every $\alpha \in [0, 1)$ the existence of an $E \in \mathcal{E}_0$ such that $|\widehat{\mathbf{1}}_{-E}(\alpha)| \geq \frac{1}{300}N$. Thus, we get for every $\alpha \in [0, 1)$

$$\sum_{E \in \mathcal{E}_0} |\widehat{\mathbf{1}}_{-E}(\alpha)|^2 \geq \frac{1}{90000}N^2.$$

Hence, we can continue the estimation of the sum of squared discrepancies as follows.

$$\sum_{E \in \mathcal{E}_0} \sum_{a \in \mathbb{Z}} |\chi(E_a)|^2 = \int_0^1 |\hat{\chi}(\alpha)|^2 \left(\sum_{E \in \mathcal{E}_0} |\widehat{\mathbf{1}}_{-E}(\alpha)|^2 \right) d\alpha$$

$$\geq \frac{1}{90000}N^2 \|\hat{\chi}\|_2^2$$

$$= \frac{1}{90000}N^2 \|\chi\|_2^2$$

$$= \frac{1}{90000}N^3.$$

Since every $E \in \mathcal{E}_0$ satisfies $E \subseteq [N]$, we get for all $a \in \mathbb{Z} \setminus \{-N + 1, \ldots, N\}$ that $E_a \cap [N] = \emptyset$ and thus $\chi(E_a) = 0$. Therefore, $\sum_{E \in \mathcal{E}_0} \sum_{a \in \mathbb{Z}} |\chi(E_a)|^2$ is the sum of at most $2N|\mathcal{E}_0| \leq 14N^2$ non-trivial elements (Lemma 113). Hence, there exists an $E \in \mathcal{E}_0$ and an $a \in \{-N = 1, \ldots, N\}$ such that

$$|\chi(E_a)|^2 \geq \frac{1}{1260000}N.$$

Thus, we have proven

$$\mathrm{disc}(\mathcal{H}_{N,2}) \geq |\chi(E_a)| > \frac{1}{1200}N^{1/2}$$

i.e. $\mathrm{disc}(\mathcal{H}_{N,2}) = \Omega\left(N^{1/2}\right)$. □

For the proof of the Main Lemma we need the following auxiliary lemmas.

Lemma 115. *For every $\alpha \in [0, 1)$ and every $k \in \mathbb{N}$ there exists a $\delta \in [k]$ and an $a \in \mathbb{Z}$ such that*

$$|\delta\alpha - a| < \frac{1}{k}.$$

Proof. For all $j \in [k]$ we set

$$M_j := \left\{ \delta \in [k] : \delta\alpha - \lfloor \delta\alpha \rfloor \in \left[\frac{j-1}{k}, \frac{j}{k} \right) \right\}.$$

For every $\delta \in M_1$ it holds $|\delta\alpha - \lfloor \delta\alpha \rfloor| < \frac{1}{k}$. Thus, we can assume $M_1 = \emptyset$. By the pigeonhole argument there exists a $j \in [k] \setminus \{1\}$ with $|M_j| \geq 2$. Let $\delta_1, \delta_2 \in M_j$ with $\delta_1 < \delta_2$. Set $\delta := \delta_2 - \delta_1$. Using $\delta_1, \delta_2 \in M_j$, we get

$$|\delta - (\lfloor \delta_2\alpha \rfloor - \lfloor \delta_1\alpha \rfloor)| = |(\delta_2 - \lfloor \delta_2\alpha \rfloor) - (\delta_1 - \lfloor \delta_1\alpha \rfloor)| < \frac{1}{k}$$

as required.   □

Lemma 116. *Let $\alpha \in [0, 1)$, $\delta_1, \delta_2, L_1, L_2 \in \mathbb{N}$ with $L_1 \neq 1 \neq L_2$ be chosen such that for suitable $a_1, a_2 \in \mathbb{Z}$ we have*

$$|\delta_j\alpha - a_j| \leq \frac{1}{12(L_j - 1)}, \quad (j = 1, 2).$$

Set $E := \{j_1\delta_1 + j_2\delta_2 : j_1 \in [0, L_1-1], j_2 \in [0, L_2-1]\}$. For the Fourier transform of the indicator function $\mathbf{1}_{-E}$ of the set $-E$ we get

$$|\widehat{\mathbf{1}}_{-E}(\alpha)| \geq \frac{|E|}{2}.$$

Proof. The Fourier transform of a function $f : \mathbb{Z} \to \mathbb{C}$ is $\hat{f} : [0, 1) \to \mathbb{C}, \alpha \mapsto \sum_{z \in \mathbb{Z}} f(z)e^{-2\pi i z\alpha}$. Thus,

$$\widehat{\mathbf{1}}_{-E}(\alpha) = \sum_{z \in E} e^{2\pi i z\alpha}.$$

Let $z \in E$. There exists a $j_1 \in [0, L_1-1]$ and a $j_2 \in [0, L_2-1]$ with $z = j_1\delta_1 + j_2\delta_2$. Hence,

$$\begin{aligned} e^{2\pi i z\alpha} &= e^{2\pi i (j_1\delta_1 + j_2\delta_2)\alpha} \\ &= e^{2\pi i [j_1(\delta_1\alpha - a_1) + j_2(\delta_2\alpha - a_2)]} e^{2\pi i (j_1 a_1 + j_2 a_2)} \\ &= e^{2\pi i [j_1(\delta_1\alpha - a_1) + j_2(\delta_2\alpha - a_2)]}. \end{aligned}$$

Using $|j_1(\delta_1\alpha - a_1) + j_2(\delta_2\alpha - a_2)| \le \frac{L_1-1}{12(L_1-1)} + \frac{L_2-1}{12(L_2-1)} = \frac{1}{12} + \frac{1}{12} = \frac{1}{6}$, we get $Re(e^{2\pi iz\alpha}) \ge \frac{1}{2}$. This proves

$$|\widehat{\mathbf{1}}_{-E}(\alpha)| = \left|\sum_{z \in E} e^{2\pi iz\alpha}\right| \ge Re(\sum_{z \in E} e^{2\pi iz\alpha}) \ge \frac{|E|}{2}$$

as asserted.   □

We are able to demonstrate the proof idea of the Main Lemma.

Proof (Main Lemma). Using Lemma 115 we can find a $\delta_1 \in [N^{\frac{1}{2}}]$ such that for an appropriate $a_1 \in \mathbb{Z}$ it holds $|\delta_1\alpha - a_1| < N^{-\frac{1}{2}}$. Dividing by δ_1 we get

$$|\alpha - \frac{a_1}{\delta_1}| < N^{-\frac{1}{2}}\delta_1^{-1}. \tag{5.55}$$

Choose δ_1 and a_1 in such a way that $\frac{a_1}{\delta_1}$ is an irreducible fraction. We distinguish three cases, but give the proof only in the first case as the other cases are similar, but little bit more involved.

Case 1 : $|\alpha - \frac{a_1}{\delta_1}| < N^{-1}$ and $\delta_1 \le 24$.
Case 2 : $|\alpha - \frac{a_1}{\delta_1}| < N^{-1}$ and $\delta_1 > 24$.
Case 3 : $|\alpha - \frac{a_1}{\delta_1}| \ge N^{-1}$.

Proof for Case 1. Set $L_1 := \lceil \frac{N}{6\delta_1} \rceil$, $\delta_2 := 1$, and $L_2 := 1$. The set $E := E_{\delta_1,L_1,\delta_2,L_2}$ is an element of the special set of hyperedges $\mathcal{E}_0$. More precisely, $E \in \mathcal{E}_1$. Arguments similar to those in the proof of Lemma 116 show

$$|\widehat{\mathbf{1}}_{-E}(\alpha)| \ge Re\left(\sum_{z \in E} e^{2\pi iz\alpha}\right)$$

$$= \sum_{j_1=0}^{L_1-1} Re\left(e^{2\pi ij_1\delta_1\alpha}\right)$$

$$\ge L_1 Re\left(e^{\frac{2\pi i}{6}}\right) \ge \frac{N}{288}.$$
                                                                                       □

5.5 Discrepancy of Arithmetic Progressions in $\mathbb{Z}_p$

In this section, which is based on [40] and previous material treated in [38], we consider the problem of finding the c-color discrepancy, $c \ge 2$, of arithmetic progressions and centered arithmetic progressions resp. in $\mathbb{Z}_p$. We will show that its discrepancy is much larger than for arithmetic progressions in the first n integers.

It is easy to prove that for $p = 2$ the discrepancy of the arithmetic progressions in $\mathbb{Z}_p$ is exactly $(c - 1)/c$ and it is 1 for any p and $c = 2$ for the centered arithmetic progressions. But no matching upper and lower bounds were known beyond these special cases. We show that for both hypergraphs the lower bound is tight up to a logarithmic factor in p: the lower bound is $\Omega\left(\sqrt{\frac{p}{c}}\right)$ and the upper bound is $O\left(\sqrt{\frac{p}{c}\ln p}\right)$. The main work is the proof of the lower bounds with Fourier analysis on $\mathbb{Z}_p$. The result for the centered arithmetic progressions is the interesting result, because here due to the lack of translation-invariance of the hypergraph, Fourier analysis alone does not work and has to be combined with a suitable decomposition of such progressions. Techniques of this type may be useful also in other areas where lower bounds for combinatorial functions are sought.

5.5.1 $\mathbb{Z}_p$-Invariant Hypergraphs

We define the hypergraph $\mathcal{H}_{\mathbb{Z}_p}$.

Definition 117. $\mathcal{H}_{\mathbb{Z}_p} := (\mathbb{Z}_p, \mathcal{E}_{\mathbb{Z}_p})$, where

$$\mathcal{E}_{\mathbb{Z}_p} := \{A_{a,\delta,L} : 0 \le a \le p - 1, \quad 1 \le \delta \le p - 1, \quad 1 \le L \le p\}$$

and $A_{a,\delta,L} := \{(a + j\delta) + p\mathbb{Z} : 0 \le j \le L - 1\}$ is the hypergraph of *arithmetic progressions in $\mathbb{Z}_p$*.

Unlike in the hypergraph of arithmetic progressions in $[n]$, an arithmetic progression in $\mathbb{Z}_p$ can be wrapped around (several times). Thus by Roth's lower bound, or Theorem 65 for $d = 2$, the discrepancy of $\mathcal{H}_{\mathbb{Z}_p}$ is at least $\Omega\left(p^{1/4}\right)$. According to Theorem 45 its c-color discrepancy is $O\left(\sqrt{\frac{p}{c}\ln p}\right)$. So there is a significant gap between lower and upper bound. Furthermore, the central object in this chapter is the hypergraph of centered arithmetic progressions.

Definition 118. Define $\mathcal{H}_{\mathbb{Z}_p,0} := \left(X, \mathcal{E}_{\mathbb{Z}_p,0}\right)$, where the set of hyperedges $\mathcal{E}_{\mathbb{Z}_p,0}$ is the set of all arithmetic progressions of the form $C_{\delta,L} := \{j\delta : -L \le j \le L\}$ with $\delta \in \mathbb{Z}_p \setminus \{0\}$ and $0 \le L \le \frac{p-1}{2}$. $\mathcal{H}_{\mathbb{Z}_p,0}$ is called the hypergraph of *centered arithmetic progressions* in $\mathbb{Z}_p$.

Again by Theorem 45 an upper bound for its discrepancy is $O\left(\sqrt{\frac{p}{c}\ln p}\right)$, but at the moment we do not have a lower bound. There is an important structural difference between arithmetic progressions and centered arithmetic progressions. The translate of an arithmetic progression is again an arithmetic progression, but this is obviously not true for centered arithmetic progressions. The translation-invariance is essential for a lower bound proof based on Fourier analysis as in the paper of Roth [67]. For hypergraphs missing the translation-invariance, lower bound proofs are a challenging task requiring new combinatorial ideas.

Let $G = \mathbb{Z}_p$, p a prime. Let us briefly recall some facts about the Fourier transform from Sect. 5.3.2. It is well-known that the dual group $\hat{G}$ of G is G itself. The Fourier transform of a function $f : \mathbb{Z}_p \to \mathbb{C}$ can be written as

$$\hat{f} : \hat{\mathbb{Z}}_p \to \mathbb{C}, \quad r \mapsto \sum_{x \in \mathbb{Z}_p} f(x) e^{-2\pi i \, xr/p}.$$

For $f, g : \mathbb{Z}_p \to \mathbb{C}$ the convolution $f * g : G \to \mathbb{C}$ is defined by $(f * g)(y) := \sum_{x \in G} f(x) g(y - x)$, $y \in G$. The Fourier transform of a convolution is multiplicative, i.e., $\widehat{f * g} = \hat{f} \hat{g}$. The Plancherel theorem says $\|\hat{f}\|_2^2 = p \|f\|_2^2$ for all $f : G \to \mathbb{C}$, where $\|f\|_2 := \left(\frac{1}{p} \sum_{x \in G} |f(x)|^2 \right)^{1/2}$ is the usual 2-norm of f.

Let $\mathscr{H} = (\mathbb{Z}_p, \mathscr{E})$ be a hypergraph with the property that there is an $E \in \mathscr{E}$ such that for all $a \in \mathbb{Z}_p \setminus \{0\}$ and $b \in \mathbb{Z}_p$ it holds that $b + aE \in \mathscr{E}$. We call such hypergraphs $\mathbb{Z}_p$-*invariant* with respect to the hyperedge E. The hypergraph of arithmetic progressions in $\mathbb{Z}_p$ has this property: with $E := \{0, 1, \ldots, \frac{p-1}{2}\}$, $b + aE$ is the arithmetic progression with starting point b, difference a and length $\frac{p+1}{2}$. The next result on functions on $\mathbb{Z}_p$ is the basis for the investigation of the coloring function.

Theorem 119. *Let $\mathscr{H} = (\mathbb{Z}_p, \mathscr{E})$ be a hypergraph that is $\mathbb{Z}_p$-invariant with respect to $E \in \mathscr{E}$. Let $\delta_E := \frac{1}{p} |E|$ be the density of E in $\mathbb{Z}_p$ and $f : \mathbb{Z}_p \to \mathbb{C}$ be a function. There are $a \in \mathbb{Z}_p \setminus \{0\}$ and $b \in \mathbb{Z}_p$ with*

$$|f(b + aE)| \geq \sqrt{\delta_E (1 - \delta_E)} \, \|f\|_2.$$

Proof. In order to apply Proposition 76, we have to find an $\gamma > 0$ such that $\sum_{a \in \mathbb{Z}_p \setminus \{0\}} |\widehat{\mathbf{1}}_{-aE}(r)|^2 \geq \gamma$ for all $r \in \mathbb{Z}_p$.

Let $r \in \mathbb{Z}_p \setminus \{0\}$. It holds for all $a \in \mathbb{Z}_p \setminus \{0\}$

$$\widehat{\mathbf{1}}_{-aE}(r) = \sum_{x \in (-aE)} e^{-2\pi i \, xr/p} = \sum_{x \in (-E)} e^{-2\pi i \, xar/p} = \widehat{\mathbf{1}}_{-E}(ar).$$

The multiplication with r is a bijection on $\mathbb{Z}_p \setminus \{0\}$. Hence

$$\sum_{a \in \mathbb{Z}_p \setminus \{0\}} |\widehat{\mathbf{1}}_{-aE}(r)|^2 = \sum_{a \in \mathbb{Z}_p \setminus \{0\}} |\widehat{\mathbf{1}}_{-E}(ar)|^2$$

$$= \sum_{a \in \mathbb{Z}_p \setminus \{0\}} |\widehat{\mathbf{1}}_{-E}(a)|^2$$

$$= \left(\sum_{a \in \mathbb{Z}_p} |\widehat{\mathbf{1}}_{-E}(a)|^2 \right) - |\widehat{\mathbf{1}}_{-E}(0)|^2$$

$$\overset{\text{Plancherel}}{=} p \sum_{a \in \mathbb{Z}_p} |\mathbf{1}_{-E}(a)|^2 - |E|^2 = |E|(p - |E|).$$

Fig. 5.5 A large color class
case

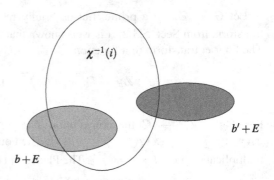

For $r = 0$ we have

$$\sum_{a\in\mathbb{Z}_p\setminus\{0\}} |\widehat{\mathbf{1}}_{-aE}(0)|^2 = \sum_{a\in\mathbb{Z}_p\setminus\{0\}} |(-aE)|^2 = (p-1)|E|^2.$$

Therefore $\sum_{a\in\mathbb{Z}_p\setminus\{0\}} |\widehat{\mathbf{1}}_{-aE}(r)|^2 \geq |E|(p-|E|)$ for all $r \in \mathbb{Z}_p$. By Proposition 76

$$\sum_{a\in\mathbb{Z}_p\setminus\{0\}} \sum_{b\in\mathbb{Z}_p} |f(b+aE)|^2 \geq |E|(p-|E|)\|f\|_2^2 = p^2\delta_E(1-\delta_E)\|f\|_2^2.$$

The right hand side is a sum of $p(p-1)$ terms. Hence there are $a \in \mathbb{Z}_p \setminus \{0\}$ and $b \in \mathbb{Z}_p$ with

$$|f(b+aE)| \geq \sqrt{\frac{p^2}{p(p-1)}\delta_E(1-\delta_E)\|f\|_2^2} \geq \sqrt{\delta_E(1-\delta_E)}\,\|f\|_2.$$

This completes the proof. $\square$

Let us informally sketch the general line of our approach. We wish to determine a lower bound for the discrepancy of a $\mathbb{Z}_p$-invariant hypergraph $\mathcal{H} = (\mathbb{Z}_p, \mathcal{E})$. For every c-coloring χ of $\mathcal{H}$, there exists a color $i \in \{1,2,\ldots,c\}$ with $|\chi^{-1}(i)| \geq \frac{p}{c}$. Now for a constant $\alpha > 0$ that we choose to optimize the lower bound, there exists a color $i \in \{1,2,\ldots,c\}$ with $|\chi^{-1}(i)| > \frac{p}{c} + \alpha\sqrt{\frac{p}{c}}$ or there is no such color. In the first case an average argument will show that for some $b \in \mathbb{Z}_p$, the translation $b + E$ has discrepancy $\Omega(\sqrt{\frac{p}{c}})$ (see Fig. 5.5). In the other case there must be a color $i \in \{1,2,\ldots,c\}$ such that for $A := \chi^{-1}(i)$ it holds $\cdot\frac{p}{c} \leq |A| \leq \frac{p}{c} + \alpha\sqrt{\frac{p}{c}}$. Let $\delta_A := \frac{1}{p}|A|$ be the density of A in $\mathbb{Z}_p$. We define the $f_A : \mathbb{Z}_p \to \mathbb{C}$ by

$$f_A(x) := \begin{cases} 1 - \delta_A & : \quad x \in A, \\ -\delta_A & : \quad x \in \mathbb{Z}_p \setminus A. \end{cases}$$

Obviously, we have $f_A(x) = \mathbf{1}_A(x) - \delta_A$ for all $x \in \mathbb{Z}_p$. This particular function is closely related to the discrepancy function: for every $E \in \mathscr{E}$

$$f_A(E) = \sum_{x \in E} f_A(x) = \sum_{x \in E \cap A} (1 - \delta_A) + \sum_{x \in E \setminus A} (-\delta_A) = |E \cap A| - \delta_A |E|.$$

So $|f_A(E)|$ is a kind of discrepancy of the hyperedge E with respect to the color class A.

In view of good lower bounds, we are interested in a large value of $|f_A(E)|$ for some $E \in \mathscr{E}$. The next corollary exhibits a lower bound of $\Omega\left(\sqrt{p}\right)$.

Corollary 120. *Let* $A, E \subseteq \mathbb{Z}_p$, $\delta_A := \frac{1}{p}|A|$ *and* $\delta_E := \frac{1}{p}|E|$. *Furthermore let* $f_A : \mathbb{Z}_p \to \mathbb{C}$, $x \mapsto \mathbf{1}_A(x) - \delta_A$. *There exist a* $\in \mathbb{Z}_p \setminus \{0\}$ *and* $b \in \mathbb{Z}_p$ *such that*

$$|f_A(b + aE)| \geq \sqrt{\delta_A(1 - \delta_A)\delta_E(1 - \delta_E)p}.$$

Proof. According to Theorem 119 there are $a \in \mathbb{Z}_p \setminus \{0\}$ and $b \in \mathbb{Z}_p$ such that

$$\begin{aligned} |f_A(b + aE)| &\geq \sqrt{\delta_E(1 - \delta_E)} \, \|f_A\|_2 \\ &= \sqrt{\delta_E(1 - \delta_E)} \sqrt{\sum_{x \in A}(1 - \delta_A)^2 + \sum_{x \in \mathbb{Z}_p \setminus A}(-\delta_A)^2} \\ &= \sqrt{\delta_E(1 - \delta_E)} \sqrt{p\delta_A(1 - \delta_A)^2 + p(1 - \delta_A)\delta_A^2} \\ &= \sqrt{\delta_A(1 - \delta_A)\delta_E(1 - \delta_E)p} \end{aligned}$$

as claimed.   □

5.5.2 Discrepancy of Arithmetic Progressions in $\mathbb{Z}_p$

The next theorem is the basis for the discrepancy result for centered arithmetic progressions, but also of independent interest as nearly tight bounds are proved.

Theorem 121 (Hebbinghaus, Srivastav [40]). *There exists a constant* $\alpha > 0$ *such that for the hypergraph of the arithmetic progressions* $\mathscr{H}_{\mathbb{Z}_p}$ *in* $\mathbb{Z}_p$ *we have*

$$\frac{1}{3}\sqrt{\frac{p}{c}} \leq \mathrm{disc}(\mathscr{H}_{\mathbb{Z}_p}, c) \leq \alpha\sqrt{\frac{p}{c}\ln p} + 1.$$

Proof. For $p \in \{2, 3\}$ it is easy to check that the claimed bounds are valid (for $p = 2$ disc$(\mathcal{H}_{\mathbb{Z}_p}, c) = \frac{c-1}{c}$). Since p is a prime, we can assume that $p \geq 5$ and p is odd.

Proof of the lower bound. Let $\chi : \mathbb{Z}_p \to \{1, 2, \ldots, c\}$ be a c-coloring. There exists a color $i \in \{1, 2, \ldots, c\}$ with $|\chi^{-1}(i)| \geq \frac{p}{c}$. If there is a color $i \in \{1, \ldots, c\}$ with $|\chi^{-1}(i)| - \frac{p}{c} > \frac{1}{3}\sqrt{\frac{p}{c}}$, then disc$(\mathcal{H}_{\mathbb{Z}_p}, c, \chi) > \frac{1}{3}\sqrt{\frac{p}{c}}$, because $\mathbb{Z}_p$ itself is an arithmetic progression in $\mathbb{Z}_p$. Thus, we can assume that there is a color $i \in \{1, 2, \ldots, c\}$ such that for $A := \chi^{-1}(i)$ it holds $0 \leq |A| - \frac{p}{c} \leq \frac{1}{3}\sqrt{\frac{p}{c}}$.

Set $\delta_A := \frac{|A|}{p}$. We define the function $f_A : \mathbb{Z}_p \to \mathbb{C}$ by

$$
f_A(x) := \begin{cases} 1 - \delta_A & : \quad x \in A, \\ -\delta_A & : \quad \text{otherwise.} \end{cases}
$$

For every subset $X \subseteq \mathbb{Z}_p$

$$
f_A(X) = \sum_{x \in X} f_A(x) = \sum_{x \in A \cap X} (1 - \delta_A) + \sum_{x \in X \setminus A} (-\delta_A) = |A \cap X| - \delta_A |X|.
$$

Let us sketch the argumentation in the following: we wish to find an arithmetic progression P in $\mathbb{Z}_p$ such that $|f_A(P)| \geq \frac{1}{3}\sqrt{\frac{p}{c}}$. Thereafter, it is proved that the discrepancy of P or the complement of P in $\mathbb{Z}_p$ with respect to χ is at least $|f_A(P)|$. Since the complement of an arithmetic progression in $\mathbb{Z}_p$ is also an arithmetic progression in $\mathbb{Z}_p$, the lower bound is achieved. Let us define $E := A_{0,1,\frac{p+1}{2}}$. For every $a \in \mathbb{Z}_p \setminus \{0\}$ and every $b \in \mathbb{Z}_p$, $b + aE = A_{b,a,\frac{p+1}{2}}$ is the arithmetic progression with starting point b, difference a and length $\frac{p+1}{2}$ in $\mathbb{Z}_p$. By Corollary 120 there are $a \in \mathbb{Z}_p \setminus \{0\}$ and $b \in \mathbb{Z}_p$ such that

$$
\begin{aligned}
|f_A(b + aE)| &\geq \sqrt{\delta_A(1 - \delta_A)\delta_E(1 - \delta_E)p} \\
&= \sqrt{\delta_A(1 - \delta_A)\frac{p+1}{2p}\frac{p-1}{2p}p} \\
&\geq \sqrt{\frac{p^2 - 1}{4p^2}}\sqrt{\delta_A(1 - \delta_A)p} \\
&\geq \sqrt{\frac{24}{100}}\sqrt{\delta_A(1 - \delta_A)p} \\
&\geq \frac{\sqrt{6}}{5}\sqrt{\delta_A(1 - \delta_A)p}.
\end{aligned}
$$

Let $x := \delta_A - \frac{1}{c}$ then $0 \le x \le \frac{1}{3\sqrt{pc}}$. Thus, we have

$$\delta_A(1 - \delta_A) = \left(\frac{1}{c} + x\right)\left(\frac{c-1}{c} - x\right) = \frac{c-1}{c^2} + \frac{c-2}{c}x - x^2 \ge \frac{c-1}{c^2} - \frac{1}{9pc}.$$

For $p \ge 5$ we conclude

$$|f_A(b + aE)| \ge \frac{\sqrt{6}}{5}\sqrt{\frac{c-1}{c} - \frac{1}{9p}}\sqrt{\frac{p}{c}} \ge \frac{\sqrt{6}}{5}\sqrt{\frac{1}{2} - \frac{1}{45}}\sqrt{\frac{p}{c}} > \frac{1}{3}\sqrt{\frac{p}{c}}.$$

If $f_A(b + aE) > 0$, define $Q := b + aE$. Otherwise, if $f_A(b + aE) < 0$, we set $Q := \mathbb{Z}_p \setminus (b + aE)$. Q is an arithmetic progression in $\mathbb{Z}_p$ with $f_A(Q) \ge \frac{1}{3}\sqrt{\frac{p}{c}}$, because $f_A(\mathbb{Z}_p \setminus (b + aE)) = f_A(\mathbb{Z}_p) - f_A(b + aE) = -f_A(b + aE)$. Hence

$$\text{disc}(\mathcal{H}_{\mathbb{Z}_p}, c, \chi) \ge |A \cap Q| - \frac{1}{c}|Q|$$

$$= |A \cap Q| - \delta_A|Q| + (\delta_A - \frac{1}{c})|Q| \ge f_A(Q) > \frac{1}{3}\sqrt{\frac{p}{c}}.$$

Proof of the upper bound. The number of hyperedges in $\mathcal{H}_{\mathbb{Z}_p}$ is bounded by p^3 because for every tuple (a, δ, L) with $0 \le a \le p - 1$, $1 \le \delta \le p - 1$ and $1 \le L \le p$ there is only one arithmetic progression P in $\mathbb{Z}_p$ with $P = A_{a,\delta,L}$ and every hyperedge of $\mathcal{H}_{\mathbb{Z}_p}$ is of this form. By Theorem 45 there is a constant $\alpha > 0$ such that

$$\text{disc}(\mathcal{H}_{\mathbb{Z}_p}, c) \le \alpha\sqrt{\frac{p}{c}\ln p} + 1.$$

This completes the proof. $\qquad\qquad\qquad\qquad\qquad\qquad\qquad\qquad\qquad\qquad\qquad$ □

We proceed to centered arithmetic progressions.

5.5.3 Centered Arithmetic Progressions

The main result is:

Theorem 122 (Hebbinghaus, Srivastav [40]). *Let $c \ge 3$. For the hypergraph of centered arithmetic progressions in $\mathbb{Z}_p$ there exists a constant $\alpha > 0$ such that*

$$\frac{1}{31}\sqrt{\frac{p}{c}} \le \text{disc}(\mathcal{H}_{\mathbb{Z}_p,0}, c) \le \alpha\sqrt{\frac{p}{c}\ln p} + 1.$$

We state bounds for the c-color discrepancy of $\mathcal{H}_{\mathbb{Z}_p,0}$ only for $c \geq 3$, because $\mathrm{disc}(\mathcal{H}_{\mathbb{Z}_p,0}, 2) = 1$, using the coloring

$$\chi : \mathbb{Z}_p \to \{-1, 1\}, \quad x \mapsto \begin{cases} 1 & : & x \in \{0, \ldots, \frac{p-1}{2}\}, \\ -1 & : & x \in \{\frac{p+1}{2}, \ldots, p-1\}. \end{cases}$$

For the proof of Theorem 122, we need a couple of lemmas. Define for a subset $A \subseteq \mathbb{Z}_p$ of density $\delta_A := \frac{1}{p}|A| < \frac{1}{2}$ the function

$$g_A : \mathbb{Z}_p \to \mathbb{C}, \quad x \mapsto \begin{cases} 2 - 2\delta_A & : & \{x, -x\} \subseteq A, \\ 1 - 2\delta_A & : & \{x, -x\} \neq \{x, -x\} \cap A \neq \emptyset, \\ -2\delta_A & : & \{x, -x\} \cap A = \emptyset. \end{cases} \quad (5.56)$$

Let $f_A : \mathbb{Z}_p \to \mathbb{C}$ be defined as in Sect. 5.5.2, i.e., $f_A(x) = 1 - \delta_A$ if $x \in A$ and $f(x) = -\delta_A$ otherwise. The function g_A is a kind of symmetrization of f_A:

$$g_A(x) = f_A(x) + f_A(-x), \quad \text{for all } x \in \mathbb{Z}_p,$$

and it turns out to be a key concept in the proof of Theorem 122. The next lemma gives an estimation for $\|g_A\|_2^2$.

Lemma 123. *Let A be a subset of $\mathbb{Z}_p$ with density $\delta_A = \frac{1}{p}|A| < \frac{1}{2}$, then*

$$\|g_A\|_2^2 \geq 4p\delta_A(\tfrac{1}{2} - \delta_A).$$

Proof. We set $\mu := \frac{1}{p}|\{x \in \mathbb{Z}_p : \{x, -x\} \neq \{x, -x\} \cap A \neq \emptyset\}|$, i.e. μ is the density of the subset of $\mathbb{Z}_p$ of all x with $\mathbf{1}_A(x) \neq \mathbf{1}_A(-x)$. For this density we get $\mu \leq 2\delta_A$, because for every $y \in \{x \in \mathbb{Z}_p : \{x, -x\} \neq \{x, -x\} \cap A \neq \emptyset\}$ either $y \in A$ or $-y \in A$. Since there are $(\delta_A - \frac{1}{2}\mu)p$ elements $x \in \mathbb{Z}_p$ with $g_A(x) = 2 - 2\delta_A$, μp elements $x \in \mathbb{Z}_p$ with $g_A(x) = 1 - 2\delta_A$ and $(1 - \delta_A - \frac{1}{2}\mu)p$ elements $x \in \mathbb{Z}_p$ with $g_A(x) = -2\delta_A$, we have

$$\|g_A\|_2^2 = \sum_{x \in \mathbb{Z}_p} |g_A(x)|^2$$

$$= (\delta_A - \tfrac{1}{2}\mu)p(2 - 2\delta_A)^2 + \mu p(1 - 2\delta_A)^2 + (1 - \delta_A - \tfrac{1}{2}\mu)p(-2\delta_A)^2$$

$$= p\left(\mu(-2 + 4\delta_A - 2\delta_A^2 + 1 - 4\delta_A + 4\delta_A^2 - 2\delta_A^2) + 4\delta_A(1 - \delta_A)\right)$$

$$= 4p\left(\delta_A(1 - \delta_A) - \tfrac{1}{4}\mu\right)$$

Using $\mu \leq 2\delta_A$ we get $\|g_A\|_2^2 \geq 4p\delta_A(\tfrac{1}{2} - \delta_A)$. $\qquad\square$

With the function g_A multi-sets come into play. For every multi-set K over $\mathbb{Z}_p$ let us denote by $K(x)$ the frequency of occurrence of $x \in \mathbb{Z}_p$ in K. For a multi-set K

over $\mathbb{Z}_p$ and a function $f : \mathbb{Z}_p \to \mathbb{C}$ we extend the definition $f(X) = \sum_{x \in X} f(x)$ for subsets $X \subseteq \mathbb{Z}_p$ to multisets by

$$f(K) := \sum_{x \in \mathbb{Z}_p} K(x) f(x).$$

Let $E \subseteq \mathbb{Z}_p$ and let M be the multi-set $E \cup (-E)$, i.e., every element of $E \cap (-E)$ occurs twice in M. Thus, we have

$$M(x) = \mathbf{1}_E(x) + \mathbf{1}_E(-x)$$

for all $x \in \mathbb{Z}_p$. With this definition we have

$$g_A(E) = \sum_{x \in E} g_A(x) = \sum_{x \in E} (f_A(x) + f_A(-x)) = \sum_{x \in \mathbb{Z}_p} M(x) f_A(x) = f_A(M).$$

$$\tag{5.57}$$

The proof of Theorem 122 will use this equation in the following way. As in Sect. 5.5.2 f_A is the color-function for a color-class A. The term $|f_A(M)|$ is a kind of discrepancy of the multi-set $M = E \cup (-E)$. Via (5.57) we can calculate a lower bound for $|f_A(M)|$ using the modified color-function g_A. The set E is an arithmetic progression with starting point 0 in $\mathbb{Z}_p$, the multi-set $M = E \cup (-E)$ can be separated into at most three sets that are either a centered arithmetic progression or complement of a centered arithmetic progression in $\mathbb{Z}_p$, as the following lemma states. Let us further define

Definition 124. Let E be an arithmetic progression with starting point 0 in $\mathbb{Z}_p$ and let M be the multi-set $M = E \cup (-E)$. We define $M_0 := \{x \in \mathbb{Z}_p : M(x) \geq 1\}$.

Note that $M_0 = E \cup (-E)$, in the *usual* sense of the union of two sets. We have

Lemma 125. M_0 *is a centered arithmetic progression.*

Proof. Let $E = \{ka : k \in \{0, 1, \ldots, L - 1\}\}$ with $a \in \mathbb{Z}_p \setminus \{0\}$ and $L \leq p$.

Case 1: There is an $x \in E \cap (-E)$. Then there are $k_1, k_2 \in \{0, 1, \ldots, L - 1\}$ with $k_1 a = x = -k_2 a$. Thus $(k_1 + k_2)a \equiv 0 \mod p$. Since p is prime, p is a divisor of $k_1 + k_2$, so $L - 1 \geq p/2$, and hence $M_0 = \mathbb{Z}_p$, which is a centered arithmetic progression.

Case 2: $E \cap (-E)$ is empty. Then $M_0 = \{ka : k \in \{-L + 1, \ldots, -1, 0, 1, \ldots, L - 1\}\}$ is a centered arithmetic progression.

$\square$

Lemma 126. *Let P be an arithmetic progression in $\mathbb{Z}_p$. The set $P \cap (-P)$ is the union of at most two sets that are either a centered arithmetic progression or the complement of a centered arithmetic progression in $\mathbb{Z}_p$.*

Proof. Let $\alpha \in \mathbb{Z}_p$, $\beta \in \mathbb{Z}_p \setminus \{0\}$ and $1 \leq L \leq p - 1$ such that $P = \{\alpha + i\beta : 0 \leq i \leq L - 1\}$. We can assume that $L < p$, because otherwise $P \cap (-P) = P = \mathbb{Z}_p$ and there is nothing left to prove. For each $x \in P \cap (-P)$ there are $i, j \in \{0, 1, \ldots, L - 1\}$ such that

$$x \equiv \alpha + i\beta \equiv -\alpha - j\beta \mod p \tag{5.58}$$

This is equivalent to $(i + j)\beta \equiv -2\alpha \mod p$. Let $k_1 \in \{0, 1, \ldots, p - 1\}$ be the unique element with $k_1\beta \equiv -2\alpha \mod p$. Set $k_2 := k_1 + p$. Then $\{k_1, k_2\}$ is the solution set for the congruence $k\beta \equiv -2\alpha$ in the set $\{0, 1, \ldots, 2p - 1\}$. Clearly, each $i + j$ is contained in this set, so $i + j \in \{k_1, k_2\}$. We make the following case distinction.

Case 1: $k_1 > 2L - 2$
Case 2: $L - 1 < k_1 \leq 2L - 2$
Case 3: $k_1 \leq L - 1$ and $k_2 > 2L - 2$
Case 4: $k_1 \leq L - 1$ and $k_2 \leq 2L - 2$

Let us show the pattern of the proof in the second case as the proof for the remaining cases is along this line. We have $k_2 = k_1 + p > L - 1 + p > 2L - 2$. Therefore the solutions $(i, j) \in \{0, 1, \ldots, L - 1\}^2$ of the congruence (5.58) are exactly the solutions of the equation $i + j = k_1$. The possible values for i and j resp. are the elements of the set $Y := \{k_1 - L + 1, k_1 - L + 2, \ldots, L - 1\}$. Thus $P \cap (-P) = \{\alpha + i\beta : i \in Y\}$.

Case 2.1: k_1 is even. Then $|P \cap (-P)| = L - 1 - (k_1 - L + 1) + 1 = 2L - 1 - k_1$ is odd. The central element in $P \cap (-P)$ is $\alpha + \frac{k_1}{2}\beta = \alpha - \alpha = 0$ and $P \cap (-P)$ is a centered arithmetic progression.

Case 2.2: k_1 is odd. Since p is a prime, we can write $X := \mathbb{Z}_p \setminus (P \cap (-P))$ in the form $X = \{\alpha + i\beta : i \in Y_0\}$ with $Y_0 = \{L, L + 1, \ldots, k_1 - L + p\}$. The cardinality of X is odd, because the cardinality of $P \cap (-P)$ is even. The central element in X is

$$\alpha + \frac{k_1 + p}{2}\beta = \alpha - \frac{2\alpha - p\beta}{2} = \alpha - \alpha = 0 \mod p.$$

Hence X is a centered arithmetic progression.

$\square$

Proof (of Theorem 122). The upper bound for the discrepancy of the hypergraph $\mathcal{H}_{\mathbb{Z}_p}$ of all arithmetic progression in $\mathbb{Z}_p$ from Theorem 121 is also an upper bound for the discrepancy of $\mathcal{H}_{\mathbb{Z}_p,0}$, since $\mathcal{E}_{\mathbb{Z}_p,0} \subseteq \mathcal{E}_{\mathbb{Z}_p}$. Thus, only the lower bound is left to prove. The assertion is trivial for $p \in \{2, 3\}$. Hence, we can assume $p \geq 5$. We consider the case $c \geq 4$ and refer for $c = 3$ (the easier case) to [40]. Let be $c \geq 4$ and fix an arbitrary c-coloring $\chi : \mathbb{Z}_p \to \{1, 2, \ldots, c\}$ of $\mathcal{H}_{\mathbb{Z}_p,0}$. There exists at least one color $i \in \{1, 2, \ldots, c\}$ with $|\chi^{-1}(i)| \geq \frac{p}{c}$. If there is a color $i \in \{1, 2, \ldots, c\}$ such that $|\chi^{-1}(i)| > \frac{p}{c} + \frac{1}{25}\sqrt{\frac{p}{c}}$, then

$$\operatorname{disc}(\mathcal{H}_{\mathbb{Z}_p,0}, c, \chi) \geq \left| |\mathbb{Z}_p \cap \chi^{-1}(i)| - \frac{1}{c}|\mathbb{Z}_p| \right| > \frac{1}{25}\sqrt{\frac{p}{c}},$$

because $\mathbb{Z}_p$ itself is a centered arithmetic progression in $\mathbb{Z}_p$. Thus, we can assume that this is not the case and get the existence of a color $i \in \{1, 2, \ldots, c\}$ such that it holds for $A := \chi^{-1}(i)$

$$\frac{p}{c} \leq |A| \leq \frac{p}{c} + \frac{1}{25}\sqrt{\frac{p}{c}}.$$

For the density $\delta_A := \frac{1}{p}|A|$ of A in $\mathbb{Z}_p$ we have

$$0 \leq \delta_A - \frac{1}{c} \leq \frac{1}{25}\sqrt{\frac{1}{pc}}.$$

Set $E := \{0, 1, \ldots, \frac{p-1}{2}\}$ and $\delta_E := |E| = \frac{p+1}{2}$. For the function g_A defined in (5.56) we get by Theorem 119 the existence of an $a \in \mathbb{Z}_p \setminus \{0\}$ and a $b \in \mathbb{Z}_p$ such that

$$|g_A(b + aE)| \geq \sqrt{\delta_E(1 - \delta_E)}\|g_A\|_2 = \frac{1}{2}\sqrt{\frac{p^2 - 1}{p^2}}\|g_A\|_2.$$

Using Lemma 123 we have

$$|g_A(b + aE)| \geq \sqrt{\frac{p^2 - 1}{p^2}}\sqrt{\delta_A\left(\frac{1}{2} - \delta_A\right)}p.$$

The set $b + aE$ is an arithmetic progression with starting point b, difference a and length $\frac{p+1}{2}$. Setting $x := \delta_A - \frac{1}{c}$ we get $0 \leq x \leq \frac{1}{25}\sqrt{\frac{1}{pc}}$. Therefore,

$$\delta_A\left(\frac{1}{2} - \delta_A\right) = \left(\frac{1}{c} + x\right)\left(\frac{1}{2} - \frac{1}{c} - x\right) = \frac{c - 2}{2c^2} + \frac{c - 4}{2c}x - x^2 \geq \frac{c - 2}{2c^2} - \frac{1}{625pc}$$

and hence (using $p \geq 5$ and $c \geq 4$)

$$|g_A(b + aE)| \geq \sqrt{\frac{p^2 - 1}{p^2}}\sqrt{\left(\frac{c - 2}{2c^2} - \frac{1}{625pc}\right)}p$$

$$\geq \sqrt{\frac{24}{25}\left(\frac{1}{4} - \frac{1}{3125}\right)}\sqrt{\frac{p}{c}}$$

$$> \frac{12}{25}\sqrt{\frac{p}{c}}.$$

Since $|g_A(\mathbb{Z}_p \setminus (b + aE))| = |g_A(b + aE)|$, we can assume that $0 \notin (b + aE)$. Every arithmetic progression in $\mathbb{Z}_p$ can be supplemented step by step until it is the whole set $\mathbb{Z}_p$, because p is a prime. Thus, we can supplement $b + aE$ to an arithmetic progression P_1 with starting point 0. Using the triangle-inequality P_1 or $P_1 \setminus (b + aE)$ is an arithmetic progression P with starting point 0 and

$$|g_A(P)| \geq \frac{1}{2}|g_A(b + aE)| > \frac{6}{25}\sqrt{\frac{p}{c}}.$$

We define the multi-set $M := P \cup (-P)$ and get according to Eq. (5.57),

$$|f_A(M)| = |g_A(P)|.$$

Lemma 126 states that $P \cap (-P)$ is the union of at most two sets that are either a centered arithmetic progression or the complement of a centered arithmetic progression in $\mathbb{Z}_p$. Since the set $M_0 := \{x \in \mathbb{Z}_p : M(x) \geq 1\} = P \cup (-P)$ is a centered arithmetic progression, the multi-set M is the union of at most three sets that are either a centered arithmetic progression or the complement of a centered arithmetic progression in $\mathbb{Z}_p$. Using the triangle-inequality at least one of these sets, which we denote by P_0, satisfies

$$|f_A(P_0)| \geq \frac{1}{3}|f_A(M)| = \frac{1}{3}|g_A(P)| > \frac{2}{25}\sqrt{\frac{p}{c}}.$$

It holds $|f_A(\mathbb{Z}_p \setminus P_0)| = |f_A(P_0)| > \frac{2}{25}\sqrt{\frac{p}{c}}$. Therefore we can assume that P_0 is a centered arithmetic progression in $\mathbb{Z}_p$. The triangle-inequality yields

$$\mathrm{disc}(\mathcal{H}_{\mathbb{Z}_p,0}, c, \chi) \geq \left| |P_0 \cap A| - \frac{1}{c}|P_0| \right|$$

$$= \left| |P_0 \cap A| - \delta_A|P_0| + \left(\delta_A - \frac{1}{c}\right)|P_0| \right|$$

$$\geq \left| |P_0 \cap A| - \delta_A|P_0| \right| - \left| \left(\delta_A - \frac{1}{c}\right)p \right|$$

$$= |f_A(P_0)| - \left| |A| - \frac{p}{c} \right|$$

$$> \frac{2}{25}\sqrt{\frac{p}{c}} - \frac{1}{25}\sqrt{\frac{p}{c}}$$

$$= \frac{1}{25}\sqrt{\frac{p}{c}}$$

as desired.   □

5.6   One-Sided Discrepancy of Linear Hyperplanes

In the previous sections we were concerned with the c-color discrepancy of certain arithmetic hypergraphs. A new aspect of combinatorial discrepancy theory is the notion of the one-sided discrepancy. Such a discrepancy function appears for example in the context of the declustering problem [28]. The one-sided discrepancy function cannot be bounded with discrepancy bounds, and has its own structure, demanding for a specific theory. Hebbinghaus, Schoen and Srivastav [41] studied the one-sided c-color discrepancy ($c \geq 2$) of linear hyperplanes in the finite vector space $\mathbb{F}_q^r$. Let $\mathcal{H}_{q,r}$ denote the hypergraph with vertex set $\mathbb{F}_q^r$ and all linear hyperplanes, i.e., subspaces of codimension one in $\mathbb{F}_q^r$ as hyperedges. It was shown that the one-sided discrepancy of $\mathcal{H}_{q,r}$ is bounded from below by $\Omega_q\left(\sqrt{nz(1-z)/c}\right)$ using Fourier analysis on $\mathbb{F}_q^r$. For the discrepancy of $\mathcal{H}_{q,r}$ there is a lower bound $\Omega_q\left(\sqrt{nz(1-z)}\right)$ and an upper bound of $O_q\left(\sqrt{nz(1-z)\log c}\right)$. The upper bound is derived by the c-color extension of Spencer's six standard deviation theorem (Theorem 44) and is also valid for the one-sided discrepancy. Thus, the gap between the upper and lower bound for the one-sided discrepancy is a factor of $\sqrt{c\log c}$ and the bounds are tight for any constant c and q. In the following we will present the techniques and main results with proofs.

5.6.1   Discrepancy of $\mathcal{H}_{q,r}$

The c-color discrepancy measures deviations from the average value $\frac{|E|}{c}$ in both directions. Therefore one cannot decide whether the discrepancy is caused by a lack or an excess of vertices in a hyperedge in one color. To measure the guaranteed excess of vertices in one hyperedge and color, we define another discrepancy notion. The one-sided c-color discrepancy of $\mathcal{H}$ with respect to χ is defined by

$$\mathrm{disc}^+(\mathcal{H}, \chi, c) = \max_{i \in [c]} \max_{E \in \mathcal{E}} \left(|A_i \cap E| - \frac{|E|}{c} \right)$$

and the one-sided c-color discrepancy of $\mathcal{H}$ by

$$\mathrm{disc}^+(\mathcal{H}, c) = \min_{\chi: X \to [c]} \mathrm{disc}^+(\mathcal{H}, \chi, c).$$

Trivially, $\mathrm{disc}^+(\mathcal{H}, c) \leq \mathrm{disc}(\mathcal{H}, c)$, and since $\sum_{i \in [c]} \left(|A_i \cap E| - \frac{|E|}{c} \right) = 0$ for every $E \in \mathcal{E}$, we get

$$\mathrm{disc}(\mathcal{H}, c) \leq (c-1)\,\mathrm{disc}^+(\mathcal{H}, c). \qquad (5.59)$$

Let us define the hypergraph under consideration in this section. Let $\mathbb{F}_q$ be the field with q elements, where $q = p^k$ is a prime power and let $X = \mathbb{F}_q^r$ be the r-dimensional vector space over $\mathbb{F}_q$. Furthermore, let $\mathscr{E}_{q,r}$ be the set of linear hyperplanes of X. This means that $\mathscr{E}_{q,r}$ is the set of all subspaces of codimension 1. For a set $S \subseteq \mathbb{F}_q^r$ we define $S^\sharp = S \setminus \{0\}$. Set $n = |X| = q^r$ and $\mathscr{H}_{q,r} = (X, \mathscr{E}_{q,r})$. $\mathscr{H}_{q,r}$ is an $\frac{n}{q}$-uniform hypergraph with n vertices and $|\mathscr{E}_{q,r}| = \frac{n-1}{q-1}$ hyperedges.

First we consider the case $c|(q-1)$.

Proposition 127. *If $c|(q-1)$ we have*

$$\mathrm{disc}^+(\mathscr{H}_{q,r}, c) = \mathrm{disc}(\mathscr{H}_{q,r}, c) = \frac{c-1}{c}.$$

Proof. Let $\mathscr{E}_1$ be the set of all one-dimensional subspaces of X. Then

$$X = \{0\} \cup \bigcup_{W \in \mathscr{E}_1} (W \setminus \{0\}). \tag{5.60}$$

For every $W \in \mathscr{E}_1$ the set $W \setminus \{0\}$ contains $q-1$ elements. Since $c|(q-1)$, we can color $W \setminus \{0\}$ in an exactly balanced way: we just color $\frac{q-1}{c}$ vertices of $W \setminus \{0\}$ in each color. Doing this for every $W \in \mathscr{E}_1$ and coloring the origin with the color 1 we get a c-coloring $\chi: X \to [c]$. This is a coloring with the least possible discrepancy

$$\mathrm{disc}^+(\mathscr{H}_{q,r}, \chi, c) = \mathrm{disc}(\mathscr{H}_{q,r}, \chi, c) = \frac{c-1}{c},$$

which proves the claim.   □

For the remainder we assume that $c \nmid (q-1)$.

Theorem 128. *Let $z = \frac{(q-1) \bmod c}{c}$. There exists a constant $\alpha > 0$ such that*

$$\frac{1}{q}\sqrt{nz(1-z)} - 1 \leq \mathrm{disc}(\mathscr{H}_{q,r}, c) \leq \alpha\sqrt{nz(1-z)\log c}.$$

As in Sect. 5.1, define for a color $i \in [c]$ the vector $m^{(i)} \in \mathbb{R}^c$ with

$$m_j^{(i)} = \begin{cases} \frac{c-1}{c} & : \quad i = j, \\ -\frac{1}{c} & : \quad \text{otherwise.} \end{cases}$$

and set $M_c = \{m^{(i)} : i \in [c]\}$. Let $\chi: X \to M_c$ be a c-coloring of $\mathscr{H}$. Further, we define $\bar{\chi} \in \mathbb{R}^{nc}$ by $\bar{\chi}_{i+(j-1)c} = (\chi(v_j))_i$ for all $i \in [c]$ and $j \in [n]$, where $X = \{v_1, v_2, \ldots, v_n\}$. Let I_c denote the c-dimensional identity matrix. Then, according to (5.18) in Sect. 5.1, we have

$$\text{disc}(\mathcal{H}, \chi, c) = \max_{E \in \mathcal{E}} \left\| \sum_{x \in E} \chi(x) \right\|_{\infty} = \| (M \otimes I_c) \bar{\chi} \|_{\infty}.$$

If we delete the origin from the vector space, the lower bound proof becomes easier. Thus define $X' = X \setminus \{0\}$, $\mathcal{E}'_{q,r} = \{E \cap X' : E \in \mathcal{E}_{q,r}\}$, and $\mathcal{H}'_{q,r} = (X', \mathcal{E}'_{q,r})$. It is obvious that $\text{disc}(\mathcal{H}_{q,r}, c) \geq \text{disc}(\mathcal{H}'_{q,r}, c) - 1$, thus we can focus on $\mathcal{H}'_{q,r}$. The hypergraph $\mathcal{H}'_{q,r}$ has an interesting property. For two arbitrary elements $x, y \in X'$ let us denote by $d(x, y)$ the number of hyperedges $E \in \mathcal{E}'_{q,r}$ which contain x and y and call it the pair-degree of x and y. One can check that

$$d(x, y) = \begin{cases} \frac{q^{r-1}-1}{q-1} & : \text{ if } x \text{ and } y \text{ are linearly dependent,} \\ \frac{q^{r-2}-1}{q-1} & : \text{ otherwise.} \end{cases}$$

Let M be the incidence matrix of $\mathcal{H}'_{q,r}$: the rows of M correspond to the hyperedges of $\mathcal{H}'_{q,r}$, the columns to the vertices of $\mathcal{H}'_{q,r}$ and the component $M_{i,j}$ of M is 1 if $v_j \in E_i$ and is 0 otherwise. W.l.o.g. we can assume that the columns of M are arranged according to the one-dimensional subspaces of X. Let us denote by $(M^T M)_{x,y}$ the component of the matrix $M^T M$ in the row that corresponds to the vertex $x \in X'$ and the column that corresponds to the vertex $y \in X'$. Then we have $(M^T M)_{x,y} = d(x, y)$, and the matrix $M^T M$ is of the following form: on the diagonal there are $\frac{n-1}{q-1}$ copies of the $(q-1) \times (q-1)$-matrix with all components equal to $\frac{q^{r-1}-1}{q-1}$, and all other components of M are equal to $\frac{q^{r-2}-1}{q-1}$.

We need the following lemma, which extends the lemma of Beck and Sos [13, Theorem 2.8 and Corollary 2.9] to block diagonal matrices and the c-color discrepancy. For simplicity we focus on block diagonal matrices consisting of blocks of same size and identical elements.

Lemma 129. *Let $\mathcal{H} = (X, \mathcal{E})$ be a finite hypergraph. Let $n = |X|$, $m = |\mathcal{E}|$ and $k, l \in \mathbb{N}$ such that $n = kl$. Set $z = \frac{l \bmod c}{c}$ and let M be the incidence matrix of $\mathcal{H}$. If for some block diagonal matrix D consisting of k copies of an $l \times l$-matrix Y with all entries equal to $\gamma > 0$, the matrix $M^T M - D$ is positive semidefinite, then*

$$\text{disc}(\mathcal{H}, c) \geq \left(\frac{k \gamma z (1-z)}{m} \right)^{\frac{1}{2}}.$$

Proof. Let $\chi: X \to M_c$ be a c-coloring such that $\text{disc}(\mathcal{H}, c) = \text{disc}(\mathcal{H}, \chi, c)$. For all $i \in [l]$ let us denote the coloring of the set X_i of vertices corresponding to the i-th block of D by $\chi_i: X_i \to M_c$. Using the discrepancy notion and the color vector $\bar{\chi}$ introduced at the beginning of this subsection, we obtain

$$\operatorname{disc}(\mathcal{H}, c) = \|(M \otimes I_c)\bar{\chi}\|_\infty \geq \frac{1}{\sqrt{cm}}\|(M \otimes I_c)\bar{\chi}\|_2$$

$$= \left(\frac{1}{cm}\bar{\chi}^T[(M^TM - D) \otimes I_c + D \otimes I_c]\bar{\chi}\right)^{\frac{1}{2}} \geq \left(\frac{1}{cm}\bar{\chi}^T(D \otimes I_c)\bar{\chi}\right)^{\frac{1}{2}}$$

$$\geq \left(\frac{k\gamma}{cm}\left(cz(1 - z)^2 + c(1 - z)z^2\right)\right)^{\frac{1}{2}} = \left(\frac{k\gamma z(1 - z)}{m}\right)^{\frac{1}{2}}$$

which finishes the proof.   □

Proof (of Theorem 128). Lower Bound. Consider the hypergraph $\mathcal{H}'_{q,r}$, where the origin has been deleted. Let D be the block diagonal matrix consisting of $\frac{n-1}{q-1}$ blocks on the diagonal, where each block is a $(q - 1) \times (q - 1)$-matrix with all components equal to $\frac{q^{r-1}-1}{q-1} - \frac{q^{r-2}-1}{q-1} = q^{r-2}$ and all other components of D are 0. Since we have for all $x, y \in X'$

$$(M^TM)_{x,y} = d(x, y) = \begin{cases} \frac{q^{r-1}-1}{q-1} & : \text{ if } x \text{ and } y \text{ are linearly dependent,} \\ \frac{q^{r-2}-1}{q-1} & : \text{ otherwise,} \end{cases}$$

all components of the matrix $M^TM - D$ are $\frac{q^{r-2}-1}{q-1}$. Therefore, the matrix $M^TM - D$ is positive semidefinite, and we can apply Lemma 129. The constants for Lemma 129 in this situation are $k = \frac{n-1}{q-1}$, $\gamma = q^{r-2}$, and $m = \frac{n-1}{q-1}$. Thus, we get

$$\operatorname{disc}(\mathcal{H}'_{q,r}, c) \geq \left(q^{r-2}z(1 - z)\right)^{\frac{1}{2}} = \tfrac{1}{q}\sqrt{nz(1 - z)}.$$

And for the lower bound for the c-color discrepancy of $\mathcal{H}_{q,r}$ we have

$$\operatorname{disc}(\mathcal{H}_{q,r}, c) \geq \operatorname{disc}(\mathcal{H}'_{q,r}, c) - 1 \geq \tfrac{1}{q}\sqrt{nz(1 - z)}.$$

Upper Bound. For the proof of the upper bound recall our assumption $z \neq 0$, i.e. $(q - 1) \mod c \neq 0$. Let W be an arbitrary one-dimensional subspace of X. Then for every $E \in \mathcal{E}_{q,r}$ either $W \subseteq E$ or $W \cap E = \{0\}$. So for any two non-trivial $x, y \in W$ there is no $E \in \mathcal{E}_{q,r}$ with $x \in E$ and $y \notin E$. Thus we can color all but $s = (q - 1) \mod c$ non-trivial elements of W in such a way that every color is used in the same amount, for every one-dimensional subspace of X. One can check that the sub-hypergraph induced by the pre-colored vertices has discrepancy 0. Let $\mathcal{H}'_{q,r} = (X', \mathcal{E}'_{q,r})$ be the sub-hypergraph induced by the other $n' = \frac{n-1}{q-1}s + 1$ vertices of X. By construction it is clear that

$$\operatorname{disc}(\mathcal{H}_{q,r}, c) \leq \operatorname{disc}(\mathcal{H}'_{q,r}, c).$$

Applying Theorem 43 to the hypergraph $\mathscr{H}'_{q,r}$ we get

$$\mathrm{disc}(\mathscr{H}_{q,r}, c) \leq \mathrm{disc}(\mathscr{H}'_{q,r}, c) = O\left(\sqrt{\frac{n'}{c}} \log c\right) = O\left(\sqrt{\frac{nz}{q}} \log c\right).$$

The only thing left to prove is $\frac{1}{q} \leq (1-z)$. If $c < q$ then $(1-z) \geq \frac{1}{c} > \frac{1}{q}$ is obvious. Hence, we can assume $c \geq q$. Using $q - 1 \leq \frac{q-1}{q}c$, we get

$$1 - z = \frac{c - (q-1)}{c} \geq \frac{c - \frac{(q-1)c}{q}}{c} = \frac{1}{q},$$

and are done. $\square$

5.6.2 One-Sided Discrepancy of $\mathscr{H}_{q,r}$

Obviously the upper bound in Theorem 128 for the c-color discrepancy is also an upper bound for the one-sided c-color discrepancy $\mathrm{disc}^+(\mathscr{H}_{q,r}, c)$. But the lower bound needs to be investigated separately. Let $r_0 : \mathbb{N} \to \mathbb{N}$ be the function defined by $r_0(2) = 6$, $r_0(q) = 5$ for $3 \leq q \leq 5$, and $r_0(q) = 4$ otherwise. The main result is the following.

Theorem 130. *Let* $z = \frac{(q-1) \bmod c}{c}$, $z \neq 0$. *If* $r \geq r_0(q)$, *we have*

$$\mathrm{disc}^+(\mathscr{H}_{q,r}, c) \geq \frac{\sqrt{z(1-z)}}{4q(q-1)} \sqrt{\frac{n}{c}} - 1.$$

Proof Idea. We will first observe that for a c-coloring χ of X the one-sided discrepancy $\mathrm{disc}^+(\mathscr{H}_{q,r}, \chi, c)$ can be written as

$$\mathrm{disc}^+(\mathscr{H}_{q,r}, \chi, c) = \max_{i \in [c]} \max_{E \in \mathscr{E}_{q,r}} \left(\frac{\hat{A}_i(E^\perp)}{q} + \frac{|A_i|}{q} - \frac{n}{qc}\right), \qquad (5.61)$$

where $\hat{A}_i$ is the Fourier transform of the indicator function $\mathbf{1}_{A_i}$ in $\mathbb{F}_q^r$ and for all $E \in \mathscr{E}_{q,r}$ we set $\hat{A}_i(E^\perp) = \sum_{x \in E^\perp \setminus \{0\}} \hat{A}_i(x)$. Then we will prove that for any set $A \subseteq \mathbb{F}_q^r$ there exists an $E \in \mathscr{E}_{q,r}$ with $\hat{A}(E^\perp) \geq -1$. And finally a certain tradeoff between the size of A_i and $\hat{A}_i(E^\perp)$ in (5.61) can be shown that leads to the stated lower bound. As a first step towards the proof of Theorem 130 we prove some lemmas for the Fourier transform on $\mathbb{F}_q^r$.

5.6.3   The Fourier Transform and Some Consequences

Basic facts about the Fourier transform on finite groups can be found in the book of
R. Lidl and H. Niederreiter [46]. For convenience, let us recall what we need. We
have defined $X = \mathbb{F}_q^r$. Let $f : X \to \mathbb{C}$ be a function. $\mathrm{Tr}_{\mathbb{F}_q/\mathbb{F}_p} : \mathbb{F}_q$ is the absolute trace
function defined by

$$\mathrm{Tr}_{\mathbb{F}_q/\mathbb{F}_p} : \mathbb{F}_q \to \mathbb{F}_p, \ \alpha \mapsto \alpha + \alpha^p + \alpha^{p^2} + \ldots + \alpha^{p^{k-1}}.$$

When there is no danger of confusion, we write $\mathrm{Tr}(.)$. Theorem 2.23 (iii) in the book
of Niederreiter [46] states that the function $\mathrm{Tr}_{\mathbb{F}_q/\mathbb{F}_p}$ is linear and onto. The Fourier
Transform $\hat{f}$ is defined by

$$\hat{f} : X \to \mathbb{C}, \ z \mapsto \sum_{x \in X} f(x) e^{\frac{2\pi i}{p} \mathrm{Tr}(\langle x, z \rangle)},$$

where $\langle x, z \rangle = x_1 z_1 + x_2 z_2 + \ldots + x_r z_r$ is the inner product in $\mathbb{F}_q^r$. Furthermore we
set for all $W \leq X$:

$$\hat{f}(W) = \sum_{z \in W^\sharp} \hat{f}(z),$$

where $W^\sharp = W \setminus \{0\}$. Let us denote by $\hat{A} = \hat{\mathbf{1}}_A$ the Fourier transform of the
indicator function $\mathbf{1}_A$ ($A \subseteq X$). Note that for all $E \in \mathscr{E}_{q,r}$ there is a unique subspace
$E^\perp$ of X of dimension one, which is orthogonal to E. To shorten notation, define

Definition 131. For every $A \subseteq X$ and every hyperedge $E \in \mathscr{E}_{q,r}$

$$d^+(A, E) := |A \cap E| - \frac{|A|}{q}.$$

We have the following orthogonality relation.

Lemma 132. *For each $z \in X^\sharp$ and $a \in X$ we have*

$$\frac{1}{q} \sum_{x \in \langle z \rangle} e^{\frac{2\pi i}{p} \mathrm{Tr}(\langle a, x \rangle)} = \begin{cases} 0 & : \ \langle a, z \rangle \neq 0, \\ 1 & : \ \langle a, z \rangle = 0. \end{cases}$$

Proof. Let $\langle z \rangle$ be the subspace generated by z. If $\langle a, z \rangle = 0$, then $\langle a, x \rangle = 0$ for
every $x \in \langle z \rangle$. Hence,

$$\frac{1}{q} \sum_{x \in \langle z \rangle} e^{\frac{2\pi i}{p} \mathrm{Tr}(\langle a, x \rangle)} = \frac{1}{q} \sum_{x \in \langle z \rangle} e^{\frac{2\pi i}{p} \mathrm{Tr}(0)} = \frac{1}{q} \sum_{x \in \langle z \rangle} 1 = \frac{1}{q} |\langle z \rangle| = 1.$$

Now we assume $\langle a, z \rangle \neq 0$. Then $\langle a, x \rangle$ runs through the whole field $\mathbb{F}_q$, if x runs through the whole subspace $\langle z \rangle$. Since the trace function Tr is onto and non-trivial, there exists a $y \in \langle z \rangle$ with $e^{\frac{2\pi i}{p} \text{Tr}(\langle a, y \rangle)} \neq 1$. Moreover, the function $f : \langle z \rangle \rightarrow \langle z \rangle, x \mapsto x + y$ is bijective. Thus,

$$e^{\frac{2\pi i}{p} \text{Tr}(\langle a, y \rangle)} \left(\frac{1}{q} \sum_{x \in \langle z \rangle} e^{\frac{2\pi i}{p} \text{Tr}(\langle a, x \rangle)} \right) = \frac{1}{q} \sum_{x \in \langle z \rangle} e^{\frac{2\pi i}{p} \text{Tr}(\langle a, x+y \rangle)}$$

$$= \frac{1}{q} \sum_{x \in \langle z \rangle} e^{\frac{2\pi i}{p} \text{Tr}(\langle a, x \rangle)}.$$

Using $e^{\frac{2\pi i}{p} \text{Tr}(\langle a, y \rangle)} \neq 1$, we get $\frac{1}{q} \sum_{x \in \langle z \rangle} e^{\frac{2\pi i}{p} \text{Tr}(\langle a, x \rangle)} = 0$. $\square$

Lemma 133 establishes a first link between one-sided discrepancy and Fourier analysis.

Lemma 133. *For every subset $A \subseteq X$ and every hyperedge $E \in \mathscr{E}_{q,r}$*

$$d^+(A, E) = \frac{1}{q} \hat{A}(E^\perp).$$

Proof. Let $z \in E^\perp \setminus \{0\}$. Using Lemma 132 we have

$$|A \cap E| = \sum_{a \in A} \delta_{\langle a, z \rangle, 0} = \sum_{a \in A} \frac{1}{q} \sum_{x \in \langle z \rangle} e^{\frac{2\pi i}{p} \text{Tr}(\langle a, x \rangle)} = \frac{1}{q} \sum_{x \in E^\perp} \sum_{a \in A} e^{\frac{2\pi i}{p} \text{Tr}(\langle a, x \rangle)}$$

$$= \frac{1}{q} \sum_{x \in E^\perp} \hat{A}(x) = \frac{1}{q} \hat{A}(E^\perp) + \frac{1}{q} \hat{A}(0) = \frac{1}{q} \hat{A}(E^\perp) + \frac{1}{q} |A|.$$

Thus, we have $d^+(A, E) = |A \cap E| - \frac{1}{q}|A| = \frac{1}{q} \hat{A}(E^\perp)$. $\square$

Note that Lemma 133 immediately implies the statement (5.61). Let us define $\alpha := |\mathscr{E}_{q,r}|$ and $\beta := |\{E \in \mathscr{E}_{q,r} : v \in E\}|$ for an arbitrary $v \in X^\sharp = X \setminus \{0\}$. Note that the definition of β does not depend on the choice of v.

Lemma 134. *We have $\alpha = \frac{q^r - 1}{q - 1}$ and $\beta = \frac{q^{r-1} - 1}{q - 1}$.*

Proof. Due to orthogonality, there is a one-to-one correspondence between the subspaces of dimension one and codimension one in X. Thus the number of linear hyperplanes in X is the number of basis of one-dimensional subspace of X divided by the number of basis of a fixed one-dimensional subspace of X. So we have $\alpha = \frac{q^r - 1}{q - 1}$. Double-counting yields

$$\beta(q^r - 1) + \alpha = \left(\sum_{v \in X^\sharp} |\{E \in \mathscr{E}_{q,r} : v \in E\}|\right) + |\{E \in \mathscr{E}_{q,r} : 0 \in E\}|$$

$$= \sum_{v \in X} |\{E \in \mathscr{E}_{q,r} : v \in E\}| = \sum_{E \in \mathscr{E}_{q,r}} |E| = \alpha q^{r-1}.$$

Thus, we have $\beta = \alpha \frac{q^{r-1}-1}{q^r-1} = \frac{q^{r-1}-1}{q-1}$. $\qquad\qquad\qquad\qquad\qquad\qquad\qquad\qquad\square$

The three statements in the next lemma will be useful for our calculations. The proof is just an rearrangement of sums [41]. For convenience we define

$$M_2 = \left\{(a_1, a_2, k_1, k_2) \in A \times A \times \mathbb{F}_q^\sharp \times \mathbb{F}_q^\sharp : k_1 a_1 + k_2 a_2 = 0\right\},$$

$$M_3 = \left\{(a_1, a_2, a_3, k_1, k_2, k_3) \in A^3 \times \left(\mathbb{F}_q^\sharp\right)^3 : k_1 a_1 + k_2 a_2 + k_3 a_3 = 0\right\}.$$

Let $\mathscr{E}_1$ denote the set of all one-dimensional subspaces of X.

Lemma 135. *Let* $A \subseteq X$. *We have*

(i) $\displaystyle\sum_{E \in \mathscr{E}_{q,r}} \hat{A}(E^\perp) = n\, \mathbf{1}_A(0) - |A|,$

(ii) $\displaystyle\sum_{E \in \mathscr{E}_{q,r}} \hat{A}(E^\perp)^2 = \frac{n}{q-1}|M_2| - (q-1)|A|^2,$

(iii) $\displaystyle\sum_{E \in \mathscr{E}_{q,r}} \hat{A}(E^\perp)^3 = \frac{n}{q-1}|M_3| - (q-1)^2|A|^3.$

Calculating $|M_2|$ gives a Parseval-type equation.

Corollary 136. *For every* $A \subseteq X$

$$\sum_{E \in \mathscr{E}_{q,r}} |\hat{A}(E^\perp)|^2 = n(q-1)\mathbf{1}_A(0) - (q-1)|A|^2 + n \sum_{W \in \mathscr{E}_1} |A \cap W^\sharp|^2$$

and in particular

$$\sum_{E \in \mathscr{E}_{q,r}} |\hat{A}(E^\perp)|^2 \geq |A|\,(n - (q-1)|A|).$$

For some hyperedge the one-sided discrepancy is not too small:

Lemma 137. *Let* $A \subseteq X$. *There exists an* $E \in \mathscr{E}_{q,r}$ *with* $d^+(A, E) \geq -1$.

Proof. With Lemma 133, we have

$$\sum_{E \in \mathscr{E}_{q,r}} d^+(A, E) = \frac{1}{q} \sum_{E \in \mathscr{E}_{q,r}} \hat{A}(E^\perp) = \frac{1}{q} \sum_{E \in \mathscr{E}_{q,r}} \sum_{z \in (E^\perp)^\sharp} \sum_{a \in A} e^{\frac{2\pi i}{p} \operatorname{Tr}(\langle a, z \rangle)}$$

$$= \frac{1}{q} \sum_{a \in A} \sum_{z \in X^\sharp} \sum_{\substack{E \in \mathscr{E}_{q,r} \\ z \in E^\perp}} e^{\frac{2\pi i}{p} \mathrm{Tr}(\langle a, z \rangle)}$$

$$= \frac{1}{q} \left(\sum_{a \in A^\sharp} \sum_{z \in X^\sharp} e^{\frac{2\pi i}{p} \mathrm{Tr}(\langle a, z \rangle)} + \mathbf{1}_A(0) \sum_{z \in X^\sharp} e^{\frac{2\pi i}{p} \mathrm{Tr}(\langle 0, z \rangle)} \right)$$

$$= \frac{1}{q} \left(\sum_{a \in A^\sharp} (-1) + \mathbf{1}_A(0)(n-1) \right)$$

$$= \frac{1}{q} \left(\mathbf{1}_A(0)(n-1) - |A^\sharp| \right)$$

$$= \frac{1}{q} \left(\mathbf{1}_A(0)(n) - |A| \right).$$

We already showed $|\mathscr{E}_{q,r}| = \frac{n-1}{q-1}$. There exists an $E \in \mathscr{E}_{q,r}$ with

$$d^+(A, E) \geq \frac{q-1}{n-1} \frac{1}{q} \left(\mathbf{1}_A(0)n - |A| \right).$$

If $0 \in A$, we get $d^+(A, E) \geq 0$. Thus, we can assume $0 \notin A$ and get

$$d^+(A, E) \geq -\frac{q-1}{n-1} \frac{|A|}{q} \geq -\frac{q-1}{n-1} \frac{n-1}{q} = -\frac{q-1}{q}$$

completing the proof. $\square$

5.6.4 Proof of the Main Result

As already explained, our strategy for the proof of Theorem 130 is to find either a large color-class A_i, or to show that the sum of squared Fourier coefficients is very large. In the first case, for the large color-class A_i the term $\frac{|A_i|}{q} - \frac{|E|}{c}$ is large. And using Lemma 137, we get a hyperedge $E \in \mathscr{E}_{q,r}$ with $\frac{\hat{A}_i(E^\perp)}{q} \geq -1$. This results in a large discrepancy. In the latter case we will use the sum of squared Fourier coefficients to deduce large discrepancy. The next lemma is a technical result, and for a proof we refer again to [41].

Lemma 138. *Let $k \in \mathbb{R}^+$ with $\frac{1}{q^{r-1}-q^{\frac{r}{2}}-1} \leq k \leq \frac{1}{3q}$ and $z = \frac{(q-1) \bmod c}{c}$. Either there exists a color $i \in [c]$ and a hyperedge $E \in \mathcal{E}_{q,r}$ with*

$$|A_i \cap E| - \frac{|E|}{c} > k\frac{\sqrt{n}}{c}$$

or we have

$$\sum_{i\in[c]} \sum_{E\in\mathcal{E}_{q,r}} |\hat{A}_i(E^{\perp})|^2 \geq \frac{n(n-1)c}{q-1}\left(z(1-z) - \frac{k}{c^2}(q-1)^2\right).$$

We need some definitions. For $E \in \mathcal{E}_{q,r}$ set

$$I^+(E) := \{i \in [c] : \hat{A}_i(E^{\perp}) \geq 0\},$$
$$I^-(E) := [c] \setminus I^+(E),$$
$$M = \max_{i\in[c]} \max_{E\in\mathcal{E}_{q,r}} |\hat{A}_i(E^{\perp})|.$$

Proposition 139. *There is an $E \in \mathcal{E}_{q,r}$ with*

$$\sum_{i\in I^+(E)} \hat{A}_i(E^{\perp}) = -\sum_{i\in I^-(E)} \hat{A}_i(E^{\perp}) \geq M. \qquad (5.62)$$

Proof. By Lemma 133, for any $E \in \mathcal{E}_{q,r}$

$$\sum_{i\in[c]} \hat{A}_i(E^{\perp}) = q\sum_{i\in[c]} d^+(A_i, E) = 0,$$

so

$$\sum_{i\in I^+(E)} \hat{A}_i(E^{\perp}) = -\sum_{i\in I^-(E)} \hat{A}_i(E^{\perp}) = \sum_{i\in I^-(E)} \left|\hat{A}_i(E^{\perp})\right|,$$

and the last sum is at least M for some E.   □

Proof (of Theorem 130). Let $r \geq r_0(q)$ and $\chi: X \to [c]$ be a c-coloring of $\mathcal{H}_{q,r}$. Set $k = \frac{\sqrt{cz(1-z)}}{4q(q-1)}$. We have to show

$$\mathrm{disc}^+(\mathcal{H}_{q,r}, c) \geq k\frac{\sqrt{n}}{c}.$$

We fix the constant $\xi = \sqrt{\frac{nc}{8q}\left(z(1-z) - \frac{k}{c^2}(q-1)^2\right)}$. A straightforward calculation shows

$$\frac{\xi}{c} \geq k\frac{\sqrt{n}}{c}. \tag{5.63}$$

which we will use later on. Both numbers are positive, therefore it suffice to prove $\frac{\xi^2}{n} - k^2 \geq 0$. We have

$$\begin{aligned}
\frac{\xi^2}{n} - k^2 &= \frac{c}{8q}\left(z(1-z) - \frac{\sqrt{cz(1-z)}(q-1)}{4qc^2}\right) - \frac{cz(1-z)}{16q^2(q-1)^2} \\
&= \frac{c}{8q}\left(z(1-z) - \frac{\sqrt{cz(1-z)}(q-1)}{4qc^2} - \frac{z(1-z)}{2q(q-1)^2}\right) \\
&\geq \frac{c}{8q}\left(z(1-z) - \frac{z(1-z)}{4} - \frac{z(1-z)}{4}\right) \\
&= \frac{cz(1-z)}{16q} \geq 0.
\end{aligned}$$

We make the following nested case distinctions.

Case 1:
$\sum_{i \in I^+(E)} \left(|A_i \cap E| - \frac{|E|}{c}\right) \geq \xi$ for an $E \in \mathscr{E}_{q,r}$.
In this case we have

$$c\, \mathrm{disc}^+(\mathscr{H}_{q,r}, \chi, c) \geq \sum_{i \in I^+(E)} \left(|A_i \cap E| - \frac{|E|}{c}\right) \geq \xi$$

and hence,

$$\mathrm{disc}^+(\mathscr{H}_{q,r}, \chi, c) \geq \frac{\xi}{c} \overset{(5.63)}{\geq} k\frac{\sqrt{n}}{c}. \tag{5.64}$$

Case 2: $\sum_{i \in I^+(E)} \left(|A_i \cap E| - \frac{|E|}{c}\right) < \xi$ for all $E \in \mathscr{E}_{q,r}$.
Thus,

$$\sum_{i \in I^-(E)} \left(|A_i \cap E| - \frac{|E|}{c}\right) = -\sum_{i \in I^+(E)} \left(|A_i \cap E| - \frac{|E|}{c}\right) > -\xi \tag{5.65}$$

for all $E \in \mathscr{E}_{q,r}$.

Case 2.1: $M \geq 2q\xi$.
By (5.62), we have for an appropriate $E \in \mathscr{E}_{q,r}$:

$$\frac{1}{q}\sum_{i \in I^-(E)} |A_i| - |I^-(E)|\frac{|E|}{c}$$

$$\overset{\text{Lemma 133}}{=} \sum_{i \in I^-(E)} \left(|A_i \cap E| - \frac{|E|}{c} \right) - \frac{1}{q} \sum_{i \in I^-(E)} \hat{A}_i(E^\perp)$$

$$\overset{(5.62, 5.65)}{>} -\xi + \frac{1}{q} M \geq -\xi + \frac{2q\xi}{q} = \xi.$$

Hence, there exists an $i_0 \in I^-(X)$ with

$$\frac{1}{q}|A_{i_0}| - \frac{|E|}{c} \geq \frac{\xi}{c} \overset{(5.63)}{\geq} k\frac{\sqrt{n}}{c}.$$

Lemma 137 ensures the existence of an $E_0 \in \mathscr{E}_{q,r}$ with

$$\frac{1}{q} \hat{A}_{i_0}(E_0^\perp) \geq -1$$

and thus,

$$\text{disc}^+(\mathscr{H}_{q,r}, \chi, c) \geq |A_{i_0} \cap E_0| - \frac{|E_0|}{c} \tag{5.66}$$

$$= \frac{1}{q} \hat{A}_{i_0}(E_0^\perp) + \frac{1}{q}|A_{i_0}| - \frac{|E_0|}{c}$$

$$\geq k\frac{\sqrt{n}}{c} - 1. \tag{5.67}$$

Case 2.2: $M < 2q\xi$.

To use Lemma 138 we have to verify $\frac{1}{q^{r-1}-q^{\frac{r}{2}-1}} \leq k \leq \frac{1}{3q}$, which can be done with a little algebraic manipulations. By Lemma 138 either there exists a color $i \in [c]$ and an $E \in \mathscr{E}_{q,r}$ such that

$$|A_i \cap E| - \frac{|E|}{c} > k\frac{\sqrt{n}}{c} \tag{5.68}$$

or we have

$$\sum_{i \in [c]} \sum_{E \in \mathscr{E}_{q,r}} |\hat{A}_i(E^\perp)|^2 \geq \frac{n(n-1)c}{q-1} \left(z(1-z) - \frac{k}{c^2}(q-1)^2 \right) \tag{5.69}$$

In the first case we get a lower bound for the one-sided discrepancy as desired. Thus we can assume that (5.69) is satisfied. There exists an $E \in \mathscr{E}_{q,r}$ with

$$\sum_{i \in [c]} |\hat{A}_i(E^\perp)|^2 \geq nc \left(z(1-z) - \frac{k}{c^2}(q-1)^2 \right).$$

And we get

$$2M \left(\sum_{i \in I^-(E)} |\hat{A}^i(E^\perp)| \right) = M \left(\sum_{i \in I^-(E)} |\hat{A}^i(E^\perp)| + \sum_{i \in I^+(E)} |\hat{A}^i(E^\perp)| \right)$$

$$= M \sum_{i \in [c]} |\hat{A}^i(E^\perp)| \geq \sum_{i \in [c]} |\hat{A}^i(E^\perp)|^2$$

$$\geq nc \left(z(1-z) - \frac{k}{c^2}(q-1)^2 \right).$$

Furthermore

$$\sum_{i \in I^-(E)} |\hat{A}^i(E^\perp)| \geq \frac{nc}{2M} \left(z(1-z) - \frac{k}{c^2}(q-1)^2 \right)$$

$$> \frac{nc}{4q\xi} \left(z(1-z) - \frac{k}{c^2}(q-1)^2 \right) = 2\xi$$

and (5.61) gives

$$\frac{1}{q} \sum_{i \in I^-(E)} |A_i| - |I^-(E)| \frac{|E|}{c}$$

$$= \sum_{i \in I^-(E)} \left(|A_i \cap E| - \frac{|E|}{c} \right) - \frac{1}{q} \sum_{i \in I^-(E)} \hat{A}_i(E^\perp)$$

$$\overset{(5.65)}{>} -\xi + 2\xi = \xi.$$

So, there exists an $i_0 \in I^-(X)$ such that

$$\frac{1}{q}|A_{i_0}| - \frac{|E|}{c} > \frac{\xi}{c} \overset{(5.63)}{\geq} k\frac{\sqrt{n}}{c}.$$

Lemma 137 implies the existence $E_0 \in \mathscr{E}_{q,r}$ with

$$\frac{1}{q}\hat{A}_{i_0}(E_0^\perp) \geq -1.$$

It follows that

$$\mathrm{disc}^+(\mathscr{H}_{q,r}, \chi, c) \geq |A_{i_0} \cap E_0| - \frac{|E_0|}{c}$$

$$= \frac{1}{q} \hat{A}_{i_0}(E_0^\perp) + \frac{1}{q} |A_{i_0}| - \frac{|E|}{c}$$

$$\geq k \frac{\sqrt{n}}{c} - 1. \tag{5.70}$$

We have shown

$$\mathrm{disc}^+(\mathcal{H}_{q,r}, \chi, c) \geq k \frac{\sqrt{n}}{c} - 1 = \frac{\sqrt{z(1-z)n}}{4q(q-1)\sqrt{c}} - 1.$$

This completes the proof. $\square$

5.6.5 Large Number of Colors

So far, our upper and lower bounds for the one-sided c-color discrepancy differ by a factor of $\sqrt{c \log c}$. Interestingly, the gap can be reduced to a factor of $\sqrt{\log c}$ for a large number of colors, namely if $c \geq q n^{1/3}$. The upper bound in (128) then is

$$\mathrm{disc}^+(\mathcal{H}_{q,r}, c) = O_q\left(\sqrt{\frac{n}{c}} \log c\right).$$

Let us proceed to the proof of the lower bound.

Theorem 140. *Let $c \geq q n^{1/3}$ and $r \geq 4$. We have*

$$\mathrm{disc}^+(\mathcal{H}_{q,r}, c) \geq \frac{1}{22\sqrt{q}} \sqrt{\frac{n}{c}} - 1.$$

In particular,

$$\mathrm{disc}^+(\mathcal{H}_{q,r}, c) = \Omega_q\left(\sqrt{\frac{n}{c}}\right).$$

The key for the proof of Theorem 140 is the following lemma.

Lemma 141. *Let $A \subseteq X$ with $|A| \leq \frac{1}{2} q^{r-1}$ and $0 \in A$. There exists an $E \in \mathcal{E}_{q,r}$ with*

$$d^+(A, E) \geq \min\left\{\frac{1}{16(q-1)^2} \frac{n}{|A|}, \frac{1}{\sqrt{10q}} \sqrt{|A|}\right\}.$$

Proof. Corollary 136 yields

$$\sum_{E \in \mathcal{E}_{q,r}} |\hat{A}(E^\perp)|^2 \geq |A|(n - (q-1)|A|) \geq \frac{1}{2} n |A|.$$

Let us denote by $\mathscr{E}^+$ the set of all $E \in \mathscr{E}_{q,r}$ with $\hat{A}(E^\perp) \geq 0$ and by $\mathscr{E}^-$ the set of all $E \in \mathscr{E}_{q,r}$ with $\hat{A}(E^\perp) < 0$. Furthermore define $M = \max_{E \in \mathscr{E}_{q,r}} \hat{A}(E^\perp)$. Recall that we have defined $\alpha = |\mathscr{E}_{q,r}| = \frac{n-1}{q-1} \leq 2q^{r-1}$. Lemma 135 (i) yields

$$\sum_{E \in \mathscr{E}^-} \left| \hat{A}(E^\perp) \right| = - \sum_{E \in \mathscr{E}_{q,r}} \hat{A}(E^\perp) + \sum_{E \in \mathscr{E}^+} \hat{A}(E^\perp)$$

$$= |A| - n\chi_A(0) + \sum_{E \in \mathscr{E}^+} \hat{A}(E^\perp)$$

$$\leq \alpha M.$$

Using Lemma 135 (iii) in the same way, we get

$$\sum_{E \in \mathscr{E}^-} \left| \hat{A}(E^\perp) \right|^3 = \alpha M^3 + (q-1)^2 |A|^2$$

Thus, with the Cauchy–Schwarz inequality

$$\tfrac{1}{2} n |A| \leq \sum_{E \in \mathscr{E}^+} |\hat{A}(E^\perp)|^2 + \sum_{E \in \mathscr{E}^-} |\hat{A}(E^\perp)|^2$$

$$\leq \alpha M^2 + \sum_{E \in \mathscr{E}^-} |\hat{A}(E^\perp)|^{\frac{1}{2}} |\hat{A}(E^\perp)|^{\frac{3}{2}}$$

$$\leq \alpha M^2 + \left(\sum_{E \in \mathscr{E}^-} |\hat{A}(E^\perp)| \right)^{\frac{1}{2}} \left(\sum_{E \in \mathscr{E}^-} |\hat{A}(E^\perp)|^3 \right)^{\frac{1}{2}}$$

$$\leq \alpha M^2 + (\alpha M)^{\frac{1}{2}} \left((q-1)^2 |A|^3 + \alpha M^3 \right)^{\frac{1}{2}}.$$

The first case is $(q-1)^2 |A|^3 < \alpha M^3$. Here we have

$$\frac{1}{2} n |A| \leq \alpha M^2 + (\alpha M)^{\frac{1}{2}} (2\alpha M^3)^{\frac{1}{2}}$$

$$\leq (1 + \sqrt{2}) \alpha M^2$$

and thus, $M \geq \sqrt{\frac{q|A|}{10}}$. For the $E \in \mathscr{E}_{q,r}$ corresponding to M we have

$$d^+(A, E) = \frac{1}{q} \hat{A}(E^\perp) \geq \sqrt{\frac{|A|}{10q}}.$$

The other case, $(q-1)^2 |A|^3 \geq \alpha M^3$, can be treated in a similar way. $\square$

Proof (of Theorem 140). Consider an arbitrary c-coloring $\chi: X \to [c]$ of $\mathcal{H}_{q,r}$. Let $A_i := \chi^{-1}(i)$ for all $i \in [c]$. There exists at least one color-class A_i with $|A_i| \geq \frac{n}{c}$. In the case that there is no color-class A_i with $\frac{n}{c} \leq |A_i \cup \{0\}| \leq \frac{n}{c} + \frac{1}{3}\sqrt{\frac{n}{c}}$ there must be a color-class A_{i_0} with $|A_{i_0} \cup \{0\}| > \frac{n}{c} + \frac{1}{3}\sqrt{\frac{n}{c}}$. Using Lemma 137 we get

$$\mathrm{disc}^+(\mathcal{H}_{q,r}, \chi, c) \geq \max_{E \in \mathcal{E}_{q,r}} \left(d^+(A_{i_0} \cup \{0\}, E) + \frac{|A_{i_0} \cup \{0\}|}{q} - \frac{|E|}{c} \right) - 1$$

$$\geq \frac{1}{3q}\sqrt{\frac{n}{c}} - 1.$$

Now, there is a color-class A_{i_0} with $\frac{n}{c} \leq |A_{i_0} \cup \{0\}| \leq \frac{n}{c} + \frac{1}{3}\sqrt{\frac{n}{c}}$, because otherwise we are done. In order to invoke Lemma 141 we have to ensure $|A_{i_0} \cup \{0\}| < \frac{1}{2}q^{r-1}$. This can be seen as follows:

$$|A_{i_0} \cup \{0\}| \leq \frac{n}{c} + \frac{1}{3}\sqrt{\frac{n}{c}} \leq q^{r-\frac{4}{3}-1} + \frac{1}{3}q^{\frac{r}{2}-\frac{2}{3}-\frac{1}{2}}$$

$$\leq q^{r-1}\left(2^{-\frac{4}{3}} + \frac{1}{3}2^{-\frac{13}{6}}\right) < \frac{1}{2}q^{r-1}.$$

Hence, there exists an $E \in \mathcal{E}_{q,r}$ with

$$d^+(A_{i_0} \cup \{0\}, E) \geq \min\left\{ \frac{1}{16(q-1)^2}\frac{n}{|A_{i_0} \cup \{0\}|}, \frac{1}{\sqrt{10q}}\sqrt{|A_{i_0} \cup \{0\}|} \right\}$$

$$\geq \frac{1}{22\sqrt{q}}\sqrt{\frac{n}{c}}.$$

Thus, we have

$$\mathrm{disc}^+(\mathcal{H}_{q,r}, \chi, c)) \geq d^+(A_{i_0} \cup \{0\}, E) + \frac{|A_{i_0}|}{q} - \frac{|E|}{c} - 1 \geq \frac{1}{22\sqrt{q}}\sqrt{\frac{n}{c}} - 1$$

as asserted.   □

5.7 Some Open Problems

It is still an interesting problem to present algorithms that compute colorings with optimal or nearly optimal discrepancy which are both, *practically and theoretically* efficient.

We have proved in Sect. 5.3.5 an upper bound for the 2-color discrepancy of the hypergraph of Cartesian products of symmetric arithmetic progressions, but a lower bound is not known. In general, good lower and upper bounds would be of interest, both in two and more colors, for products of hypergraphs.

We discussed arithmetic progressions in the integers and sums of such progressions (Sects. 5.3 and 5.4). A natural generalization are quadratic arithmetic progressions, with a linear and a quadratic term. The analysis of the discrepancy function for such progressions seems to be difficult, because here exponential sums with quadratic terms in the exponent, special cases of Weyl sums, appear demanding for techniques able to master this kind of non-linearity in Fourier analysis.

Furthermore, in Sect. 5.5 we analyzed the discrepancy of arithmetic progressions in $\mathbb{Z}_p$, where p is a prime. The proof methods were using this fact. Can we develop also theory for non-prime p's?

In Sect. 5.6 we have established almost tight bounds for the c-color discrepancy of the hypergraph of linear hyperplanes in $\mathbb{F}_q^r$. For the one-sided c-color discrepancy of the same hypergraph we have given a lower and an upper bound which differ by a factor $\sqrt{c \log c}$. In particular, for any constant c and q upper and lower bound are tight up to a constant factor. Moreover, in the case that the number of colors c is large ($c \geq q n^{\frac{1}{3}}$) we have reduced this gap to a factor of $\sqrt{\log c}$. A challenging algorithmic problem still is to construct an $O_q(\sqrt{n})$-discrepancy coloring for the linear hyperplanes in $\mathbb{F}_q^r$.

Acknowledgements We thank Mayank Singhal, MSc, for reading this chapter and Dr. Volkmar Sauerland for his assistance in Latex.

References

1. N. Alon, J.H. Spencer, P. Erdős, *The Probabilistic Method* (Wiley, New York, 1992)
2. K. Appel, W. Haken, Every planar map is four colorable. I: Discharging. Ill. J. Math. **21**, 429–490 (1977)
3. K. Appel, W. Haken, J. Koch, Every planar map is four colorable. II: Reducibility. Ill. J. Math. **21**, 491–567 (1977)
4. L. Babai, T.P. Hayes, P.G. Kimmel, The cost of the missing bit: Communication complexity with help, in *Proceedings of the 30th Annual ACM Symposium on Theory of Computing, Dallas, Texas, USA, May 1998 (STOC 1998)*, 1998, pp. 673–682
5. W. Banaszczyk, Balancing vectors and Gaussian measures of n-dimensional convex bodies. Random Struct. Algorithm **12**(4), 351–360 (1998)
6. N. Bansal, Constructive algorithms for discrepancy minimization, in *Proceedings of the 51st Annual IEEE Symposium on Foundations of Computer Science, Las Vegas, Nevada, USA, October 2010 (FOCS 2010)*, 2010, pp. 3–10
7. N. Bansal, J.H. Spencer, *Deterministic discrepancy minimization*, Demetrescu, C. (ed.) et al., Algorithms – ESA 2011, 19th annual European symposium, Saarbrücken, Germany, September 5–9, 2011, Proceedings. (Springer, Berlin). Lecture Notes in Computer Science 6942, 408–420, 2011. doi:10.1007/978-3-642-23719-5_35
8. I. Bárány, B. Doerr, Balanced partitions of vector sequences. Lin. Algebra Appl. **414**(2–3), 464–469 (2006). doi:10.1016/j.laa.2005.10.024
9. I. Bárány, V.S. Grinberg, On some combinatorial questions in finite dimensional spaces. Lin. Algebra Appl. **41**, 1–9 (1981)
10. J. Beck, Roth's estimate of the discrepancy of integer sequences is nearly sharp. Combinatorica **1**, 319–325 (1981)

11. J. Beck, W.W.L. Chen, *Irregularities of distribution*. Cambridge Tracts in Mathematics, vol. 89 (Cambridge University Press, Cambridge, 1987)
12. J. Beck, T. Fiala, "Integer making" theorems. Discrete Appl. Math. **3**(1), 1–8 (1981)
13. J. Beck, V.T. Sós, Discrepancy Theory, in *Handbook of Combinatorics*, ed. by R.L. Graham, M. Grötschel, L. Lovász (Elsevier, Amsterdam, 1995), pp. 1405–1446. Chap. 26
14. J. Beck, J.H. Spencer, Integral approximation sequences. Math. Program. **30**, 88–98 (1984)
15. J. Beck, J.H. Spencer, Well distributed 2-colorings of integers relative to long arithmetic progressions. Acta Arithmetica **43**, 287–298 (1984)
16. T. Bohman, R. Holzman, Linear versus hereditary discrepancy. Combinatorica **25**(1), 39–47 (2005). doi:10.1007/s00493-005-0003-9
17. S.A. Cook, The complexity of theorem-proving procedures, *ACM, Proc. 3rd ann. ACM Sympos. Theory Computing, Shaker Heights, Ohio 1971*, 1971, pp. 151–158
18. M. Charikar, A. Newman, A. Nikolov, Tight hardness results for minimizing discrepancy, in *Proceedings of the Twenty-Second Annual ACM-SIAM Symposium on Discrete Algorithms*. SODA '11 (SIAM, Philadelphia, 2011), pp. 1607–1614
19. B. Chazelle, *The Discrepancy Method* (Cambridge University Press, Cambridge, 2000)
20. J. Cilleruelo, N. Hebbinghaus, Discrepancy in generalized arithmetic progressions. Eur. J. Combinator. **30**(7), 1607–1611 (2009). doi:10.1016/j.ejc.2009.03.006
21. B. Doerr, Linear discrepancy of totally unimodular matrices, in *Proceedings of the 12th Annual ACM-SIAM Symposium on Discrete Algorithms, Washington, DC, USA, January 2001 (SODA 2001)*, 2001, pp. 119–125
22. B. Doerr, Balanced coloring: Equally easy for all numbers of colors?, in *Proceedings of the 19th Annual Symposium on Theoretical Aspects of Computer Science, Juan les Pins, France, March 2002 (STACS 2002)*, ed. by H. Alt, A. Ferreira, Lecture Notes in Computer Science, vol. 2285 (Springer, Berlin-Heidelberg, 2002), pp. 112–120
23. B. Doerr, Discrepancy in different numbers of colors. Discrete Math. **250**, 63–70 (2002)
24. B. Doerr, Linear discrepancy of totally unimodular matrices. Combinatorica **24**, 117–125 (2004)
25. B. Doerr, l_p linear discrepancy of totally unimodular matrices. Lin. Algebra Appl. **420**(2–3), 663–666 (2007). doi:10.1016/j.laa.2006.08.019
26. B. Doerr, A. Srivastav, Multi-color discrepancies. Combinator. Probab. Comput. **12**, 365–399 (2003). doi:10.1017/S0963548303005662
27. B. Doerr, M. Gnewuch, N. Hebbinghaus, Discrepancy of symmetric products of hypergraphs. Electron. J. Combinator. **13**(1) (2006)
28. B. Doerr, N. Hebbinghaus, S. Werth, Improved bounds and schemes for the declustering problem. Theor. Comput. Sci. **359**, 123–132 (2006)
29. B. Doerr, A. Srivastav, P. Wehr, Discrepancy of cartesian products of arithmetic progressions. Electron. J. Combinator. **11**(1), 16 (2004). http://www.informatik.uni-kiel.de/~discopt/person/asr/publication/ap-electr04asrbedpetr.ps
30. P. Erdős, Some unsolved problems. Mich. Math. J. **4**(3), 291–300 (1957). doi:10.1307/mmj/1028997963
31. P. Erdős, J.H. Spencer, *Probabilistic Methods in Combinatorics* (Akadémia Kiadó, Budapest, 1974)
32. G.A. Freiman, *Elements of a structural theory of sumsets*. (Kazan': Kazan. Gosudarstv. Ped. Inst; Elabuzh. Gosudarstv. Ped. Inst., n.A., 1966)
33. A. Giannopoulos, On some vector balancing problems. Studia Math. **122**(3), 225–234 (1997)
34. E.D. Gluskin, Extremal properties of orthogonal parallelepipeds and their applications to the geometry of Banach spaces. Math. USSR Sbornik **64**(1), 85–96 (1989). doi:10.1070/SM1989v064n01ABEH003295
35. G. Gonthier, Formal proof - the four color theorem. Not. Am. Math. Soc. **55**(11), 1382–1393 (2008)
36. R.L. Graham, B.L. Rothschild, J.H. Spencer, *Ramsey Theory* (Wiley, New York, 1990)
37. D. Haussler, E. Welzl, ε-nets and simplex range queries. Discrete Comput. Geometry **2**, 127–151 (1987)

38. N. Hebbinghaus, Discrepancy of arithmetic structures, PhD thesis, Christian-Albrechts-Universität Kiel, Technische Fakultät, 2005. http://eldiss.uni-kiel.de/macau/receive/dissertation_diss_00001851
39. N. Hebbinghaus, Discrepancy of sums of two arithmetic progressions. Electron. Notes Discrete Math. **29**, 547–551 (2007). doi:10.1016/j.endm.2007.07.087
40. N. Hebbinghaus, A. Srivastav, Discrepancy of (centered) arithmetic progressions in $\mathbb{Z}_p$. Eur. J. Combinator. **35**, 324–334 (2014). doi:10.1016/j.ejc.2013.06.039
41. N. Hebbinghaus, T. Schoen, A. Srivastav, One-Sided Discrepancy of linear hyperplanes in finite vector spaces, in *Analytic Number Theory, Essays in Honour of Klaus Roth*, ed. by W.W.L. Chen, W.T. Gowers, H. Halberstam, W.M. Schmidt, R.C. Vaughan (Cambridge University Press, Cambridge, 2009), pp. 205–223. https://www.informatik.uni-kiel.de/~discopt/person/asr/publication/fq2rhneasrtos05.ps
42. M. Helm, On the Beck-Fiala theorem. Discrete Math. **207**(1–3), 73–87 (1999)
43. R.M. Karp, Reducibility among combinatorial problems, *(Russian. English original). Kibern. Sb., Nov. Ser. 12, 16-38 (1975); translation from Complexity of Computer Computations 1972, Plenum Press, New York*, 1973, pp. 85–103
44. L. Kliemann, O. Kliemann, C. Patvardhan, V. Sauerland, A. Srivastav, A New QEA Computing Near-Optimal Low-Discrepancy Colorings in the Hypergraph of Arithmetic Progressions, in *Proceedings of the 12th International Symposium on Experimental and Efficient Algorithms, Rome, Italy, June 2013 (SEA 2013)*, ed. by V. Bonifaci, C. Demetrescu, A. Marchetti-Spaccamela, Lecture Notes in Computer Science (Springer, Heidelberg, 2013), pp. 67–78. doi:10.1007/978-3-642-38527-8_8
45. P. Knieper, Discrepancy of Arithmetic Progressions, PhD thesis, Institut für Informatik, Humboldt-Universität zu Berlin, 1997
46. R. Lidl, H. Niederreiter, *Introduction to Finite Fields and Their Applications*, 2nd edn. (Cambridge University Press, Cambridge, 1994)
47. L. Lovász, *Coverings and colorings of hypergraphs*, Proceedings of the 4th Southeastern Conference on Combinatorics, Graph Theory and Computing, Boca Raton, FL, 1973
48. L. Lovász, J.H. Spencer, K. Vesztergombi, Discrepancies of set-systems and matrices. Eur. J. Combinator. **7**(2), 151–160 (1986)
49. S. Lovett, R. Meka, Constructive discrepancy minimization by walking on the edges. CoRR **abs/1203.5747** (2012)
50. J. Matoušek, Tight upper bound for the discrepancy of half-spaces. Discrete Comput. Geometry **13**, 593–601 (1995)
51. J. Matoušek, *Geometric Discrepancy* (Springer, Heidelberg, New York, 1999)
52. J. Matoušek, On the linear and hereditary discrepancies. Eur. J. Combinator. **21**(4), 519–521 (2000)
53. J. Matoušek, *Geometric Discrepancy* (Springer, Heidelberg, New York, 2010)
54. J. Matoušek, The determinant bound for discrepancy is almost tight. Proc. Am. Math. Soc. **141**(2), 451–460 (2013). doi:10.1090/S0002-9939-2012-11334-6
55. J. Matoušek, A. Nikolov, Combinatorial Discrepancy for Boxes via the Ellipsoid-Infinity Norm. **arXiv: 1408.1376** (2014)
56. J. Matoušek, J.H. Spencer, Discrepancy in arithmetic progressions. J. Am. Math. Soc. **9**(1), 195–204 (1996). http://www.jstor.org/stable/2152845
57. J. Matoušek, E. Welzl, L. Wernisch, Discrepancy and approximations for bounded VC-dimension. Combinatorica **13**(4), 455–466 (1993). doi:10.1007/BF01303517
58. K. Mulmuley, *Computational Geometry: An Introduction through Randomized Algorithms* (Prentice-Hall, Englewood Cliffs, 1994)
59. M.B. Nathanson, *Long arithmetic progressions and powers of 2*, Théorie des nombres, C. R. Conf. Int., Québec/Can. 1987, 735–739 (1989)
60. A. Nikolov, The Komlós conjecture holds for vector colorings. CoRR **abs/1301.4039** (2013). http://arxiv.org/abs/1301.4039
61. A. Nikolov, K. Talwar, L. Zhang, The geometry of differential privacy: the sparse and approximate cases, in *Proceedings of the 45th ACM Symposium on Theory of Computing*

Conference, Palo Alto, CA, USA, June 2013 (STOC 2013), ed. by D. Boneh, T. Roughgarden, J. Feigenbaum, 2013, pp. 351–360

62. A. Přívětivý, Discrepancy of sums of three arithmetic progressions. Electron. J. Combinator. **13**(1), 49 (2006)

63. A. Přívětivý, Coloring Problems in Geometric Context, PhD thesis, Charles University Prague, Faculty of Mathematics and Physics, 2007

64. R. Rado, Verallgemeinerung eines Satzes von van der Waerden mit Anwendungen auf ein Problem der Zahlentheorie. Sitzungsberichte der Preußischen Akademie der Wissenschaften, Physikalisch-Mathematische Klasse **16/17**, 589–596 (1933)

65. J. Riordan, *An Introduction to Combinatorial Analysis* (Wiley, New York, 1958)

66. N. Robertson, D.P. Sanders, P. Seymour, R. Thomas, The four-colour theorem. J. Combinator. Theory Ser. B **70**(1), 2–44 (1997). doi:10.1006/jctb.1997.1750

67. K.F. Roth, Remark concerning integer sequences. Acta Arithmetica **9**, 257–260 (1964)

68. W. Rudin, *Fourier Analysis on Groups* (Wiley, New York, 1962)

69. V. Sauerland, Algorithm Engineering for some Complex Practice Problems, PhD thesis, Christian-Albrechts-Universität Kiel, Technische Fakultät, 2012. http://eldiss.uni-kiel.de/macau/receive/dissertation_diss_00009452

70. J.H. Spencer, Six standard deviations suffice. Trans. Am. Math. Soc. **289**(2), 679–706 (1985)

71. J.H. Spencer, *Ten Lectures on the Probabilistic Method*, 2nd edn. (CBMS-NSF Regional Conference Series in Applied Mathematics. SIAM, Society for Industrial and Applied Mathematics, Philadelphia, 1994)

72. A. Srinivasan, Improving the discrepancy bound for sparse matrices: better approximations for sparse lattice approximation problems, in *Proceedings of the 8th Annual ACM-SIAM Symposium on Discrete Algorithms, New Orleans, LA, USA, January 1997 (SODA 1997)*, 1997, pp. 692–701

73. A. Srivastav, Derandomization in Combinatorial Optimization, in *Handbook of Randomized Computing, Vol. II*, ed. by S. Rajasekaran, P.M. Pardalos, J.H. Reif, J.D.P. Rolim (Academic Publisher, Dordrecht, 2001), pp. 731–842. http://www.informatik.uni-kiel.de/~discopt/person/asr/publication/kluwerartikel.ps

74. A. Srivastav, P. Stangier, Algorithmic Chernoff-Hoeffding inequalities in integer programming. Random Struct. Algorithm **1**(8), 27–58 (1996). doi:10.1002/(SICI)1098-2418(199601)8:1<27::AID-RSA2>3.0.CO;2-T

75. B.L. van der Waerden, Beweis einer Baudetschen Vermutung. Nieuw Archief voor Wiskunde **15**, 212–216 (1927)

Chapter 6
Algorithmic Aspects of Combinatorial Discrepancy

Nikhil Bansal

Abstract This chapter describes some recent results in combinatorial discrepancy theory motivated by designing efficient polynomial time algorithms for finding low discrepancy colorings. Until recently, the best known results for several combinatorial discrepancy problems were based on counting arguments, most notably the entropy method, and were widely believed to be non-algorithmic. We describe some algorithms based on semidefinite programming that can achieve many of these bounds. Interestingly, the new connections between semidefinite optimization and discrepancy have lead to several new structural results in discrepancy itself, such as tightness of the so-called determinant lower bound and improved bounds on the discrepancy of union of set systems. We will also see a surprising new algorithmic proof of Spencer's celebrated six standard deviations result due to Lovett and Meka, that does not rely on any semidefinite programming or counting argument.

6.1 Introduction

In this chapter we consider the algorithmic aspects of several problems arising in discrepancy theory. In particular, we will focus on combinatorial discrepancy, which deals with the following type of question. There is a set-system (V, C) specified by the elements $V = \{1, \ldots, n\}$ and a collection of subsets $C = \{S_1, \ldots, S_m\}$ of V. Find a red-blue coloring of V such that each set in C is colored as evenly as possible.

Formally, let us use -1 and $+1$ to denote the colors red and blue. Given a coloring $\chi : V \to \{-1, 1\}$, let us define the discrepancy of χ for a set S as $\mathrm{disc}(\chi, S) := |\chi(S)|$ where $\chi(S) := \sum_{i \in S} \chi(i)$. Note that $\chi(S)$ measures the imbalance between the number of elements in S that are colored -1 and 1. The discrepancy of the

N. Bansal (✉)
Eindhoven University of Technology, 5600 MB Eindhoven, The Netherlands
e-mail: n.bansal@tue.nl

W. Chen et al. (eds.), *A Panorama of Discrepancy Theory*, Lecture Notes in Mathematics 2107, DOI 10.1007/978-3-319-04696-9_6,
© Springer International Publishing Switzerland 2014

coloring χ for the system (V, C) is defined as the maximum discrepancy over all sets $S \in C$,

$$\text{disc}(\chi, C) := \max_{S \in C} |\chi(S)|.$$

The discrepancy of the system (V, C) is the minimum discrepancy over all possible colorings, i.e.

$$\text{disc}(C) = \min_{\chi} \max_{j \in [m]} |\chi(S_j)|.$$

More generally, one can define the discrepancy of a $m \times n$ matrix A as $\text{disc}(A) := \min_{x \in \{-1,1\}^n} \|Ax\|_\infty$. Clearly, if A is the incidence matrix of a set system, then this is precisely the discrepancy of the set system as defined above. Almost all of the results that we consider in this chapter generalize to arbitrary matrices in a straightforward way. However we will focus on the case of set systems to keep the notation simple, and also for historical reasons, as most results in combinatorial discrepancy are stated for set systems.

Roughly speaking, the discrepancy of a set system is a useful measure of its inherent complexity, and hence its understanding it is related to several areas in mathematics and theoretical computer science. In computer science for example, discrepancy is useful in topics such as probabilistic and approximation algorithms, computational geometry, numerical integration, derandomization, complexity theory, data structures and so on. We discuss one such application in Sect. 6.1.1, but for more details we refer the reader to [10, 13, 20].

Algorithmic Aspects. Motivated by these applications, several interesting and non-trivial techniques have been developed for both upper bounding and lower bounding the discrepancy of various classes of set systems. As we shall see, many of these techniques are non-constructive in the sense that they prove the existence of a low discrepancy coloring, but give no clue about how to find one *efficiently*. As usual, an algorithm is called efficient if its running time on every input instance is polynomial in the size of the description of the instance. For us, this means that the algorithm must run in time polynomial in n and m. In particular, the algorithm that enumerates over all possible 2^n colorings and picks the best one is not efficient.

Designing efficient algorithms for finding low discrepancy colorings has several motivations. First, in many applications (see Sect. 6.1.1) one actually needs to find such a coloring efficiently. Another motivation is from theoretical computer science where one wishes to understand which problems admit efficient algorithms and which ones do not. Until recently, many techniques for upper-bounding discrepancy were believed to be inherently non-algorithmic.

Approximating Discrepancy. Perhaps the most natural question is the following. Given a set system (V, C) can we determine its discrepancy exactly, or perhaps even approximately, in polynomial time? Unfortunately, it turns out that this general question is essentially hopeless in a very strong sense. In particular, recently Charikar, Newman and Nikolov [9] showed the following.

Theorem 1. *Given a set system on n elements and m = O(n) sets, it is NP-hard to distinguish whether the system has discrepancy 0 or $\Omega(\sqrt{n})$.*

In particular, Theorem 1 says that assuming P≠NP, no polynomial time algorithm can distinguish even among the following two extreme cases (i) whether a given set system with $O(n)$ sets has discrepancy 0 or (ii) has discrepancy at least $c\sqrt{n}$ for some universal constant c.

We assume that the reader is familiar with basic notions of computational complexity such as P≠ NP. For more details, an excellent reference is [26].

Theorem 1 is surprisingly tight, as the celebrated "six standard deviations suffice" result of Spencer [27] shows that

Theorem 2. *Every set system with m = n sets has discrepancy at most $6\sqrt{n}$. More generally, for m > n, the discrepancy is $O((n\log(m/n))^{1/2})$.*

We shall prove Theorems 1 and 2 in Sects. 6.6 and 6.2.2.

Hereditary Discrepancy. One reason why discrepancy is so hard to estimate (in the sense of Theorem 1) is that it is a very *fragile* quantity. The following example is instructive.

Let (V, C) be a set system with high discrepancy, where $V = \{1, \ldots, n\}$ and $C = \{S_1, \ldots, S_m\}$. Let (V', C') be a copy of (V, C) on a different ground set. That is, let $V' = \{n + 1, \ldots, 2n\}$ and $C' = \{S'_1, \ldots, S'_m\}$ where each $S'_i = \{n + j : j \in S_i\}$ is a copy of S_i on V'. Consider the system $(W, D) := ((V \cup V', \{S_1 \cup S'_1, \ldots, S_m \cup S'_m\})$, that is, with elements $V \cup V'$ and with i-th set $S_i \cup S'_i$. As the system (W, D) contains the system (V, C) (restricting (W, D) to V gives (V, C)), it is at least as complex as C, and hence one would expect its discrepancy to be no less than that of (V, C). However, (W, D) has discrepancy zero for trivial reasons, as one can color all the elements in V by 1 and those in V' by -1.

To get around these kinds of anomalies, it is very useful to define the *hereditary* discrepancy of a set system (V, C). Specifically, given $V' \subseteq V$, let $C|_{V'}$ denote the collection $\{S \cap V' : S \in C\}$. Then, the hereditary discrepancy of (V, C) is defined as

$$\text{herdisc}(C) = \max_{V' \subseteq V} \text{disc}(C|_{V'}).$$

Most upper bounds stated in terms of discrepancy also imply the same bound for hereditary discrepancy (for example if a class of systems is closed under taking restrictions of the ground set, then bounding the discrepancy of systems in this class, also implies bounds on hereditary discrepancy).

Let us consider an application to see how the concepts of discrepancy and hereditary discrepancy can be useful.

6.1.1 An Application

Suppose we are given a fractional solution $x \in \mathbb{R}^n$, to a linear system $Ax = b$ on n variables and m constraints. We would like to *round* x to an integral solution $\tilde{x}$ such that the error in each of the m equations is as low as possible, i.e. find $\tilde{x} \in \mathbb{Z}^n$ that minimizes $\|A(x - \tilde{x})\|_\infty$.

The answer to this question turns out to be closely related to the hereditary discrepancy of the matrix A.

Theorem 3 (Lovász, Spencer, and Vesztergombi [18]). *For any $x \in \mathbb{R}^n$ satisfying $Ax = b$, there is a $\tilde{x} \in \mathbb{Z}^n$ with $\|\tilde{x} - x\|_\infty < 1$, such that $\|A(x - \tilde{x})\|_\infty \leq$* herdisc($A$).

Proof. Let $x = (x_1, \ldots, x_n)$, and consider the binary expansion of each x_i. That is, we write x_i as $\lfloor x_i \rfloor + \sum_{j \geq 1} q_{ij} 2^{-j}$, where $q_{ij} \in \{0, 1\}$ denotes the j-th bit of x_i after the decimal point. The idea of the proof is to round the bits q_{ij} to 0, while introducing low error in each of the m equations.

Let k be a fixed positive integer. Consider the k-th bit q_{ik} of each x_i. Let $A^{(k)}$ be the sub-matrix of A restricted to those columns i for which $q_{ik} = 1$. By the definition of hereditary discrepancy, there is a $\{-1, 1\}$ coloring $\chi^{(k)}$ of the columns of $A^{(k)}$ such that the discrepancy of each row of $A^{(k)}$ is at most herdisc(A). Viewing $\chi^{(k)}$ as a vector in $\mathbb{R}^n$ with entries $\{-1, 0, 1\}$ (where the 0 entries correspond to the columns not in $A^{(k)}$), consider the vector $x' = x + 2^{-k}\chi^{(k)}$. Now, the k-th bit of each x_i' is 0, as $\chi^{(k)}(i) \in \{-1, 1\}$ if $q_{ik} = 1$ and 0 if $q_{ik} = 0$. Moreover, $Ax' - Ax = A(x' - x) = 2^{-k} A\chi^{(k)}$ and hence $\|Ax - Ax'\|_\infty \leq 2^{-k} \cdot$ herdisc(A).

We now iterate this process (treating x' as x) on the bits at position $k - 1, k - 2, \ldots, 1$. This produces a vector $x' = x + 2^{-k}\chi^{(k)} + 2^{-k+1}\chi^{(k-1)} + \ldots + 2^{-1}\chi^{(1)}$, where the first k bits of each x_i' are 0, and $|Ax - Ax'| \leq$ herdisc(A)$(2^{-k} + 2^{-k+1} + \ldots 2^{-1}) <$ herdisc(A). Making k arbitrarily large implies the final result. $\square$

Let us note a few things about the proof, as these ideas will also be useful later. (i) For each bit position j, the coloring $\chi^{(j)}$ is used as a guide whether to round the bit q_{ij} up or down. (ii) After the update is applied to the j-th bit, the bits in positions $j - 1$ or earlier might change due to carry-overs, and hence the submatrix $A^{(j-1)}$ and the coloring $\chi^{(j-1)}$ at the next step depend on the previous choices $\chi^j, \chi^{(j+1)}, \ldots$ (iii) The proof does not give an efficient algorithm to find $\tilde{x}$, as the low hereditary discrepancy property only shows the existence of some good χ^j.

We remark that Doerr [12] gives an improved error bound of $(1 - 1/2m)$herdisc(A) in Theorem 3.

6.1.2 Chapter Overview

The chapter is organized as follows. We first describe some classical methods for both upper bounding and lower bounding the discrepancy of various types of set systems. We then discuss the more recent results, that will build upon these previous ideas. To keep our discussion manageable, in this chapter we mostly focus on two problems which will suffice to convey the main ideas.

Arbitrary Set Systems. Given n elements and an arbitrary collection of $m = n$ sets, find a minimum discrepancy coloring.

Bounded Degree Set Systems. Given n elements and an arbitrary collection of sets such that each element lies in at most t sets, find a minimum discrepancy coloring.

We now give a brief overview of the topics and the results that we will consider.

6.1.2.1 Classical Upper Bound Methods

Random Coloring. Given a set system (V, C), perhaps the simplest idea is to color each element randomly and independently ± 1 with probability $1/2$ each.

Lemma 4. *For any set system (V, C) on n elements and m sets, a random coloring has discrepancy $O(\sqrt{n \log m})$ with high probability.*

Proof. For any set $S \in C$, $\chi(S)$ is a random variable with mean 0 and variance $|S|$. Thus, by standard Chernoff bounds (see for e.g. [1]), $\Pr[|\chi(S)| \geq c\sqrt{|S|}] \leq 2\exp(-c^2/2)$. Choosing, say $c = 2\sqrt{\log m}$, this probability is at most $2/m^2$, and hence by a union bound over the m sets in C, $\max_S |\chi(S)| \leq 2\sqrt{n \log m}$ with probability at least $1 - 2/m$. □

In particular, for $m = n$ this gives a discrepancy of $O(\sqrt{n \log n})$. This is reasonably good as there exist set systems on n sets with discrepancy at least $\Omega(\sqrt{n})$ (we will see this in Sect. 6.6). However, a random coloring is not very interesting as it is completely oblivious to the problem structure. For example, even for bounded degree set systems with $t = O(1)$, a random coloring only gives $\Omega(\sqrt{n})$ discrepancy, while the actual discrepancy is $O(1)$ (see Theorem 5 below). Moreover, note that Theorem 2 always beats random coloring for any set system.

Linear Algebraic Method. A simple but powerful approach for bounding discrepancy is based on basic linear algebra. An interesting application of this method is the following result for bounded degree set systems.

Theorem 5 ([6]). *If A is a set system where each element lies in at most t sets, then $\mathrm{disc}(A) \leq 2t - 1$.*

We shall see the proof of this theorem is Sect. 6.2.1. While the idea itself is quite simple, the underlying idea will play a key role in later results.

The Entropy Method. One of the most powerful and widely used tools in combinatorial discrepancy is the so-called partial coloring method due to [5] and its refinements [27] based on the so-called entropy method. For example, Spencer's original proof of Theorem 2 was based on the entropy method. It can also be used to prove a $O(\sqrt{t} \log n)$ bound [28] for bounded degree systems,[1] which can sometimes be better than the bound in Theorem 5.

As we shall see in Sect. 6.2.2, the entropy method is based on a clever application of the pigeonhole principle to the space of all 2^n possible colorings, and only proves the existence of a low discrepancy coloring, without giving any algorithmic insight on how to find it efficiently. For example, even though Theorem 2 was long known, no polynomial time algorithm better than random coloring was known until recently for general set systems.

6.1.2.2   A Lower Bound Method

A variety of techniques have also been developed for proving lower bounds on discrepancy. Many of these are based on deep results and connections to various areas of mathematics (see for e.g. [10, 20]). One of the strongest known results in this direction is the following *determinant lower bound* due to [18]. For a real matrix A, define

$$\mathrm{detlb}(A) := \max_k \max_B |\det B|^{1/k},$$

where the maximum is over all $k \times k$ submatrices B of A. For a set system (V, C), let $\mathrm{detlb}(C)$ denote $\mathrm{detlb}(A)$ where A is the incidence matrix of (V, C).

Theorem 6 ([18]). *For every set system C,* $\mathrm{herdisc}(C) \geq \frac{1}{2}\mathrm{detlb}(C)$.

We shall prove Theorem 6 in Sect. 6.7.1. The proof is based on an interesting geometric interpretation of hereditary discrepancy which will be useful later.

6.1.2.3   Some Recent Results

Recently, several advances have been made in combinatorial discrepancy theory. First, many previous (non-constructive) results based on the entropy method, can now be made algorithmic. In particular, Bansal [3] showed the following algorithmic version of Spencer's result.

Theorem 7 ([3]). *For any set system on $m = O(n)$ sets, there is a randomized polynomial time algorithm to find an $O(\sqrt{n})$ discrepancy coloring.*

This result is based on semidefinite programming (SDPs) and in particular a new method to round SDP solutions based on designing correlated gaussian random walks. This method has several other applications. For example, it gives an

---

[1]Interestingly, an improved $O(\sqrt{t} \log n)$ bound is also known [2] using a different method based on convex geometry.

algorithm to find $O(\sqrt{t} \log n)$ discrepancy coloring for bounded degree systems, matching the bound of [28]. It also gives a good approximation algorithm to find a low discrepancy coloring for systems with low hereditary discrepancy.

Theorem 8 ([3]). *For any set system (V, C) on n elements and m sets, there is a randomized polynomial time algorithm to find a coloring with discrepancy at most $O((\log m \log n)^{1/2}) \cdot \text{herdisc}(C)$.*

Theorem 8 directly implies the following algorithmic version of Theorem 3.

Corollary 9. *Given any fractional solution $x \in \mathbb{R}^n$ to the system $Ax = b$ on m equations, there is a polynomial time algorithm to round x to $\tilde{x} \in \mathbb{Z}^n$ such that $\|A(x - \tilde{x})\|_\infty = O((\log m \log n)^{1/2}) \cdot \text{herdisc}(A)$.*

In Sect. 6.3 we give the relevant background on semidefinite programming and prove Theorem 8. Then we show how these ideas can be refined and combined with the entropy method to obtain Theorem 7. However, instead of proving the $O(\sqrt{n})$ bound, we show a weaker $O((n \log \log \log n)^{1/2})$ bound (which is already much stronger than random coloring). This weaker bound illustrates all the ideas involved without the tedious calculations needed for the $O(\sqrt{n})$ bound.

Somewhat surprisingly, even though the algorithm in Theorem 7 is polynomial time, it crucially relies on the non-constructive entropy method in its design, and in particular does not give a truly constructive proof of Spencer's result. This unsatisfying situation was resolved very recently by Lovett and Meka [19] who gave a simpler and completely constructive proof of Spencer's result based on gaussian random walks and linear algebraic ideas. In particular, their proof does not use entropy method. We will see their proof in Sect. 6.5. Interestingly, their main result implies a variant of the partial coloring that is quantitatively stronger (in a certain regime) than the one obtained by the entropy method. This variant can also be viewed as a "robust" version of the so-called iterated rounding technique in approximation algorithms, and has found some other very interesting algorithmic uses recently [25].

The connections between discrepancy and semidefinite programming have also been useful in other ways besides algorithm design. In a very interesting result, Matoušek [21] gave the first non-trivial upper bound on the gap between the determinant lower bound and hereditary discrepancy.

Theorem 10 ([21]). *For any set system (V, C), $\text{herdisc}(C) \leq O(\log n \sqrt{\log m}) \cdot \text{detlb}(C)$.*

This result is remarkably tight, as there exist set systems for which $\text{herdisc}(C) = \Omega(\log n) \cdot \text{detlb}(C)$ [23]. The proof of Theorem 10 is based on SDP duality, and relating the dual SDP solution to sub-determinants of A via eigenvalues. We will prove Theorem 10 in Sect. 6.7.1.

Among other things, Theorem 10 also implies the following new structural result in discrepancy [21]: For any two set systems (V, C_1) and (V, C_2),

$$\text{herdisc}(C_1 \cup C_2) \leq \max(\text{herdisc}(C_1), \text{herdisc}(C_2)) \cdot O(\log n \sqrt{\log m}).$$

Previously such a result was known only for the special case when C_2 consists of a single set [16]! A further extension of this result to the union of t set-system can also be found in [21].

Some Related Results that We Do not Discuss. A very recent and surprising result that we do not discuss in this chapter, is the first algorithm for approximating hereditary discrepancy to within poly-logarithmic factors [22]. Note that apriori it is not even clear whether the minimum hereditary discrepancy problem is in NP.[2] For this to hold, given a set system (V, C) and a target hereditary discrepancy λ, there must exist a (short) polynomial time verifiable witness that certifies that $\mathrm{disc}(C|_J) \leq \lambda$ for each of the 2^n subsets $J \subseteq V$. The result of [22] is based on combining the geometric interpretation of hereditary discrepancy together with powerful results and ideas from convex geometry, most notably the restricted invertibility principle by Bourgain and Tzafriri [7] and its refinements due to Vershynin [30].

Finally, the recent algorithmic ideas developed for addressing discrepancy related questions have also led to some very interesting results in optimization and algorithms (e.g. [8, 14, 24, 25]). However, a discussion of these is beyond the scope of this chapter.

6.2 Some Classic Results

We describe some classical results and techniques which will be very useful later.

6.2.1 Linear Algebraic Method

Beck and Fiala proved the following bound on the discrepancy of bounded degree set systems, based on the linear algebraic method. Even though very elementary, this will be an important proof for us, as its high level idea will be used many times later.

Theorem 11 ([6]). *If (V, C) is a set system such that each element lies in at most d sets, then $\mathrm{disc}(S) \leq 2d - 1$. Moreover, such a coloring can be found in polynomial time.*

Proof. We will give an algorithm that starts with the all 0-coloring χ_0, i.e. $\chi_0(i) = 0$ for all i, and iteratively updates the colors over time until they all reach -1 or 1. During the intermediate steps, the elements may be assigned a fractional coloring, i.e. in $[-1, 1]$ instead of $\{-1, 1\}$. We now describe how the algorithm proceeds.

---

[2]Observe that proving a c-approximation for a problem, implies that the (approximate) problem is both in NP and co-NP. Note that both Theorems 6 and 8 can be used to give a co-NP witness.

Let $\chi_t = (\chi_t(1), \ldots, \chi_t(n))$ denote the coloring after the t-th iteration. Initially, $\chi_0(i) = 0$ for all i. We say that i is *alive* at time t, if $|\chi_{t-1}(i)| < 1$. Otherwise, if $\chi_{t-1}(i) \in \{-1, 1\}$, it is considered *fixed* and is never updated again. Call a set S *safe* at time t if at most d elements of S are alive. Otherwise, we call S *dangerous*.

The algorithm will ensure that at least one alive variable becomes fixed in each iteration, and hence it terminates in at most n steps. As an element's color is not updated once it is fixed, once a set becomes safe it stays safe henceforth. The coloring is updated in each iteration as follows. At iteration t, let C^t denote the collection of dangerous sets. The crucial observation is that the number of dangerous sets is strictly less than the number of alive elements. This follows as each (alive) element lies in at most d sets, and each dangerous set contains strictly more than d alive elements.

Let $v = (v_1, \ldots, v_n)$, and consider the linear system defined by $S_j \cdot v = 0$ ($\sum_{i \in S_j} v_i = 0$) for each dangerous set S_j and $v_i = 0$ for each element i that is fixed. By the observation above, there are strictly less than n constraints and hence by basic linear algebra, there must exist a non-zero update direction $v = (v(1), \ldots, v(n))$ such that $v(i) = 0$ if i is fixed, and $S_j \cdot v = 0$ if j is dangerous. We set $\chi_t = \chi_{t-1} + \delta v$ where $\delta > 0$ is the smallest real such that some alive variable reaches -1 or 1 and gets fixed.

To see that the discrepancy of any set S is at most $2d - 1$, observe that as long as a set is dangerous, the update rule ensures that $S \cdot \chi_t = 0$. However, once S becomes safe, it has at most d alive variables and thus no matter how these variables will be subsequently updated, the discrepancy added with be strictly less than $2d$. As the discrepancy is an integer, it can be at most $2d - 1$. $\square$

Let us note some key points about the proof. (i) The algorithms works with fractional colorings at intermediate steps. (ii) At each step it makes progress towards becoming an integral coloring. (iii) During the algorithm a set is protected as long as it is dangerous.

6.2.2   Entropy Method

The following result is known as the partial coloring lemma and is one of the most widely used techniques in combinatorial discrepancy. Its proof is based on a refined counting approach called the entropy method, and a clever pigeonhole principle argument. We closely follow the exposition in [20].

Theorem 12 (Partial Coloring Lemma via the Entropy Method). *Let (V, C) be a set system on n elements, and let a number $\Delta_S > 0$ be given for each set $S \in C$. Suppose the Δ_S satisfy the condition*

$$\sum_{S \in C} g\left(\frac{\Delta_S}{\sqrt{|S|}}\right) \le \frac{n}{5} \tag{6.1}$$

where

$$g(\lambda) = \begin{cases} Ke^{-\lambda^2/9} \ if \ \lambda > 0.1 \\ K\ln(\lambda^{-1}) \ if \ \lambda \le 0.1 \end{cases}$$

and K is some absolute constant. Then there is a partial coloring χ that assigns ± 1 to at least $n/10$ variables (and 0 to the rest), satisfying $|\chi(S)| \le \Delta_S$ for each $S \in C$.

We begin with some standard results that we will need to prove Theorem 12.

Entropy. Let Y be a discrete random variable that takes value y with probability p_y. Then, its *entropy* is defined as $H(Y) := \sum_y p_y \log_2(1/p_y)$.

Entropy satisfies the following properties (see e.g. [11]).

1. If $H(Y) \le k$, then $p_y \ge 2^{-k}$ for some value y.
2. If Y attains ℓ different values, then $H(Y) \le \log_2 \ell$. The equality is attained iff $Y = U_\ell$, the uniform distribution on ℓ values.
3. Subadditivity: If $Y_1, \ldots, Y_m$ are arbitrary (correlated) random variables, and $X = (Y_1, \ldots, Y_m)$ is a random vector with components $Y_1, \ldots, Y_m$, then $H(Y) \le \sum_i H(Y_i)$.

Roughly speaking, the entropy $H(X)$ measures the amount of randomness in X.

Lemma 13. *Let $k = 2^{4n/5}$ and suppose $\chi_1, \ldots, \chi_k$ are k distinct ± 1 colorings of $[n]$, such that for every two colorings χ_i and χ_j, with $i, j \in [k]$, it holds that $|\chi_i(S) - \chi_j(S)| \le 2\Delta_S$ for each set $S \in C$. Then, there exists a partial coloring χ that assigns ± 1 to at least $n/10$ elements (and 0 to the rest) satisfying $\chi(S) \le \Delta_S$ for each $S \in C$.*

Proof. This follows from the standard isoperimetric inequality for the hamming cube [17], which states that any subset $C \subset \{-1, 1\}^n$ of the cube with $|C| > \sum_{j=0}^{\ell} \binom{n}{j}$ contains two points in C with hamming distance at least 2ℓ.

As $2^{4n/5} > \sum_{h=0}^{n/20} \binom{n}{h}$ this implies that there must exist two colorings χ_i, χ_j in $\{\chi_1, \ldots, \chi_k\}$ with hamming distance at least $n/10$. Now, consider the vector $\chi = (\chi_i - \chi_j)/2$. Note that $\chi(\ell) = \pm 1$ whenever $\chi_i(\ell) \ne \chi_j(\ell)$, and is 0 otherwise. Moreover, for any set $S \in C$

$$|\chi(S)| = |\sum_{\ell \in S} \chi(\ell)| = |\frac{1}{2} \sum_{\ell \in S} (\chi_i(\ell) - \chi_j(\ell))| = |\frac{1}{2}(\chi_i(S) - \chi_j(S))| \le \Delta_S.$$

Here the last inequality follows from the property of the colorings $\chi_1, \ldots, \chi_k$. □

We now prove Theorem 12.

Proof. (Theorem 12) For a ± 1 coloring χ and a set $S \in C$, let $Y_\chi(S) :=$ round $\left(\frac{\chi(S)}{2\Delta_S}\right)$, where round$(x) = \lfloor x + 1/2 \rfloor$ is the rounding function to the nearest integer. Thus $Y_\chi(S)$ simply indicates which bucket of size $2\Delta_S$ the discrepancy $\chi(S)$ lies in.

Let $Y(S)$ be the random variable that takes value $Y_\chi(S)$ where the coloring χ is chosen uniformly at random among the 2^n possible colorings of $[n]$. We can determine $Y(S)$ exactly. In particular, if χ is a random coloring, then $\Pr[\chi(S) = k] = \binom{|S|}{(|S|-k)/2} 2^{-|S|}$ which is roughly $\exp(-k^2/(2|S|))$. In particular, $\chi(S)$ is distributed approximately uniformly in $[-\sqrt{S}, \sqrt{S}]$ and then decays super-exponentially in subsequent intervals of size $\sqrt{|S|}$. Let $\lambda_S := \Delta_S / \sqrt{|S|}$.

Claim. The entropy of Y_S satisfies $H(Y_S) \le g(\lambda_S)$, for g as defined in Theorem 12.

Proof (Sketch). We refer the reader to [20] for the precise calculation, but the idea is the following. For $\lambda_S \le 0.1$, Y_S is distributed essentially uniformly in $[-1/\lambda_S, 1/\lambda_S]$ and the probability that Y_S takes values outside this range decreases super-exponentially. Thus $H(Y_S) \approx H(U_{2\lambda_S}) = O(\log(1/\lambda_S))$.

On the other hand if λ_S is large (say $\lambda_S = 10$), then $p_0 = \Pr[Y_S = 0] \approx 1 - \exp(-\lambda_S^2/2)$ (and hence very close to 1), and for $\ell > 1$, $\Pr[|Y_S| = \ell] \le 2\exp(-\lambda_S^2\ell^2/2)$. Thus Y_S is essentially always 0 and has entropy roughly $p_0 \log(1/p_0) \approx \log(1/p_0) = O(\exp(-\lambda_S^2/2))$. Together this gives, $H(Y_S) \le g(\lambda_S)$ for any set S.   □

For $Y(S)$ as defined above, let Y denote the random vector $Y = (Y(S_1), \dots, Y(S_m))$ where $S_1, \dots, S_m$ are the sets in C. By sub-additivity of entropy and the claim above $H(Y) \le \sum_j H(Y_j) \le \sum_{S\in C} g(\lambda_S)$. As $\sum_S g(\lambda_S) \le n/5$, this gives $H(Y) \le n/5$ and hence Y attains some value $b = (b_1, \dots, b_m)$ with probability at least $2^{-n/5}$. Equivalently, there exist $k \ge 2^{4n/5}$ different colorings $\chi_i, i = 1, \dots, k$ such that $Y(\chi_i(S_j)) = b_j$ for each $i = 1, \dots, k$ and $j = 1, \dots, m$, and thus for every $1 \le i, i' \le k$ and $j = 1, \dots, m$ it holds that $|\chi_i(S_j) - \chi_{i'}(S_j)| \le 2\Delta_S$. Applying Lemma 13 now gives the result.   □

Let us see how Theorem 12 implies Spencer's result and the $O(\sqrt{t} \log n)$ bound for bounded-degree set systems.

Proof of Theorem 2. The coloring is constructed in phases. Let $n_0 = n$ and let n_i denote number of uncolored elements left at the beginning of phase i, for $i = 0, 1, \dots$. In phase i, we apply Theorem 12 to these n_i elements with $\Delta_S^i = c(n_i \log(2n/n_i))^{1/2}$ and verify that (6.1) holds when c is a large enough constant. This gives a partial coloring on at least $n_i/10$ elements, with discrepancy for any set S at most Δ_S^i. This gives that $n_{i+1} \le (0.9)n_i$ and hence $n_i \le (0.9)^i n$. Summing up over the phases, the discrepancy for any set is at most

$$\sum_i \Delta_S^i \le \sum_i c\left(n(0.9)^i \log\left(\frac{2n}{n(0.9)^i}\right)\right)^{1/2} = O(n^{1/2}).$$

□

Proof of $O(\sqrt{t} \log n)$ Discrepancy for Bounded Degree Systems [28]. The coloring is constructed in phases where at most $n_i \le n(0.9)^i$ elements are left uncolored in phase i. In phase i, let $s_{i,j}$ denote the number of sets with the number of uncolored elements in the range $[2^j, 2^{j+1})$. As the degree of the set system is at most t, we have that $s_{i,j} \leq \min(m, n_i t / 2^j)$. Using this fact, it can be verified that (6.1) holds if $\Delta_S = ct^{1/2}$ for some large enough constant c. Thus each set incurs $O(\sqrt{t})$ discrepancy in each phase and hence the total discrepancy incurred is $O(t^{1/2} \log n)$.

6.3   Systems with Low Hereditary Discrepancy

In this section we prove Theorem 8.

Theorem 8. *For any set system (V, C) on n elements and m sets, there is a randomized polynomial time algorithm to find a coloring with discrepancy $O((\log m \log n)^{1/2} \cdot \mathrm{herdisc}(C))$.*

The algorithm will be based on semidefinite programming, so we first give a brief overview of semidefinite programming.

Semidefinite Programming. Recall that a linear program (LP) consists of some collection of variables $x_1, \ldots, x_n$ where each x_i takes values in $\mathbb{R}$, and the goal is to optimize some linear objective function $\sum_i c_i x_i$, subject to some linear constraints $\sum_i a_{ji} x_i \leq b_j$ for $j = 1, \ldots, m$. Linear programs can be solved optimally in time that is polynomial in n, m, and the bit length required to describe the entries a_{ji}, b_j, c_i.

A semidefinite program (SDP) can be viewed as the following linear program. The variables are written in the form x_{ij} where $1 \leq i, j \leq n$ (the reason for writing variables in this form will become clear soon). There are m arbitrary linear constraints on x_{ij} of the type $\sum_{ij} a_{ij}^k x_{ij} \leq b^k$ for $k = 1, \ldots, m$ and there is some linear objective function $\sum c_{ij} x_{ij}$. Moreover, we require the symmetry condition $x_{ij} = x_{ji}$. Finally, one imposes the constraint that the $n \times n$ matrix $X = (x_{ij})$, consisting of entries x_{ij}, be positive semidefinite. That is, all its eigenvalues must be non-negative (this is well defined as X is symmetric and hence has only real eigenvalues), and we denote this by $X \succeq 0$. To summarize, a general SDP has the following form.

$$\min \sum c_{ij} x_{ij}$$

$$\text{s.t.} \quad \sum_{ij} a_{ij}^k x_{ij} \leq b^k, \qquad 1 \leq k \leq m$$

$$X \succeq 0$$

$$x_{ij} = x_{ji}, \qquad 1 \leq i, j \leq n$$

where a_{ij}^k, b^k, c_{ij} are arbitrary real numbers.

While the condition $X \succeq 0$ may appear non-linear, note that it can be enforced by adding (infinitely many) linear constraints of the form $a^T X a \geq 0$, one for each vector $a \in \mathbb{R}^n$. Despite the infinitely many constraints, by standard optimization theory and in particular the Ellipsoid method, this program can be solved to any desired level of accuracy in polynomial time. In particular, given a candidate solution X there exists an efficient separation procedure as one can determine in polynomial time whether $a^T X a < 0$ for some a, by computing the least eigenvalue of X and checking if it is negative. For more details about solving semidefinite programs, we refer the reader to [29].

Vector Program View. Recall that a symmetric $n \times n$ matrix X is positive semidefinite if and only if it is the Gram matrix of some vectors $v_1, \ldots, v_n \in \mathbb{R}^n$. That is, each entry x_{ij} can be written as $x_{ij} = \langle v_i, v_j \rangle$ where $\langle, \rangle$ denotes the standard inner product. Moreover, given X, the vectors v_i can be computed in polynomial time using the Cholesky decomposition procedure.

This implies that an SDP can equivalently be viewed as an arbitrary linear program where the variables correspond to inner product of vectors. This is referred to as the vector program view of an SDP, and it will be extremely useful for our purposes. Note that one can only impose constraints on the dot products of v_i's and not on the vectors v_i's themselves. Let us see how this is useful for discrepancy.

SDP Relaxation for Discrepancy. The natural SDP relaxation for the problem of finding a ± 1 coloring with discrepancy at most λ is the following.

$$\left\| \sum_{i \in S_j} v_i \right\|_2^2 \leq \lambda^2 \qquad \text{for each set } S_j \in C \tag{6.2}$$

$$\|v_i\|_2^2 = 1 \qquad \text{for each element } i \in V. \tag{6.3}$$

Here, as usual, $\|v\|_2 = (\langle v, v \rangle)^{1/2}$ denotes the length of v. The first constraint (6.2) says that the discrepancy of each set S_j must be at most λ. Observe that this is a linear constraint on the dot product of variables as the left hand side of (6.2) can be written as $\sum_{i \in S_j, i' \in S_j} \langle v_i, v_{i'} \rangle$. The second constraint says that each $\langle v_i, v_i \rangle = 1$, i.e. each v_i must be a unit vector.

This is a valid relaxation as any ± 1 coloring with discrepancy at most λ is a feasible solution to the above program (corresponding to the solution $v_i = (1, 0, \ldots, 0)$ or $v_i = (-1, 0, \ldots, 0)$ depending on whether i is colored 1 or -1). We will call any feasible solution to this SDP, a *vector-coloring* for C, and the smallest value λ for which this SDP is feasible as the *vector discrepancy* of C, denoted by $\text{vecdisc}(C)$. Clearly, $\text{vecdisc}(C) \leq \text{disc}(C)$.

Using this SDP. Before we describe our algorithm, we describe a natural approach for using this SDP that does not work, but it will give important insights.

Let A be a matrix with discrepancy λ (the reader should think of λ as being very small, say 0). First, we can assume that the algorithm knows λ, as it can try all values $0, 1, \ldots, n$ and pick the smallest λ for which the SDP given by (6.2)–(6.3) is feasible. Now let us consider some vector-coloring v_i obtained by solving this SDP. In this solution, the unit vectors v_i will be nicely correlated such that for every set S_j, the vector $\sum_{i \in S_j} v_i$ has length at most λ (say 0).

Our goal then is to convert these vectors v_i into the numbers ± 1 without increasing $\sum_{i \in S_j} v_j$ too much. A natural first step is to try to convert v_i into real numbers (hopefully close to ± 1) without substantially violating the sums $\sum_{i \in S_j} v_j$. So we project the vectors v_i on some vector $g \in \mathbb{R}^n$ to get real numbers $y_i = \langle g, v_i \rangle$. This seems reasonable as this maintains the correlations among v_i. In particular, these y_i satisfy

$$\sum_{i \in S_j} y_i = \sum_{i \in S_j} \langle g, v_i \rangle = \langle g, \sum_{i \in S_j} v_i \rangle \leq \|g\|_2 \cdot \| \sum_{i \in S_j} v_i \|_2,$$

implying that $\sum_{i \in S_j} y_i$ is also small if $\| \sum_{i \in S_j} v_i \|$ is small. To this end, it will be very convenient to project the v_i on to random gaussian vectors in $\mathbb{R}^n$.

Gaussian Random Variables. We recall the following standard facts about gaussian distributions.

1. The gaussian distribution $N(\mu, \sigma^2)$ with mean μ and variance σ^2 has probability distribution function

$$f(x) = \frac{1}{(2\pi)^{1/2}\sigma} e^{-(x-\mu)^2/2\sigma^2}.$$

2. If X is distributed as $N(0, \sigma^2)$, then $\Pr[|X| \geq t\sigma] \leq 2e^{-t^2/2}$ for any $t \geq 1$.
3. *Additivity:* If $g_1 \sim N(\mu_1, \sigma^2)$ and $g_2 \sim N(\mu_2, \sigma_2^2)$ are independent gaussian random variables, then for any $t_1, t_2 \in \mathbb{R}$, the random variable $t_1 g_1 + t_2 g_2$ is distributed as $N(t_1\mu_1 + t_2\mu_2, t_1^2\sigma_1^2 + t_2^2\sigma_2^2)$.

The additivity property of gaussians implies the following useful property.

Lemma 14. *Let $g \in \mathbb{R}^n$ be a random gaussian, i.e. each coordinate is chosen independently according to distribution $N(0, 1)$. Then for any arbitrary vector $v \in \mathbb{R}^n$, the random variable $\langle g, v \rangle \sim N(0, \|v\|_2^2)$.*

Proof. As $\langle g, v \rangle = \sum_i g(i)v(i)$, where $g(i)$ and $v(i)$ denote the i-th coordinate of g and v, and as the $g(i)'s$ are independent, the additivity property implies that $\langle g, v \rangle$ is distributed as $N(0, \sum_i v(i)^2) = N(0, \|v\|_2^2)$. $\qquad\qquad\square$

If the vectors v_i satisfy (6.2)–(6.3), then if we choose a random gaussian g and let $y_i = \langle g, v_i \rangle$, Lemma 14 implies that

1. Each y_i is distributed as $N(0, 1)$. This follows as $\|v_i\|_2^2 = 1$.
2. For each j, the discrepancy $\sum_{i \in S_j} y_i$ is distributed as $N(0, \leq \lambda^2)$, i.e. as a gaussian with mean 0 and variance at most λ^2. This follows as $\| \sum_{i \in S_j} v_i \|^2 \leq \lambda^2$ by the SDP constraint.

This seems quite close to what we would like. As $y_i \sim N(0,1)$, we have that $y_i/(c(\log n)^{1/2}) \in [-1, 1]$ with high probability (for some large enough constant c). Moreover, for any j, the discrepancy $|\sum_{i \in S_j} y_i| = O(\lambda(\log n)^{1/2})$ with high probability. Perhaps, one could now hope to round these y_i's to ± 1 without increasing the discrepancy substantially.

However, this possibility is ruled out by the hardness result in Theorem 1. In particular, this result implies that there must exist systems with discrepancy $\Omega(\sqrt{n})$ but vector-discrepancy 0 (if such systems did not exist, then solving the discrepancy SDP with $\lambda = 0$ would give an algorithm to distinguish between discrepancy 0 and $\Omega(\sqrt{n})$).

So, we adopt a different approach. Instead of trying to round the y_i's directly into ± 1, we will obtain a ± 1 solution gradually by combining several different collections of correlated $y_i's$ and solving several SDPs. This is also where we will really use that the guarantee is Theorem 8 is with respect to the *hereditary* discrepancy.

Algorithm Overview. As mentioned above, instead of trying to obtain a coloring using a single SDP solution, we will gradually produce a solution by using several SDPs over time. At time 0, we start with the "empty" coloring $x^0 = (0, \dots, 0)$ where each element is colored 0. We slowly modify it over time as follows: Suppose x^{t-1} denotes the coloring of elements at time $t - 1$, we obtain the coloring x^t by adding a small perturbation vector u^t (how this is chosen will be described later) to x^{t-1}, i.e. $x^t(i) = x^{t-1}(i) + u^t(i)$ for each element i. As the perturbations are added, the color of the elements will evolve over time. Whenever an element's color reaches -1 or $+1$, we freeze that element's color and it is not longer updated.

It remains to specify how to generate the updates u^t. This is done using the gaussian rounding idea described above. In particular, at time t, we consider the SDP given by (6.2)–(6.3) with $\lambda = \text{herdisc}(C)$ (but only restricted to elements i that are still *alive*, i.e. not frozen yet). As $\lambda = \text{herdisc}(C)$, no matter which variables are alive at time t, the SDP is always feasible. We take the vectors v_i^t corresponding to some feasible SDP solution and set $u_i^t = \gamma \langle g, v_i^t \rangle$, where g is a random gaussian in $\mathbb{R}^n$ and γ is some polynomially small scaling factor ($\gamma = 1/n^c$ for any $c \geq 1$ suffices).

6.3.1   The Algorithm

We now state the algorithm formally.

1. Let x^t denote the coloring at time t. Let $\gamma = 1/n$ and $\ell = 8 \log n/\gamma^2$. We initialize, $x^0(i) = 0$ for all $i \in [n]$. The F^t denote the set of frozen variables by time t, where we initialize $F^0 = \emptyset$.

2. For each time step $t = 1, 2, \ldots, \ell$ repeat the following steps:

   a. Find a feasible solution to the following semidefinite program:

$$\| \sum_{i \in S_j} v_i \|_2^2 \leq \lambda^2 \qquad \text{for each set } S_j$$

$$\|v_i\|_2^2 = 1 \qquad \forall i \notin F^{t-1}$$

$$\|v_i\|_2^2 = 0 \qquad \forall i \in F^{t-1}$$

   b. Pick a random gaussian vector $g^t \in \mathbb{R}^n$.
   c. For each $i \in [n]$, update $x^t(i) = x^{t-1}(i) + \gamma \langle g^t, v_i^t \rangle$.
   d. Set $F^t = F^{t-1}$. For each i, freeze i if $|x^t(i)| > 1$, and update $F^t = F^t \cup \{i\}$.

3. After time $t = \ell$, if some $|x^\ell(i)| < 1$ (i.e. some element is alive), return fail.
4. For each i, set $x_i^\ell = -1$ if $x_i^\ell < -1$, and $x_i^\ell = 1$ if $x_i^\ell > 1$. Output the coloring x_ℓ.

Remark. The SDP in step 2(a) changes only when the set of frozen variables F changes. Moreover, the SDP is always feasible irrespective of the set F^t as herdisc$(C) \leq \lambda$.

6.3.2 Analysis

We will prove that the algorithm above produces a coloring with discrepancy $O((\log m \log n)^{1/2} \lambda)$ with probability at least $1/2$, where λ is the hereditary discrepancy of A. The proof relies on two simple ideas, that we first describe informally.

The Proof Sketch. First we show that all elements are frozen by time ℓ with high probability. Let us consider some element i. Observe that its color $x^t(i)$ starts at 0, and evolves over time until it crosses ± 1 and is frozen. At each step the update $u^t(i) = \gamma \langle v_i^t, g^t \rangle$ is added. As $\|v_i^t\| = 1$, by Lemma 14, $u_i^t \sim N(0, \gamma^2)$, and as g^t is chosen independently at each time step, $u^t(i)$ is independent of the previous updates $u^{t-1}(i), u^{t-2}(i), \ldots$ for $x(i)$ (this is not strictly true, but let us ignore this technicality here). As the increments are $N(0, \gamma^2)$, $x^t(i)$ will reach ± 1 in $O(1/\gamma^2)$ steps with constant probability, and thus the probability that it does not reach ± 1 until $\ell = O(\log n / \gamma^2)$ steps is at most $1/n^2$. So, with probability at least $1 - 1/n$, all elements will be frozen by time ℓ.

We show now that the discrepancy is bounded with high probability. Let us consider how the discrepancy $x^t(S_j) = \sum_{i \in S_j} x^t(i)$ of a set S_j evolves over time. It is 0 initially at $t = 0$, at each step t, it is updated by $u^t(S_j) = \sum_{i \in S_j} \gamma \langle g^t, v_i^t \rangle$. Let $\lambda_t := \| \sum_{i \in S_j} v_j^t \|_2$. The SDP constraint ensures that $\lambda_t \leq \lambda$. Thus (roughly speaking) $x^t(S_j)$ evolves as a random walk with steps $N(0, \leq \gamma\lambda)$. By standard tail bounds, $\Pr[x^\ell(S_j) > c(\log m)^{1/2} \cdot \sqrt{\ell} \cdot \gamma\lambda] \leq 1/m^2$ for some suitable constant c.

By union bound over the sets implies that

$$\text{disc}(C) = O((\log m)^{1/2} \cdot \sqrt{\ell} \cdot \gamma\lambda) = O(\lambda \cdot (\log m \log n)^{1/2}).$$

Finally, note that truncating $x^\ell(i)$ to ± 1 in the last step introduces very low error. As $|x_i^t| < 1$ holds just before it is frozen and the next increment is a $N(0, \gamma^2)$ update, it must hold that $|x_i^t| < 1 + \gamma \cdot O((\log n)^{1/2})$ with high probability when it freezes. Thus, truncation adds at most $n \cdot \gamma \cdot O((\log n)^{1/2}) = O((\log n)^{1/2})$ error to any set.

The Formal Proof. Recall that a sequence of random variables $X_0, \ldots, X_t$ forms a martingale, if $\mathbb{E}[X_t | X_{t-1}, \ldots, X_0] = X_{t-1}$. We first need the following tail bound on martingales with gaussian increments.

Lemma 15. *Let $0 = X_0 = X_1, \ldots, X_n$ be a martingale with increments $Y_i = X_i - X_{i-1}$. Suppose for $1 \leq i \leq n$, we have that $Y_i | (X_{i-1}, \ldots, X_0)$ is distributed as $\eta_i G$, where G is a standard gaussian $N(0, 1)$ and η_i is a constant such that $|\eta_i| \leq 1$ (note that η_i may depend on $X_0, \ldots, X_{i-1}$). Then,*

$$\Pr[|X_n| \geq \lambda \sqrt{n}] \leq 2e^{-\lambda^2/2}.$$

Proof. Let α be a parameter to be optimized later. We have,

$$\mathbb{E}[e^{\alpha Y_i} | X_{i-1}, \ldots, X_0] \leq \int_{-\infty}^{\infty} e^{\alpha y} \cdot \left(\frac{1}{(2\pi)^{1/2}\eta_i} e^{-y^2/2\eta_i^2} \right) dy$$

$$= e^{\alpha^2 \eta_i^2/2} \cdot \int_{-\infty}^{\infty} \left(\frac{1}{(2\pi)^{1/2}\eta_i} e^{-(y-\alpha\eta_i^2)^2/2\eta_i^2} \right) dy$$

$$= e^{\alpha^2 \eta_i^2/2} \leq e^{\alpha^2/2}.$$

Now,

$$\mathbb{E}[e^{\alpha X_n}] = \mathbb{E}[e^{\alpha X_{n-1}} e^{\alpha Y_n}] = \mathbb{E}[e^{\alpha X_{n-1}} \mathbb{E}[e^{\alpha Y_n} | X_{n-1}, \ldots, X_0]] \leq e^{\alpha^2/2} \mathbb{E}[e^{\alpha X_{n-1}}].$$

Thus it follows by induction that $\mathbb{E}[e^{\alpha X_n}] \leq e^{\alpha^2 n/2}$. Finally by Markov's inequality,

$$\Pr[X_n \geq \lambda \sqrt{n}] = \Pr[e^{\alpha X_n} \geq e^{\alpha\lambda\sqrt{n}}] \leq e^{-\alpha\lambda\sqrt{n}} \mathbb{E}[e^{\alpha X_n}] \leq e^{-\alpha\lambda\sqrt{n}+\alpha^2 n/2}.$$

Setting $\alpha = \lambda/\sqrt{n}$ and noting that $\Pr[X_n \geq \lambda \sqrt{n}] = \Pr[X_n \leq -\lambda \sqrt{n}]$ implies the claim. $\qquad\square$

Lemma 16. *Let $X_t = g_1 + \ldots + g_t$ where g_i are iid $N(0, 1)$ random variables. Then there is some universal constant $c > 0$ such that for any integer $k \geq 1$,*

$$\Pr[|X_t| < \sqrt{n} \text{ for } t = 1, \ldots, k\sqrt{n}] < (1-c)^{-k+1}.$$

While more precise estimates known for the above quantity, the following simple proof suffices for our purposes.

Proof. By additivity of gaussians, X_n is distributed as $N(0, n)$. Thus $\Pr[|X_n| > 2\sqrt{n}] \geq c$ for some constant $c > 0$. Now, if $|X_t| < \sqrt{n}$ holds for each $t = 1, \ldots, k\sqrt{n}$, it must necessarily hold that $|X_n| < \sqrt{n}$ and that $|X_{jn} - X_{(j-1)n}| < 2\sqrt{n}$ for each $j = 2, \ldots, k$. As each of these events is independent and the probability of each latter event is at most $1 - c$. This implies the claimed result. □

We can now prove Theorem 8.

Lemma 17. *With probability at least $1 - 1/n$, all elements will be frozen by time $\ell = O(\log n/\gamma^2)$.*

Proof. Let us consider an element i. We bound the probability that its color $x^t(i)$ never crosses ± 1 until time ℓ. Starting from 0, at each step t, the update $u^t(i) = \gamma \langle v_i^t, g^t \rangle$ is added to $x^{t-1}(i)$. Now, while the vector v_i^t may depend on previous choices of the gaussian vectors $g^{t'}$ for $t' < t$ (as they determine the SDP at time t), note that u_i^t is distributed as $N(0, \gamma^2)$ and is independent of $u_i^{t'}$ for $t' < t$, whenever $\|v_i^t\| = 1$. By Lemma 16 there is some constant c' such that the probability that x_i does not each ± 1 by $c' \log n/\gamma^2$ steps is at most $1/n^2$. The result follows by a union bound over the n elements. □

Lemma 18. *With probability at least $1 - 1/m$, for each set S, it holds that $disc(S) = O(\lambda \cdot (\log m \log n)^{1/2})$.*

Proof. The discrepancy of a set S_j at time t is $x^t(S_j) = \sum_{i \in S_j} x^t(i)$. It is 0 at $t = 0$, and at each step t it gets updated by $u^t(S_j) = \sum_{i \in S_j} \gamma \langle g^t, v_i^t \rangle$. Let $\lambda_t := \|\sum_{i \in S_j} v_j^t\|_2$. The SDP constraint ensures that $\lambda_t \leq \lambda$. Now, λ_t may depend on the previous choices $g^{t'}$ for $t' \leq t - 1$ (as these choices affect the SDP at time t), but as g^t is chosen independently at time t, conditioned on the previous random choices $g^{t'}$, the update $u^t(S_j) \sim N(0, \gamma^2 \lambda_t^2)$. Thus, $x^t(S_j)$ forms a martingale with increments $N(0, \leq \gamma^2 \lambda^2)$. So by Lemma 15,

$$\Pr[x^\ell(S_j) > c(\log m)^{1/2} \cdot \sqrt{\ell} \cdot \gamma\lambda] \leq 1/m^2$$

for some suitable constant c. By a union bound over the m sets, $disc(S) = O((\log m)^{1/2} \cdot \sqrt{\ell} \cdot \gamma\lambda) = O(\lambda \cdot (\log m \log n)^{1/2})$ with probability at least $1 - 1/m$. □

Finally, as discussed previously, truncating a frozen variable $x^\ell(i)$ to ± 1 introduces only negligible error. Combining Lemmas 17 and 18 gives Theorem 8.

6.3.3 Bounds Based on Partial Hereditary Discrepancy

Let us define the *partial hereditary discrepancy* of a system as the smallest number λ such that for any sub-system, there exists a partial coloring (say that colors at least half the elements of that sub-system) with discrepancy at most λ. The result above can be refined to show that

Theorem 19. *There is a polynomial time algorithm that finds a coloring with discrepancy $O(\lambda(\log m \log n)^{1/2})$, where λ is the partial hereditary discrepancy of A.*

This is useful because for many problems better bounds are known on partial hereditary discrepancy than for hereditary discrepancy. For example, for bounded degree systems the best bound we know on hereditary discrepancy is $O((t \log n)^{1/2})$ [2], while the partial hereditary discrepancy is $O(\sqrt{t})$ (as we saw in Sect. 6.2.2).

The algorithm is a direct modification of the one in Sect. 6.3.1. We replace the SDP constraint (6.3) that $\|v_i\|^2 = 1$, by the conditions $\sum_i \|v_i\|^2 \geq n/2$ and $\|v_i\|^2 \leq 1$ for each i. In particular, we consider the following SDP.

$$\| \sum_{i \in S_j} v_i \|_2^2 \leq \lambda^2 \qquad \text{for each set } S_j \qquad (6.4)$$

$$\sum_{i \notin F} \|v_i\|_2^2 \geq |F^c|/2 \qquad (6.5)$$

$$\|v_i\|_2^2 \leq 1 \qquad \forall i \in F^c \qquad (6.6)$$

$$\|v_i\|_2^2 = 0 \qquad \forall i \in F \qquad (6.7)$$

Here F^c denotes the complement of F (i.e. alive variables) and λ is the partial hereditary discrepancy. Note that the constraints (6.5) and (6.6) only require that at least half of the alive variables must be colored.

Analysis. While the algorithm is same as before, the analysis needs some more care. The problem is that the alive variables do not necessarily satisfy $\|v_i\|_2 = 1$, but only the weaker condition (6.5). So, a priori it is possible that some variable always has $\|v_i\| \approx 0$ and hence never makes progress towards reaching ± 1. To get around this, one needs a more careful "energy increment" argument to show that after every $1/\gamma^2$ time steps, a constant fraction of the variables do reach ± 1 in expectation. One can then show that all elements are eventually colored ± 1 in $O(\log n/\gamma^2)$ time steps with probability at least $1/n^{O(1)}$.

The key result is the following.

Theorem 20. *Let $x \in [-1, +1]^n$ be an arbitrary fractional coloring with at most k alive variables. Starting from the coloring x, let z be the coloring obtained after applying the steps (2a)–(2d) of the algorithm for the SDP given by (6.4)–(6.7), for $16/\gamma^2$ time units. Then the probability that z has more than $k/2$ alive variables is at most $1/4$.*

Proof. Let $u = 16/\gamma^2$ and for each $1 \leq t \leq u$, let x_t denote the coloring at the end of time t starting from the initial coloring $x_0 = x$, i.e. after t applications of steps (2a)–(2d). Let K denote the set of alive variables at time $t = 0$. Let k_t denote the number of variables alive at the end of time t, and let $k = k_0 = |K|$. We would like to show that $k_u \leq k_0/2$ with probability at least $3/4$. To do this, we track how the "energy" $\sum_i x_t(i)^2$ of the coloring x^t evolves as the algorithm proceeds.

For each time $t = 1, \ldots, u$, let us define

$$r_t = \begin{cases} \sum_{i \in K} x_t(i)^2 & \text{if } k_{t-1} \geq k/2, \\ r_{t-1} + \gamma^2 k/4 & \text{otherwise.} \end{cases}$$

Lemma 21. *Conditioned on any coloring x_{t-1} at the end of time $t-1$, the expected increment $r_t - r_{t-1}$ at time step t is at least $\gamma^2 k/4$, where the expectation is over the choice of the random gaussian $g \in R^n$ at time t. That is, $\mathbb{E}[r_t - r_{t-1}|x_{t-1}] \geq \gamma^2 k/4$ for any x_{t-1}.*

Proof. If $k_{t-1} < k/2$, this follows trivially from the definition of r_t, as r_t is deterministically set to $r_{t-1} + \gamma^2 k/4$. So it suffices to consider the case when $k_{t-1} \geq k/2$. Let v^t the vector solution to the SDP (6.4)–(6.7). Then,

$$\mathbb{E}[r_t - r_{t-1}|x_{t-1}] = \mathbb{E}[r_t|x_{t-1}] - r_{t-1}$$

$$= \mathbb{E}_g\left[\sum_i (x_{t-1}(i) + \gamma\langle g, v_i^t\rangle)^2\right] - \sum_i x_{t-1}(i)^2$$

$$= \sum_i \left(2\gamma x_{t-1}\mathbb{E}[\langle g, v_i^t\rangle] + \gamma^2\mathbb{E}[(\langle g, v_i^t\rangle)^2]\right) \geq \gamma^2 k_{t-1} \geq \gamma^2 k/4.$$

The first step follows as r_{t-1} is completely determined by x_{t-1}. The last step follows for the following reason. By Lemma 14, $\langle g, v_i^t\rangle$ is distributed as $N(0, \|v_i^t\|^2)$ which implies that $\mathbb{E}_g[\langle g, v_i^t\rangle] = 0$ and $\mathbb{E}_g[(\langle g, v_i^t\rangle)^2] = \|v_i^t\|^2$. Now, the SDP constraint (6.5) ensures that $\sum_i \|v_i^t\|^2 \geq k_{t-1}/2$ which is at least $k/4$. $\qquad\square$

We now use this lemma together with Markov's inequality to finish off the proof.

The crucial observation is the following. Consider time $t = u$. If $k_u \geq k/2$, then by definition $r_u = \sum_{i \in K} x_u(i)^2 \leq k$. On the other hand, it always holds that $r_u \leq k + u\gamma^2 k/4$. This is because in any run of the algorithm, $r_t = \sum_{i \in K} x_t(i)^2 \leq k$ as long as $k_t \geq k/2$, and when k_t falls below $k/2$, then r_t increases deterministically by $\gamma^2 k/4$ at each subsequent time step.

This gives us that

$$\mathbb{E}[r_u] = \mathbb{E}[r_u|k_u \geq k/2]\Pr[k_u \geq k/2] + \mathbb{E}[r_u|k_u < k/2]\Pr[k_u < k/2]$$

$$\leq k \cdot \Pr[k_u \geq k/2] + (k + \gamma^2 uk/4) \cdot (1 - \Pr[k_u \geq k/2])$$

$$= k + (1 - \Pr[k_u \geq k/2]) \cdot (\gamma^2 uk/4).$$

As $\mathbb{E}[r_u] \geq \gamma^2 uk/4$ by Lemma 21, together this gives that $\gamma^2 uk/4 \leq k + (1 - \Pr[k_u \geq k/2]) \cdot (\gamma^2 uk/4)$, and hence that $\Pr[k_u \geq k/2] \leq k/(\gamma^2 uk/4) = 1/4$ as claimed. $\qquad\square$

Lemma 22. *Let* $\ell = 16 \log n / \gamma^2$. *The probability that every element is colored* ± 1 *by time* ℓ, *is at least* $1/n$.

Proof. We apply Theorem 20 repeatedly with the starting coloring $x = x_t$ at the time steps $t = 0, 16/s^2, 32/s^2, \ldots, (16 \log n)/s^2 = \ell$. With probability at least $(1 - 1/4)^{\log n} \geq 1/n$, the number of alive variables at least halves at each epoch and hence reaches 0.   □

The proof of Theorem 19 now follows directly by combining Lemma 22 together with the argument in Lemma 18.

6.4   Algorithmic Version of Spencer's Result

In this section we consider Theorem 7. This result turns out to be much more tricky that the algorithm in the previous section.

To keep the focus on the main ideas, we will first describe a weaker version of Theorem 7, that gives an $O((n \log \log \log n)^{1/2})$ discrepancy coloring (note that is still much better than randomized rounding). Later, in Sect. 6.5 we will describe the recent and much simpler algorithm due to Lovett and Meka [19] to find an $O(\sqrt{n})$ discrepancy coloring.

Some Problematic Issues.   The reason that proving Theorem 7 is much more tricky is the following. First, it is not at all clear whether semidefinite programming is useful. In particular, consider the SDP given by (6.2)–(6.3). The natural thing is to set $\lambda = O(\sqrt{n})$, and try to use this SDP solution.[3] However, if we set $\lambda \geq \sqrt{n}$, then this SDP can always return the trivially feasible solution $v_i = e_i$ for $i \in [n]$, where e_i denotes the unit vector in the ith direction. This SDP solution is feasible as the v_i are unit vectors $\|v_i\| = 1$ and their orthogonality implies that $\| \sum_{i \in S} v_i \|_2 = (|S|)^{1/2} \leq n^{1/2}$ for any S. Thus, the SDP does not seems to reveal any useful information.

A second problem is that in the previous algorithm the discrepancy of a set performs a random walk over time. So, even if we the expected discrepancy of a set is $O(\sqrt{n})$, some of them are very likely to deviate from the expectation by factor $\Omega(\sqrt{\log n})$. So we do not seem to get anything better than the $O(\sqrt{n \log n})$ bound that random coloring would have given us anyway.

The Additional Idea.   To get around these issues, the idea is to let the discrepancy bound λ_S for set S (in the SDP) vary over time depending on how the discrepancy of S evolves. If a set S gets dangerously closely to violating the target $c\sqrt{n}$ discrepancy bound, we set its λ_S (denoted by Δ_S in Lemma 12) in the SDP to be much smaller than $\sqrt{n}$, thereby ensuring that its discrepancy is extremely unlikely

---

[3]Note that one cannot set $\lambda = o(\sqrt{n})$ in our setting, there are set systems on $m = O(n)$ sets (e.g. the Hadamard set system, that we will see in Sect. 6.6) with vector discrepancy $\Omega(\sqrt{n})$.

to exceed in $c\sqrt{n}$ is subsequent iterations. The point is that the entropy method (Lemma 12) guarantees a partial coloring provided the discrepancy bounds satisfy the condition (6.1). So if we can argue that not too many sets become dangerous, then the argument still goes through.

Before we describe the algorithm and its analysis, we state the following corollary of Theorem 12 that we will need.

Corollary 23. *Let (V, C) be any set system on $m = O(n)$ sets, and $C' \subset C$ be any sub-collection of $O(n/(\log\log n)^2)$ sets. Then there exists a partial coloring where the discrepancy of sets in C' is at most $\sqrt{n}/\log n$ and the ones in $C \setminus C'$ is $O(\sqrt{n})$.*

Proof. For each set $S \in C'$, setting $\lambda_S = \sqrt{n}/\log n$ contributes at most $g(1/\log n) = O(\log\log n)$ to the left hand side of (6.1). As $|C'| = O(n/(\log\log)^2 n)$, the overall contribution due to C' is $o(n)$. For the sets in $C \setminus C'$ we set $\lambda_S = c\sqrt{n}$ for c large enough, so that their total contribution to (6.1) is at most $n/10$. Thus the claimed partial coloring exists by Lemma 12. □

6.4.1 Algorithmic Subroutine and Analysis

We now describe the algorithm. We only consider the first phase when the number of uncolored variables reduces from n to $n/2$. This is the hardest phase and contains all the main ideas.

Algorithm for the First Phase. We start with the all 0 coloring, and consider the following partial coloring SDP.

$$\left\| \sum_{i \in S} v_i \right\|_2^2 \leq \lambda_S^2 \qquad \text{for each set } S = S_1, \ldots, S_m \tag{6.8}$$

$$\sum_{i \notin F} \|v_i\|_2^2 \geq |F^c|/2 \tag{6.9}$$

$$\|v_i\|_2^2 \leq 1 \qquad \forall i \notin F \tag{6.10}$$

$$\|v_i\|_2^2 = 0 \qquad \forall i \in F \tag{6.11}$$

Initially we set $\lambda_S = cn^{1/2}$ for each set S, where c is a large enough constant such that (6.1) is satisfied easily with some slack. As previously, for each time step t, we obtain the update $u_i^t = \gamma \langle g^t, v_i^t \rangle$ for $i = 1, \ldots, n$ and add it to the coloring thus far. We repeat this for $O(1/\gamma^2)$ steps, at which point we expect half the colors to reach ± 1.

During these steps, if the discrepancy $|x^t(S)|$ of S ever exceeds $2(n \log\log \log n)^{1/2}$, we label S *dangerous* and set $\lambda_S = n^{1/2}/\log n$ in all the subsequent SDPs. This ensures that its discrepancy increment $u^t(S)$ will have standard deviation at most $O(\gamma \cdot (n^{1/2}/\log n))$ henceforth, making S extremely unlikely to incur an additional $\Omega(n^{1/2})$ discrepancy over the remaining $O(1/\gamma^2)$ steps.

Analysis. By design, the reduction of λ_S for dangerous sets ensures that after $O(1/\gamma^2)$ steps, the discrepancy of every set is $O((n \log \log \log n)^{1/2})$ with high probability. Moreover, Lemma 22 implies that with probability at least $3/4$, at least $n/2$ elements are colored ± 1 by the end of the phase.

It suffices to show that with probability at least $3/4$, the SDP never becomes infeasible. Indeed, as the discrepancy of any set forms a martingale with gaussian increments with standard deviation $O(\sqrt{n})$, thus by Lemma 15 the probability of a set ever becoming dangerous is $O(\exp(-2 \log \log \log n)) = O(\log \log n)^{-2}$. Thus the total expected number of dangerous sets is $O(n/(\log \log n)^2))$, and by Markov's inequality, with probability at least $3/4$, this number does not exceed $O(n(\log \log n)^{-2})$. Corollary 23 now gives the claimed result.

Remarks.

1. The $O(n^{1/2})$ bound in Theorem 7 follows by refining this idea by having multiple danger levels and by setting the bounds λ_S for a set S appropriately for each danger level.
2. Even though Theorem 7 gives a polynomial time algorithm, it crucially uses the entropy method to argue about the feasibility of the SDP, and hence is not truly constructive.

Recently, Lovett and Meka [19] discovered a much simpler argument to obtain an algorithmic version of Spencer's result. Their idea was to combine the gaussian random walk approach above together with the linear algebraic ideas that we saw in Sect. 6.2.1. We now describe their algorithm and analysis.

6.5   The Result of Lovett and Meka

As usual, let (V, C) be a set system with n elements and m sets $S_1, \ldots, S_m$. Lovett and Meka [19] gave the following algorithmic version of the partial coloring lemma.

Theorem 24. *Let $x \in [-1, 1]^n$ be some fractional coloring, with k alive elements (i.e. k elements i such that $x(i) \neq \pm 1$). For $j = 1, \ldots, m$, let λ_j be such that*

$$\sum_j \exp(-\lambda_j^2/16) \leq k/16. \tag{6.12}$$

Then there is a randomized polynomial time algorithm to find a coloring x' with at most $k/2$ alive variables such that the additional discrepancy added to each set S_j is at most $\Delta_{S_j} = \lambda_j \sqrt{|S_j|}$. That is, $|x'(S_j) - x(S_j)| \leq \Delta_{S_j}$ for each $j = 1, \ldots, m$.

Remark. The theorem is stated in the form above so that it can be applied repeatedly to obtain a complete coloring after $O(\log n)$ rounds. The reader may assume that the starting coloring x is $(0, \ldots, 0)$.

An interesting aspect of Theorem 24 is that it is quantitatively stronger than the entropy method (Theorem 12). In particular, if we require $\lambda \ll 1$ for the sets, then using (6.1) this can only be done for $O(n/\log n)$ sets, while Theorem 24 allows this for $\Omega(n)$ sets. For more discussion on the comparison with the entropy method we refer the reader to [19].

We now prove Theorem 24. Let K denote the set of alive elements at $t = 0$. Without loss of generality let us assume that $K = \{1, \ldots, k\}$. As in the previous algorithms, the algorithm will start from the initial coloring $x_0 = x$ and update it over time as $x_t = x_{t-1} + u^t$ by adding a suitably chosen tiny update vector u^t at time t.

6.5.1 The Algorithm

Let $\gamma \leq 1/n^2$ be a tiny parameter as before, and let $\delta = (10 \log n)\gamma$. Let us call a set S_j dangerous after time $t - 1$, if $|x_{t-1}(S_j) - x_0(S_j)| \geq \lambda_j - \delta$. Also, we call a variable i frozen after time $t - 1$ if $|x_{t-1}(i)| \geq 1 - \delta$. The algorithm will ensure that the color $x_t(i)$ of any element i does not change once it freezes, and moreover the discrepancy of a set S also never changes once it becomes dangerous.

To achieve this the algorithm makes the update u^t at time t in an appropriate linear subspace defined as follows. Given the coloring x_{t-1} after time $t - 1$, let $V^t \subseteq \mathbb{R}^k$ be the subspace of points $(v(1), \ldots, v(k))$ satisfying:

1. If the element i is frozen, then $v(i) = 0$.
2. If the set S_j is dangerous, then $\sum_{i \in S_j \cap [k]} v(i) = 0$.

Algorithm. Initialize $x_0 = x$ and $V^0 = \mathbb{R}^k$.
  For $t = 1, \ldots, T$, where $T = 16/(3\gamma^2)$, repeat the following steps.

1. Pick $g \sim N(V^t)$. Update the coloring as $x_t = x_{t-1} + \gamma g$.
2. If some element i freezes or if some set S_j becomes dangerous at time t, update V^{t+1} accordingly.

Note that as the algorithm proceeds $\mathbb{R}^k = V^0 \supseteq V^1 \supseteq \cdots \supseteq V^t$. Moreover, as the update γg lies in V^t, it follows that for any element i, we have that $|x_t(i)|$ is at most $1 - \delta + O(\sqrt{\log(T)})\gamma \leq 1$ with high probability. Similarly, the additional discrepancy $|x_{t-1}(S) - x_0(S)|$ of any set does not exceed $\lambda_j - \delta + O(\gamma\sqrt{|S_j|}\sqrt{\log T}) \leq \lambda_j + 1/n$.

Thus one only needs to show that the algorithm does not get stuck (i.e. $V^t = \emptyset$, while more than half the variables are still alive).

6.5.2   Analysis

We begin with two simple properties of random gaussian vectors that play a key role in the analysis.

For a linear subspace $V \subseteq \mathbb{R}^n$, let $N(V)$ denote the standard multi-dimensional gaussian distribution supported on V. A random vector g is distributed according to $N(V)$ if $g = g(1)v_1 + \ldots + g(d)v_d$ where $\{v_1, \ldots, v_d\}$ is an orthonormal basis for V, and $g(1), \ldots, g(d)$ are independent $N(0, 1)$ random variables. By the rotational invariance of the multi-dimensional gaussian distribution, it is easily seen that g is invariant of the choice of the basis $\{v_1, \ldots, v_d\}$.

Lemma 25. *If $g \sim N(V)$, then for all $u \in \mathbb{R}^n$, $\langle g, u \rangle \sim N(0, \sigma^2)$ where $\sigma^2 \le \|u\|^2$.*

Proof. Let u' denote the projection of u onto V. Clearly, $\|u'\| \le \|u\|$. As $g \in V$, $\langle g, u \rangle = \langle g, u' \rangle$, which by Lemma 14 is distributed as $N(0, \|u'\|^2)$.   □

Let e_i denote the unit vector in the i-th direction.

Lemma 26. *Let V be a d-dimensional subspace of $\mathbb{R}^n$ and $g \sim N(V)$. For $i = 1, \ldots, n$ let σ_i be such that $\langle g, e_i \rangle \sim N(0, \sigma_i^2)$. Then $\sum_{i=1}^{n} \sigma_i^2 = d$.*

Proof. If $v_1, \ldots, v_d$ is an orthogonal basis for V, then by Lemma 14, $\sigma_i^2 = \sum_{j=1}^{d} \langle e_i, v_j \rangle^2$. Thus $\sum_{i=1}^{n} \sigma_i^2 = \sum_{i=1}^{n} \sum_{j=1}^{d} \langle e_i, v_j \rangle^2 = \sum_{j=1}^{d} \sum_{i=1}^{n} \langle v_j, e_i \rangle^2 = \sum_{j=1}^{d} \|v_j\|^2 = d$, where the second last equality follows as $\{e_1, \ldots, e_n\}$ is an orthogonal basis for $\mathbb{R}^n$ and hence $\|v\|^2 = \sum_{i=1}^{n} \langle v, e_i \rangle^2$ for every v.   □

The proof of theorem now follows by arguments similar to the ones in Sect. 6.3.3. We sketch the main idea here and refer the reader to [19] for the detailed computations.

First we claim that not many sets become dangerous in expectation. This follows as at each time step t, Lemma 25 implies that irrespective of the choice of V^t, the discrepancy increment of each set S_j is distributed as $N(0, \le \gamma^2 |S_j|)$. Thus, by Lemma 15,

$$\Pr[|x_T(S_j) - x_0(S_j)| \ge \Delta_{S_j}] \le 2 \exp(-\Delta_{S_j}^2 / (T\gamma^2 |S_j|)) = 2 \exp(-3\lambda_j^2/16).$$
(6.13)

As λ_j satisfy (6.12), by a standard calculation that we skip, (6.13) implies that the probability that more than $k/8$ sets become dangerous is at most $1/8$.

Let us condition on the event that no more than $k/8$ sets become dangerous. Then we claim that with probability at least $3/4$, at least $k/2$ elements become frozen. This follows by the argument in Lemma 20. In particular, if fewer than $k/2$ elements are frozen, then at any time during the algorithm the dimension of the subspace V^t is at least $k - k/2 - k/8 \ge 3k/8$. Now, Lemma 26 implies that the expected energy increment $\sum_i x_t(i)^2 - x_{t-1}(i)^2 \ge (3k/8)\gamma^2$. So, if fewer than $k/2$ elements are frozen, the total energy increment will be $T(3k/8)\gamma^2 = 2k$. But as $x_t(i)^2$ can never exceed k, we can use the argument in the proof of Lemma 20 (with constants modified) to upper bound the probability that fewer than $k/2$ elements are fixed. Together, these two claims imply the result.

6.6 Inapproximability of Discrepancy

In this section we prove Theorem 1, which shows that discrepancy is essentially hopeless to approximate. The proof has two main ingredients. One starts with a very weak hardness result for discrepancy, which states that it is NP-hard to determine if a set system has discrepancy 0 or if a constant fraction of sets have discrepancy at least 1. The next step is to amplify this hardness by composing it with the Hadamard set system. The Hadamard system is a classic example that shows that Theorem 2 is tight up to constant factors.

Hadamard Set System. We recall the construction of the $2^k \times 2^k$ Hadamard matrix $H^{(k)}$, which is defined as the k-fold tensor product $H^{(k)} := H^{(1)} \otimes \cdots \otimes H^{(1)}$, where $H^{(1)}$ is the 2×2 matrix with entries $h(1,1) = h(1,2) = h(2,1) = 1$ and $h(2,2) = -1$.

Let us denote $n = 2^k$ and $H = H^{(k)}$. The only property we need of the H is that (i) it is symmetric and consists of ± 1 entries, (ii) its first row consists entirely of 1's and (iii) its columns are mutually orthogonal, i.e. $H^T H = nI$.

Let J denote the $n \times n$ matrix with all 1's, and let Z be the 0–1 matrix $Z := \frac{1}{2}(H + J)$. Then Z satisfies the following.

Lemma 27. *Let* $x \in \mathbb{R}^n$ *be such that* $\sum_{i>1} x_i^2 = \Omega(n)$. *Then* $\|Zx\|_2^2 = \Omega(n^2)$.

Proof. Let Z_i (resp. J_i, H_i) denote the i-th column of Z (resp. J, H). We have

$$\|Zx\|_2^2 = (Zx)^T(Zx) = \sum_{i,j} x_i Z_i^T Z_j x_j = \frac{1}{4} \sum_{i,j} x_i (H_i + J_i)^T (H_j + J_j) x_j.$$

As the columns of H are orthogonal, $\sum_{i,j} x_i x_j H_i^T H_j = \sum_i x_i^2 H_i^T H_i = n\|x\|_2^2$. Moreover, as $J_j = H_1$ for all j, $\sum_{i,j} x_i x_j H_i^T J_j = \sum_{i,j} x_i x_j H_i^T H_1 = (\sum_j x_j)(nx_1)$. Similarly, $\sum_{i,j} x_i x_j J_i^T H_j = (\sum_i x_i)(nx_1)$. Finally, the last term $\sum_{i,j} x_i x_j J_i^T J_j = n(\sum_i x_i)^2$. Combining these terms gives

$$\|Zx\|_2^2 = n(\|x\|_2^2 + 2x_1(\sum_i x_i) + (\sum_i x_i)^2) = n(\|x\|_2^2 + (x_1 + \sum_i x_i)^2 - x_1^2) \geq n(\sum_{i>1} x_i^2).$$

□

Max-2-2-Set-Splitting Problem. We will use the following max-2-2-set-splitting problem (henceforth referred to as MSS). An instance of MSS consists of m sets, $C = \{C_1, \ldots, C_m\}$ on n elements, each set consisting of exactly four distinct elements. The objective is to assign each element either $\{-1, +1\}$ such that number of sets for which the values sum to exactly 0 is maximized. Clearly, for any assignment, the sets can have values $\{0, \pm2, \pm4\}$. The sets that have value 0 are called *split*, and otherwise they are *unsplit*.

Theorem 28 ([15]). *There exists a universal constant $\phi > 0$ such that it is NP-hard to distinguish between instances of MSS whether (i) there is an assignment such that all the sets are split, or (ii) any assignment will result in at least ϕm unsplit sets.*

Moreover, one can assume that in these instances no element appears in more than b sets for some universal constant b.

If C denotes the $m \times n$ incidence matrix of an MSS instance, the result above implies that it is NP-hard to distinguish whether (i) the discrepancy of C is 0 (ii) for any $y \in \{-1, 1\}^n$, at least $\Omega(m)$ sets have non-zero discrepancy. We amplify this gap by composing the MSS instance with the Hadamard system. At first, we obtain a matrix whose entries are not necessarily $\{0, 1\}$, but lie in the range $[0, b]$ for some constant b. Later, we show how to modify the argument to obtain a $\{0, 1\}$ matrix.

Theorem 29. *Given a $m \times n$ matrix B with $m = O(n)$ and integer entries in the range $[0, b]$, where b is a constant, it is NP-hard to distinguish between the cases: (i) the discrepancy of B is 0 (ii) for any $y \in \{-1, 1\}^n$, $\|By\|_2^2 = \Omega(n^2)$.*

Proof. Let us take an instance of MSS on n elements and $m = O(n)$ sets, and let C be its incidence matrix. By duplicating some sets if necessary, we can assume without loss of generality that m is an integer power of 2. Consider the matrix $B = ZC$, where Z is the $m \times m$ set system in Lemma 27. As each column of C contains at most b 1's and Z is a $0 - 1$ matrix, each entry of B is an integer in $[0, b]$.

Now, if the MSS instance corresponding to C has an assignment y that splits each set, then $Cy = 0$ and hence $By = Z(Cy) = 0$. On the other hand if $\|Cy\|_2^2 = \Omega(m)$, then by Lemma 27 and noting that the first entry of $Cy \in \{0, \pm2, \pm4\}$, it follows that $\|By\|_2^2 = \|Z(CY)\|_2^2 = \Omega(m^2) = \Omega(n^2)$. □

We now modify the argument so that B only has $\{0, 1\}$ entries. Let us partition the sets in the MSS instance into $h \leq 4b + 1$ parts $T^1, \ldots, T^h$ such that an element j appears in at most one set from any collection T^i. Such a partition exists as each set contains four elements and each elements lies in at most b sets, and hence each set shares an element with at most $4b$ other sets. Duplicating sets if necessary, we assume that the number of sets m_i in T^i is an integer power of 2. For each $i = 1, \ldots, h$, let H^i denote the $m_i \times m_i$ Hadamard set system, and let $B^i = H^i T^i$. Note that B^i has $\{0, 1\}$ entries as each element appears in at most one set in T^i. Let B be the $(\sum_i m_i) \times n$ matrix obtained by placing the rows of $B^1, \ldots, B^h$ one after the other.

Now if $Cy = 0$ then $T^i y = 0$ for each i, and hence each $B^i y = H^i T^i y = 0$ which implies that $By = 0$. On the other hand, if $\|Cy\|_2^2 = \Omega(m)$, then $\|T^i y\|_2^2 = \Omega(m/b) = \Omega(m)$ for some i, and hence $\|By\|_2^2 \geq \|B^i y\|_2^2 = \|H^i (T^i y)\|_2^2 = \Omega(m^2)$, where the last step follows from Lemma 27.

6.7 Tightness of the Determinant Lower Bound

In this section, we see how the connection between discrepancy and semidefinite programming can be used to show that the determinant lower bound characterizes hereditary discrepancy up to poly-logarithmic factors. We begin by describing the determinant lower bound and proving Theorem 6. Then in Sect. 6.7.2, we prove Theorem 10.

6.7.1 Determinant Lower Bound

Recall that for a real matrix A, we denote $\text{detlb}(A) := \max_k \max_B |\det B|^{1/k}$, where the maximum is over all $k \times k$ submatrices B of A. We will show that

Theorem 6. *For any real matrix A, $\text{herdisc}(A) \geq \frac{1}{2}\text{detlb}(A)$.*

Let us define the linear discrepancy of a matrix A as

$$\text{lindisc}(A) := \max_{x \in [-1,1]^n} \min_{y \in \{-1,1\}^n} \|Ax - Ay\|_\infty.$$

That is, it is the worst case error over all points $x \in [-1, 1]^n$, when x is rounded to the "best" integral $y \in \{-1, 1\}^n$. Theorem 3 directly implies that

$$\text{lindisc}(A) \leq 2\,\text{herdisc}(A) \tag{6.14}$$

(we lose the extra factor 2 here as $y \in \{-1, 1\}^n$, instead of $y \in \{0, 1\}^n$).

Lemma 30. *To prove Theorem 6, it suffices to show that for any $k \times k$ matrix B, $\text{lindisc}(B) \geq \det(B)^{1/k}$.*

Proof. Let A be any $m \times n$ matrix. By the definition of hereditary discrepancy and (6.14)

$$\text{herdisc}(A) \geq \max_k \max_B \text{herdisc}(B) \geq \frac{1}{2} \max_k \max_B \text{lindisc}(B),$$

where B ranges over the $k \times k$ submatrices of A. By the claim, this is at least $\frac{1}{2} \max_k \max_B \det(B)^{1/k}$ which is exactly $\frac{1}{2}\text{detlb}(A)$. □

Thus we focus on proving Lemma 30 for square matrices. The proof is based on a geometric interpretation of lindisc.

Geometric Interpretation of Linear Discrepancy. Let B be an invertible $k \times k$ matrix (if B is non-invertible, then $\det(B) = 0$ and Lemma 30 holds trivially). Let $P \in \mathbb{R}^k$ be the set of points x satisfying $-\mathbf{1} \leq Bx \leq \mathbf{1}$, where $\mathbf{1}$ is the all 1's vector in $\mathbb{R}^k$. Clearly, if $x \in P$ then $-x$ also lies in P and hence P is symmetric about the origin. Note that P is the inverse image of the unit cube $[-1, 1]^k$ under B^{-1}, i.e. $P = \{B^{-1}z : z \in [-1, 1]^k\}$. We claim the following.

Lemma 31. *Linear discrepancy is the smallest real number t such that placing a copy of tP at every point $\{-1, 1\}^k$ covers the cube $[-1, 1]^k$ completely.*

Proof. By definition of linear discrepancy and P, lindisc(B) is smallest t such that for each point $x \in [-1, 1]^k$, the polytope $tP + x$ (i.e. P scaled t times and shifted by x) contains some point y in $\{-1, 1\}^k$. Let x be any point in $[-1, 1]^k$. Now if $y \in \{-1, 1\}^k$ is such that $y \in tP + x$, then $y - x \in tP$. But then, $x - y \in tP$ by the symmetry of P which implies that $x \in tP + y$. □

We now finish the proof of Theorem 6. For each $y \in \{-1, 1\}^k$, let $R_y = [-1, 1]^k \cap (tP + y)$ denote the intersection of the unit cube with the copy of tP at y. Observe that R_y is identical to $tP \cap I(-y)$ where $I(y)$ is the orthant $\Pi_i[0, y_i]$ formed by the origin and y. As $tP = \cup_{y=\{-1,1\}^k}(tP \cap I(y))$, this implies that R_y's give a partitioning of tP and hence

$$\text{Vol}(tP) = t^k \text{Vol}(P) = \sum_{y \in \{-1,1\}^k} \text{Vol}(R_y) \geq \text{Vol}([-1, 1]^k) = 2^k \qquad (6.15)$$

where the inequality above follows by Lemma 31 and the definition of R_y.

As $P = \{B^{-1}z : z \in [-1, 1]^k\}$ we have that

$$\text{Vol}(P) = \det(B^{-1})\text{Vol}([-1, 1]^k) = 2^k \det(B^{-1}) = 2^k / \det(B). \qquad (6.16)$$

Taking logarithms in (6.15) and using (6.16) this gives $t \geq 2/\text{Vol}(P)^{1/k} = \det(B)^{1/k}$, which implies Lemma 30 and hence Theorem 10.

6.7.2 Matoušek's Upper Bound

A natural question is whether the determinant lower bound essentially determines the hereditary discrepancy of a system. Hoffman gave an elegant example (see [20]) of a set system with detlb(C) $= O(1)$ and disc(C) $\approx (\log n)/(\log \log n)$, and hence implies that the gap between the two can be $\Omega(\log n/ \log \log n)$. Recently, Pálvölgyi [23] gave an improved example with an $\Omega(\log n)$ gap. However, there was no non-trivial result upper bound known on this gap until the following recent result [21] of Matoušek.

Theorem 10. [21] *For any set system (V, C), herdisc(C) $\leq O(\log n \sqrt{\log m}) \cdot$ detlb(C).*

The proof uses several interesting ideas, and in particular the duality for semidefinite programming.

SDP Dual for Discrepancy. Let us recall the strong duality for convex programs which states the following. Let P be a convex programming problem (say, with minimization objective). Then any feasible solution to the dual program of P is a lower bound on the optimum solution for P. Moreover under some mild technical conditions (see [29] for details), the value of the optimum feasible dual solution is equal to the value of the optimum feasible solution for P. Let us apply this to the following discrepancy minimization SDP for the system (V, C) (having the optimum value vecdisc(C)).

$$\min \lambda^2, \text{ s.t. } (i) \quad \| \sum_{i \in S_j} v_i \|_2^2 \le \lambda^2 \quad \forall S_j \in C, \text{ and } (ii) \quad \|v_i\|_2^2 = 1, \quad \forall i \in V.$$

Lemma 32. *For any set system (V, C) with n elements and m sets, we have that* vecdisc$(C) \ge D$ *if and only if there are nonnegative reals $w_1, \ldots, w_m$ with $\sum_{j=1}^m w_j \le 1$ and reals $z_1, \ldots, z_n$ with $\sum_{i=1}^n z_i \ge D^2$ such that for all $\mathbf{x} \in \mathbb{R}^n$,*

$$\sum_{j=1}^m w_j \left(\sum_{i \in S_j} x_i \right)^2 \ge \sum_{i=1}^n z_i x_i^2. \tag{6.17}$$

While computing the dual formally needs some work (and we refer the reader to [21] for details), this dual has a rather intuitive interpretation. Suppose there exists a convex combination with coefficients w_i, of the set discrepancy constraints $(\sum_{i \in S_j} x_i)^2$ such that this sum always exceeds $\sum_{i=1}^n z_i x_i^2$ no matter what real values are assigned to x_i's. Or in other words, $\sum_{j=1}^m w_j (\sum_{i \in S_j} x_i)^2 - \sum_{i=1}^n z_i x_i^2$ is a positive definite quadratic form. Then, indeed $\sum_{i=1}^n z_i x_i^2 = \sum_i z_i$ for any ± 1 assignment to the x_i's and hence this certifies that $\sum_{i=1}^n z_i$ is a lower bound on the discrepancy. The duality says that if the vector discrepancy is D, then there always exists a choice of witnesses w_j's and z_i's of this form.

We now show the following result, which directly implies Theorem 10.

Theorem 33. *Let $C = \{S_1, \ldots, S_m\}$ be a set system on $[n]$ with* vecdisc$(C) = D$. *Then* detlb$(C) = \Omega(D/\sqrt{\log n})$.

Before proving Theorem 33, we show how it implies Theorem 10.

Proof (of Theorem 10). Let $J \subset [n]$ be such that vecdisc$(C|_J)$ is maximized. Then, it must hold that

$$\text{vecdisc}(C|_J) = \Omega(\text{herdisc}(C)/(\log m \log n)^{1/2}) \tag{6.18}$$

otherwise, one could use the algorithm in Theorem 8 to color any subsystem $(J', C|_{J'})$ of (V, C) with discrepancy strictly less then herdisc(C), contradicting the definition of hereditary discrepancy.

Applying Theorem 33 to $C|_J$ we obtain detlb$(C|_J) = \Omega(\text{vecdisc}(C|_J)/\sqrt{\log n})$. Together with (6.18), this gives that

$$\text{detlb}(C) \geq \text{detlb}(C|_J) \geq \Omega(\text{herdisc}(C)/(\sqrt{\log n}(\log m \log n)^{1/2})),$$

where the first inequality follows as any sub-matrix of $C|_J$ is also a sub-matrix of C.   □

Proof (of Theorem 33). Let us consider the dual formulation of vector discrepancy from Lemma 32. For more convenient notation, let us write the nonnegative weight w_j as β_j^2. Moreover, let $L \subseteq [n]$ consist of the indices i with $z_i > 0$. By Lemma 32 applied with $x_i = 0$ for $i \notin L$, and writing $z_i = \gamma_i^2$ for $i \in L$, we obtain

$$\sum_{i=j}^{m} \beta_j^2 \left(\sum_{i \in S_j \cap L} x_i \right)^2 \geq \sum_{i \in L} \gamma_i^2 x_i^2 \tag{6.19}$$

for all $x \in \mathbb{R}^L$, where $\|\beta\|^2 \leq 1$ and $\|\gamma\|^2 \geq D$.

Next, we select $K \subseteq L$ with $\|\gamma[K]\|^2 = \Omega(D/\sqrt{\log n})$ and such that all entries of $\gamma[K]$ are within a factor of 2 of each other (here $\gamma[K]$ denotes the vector γ restricted to coordinates in K). Such a subset K exists for the following reason: Let $\gamma_{\max} = \max_i |\gamma_i|$. For $\ell = 0, 1, 2, \ldots, 2 \log n - 1$, let $K_\ell = \{i : |\gamma_i| \in (2^{-\ell-1}\gamma_{\max}, 2^{-\ell}\gamma_{\max}]\}$. The contribution to $\|\gamma\|$ of the components of γ with $\gamma_i \leq \gamma_{\max}/n^2$ is negligible, and so there exists some ℓ_0 for which $\sum_{i \in K_{\ell_0}} \gamma_i^2 = \Omega(\|\gamma\|^2/\log n)$.

Let us denote $k = |K|$ and $\tilde{D} = \frac{1}{2}\|\gamma[K]\|$. As $\sum_{i \in K} \gamma_i^2 = 4\tilde{D}^2$, and all these γ_i are within twice of each other, we have that $\gamma_i \geq \tilde{D}/\sqrt{k}$ for all $i \in K$. So, restricting (6.19) to vectors x with $x_i = 0$ for $i \notin K$, we have that

$$\sum_{j=1}^{m} \beta_j^2 \left(\sum_{i \in S_j \cap K} x_i \right)^2 \geq \frac{\tilde{D}^2}{k} \sum_{i \in K} x_i^2. \tag{6.20}$$

Let $C = A[*, K]$ be the $m \times k$ incidence matrix of the system $A|_K$ and let $\check{C}$ be the $m \times k$ matrix obtained from C by multiplying the jth row by β_j. Then (6.20) can be rewritten as

$$x^T \check{C}^T \check{C} x = \|\check{C}x\|^2 \geq \frac{\tilde{D}^2}{k} \|x\| \qquad \text{for all } x \in R^k.$$

This, by the usual variational characterization of eigenvalues, tells us that the smallest eigenvalue of the $k \times k$ matrix $\check{C}^T \check{C}$ is at least $\tilde{D}^2/k$. Now, as the determinant is the product of eigenvalues, this implies that $\det(\check{C}^T \check{C}) \geq (\tilde{D}^2/k)^k$. By the Binet–Cauchy formula we obtain

$$\det(\check{C}^T \check{C}) = \sum_I \det(\check{C}[I, *])^2, \tag{6.21}$$

where the summation is over all k-element subsets $I \subseteq [m]$ and $\check{C}[I, *]$ consists
of the rows of $\check{C}$ whose indices lie in I. Setting $M := \max_I |\det(C[I, *])|$ and
noting that $\det(\check{C}[I, *]) = \det(C[I, *]) \prod_{j \in I} \beta_j$, we can bound the right-hand side
of (6.21) as

$$\sum_I \det(\check{C}[I, *])^2 = \sum_I \det(C[I, *])^2 \prod_{j \in I} \beta_j^2 \le M^2 \sum_I \prod_{j \in I} \beta_j^2$$

$$\le M^2 \frac{\left(\sum_{j=1}^m \beta_j^2\right)^k}{k!} \le \frac{M^2}{k!},$$

where the second inequality follows as every term $\prod_{j \in I} \beta_j^2$ occurs $k!$ times in
the multinomial expansion of $(\beta_1^2 + \cdots + \beta_m^2)^k$. Letting $B := C[I, *]$ for an I
maximizing $|\det C[I, *]|$, we have

$$\det(B)^2 \ge k! \det(\check{C}^T \check{C}) \ge k!(\tilde{D}^2/k)^k \ge (k/e)^k(\tilde{D}^2/k)^k = \Omega(\tilde{D})^{2k} = \Omega(D/\sqrt{\log n})^{2k}.$$

So the $k \times k$ matrix B witnesses that $\mathrm{detlb}(C) = \Omega(D/\sqrt{\log n})$, which implies the
claimed result. $\square$

References

1. N. Alon, J.H. Spencer, *The probabilistic method. With an appendix on the life and work of Paul Erdős. 2nd ed.* (Wiley-Interscience Series in Discrete Mathematics and Optimization. Wiley, Chichester, xvi, 301 p., New York, 2000)
2. W. Banaszczyk, Balancing vectors and Gaussian measures of n-dimensional convex bodies. Random Struct. Algorithm **12**(4), 351–360 (1998). doi:10.1002/(SICI)1098-2418(199807)12: 4<351::AID-RSA3>3.0.CO;2-S
3. N. Bansal, Constructive Algorithms for Discrepancy Minimization, in *Foundations of Computer Science (FOCS)*, 2010, pp. 3–10
4. N. Bansal, J.H. Spencer, Deterministic discrepancy minimization., in *19th annual European symposium, Saarbrücken, Germany, September 5–9, 2011. Proceedings, Algorithms – ESA 2011*, ed. by C. Demetrescu, et al. Lecture Notes in Computer Science, vol. 6942 (Springer, Berlin, 2011), pp. 408–420. doi:10.1007/978-3-642-23719-5_35
5. J. Beck, Roth's estimate of the discrepancy of integer sequences is nearly sharp. Combinatorica **1**, 319–325 (1981). doi:10.1007/BF02579452
6. J. Beck, T. Fiala, Integer-making theorems. Discrete Appl. Math. **3**, 1–8 (1981)
7. J. Bourgain, L. Tzafriri, Invertibility of "large" submatrices with applications to the geometry of Banach spaces and harmonic analysis. Isr. J. Math. **57**(2), 137–224 (1987). doi:10.1007/BF02772174
8. K. Chandrasekaran, S. Vempala, A discrepancy based approach to integer programming. CoRR **abs/1111.4649** (2011)
9. M. Charikar, A. Newman, A. Nikolov, Tight Hardness Results for Minimizing Discrepancy, in *Proceedings of the Twenty-Second Annual ACM-SIAM Symposium on Discrete Algorithms, SODA 2011, San Francisco, California, USA, January 23–25, 2011*, ed. by D. Randall, 2011, pp. 1607–1614

10. B. Chazelle, *The discrepancy method. Randomness and complexity*. (Cambridge University Press, Cambridge, 2000), p. 463
11. T.M. Cover, J.A. Thomas, *Elements of information theory. 2nd ed.* (Wiley-Interscience. Wiley, Hoboken, NJ, 2006), p. 748. doi:10.1002/047174882X
12. B. Doerr, Linear and hereditary discrepancy. Combinator. Probab. Comput. **9**(4), 349–354 (2000). doi:10.1017/S0963548300004272
13. M. Drmota, R.F. Tichy, *Sequences, discrepancies and applications*. Lecture Notes in Mathematics, vol. 1651 (Springer, Berlin, 1997), p. 503. doi:10.1007/BFb0093404
14. F. Eisenbrand, D. Pálvölgyi, T. Rothvoß, Bin Packing via Discrepancy of Permutations, in *Proceedings of the Twenty-Second Annual ACM-SIAM Symposium on Discrete Algorithms, SODA 2011, San Francisco, California, USA, January 23–25, 2011*, ed. by D. Randall, 2011, pp. 476–481
15. V. Guruswami, Inapproximability results for set splitting and satisfiability problems with no mixed clauses. Algorithmica **38**(3), 451–469 (2004). doi:10.1007/s00453-003-1072-z
16. J.H. Kim, J. Matoušek, V.H. Vu, Discrepancy after adding a single set. Combinatorica **25**(4), 499–501 (2005). doi:10.1007/s00493-005-0030-x
17. D.J. Kleitman, On a combinatorial conjecture of Erdős. J. Combin. Theory **1**, 209–214 (1966). doi:10.1016/S0021-9800(66)80027-3
18. L. Lovász, J.H. Spencer, K. Vesztergombi, Discrepancy of set-systems and matrices. Eur. J. Combinator. **7**, 151–160 (1986)
19. S. Lovett, R. Meka, Constructive Discrepancy Minimization by Walking on the Edges, in *Foundation of Computer Science (FOCS)*, 2012, pp. 61–67
20. J. Matoušek, *Geometric discrepancy. An illustrated guide. Revised paperback reprint of the 1999 original*. Algorithms and Combinatorics, vol. 18 (Springer, Dordrecht, 2010), p. 296. doi:10.1007/978-3-642-03942-3
21. J. Matoušek, The determinant bound for discrepancy is almost tight. Proc. Am. Math. Soc. **141**(2), 451–460 (2013). doi:10.1090/S0002-9939-2012-11334-6
22. A. Nikolov, K. Talwar, L. Zhang, The Geometry of Differential Privacy: the Approximate and Sparse Cases., in *Symposium on Theory of Computing (STOC)*, 2013. to appear.
23. D. Pálvölgyi, Indecomposable coverings with concave polygons. Discrete Comput. Geom. **44**(3), 577–588 (2010). doi:10.1007/s00454-009-9194-y
24. T. Rothvoß, The entropy rounding method in approximation algorithms, in *Symposium on Discrete Algorithms (SODA)*, 2012, pp. 356–372
25. T. Rothvoß, Approximating bin packing within o(log opt * log log opt) bins. CoRR **abs/1301.4010** (2013)
26. M. Sipser, *Introduction to the theory of computation*. (PWS Publishing, Boston, MA, 1997), p. 396
27. J.H. Spencer, Six standard deviations suffice. Trans. Am. Math. Soc. **289**(2), 679–706 (1985). doi:10.2307/2000258
28. A. Srinivasan, Improving the Discrepancy Bound for Sparse Matrices: Better Approximations for Sparse Lattice Approximation Problems, in *Symposium on Discrete Algorithms (SODA)*, 1997, pp. 692–701
29. L. Vandenberghe, S. Boyd, Semidefinite programming. SIAM Rev. **38**(1), 49–95 (1996). doi:10.1137/1038003
30. R. Vershynin, John's decompositions: Selecting a large part. Isr. J. Math. **122**, 253–277 (2001). doi:10.1007/BF02809903

Chapter 7
Practical Algorithms for Low-Discrepancy 2-Colorings

Lasse Kliemann

Abstract We present practical approaches for low-discrepancy 2-colorings in the hypergraph of arithmetic progressions. A simple randomized algorithm, a deterministic combinatorial algorithm (Sárközy 1974), and three estimation of distribution algorithms are compared. The best of them experimentally achieves a constant-factor approximation.

We consider practical approaches for computing low-discrepancy 2-colorings in hypergraphs, using modern parallel computer systems. The hypergraph $\mathscr{A}_n$ of arithmetic progressions in the first n integers will be our benchmark. Our interest is in the range $100{,}000 \leq n \leq 250{,}000$. We have chosen $\mathscr{A}_n$ for two reasons. First, its discrepancy is well understood and known to be $\Theta(\sqrt[4]{n})$. Second, $\mathscr{A}_n$ is a massive hypergraph with approximately $n^2 \ln(n)/2$ hyperedges, but at the same time it can be represented very succinctly in a computer program. The highlight is an estimation of distribution algorithm (EDA) that experimentally achieves $O(\sqrt[4]{n})$ for up to $n = 250{,}000$. For this n, the number of hyperedges in $\mathscr{A}_n$ is more than $377 \cdot 10^9$. We will explain the idea behind EDAs and their technical features in detail.

7.1 Facts on $\mathscr{A}_n$

Given $a \in \mathbb{N}_0 = \{0, 1, 2, 3, \ldots\}$ and $d, \ell \in \mathbb{N} = \{1, 2, 3, \ldots\}$, the set

$$A_{a,d,\ell} := \{a + id;\ 0 \leq i < \ell\} = \{a, a+d, a+2d, \ldots, a+(\ell-1)d\}$$

L. Kliemann (✉)
Department of Computer Science, Kiel University, Christian-Albrechts-Platz 4, 24118 Kiel, Germany
e-mail: lki@informatik.uni-kiel.de

W. Chen et al. (eds.), *A Panorama of Discrepancy Theory*, Lecture Notes in Mathematics 2107, DOI 10.1007/978-3-319-04696-9__7, © Springer International Publishing Switzerland 2014

Table 7.1 Approximating the number of APs

n	$n^2 \ln(n)/2$	$|\mathscr{A}_n|$	Relative error (rounded)
$100 \cdot 10^3$	57564627325	55836909328	3.1 %
$125 \cdot 10^3$	91688039190	88988439957	3.0 %
$150 \cdot 10^3$	134081893947	130194434212	3.0 %
$175 \cdot 10^3$	184860787935	179569486783	3.0 %
$200 \cdot 10^3$	244121452911	237210329226	2.9 %
$225 \cdot 10^3$	311947596930	303200671205	2.9 %
$250 \cdot 10^3$	388413006151	377614297340	2.9 %

is the *arithmetic progression* (AP) with starting point a, difference d, and length ℓ. Its cardinality is $|A_{a,d,\ell}| = \ell$. For $n \in \mathbb{N}$, denote $[\![n]\!] := \{0, \ldots, n-1\}$ and call $\mathscr{A}_n := \{A_{a,d,\ell} \cap [\![n]\!]; \; a \in \mathbb{N}_0 \wedge d, \ell \in \mathbb{N}\}$ the *hypergraph of arithmetic progressions* in the first n integers. The set of vertices is implicitly assumed to be $[\![n]\!]$. In the literature, it is more common to use the set $[n] = \{1, \ldots, n\}$ as the vertex set, but for us, starting from 0 is more convenient. Likewise, we make the convention that for any set X and $k \in \mathbb{N}$, vectors from X^k are indexed by $0, \ldots, k-1$.

$\mathscr{A}_n$ is a massive hypergraph. Its number of hyperedges is

$$|\mathscr{A}_n| = n + \sum_{a=0}^{n-2} \sum_{d=1}^{n-1-a} \left\lfloor \frac{n-1-a}{d} \right\rfloor = n + \sum_{a=0}^{n-2} \sum_{\ell=2}^{n-a} \left\lfloor \frac{n-1-a}{\ell-1} \right\rfloor , \qquad (7.1)$$

which can be estimated using integrals [22] as

$$\frac{(n-1)^2 \ln(n-1)}{2} - \frac{3n^2 - 8n}{4} \leq |\mathscr{A}_n| \leq \frac{n^2 \ln(n)}{2} + \frac{n^2 + 2n + 1}{4} .$$

It follows $|\mathscr{A}_n| = \Theta(n^2 \ln(n))$. For large n, by which we mean $100 \cdot 10^3 \leq n \leq 250 \cdot 10^3$, a pretty good estimate is $|\mathscr{A}_n| \approx n^2 \ln(n)/2$ with an error around 3 %, shown by Table 7.1.

For $n = 100 \cdot 10^3$, storing this hypergraph explicitly as an hyperedge-vertex incidence matrix requires about 635 TiB provided we only use 1 bit for each entry. For $n = 250 \cdot 10^3$ it is more than 10 PiB.[1] Using lists of 32 bit integers for the hyperedges is even more costly. Based on (7.1), we get the expression

$$n + \sum_{a=0}^{n-2} \sum_{\ell=2}^{n-a} \ell \cdot \left\lfloor \frac{n-1-a}{\ell-1} \right\rfloor$$

for the number of required list entries. For $n = 100 \cdot 10^3$, this gives about 998 TiB and for $n = 250 \cdot 10^3$ more than 15 PiB. Fortunately, there is no need to store the hypergraph explicitly for the algorithmic approaches we are about to present.

---

[1] We have 1 TiB $= 2^{40}$ byte and 1 PiB $= 2^{50}$ byte.

When we speak of *discrepancy* in the following, we always mean the (combinatorial) 2-color discrepancy. We denote it with "disc", so if $\mathcal{H} \subseteq 2^V$ is any hypergraph on vertex set V, e.g. $\mathscr{A}_n$ on $V = [\![n]\!]$, the *discrepancy* of $\mathcal{H}$ is

$$\text{disc}(\mathcal{H}) := \min_{\chi \in \{-1,+1\}^V} \text{disc}_{\mathcal{H}}(\chi) ,$$

where $\text{disc}_{\mathcal{H}}(\chi) := \max_{E \in \mathcal{H}} |\chi(E)|$ and $\chi(E) := \sum_{v \in E} \chi_v$ for a coloring χ. We call $\text{disc}_{\mathcal{H}}(\chi)$ the *discrepancy* of $\mathcal{H}$ with respect to χ and $|\chi(E)|$ the *discrepancy* or *imbalance* of hyperedge E with respect to χ. In total the expression is:

$$\text{disc}(\mathcal{H}) = \min_{\chi \in \{-1,+1\}^V} \max_{E \in \mathcal{H}} \left| \sum_{v \in E} \chi_v \right| .$$

We denote colorings as functions, i.e. from $\{-1, +1\}^V$, or as vectors, i.e. from $\{-1, +1\}^n$, but always use vector notation χ_v when referring to the color of a particular vertex v.

In 1964, it was shown by Roth [20] that $\text{disc}(\mathscr{A}_n) = \Omega(\sqrt[4]{n})$. More than 30 years later, in 1996, it was shown by Matoušek and Spencer [14] that $\text{disc}(\mathscr{A}_n) = O(\sqrt[4]{n})$, so together we have $\text{disc}(\mathscr{A}_n) = \Theta(\sqrt[4]{n})$. The proofs do not yield an efficient algorithm to actually construct a coloring with discrepancy $\Theta(\sqrt[4]{n})$.

7.2 Overview of Algorithmic Approaches

Standard probabilistic arguments show that discrepancy $O(\sqrt{n \ln(m)})$ can be reached in general hypergraphs with n vertices and m hyperedges with a simple randomized algorithm with constant probability. For $\mathscr{A}_n$, this means $O(\sqrt{n \ln(n)})$, far away from $O(\sqrt[4]{n})$. In 1974, Sárközy invented a method which we call *modulo coloring*: for an appropriately chosen prime p, the first p numbers are colored in a certain randomized way and then this coloring is simply repeated for the rest of the numbers, up to n. The algorithm achieves discrepancy $O(\sqrt[3]{n \ln(n)})$ with constant probability.

In 2010, in a pioneering work, the algorithmic problem was solved (up to logarithmic factors) by Bansal [5], using semi-definite programs (SDP). However, Bansal's algorithm requires solving a series of SDPs that grow in the number of hyperedges, making it practically problematic given the enormous size of $\mathscr{A}_n$. In our experiments with a sequential implementation of Bansal's algorithm, even for $n < 100$, it requires several hours to complete. We will not consider details of Bansal's algorithm here and instead point to his Chap. 6. However, I believe that it would be worthwhile considering and perhaps refining his techniques from a practical point of view for arithmetic progressions in future work.

In 2012, Sauerland [22] evaluated a univariate EDA based on the "quantum-inspired evolutionary algorithm" by Han and Kim [7] in the range up to

Table 7.2 Results for $\mathscr{A}_n$ obtained with the EDA from [11], which is called *apQEA* and will be explained in detail later. Running times are given in minutes. Each row, except the last, is based on 30 runs with the same parameters. The last row is based on 5 runs. For $n \leq 200 \cdot 10^3$, a number of 96 processor cores is used in parallel. For $n = 250 \cdot 10^3$, a number of 192 processor cores is used

n	$3 \lfloor \sqrt[4]{n} \rfloor$	Running time
$100 \cdot 10^3$	53	$7_{\sigma=01}$
$125 \cdot 10^3$	56	$11_{\sigma=01}$
$150 \cdot 10^3$	59	$20_{\sigma=02}$
$175 \cdot 10^3$	61	$39_{\sigma=05}$
$200 \cdot 10^3$	63	$74_{\sigma=15}$
$250 \cdot 10^3$	67	$83_{\sigma=16}$

$n = 100 \cdot 10^3$. He showed that this EDA beats other randomized algorithms consistently, however it remained unclear whether the asymptotic $\Theta(\sqrt[4]{n})$ could be achieved by this approach.

In 2013, Kliemann et al. [11] presented a new univariate EDA, capable of treating the cases of larger n, up to $n = 250 \cdot 10^3$ and perhaps even beyond, reliably attaining discrepancy $3 \lfloor \sqrt[4]{n} \rfloor$ in experiments. This was the first practical constant-factor approximation algorithm for the discrepancy problem in $\mathscr{A}_n$. The achieved discrepancies are given in Table 7.2 together with mean running times and their standard deviations over a number of runs for each n. Computations were conducted on 96 processor cores in parallel for $n \leq 200 \cdot 10^3$ and on 192 cores for $n = 250 \cdot 10^3$. All those computations as well as those described later in this chapter were carried out on the NEC™ Linux Cluster of the Rechenzentrum at Kiel University, with SandyBridge-EP™ processors. The numbers given in Table 7.2 will serve as a comparison in the discussion of other approaches in the rest of this text.

7.3   Discrepancy Computation: Shortcutting, Parallelization

Given a coloring $\chi \in \{-1, +1\}^n$, the discrepancy of $\mathscr{A}_n$ with respect to χ will be denoted by $\mathrm{disc}_{\mathscr{A}}(\chi)$ in the following. Computing $\mathrm{disc}_{\mathscr{A}}(\chi)$ naively requires checking all the approximately $n^2 \ln(n)/2$ APs in $\mathscr{A}_n$. Using the following obvious relation, this can be done more efficiently than treating each AP on its own:

$$\chi(A_{a,d,\ell}) = \chi(A_{a,d,\ell-1}) + \chi_{a+(\ell-1)d} \ .$$

So we have the algorithm given in Algorithm 1. We initialize the discrepancy computed so far to $\delta := 1$ in line 1, the discrepancy of the singleton APs, so we have the condition $a + d \leq n - 1$ on d. Note that still, lines 6 and 7 have to be executed approximately $n^2 \ln(n)/2$ times.

Sometimes, we are only interested in the exact discrepancy provided that it is on or below some δ_0. Then we can initialize $\delta := \delta_0$ in line 1 and have the condition $a + \delta_0 d \leq n - 1$ on d, since we are not interested in the discrepancy of APs of

---

Algorithm 1: Simple discrepancy computation. For-loops use inclusive ranges, so e.g. a runs through the set $[\![n]\!] = \{0, \dots, n-1\}$

Input: $\chi \in \{-1, +1\}^n$
Output: $\mathrm{disc}_{\mathscr{A}}(\chi)$

1 $\delta := 1$;
2 **for** $a := 0$ **to** $n-1$ **do**
3   **for** $d := 1$ **to** $n-1-a$ **do**
4          $s := 0$;
5   **for** $\ell := 1$ **to** $\left\lfloor \frac{n-1-a}{d} \right\rfloor + 1$ **do**
6              $s := s + \chi_{a+(\ell-1)d}$;
7   **if** $|s| > \delta$ **then** $\delta := |s|$;

8 **return** δ;

Algorithm 2: Discrepancy computation with additional parameters for speedup

---

Input: $\chi \in \{-1, +1\}^n$ and $\delta_0, \delta_1 \in \mathbb{N}$
Output: $\max\{\mathrm{disc}_{\mathscr{A}}(\chi), \delta_0\}$ or report that $\mathrm{disc}_{\mathscr{A}}(\chi) \geq \delta_1$

1 $\delta := \delta_0$;
2 **for** $a := 0$ **to** $n-1$ **do**
3   **for** $d := 1$ **to** $\left\lfloor \frac{n-1-a}{\delta_0} \right\rfloor$ **do**
4          $s := 0$;
5   **for** $\ell := 1$ **to** $\left\lfloor \frac{n-1-a}{d} \right\rfloor + 1$ **do**
6              $s := s + \chi_{a+(\ell-1)d}$;
7   **if** $|s| > \delta$ **then**
8                  $\delta := |s|$;
9   **if** $\delta \geq \delta_1$ **then return** "$\mathrm{disc}_{\mathscr{A}}(\chi) \geq \delta_1$";

10 **return** δ;

---

length δ_0 or shorter. Hence the loop for d only runs from 1 to $\left\lfloor \frac{n-1-a}{\delta_0} \right\rfloor$. We will use this for $\delta_0 := \lfloor 3\sqrt[4]{n} \rfloor$ in some of our algorithms when we are just aiming to achieve this discrepancy, providing a substantial speedup.

Moreover, sometimes we are only interested in the exact value of $\mathrm{disc}_{\mathscr{A}}(\chi)$ provided it is better than some δ_1. If $\mathrm{disc}_{\mathscr{A}}(\chi) \geq \delta_1$, then the exact value is uninteresting since χ will be discarded anyway. This happens in the EDAs presented later, but also in the random coloring algorithm presented in the next section. Since $\mathrm{disc}_{\mathscr{A}}(\chi)$ is a maximum over all APs, we can abort computation once an AP E is found with $|\chi(E)| \geq \delta_1$. We call this *shortcutting* and speak of a *shortcut* discrepancy computation, as opposed to a *full* one. Shortcutting has shown to be an enormous time saver in practice. In total, the algorithm looks like given in Algorithm 2.

The discrepancy computation can be thread-parallelized. The range $[\![n]\!]$ for starting points a is split into chunks, and these are assigned to different threads.

Communication is light: when a thread has found the shortcutting condition to be true, this must be communicated so that all threads terminate. In case of no shortcut, the maximum discrepancy found over all threads must be determined in the end. However, we must be aware that although communication is light, there is an overhead involved with using multiple threads. When there are many shortcuts during the run of an algorithm, it should therefore be considered using only a moderate number of threads for discrepancy computation. If unsure, a good value for the number of threads should be determined in preliminary experiments.

7.4   Random Coloring

This approach works for general hypergraphs. Consider the simple algorithm of flipping a fair coin for each vertex to decide its color, in other words we endow $\{-1, +1\}^V$ with the uniform distribution and sample our colorings from this probability space. Denote m the number of hyperedges and let $\varepsilon > 0$. Denote $\alpha := \alpha(n, m, \varepsilon) := \sqrt{2n \ln(2m/\varepsilon)}$. By Chernoff bounds, see, e.g. [15, Thm. 7.1.1] or [1, Cor. A.1.2], we easily derive that for each hyperedge E:

$$\mathbb{P}\big(|\chi(E)| > \alpha\big) < 2\,e^{-\frac{\alpha^2}{2|E|}} \le 2\,e^{-\frac{\alpha^2}{2n}} = \varepsilon/m \ .$$

By the union bound, it follows that the probability of having a hyperedge E violating the α bound is at most ε. Hence with probability at least $1 - \varepsilon$ we attain discrepancy $\alpha = O(\sqrt{n \ln(m)})$.

How does this approach perform in practice for $\mathscr{A}_n$? We test it on $n = 100 \cdot 10^3$ with 96 parallel processes, each using one core. Since there is only a handful of full discrepancy computations, we decided against thread-parallelizing this part. Repeatedly, new colorings are generated randomly and their discrepancy computed, using the best discrepancy so far for shortcutting. The parallel processes communicate the best solution among themselves, so that each can make best use of shortcutting. It turns out that after 60 min of computation, no better than discrepancy 220 is reached. The number of trials conducted in this time was $166 \cdot 10^6$. There was no improvement during the last 25 of the 60 min. For comparison: with the right algorithm, discrepancy 53 is possible within 7 min as per Table 7.2. On the other hand, it should be noted that discrepancy 220 is much better than the theoretical guarantee, which is $2\,286$ for $\varepsilon = 1/2$.

7.5   Modulo Colorings: Sárközy's Technique

Let $p \le n$ be a prime number and $\pi \in \{-1, +1\}^p$ a coloring of the first p integers $[\![p]\!] = \{0, \ldots, p-1\}$. Then a coloring χ of the first n integers $[\![n]\!] = \{0, \ldots, n-1\}$ can be defined by $\chi_v := \pi_{v \bmod p}$ for each $v \in [\![n]\!]$. In other words, we repeat π until all n integers are colored; we do not require an integral number of repetitions, so the last repetition of π may be cut off, just when $n - 1$ has been colored. We call π a *generating coloring* and χ a *p-modulo coloring*.

Now let π be *balanced*, i.e. $|\pi(\llbracket p \rrbracket)| = |\sum_{v \in \llbracket p \rrbracket} \pi_v| \leq 1$. Let $A_{a,d,\ell}$ be any AP and $\ell = qp + r$ with integers q, r and $r < p$. Then we have the decomposition:

$$A_{a,d,\ell} = \underbrace{\bigcup_{i=0}^{q-1} A_{a+ipd,d,p}}_{B_i :=} \cup \underbrace{A_{a+qpd,d,r}}_{B_q :=} , \qquad (7.2)$$

where the unions are disjoint. For any set $X \subseteq \mathbb{N}$ denote $(X \bmod p) := \{x \bmod p; \ x \in X\}$. Since p is prime, each of the p-element sets $B_0, \dots, B_{q-1}$ is mapped exactly onto $\llbracket p \rrbracket$ by modulo p, so $(B_i \bmod p) = \llbracket p \rrbracket$, thus $\chi(B_i) = \pi(\llbracket p \rrbracket)$ for each $i \in \llbracket q \rrbracket$. By π being balanced, we arrive at

$$|\chi(A_{a,d,\ell})| \leq \sum_{i=0}^{q-1} |\pi(B_i)| + |\chi(B_q)| \leq q + |\chi(B_q)| \leq \frac{n}{p} + |\chi(B_q)| , \qquad (7.3)$$

where for the last bound we use $q = \frac{\ell - r}{p} \leq \frac{n}{p}$.

This observation suggests that if p and π are appropriately chosen, modulo colorings will tend to have small discrepancy. Moreover, when actually computing discrepancy, a p-modulo coloring allows us restricting to starting points $a \in \llbracket p \rrbracket$, which reduces computation time drastically. We will give more concrete figures on this later in Sect. 7.8.

7.5.1   Analysis

It may be tempting to choose p very small, but the best discrepancy we can hope for in a p-modulo coloring is $\lceil n/p \rceil$. To see this, consider $A_{0,p,\lceil n/p \rceil}$ noting that $\lceil n/p \rceil \leq n/p + 1 - 1/p$ and so $0 + (\lceil n/p \rceil - 1) p \leq n - 1$, hence this is indeed an AP in $\llbracket n \rrbracket$, and it is monochromatic. However, with p chosen such that $n \approx \sqrt{p^3 \ln(p)}$, we get discrepancy $O(\sqrt[3]{n \ln(n)})$ by a randomized algorithm with probability $1 - 4/p$; the choice of p will be made precise later. This result is due to Sárközy [6, p. 39]; the proofs given here that lead up to Theorem 5 follow Sauerland [22].

First we have to discuss how to create a balanced coloring of $\llbracket p \rrbracket$. Denote $h := \frac{p+1}{2}$. One way is to randomly color the first $h - 1$ numbers with colors $\eta_0, \dots, \eta_{h-2}$ and the next $h - 1$ ones based on that with the opposite sign. The one remaining number $p - 1$ is colored arbitrarily, say with η_{h-1}. So we have $\eta \in \{-1, +1\}^h$. There is some freedom of choice how to color $h, \dots, p - 2$ based on η. The following *mirror* construction (with an additional feature explained later in Sect. 7.5.2) has shown to be good in experiments. The coloring of $\llbracket p \rrbracket$ obtained by the mirror construction looks like this:

Algorithm 3: Constructing a random balanced coloring. This algorithm has to be seen in the context of Lemma 1. Note that we will use this algorithm mainly to construct generating colorings of $[\![p]\!]$ for some prime p, in which case the coloring will be called "π" and not "χ"

---

1 foreach $v \in V_1$ **do**
2 $\quad\big|\quad \chi_v :=_{\text{unif}} \{-1, +1\}$;
3 $\quad\big\lfloor\quad \chi_{\varphi(v)} := -\chi_v$;
4 if $V \neq V_1 \cup V_2$ **then** color $v \in V \setminus V_0$ arbitrarily;
5 return χ;

$$(\eta_0, \ldots, \eta_{h-2}, -\eta_{h-2}, \ldots, -\eta_0, \eta_{h-1}) \in \{-1, +1\}^p . \qquad (7.4)$$

The following lemma and algorithm work for general hypergraphs. The operator $:=_{\text{unif}}$ means drawing uniformly at random; when multiple such statements appear in an algorithm, they are assumed to describe independent random experiments.

Lemma 1. *Let $\varepsilon > 0$. Let $\mathscr{H}$ be a hypergraph with an n-element vertex set V and m hyperedges. Let $V_1, V_2 \subset V$ with $|V_1| = |V_2|$, bijection $\varphi : V_1 \longrightarrow V_2$, and $|V \setminus V_0| \leq 1$ with $V_0 := V_1 \cup V_2$. Then with probability at least $1 - \varepsilon$, Algorithm 3 on this page constructs a coloring χ with $\text{disc}_{\mathscr{H}}(\chi) \leq 2\sqrt{n \ln(4m/\varepsilon)} + 1$.*

Proof. Denote $\alpha := \sqrt{n \ln(4m/\varepsilon)}$, which is half of the main term in the stated bound. Each of the families $(\chi_v)_{v \in V_1}$ and $(\chi_v)_{v \in V_2} = (\chi_{\varphi(v)})_{v \in V_1}$ is a family of independent random variables. Hence by Chernoff bounds, similar to Sect. 7.4, for each $i \in \{1, 2\}$ and $E \in \mathscr{H}$ we have

$$\mathbb{P}\big(|\chi(E \cap V_i)| > \alpha\big) < 2\,e^{-\frac{\alpha^2}{2|E \cap V_i|}} \leq 2\,e^{-\frac{\alpha^2}{n}} = \frac{\varepsilon}{2m} .$$

We use $|E \cap V_i| \leq n/2$ here. By the union bound and the triangle inequality it follows

$$\mathbb{P}\big(\exists E \in \mathscr{H} : |\chi(E \cap V_0)| > 2\alpha\big) \leq \sum_{E \in \mathscr{H}} \mathbb{P}\big(|\chi(E \cap V_0)| > 2\alpha\big)$$

$$\leq \sum_{E \in \mathscr{H}} \mathbb{P}\big((|\chi(E \cap V_1)| > \alpha) \vee (|\chi(E \cap V_2)| > \alpha)\big) < \sum_{E \in \mathscr{H}} 2\frac{\varepsilon}{2m} \leq \varepsilon .$$

Since the at most one vertex in $V \setminus V_0$ can worsen discrepancy of a hyperedge by at most 1, we are done. $\qquad\qquad\square$

This result is particularly useful for hypergraphs with few hyperedges. We turn our attention to the group $\mathbb{Z}_p$, by which we understand the set $[\![p]\!] = \{0, \ldots, p - 1\}$ with the operation $x \oplus y := ((x + y) \bmod p)$ for $x, y \in \mathbb{N}_0$. Although we define the operation also outside of $[\![p]\!]$, the result is always in $[\![p]\!]$. For $a \in \mathbb{N}_0$ and $d, \ell \in \mathbb{N}$ we denote

$$A_{a,d,\ell}^p := \{a \oplus id; \ 0 \le i < \ell\} = \{((a + id) \bmod p); \ 0 \le i < \ell\}$$

the *arithmetic progression* in $\mathbb{Z}_p$ with starting point a, difference d and length ℓ. Denote $\mathscr{A}^p := \{A_{a,d,\ell}^p; \ a \in \mathbb{N}_0 \wedge d, \ell \in \mathbb{N}\}$ the *hypergraph of arithmetic progressions* in $\mathbb{Z}_p$. It is easy to see that

$$A_{a,d,\ell}^p = A_{(a \bmod p),(d \bmod p),\min\{\ell,p\}}^p .$$

Hence $|\mathscr{A}^p| \le p^3$. It follows:

Corollary 2. *By Algorithm 3, we find with probability at least $1 - \varepsilon$ a balanced coloring π of $\mathscr{A}^p$ with $\mathrm{disc}_{\mathscr{A}^p}(\pi) \le 2\sqrt{p \ \ln(4p^3/\varepsilon)} + 1 = O(\sqrt{p} \ \ln(p))$.*

In [6, p. 39], it is claimed that a balanced coloring with discrepancy $O(\sqrt{p} \ \ln(p))$ can be constructed deterministically[2] by coloring $p - 1$ arbitrarily, e.g. with $+1$, and coloring $v < p - 1$ using the *Legendre symbol* $\left(\frac{v+1}{p}\right)$, i.e. in total:

$$\pi_v := \begin{cases} +1 & \text{if } v = p - 1 \\ +1 & \text{if } v + 1 \text{ is quadratic residue modulo } p \qquad \text{for } v \in [\![p]\!]. \\ -1 & \text{else} \end{cases}$$

The proof of the $O(\sqrt{p} \ \ln(p))$ bound allegedly works via number-theoretic arguments, but to the best of my knowledge has never been published. We give practical results for the Legendre construction at the end of Sect. 7.5.2.

The next lemma shows how discrepancy of $\mathscr{A}^p$ propagates under a p-modulo coloring.

Lemma 3. *There exists a prime p such that $\sqrt{p^3 \ \ln(p)} \in [\frac{n}{4}, n]$. Furthermore, with such a choice of p, let $\pi \in \{-1, +1\}^p$ be balanced. Then*

$$\mathrm{disc}_{\mathscr{A}}(\chi) \le 4\sqrt{p \ \ln(p)} + \mathrm{disc}_{\mathscr{A}^p}(\pi) + 1 ,$$

where χ is the p-modulo coloring constructed from π.

Proof. Choose $x \in \mathbb{R}$ so that $n = \sqrt{x^3 \ \ln(x)}$ and p as the largest prime $p \le x$. Then by Bertrand's postulate, $\frac{x}{2} \le p \le x$. Clearly $\sqrt{p^3 \ \ln(p)} \le \sqrt{x^3 \ \ln(x)} = n$ and moreover

$$n = \sqrt{x^3 \ \ln(x)} \le \sqrt{8p^3 \ \ln(2p)} \le \sqrt{16p^3 \ \ln(p)} = 4\sqrt{p^3 \ \ln(p)} .$$

---

[2]The description given in [6] reads differently to what we present here, since there, APs live on $[n] = \{1, \ldots, n\}$ and not $[\![n]\!] = \{0, \ldots, n - 1\}$, as here. Still, the generating coloring is defined on $[\![p]\!]$ in [6].

It follows

$$\frac{n}{p} \le \frac{4\sqrt{p^3 \ln(p)}}{p} = 4\sqrt{p \ln(p)} \ . \tag{7.5}$$

Fix a, d, ℓ. We consider two cases for d. The first case is $d \ge p$. Then $n - 1 \ge a + (\ell - 1)d \ge a + (\ell - 1)p$, so $\frac{n-1-a}{p} + 1 \ge \ell$ and with (7.5) we have $\ell \le 4\sqrt{p \ln(p)} + 1$. In the worst case, $|\chi(A_{a,d,\ell})| = \ell$, so we conclude in the case $d \ge p$ that

$$|\chi(A_{a,d,\ell})| \le 4\sqrt{p \ln(p)} + 1 \ .$$

Now consider the case $d < p$. We will use (7.3) and bound $|\chi(B_q)|$. Since d is not a multiple of p, each number in B_q is mapped to a different number in $[\![p]\!]$ by the modulo operation. Thus B_q is one-to-one with an AP in $\mathbb{Z}_p$ and so $|\chi(A_{a,d,\ell})| \le n/p + \mathrm{disc}_{\mathscr{A}^p}(\pi) \le 4\sqrt{p \ln(p)} + 1 + \mathrm{disc}_{\mathscr{A}^p}(\pi)$, where the final inequality follows from (7.5).   □

We have immediately:

Corollary 4. *By Algorithm 3, we find with probability at least* $1 - \varepsilon$ *a balanced coloring* π *of* $\mathscr{A}^p$, *where*

$$\mathrm{disc}_{\mathscr{A}}(\chi) \le 4\sqrt{p \ln(p)} + 2\sqrt{p \ln(4p^3/\varepsilon)} + 1$$

holds for the p-modulo coloring χ *constructed from* π. *If restricting to* $\varepsilon \ge 4/p$, *this bound simplifies to*

$$\mathrm{disc}_{\mathscr{A}}(\chi) \le 6\sqrt{p \ln(p)} + 1 \ .$$

We combine these results to get:

Theorem 5. *By using the coloring constructed by Algorithm 3 as a generating coloring for a p-modulo coloring* χ, *we have with probability at least* $1 - 4/p$ *that*

$$\mathrm{disc}_{\mathscr{A}}(\chi) \le 6\sqrt[3]{2/3 \cdot n \ln(n)} + 1 = O(\sqrt[3]{n \ln(n)}) \ .$$

Proof. Let p as in Lemma 1, thus in particular $\sqrt{p^3 \ln(p)} \le n$. We have

$$\sqrt{p^3 \ln^3(p)} = \sqrt{p^3 \ln(p)} \ln(p) = \sqrt{p^3 \ln(p)} \cdot \frac{2}{3} \cdot \ln\left(p^{3/2}\right)$$

$$\le \sqrt{p^3 \ln(p)} \cdot \frac{2}{3} \cdot \ln\left(\sqrt{p^3 \ln(p)}\right) \le n \cdot \frac{2}{3} \cdot \ln(n) \ ,$$

thus $\sqrt{p \ln(p)} \le \sqrt[3]{2/3 \cdot n \ln(n)}$. The theorem follows with Corollary 4.   □

As the last step of the theoretical analysis, we provide an asymptotically almost matching lower bound. Recall that discrepancy is lower-bounded by $\lceil n/p \rceil$. For our choice of p we have $\sqrt{p^3 \ln(p)} \leq n$ hence $p \leq \frac{n^{2/3}}{\ln(p)^{1/3}} \leq n^{2/3}$ and so $n/p \geq \sqrt[3]{n}$.

7.5.2 Experiments

Before putting this algorithm into practice, we do one more refinement to the way we construct a coloring χ of $[\![n]\!]$ out of a coloring η of $[\![h]\!]$ for $h = \frac{p+1}{2}$. Recall the mirror construction from (7.4), which gives us a coloring π of $[\![p]\!]$, which is then repeated to color $[\![n]\!]$. In the experimental studies in [11], an *alternating mirror* technique is used, which alternately uses π_{p-1} and $-\pi_{p-1}$ for numbers of the form $kp - 1$ with $k \in \mathbb{N}$. This means that we repeat the following coloring of $[\![2p]\!]$ via $v \mapsto (v \bmod 2p)$:

$$(\eta_0, \ldots, \eta_{h-2}, -\eta_{h-2}, \ldots, -\eta_0, \eta_{h-1}, \eta_0, \ldots, \eta_{h-2}, -\eta_{h-2}, \ldots, -\eta_0, -\eta_{h-1}).$$

A review of the proofs from Sect. 7.5.1 yields that the discrepancy bounds of Theorem 5 still hold. Instead of thinking modulo $2p$ let us stick to modulo p but consider the color of $p-1$ as not being determined and appearing different each time we look at it. We review the proof of Lemma 3 for this case. Fix a, d, ℓ. The case of $d \geq p$ uses the roughest estimation possible, namely the length of the progression. It is independent of the actual coloring. For the case $d < p$, realize that each of the sub-progressions $B_0, \ldots, B_q$ when mapped to $[\![p]\!]$ by the modulo operation contains each number from $[\![p]\!]$ at most once. Due to our construction, the coloring of $[\![p]\!]$ is balanced regardless of the color that $p - 1$ receives. This handles $B_0, \ldots, B_{q-1}$. For B_q, we look more precisely what V_1 and V_2 in Lemma 1 are for the mirror construction, namely $V_1 = [\![h]\!] = \{0, \ldots, h - 1\}$ and $V_2 = \{h, \ldots, p - 2\}$. The proof of that lemma works with any color being assigned to the last vertex $p - 1$ (cf. line 4 in Algorithm 3), it does not have to be in any accordance with the rest of the colors, nor does it have to be probabilistically chosen. The "$+1$" in the bound takes care of it.

The alternating mirror technique gives rise to the hope that we could attain a better discrepancy than $\lceil n/p \rceil$. But considering $A_{1,p,\ell}$ with ℓ as large as possible, we get $\ell = \lfloor n/p - 2/p + 1 \rfloor \in \{\lfloor n/p \rfloor, \lceil n/p \rceil\}$, which is no big improvement.

Table 7.3 gives results for $n = 100 \cdot 10^3$ and different primes. For each prime, we run for 60 min in the same framework that we used in Sect. 7.4. The result is given in the third column. The number of trials is rounded to the nearest 10 millions.

The best result 66 is attained after 36 min with no improvement during the remaining 24 min. The smallest prime, 1,123, is the largest prime p such that

Table 7.3 Sampling from the uniform distribution with alternating mirror technique

p	$\left\lfloor \frac{n}{p} - \frac{2}{p} + 1 \right\rfloor$	Result	Number of trials
2803	36	76	$400 \cdot 10^6$
2243	45	73	$430 \cdot 10^6$
1907	53	69	$410 \cdot 10^6$
1873	54	69	$430 \cdot 10^6$
1607	63	66	$390 \cdot 10^6$
1399	72	72	$300 \cdot 10^6$
1249	81	81	$230 \cdot 10^6$
1123	90	90	$220 \cdot 10^6$

$\sqrt{p^3 \ln(p)} \leq n$, which is the one suggested by Sárközy and used in the proof of Theorem 5. It is noteworthy that for the three smallest primes, the best possible result was found. Moreover, this happened virtually immediately, within the first minute.

Finally, we test the Legendre construction. Since it deterministically determines a coloring of $[\![p]\!]$, there is no need for a mirror construction. However, we use the alternating technique. A vast range of primes is tested, starting with 563 (which is approximately $\frac{1123}{2}$) and then all primes in ascending order until 60 min are up. A single-threaded implementation is used, but this is enough to reach prime 11,059. The best discrepancy of 71 is obtained for prime 1,907 after 11 min; the randomized version attained discrepancy 69 for this prime, however using much more computing power. For the primes on or below 1,123, we only get discrepancy 90 with the Legendre construction. Without the alternating technique, best discrepancy we obtain with the Legendre construction in the range 563–10,429 of primes during 60 min is 88 for prime 3,083, which is clearly inferior to the 71.

7.6 EDA: Beyond the Uniform Distribution

Using modulo colorings, discrepancy can be brought to much smaller values than what we have seen for fully random colorings in Sect. 7.4. However, this is still some distance from the 53 which allegedly can be achieved in only 7 min. How can we do this? The idea is not sticking to the uniform distribution to generate the random portion of the colorings, which is $(\eta_0, \ldots, \eta_{h-1}) \in \{-1, +1\}^h$. Instead, we gradually modify the distribution on $\{-1, +1\}^h$ with the intention of concentrating it in promising regions of this search space. Estimation of distribution algorithms (EDA) follow this approach, more precisely the idea of gradually modifying a distribution characterizes *incremental* EDAs. We describe the idea of EDAs in more detail, starting with the non-incremental version.

7.6.1 Population-Based EDAs

Assume the search space is $\{0, 1\}^k$ for some $k \in \mathbb{N}$, and the task is to maximize a function $f : \{0, 1\}^k \longrightarrow \mathbb{Q}$. The function f is typically called the *fitness function*, and a solution $x \in \{0, 1\}^k$ is sometimes called an *individual*. For the discrepancy problem, we would use $\{-1, +1\}^h$ as the search space (which is essentially the same as $\{0, 1\}^h$) and the negative of the discrepancy as fitness, so larger fitness is better.

In a population-based EDA, we start by sampling a number of, say, N, solutions using the uniform distribution. This set $P \subset \{0, 1\}^k$ of N solutions is called a *population*. From the population, we select the, say, 50 % solutions with highest fitness, denote them $A \subset P \subset \{0, 1\}^k$. We ask: what probability distribution could have produced this set A of relatively good solutions? If we knew this distribution, we could sample it some more and hope for even better solutions. An important characteristic of an EDA is what types of distributions it considers. In the simplest case, the k coordinates in the solution are treated as a family of independent random variables. All such distributions look like $Q = (Q_0, \ldots, Q_{k-1}) \in [0, 1]^k$ with Q_i stating the probability of sampling a 1 in position i. An EDA assuming this independence is called a *univariate* EDA. One of the earliest EDAs, the *univariate marginal distribution algorithm* (UMDA) by Mühlenbein and Paaß [16], is of this type. In UMDA, given the selection $A \subset P$ of promising solutions, we build a distribution, also called a *(probabilistic) model*, by defining

$$Q_i := \frac{|\{x \in A; \ x_i = 1\}|}{|A|} \quad \text{for each } i \in [\![k]\!].$$

Then we use Q to sample N new solutions, again select promising solutions out of those (e.g. the 50 % best) and build the next model, and so on.

This continues until we are satisfied or we run out of time or some other termination criterion holds. One other possible termination criterion is that the models have low entropy. *Entropy* is a measure of randomness, defined as $-\sum_{i=0}^{k-1} Q_i \log(Q_i)$, which is at its maximum k when $Q_i = \frac{1}{2}$ for all i, and at its minimum 0 if $Q_i \in \{0, 1\}$ for all i. When entropy is low, sampling will yield similar solutions.

Population-based EDAs are considered *evolutionary algorithms* since they allow the fittest individuals in the population P to reproduce via the distribution Q, which is build upon them. Each iteration is therefore also called a *generation*.

7.6.2 Incremental EDAs

Instead of building the model from scratch in each generation, it is possible to update an existing model. Then we speak of an *incremental* EDA. One of the earliest incremental EDAs was given in 1994 and 1995 by Baluja [3] and Baluja and Caruana [4], known as *population-based incremental learning* (PBIL). It starts with $Q = (1/2, \ldots, 1/2) \in [0, 1]^k$, the uniform distribution. In each generation, a population P is sampled from Q. Then an individual $x^* \in P$ with maximal fitness among all the individuals in P is determined and Q slightly adjusted towards it:

$$Q_i := \begin{cases} Q_i - \lambda Q_i & \text{if } x_i^* = 0 \\ Q_i + \lambda(1 - Q_i) & \text{if } x_i^* = 1 \end{cases} \quad \text{for each } i \in [\![k]\!].$$

$\lambda \in [0, 1]$ is a parameter, called *learning rate*. Variations include using the best N individuals $x^{(1)}, \ldots, x^{(N)} \in P$ for some $N < |P|$ and update the model with each of them, starting with the best.

The *compact genetic algorithm* (cGA) given by Harik et al. [9] in 1997 is similar, but only two solutions are sampled each generation. Let x^* and x be the two solutions and $f(x^*) \geq f(x)$. Then the model update works as follows:

$$Q_i := \begin{cases} \max\{0, Q_i - \lambda\} & \text{if } x_i^* = 0 \\ \min\{1, Q_i + \lambda\} & \text{if } x_i^* = 1 \end{cases} \quad \text{for each } i \in [\![k]\!] \text{ such that } x_i^* \neq x_i.$$

So only in coordinates where the better solution differs from the inferior one, an update is performed. Another difference to PBIL is that updates are in absolute terms. Variations include sampling N solutions $x^{(1)}, \ldots, x^{(N)}$ instead of just 2, and then for each unordered pair $\{j, j'\} \in \binom{[N]}{2}$ we let $x^{(j)}$ and $x^{(j')}$ take the roles of two solutions in the simple version. The cGA algorithm was evaluated on large-scale optimization problems by Sastry et al. [21].

Incremental EDAs are also considered evolutionary algorithms, in the sense that the probabilistic model gives a compact representation of a population of individuals. Over time, the frequency of fit individuals increases, as the model is adjusted towards producing fitter individuals upon being sampled.

In 2002, Han and Kim [7] gave an incremental EDA using an *attractor*. Their algorithm is called the *quantum-inspired evolutionary algorithm* (QEA); we will comment on this name later. Since we will discuss variants of this algorithm, we call the version given by Han and Kim the *standard QEA* (sQEA). sQEA also starts with $Q = (1/2, \ldots, 1/2)$. Additionally, there is a solution $a \in \{0, 1\}^k$ called the *attractor*. In the simplest case, it is initialized by drawing from $\{0, 1\}^k$ uniformly at random. In each generation, Q is sampled yielding a solution x and then x is compared to a. If $f(x) \geq f(a)$, i.e. x is at least as good as a, then we update $a := x$ and continue with the next generation; we also say that the sample x is *accepted*. If, on the other hand, $f(x) < f(a)$, i.e. the sample x is inferior to a, then Q is updated in a similar way as in PBIL or cGA, namely moved towards a where x and a differ. Then the next generation starts, either with the updated attractor (if $f(x) \geq f(a)$) or with the same attractor as before (if $f(x) < f(a)$). Since in case $f(x) < f(a)$ the sample x is discarded, sQEA can make use of shortcutting when used for the discrepancy problem. As an extension, multiple samples can be taken in each generation. The model is only adjusted if none of them satisfies $f(x) \geq f(a)$, and then an arbitrary one, say the first, is used as reference for the test $x_i \neq a_i$.

The original description of sQEA carries out the model update using the following procedure: the point $(\sqrt{1 - Q_i}, \sqrt{Q_i})$ in the plane is rotated by a certain angle ϑ either clockwise (if $a_i = 0$) or counter-clockwise (if $a_i = 1$), and the new value of Q_i becomes the square root of the new ordinate. This is inspired by quantum computing; an actual benefit could be that towards the extremes (0 and 1) shifts become smaller. We call this *rotation learning*, opposed to *linear learning* as in cGA. The angle ϑ in rotation learning corresponds to λ in linear learning. Instead of giving ϑ or λ, we will often prefer to specify how many different values each Q_i can attain in the interval $[0, 1]$, called *learning resolution*. Learning resolution is $R = \frac{\pi}{2\vartheta}$ and $R = \frac{1}{\lambda}$, respectively. Of course, the attractor concept of sQEA can also be used with linear learning, not only with rotation learning.

The term "quantum-inspired" is due to recognizing that the $Q_0, \ldots, Q_{k-1}$ behave similar to k qubits in a quantum computer: each is in a state between 0 and 1, and only upon observation takes on states 0 or 1 with certain probabilities. Hence what we call "sampling" is also called "observing" in the literature. Clearly, any univariate EDA is "quantum-inspired" in this sense, including UMDA and PBIL from the 1990s. However, it appears that "quantum-inspired" has become the attribute to describe incremental EDAs that work with an attractor. In [19], Platel et al. formally state and explain that QEAs indeed belong to the class of EDAs.

Although EDAs are often better than just sampling uniformly over and over again, they can suffer from *premature convergence*, i.e. the probabilities $Q_0, \ldots, Q_{k-1}$ can quickly move to one of the extremes (0 or 1) and the algorithm does not provide means to break out of this. Premature convergence can also be seen as a *lack of exploration* of the search space.

sQEA can be implemented treating multiple models in parallel, typically up to 100, each having an attractor attached to it. Periodically, all models replace their attractor with the best one over all models. This is called *migration* or *synchronization*. As a refinement, models are divided into groups, and we have *local migration* and *global migration*. In each generation, we have local migration: the attractor of a model is replaced by the best attractor in its group. Only every T_g generations, we have global migration: each attractor is replaced by the best attractor over all models, regardless of groups. The number T_g is called the *global migration period*. Not conducting global migration in each generation is intended to prevent premature convergence. Indeed, it can be observed sometimes that entropy drastically rises after a global migration when using a large enough global migration period. This must be considered when using low entropy as a termination criterion.

Given an appropriate implementation, it can be considered treating the different models asynchronously in parallel in order to prevent idle time. This is particularly interesting if the fitness function provides a shortcutting mechanism, as in the discrepancy problem. A full description of sQEA is given in Algorithm 4.

In 2007, Platel, et al. [18] gave the *versatile QEA* (vQEA). It works similar to sQEA with the exception that the attractor is replaced by the sample x in each generation *unconditionally*, even if $f(x) < f(a)$. This feature is also intended to prevent premature convergence. Experimental evidence suggests this being a very promising approach. When using multiple models in parallel, migration is

---

Algorithm 4: The standard QEA (sQEA). By $\mathcal{M}$ we denote the set of all models. The shift $\Delta(Q_i)$ is either determined by rotation when using rotation learning, then it depends on Q_i, or it is just a constant if using linear learning. The termination criterion "hopeless" could be low entropy

---

1 **in parallel for each** *model* $Q = (Q_0, \ldots, Q_{k-1}) \in \mathcal{M}$ **do**
2   initialize generation counter $t := 0$;
3   initialize model $Q := (1/2, \ldots, 1/2)$;
4   initialize attractor $a :=_{\text{unif}} \{0, 1\}^k$;
5   **repeat**
6   **if** $t \bmod T_g = 0$ **then** $a :=$ best attractor over all models;
7   **else** $a :=$ best attractor over all models in Q's group;
8   sample Q for N times yielding $x^{(1)}, \ldots, x^{(N)} \in \{0, 1\}^k$;
9   **if** $\exists j \in [N] : f(x^{(j)}) \geq f(a)$ **then** $a :=$ best of $x^{(1)}, \ldots, x^{(N)}$;
10   **else for** $i = 0, \ldots, k - 1$ **do**
11   **if** $x_i^{(1)} \neq a_i$ **then** $Q_i := \begin{cases} \max\{0, \ Q_i - \Delta(Q_i)\} & \text{if } a_i = 0 \\ \min\{1, \ Q_i + \Delta(Q_i)\} & \text{if } a_i = 1 \end{cases}$
12         $t := t + 1$;
13   **until** *satisfied or hopeless or out of time*;

---

performed in each generation such that the attractor of each model in generation $t + 1$ is the best sample taken in generation t. This reduces the possibilities for shortcutting: each generation, at least one full fitness function evaluation must be conducted. A similar observation holds for PBIL and cGA.

In [11], sQEA, vQEA, and a novel variant called *attractor population QEA* (apQEA) are compared against each other on the low-discrepancy problem for APs. The next section will be devoted to apQEA, and in the section that follows we will summarize the experimental comparison of the three.

For more on EDAs in general, the reader is referred to the survey by Hauschild and Pelikan [10] and the references therein. That survey also points out and explains several EDAs which are not univariate, i.e. they do not necessarily treat the k coordinates of a solution as independent random variables. For experimental work see, e.g., [2, 8, 12, 13, 17].

7.7 Attractor Population QEA (apQEA)

This and the following section is based on [11], where the attractor population QEA (apQEA) was invented and tested on the discrepancy problem in $\mathscr{A}_n$. apQEA strikes a balance between the approaches of sQEA and vQEA. In sQEA, attractors basically follow the best solution found so far, with some delay if using global migration period. In vQEA, attractors change frequently even if this means that the new attractor is less fit. In apQEA, an *attractor population* $\mathscr{P} \subseteq \{0, 1\}^k$ of a

---

Algorithm 5: apQEA

1 randomly initialize attractor population $\mathscr{P} \subseteq \{0, 1\}^k$ of cardinality S;
2 **in parallel for each** *model* $Q = (Q_0, \ldots, Q_{k-1}) \in \mathscr{M}$ **do**
3   initialize model $Q := (1/2, \ldots, 1/2)$;
4   **repeat**
5          $a :=$ select from $\mathscr{P}$;
6   **do** P **times**
7              $f_0 :=$ worst fitness in $\mathscr{P}$;
8   sample Q for N times yielding $x^{(1)}, \ldots, x^{(N)} \in \{0, 1\}^k$;
9   **if** $\exists j \in [N] : f(x^{(j)}) > f_0$ **then**
10   inject each $x^{(j)}$ into $\mathscr{P}$ if $f(x^{(j)}) > f_0$;
11   trim $\mathscr{P}$ to the size of S, removing worst solutions;
12   **else for** $i = 0, \ldots, k-1$ **do**
13   **if** $x_i^{(1)} \neq a_i$ **then** $Q_i := \begin{cases} \max\{0, \, Q_i - \Delta(Q_i)\} & \text{if } a_i = 0 \\ \min\{1, \, Q_i + \Delta(Q_i)\} & \text{if } a_i = 1 \end{cases}$
14   **until** *satisfied or hopeless or out of time*;

---

fixed size S is maintained, which is a set of solutions from which attractors are drawn. For each model, every $P \in \mathbb{N}$ generations a new attractor is selected from $\mathscr{P}$. The parameter P, which is the number of generations that an attractor stays in function, is called the *attractor persistence*. We made good experiences with $P = 10$. Selection can for example be done by a tournament: draw two solutions from $\mathscr{P}$ uniformly at random and use the better of the two with probability 60 % and the inferior of the two with probability 40 %.

In each generation, the sampled solution x is compared to the worst solution in $\mathscr{P}$. Denote f_0 the worst fitness in $\mathscr{P}$ at that point in time. Then the sample is accepted if $f(x) > f_0$, in which case it is injected into $\mathscr{P}$ and then $\mathscr{P}$ trimmed to its original size S, removing worst solutions. If $f(x) \leq f_0$, then the model is adjusted using rotation learning or linear learning, then the sample x is discarded. It is also conceivable to use $f(x) \geq f_0$ as the acceptance criterion, similar to sQEA. For the discrepancy problem, $f(x) > f_0$ has shown to be superior, but this could be different for other problems. It is also conceivable to use the Nth percentile in $\mathscr{P}$ for comparison instead of f_0, for a parameter N. As in sQEA, multiple samples can be taken per generation.

A very important parameter is the size $S = |\mathscr{P}|$ of the attractor population. We will see in experiments that larger S means better exploration abilities. For the discrepancy problem, we will have to increase S (moderately) when n increases. A complete description of apQEA is given as Algorithm 5 on the current page.

apQEA will benefit from shortcutting since $f(x)$ has only to be computed exactly when $f(x) > f_0$. It is also appropriate to treat the models *asynchronously* in apQEA, hence preventing idle time: the attractor population is there, any process may inject into it or select from it at any time—given an appropriate implementation.

7.7.1 Flatline Termination Criterion

Since the attractor changes often in apQEA, entropy oftentimes never reaches near zero but instead oscillates around values like 20 or 30. A more stable measure is the mean Hamming distance in the attractor population, i.e.

$$\frac{1}{\binom{S}{2}} \cdot \sum_{\{x,x'\} \in \binom{\mathscr{P}}{2}} |\{i; \ x_i \neq x_i'\}| \ .$$

However, it also can get stuck well above zero. To determine a hopeless situation, we instead developed the concept of a *flatline*. A flatline is a period of time in which neither the mean Hamming distance reaches a new minimum nor the best-so-far solution improves. When we encounter a flatline stretching over 25 % of the computation time so far, we declare the situation hopeless. To avoid erroneously aborting in early stages, we additionally demand that the relative mean Hamming distance, which is the mean Hamming distance divided by k, falls below $1/10$. Those thresholds were found to be appropriate (for the discrepancy problem) in preliminary experiments. For other problems, those values might have to be adjusted.

7.7.2 Implementation

To fully benefit from the features of apQEA, we need an implementation which allows *asynchronous* communication between processes. Our MPI-based implementations (version 1.2.4) exhibited unacceptable idle times when used for asynchronous communication, so we wrote our own client-server-based parallel framework. It provides a server that manages the attractor population. Clients can connect to it at any time via TCP/IP and do selection and injection. The server takes care of trimming the population after injection. Great care was put into making the implementation well-performing and free of race conditions. The server uses a simple filesystem-based database. Any number of server processes (connected to client processes over the network) can access the database at the same time and do selection and place injection requests. Periodically, every few seconds, a single process treats the injection requests and performs the actual injection and trimming of the attractor population. This way, the main data structure is never modified by more than one process at a time. The system gets by without any locking mechanisms.

Most parts of the software is written in Bigloo,[3] an implementation of the Scheme programming language, which, quoting its website, is "devoted to one goal:

---

[3]http://www-sop.inria.fr/indes/fp/Bigloo/.

enabling Scheme based programming style where C(++) is usually required." In our experience, Scheme allows us to quickly arrive at a working implementation, which can easily be adapted afterwards. Scheme code usually is very concise. Nonetheless, the fitness function (discrepancy computation) and a few other parts were written in C, for performance reasons and to have OpenMP available. This is possible since Bigloo provides a C interface. OpenMP is used to distribute fitness function evaluation across multiple processor cores. So we have a two-level parallelization: on the higher level, we have multiple processes treating multiple models and communicating via the attractor population server. On the lower level, we have thread parallelization for the fitness function. The framework provides also means to run sQEA and vQEA, although in the case of vQEA perhaps not in the most efficient way.

We always use 8 threads (on 8 processor cores, one for each thread) for the fitness function. If not stated otherwise, a total of 96 processor cores is used. This allows us to have 12 models fully in parallel; if we use more models on the same number of processors, then the set of models is partitioned and the models from each partition are treated sequentially.

7.8 QEA Results for the Discrepancy Problem in $\mathscr{A}_n$

We prepare for the three algorithms sQEA, vQEA, and apQEA to be tested on $\mathscr{A}_n$. We will use modulo colorings with the alternating mirror construction as introduced in Sect. 7.5. Recall that the best discrepancy to hope for is $\lfloor n/p \rfloor$. Our aim is to attain the asymptotic $\Theta(\sqrt[4]{n})$, therefore we choose p depending on n so that $n/p = \Theta(\sqrt[4]{n})$. Call $\lfloor 3\sqrt[4]{n} \rfloor$ the *target discrepancy*; preliminary experiments had shown this to be an attainable goal for apQEA in reasonable time. Better approximations may be possible with other parameters and more processors and/or more time, but we do not investigate this here. Since we wish to attain the target discrepancy, we choose p large enough so that we have some room relative to the target discrepancy, say $n/p \approx 2.5 \cdot \sqrt[4]{n}$, i.e. $p \approx 2/5 \cdot n^{3/4}$. Precisely, we choose p to be the largest prime subject to $p \leq \lceil 2/5 \cdot n^{3/4} \rceil$.

Recall also that modulo colorings allow restricting to small starting points for APs, namely with alternating modulo coloring we can restrict to starting points in $[\![2p]\!]$. Denote $m(n) := |\mathscr{A}_n|$, and also denote $m(n, p) := |\{A_{a,d,\ell} \cap [\![n]\!]; a \in [\![2p]\!] \wedge d, \ell \in \mathbb{N}_0\}|$ the number of APs in $\mathscr{A}_n$ with starting points in $[\![2p]\!]$. Table 7.4 for each n gives the prime p as explained above and the corresponding values for $m(n)$ and $m(n, p)$ as well as to which percent the number of APs to test is reduced (rounded to full percents).

Of course, where applicable, we do shortcutting by setting δ_1 appropriately in Algorithm 2. Parameter δ_0 is set to the target discrepancy. In fact, in our implementation, we use some additions to Algorithm 2, exploiting further the structure of APs and making it substantially more technically involved. We do not give the details here; they are not necessary for understanding this chapter.

Table 7.4 Reducing the number of APs to check

n	p	$m(n)$	$m(n, p)$	$\frac{m(n,p)}{m(n)}$
$100 \cdot 10^3$	2243	55836909328	5105563932	9 %
$125 \cdot 10^3$	2659	88988439957	7720722747	9 %
$150 \cdot 10^3$	3049	130194434212	10798481037	8 %
$175 \cdot 10^3$	3413	179569486783	14295844944	8 %
$200 \cdot 10^3$	3779	237210329226	18300540427	8 %
$225 \cdot 10^3$	4133	303200671205	22745014549	8 %
$250 \cdot 10^3$	4463	377614297340	27536746469	7 %

Table 7.5 Results for sQEA

R	$M = 12$		$M = 24$		$M = 48$		$M = 96$	
50	60	61	59	59	58	57	58	58
100	59	59	57	59	59	56	56	56
500	57	56	57	57	55	56	56	57
1 000	57	57	56	55	57	56	58	59
2 000	57	57	57	58	59	60	63	62
3 000	56	57	59	58	61	61	64	65

7.8.1 sQEA

The important parameters of sQEA are number of models M, number of samples N taken in each generation, learning resolution R, global migration period T_g, and number of groups. We stick to $N = 10$ samples in each generation for all experiments. For R, we use 50, 100 and 500, which are common settings, and also try 1 000, 2 000, and 3 000. In [8], it is proposed to choose T_g in linear dependence on R, which in our notation and neglecting a small additive constant reads $T_g = 2 \cdot R\alpha$ with $1.15 \leq \alpha \leq 1.34$. We use $\alpha = 1.25$ and $\alpha = 1.5$, so $T_g = 2.5 \cdot R$ and $T_g = 3 \cdot R$. In [8], the number of groups is fixed to 5 and the number of models ranges up to 100. We use 6 groups and up to 96 models. We use rotation learning, but made similar observations with linear learning.

We fix $n = 100 \cdot 10^3$, which implies $p = 2\,243$ as per our way to choose the prime, and do 3 runs for each set of parameters. Computation is aborted after 15 min, which is roughly double the time apQEA needs to reach the target discrepancy of 53. Table 7.5 (taken from [11]) gives mean discrepancies over the 3 runs. The left number is for smaller T_g and the right for higher T_g, e.g. for $M = 12$ and $R = 50$ we have 60 for $T_g = 2.5 \cdot 50 = 125$ and 61 for $T_g = 3 \cdot 50 = 150$.

Although those discrepancies look better than those we computed in Sect. 7.5.2—in particular compare with the 73 we got for $p = 2\,243$ there— target discrepancy 53 is never reached. For two runs we reach 54, namely for $(M, R, T_g) = (24, 1\,000, 3\,000)$ and $(96, 500, 1\,250)$. But for each of the 2 settings, only 1 of the 3 runs reached 54. There is no clear indication whether smaller or larger T_g is better. Entropy left in the end generally increases with M and R.

Table 7.6 Results for sQEA for larger n

$n =$	$100 \cdot 10^3$	$125 \cdot 10^3$	$150 \cdot 10^3$	$175 \cdot 10^3$	$200 \cdot 10^3$
Time limit in minutes	15	25	40	80	150
Best sQEA result	54	60	63	68	69
Target	53	56	59	61	63

We pick the setting $(M, R, T_g) = (24, 1\,000, 3\,000)$, which attained 54 in 1 run and also has lowest mean value of 55, for a 5 h run. In the 15 min runs with this setting, entropy in the end was 48 on average. What happens if we let the algorithm use up more of its entropy? As it turns out while entropy is brought down to 16 during the 5 h, only discrepancy 56 is attained.

The main problem with sQEA here is that there is no clear indication which parameter to tune in order to get higher quality solutions—at least not within reasonable time (compared to what apQEA can do). In preliminary experiments, we reached target discrepancy 53 on some occasions, but with long running times. We found no way to *reliably* reach discrepancy 55 or better with sQEA.

Finally, for $(M, R, T_g) = (24, 1\,000, 3\,000)$, we do experiments for up to $n = 200 \cdot 10^3$, with 3 runs for each n. The time limit is twice what apQEA requires on average, rounded to the next multiple of 5. Table 7.6 for each n gives the best result obtained over the 3 runs and the target for comparison.

Although we do multiple runs and allocate twice the time apQEA would need to attain the target, the best result for sQEA stays clearly away from the target.

In apQEA, we use $f(x) > f_0$ as the criterion for accepting the sample x. In sQEA, the non-strict $f(x) \geq f(a)$ is used. Changing this to $f(x) > f(a)$ in sQEA has shown to be detrimental for the discrepancy problem: the best discrepancy we reached with this modification in similar experiments was 57.

7.8.2 vQEA

Recall that vQEA in generation $t + 1$ unconditionally replaces the attractor for each model with the best sample found during generation t. vQEA does not use groups nor global migration period, instead all models form a single group "to ensure convergence" [18].[4] Indeed, our experiments confirm that vQEA has no problem keeping entropy high. We conducted 5 runs with $N = 10$ samples each generation, $M = 12$ models in parallel, learning resolution $R = 50$, and rotation learning for $n = 100 \cdot 10^3$. In all of the runs, the target of 53 was hit with about 100 of

---

[4]However, in [18] it is suggested to use groups with vQEA in future work.

entropy left. However, the time required was almost 2 h.[5] We also conducted 5 runs with linear learning, yielding the same solution quality at a 12 % higher running time. We also did a run for $n = 125 \cdot 10^3$; there vQEA attained the target discrepancy after 3.5 h.

The high running times were to be expected since vQEA can only make limited use of shortcutting. Since we take multiple samples in each generation (10 for each model), fitness function evaluation from the second sample on can make use of shortcutting. However, necessarily each generation takes at least the time of one full fitness function evaluation. We conclude that while vQEA has impressive exploration capabilities and delivers high solution quality "out of the box", i.e. without any particular parameter tuning, it is not well suited for the discrepancy problem.

7.8.3 apQEA

Recall that the most important parameter for apQEA is the attractor population size S. We fix attractor persistence to $P = 10$, number of models to $M = 12$, number of samples in each generation to $N = 10$, and learning resolution to $R = 100$ with linear learning. The main parameter S is varied in steps of 10. Computation is aborted when the target of $\lfloor 3\sqrt[4]{n} \rfloor$ is hit (a *success*) or a long flatline is observed (a *failure*), as explained in Sect. 7.7.1. For selecting attractors, we use tournament selection: draw two solutions randomly from $\mathscr{P}$ and use the better one with 60 % probability and the inferior one with 40 % probability. For each n and appropriate choices of S, we do 30 runs and record the following: whether it is a success or a failure, final discrepancy (equals target discrepancy for successes), running time (in minutes), mean final entropy. For failures, we also record the time at which the last discrepancy improvement took place (for successes, this value is equal to the running time). Table 7.7 (taken from [11]) gives results grouped into successes and failures, all numbers are mean values over the 30 runs together with their standard deviation.

We observe that by increasing S, we can guarantee the target to be hit. Dependence of S on n for freeness of failure appears to be approximately linear or slightly super-linear: ratios of S to $n/1\,000$ for no failures are 0.30, 0.24, 0.27, 0.29, and 0.30. But even if S is one step below the required size, discrepancy is only 1 away from the target (with an exception for $n = 125 \cdot 10^3$ and $S = 20$, where we recorded discrepancy 58 in 1 of the 30 runs). Running times for failures tend to be longer than for successes, even if the failure is for a smaller S. This is because it takes some time to detect a failure by the flatline criterion. Larger S effects

---

[5]Even more, these 2 h is only the time spent in fitness function evaluation. Total running time was about 4 h, but we suspect this to be partly due to our implementation being not particularly suited for vQEA resulting in communication overhead.

Table 7.7 Results for apQEA

		Successes				Failures				
n	S	#	Result	Time	Entropy	#	Result	Time	Entropy	Last imp.
$100 \cdot 10^3$	20	28	$53_{\sigma=00}$	$05_{\sigma=01}$	$38_{\sigma=09}$	02	$54_{\sigma=00}$	$08_{\sigma=01}$	$22_{\sigma=01}$	$06_{\sigma=01}$
$100 \cdot 10^3$	30	30	$53_{\sigma=00}$	$07_{\sigma=01}$	$63_{\sigma=12}$	00	na	na	na	na
$125 \cdot 10^3$	20	17	$56_{\sigma=00}$	$09_{\sigma=02}$	$31_{\sigma=08}$	13	$57_{\sigma=00}$	$12_{\sigma=01}$	$19_{\sigma=09}$	$08_{\sigma=01}$
$125 \cdot 10^3$	30	30	$56_{\sigma=00}$	$11_{\sigma=01}$	$48_{\sigma=11}$	00	na	na	na	na
$150 \cdot 10^3$	30	26	$59_{\sigma=00}$	$16_{\sigma=02}$	$46_{\sigma=11}$	04	$60_{\sigma=00}$	$23_{\sigma=02}$	$32_{\sigma=09}$	$16_{\sigma=02}$
$150 \cdot 10^3$	40	30	$59_{\sigma=00}$	$20_{\sigma=02}$	$62_{\sigma=10}$	00	na	na	na	na
$175 \cdot 10^3$	30	16	$61_{\sigma=00}$	$24_{\sigma=04}$	$31_{\sigma=08}$	14	$62_{\sigma=00}$	$38_{\sigma=05}$	$21_{\sigma=08}$	$24_{\sigma=03}$
$175 \cdot 10^3$	40	27	$61_{\sigma=00}$	$32_{\sigma=05}$	$42_{\sigma=09}$	03	$62_{\sigma=00}$	$47_{\sigma=02}$	$26_{\sigma=05}$	$30_{\sigma=01}$
$175 \cdot 10^3$	50	30	$61_{\sigma=00}$	$39_{\sigma=05}$	$57_{\sigma=11}$	00	na	na	na	na
$200 \cdot 10^3$	30	02	$63_{\sigma=00}$	$38_{\sigma=02}$	$22_{\sigma=06}$	28	$65_{\sigma=01}$	$47_{\sigma=08}$	$24_{\sigma=17}$	$30_{\sigma=05}$
$200 \cdot 10^3$	50	28	$63_{\sigma=00}$	$53_{\sigma=08}$	$44_{\sigma=08}$	02	$64_{\sigma=00}$	$76_{\sigma=06}$	$28_{\sigma=02}$	$53_{\sigma=03}$
$200 \cdot 10^3$	60	30	$63_{\sigma=00}$	$74_{\sigma=15}$	$53_{\sigma=12}$	00	na	na	na	na

larger entropy; failures tend to have lowest entropy, indicating that the problem is the models having locked onto an inferior solution. The table also shows what happens if we do *lazy S tuning*, i.e. fixing S to the first successful value $S = 30$ and then increasing n: failure rate increases and solution quality for failures deteriorates moderately. The largest difference to the target is observed for $n = 200 \cdot 10^3$ and $S = 30$, namely we got discrepancy 67 in 1 of the 30 runs; target is 63. For comparison, the best discrepancy we found via sQEA in 3 runs for such n was 69 and the worst was 71. We conclude that a mistuned S does not necessarily have catastrophic implications, and apQEA can still beat sQEA.

To see the effect of parallelization, we double number of cores from 96 to 192 and increase number of models to $M = 24$, so that they can be treated in parallel with 8 cores each. Table 7.8 (taken from [11]) shows results for 5 runs for each set of parameters. First consider $n = 175 \cdot 10^3$ and $200 \cdot 10^3$. The best failure-free settings for S are 30 and 50. In comparison with the best failure-free settings for 96 cores, running time is reduced to $17/39 = 44\%$ and $42/74 = 57\%$, respectively. 50 % or less would mean a perfect scaling. When we do not adjust S, i.e. we use 50 and 60, running time is reduced to $26/39 = 67\%$ and $53/74 = 72\%$, respectively. We also compute for $n = 250 \cdot 10^3$.

Colorings found by apQEA can be downloaded[6] and verified. A small Bigloo Scheme program is included, but a similar program can be implemented quickly from scratch in any programming language.

---

[6]http://www.informatik.uni-kiel.de/~lki/discap-results.html.

Table 7.8 Results for apQEA on 192 processor cores

		Successes				Failures				
n	S	#	Result	Time	Entropy	#	Result	Time	Entropy	Last imp.
$175 \cdot 10^3$	20	02	$61_{\sigma=00}$	$14_{\sigma=02}$	$22_{\sigma=02}$	03	$62_{\sigma=00}$	$20_{\sigma=02}$	$31_{\sigma=12}$	$13_{\sigma=00}$
$175 \cdot 10^3$	30	05	$61_{\sigma=00}$	$17_{\sigma=04}$	$43_{\sigma=07}$	00	na	na	na	na
$175 \cdot 10^3$	50	05	$61_{\sigma=00}$	$26_{\sigma=03}$	$72_{\sigma=12}$	00	na	na	na	na
$200 \cdot 10^3$	40	04	$63_{\sigma=00}$	$30_{\sigma=03}$	$42_{\sigma=12}$	01	$64_{\sigma=00}$	$42_{\sigma=00}$	$32_{\sigma=02}$	$28_{\sigma=00}$
$200 \cdot 10^3$	50	05	$63_{\sigma=00}$	$42_{\sigma=10}$	$52_{\sigma=05}$	00	na	na	na	na
$200 \cdot 10^3$	60	05	$63_{\sigma=00}$	$53_{\sigma=08}$	$60_{\sigma=11}$	00	na	na	na	na
$250 \cdot 10^3$	50	04	$67_{\sigma=00}$	$58_{\sigma=07}$	$42_{\sigma=08}$	01	$68_{\sigma=00}$	$110_{\sigma=00}$	$29_{\sigma=02}$	$70_{\sigma=00}$
$250 \cdot 10^3$	60	05	$67_{\sigma=00}$	$83_{\sigma=16}$	$61_{\sigma=14}$	00	na	na	na	na

7.9 Conclusion and Future Work

Different randomized approaches were presented for the discrepancy problem in the hypergraph of arithmetic progressions in the first n integers. The first important step was to use modulo colorings for certain prime numbers, as suggested by Sárközy in 1974. The second step was to look beyond the uniform distribution and use estimation of distribution algorithms instead. The third step was to refine the QEA procedure into what became apQEA. This new form of QEA is able to attain the asymptotic $\Theta(\sqrt[4]{n})$ and beat similar previous EDAs, namely sQEA and vQEA.

At the time of writing, apQEA has only been tested on this particular discrepancy problem. It should be determined in the future whether apQEA is also applicable for computing low-discrepancy colorings in other hypergraphs and also whether it can improve on known results for other hard problems, like packing and covering in graphs and hypergraphs. At least for the discrepancy problem in $\mathscr{A}_n$, apQEA is easy to tune, since a single parameter, the size S of the attractor population, has a clear and foreseeable effect: it improves solution quality (if possible) at the price of an acceptable increase in running time. Will this hold for other challenging problems as well?

Experimentation is emerging as a tool in Combinatorics. For example, experimentation is used in a Polymath project[7] on one of the most challenging open problems of Paul Erdős on homogeneous arithmetic progressions. Our problem sizes are far beyond those of the Polymath project. Of course, our problem is different, but related and in future work it should be attempted to apply apQEA to homogeneous arithmetic progressions.

Acknowledgements I thank the editors for inviting me to contribute this chapter to "A Panorama of Discrepancy Theory". I thank Volkmar Sauerland for proofreading. I thank my co-authors from our SEA 2013 publication [11] for joint work. Financial support through DFG Priority Program "Algorithm Engineering" (Grant Sr7/12-3) is also gratefully acknowledged.

---

[7]http://michaelnielsen.org/polymath1/index.php?title=The_Erd%C5%91s_discrepancy_problem.

References

1. N. Alon, J.H. Spencer, *The Probabilistic Method* (Wiley, New York, 2000)
2. G.S.S. Babu, D.B. Das, C. Patvardhan, Solution of Real-parameter Optimization problems using novel Quantum Evolutionary Algorithm with applications in Power Dispatch, in *Proceedings of the IEEE Congress on Evolutionary Computation, Trondheim, Norway, May 2009 (CEC 2009)*, 2009, pp. 1927–1920
3. S. Baluja, Population-based incremental learning: A method for integrating genetic search based function optimization and competitive learning, Technical report, Carnegie Mellon University, Pittsburgh, PA, 1994
4. S. Baluja, R. Caruana, Removing the genetics from the standard genetic algorithm, Technical report, Carnegie Mellon University, Pittsburgh, PA, 1995
5. N. Bansal, Constructive algorithms for discrepancy minimization, in *Proceedings of the 51st Annual IEEE Symposium on Foundations of Computer Science, Las Vegas, Nevada, USA, October 2010 (FOCS 2010)*, 2010, pp. 3–10
6. P. Erdős, J.H. Spencer, *Probabilistic Methods in Combinatorics* (Akadémia Kiadó, Budapest, 1974)
7. K.-H. Han, J.-H. Kim, Quantum-inspired evolutionary algorithm for a class of combinatorial optimization. IEEE Trans. Evol. Comput. **6**(6), 580–593 (2002). doi:10.1109/TEVC.2002. 804320
8. K.-H. Han, J.-H. Kim, On Setting the Parameters of Quantum-inspired Evolutionary Algorithm for Practical Applications, in *Proceedings of the IEEE Congress on Evolutionary Computation, Canberra, Australia, December 2003 (CEC 2003)*, 2003, pp. 178–184
9. G.R. Harik, F.G. Lobo, D.E. Goldberg, The compact genetic algorithm, Technical report, Urbana, IL: University of Illinois at Urbana-Champaign, Illinois Genetic Algorithms Laboratory, 1997
10. M. Hauschild, M. Pelikan, *An Introduction and Survey of Estimation of Distribution Algorithms*, 2011. http://medal-lab.org/files/2011004_rev1.pdf
11. L. Kliemann, O. Kliemann, C. Patvardhan, V. Sauerland, A. Srivastav, A New QEA Computing Near-Optimal Low-Discrepancy Colorings in the Hypergraph of Arithmetic Progressions, in *Proceedings of the 12th International Symposium on Experimental and Efficient Algorithms, Rome, Italy, June 2013 (SEA 2013)*, ed. by V. Bonifaci, C. Demetrescu, A. Marchetti-Spaccamela Lecture Notes in Computer Science (Springer, Heidelberg, 2013), pp. 67–78. doi:10.1007/978-3-642-38527-8_8
12. A. Mani, C. Patvardhan, An Adaptive Quantum inspired Evolutionary Algorithm with Two Populations for Engineering Optimization Problems, in *Proceedings of the International Conference on Applied Systems Research, Dayalbagh Educational Institute, Agra, India, 2009 (NSC 2009)*, 2009
13. A. Mani, C. Patvardhan, A hybrid quantum evolutionary algorithm for solving engineering optimization problems. Int. J. Hybrid Intell. Syst. **7**, 225–235 (2010)
14. J. Matoušek, J.H. Spencer, Discrepancy in arithmetic progressions. J. Am. Math. Soc. **9**, 195–204 (1996). http://www.jstor.org/stable/2152845
15. J. Matoušek, J. Vondrák, *The Probabilistic Method*, 2008. Lecture notes. Revision March 2008. http://kam.mff.cuni.cz/\char126\relaxmatousek/prob-ln.ps.gz
16. H. Mühlenbein, G. Paaß, From recombination of genes to the estimation of distributions I. Binary parameters, in *Proceedings of the 4th International Conference on Parallel Problem Solving from Nature, Berlin, Germany, September 1996 (PPSN 1996)*, 1996, pp. 178–187
17. C. Patvardhan, P. Prakash, A. Srivastav, A Novel Quantum-inspired Evolutionary Algorithm for the Quadratic Knapsack Problem, in *Proceedings of the International Conference on Operations Research Applications In Engineering And Management, Tiruchirappalli, India, May 2009 (ICOREM 2009)*, 2009, pp. 2061–2064

484 L. Kliemann

18. M.D. Platel, S. Schliebs, N. Kasabov, A Versatile Quantum-inspired Evolutionary Algorithm, in *Proceedings of the IEEE Congress on Evolutionary Computation, Singapore, September 2007 (CEC 2007)*, 2007, pp. 423–430. doi:10.1109/CEC.2007.4424502
19. M.D. Platel, S. Schliebs, N. Kasabov, Quantum-inspired evolutionary algorithm: A multimodel EDA. IEEE Trans. Evol. Comput. **13**(6), 1218–1232 (2009). doi:10.1109/TEVC.2008.2003010
20. K.F. Roth, Remark concerning integer sequences. Acta Arithmetica **9**, 257–260 (1964)
21. K. Sastry, D.E. Goldberg, X. Llorà, Towards billion bit optimization via parallel estimation of distribution algorithm, in *Proceedings of the Genetic and Evolutionary Computation Conference, London, England, UK, July 2007 (GECCO 2007)*, 2007, pp. 551–558
22. V. Sauerland, Algorithm Engineering for some Complex Practice Problems, PhD thesis, Christian-Albrechts-Universität Kiel, Technische Fakultät, 2012. http://eldiss.uni-kiel.de/macau/receive/dissertation_diss_00009452

Part III
Applications and Constructions

Chapter 8
On the Distribution of Solutions to Diophantine Equations

Ákos Magyar

Abstract Let P be a positive homogeneous polynomial of degree d, with integer coefficients, and for natural numbers λ consider the solution sets

$$Z_{P,\lambda} = \{m \in \mathbf{Z}^n : P(m) = \lambda\}.$$

We'll study the asymptotic distribution of the images of these sets when projected onto the unit level surface $\{P = 1\}$ via the dilations, and also when mapped to the flat torus $\mathbf{T}^n$. Assuming the number of variables n is large enough with respect to the degree d we will obtain quantitative estimates on the rate of equi-distribution in terms of upper bounds on the associated discrepancy. Our main tool will be the Hardy-Littlewood method of exponential sums, which will be utilized to obtain asymptotic expansions of the Fourier transform of the solution sets

$$\omega_{P,\lambda}(\xi) = \sum_{m \in \mathbf{Z}^n, \, P(m)=\lambda} e^{2\pi i m \cdot \xi},$$

relating these exponential sums to Fourier transforms of surface carried measures. This will allow us to compare the discrete and continuous case and will be crucial in our estimates on the discrepancy.

8.1 Introduction

A fundamental problem in number theory is to find integer solutions of diophantine equations, that is equations of the form

Á. Magyar (✉)
UBC, 1984 Mathematics Road, Vancouver, BC, Canada V6T 1Z2
e-mail: magyar@math.ubc.ca

W. Chen et al. (eds.), *A Panorama of Discrepancy Theory*, Lecture Notes
in Mathematics 2107, DOI 10.1007/978-3-319-04696-9_8,

$$P(m_1, \ldots, m_n) = \lambda$$

where P is a polynomial with integer coefficients. The approaches fall into two broad categories, algebraic and analytic, the latter being especially useful when the number of variables is large with respect to the degree of the polynomial.

If the polynomial P is also positive and homogeneous of (even) degree d, then for each natural number λ, there is a finite solution set

$$Z_{P,\lambda} = \{m \in \mathbf{Z}^n : P(m) = \lambda\}. \tag{8.1}$$

One may view these sets as the set of lattice points on the level surfaces $\{P = \lambda\}$ and by homogeneity they can be projected onto the unit level surface $S_P := \{P = 1\}$ via the dilations $m \to \lambda^{-1/d} m$. We will study the rate of equi-distribution of the images of the solution sets $Z'_{P,\lambda}$ on the unit level surface S_P as $\lambda \to \infty$. Of course one needs some more conditions on the polynomial P in order to have solutions at all of the diophantine equation $P(m) = \lambda$. For example if $P(m) = m_1^8 + (m_2^2 + \ldots + m_n^2)^4$ then even for large n there are only a sparse set of λ's (namely which can be written as a sum of an 8-th and a 4-th power) for which there are solutions, and even for those values of λ one cannot have equi-distribution as the first coordinate m_1 can take very few values. A natural condition on the polynomial P, introduced by Birch [4], is that of P being non-singular in the sense that

$$\nabla P(z) = (\partial_1 P(z), \ldots, \partial_n P(z)) \neq 0, \quad \text{for all} \quad z \in \mathbf{C}^n, \ z \neq 0.$$

Also, there are local or congruence obstructions. For example, the polynomial $P(m) = m_1^d + p(m_2^d + \ldots + m_n^d)$ is non-singular, but the equation $P(m) = \lambda$ can only have an integer solution if λ is congruent to a d-th power modulo p. Nevertheless, as it is implicit in the work of Birch [4], that if P is non-singular and if the number of variables n is large enough with respect to the degree d, then there is an infinite arithmetic progression Λ depending on P, which can be explicitly determined, such that for each $\lambda \in \Lambda$ the equation $P(m) = \lambda$ has the expected number of solutions $\approx \lambda^{n/d-1}$, in fact the number of solutions can be asymptotically determined. We will refer to such a set Λ as a set of *regular values* of the polynomial P.

As mentioned earlier, one of the problems we will be interested in is the asymptotic distribution of the images of the solution sets $Z'_{P,\lambda} = \{\lambda^{-1/d} m; \ P(m) = \lambda\}$ as $\lambda \to \infty$ ($\lambda \in \Lambda$), on the unit level surface S_P. First, one can show that there is a natural measure σ_P on the surface S_P, such that the sets $Z'_{P,\lambda}$ become weakly equi-distributed with respect to the measure $d\sigma_P$. That is for any smooth function ϕ one has that

$$\frac{1}{N_\lambda} \sum_{x \in Z'_{P,\lambda}} \phi(x) \to \int_{S_P} \phi(x) \, d\sigma_P(x), \quad \text{as} \quad \lambda \to \infty, \ \lambda \in \Lambda,$$

where N_λ is the number of solutions of the equation $P(m) = \lambda$. To get quantitative information on the rate of equi-distribution, we define below the discrepancy of a finite set $Z \subset S_P$ with respect to *caps*. For a unit vector $\xi \in \mathbf{R}^n$ and positive number a, define the cap

$$C_{a,\xi} := \{x \in S_P : x \cdot \xi \geq a\},$$

where $x \cdot \xi$ is the dot product of the vectors x and ξ. Note that $C_{a,\xi}$ is the intersection of the surface S_P with the half-space defined by $x \cdot \xi \geq a$, and we will refer to ξ as the *direction* of the cap. The associated discrepancy of a finite set $Z \subset S_P$, consisting of N points, with respect to caps of a given direction ξ is given by

$$D(Z,\xi) = \sup_{a>0} ||Z \cap C_{a,\xi}| - N\,\sigma_P(C_{a,\xi})|, \tag{8.2}$$

where $|A|$ denotes the size of a set A.

It turns out that for the solution sets $Z'_{P,\lambda}$ the discrepancy depends heavily on the direction of the cap. To see this consider the polynomial $P(m) = m_1^2 + \ldots + m_n^2$, so that one is interested in the distribution of lattice points on spheres, projected back to the unit sphere. It is well-known that for $n \geq 5$, the size of the solution sets are $N_\lambda \approx \lambda^{\frac{n}{2}-1}$. If $\xi = (0,\ldots,0,1)$ then for certain values of a, the boundary of the cap contain as many as $\approx \lambda^{\frac{n-3}{2}}$ points from the set $Z'_{P,\lambda}$. Indeed, after scaling back with a factor of $\lambda^{1/2}$, the boundary of the cap is given by the equation.

$$m_1^2 + \ldots + m_{n-1}^2 = \mu \text{ for some } \mu \text{ depending on } \lambda \text{ and } a. \text{ Thus the discrepancy}$$

cannot be smaller than $\lambda^{\frac{n-3}{2}} \approx N_\lambda^{1-\frac{1}{n-2}}$. In contrast, we will show that if the direction of the cap points away from rational points as much as possible, then one can obtain much better bounds on the discrepancy. To be more precise let us call a point $\alpha \in \mathbf{R}^{n-1}$ *diophantine*, if for every $\varepsilon > 0$ there exists a constant $C_\epsilon > 0$ such that for all $q \in \mathbf{N}$

$$\|q\alpha\| = \min_{m \in \mathbf{Z}^{n-1}} |q\alpha - m| \geq C_\varepsilon\, q^{-\frac{1}{n-1}-\varepsilon}. \tag{8.3}$$

Correspondingly a point $\xi \in S^{n-1}$ is called *diophantine* if for every $1 \leq i \leq n$ for which $\xi_i \neq 0$, the point $\alpha^i \in \mathbf{R}^{n-1}$ is diophantine, where the coordinates of α^i are obtained by dividing each coordinate of ξ by ξ_i and deleting the i-th coordinate. It is easy to see that the complement of diophantine points has measure 0 in $\mathbf{R}^{n-1}$ and hence in S^{n-1} as well, see [Lemma 3, Sec. 2.2]. We will show, see also [13], in dimensions $n \geq 4$, that the discrepancy is bounded by above by

$$D(Z'_{P,\lambda},\xi) \leq C_{\xi,\varepsilon}\, N_\lambda^{\frac{1}{2}+\frac{1}{2(n-2)}} \tag{8.4}$$

for all $\varepsilon > 0$, when ξ is diophantine. This is especially significant in large dimensions as it is known from the works of Beck and Schmidt [3, 15], see also [12], that for any set of N points on the unit sphere S^{n-1}, the L^2 average of the discrepancy with respect to spherical caps is at least $N^{\frac{1}{2}-\frac{1}{2(n-1)}}$. For general non-singular, positive and homogeneous polynomials P, the same observation shows that for rational directions (p.e. when $\xi = (0, \ldots, 0, 1)$), the discrepancy is at least $N_\lambda^{1-\frac{1}{n-d}}$, while we'll show that in diophantine directions it is bounded by $N_\lambda^{1-\gamma_d}$ with $\gamma_d = \frac{1}{(d-1)2^d+1}$, in large enough dimensions.

We will also study the equi-distribution of the solutions when mapped to the flat torus $\mathbf{T}^n = \mathbf{R}^n/\mathbf{Z}^n$. Let $\alpha = (\alpha_1, \ldots, \alpha_n) \in \mathbf{R}^n$ and consider the map $T_\alpha : \mathbf{Z}^n \to \mathbf{T}^n$, defined by $T_\alpha(m) = (m_1\alpha_1, \ldots, m_n\alpha_n)$ (mod 1). Then the images of the solution sets take the form

$$\Omega_{\lambda,\alpha} = \{(m_1\alpha_1, \ldots, m_n\alpha_n); \ P(m_1, \ldots, m_n) = \lambda\} \subseteq \mathbf{T}^n.$$

It is clear that if one of the coordinates of the point α is rational then the corresponding coordinate of the points in the image set can take only finitely many different values and the sets $\Omega_{\lambda,\alpha}$ cannot become equi-distributed as $\lambda \to \infty$. It turns out that this is the only obstruction for non-singular polynomials P in sufficiently many variables. Indeed we will see that if $\alpha \in (\mathbf{R}\backslash\mathbf{Q})^n$, then for any $\phi \in C^\infty(T^n)$ we have that

$$N_\lambda^{-1} \sum_{P(m)=\lambda} \phi(m_1\alpha_1, \ldots, m_n\alpha_n) \to \int_{\mathbf{T}^n} \phi(x)\, dx, \qquad (8.5)$$

as $\lambda \to \infty$ through regular values of the polynomial P. To obtain quantitative bounds on the rate of equi-distribution, we will assume that each coordinate of the point α is diophantine, that is $\|q\alpha_i\| \geq C_\varepsilon q^{-1-\varepsilon}$, for all $\varepsilon > 0$ and for all $q \in \mathbf{N}$ with an appropriate constant $C_\varepsilon > 0$. Identify the torus with the set $[-\frac{1}{2}, \frac{1}{2})^n$ and let $K \subseteq (-\frac{1}{2}, \frac{1}{2})^n$ be a compact, convex set with nonempty interior. The discrepancy of the image set $\Omega_{\lambda,\alpha}$ with respect to the convex body K is defined by

$$D(K, \alpha, \lambda) = \sum_{P(m)=\lambda} \chi_K(m_1\alpha_1, \ldots, m_n\alpha_n) - N_\lambda \, vol_n(K),$$

where χ_K is the indicator function of the set K. We will show that for diophantine points α one has the upper bound

$$|D(K, \alpha, \lambda)| \leq C_P \lambda^{\frac{n}{d}-1-\gamma_d}, \qquad (8.6)$$

for some constant $\gamma_d > 0$ depending only on the degree d.

Let us remark that the above is a special case of a more general phenomenon; namely if (X, μ) is a probability measure space, and if $T = (T_1, \ldots, T_n)$ is a commuting, fully ergodic family of measure-preserving transformations, then the images of the solution sets

$$\Omega_{\lambda,x} := \{T_1^{m_1} \ldots T_n^{m_n}(x); \ P(m_1, \ldots, m_n) = \lambda\} \subseteq X,$$

become equi-distributed as $\lambda \to \infty$, $(\lambda \in \Lambda)$, for almost every $x \in X$ [11]. To prove such results one needs estimate certain maximal operators associated to averages over the solution sets $P(m) = \lambda$, however, as in this generality one cannot hope for quantitative bounds on the rate of equi-distribution, we will not discuss such results below.

Crucial to all these results is the structure of the Fourier transform of the indicator function of the set of lattice points on the level surface $\{P = \lambda\}$. This is an exponential sum of the form

$$\hat{\omega}_\lambda(\xi) = \sum_{m \in \mathbf{Z}^n, \, P(m)=\lambda} e^{-2\pi i \, m \cdot \xi}. \tag{8.7}$$

Note that $\hat{\omega}_\lambda(0) = N_\lambda$, that is the number of solutions to the equation $P(m) = \lambda$, a quantity which has been extensively studied in analytic number theory. Indeed for the special case $P(m) = m_1^2 + \ldots + m_n^2$ asymptotic formulae for the number of solutions were obtained by Hardy and Littlewood, by developing the so-called "circle method" of exponential sums. Their methods were later further extended by Birch and Davenport [4, 5], to treat higher degree non-diagonal forms; in fact they have shown that

$$N_\lambda = \hat{\omega}_\lambda(0) = c_p \lambda^{\frac{n}{d}-1} \sum_{q=1}^{\infty} K(q, 0, \lambda) + O(\lambda^{\frac{n}{d}-1-\delta}), \tag{8.8}$$

for some $\delta > 0$. The expression $K(\lambda) = \sum_{q=1}^{\infty} K(q, 0, \lambda)$ is called the singular series, and it capturers all the local arithmetic information about the polynomial P. Without recalling the precise definition of the terms $K(q, 0, \lambda)$ here (see Sect. 8.3.1.4), it is enough to note here that for regular values $\lambda \in \Lambda$ the singular series $K(\lambda)$ bounded below by a fixed constant $A_P > 0$. It turns out that one can derive similar asymptotic formulas for the exponential sums $\hat{\omega}_\lambda(\xi)$, which are uniform in the phase variable ξ. Namely, we will show that

$$\hat{\omega}_\lambda(\xi) = c_P \, \lambda^{\frac{n}{d}-1} \sum_{q=1}^{\infty} m_{q,\lambda}(\xi) + \mathscr{E}_\lambda(\xi), \tag{8.9}$$

where

$$\sup_{\xi} |\mathscr{E}(\xi)| \leq C \, \lambda^{(\frac{n}{d}-1)(1-\gamma)}.$$

Moreover

$$m_{q,\lambda}(\xi) = \sum_{l \in \mathbf{Z}^n} K(q,l,\lambda) \, \psi(q\xi - l) \, \tilde{\sigma}_P(\lambda^{\frac{1}{d}}(\xi - l/q)),$$

where ψ is a smooth cut-off function supported near the origin, and $\tilde{\sigma}_P$ is the Euclidean Fourier transform of the surface measure σ_P

$$\tilde{\sigma}_P(\xi) = \int_{S_P} e^{-2\pi i \, x \cdot \xi} \, d\sigma_P(x).$$

To describe the meaning of this formula, note first that for ξ near the origin

$$m_{1,\lambda}(\xi) = \psi(\xi) \, \tilde{\sigma}_P(\lambda^{\frac{1}{d}}(\xi)),$$

since $K(1,0,\lambda) = 1$. The term $c_P \, \lambda^{\frac{n}{d}-1} \tilde{\sigma}_P(\lambda^{\frac{1}{d}}\xi)$ can be interpreted as the Fourier transform of a smooth density supported on the level surface $\{P = \lambda\}$. Thus the first term in the approximation formula may be viewed as an approximation near the origin via the Fourier transform of a surface carried measure. Notice also that all other terms are similar involving the Fourier transform $\tilde{\sigma}_P(\lambda^{\frac{1}{d}}(\xi - l/q))$, and may be viewed as higher order approximations near the rational points l/q. In fact if ξ is near a rational point l/q, then the sum expressing $m_{q,\lambda}(\xi)$ has only one nonzero term taken at $l = [q\xi]$, the nearest integer point to $q\xi$.

Let us sketch below how this formula will allow us to compare the discrete and the continuous case and to estimate the rate of equi-distribution of the solution sets in terms of the discrepancy.

Let χ_a be the indicator function of the interval $[a, b]$ (b being a fixed number depending on P), then by taking the inverse Fourier transform $\chi_a = \int \hat{\chi}_a(t) e^{2\pi i t \cdot} dt$, and by making a change of variables $t \to \lambda^{-1/d} t$, one may write

$$|Z'_{P,\lambda} \cap C_{a,\xi}| = \sum_{P(m)=\lambda} \chi_a(\lambda^{-\frac{1}{d}} m \cdot \xi) = \int_{\mathbf{R}} \lambda^{\frac{1}{d}} \hat{\chi}_a(t\lambda^{\frac{1}{d}}) \, \hat{\omega}_\lambda(t\xi) \, dt,$$

and also

$$\sigma_P(C_{a,\xi}) = \int_{S_P} \chi_a(x \cdot \xi) \, d\sigma_P(x) = \int_{\mathbf{R}} \hat{\chi}_a(t) \, \tilde{\sigma}_P(t\xi) \, dt.$$

Substituting the asymptotic formula (8.9) into this expression one may study the contribution of each term separately

$$I_{q,\lambda}(\xi) := \int_{\mathbf{R}} \lambda^{\frac{1}{d}} \hat{\chi}_a(t\lambda^{\frac{1}{d}}) \, m_{q,\lambda}(t\xi) \, dt.$$

A crucial point is that if $|t| \ll q^{-1}$ then $\psi(qt\xi - l) = 0$ unless $l = 0$, moreover $\psi(qt\xi) = 1$, hence

$$m_{q,\lambda}(t\xi) = K(q,0,\lambda)\, \tilde{\sigma}_P(\lambda^{\frac{1}{d}} t\xi).$$

Writing

$$I_{q,\lambda}(\xi) = \int_{|t| \ll q^{-1}} + \int_{|t| \gg q^{-1}} = I_{q,\lambda}^1(\xi) + I_{q,\lambda}^2(\xi),$$

one has, after a change of variables $t := \lambda^{1/d}\, t$, that

$$I_{q,\lambda}^1(\xi) = K(q,0,\lambda) \int_{|t| \ll \lambda^{\frac{1}{d}} q^{-1}} \hat{\chi}_a(t)\tilde{\sigma}_P(t\xi)\, dt.$$

At this point one exploits the cancelation in the normalized exponential sums $K(q,l,\lambda)$ and oscillatory integrals $\tilde{\sigma}_P(\xi)$, expressed in estimates roughly of the form

$$|K(q,l,\lambda)| \ll q^{-cn}$$

$$|\tilde{\sigma}_P(\xi)| \ll (1 + |\xi|)^{-c'n}.$$

Then, for $|\xi| = 1$, one can extend the integral to the whole real line by making a small error. This gives

$$I_{q,\lambda}^1(\xi) \approx K(q,0,\lambda) \int_{\mathbf{R}} \hat{\chi}_a(t)\tilde{\sigma}_P(t\xi)\, dt = K(q,0,\lambda)\, \sigma_P(C_{a,\xi}).$$

Thus by formula (8.8), we have that

$$c_P \lambda^{\frac{n}{d}-1} \sum_{q=1}^{\infty} I_{q,\lambda}^1(\xi) \approx N_\lambda\, \sigma_P(C_{a,\xi}). \tag{8.10}$$

To get upper bounds for the discrepancy one needs to estimate the total contribution of the rest of terms $I_{q,\lambda}^2(\xi)$, exploiting the diophantine properties of the point ξ. In fact by making a change of variables $t := qt$, and noting that the only nonzero term of the sum expressing $m_{q,\lambda}(\frac{l}{q}\xi)$ is taken at $l = [t\xi]$, one may write

$$I_{q,\lambda}^2(\xi) = \frac{\lambda^{\frac{1}{d}}}{q} \int_{|t| \gg 1} \hat{\chi}_a\left(t\frac{\lambda^{\frac{1}{d}}}{q}\right) K(q,[t\xi],\lambda)\, \psi(\{t\xi\})\, \tilde{\sigma}_P\left(\frac{\lambda^{\frac{1}{d}}}{q}\{t\xi\}\right) dt,$$

where $\{t\xi\} = t\xi - [t\xi]$ denotes the fractional part of the point $t\xi$. At diophantine points it is not hard to show that on average $|\{t\xi\}| = \|t\xi\| \geq c_\varepsilon t^{-\varepsilon}$ (see Lemma 6 in Sect. 8.2.2), thus using the cancelation estimates for $K(q, l, \lambda)$ and $\tilde{\sigma}(\xi)$ again, the terms $I_{q,\lambda}^2(\xi)$ add up only to a small error.

The organization of the rest of this chapter is as follows. In the next section we will derive the asymptotic expansion (8.9) for the polynomial $P(m) = m_1^2 + \ldots + m_d^2$, and prove upper bounds on the discrepancy of lattice points on spheres. Next, we will extend our approach to general non-singular forms, using the Birch-Davenport method of exponential sums. Finally, in the last section we will study the equi-distribution of the images of the solution sets $\{P(m) = \lambda\}$ modulo 1, when mapped to the flat torus $\mathbf{T}^n$ via the map T_α.

As for our notations, we will think of the polynomial P hence the parameters n, d being fixed, and write $f = O(g)$ or alternatively $f \ll g$ if $|f(m)| \le C g(m)$ for all $m \in \mathbf{N}$ with a constant $C > 0$ depending only on the polynomial P or the parameters n, d. We will also write, $f \gg g$ if $g \ll f$ and $f \approx g$ if both $f \ll g$ and $f \gg g$. If the implicit constant in our estimates depend on additional parameters $\varepsilon, \delta, \ldots$ then we may write $f = O_{\varepsilon,\delta\ldots}(g)$ or $f \ll_{\varepsilon,\delta,\ldots} g$. The Fourier transform of a function f defined on $\mathbf{Z}^n$ will be denoted by $\hat{f}$, as opposed, somewhat unconventionally, we will denote the Euclidean Fourier transform of a function ϕ defined on $\mathbf{R}^n$ by $\tilde{\phi}$. This is to avoid confusion as we will often move between the discrete and continuous settings.

8.2 The Discrepancy of Lattice Points on Spheres

The uniformity of the distribution of lattice points on spheres has been extensively studied and proved in dimension at least 4, see [7], and later in dimension 3 [6] using difficult estimates for the Fourier coefficients of modular forms. These methods, however, do not take into consideration the direction of the caps, and hence the bounds obtained are subject to the limitations described in the introduction, arising from caps whose direction has rational coordinates.

We will assume that the direction ξ of the caps is diophantine in the sense that $\xi^i = \xi/\xi_i$ satisfies condition (8.3) for each $1 \le i \le n$ such that $\xi_i \ne 0$. In this case, when $Z = \{\lambda^{-1/2}m; |m|^2 = \lambda\}$, we will obtain the following upper bound on the discrepancy, defined in (8.2), see also [13].

Theorem 1. Let $n \ge 4$ and let $\xi \in S^{n-1}$ be a diophantine point. Then for every $\varepsilon > 0$, one has

$$|D_n(\xi, \lambda)| \le C_{\xi,\varepsilon} \lambda^{\frac{n-1}{4}+\varepsilon} \tag{8.11}$$

We note that for $n \ge 4$, and if $n = 4$ assuming that 4 does not divide λ, one has that $N_\lambda \gg \lambda^{\frac{n}{2}-1}$, thus (8.11) implies that

$$|D_n(\xi, \lambda)| \le C_{\xi,\varepsilon} N_\lambda^{\frac{1}{2}+\frac{1}{2(n-2)}+\varepsilon}.$$

In dimension $n = 4$, the best previous estimate for the normalized discrepancy $D(\xi, \lambda)/N_\lambda$ was given in [7] of the order of $\lambda^{-1/5+\varepsilon}$ while we get the improvement $\lambda^{-1/4+\varepsilon}$. In case $n = 4$ and $\lambda = 4^k$ there are only 24 lattice points of length $\lambda^{1/2}$, estimates for the discrepancy become trivial in such degenerate cases.

8.2.1 The Fourier Transform of Lattice Points on Spheres

Our first task will be to derive the asymptotic formula (8.9) for the special case when $P(m) = |m|^2 = m_1^2 + \ldots m_n^2$. As we have mentioned this can be viewed as an extension of the asymptotic formula for the number of representations of a positive integer λ as sum of n squares, and as such our main tool will be the Hardy-Littlewood method of exponential sums. Because of the quadratic nature of the problem, there are special tools available this case, most notably the transformation properties of certain theta functions. Also, we will use the so-called Kloosterman refinement, mainly to include the case $n = 4$. For a fixed $\lambda \in \mathbf{N}$ and $\xi \in \mathbf{T}^n$, set $\delta = \lambda^{-1}$ and write

$$e^{-2\pi} \hat{\omega}_\lambda(\xi) = \sum_{|m|^2=\lambda} e^{-2\pi\delta|m|^2} e^{2\pi i m \cdot \xi} = \sum_{|m|^2=\lambda} w(m), \qquad (8.12)$$

where the weight function $w(x) = e^{-2\pi\delta|m|^2} e^{2\pi i m \cdot \xi}$ is bounded and absolute summable. Using the fact that $\int_0^1 e^{2\pi i(|m|^2-\lambda)\alpha}\, d\alpha = 1$ if $|m|^2 = \lambda$ and is equal to 0 otherwise, one may write

$$\hat{\omega}_\lambda(\xi) = e^{2\pi} \int_0^1 S(\alpha, \xi)\, e^{-2\pi i \alpha\lambda}\, d\alpha,$$

where

$$S(\alpha, \xi) = \sum_{m \in \mathbf{Z}^n} e^{2\pi i |m|^2 \alpha}\, w(m) = \sum_{m \in \mathbf{Z}^n} e^{2\pi i\,((\alpha+i\delta)|m|^2+m\cdot\xi)} \qquad (8.13)$$

is a theta function. It is well-known, at least when $\xi = 0$, that it is concentrated near rational points a/q with small denominator. To exploit this, one dissects the interval $[0, 1]$ into small neighborhoods of the set of rational points $\mathscr{R}_N = \{a/q;\ (a,q) = 1,\ q \leq N\}$ for some specific choice of the parameter N. It is easy to see, using Dirichlet's principle, that one can choose intervals around the rational points a/q of length $|I_{a/q}| \approx 1/Nq$. This suggests that

$$\hat{\omega}_\lambda(\xi) \approx c \sum_{q \leq N} \sum_{(a,q)=1} e^{-\pi i \lambda \frac{a}{q}} \int_{-\frac{1}{Nq}}^{\frac{1}{Nq}} S\left(\frac{a}{q} + \tau, \xi\right) e^{-2\pi i \lambda \tau}\, d\tau.$$

The idea behind the Kloosterman refinement is to make a specific choice of this partition (the so-called Farey dissection) and to estimate carefully the errors arising from the fact that the length of the intervals corresponding to a fixed denominator are not quite the same. We will use the following general result

Theorem A (Heath-Brown [9]). *Let $P : \mathbf{Z}^n \to \mathbf{Z}$ be a polynomial with integral coefficients, let λ, N be natural numbers and let $w \in L^1(\mathbf{Z}^n)$. Then one has*

$$\sum_{P(m)=\lambda} w(m) = \sum_{q \leq N} \int_{-\frac{1}{qN}}^{\frac{1}{qN}} e^{-2\pi i \lambda \tau} S_0(q, \tau) d\tau + E_1(\lambda) \tag{8.14}$$

where

$$|E_1(\lambda)| \leq C \, N^{-2} \sum_{q \leq N} \sum_{|u| \leq q/2} (1 + |u|)^{-1} \max_{\tau \approx \frac{1}{qN}} |S_u(q, \tau)| \tag{8.15}$$

Here $C > 0$ is an absolute constant and

$$S_u(q, \tau) = \sum_{(a,q)=1} e^{2\pi i \frac{\bar{a}u - a\lambda}{q}} S(a/q + \tau), \quad S(\alpha) = \sum_{m \in \mathbf{Z}^n} e^{2\pi i \alpha P(m)} w(m), \tag{8.16}$$

where $a\bar{a} \equiv 1 \pmod{q}$.

This is proved in [9] for the case $\lambda = 0$ and for a non-negative weight function w, however the proof extends without any changes to all $\lambda \in \mathbf{N}$ and $w \in L^1(\mathbf{Z}^n)$. Let us postpone the proof of the above result to the end of this section and see how it translates to our situation.

By (8.13) we have that

$$S(a/q + \tau) = \sum_{m \in \mathbf{Z}^n} e^{2\pi i \frac{a}{q} |m|^2} e^{2\pi i m \cdot \xi} h_{\tau, \delta}(m),$$

with $h_{\tau, \delta}(x) = e^{2\pi i (\tau + i\delta)|x|^2}$. Writing $m := qm_1 + s$, where $m_1 \in \mathbf{Z}^n$, $s \in (\mathbf{Z}/q\mathbf{Z})^n$, and applying Poisson summation, we have

$$S(a/q + \tau) = \sum_{s \in (\mathbf{Z}/q\mathbf{Z})^n} e^{2\pi i \frac{a}{q} |s|^2} \sum_{m_1 \in \mathbf{Z}^n} e^{2\pi i (qm_1 + s) \cdot \xi} h_{\tau, \delta}(qm_1 + s)$$

$$= \sum_{s \in (\mathbf{Z}/q\mathbf{Z})^n} e^{2\pi i \frac{a}{q} |s|^2} \sum_{l \in \mathbf{Z}^n} \int_{\mathbf{R}^n} e^{2\pi i (qx + s) \cdot \xi} h_{\tau, \delta}(qx + s) e^{-2\pi i x \cdot l} dx$$

$$= \sum_{l \in \mathbf{Z}^n} q^{-n} \sum_{s \in (\mathbf{Z}/q\mathbf{Z})^n} e^{2\pi i \frac{a|s|^2 + l \cdot s}{q}} \int_{\mathbf{R}^n} h_{\tau, \delta}(y) e^{2\pi i y \cdot (\xi - \frac{l}{q})} dy$$

$$= \sum_{l \in \mathbf{Z}^n} G(a, q, l) \, \tilde{h}_{\tau, \delta}(l/q - \xi). \tag{8.17}$$

Here $G(a, q, l)$ is a normalized Gaussian sum:

$$G(a, q, l) = q^{-n} \sum_{s \in (\mathbf{Z}/q\mathbf{Z})^n} e^{2\pi i \frac{a|s|^2 - s \cdot l}{q}}. \tag{8.18}$$

The function $h_{\tau, \delta}(x)$ is of the form $e^{-\pi z |x|^2}$ with $z = 2(\delta - i\tau)$, hence, after a change of variables $x := z^{1/2} x$, its Fourier transform can be evaluated explicitly,

$$\tilde{h}_{\tau, \delta}(l/q - \xi) = (2(\delta - i\tau))^{-\frac{n}{2}} e^{-\frac{\pi |q\xi - l|^2}{2q^2(\delta - i\tau)}}. \tag{8.19}$$

Let us first estimate the error terms $S_u(q, \tau)$ in formula (8.15). Note that on the range when $|\tau| \approx 1/qN \approx 1/q\lambda^{1/2}$, one has $Re\left(\frac{1}{q^2(\delta - i\tau)}\right) = \frac{\delta}{q^2(\delta^2 + \tau^2)} \geq c$, for some absolute constant $c > 0$. Thus

$$|\tilde{h}_{\tau, \delta}(\xi - l/q)| \leq C q^{\frac{n}{2}} \lambda^{\frac{n}{4}} e^{-c|q\xi - l|^2}. \tag{8.20}$$

Also, by (8.17)

$$S_u(q, \tau) = \sum_{l \in \mathbf{Z}^n} K(q, l, \lambda; u) \, \tilde{h}_{\tau, \delta}(\xi - l/q),$$

where

$$K(q, l, \lambda; u) = \sum_{(a, q) = 1} e^{2\pi i \frac{\bar{a}u - a\lambda}{q}} G(a, l, q), \tag{8.21}$$

These exponential sums have been extensively studied in number theory, various estimates are known in the literature, going back to the original work of Kloosterman. We will use the following estimate, which we will take for granted for now, however for the sake of completeness will include a proof later.

Theorem B. *Let $K(q, l, \lambda; u)$ be the exponential sum defined in (8.21). Then one has for every $\varepsilon > 0$,*

$$|K(q, l, \lambda; u)| \leq C_{n, \varepsilon} q^{\frac{n-1}{2} + \varepsilon} (\lambda, q_1)^{\frac{1}{2}} 2^{\frac{r}{2}}, \tag{8.22}$$

where $q = q_1 2^r$ with q_1 odd, and (λ, q_1) denotes the greatest common divisor of λ and q_1.

We remark that using only standard estimates for Gaussian sums would yield to a weaker bound of $O(q^{-n/2+1})$, thus the extra cancelation in the sum over $(a, q) = 1$ is crucial. By this and estimate (8.20) we have

$$\max_{|\tau| \approx \frac{1}{qN}} S_u(q, \tau) \leq C_{\varepsilon} q^{\frac{1}{2} + \varepsilon} (\lambda, q_1)^{\frac{1}{2}} 2^{\frac{r}{2}}. \tag{8.23}$$

The factors $(\lambda, q_1)^{\frac{1}{2}} 2^{\frac{r}{2}}$ are at most λ^ε on average for $q \leq \lambda^{\frac{1}{2}}$, hence they do not play any role in our estimates. Indeed, it is easy to see that

Lemma 2. *Let $\beta \in \mathbf{R}$. Then for every $\varepsilon > 0$, one has*

$$\sum_{q \leq \lambda^{\frac{1}{2}}} q^\beta (\lambda, q_1)^{\frac{1}{2}} 2^{\frac{r}{2}} \leq C_{\beta,\varepsilon} \lambda^{\frac{\beta+1}{2}+\varepsilon}$$

Proof. Let $1 \leq \mu \leq \lambda^{1/2}$. First, we show that

$$\sum_{q \leq \mu} (\lambda, q_1)^{\frac{1}{2}} 2^{\frac{r}{2}} \leq C_\varepsilon \lambda^\varepsilon \mu$$

To see this, write $d = (\lambda, q_1)$ and $q_1 = dt$. Then d divides λ and $d 2^r t \leq \mu$, hence the left side is majorized by

$$\sum_{d | \lambda} \sum_{r \in \mathbf{N}} d^{\frac{1}{2}} 2^{\frac{r}{2}} \frac{\mu}{d 2^r} \leq C_\varepsilon \lambda^\varepsilon \mu$$

By partial summation, we have

$$C_\varepsilon \lambda^\varepsilon (\lambda^{\frac{\beta}{2}} + \sum_{\mu \leq \lambda^{\frac{1}{2}}} \mu \mu^{\beta-1}) \leq C_\varepsilon \lambda^{\frac{\beta+1}{2}+\varepsilon}.$$

$\square$

Going back to the error term $E_1(\lambda)$ defined in (8.15), we have by estimate (8.23) and Lemma 2

$$|E_1(\lambda)| \leq C_{n,\varepsilon} \lambda^{\frac{n-1}{4}+\varepsilon}, \tag{8.24}$$

for all $\varepsilon > 0$.

The main term in (8.14) takes the form

$$M(\lambda) := \sum_{q \leq N} \int_{-\frac{1}{qN}}^{\frac{1}{qN}} e^{-2\pi i \lambda \tau} S_0(q, \tau) d\tau$$

$$= \sum_{q \leq N} \sum_{l \in \mathbf{Z}^n} K(q, l, \lambda; 0) \int_{-\frac{1}{qN}}^{\frac{1}{qN}} e^{-2\pi i \lambda \tau} \tilde{h}_{\tau,\delta}(\xi - l/q) \tag{8.25}$$

We will do now a number of transformations, to obtain the asymptotic formula (8.9), described in the introduction. First we insert the functions $\psi(q\xi - l)$, the restrict the summation in l to at most one non-zero term. Then we extend the integral to the whole real line and identify it with the Fourier transform of the normalized measure on the unit sphere.

First, let $\psi(\xi)$ be a smooth cut-off function which is constant 1 on $[-\frac{1}{8}, \frac{1}{8}]^n$ and is equal to 0 for $\xi \notin [-\frac{1}{4}, \frac{1}{4}]^n$. Then by (8.19), one estimates

$$\sum_{l \in \mathbf{Z}^n} (1 - \psi(q\xi - l)) |\tilde{h}_{\tau,\delta}(\xi - l/q)| \leq C_n \, (\tau^2 + \delta^2)^{-\frac{n}{4}} \, e^{\frac{c\delta}{q^2(\tau^2 + \delta^2)}} \ll \lambda^{\frac{n}{4}} q^{\frac{n}{2}},$$

where the last inequality follows from the fact that $e^{-u} \ll u^{-\frac{n}{4}}$ taking the special value $u = \frac{\delta}{q^2(\tau^2 + \delta^2)}$. Thus, by (8.15), the total error accumulated by inserting the cut-off functions in (8.25) is bounded by

$$|E_2(\lambda)| \leq C_\varepsilon \lambda^{\frac{n}{4} - \frac{1}{2}} \sum_{q \leq N} q^{-\frac{1}{2} + \epsilon} (\lambda, q)^{\frac{1}{2}} \leq C_\varepsilon \lambda^{\frac{n-1}{4} + \epsilon}, \tag{8.26}$$

and the main term takes the form

$$M_2(\lambda) := \sum_{q \leq N} \sum_{l \in \mathbf{Z}^n} K(q, l, \lambda; 0) \, \psi(q\xi - l) \int_{-\frac{1}{qN}}^{\frac{1}{qN}} e^{-2\pi i \lambda \tau} \, \tilde{h}_{\tau,\delta}(\xi - l/q). \tag{8.27}$$

At this point, the integration can be extended to the whole real line, exploiting the fact that now there is at most one nonzero term in the l-sum. For $|\tau| \geq \frac{1}{qN} \geq \delta$ one has $|\tilde{s}_{\tau,\delta}(\xi - l/q)| \ll \tau^{-\frac{n}{2}}$, thus the total error obtained in (8.27) by extending the integration is

$$|E_3(\lambda)| \leq C_\varepsilon \sum_{q \leq N} q^{-\frac{1}{2} + \epsilon} (\lambda, q)^{\frac{1}{2}} \int_{|\tau| \geq \frac{1}{qN}} \tau^{-\frac{n}{2}} \, d\tau \leq C_\varepsilon \lambda^{\frac{n-1}{4} + \epsilon}. \tag{8.28}$$

Finally, we identify the integrals, and show that

Lemma 3.

$$I_\lambda(\xi) := e^{2\pi} \int_{\mathbf{R}} e^{-2\pi i \lambda \tau} \, \tilde{h}_{\tau,\delta}(\xi) \, d\tau = \lambda^{\frac{n}{2} - 1} \, \tilde{\sigma}(\lambda^{\frac{1}{2}} \xi), \tag{8.29}$$

where σ is one-half of the surface area measure on the unit sphere in $\mathbf{R}^n$.

Proof. By using (8.19) and making a change of variables: $t = \lambda\tau$ to take out the dependence on λ, one has that

$$I_\lambda(\xi) = e^{2\pi} \lambda^{\frac{n}{2} - 1} \int_{\mathbf{R}} e^{-2\pi i t} (2(1 - it))^{-\frac{n}{2}} e^{-\frac{\pi \lambda |\xi|^2}{2(1 - it)}} \, dt.$$

Let $\eta := \lambda^{1/2}\xi$, then out task is to show that

$$J(\eta) := e^{2\pi} \int_{\mathbf{R}} e^{-2\pi i t} (2(1-it))^{-\frac{n}{2}} e^{-\frac{\pi|\eta|^2}{2(1-it)}} \, dt = \tilde{\sigma}(\eta).$$

We now insert an extra convergence factor $e^{-\pi\gamma t^2}$ into the integral defining $J(\eta)$. Denoting the resulting integral by J^γ we have $J^\gamma \to J$ as $\eta \to 0$; moreover for any test function ϕ in the Schwartz space

$$\int_{\mathbf{R}^n} \hat{\phi}(\eta) J(\eta) \, d\eta = \lim_{\gamma \to 0} \int_{\mathbf{R}^d} \hat{\phi}(\eta) J^\gamma(\eta) \, d\eta.$$

Also,

$$\int_{\mathbf{R}^d} \hat{\phi}(\eta) J^\gamma(\eta) \, d\eta = \int_{\mathbf{R}^d} \phi(x) J^\gamma(x) \, dx. \tag{8.30}$$

Note, that by (8.19) we have that $\tilde{h}_{t,1}(\eta) = (2(1-it))^{-n/2} e^{-\frac{\pi|\eta|^2}{2(1-it)}}$, thus

$$J^\gamma(x) = e^{2\pi} \int_{\mathbf{R}} e^{-2\pi i t} e^{-2\pi|x|^2(1-it)} e^{-\pi\gamma t^2} \, dt = \gamma^{-\frac{1}{2}} e^{-\pi(1-|x|^2)/\gamma} e^{-\pi|x|^2}.$$

Inserting this into (8.30), and letting $\gamma \to 0$, we obtain

$$\int_{\mathbf{R}^n} \hat{\phi}(\eta) J(\eta) \, d\eta = \int_{\mathbf{R}^n} \phi(x) \, d\sigma(x),$$

and thus $J(\eta) = \tilde{\sigma}(\eta)$, as we wanted to prove. Note that

$$\tilde{\sigma}(0) = J(0) = \int_{\mathbf{R}} e^{-2\pi i t} \frac{dt}{(2(1-it))^{n/2}} = \frac{\pi^{n/2}}{\Gamma(n/2)}.$$

This identifies σ as one-half of the surface area measure of the unit sphere. □

Substituting the above formula (8.29) into the expression (8.27), the main term finally takes the form

$$M_3(\lambda) := \lambda^{n/2-1} \sum_{q \leq N} \sum_{l \in \mathbf{Z}^n} K(q,l,\lambda;0) \, \psi(q\xi - l) \tilde{\sigma}(\lambda^{1/2}(\xi - l/q)). \tag{8.31}$$

Note, that all error terms (8.15), (8.24), (8.26), and (8.28), we obtained in the process of transforming the main term into the above expression is of magnitude $O_\varepsilon(\lambda^{\frac{n-1}{4}+\varepsilon})$. Summarizing we have proved

Theorem 4. *Let $n \geq 4$. Then one has*

$$\hat{\omega}_\lambda(\xi) = \lambda^{\frac{n}{2}-1} \sum_{q \leq \lambda^{\frac{1}{2}}} m_{q,\lambda}(\xi) + \mathcal{E}_\lambda(\xi),$$

where

$$|\mathcal{E}_\lambda(\xi)| \leq C_\varepsilon \lambda^{\frac{n-1}{4}+\varepsilon} \tag{8.32}$$

holds uniformly in ξ for every $\varepsilon > 0$. Moreover

$$m_{q,\lambda}(\xi) = \sum_{l \in \mathbf{Z}^n} K(q, l, \lambda)\, \psi(q\xi - l)\, \tilde{\sigma}(\lambda^{\frac{1}{2}}(\xi - l/q)) \tag{8.33}$$

where

$$K(q, l, \lambda) = q^{-n} \sum_{(a,q)=1} \sum_{s \in (\mathbf{Z}/q\mathbf{Z})^n} e^{2\pi i \frac{a(|s|^2-\lambda)+s \cdot l}{q}}.$$

Here $\tilde{\sigma}$ denotes the Fourier transform of the surface-area measure σ on S^{n-1}, and ψ is a smooth cut-off function supported on $[-\frac{1}{4}, \frac{1}{4}]^n$ which is constant 1 on $[-\frac{1}{8}, \frac{1}{8}]^n$.

8.2.2 Some Properties of Diophantine Points

We will derive here a few elementary properties of diophantine points, needed later in our estimates on the discrepancy. Crucial among them is the fact if $\xi \in S^{n-1}$ is a diophantine point, then $\|t\xi\| \gg T^{-\varepsilon}$ on average for $1 \leq t \leq T$, where $\|\xi\|$ denotes the distance of a point $\xi \in \mathbf{R}^n$ to the nearest lattice point. To start, let us call a point $\alpha \in \mathbf{R}^n$ of *type ε* if it satisfies condition (8.3) with a given $\varepsilon > 0$.

Lemma 5. *For every $\epsilon > 0$ the set of points $\alpha \in [0, 1]^{n-1}$ of type ε has measure 1.*

Proof. If a point $\alpha \in \mathbf{R}^{n-1}$ is not of *type ϵ* then there are infinitely many positive integers q such that: $\|q\xi\| \leq q^{-\frac{1}{n-1}-\epsilon}$. This means that there exists an $m \in \mathbf{Z}^n$ such that: $|\xi - m/q| \leq q^{-\frac{n}{n-1}-\epsilon}$. However the sum of the volumes of all such neighborhoods around the points $m/q \in [0, 1]^{n-1}$ is bounded by

$$\sum_{n=1}^{\infty} q^{n-1} q^{-n-\epsilon} \leq C_\varepsilon,$$

thus the set of points which belong to infinitely many of such neighborhoods has measure 0. $\square$

This shows that the set of points $\alpha \in \mathbf{R}^{n-1}$ which are not diophantine has measure 0. Indeed α is diophantine if it is of *type* $\varepsilon_k = (1/2)^k$ for all $k \in \mathbf{N}$. Next we show that $\|q\alpha\| \approx 1$ on average.

Lemma 6. *Let* $\alpha \in [0, 1]^{n-1}$ *be diophantine,* $Q > 1$ *and* $1 \le k < n - 1$. *Then for every* $\varepsilon > 0$, *we have*

$$\sum_{q \le Q} \|q\alpha\|^{-k} \le C_\varepsilon \, Q^{1+\varepsilon} \tag{8.34}$$

Proof. Let $\varepsilon > 0$. Consider the set of points $\{q\alpha\} \in [-1/2, 1/2]^{n-1}$, for $1 \le q \le Q$. If $q_1 \ne q_2$ then

$$|\{q_1\alpha\} - \{q_2\alpha\}| \ge \|(q_1 - q_2)\alpha\| \ge C_\varepsilon \, Q^{-\frac{1}{n-1} - \frac{\varepsilon}{n}},$$

thus the number of points in a dyadic annulus $2^{-j} \le \|q\alpha\| < 2^{-j+1}$ is bounded by $2^{-(n-1)j} \, Q^{1+\varepsilon}$ and the sum in (8.34) is convergent for $1 \le k < n - 1$. $\qquad\square$

Lemma 7. *Let* $\xi \in S^{n-1}$ *be diophantine, and assume that* $max_j \, |\xi_j| = |\xi_n|$. *Let* $t \ge 1$, $\alpha = (\alpha_1, \ldots \alpha_{n-1})$, $\alpha_j = \xi_j / \xi_n$ *and* $q = [t\xi_n]$. *Then one has*

$$\|t\xi\| \ge \frac{1}{n} \|q\alpha\|$$

Proof. Note that

$$t\xi_j = t\xi_n\alpha_j = [t\xi_n]\alpha_j \pm \|t\xi_n\|\alpha_j$$

hence

$$|q\alpha_j - m_j| \le |t\xi_j - m_j| + \|t\xi_n\|.$$

Thus taking $m_j = [t\xi_j]$, we have

$$\|q\alpha_j\| \le \|t\xi_j\| + \|t\xi_n\|.$$

Summing for $1 \le j \le n - 1$ proves the lemma. $\qquad\square$

Lemma 8. *Suppose* $\xi \in S^{n-1}$ *is diophantine, and let* $t \ge 1$ *and* $T \ge 1$. *Then for every* $\varepsilon > 0$, *one has*

$$\|t\xi\| \ge C_\varepsilon \, t^{-\frac{1}{n-1} - \varepsilon} \tag{8.35}$$

Moreover, for $1 \le k < n - 1$

$$\int_1^T \|t\xi\|^{-k} \le C_\varepsilon \, T^{1+\varepsilon} \tag{8.36}$$

Proof. By permuting the coordinates of ξ, one can assume that $max_j |\xi_j| = |\xi_n|$. Inequality (8.35) follows immediately from Lemma 7 and the definition of a diophantine point. Similarly (8.35) is reduced to (8.34) by observing that for a fixed q, the set of t's for which $q = [t\xi_n]$ is an interval of length at most $1/\xi_n \leq \sqrt{n}$. $\square$

8.2.3 Upper Bounds on Discrepancy

We have developed all the necessary tools to prove Theorem 8.11, our main result in this section. The argument will follow the broad outline given at the end of the introduction, in addition we will use the standard stationary phase estimate on the Fourier transform of the surface area measure on the unit sphere S^{n-1}, see for example [17]

$$|\hat{\sigma}(\xi)| \ll (1 + |\xi|)^{-\frac{n-1}{2}} \tag{8.37}$$

Now, for given $a > 0$ let χ_a denote the indicator function of the interval $[a, 1 + a]$. The discrepancy may be written as

$$D_n(\xi, \lambda) = \sum_{|m|^2 = \lambda} \chi_a(\lambda^{-\frac{1}{2}} m \cdot \xi) - N_\lambda \int_{S^{n-1}} \chi_a(x \cdot \xi) \, d\sigma(x). \tag{8.38}$$

The function χ_a can be replaced with a smooth function $\phi_{a,\delta}$ by making a small error in the discrepancy. Indeed, let $0 \leq \phi(t) \leq 1$ be smooth function supported in $[-1, 1]^n$, such that $\int \phi = 1$. For a given $\delta > 0$ let $\phi_{a,\delta}^\pm = \chi_{a\pm\delta} * \phi_\delta$, where $\phi_\delta(t) = \delta^{-1}\phi(t\,\delta^{-1})$ and define the *smoothed* discrepancy as

$$D_n(\phi_{a,\delta}^\pm, \xi, \lambda) = \sum_{|m|^2 = \lambda} \phi_{a,\delta}^\pm(\lambda^{-\frac{1}{2}} m \cdot \xi) - N_\lambda \int_{S^{n-1}} \phi_{a,\delta}^\pm(x \cdot \xi) \, d\sigma(x). \tag{8.39}$$

Lemma 9. *One has*

$$|D_n(\xi, \lambda)| \leq \max\left(|D_n(\phi_{a,\delta}^+, \xi, \lambda)|, |D_n(\phi_{a,\delta}^-, \xi, \lambda)|\right) + O(\delta N_\lambda). \tag{8.40}$$

Proof. Note that $\phi_{a,\delta}^-(t) \leq \chi_a(t) \leq \phi_{a,\delta}^+(t)$ thus

$$\sum_{|m|^2 = \lambda} \phi_{a,\delta}^-(\lambda^{-\frac{1}{2}} m \cdot \xi) \leq \sum_{|m|^2 = \lambda} \chi_a(\lambda^{-\frac{1}{2}} m \cdot \xi) \leq \sum_{|m|^2 = \lambda} \phi_{a,\delta}^+(\lambda^{-\frac{1}{2}} m \cdot \xi)$$

and

$$N_\lambda \int_{S^{n-1}} \phi_{a,\delta}^+(x \cdot \xi) \, d\sigma(x) \geq N_\lambda \int_{S^{n-1}} \chi_a(x \cdot \xi) \, d\sigma(x) \geq N_\lambda \int_{S^{n-1}} \phi_{a,\delta}^-(x \cdot \xi) \, d\sigma(x)$$

Subtracting the above inequalities, (8.40) follows from the fact that

$$\int_{S^{n-1}} (\phi_{a,\delta}^+ - \phi_{a,\delta}^-) (x \cdot \xi) \, d\sigma(x) \ll \delta$$

□

In what follows, we take $\delta = \lambda^{-n}$ and write $\phi_{a,\delta}$ for $\phi_{a,\delta}^{\pm}$, as our estimates work the same way for both choices of the sign. By taking the inverse Fourier transform of $\phi_{a,\delta}(t)$ one has

$$\sum_{|m|^2=\lambda} \phi_{a,\delta} (\lambda^{-1/2} m \cdot \xi) = \int_{\mathbf{R}} \lambda^{\frac{1}{2}} \tilde{\phi}_{a,\delta}(t\lambda^{\frac{1}{2}}) \, \hat{\omega}_\lambda(t\xi) \, dt \qquad (8.41)$$

also

$$\int_{S^{n-1}} \phi_{a,\delta} (x \cdot \xi) \, d\sigma(x) = \int_{\mathbf{R}} \tilde{\phi}_{a,\delta}(t) \, \tilde{\sigma}(t\xi) \, dt \qquad (8.42)$$

We substitute the asymptotic formula (8.9) into (8.41) and study the contribution of each term separately. Accordingly, let

$$I_{q,\lambda} := \int_{\mathbf{R}} \lambda^{\frac{1}{2}} \tilde{\phi}_{a,\delta}(t\lambda^{\frac{1}{2}}) \, m_{q,\lambda}(t\xi) \, dt, \qquad (8.43)$$

and

$$E_\lambda = \int_{\mathbf{R}} \lambda^{\frac{1}{2}} \tilde{\phi}_{a,\delta}(t\lambda^{\frac{1}{2}}) \, \mathscr{E}_\lambda(t\xi) \, dt. \qquad (8.44)$$

To estimate the error term in (8.44) note that

$$\int_{\mathbf{R}} \lambda^{\frac{1}{2}} |\tilde{\phi}_{a,\delta}(t\lambda^{\frac{1}{2}})| \, dt \leq C \int_{\mathbf{R}} (1 + |t|)^{-1}(1 + \delta|t|)^{-1} \, dt \leq C \log \lambda.$$

Thus by (8.32) one has for every $\varepsilon > 0$

$$|E_\lambda| \leq C_\varepsilon \lambda^{\frac{n-1}{4} + \varepsilon}. \qquad (8.45)$$

Next, decompose the integral in (8.43) as

$$I_{q,\lambda} = \int_{|t| < 1/8q} + \int_{|t| \geq 1/8q} = I_{q,\lambda}^1 + I_{q,\lambda}^2. \qquad (8.46)$$

Here an important observation is that if $|t| < 1/8q$ then $\psi(qt\xi - l) = 0$ unless $l = 0$, moreover $\psi(tq\xi) = 1$ since $|tq\xi_j| < 1/8q$ for each j. Hence

$$m_{q,\lambda}(t\xi) = K(q,0,\lambda)\,\tilde{\sigma}(\lambda^{\frac{1}{2}}t\xi).$$

Thus by (8.43) and a change of variables: $t := t\lambda^{1/2}$

$$I_{q,\lambda}^1 = K(q,1,\lambda)\int_{|t|<\lambda^{\frac{1}{2}}/8q}\tilde{\phi}_{a,\delta}(t)\,\tilde{\sigma}(t\xi)\,dt. \tag{8.47}$$

Lemma 10. *One has for every $\varepsilon > 0$*

$$\Big|\gamma_n\lambda^{\frac{n}{2}-1}\sum_{q\leq\lambda^{\frac{1}{2}}}I_{q,\lambda}^1 - N_\lambda\int_{S^{n-1}}\phi_{a,\delta}\,(x\cdot\xi)\,d\sigma(x)\Big| \leq C_\varepsilon\lambda^{\frac{n-1}{4}+\varepsilon} \tag{8.48}$$

Proof. Using (8.37), one has

$$\int_{|t|\geq\lambda^{\frac{1}{2}}/8q}|\tilde{\phi}_{a,\delta}(t)\,\tilde{\sigma}(t\xi)|\,dt \leq C_\varepsilon\lambda^{-\frac{n-1}{4}+\varepsilon}q^{\frac{n-1}{2}} \tag{8.49}$$

Thus by (8.42) and (8.47)

$$\Big|I_{q,\lambda}^1 - K(q,0,\lambda)\int_{S^{n-1}}\phi_{a,\delta}\,(x\cdot\xi)\,d\sigma(x)\Big| \leq C_\varepsilon\lambda^{-\frac{n-1}{4}+\varepsilon}q^{\frac{n-1}{2}}\,|K(q,0,\lambda)|$$

Substituting $\xi = 0$ in (8.33) one has

$$\Big|N_\lambda - \gamma_n\lambda^{\frac{n}{2}-1}\sum_{q\leq\lambda^{\frac{1}{2}}}K(q,0,\lambda)\Big| \leq C_\varepsilon\lambda^{\frac{n-1}{4}+\varepsilon} \tag{8.50}$$

Using (8.22) and (8.50), the left side of (8.48) is estimated by

$$C_\varepsilon\left(\lambda^{\frac{n-1}{4}+\varepsilon} + \lambda^{\frac{n-3}{4}+\varepsilon}\sum_{q\leq\lambda^{\frac{1}{2}}}q^\varepsilon\,(\lambda,q_1)^{\frac{1}{2}}2^{\frac{5}{2}}\right) \leq C_\varepsilon\lambda^{\frac{n-1}{4}+\varepsilon} \tag{8.51}$$

$\square$

To estimate the remaining error terms one needs to exploit the diophantine properties of the direction ξ.

Lemma 11. *Let $\xi \in S^{n-1}$ diophantine. Then for every $\varepsilon > 0$, we have*

$$\sum_{q\leq\lambda^{\frac{1}{2}}}|I_{q,\lambda}^2| \leq C_{\xi,\varepsilon}\lambda^{-\frac{n-3}{4}+\varepsilon} \tag{8.52}$$

Proof. First, note that $\psi(q\xi - l) = 0$ unless $l = [q\xi]$, that is the closest lattice point to the point $q\xi \in \mathbf{R}^n$. Using the notation $\{q\xi\} = q\xi - [q\xi]$ one may write

$$m_{q,\lambda}(t\xi) = K(q, [qt\xi], \lambda)\, \psi(\{qt\xi\})\, \tilde{\sigma}\left(\frac{\lambda^{\frac{1}{2}}}{q}\{qt\xi\}\right)$$

By making a change of variables $t := qt$, it follows from estimates (8.22) and (8.37) that

$$|I_{q,\lambda}^2| \le C_\varepsilon\, (\lambda^{\frac{1}{2}}/q)^{-\frac{n-3}{2}}\, q^{-\frac{n-1}{2}+\varepsilon}\, (\lambda, q_1)^{\frac{1}{2}}\, 2^{\frac{r}{2}}\, J_\lambda, \tag{8.53}$$

where

$$J_\lambda = \int_{|t| \ge 1/8} |\tilde{\phi}_{a,\delta}(t\, \lambda^{\frac{1}{2}}/q)|\, \|t\xi\|^{-\frac{n-1}{2}}\, dt,$$

and $\|t\xi\|$ denotes the distance of the point $t\xi$ to the nearest lattice point. For $q \le \lambda^{1/2}$ one has

$$|\hat{\phi}_{a,\delta}(t\, \lambda^{\frac{1}{2}}/q)| \le C\, (\lambda^{\frac{1}{2}}/q)^{-1}\, |t|^{-1}\, (1 + \delta|t|)^{-1} \tag{8.54}$$

To estimate the integral J_λ we use (8.54), and integrate over dyadic intervals $2^j \le |t| < 2^{j+1}$ $(j \ge -3)$. For a fixed j we have

$$\int_{2^j}^{2^{j+1}} t^{-1}(1 + \delta t)^{-1}\, \|t\xi\|^{-\frac{n-1}{2}}\, dt \le C_\varepsilon\, 2^{j\varepsilon}\, (1 + \delta 2^j)^{-1} \tag{8.55}$$

Summing over j this gives: $J_\lambda \le C_\varepsilon\, (\lambda^{\frac{1}{2}}/q)^{-1}\lambda^\varepsilon$. Substituting into (8.53) one estimates

$$|I_{q,\lambda}^2| \le C_\varepsilon\, \lambda^{-\frac{n-1}{4}+\varepsilon}\, q^\varepsilon\, (\lambda, q_1)^{\frac{1}{2}}\, 2^{\frac{r}{2}} \tag{8.56}$$

Summing over $q \le \lambda^{1/2}$, and using Lemma 2, estimate (8.52) follows.   □

Theorem 1 follows immediately from Lemmas 9–11, and estimate (8.45).

8.2.4 The Kloosterman Refinement

For the sake of completeness we include below the proofs of Theorems A–B. The present form of Theorem A was given by Heath-Brown [9] in his study of non-singular cubic forms, the idea going back to Kloosterman. Theorem B follows from the multiplicative properties of Kloosterman sums and Weil's estimates [14].

To start let w be an absolutely summable weight function, P be an integral polynomial, N a fixed positive integer, and write

$$I := \sum_{P(m)=\lambda} w(m) = \int_{-1/N+1}^{1-1/N+1} S(\alpha)\,d\alpha, \tag{8.57}$$

with $S(\alpha) = \sum_{m \in \mathbb{Z}^n} e^{2\pi i\, P'(m)} w(m)$, $P'(m) = P(m) - \lambda$. Breaking up the interval $[-1/N + 1, 1 - 1/N + 1]$ according to the Farey dissection of order N (see [8], Ch. 3.8), we have

$$I = \sum_{q \leq N} \sum_{(a,q)=1} \int S(a/q + \beta)\,d\beta.$$

Here for fixed a the inner integral is over the interval

$$\left[\frac{a+a'}{q+q'} - \frac{a}{q}, \frac{a+a''}{q+q''} - \frac{a}{q}\right],$$

where $a'/q', a/q, a''/q''$ are consecutive Farey fractions. Since $qa' - q'a = -1$ and $qa'' - q''a = 1$ the range of β is given by

$$-(q+q')^{-1} \leq q\beta \leq (q+q'')^{-1}.$$

Since for consecutive Farey fractions, we have $q + q', q + q'' \geq N$, one may write I as

$$\sum_{q \leq N} \int_{-1/qN}^{1/qN} \sum_a S(a/q + \beta)\,d\beta, \tag{8.58}$$

where the inner sum is restricted to $1 \leq a \leq q$, $(a,q) = 1$, and

$$q' \leq \frac{1}{q|\beta|} - q \quad (\beta < 0), \qquad q'' \leq \frac{1}{q\beta} - q \quad (\beta > 0). \tag{8.59}$$

The numbers q', q'' are completely specified by a as $q' \equiv -q'' \equiv a^{-1} \pmod{q}$ and $N - q < q', q'' \leq N$, thus (8.58) eventually restricts the summation in a. The point is that if $|\beta| \leq q^{-1}(q + N)^{-1}$, then (8.58) places no restriction on a, and otherwise $\bar{a} = a^{-1} \pmod{q}$ must lie in one of two intervals $J(q, \beta) \subseteq (0, q)$. Then one estimates

$$\sum_{\bar{a} \in J(q,\beta)} S(a/q + \beta) = \sum_{(s,q)=1} S(\bar{s}/q + \beta) \sum_{t \in J(q,\beta)} \frac{1}{q} \sum_{|u| \leq q/2} e^{2\pi i \frac{u(s-t)}{q}}$$

$$= \frac{1}{q} \sum_{|u| \le q/2} S_u(q, \beta) \sum_{t \in J(q,\beta)} e^{-2\pi i \frac{ut}{q}}$$

$$\ll \sum_{|u| \le q/2} (1 + |u|)^{-1} |S_u(q, \beta)|, \qquad (8.60)$$

where

$$S_u(q, \beta) = \sum_{(s,q)=1} e^{2\pi i \frac{us}{q}} S(\bar{s}/q + \beta),$$

using the estimate

$$\frac{1}{q} \sum_{t \in J(q,\beta)} e^{-2\pi i \frac{ut}{q}} \ll (1 + |u|)^{-1}.$$

Since

$$(qN)^{-1} - q^{-1}(q + N)^{-1} = N^{-1}(q + N)^{-1} \le N^{-2}$$

and

$$q^{-1}(q + N)^{-1} \ge (2qN)^{-1},$$

the total contribution to (8.58) arising from the ranges $|\beta| \ge q^{-1}(q + N)^{-1}$ is

$$\ll N^{-2} \sum_{q \le N} \sum_{|u| \le q/2} (1 + |u|)^{-1} \max_{\frac{1}{2} \le qN|\beta| \le 1} |S_u(q, \beta)|. \qquad (8.61)$$

The remaining range for β gives

$$\sum_{q \le N} \int_{-1/q(q+N)}^{1/q(q+N)} S_0(q, \beta) \, d\beta.$$

If one integrates for $|\beta| \le 1/qN$ instead, the resulting error is again of the form of (8.61). Thus summarizing the above estimates, we have

$$I = \sum_{q \le N} \int_{-1/q(q+N)}^{1/q(q+N)} S_0(q, \beta) \, d\beta + O\Big(N^{-2} \sum_{q \le N} \sum_{|u| \le q/2} (1 + |u|)^{-1} \max_{\frac{1}{2} \le qN|\beta| \le 1} |S_u(q, \beta)|\Big).$$

and Theorem A follows.

From the standard estimate for the Gaussian sums $G(a, l, q) \ll q^{-n/2}$, it is immediate that

$$|K(q,l,\lambda;u)| \ll q^{-n/2+1}. \tag{8.62}$$

Also, $G(a,l,q)$ is a product of one dimensional sums, thus for q odd, by completing the square in the exponent, it may be written in the form (see also [14], Ch. 4)

$$G(a,l,q) = q^{-n} \epsilon_q^n \left(\frac{q}{a}\right)^n e^{-2\pi i \frac{4\bar{a}|l|^2}{q}} G(1,0,q)^n,$$

where $\left(\frac{q}{a}\right)$ denotes the Jacobi symbol, ϵ_q is a 4th root of unity, and $\bar{a}$ denotes the multiplicative inverse of a mod q. Substituting this into (8.21) we have

$$K(q,l,\lambda;u) = \epsilon_q^n q^{-n} G(1,0,q)^n \sum_{(a,q)=1} \left(\frac{q}{a}\right)^n e^{2\pi i \frac{a\lambda+4\bar{a}(u-|l|^2)}{q}}. \tag{8.63}$$

The sum in (8.63) is a Kloosterman sum or Salie sum depending on whether n is even or odd. Weil's estimates [14, Ch. 4] imply

$$|K(q,l,\lambda;u)| \le C_\varepsilon q^{-\frac{n-1}{2}+\varepsilon} (\lambda,q)^{\frac{1}{2}}. \tag{8.64}$$

Estimate (8.22) follows by writing $q = q_1 q_2$, with q_1 odd and $q_2 = 2^r$, applying (8.64) to q_1, (8.62) to $q_2 = 2^r$ and using the multiplicative property

$$K(q,l,\lambda;u) = K(q_1,l\,\overline{q_2},\lambda;u\,\overline{q_2}^2)\, K(q_2,l\,\overline{q_1},\lambda;u\,\overline{q_1}^2), \tag{8.65}$$

where $q_1\overline{q_1} \equiv 1 \pmod{q_2}$, and $q_2\overline{q_2} \equiv 1 \pmod{q_1}$. Property (8.65) is well-known, and is an easy computation using the Chinese Remainder Theorem. This finishes the proof of Theorem B.

8.3 The Discrepancy of Lattice Points on Hypersurfaces

We will study now the uniformity of distribution of lattice points on a homogeneous, non-singular, hypersurface. We will show that if the dimension of the underlying Euclidean space is large enough with respect to the degree of the hypersurface, then there are non-trivial upper bounds on the discrepancy with respect to caps.

The analysis will be similar to what we have carried out for spheres, however in this generality we will use the Birch-Davenport method of exponential sums, which will allow us to develop uniform asymptotic formulae for the Fourier transform of the set of lattice points on the hypersurface.

To formulate our main result in this section, let $P(m)$ be a positive, homogeneous polynomial of degree d with integer coefficients, and for $\lambda \in \mathbf{N}$, define the hypersurface

$$S_\lambda = \{x \in \mathbf{R}^n;\ P(x) = \lambda\}$$

We will write S for S_1, the unit level surface. Recall that the polynomial is called non-singular if for all $z \in \mathbf{C}^n / \{0\}$

$$\nabla P(z) = (\partial_1(z), \ldots, \partial_n(z)) \neq 0 \tag{8.66}$$

Our main result in this section is the following upper bound of the set of solutions $Z'_{P,\lambda} = \{\lambda^{-1/d} m; \ P(m) = \lambda\}$ with respect to the family of caps $C_{a,\xi}$ corresponding to a given diophantine direction ξ, defined in (8.2). Similar but somewhat weaker results have been obtained in [13].

Theorem 12. *Let $n > d(d-1)2^{d+1}$, and let $P(m)$ be a positive, homogeneous non-singular polynomial of degree d with integer coefficients. If $\xi \in S^{n-1}$ is diophantine, then we have*

$$|D_P(\xi, \lambda)| \leq C_{\xi,\varepsilon} \lambda^{(\frac{n}{d}-1)(1-\gamma_d)}, \tag{8.67}$$

with $\gamma_d = \frac{1}{(d-1)2^{d+1}}$.

To see why this upper bound is non-trivial, note that as P is positive, we have that $P(x) \approx |x|^d$, thus on average for $L \leq \lambda < 2L$, the surface S_λ contains $\approx \lambda^{n/d-1}$ lattice points. Indeed there are $\approx L^{n/d}$ lattice points m in the region $L \leq P(m) < 2L$, and they lie on L hypersurfaces. As we have mentioned, because of congruence obstructions, one cannot have that $|\mathbf{Z}^n \cap S_\lambda| \approx \lambda^{n/d-1}$ for all large λ, but it can be shown that this holds all $\lambda \in \Lambda$, for an infinite arithmetic progression $\Lambda \subseteq \mathbf{N}$. Such a set Λ will be called a set of regular values. Thus one has

Corollary 13. *Let $n > d(d-1)2^{d+1}$, $P(m)$ be a positive, homogeneous non-singular polynomial of degree d with integer coefficients, and let Λ be a set of regular values for P. If $\xi \in S^{n-1}$ is diophantine, then we have*

$$|D_P(\xi, \lambda)| \leq C_{\xi,\varepsilon} N_\lambda^{1-\gamma_d}, \tag{8.68}$$

for each $\lambda \in \Lambda$, with $\gamma_d = \frac{1}{(d-1)2^{d+1}}$, where N_λ denotes the number of lattice points on the surface S_λ.

8.3.1   The Fourier Transform of the Set of Lattice Points on Hypersurfaces

We will now generalize the asymptotic formula (8.9) describing the structure of the Fourier transform of lattice points on spheres, using the Birch-Davenport [4, 5, 16] version of the Hardy-Littlewood method of exponential sums. This method was developed to count solutions of (systems of) diophantine equations, when the number of variables is large enough with respect to the degrees of the polynomials, and it is one of the most far reaching application of analytic tools in the area of diophantine equations. In spite of this there are very few accessible description of this method, so perhaps it is of interest to discuss it in detail in the case of a single non-singular homogeneous polynomial.

8.3.1.1 Minor Arcs Estimates

To start, let ϕ be a smooth cut-off function which is constant 1 on the unit level surface $S = \{P = 1\}$, and let $N = \lambda^{1/d}$. Then

$$\hat{\omega}_\lambda(\xi) = \sum_{P(m)=\lambda} e^{2\pi i\, x\cdot\xi} \phi(m) = \int_0^1 S(\alpha, \xi) e^{-2\pi i\alpha\lambda}\, d\alpha, \tag{8.69}$$

where

$$S(\alpha, \xi) = \sum_{m\in\mathbf{Z}^n} e^{2\pi i(P(m)+m\cdot\xi)}\phi(m/N) \tag{8.70}$$

As is usual in the circle-method, we will now define a family of small intervals, which we will call *major arcs* on which the exponential sum $S(\alpha, \xi)$ is concentrated. Let $0 < \theta \le 1$ be a parameter, and for a given pair of natural numbers a, q such that $(a, q) = 1$, define the corresponding *major arc* centered at a/q by

$$L_{a,q}(\theta) = \{\alpha :\ 2|\alpha - a/q| < q^{-1}N^{-d+(d-1)\theta}\},$$

moreover let

$$L(\theta) = \bigcup_{q\le N^{(d-1)\theta},\,(a,q)=1} L_{a,q}(\theta).$$

If $\alpha \notin L(\theta)$, the we say α is in a *minor arc*. The following properties of the major arcs are immediate from their definition.

Proposition 14.   (i) If $\theta_1 < \theta_2$ then $\theta_1 \subseteq L(\theta_2)$.
(ii) If $\theta < \frac{d}{2(d-1)}$ then the intervals $L_{a,q}(\theta)$ are disjoint for different values of a and q.
(iii) $|L(\theta)| \le N^{-d+(d-1)\theta}$.

We will now derive standard Weyl-type estimates, following [4], for the exponential sum $S(\alpha, \xi)$, when α is in a minor arc. It will be useful to introduce the notations

$$D_h P(m) = P(m) - P(m + h), \quad \Delta_h\phi(m) = \phi(m)\overline{\phi}(m + h),$$

and inductively

$$D_{h^1,\ldots,h^k} P = D_{h^1}(D_{h^2,\ldots,h^k} P), \qquad \Delta_{h^1,\ldots,h^k}\phi = \Delta_{h^1}(\Delta_{h^2,\ldots,h^k}\phi).$$

Note, that the above expressions are independent of the order of the vectors $h^1,\ldots,h^k$. We will also use repeatedly the expression

$$|\sum_m \phi(m)|^2 = \sum_{m,h} \phi(m)\bar{\phi}(m+h) = \sum_{m,h} \Delta_h\phi(m).$$

Writing $\phi_N(m) = \phi(m/N)$, and taking averages, we have

$$|N^{-n}S(\alpha,\xi)|^2 = N^{-2n} \sum_{h^1,m} e^{2\pi i\, \alpha D_{h^1} P(m)-h\cdot\xi} \Delta_{h^1}\phi_N(m)$$

$$\leq N^{-n} \sum_{h^1} |N^{-n} \sum_m e^{2\pi i\, \alpha D_{h^1} P(m)} \Delta_{h^1}\phi_N(m)|$$

Note that the summation is restricted to $|h^1| \ll N$ and $|m| \ll N$. Applying the Cauchy-Schwarz inequality $d - 2$ times, one has

$$|N^{-n} S(\alpha,\xi)|^{2^{d-1}} \ll N^{-n(d-1)} \sum_{h^1,\ldots,h^{d-1}} N^{-n}|\sum_m e^{2\pi i\alpha D_{h^1,\ldots,h^{d-1}} P(m)} \Delta_{h^1,\ldots,h^{d-1}}\phi_N(m)| \tag{8.71}$$

Note that the implicit constant in (8.71) depends only on the dimension n and the degree d, and the summation again is restricted to $|h^i| \ll N$ and $|m| \ll N$. The point is that after taking $d - 1$ "derivatives", the polynomial $D_{h^1,\ldots,h^{d-1}} P$ becomes linear, i.e. it is of the form

$$D_{h^1,\ldots,h^{d-1}} P(m) = \sum_{j=1}^n m_j\, \Phi_j(h^1,\ldots,h^{d-1}), \tag{8.72}$$

where the coefficients $\Phi_j : \mathbf{Z}^{n(d-1)} \to \mathbf{Z}$ are multi-linear forms. In fact writing the homogeneous polynomial P as

$$P(m) = \sum_{1\leq j_1,\ldots,j_d} a_{j_1,\ldots,j_d}\, m_{j_1}\ldots m_{j_d},$$

so that the coefficients $a_{j_1,\ldots,j_d}$ are independent of the order of the indices $j_1,\ldots,j_{d-1}$, it is not hard to see that

$$\Phi_j(h^1,\ldots,h^{d-1}) = d! \sum_{j_1,\ldots,j_{d-1}} a_{j_1,\ldots,j_{d-1},j}\, h^1_{j_1}\ldots h^{d-1}_{j_{d-1}}. \tag{8.73}$$

For simplicity let us introduce the notations

$$\underline{h} := (h^1, \ldots, h^{d-1}),$$

$$\Psi_{N,\underline{h}}(x) := \Delta_{h^1 \ldots h^{d-1}} \phi_N(x),$$

$$\Phi(\underline{h}) := (\Phi_1(\underline{h}), \ldots, \Phi_n(\underline{h})).$$

Now, by (8.72) the inner sum in (8.71) is the Fourier transform the function $\Psi_{N,\underline{h}}$ at $\xi = \alpha \Phi(\underline{h})$. To estimate it, note that

$$\left| \left(\frac{d}{dx} \right)^k \Psi_{N,\underline{h}}(x) \right| \ll N^{-k}, \quad \text{for all } k \in \mathbf{N},$$

(where the implicit constant depends only on n, d and k), and that the function $\Psi_{N,\underline{h}}$ is supported on $|x| \ll N$. Thus integrating by parts k-times we have that

$$|\tilde{\Psi}_{N,\underline{h}}(\xi)| \ll N^n (1 + N|\xi|)^{-k},$$

where $\tilde{\Psi}_{N,\underline{h}}$ denotes the Fourier transform of $\Psi_{N,\underline{h}}(x)$ considered as function on $\mathbf{R}^n$. Thus by Poisson summation

$$|\hat{\Psi}_{N,\underline{h}}(\xi)| \leq \sum_{l \in \mathbf{Z}^n} |\tilde{\Psi}_{N,\underline{h}}(\xi - l)| \ll N^n (1 + N\|\xi\|)^{-k}.$$

Here we used the notation $\|\xi\| = \max_j \|\xi_j\|$, for a point $\xi = (\xi_1, \ldots, \xi_n)$, where $\|\xi_j\|$ denotes the distance of the j-th coordinate ξ_j from the nearest integer. Plugging this, into inequality (8.71) we have

$$|N^{-n} S(\alpha, \xi)|^{2^{d-1}} \ll N^{-n(d-1)} \sum_{\underline{h} \in \mathbf{Z}^{n(d-1)}, \, |\underline{h}| \ll N} (1 + N \|\alpha \Phi(\underline{h})\|)^{-k}, \quad (8.74)$$

for all $k \in \mathbf{N}$. We will fix now $k = n + 1$, and use the multi-linearity of the forms $\Phi_j(h^1, \ldots, h^{d-1})$, to estimate the right side of inequality (8.74) by the number of $(h^1, \ldots, h^{d-1})\mathbf{Z}^{n(d-1)}, |h^j| \ll N$ such that $\|\alpha \Phi(\underline{h})\| \leq N^{-1}$. More generally, for given parameters τ, η, let us introduce the quantities

$$\mathscr{R}(N^\tau, N^{-\eta}, \alpha) = |\{\underline{h} \in \mathbf{Z}^{n(d-1)}; \ |\underline{h}| \ll N, \ \|\alpha \Phi_j(\underline{h})\| \leq N^{-\eta}, \ 1 \leq j \leq n\}|. \quad (8.75)$$

Lemma 15.

$$(N^{-n}|S(\alpha, \xi)|)^{2^{d-1}} \ll N^{-n(d-1)} \mathscr{R}(N, N^{-1}, \alpha). \quad (8.76)$$

Proof. Consider the points $\{\alpha\,\Phi(\underline{h})\} \in [-\frac{1}{2},\frac{1}{2}]^n$, where $\{\;\}$ denotes the fractional part, and divide the cube $[-\frac{1}{2},\frac{1}{2}]^n$ into N^n cubes $B_{\underline{s}}$ of size $\frac{1}{N}$. Now if $B_{\underline{0}} = [-\frac{1}{2N},\frac{1}{2N}]^n$, then for each fixed $\underline{h}' = (h^1,\ldots,h^{d-2})$, the cube $B_{\underline{0}}$ will contain at least as many points of the form $\{\alpha\,\Phi(\underline{h}',h^{d-1})\}$, as any of the other cubes $B_{\underline{s}}$. Indeed, this follows immediately from the linearity of the forms Φ_j in the variable h^{d-1}. Since the center of the cubes $B_{\underline{s}}$ are $N^{-1}\underline{s} = (\frac{s_1}{N},\ldots,\frac{s_n}{N})$ with $-N/2 \leq s_j < N/2$, the right side of (8.74) is bounded by

$$N^{-n(d-1)} \sum_{-\frac{N}{2}\leq s_1,\ldots,s_n<\frac{N}{2}} (1+|\underline{s}|)^{-n-1}\, \mathscr{R}(N,N^{-1},\alpha) \ll N^{-n(d-1)}\mathscr{R}(N,N^{-1},\alpha).$$

$\square$

Next, we will use that fact that the quantities $\mathscr{R}(N^\tau, N^{-\eta},\alpha)$ can be compared to each other for different values of the parameters τ, η, in fact we will need the following

Lemma 16. *Let* $0 < \theta < \frac{1}{d-1}$. *Then we have*

$$N^{-n(d-1)}\,\mathscr{R}(N,N^{-1},\alpha) \ll N^{-n(d-1)\theta}\,\mathscr{R}(N^\theta,N^{-d+(d-1)\theta},\alpha). \tag{8.77}$$

This is based on the following result

Lemma 17 (Davenport [5]). *Let* $L_1(u),\ldots,L_n(u)$ *be* n *real linear forms in* n *variables* $u_1,\ldots,u_n$, *say*

$$L_j(u) = \sum_k \lambda_{jk}\,u_k,$$

which are symmetric in the sense that $\lambda_{jk} = \lambda_{kj}$. *Let* $1 < K_1 < K_2$ *and for* $0 < r < 1$ *let* $U(r)$ *denote the number of integer solutions of the system*

$$|u_k| < rK_1, \qquad \|L_j(u)\| < rK_2^{-1}. \tag{8.78}$$

Then for all $0 < r \leq 1$ *we have*

$$U(1) \ll r^{-n}U(r). \tag{8.79}$$

This is Lemma 3.3 in [5] and is an application of the geometry of numbers. Let us remark here only that the solutions of (8.78) can be viewed as lattice points $(u,v) \in \mathbf{Z}^{2n}$ which are inside the convex symmetric body rB, where

$$B = \{(u,v) \in \mathbf{R}^{2n};\ |u_k| < K_1,\ |v_j - L_j(u)| < K_2^{-1},\ 1 \leq k,j \leq n\}.$$

Proof (of Lemma 16). We will apply Lemma 17 in each variable $h^1, \ldots, h^{d-1}$. Fix $\underline{h}' = (h^2, \ldots, h^{d-1})$, write $u = h^1$ and $L_j(u) = \alpha \, \Phi_j(u, h\underline{h}')$. From (8.73) it is clear that the linear forms $L_j(u)$ are symmetric, thus we can apply Lemma 17, with $K_1 = K_2 = N$, $r_1 = N^{\theta-1}$, $r_2 = 1$ for each $\underline{h}'$. Summing over $\underline{h}'$ gives

$$\mathscr{R}(N, N^{-1}\alpha) \ll N^{n(1-\theta)} |\{\underline{h} \in \mathbf{Z}^{n(d-1)}; \; |h^1| \ll N^\theta, \; |\underline{h}'| \ll N, \; \|\alpha \, \Phi(\underline{h})\| < N^{\theta-2}\}.$$

Next, set $u = h^2$, fix the remaining variables and apply Lemma 8.69 with $K_1 = N$, $K_2 = N^{2-\theta}$ and $r = N^{\theta-1}$. Continuing this procedure for all the variables $h^1, \ldots, h^{d-1}$ eventually, we have

$$\mathscr{R}(N, N^{-1}\alpha) \ll N^{n(d-1)(1-\delta)} \mathscr{R}(N^\theta, N^{-d+(d-1)\theta}, \alpha),$$

which is the same as (8.77). $\qquad\square$

Note that if there is a point $\underline{h} \in \mathbf{Z}^{n(d-1)}$, $|\underline{h}| \ll N$ such that $\|\alpha \Phi_j(\underline{h})\| < N^{-d+(d-1)\theta}$ and $\Phi_j(\underline{h}) \neq 0$, then setting $q = |\Phi_j(\underline{h})|$, we have that

$$\left| \alpha - \frac{a}{q} \right| < \frac{1}{q} N^{-d+(d-1)\theta}$$

for some $a \in \mathbf{Z}^n$ such that $(a, q) = 1$. Thus $\alpha \in L(\theta)$ by the definition of major arcs, hence if α is in a minor arc, we have

$$\mathscr{R}(N^\theta, N^{-d+(d-1)\theta}, \alpha) = |\{\underline{h} \in \mathbf{Z}^{n(d-1)}; \; |\underline{h}| \ll N, \; \Phi_1(\underline{h}) = \ldots = \Phi_n(\underline{h}) = 0\}. \tag{8.80}$$

which is the number of lattice points $\underline{h} \in \mathbf{Z}^{n(d-1)}$ of size $|\underline{h}| \ll N^\theta$ on the variety

$$S_\Phi := \{\underline{z} \in \mathbf{C}^{n(d-1)}; \; \Phi_1(\underline{z}) = \ldots = \Phi_n(\underline{z}) = 0\}.$$

By (8.73) it is easy to see that $\Phi_j(h, \ldots, h) = (d-1)!(\partial/\partial_j)P(h)$, thus if we set

$$\Delta := \{(h, \ldots, h); \; h \in \mathbf{C}^n\} \subseteq \mathbf{C}^{n(d-1)},$$

then

$$S_\Phi \cap \Delta = \{h \in \mathbf{C}^n; \; \partial_1 P(h) = \ldots = \partial_n P(h) = 0\} = \{0\},$$

by our assumption that the polynomial P is non-singular. Then by basic facts from algebraic geometry it follows that

$$D := \dim S_\Phi \leq n(d-1) - n.$$

The dimension of the algebraic set S_Φ is defined algebraically, however it is well-known, see [10], Ch. 7, that if it has dimension D then every bounded part of it can be covered by $O(\rho^{-D})$ balls of diameter ρ for any $0 < \rho < 1$. Combining this with the fact that S_Φ is homogeneous, we have

$$|\{\underline{h} \in \mathbf{Z}^{n(d-1)} \cap S_\Phi; \ |\underline{h}| \ll N^\theta\}| \ll |\{\underline{h}' \in (N^{-\theta}\mathbf{Z})^{n(d-1)} \cap S_\Phi; \ |\underline{h}'| \ll 1\}| \ll N^{D\theta}. \tag{8.81}$$

Then by (8.76), (8.77) and (8.81), we have the following estimate on the minor arcs.

Lemma 18. *Let* $0 < \theta < 1$. *If* $\alpha \notin L(\theta)$, *then we have uniformly in* ξ

$$|S(\alpha, \xi)| \ll N^{n-n\theta 2^{-(d-1)}}. \tag{8.82}$$

We will also need a variant of the above estimate when the cut-off function ϕ is replaced by the indicator function χ of a cube of side length ≈ 1 centered near the origin. The estimate below is proved in [4], however it easily follows from (8.82). Indeed, choose a cut off function ϕ such that $\chi\phi = \chi$, and let $P_1(m) = P(m) + m \cdot \xi$. Then by Plancherel's identity

$$\sum_{m \in \mathbf{Z}^n} e^{2\pi i \alpha P_1(m)} \phi(m/N) \chi(m/N) = \tag{8.83}$$

$$= \int_{\mathbf{T}^n} \left(\sum_{m \in \mathbf{Z}^n} e^{2\pi i \alpha P_1(m) - m \cdot \xi} \phi(N/P) \right) (N^n \hat{\chi}(N\xi)) \, d\xi \ll N^{n - \frac{n}{2^{d-1}}} (\log N)^n.$$

Here $\mathbf{T}^n$ is the flat torus, and the above estimate follows using (8.82) for the first term of the integral uniformly in ξ, and the fact that $\|N^n \hat{\chi}(N\xi)\|_{L^1(\mathbf{T}^n)} \ll (\log N)^n$.

Corollary 19. *Let* $1 \le a < q$ *be natural numbers s.t.* $(a, q) = 1$. *The for the exponential sum*

$$G(a, q, l) = q^{-n} \sum_{s \in (\mathbf{Z}/q\mathbf{Z})^n} e^{2\pi i \frac{aP(m) - l \cdot s}{q}},$$

one has

$$|G(a, q, l)| \ll q^{-\frac{n}{(d-1)2^d}} (\log q)^n. \tag{8.84}$$

Proof. Set $N = q$, $\alpha = a/q$, $\xi = l/q$, $\theta = 1/2(d-1)$ and notice that $\alpha \notin L(\theta)$. Indeed, for $q_1 \le q^{(d-1)\theta}$ we have

$$\left| \frac{a}{q} - \frac{a_1}{q_1} \right| \ge \frac{1}{q_1 q} \ge \frac{1}{q_1} q^{-d + (d-1)\theta}.$$

Then (8.84) follows from (8.83), choosing χ to be the indicator function of $[0, 1)^n$, and identifying $(\mathbf{Z}/q\mathbf{Z})^n$ with $[0, q)^n \cap \mathbf{Z}^n$. $\Box$

Corollary 20. *If $|\alpha| < P^{-d/2}$ then one has*

$$|S(\alpha, \xi)| \ll N^n \, (N^d |\alpha|)^{-\frac{n}{(d-1)2^{d}-1}}.$$

Proof. Choose θ such that $|\alpha| = N^{-d+(d-1)\theta}$, that is $(N^d |\alpha|)^{\frac{1}{d-1}} = N^\theta$. The major arcs $L_{a,q}(\theta)$ are disjoint since $(d-1)\theta < d/2$, moreover α is an endpoint of the interval $L_{0,1}(\theta)$ hence $\alpha \notin L_{a,q}(\theta)$. By (8.82) this gives

$$|S(\alpha, \xi)| \ll N^{n-n2^{-(d-1)}\theta} = N^n \, (N^d |\alpha|)^{-\frac{n}{(d-1)2^{(d-1)}}}.$$

$\Box$

8.3.1.2   Approximations on the Major Arcs

We will now derive an asymptotic expansion for the Fourier transform of the lattice points on the hypersurface $S_\lambda = \{P = \lambda\}$ along the lines as in Sect. 8.2. Throughout this section we will assume that n is sufficiently large, in particular that $n > n_d := d(d-1)2^{d+1}$, set $\gamma_d := \frac{1}{(d-1)2^d+1}$, and for simplicity of notation introduce the quantity $D := (d-1)2^{d-1}$.

Going back to the integral defined in (8.69), for a given θ, write

$$\hat{\omega}_\lambda(\xi) = \int_{\alpha \in L(\theta)} S(\alpha, \xi) \, d\alpha + \int_{\alpha \notin L(\theta)} S(\alpha, \xi) \, d\alpha = A_\lambda(\xi) + E_\lambda^1(\xi). \quad (8.85)$$

It follows from our assumptions on n, that there is a $\theta < \frac{1}{2(d-1)}$, such that

$$n\theta 2^{-(d-1)} > d + n\gamma_d$$

thus (8.71) implies that $S(\lambda, \xi) \ll N^{n-d-n\gamma_d}$ for $\lambda \notin L(\theta)$. Thus we have the estimate, uniformly in ξ

$$|E_\lambda^1(\xi)| \ll N^{n-d-n\gamma_d}. \quad (8.86)$$

We will fix a $\theta < \frac{1}{2(d-1)}$ so that (8.86) holds, and will do a number of transformations on the main term $A_\lambda(\xi)$ which are similar the ones we have used in the special case of the spheres. For a given $\alpha \in L_{a,q}(\theta)$ for some $(a, q) = 1$, $q \leq N^{(d-1)\theta}$, write $\alpha = a/q + \beta$, with $|\beta| \leq N^{-d+(d-1)\theta}$ and $m = qm_1 + s$ with $m_1 \in \mathbf{Z}^n$, $s \in (\mathbf{Z}/q\mathbf{Z})^n$. Applying Poisson summation as in (8.17), we have

$$S(a/q + \beta, \xi) = \sum_{m \in \mathbf{Z}^n} e^{2\pi i \frac{a}{q} P(m)} e^{2\pi i m \cdot \xi} H_{\beta,N}(m)$$

$$= \sum_{s \in (\mathbf{Z}/q\mathbf{Z})^n} G(a, q, l) \, \tilde{H}_{\beta,N}(l/q - \xi), \qquad (8.87)$$

where $\tilde{H}_{\beta,N}$ is the Fourier transform of the function $H_{\beta,N}(x) = e^{2\pi i \beta P(m)} \phi(m/N)$, and $G(a, q, l)$ is the exponential sum defined in (8.84). Thus we have

$$A_\lambda(\xi) = \sum_{q \le N^{(d-1)\theta}} \sum_{(a,q)=1} \sum_{l \in \mathbf{Z}^n} G(a, l, q) \, J_\lambda(\xi - l/q), \qquad (8.88)$$

where

$$J_\lambda(\xi - l/q) = \int_{|\beta| \le N^{-d+(d-1)\theta}} \tilde{H}(l/q - \xi, \beta) e^{-2\pi i \lambda \beta} \, d\beta$$

We shall approximate the functions $A_\lambda(\xi)$ with functions $B_\lambda(\xi)$ where the cut-off function $\psi(q\xi - l)$ have been inserted in (8.88), that is let

$$B_\lambda(\xi) = \sum_{a,q} \sum_{l \in \mathbf{Z}^n} G(a, l, q) \, \psi(q\xi - l) \, J_\lambda(\xi - l/q)$$

Next, we extend the integration in β and define

$$M_\lambda(\xi) = \sum_{a,q} \sum_{l \in \mathbf{Z}^n} G(a, l, q) \, \psi(q\xi - l) \, I_\lambda(\xi - l/q)$$

with

$$I_\lambda(\xi - l/q) = \int_{\mathbf{R}} \tilde{H}(\xi - l/q, \beta) e^{-2\pi i \lambda \beta} \, d\beta. \qquad (8.89)$$

A crucial point is to identify the integrals $I_\lambda(\eta)$, in fact we will show that

$$I_\lambda(\eta) = \tilde{\sigma}_\lambda(\eta).$$

First we estimate the errors obtained.

Lemma 21. *If* $0 < \theta < \frac{1}{2(d-1)}$ *then one has uniformly in* ξ

$$|A_\lambda(\xi) - B_\lambda(\xi)| \ll N^{n-d-n\gamma_d}.$$

Proof. If we set

$$\mu_\beta(\xi) = \sum_l G(a, q, l) \, (1 - \psi(q\xi - l)) \, \tilde{H}_{N,\beta}(\xi - l/q),$$

then it is enough to show that $|\mu_\beta(\xi)| \ll N^{n-d-n\gamma_d}$ uniformly for $|\beta| \leq N^{-d+(d-1)\theta}$ and $\xi \in \mathbf{T}^n$. Let $\eta = \xi - l/q$, and estimate $\tilde{H}_{N,\beta}(\eta)$ by partial integration:

$$\tilde{H}_{N,\beta}(\eta) \leq N^n \left| \int_{\mathbf{R}^n} e^{2\pi i N^d \beta P(x)} \phi(x) e^{2\pi i N x \cdot \eta} dx \right|$$

$$\ll N^n |N\eta|^{-K} \left| \int_{\mathbf{R}^n} (d/d\eta)^K \left(e^{2\pi i N^d \beta P(x)} \phi(x) e^{2\pi i N x \cdot \eta} dx \right) \right|$$

$$\ll N^n |N\eta|^{-K} (1 + N^d |\beta|)^K.$$

Now, on the support of $1 - \psi(q\xi - l)$ we have that

$$N|\eta| = N|\xi - l/q| \gg N^{1-(d-1)\theta},$$

hence for $|\beta| \leq N^{-d+(d-1)\theta}$ and $\theta < 1/2(d-1)$, choosing $0 < \tau < \frac{1}{2} - (d-1)\theta$ we have

$$|\mu_\beta(\xi)| \ll N^n (N/q)^{-\tau K} \sum_{l \in \mathbf{Z}^n} (1 + |q\xi - l|)^{-\tau K} \ll N^{n-\tau K(1-(d-1)\theta)}.$$

The Lemma follows by choosing K sufficiently large. □

In order to estimate the error obtained by extending the integration in β, we will need the following

Lemma 22. *For given η, $L > 0$ let*

$$I(L, \eta) = \int e^{2\pi i L(P(x) + x \cdot \eta)} \phi(x) dx.$$

Then one has

$$I(L, \eta) \ll (1 + L)^{-\frac{n}{D}}, \tag{8.90}$$

with $D = (d-1)2^{d-1}$.

Proof. The estimate is obvious for $L < 1$, so let $L \geq 1$. If $|\eta| \geq C$ with a large enough constant C, then the gradient of the phase $L|P'(x) + \eta| \geq L$ on the support of ϕ and (8.90) follows by partial integration.

Suppose $|\eta| \leq C$ and introduce the parameters θ, N, α such that $L = N^{(d-1)\theta}$, $\alpha = N^{-d} L$. Note that if $\theta < 2^{d-1}/n$, then we have $N > L^{\frac{2n}{D}}$. Changing variables $y = Nx$ yields

$$I(L, \eta) = N^{-n} \int e^{2\pi i \alpha (P(y) + N^{d-1} y \cdot \eta)} \phi(y/N) dy.$$

We compare the integral to a corresponding exponential sum

$$N^{-n} S(\alpha, \eta) = N^{-n} \sum_{m \in \mathbf{Z}^n} e^{2\pi i \alpha (P(m) + N^{d-1} m \cdot \eta)} \phi(m/N).$$

If $y = m + z$ where $m \in \mathbf{Z}^n$ and $z \in [0, 1]^n$, then it is easy to see that

$$\left| e^{2\pi i \alpha (P(y) + N^{d-1} y \cdot \eta)} - e^{2\pi i \alpha (P(m) + N^{d-1} m \cdot \eta)} \right| \ll N^{-1 + (d-1)\theta},$$

since $|\alpha| = N^{-d + (d-1)\theta}$ and $|\eta| \le C$. Thus

$$|I(L, \eta) - N^{-n} S(\alpha)| \ll N^{-1 + 2(d-1)\theta} \ll N^{-\frac{1}{2}} \le L^{-\frac{n}{b}}.$$

Also, by Corollary 20

$$|N^{-n} S(\alpha, \eta)| \ll |N^d \alpha|^{-\frac{n}{b}} = L^{-\frac{n}{b}}$$

and (8.90) follows. $\square$

We remark that a better uniform estimate can be obtained by using real variable methods, exploiting the fact that $P(x) \approx |x|^d$. However we have chosen to estimate integral using exponential sums as this method works also for indefinite forms P. Now, it is easy to prove.

Lemma 23. *We have, uniformly in* ξ

$$|B_\lambda(\xi) - M_\lambda(\xi)| \ll N^{n - d - n\gamma_d}$$

Proof. One has by (8.90)

$$\int_{|\beta| \ge N^{-d + (d-1)\theta}} |\tilde{H}(\xi - l/q)| \, d\beta \ll N^{n - \frac{n}{b}} \ll N^{n - d - n\gamma_d}.$$

The factors $\psi(q\xi - l)$ restrict the sum in l to at most one non-zero term, moreover by (8.84) we have $|G(a, q, l)| \ll q^{-\frac{n}{b} + \varepsilon} \ll q^{-3}$, say. Thus

$$|B_\lambda(\xi) - M_\lambda(\xi)| \ll \left(\sum_{q \le N^{(d-1)\theta}} \sum_{(a,q)=1} q^{-3} \right) N^{n - d - n\gamma_d} \ll N^{n - d - n\gamma_d}.$$

$\square$

Summarizing, we have the asymptotic formula

$$\hat{\omega}_\lambda(\xi) = M_\lambda(\xi) + E_\lambda(\xi),$$

where

$$M_\lambda(\xi) = \sum_{q \le N^{(d-1)\theta}} \sum_{(a,q)=1} \sum_{l \in \mathbf{Z}^n} G(a,q,l)\,\psi(q\xi - l)\,I_\lambda(\xi - l/q),$$

and

$$|E_\lambda(\xi)| \ll N^{n-d-n\gamma_d},$$

uniformly in $\xi \in \mathbf{T}^n$.

8.3.1.3 The Singular Integral

We will now identify the integrals $I_\lambda(\eta)$ with the Fourier transform of a certain natural measure supported on the surface $S_\lambda = \{P = \lambda\}$. Note that by assumption that the polynomial P is non-singular and positive, S_λ is a smooth, compact hypersurface in $\mathbf{R}^n$.

There is a unique $n - 1$-form $d\sigma_P(x)$ on $\mathbf{R}^n \backslash \{0\}$ such that

$$dP \wedge d\sigma_P = dx_1 \wedge \ldots \wedge dx_n, \tag{8.91}$$

called the Gelfand-Leray form (see [1, 2], Sec.7.1). To see this, suppose that say $\partial_1 P(x) \ne 0$ on some open set U. By a change of coordinates: $y_1 = P(x), y_j = x_j$ for $2 \le j \le n$, Eq. (8.91) takes the form

$$dy_1 \wedge d\sigma_P(y) = \partial_1 H(y)\,dy_1 \wedge \ldots \wedge dy_n$$

where $x_1 = H(y), x_j = y_j$ is the inverse map. Thus the form $d\sigma_P(y) = \partial_1 H(y)\,dy_2 \wedge \ldots \wedge dy_n$ satisfies (8.91).

We define the measure σ_λ as the restriction of the $n - 1$ form $d\sigma_P$ to the level surface S_λ. This measure is absolutely continuous with respect to the Euclidean surface area measure $dS_{P,\lambda}$, more precisely one has

Proposition 24.

$$d\sigma_\lambda(x) = \frac{dS_\lambda(x)}{|P'(x)|}, \tag{8.92}$$

where dS_λ denotes the Euclidean surface area measure on the level surface $\{P = \lambda\}$.

Proof. Choose local coordinates y as before; in coordinates y level surface S_λ and surface area measure dS_λ takes the form

$$S_\lambda = \{x_1 = H(\lambda, y_2, \ldots, y_n),\ x_j = y_j;\ 2 \le j \le n\},$$

and

$$dS_\lambda(y) = (1 + \sum_{j=2}^{n} \partial_j^2 H(\lambda, y))^{1/2} \, dy_2 \wedge \cdots \wedge dy_n.$$

Using the identity $P(H(y), y_2, \ldots, y_n) = y_1$, one has

$$\partial_1 P(x)\partial_1 H(y) = 1, \quad \partial_1 P(x)\partial_j H(y) + \partial_j P(x) = 0,$$

This implies that

$$\partial_1 H(y) = (1 + \sum_{j=2}^{n} \partial_j^2 H(y))^{1/2} \cdot |P'(x)|^{-1},$$

and (8.92) follows by taking $y_1 = \lambda$.   □

A crucial observation is that the measure $d\sigma_\lambda$, considered as a distribution on $\mathbf{R}^n$, has a simple oscillatory integral representation.

Lemma 25. *Let $P(x)$ be a non-singular, homogeneous polynomial, and let λ be a real number. Then in the sense of distributions*

$$\sigma_\lambda(x) = \int_{\mathbf{R}} e^{2\pi i \, (P(x)-\lambda)t} \, dt. \tag{8.93}$$

This means that for any smooth cut-off function $\chi(t)$ and test function $\phi(x)$ one has

$$\lim_{\varepsilon \to 0} \int \int e^{2\pi i(P(x)-\lambda)t} \chi(\epsilon t)\phi(x) \, dxdt = \int \phi(x)d\sigma_\lambda(x). \tag{8.94}$$

Proof. Let U be an open set on which $\partial_1 P \neq 0$, and by a partition of unity we can assume that $\mathrm{supp} \, \phi \subseteq U$. Changing variables $y_1 = P(x)$, $y_j = x_j$ the left side of (8.94) becomes

$$\lim_{\epsilon \to 0} \int \int e^{2\pi i(y_1-\lambda)t} \chi(\epsilon t)\tilde\phi(y)|\partial_1 H(y)| \, dydt = \int \tilde\phi(\lambda, y')|\partial_1 H(\lambda, y')|dy',$$

where $y' = (y_2, \ldots y_n)$.

The last equality can be seen by integrating in t and in y_1 first, and using the Fourier inversion formula:

$$\lim_{\epsilon \to 0} \int \int e^{2\pi i(y_1-\lambda)t} \chi(\epsilon t)g(y_1) \, dy_1dt = g(\lambda).$$

On the other hand $S_\lambda \cap U = \{x_1 = H(\lambda, y_2, \ldots y_n), x_j = y_j\}$, and $\sigma_\lambda(y) = |\partial_1 H(\lambda, y')| \, dy'$ in parameters $y' = (y_2, \ldots, y_n)$.   □

Now it is easy to identity the integrals $I_\lambda(\eta)$ defined in (8.89). Indeed by (8.94), we have

$$I_\lambda(\eta) = \int_{\mathbf{R}^n} \int_{\mathbf{R}} e^{-2\pi i\,(P(x)-\lambda)\beta}\, e^{2\pi i x\cdot\eta} \phi(x/P)\, d\beta\, d\eta$$

$$= \int_{\mathbf{R}^n} \sigma_\lambda(x) e^{2\pi i x\cdot\eta} \phi(x/P)\, d\eta = \tilde{\sigma}_\lambda(\eta)$$

Also, by homogeneity, $\tilde{\sigma}_\lambda(\eta) = \lambda^{n/d-1}\tilde{\sigma}(\lambda^{1/d}\eta)$, where σ is the Gelfand-Leray measure restricted the unit level surface $S = \{P = 1\}$. Thus we have shown

Theorem 26. *Let $d \geq 2$, $n \geq d(d-1)2^{d+1}$, and let P be a positive, homogeneous, non-singular polynomial of degree d. Then we have*

$$\hat{\omega}_\lambda(\xi) = M_\lambda(\xi) + E_\lambda(\xi), \tag{8.95}$$

where

$$M_\lambda(\xi) = \lambda^{\frac{n}{d}-1} \sum_{q\leq N^{d-1}} \sum_{\theta\,(a,q)=1} \sum_{l\in\mathbf{Z}^n} G(a,q,l)\, \psi(q\xi-l)\, \tilde{\sigma}(\lambda^{\frac{1}{d}}(\xi-l/q)), \tag{8.96}$$

and

$$|E_\lambda(\xi)| \ll N^{n-d-n\gamma_d} \tag{8.97}$$

uniformly in $\xi \in \mathbf{T}^n$, where $\gamma_d = \frac{1}{(d-1)2^{d+1}}$.

Let us remark that following the error estimates carefully, in fact it was shown that

$$|E_\lambda(\xi)| \ll N^{n-d-\frac{n}{b}+2} = N^{n-d-n\gamma_d'}$$

with some constant $\gamma_d' > \gamma_d$ for $n > d(d-1)2^{d+1}$. This will be utilized in our estimates on the discrepancy, to swallow certain small factors of size N^ε.

We will also need an estimate on the decay of the Fourier transform of the measure σ, later in our upper bounds on the discrepancy.

Lemma 27. *One has*

$$|\tilde{\sigma}(\xi)| \ll (1+|\xi|)^{-\frac{n}{b}+1}$$

Proof. Suppose $|\xi| > 1$, and choose a cut-off ϕ such that $\phi\sigma = \sigma$. Then by (8.94), we have

$$\tilde{\sigma}(\xi) = \int e^{-2\pi i\,x\cdot\xi}\phi(x)\, d\sigma(x)$$

$$= \lim_{\delta\to0} \int\int e^{-2\pi i\,x\cdot\xi} e^{2\pi i(P(x)-1)t}\phi(x)\chi(\delta t)\, dx dt$$

We decompose the range of integration into two parts

$$\tilde{\sigma}(\xi) = \int_{|t| \geq c|\xi|} \int_{\mathbf{R}^n} \quad + \int_{|t| \leq c|\xi|} \int_{\mathbf{R}^n} \quad = I_1 + I_2$$

Note that if $|t| \leq C|\xi|$, with a sufficiently small constant $c > 0$, then one has for the gradient of the phase

$$|(tP(x) - x \cdot \xi)'| = |P'(x) - \xi| \geq |\xi|/2,$$

thus integrating by parts K times yields

$$|I_2| \leq C_N (1 + |\xi|)^{-K+1}.$$

For $|t| \geq C|\xi|$ we have by (8.90)

$$\left| \int e^{2\pi i (tP(x) - x \cdot \xi)} \phi(x) \, dx \right| \ll |t|^{-\frac{n}{D}},$$

hence

$$I_1 \ll \int_{|t| \geq C|\xi|} |t|^{-\frac{n}{D}} \, dt \ll |\xi|^{-\frac{n}{D}+1},$$

with $D = (d-1)2^{d-1}$.   □

8.3.1.4   The Singular Series

In order to get nontrivial upper bounds on the discrepancy for the set of lattice points on hypersurfaces, one needs to ensure that there are many lattice points on the surface. We will do this, by showing the existence of a regular set of values Λ corresponding to a non-singular polynomial P. Most of what we discuss below is standard, for example it is implicit in [4], so we only include the details for the sake of completeness.

Recall that we have a fixed θ slightly smaller than $\frac{1}{2(d-1)}$, so that the asymptotic expansion (8.96) holds with an error term of size $O(N^{n-d-n\gamma_d})$, where $N = \lambda^{1/d}$ and $\gamma_d = \frac{1}{(d-1)2^{d+1}}$. Taking $\xi = 0$ this means that

$$\hat{\omega}_\lambda(0) = \lambda^{\frac{n}{d}-1} \sum_{1 \leq N^{(d-1)\theta}} K(q, 0, \lambda) + O(N^{n-d-n\gamma_d}),$$

where

$$K(q,0,\lambda) = \sum_{(a,q)=1} G(a,q,l) = q^{-n} \sum_{(a,q)=1} \sum_{s\in(\mathbf{Z}/q\mathbf{Z})^n} e^{2\pi i \frac{a(P(s)-\lambda)-s\cdot l}{q}}.$$

To exploit the multiplicativity of the terms $K(q,0,\lambda)$ we need to extend the summation for all $q \in \mathbf{N}$, and estimate the error obtained. This can be done by using (8.84) which yields

$$|K(q,0,\lambda)| \ll (\log q)^n q^{-\frac{n}{D}+1},$$

thus for a sufficiently small $\varepsilon > 0$

$$\sum_{q \geq N^{(d-1)\theta}} |K(q,0,\lambda)| \ll_\varepsilon N^{-(d-1)\theta(\frac{n}{D}-2-\varepsilon)} \ll N^{-n\gamma_d}$$

if $n > d(d-1)2^{d+1}$, by our choice of the parameters, D and γ_d. Indeed, we have that $(n/D-2) > 2n\gamma_d$, thus choosing θ sufficiently close to (but smaller than) $1/2(d-1)$, the above estimate holds. It is well-known, and easy to see from the Chinese Remainder Theorem, that $K(q_1,0,\lambda)K(q_2,0,\lambda) = K(q_1q_2,0,\lambda)$ for q_1 and q_2 being relative primes, which implies that

$$\sum_{q=1}^\infty K(q,0,\lambda) = \prod_{p \text{ prime}} (\sum_{r=0}^\infty K(p^r,0,\lambda)) =: \prod_{p \text{ prime}} K_p(\lambda),$$

where the last equality is used to define the arithmetic factors $K_p(\lambda) = \sum_{r=0}^\infty K(p^r,0,\lambda)$. Note that $K(1,0,\lambda) = 1$ and by estimate (8.84) we have that $K_p(\lambda) = 1 + O(p^{-\frac{n}{D}+2}) = 1 + O(p^{-2})$. Thus choosing $R = R_P$ sufficiently large, we have that

$$1/2 \leq \prod_{p>R \ p \text{ prime}} |K_p(\lambda)| \leq 2 \qquad (8.98)$$

An important and well-known fact, which we will explain below, is that the arithmetic factors $K_p(\lambda)$ can be interpreted as the density of solutions of the equation $P(m) = \lambda$ among the p-adic integers (see [4]). Thus the main term in the asymptotic formula (8.8) is the product of the densities of the solutions in the p-adic integers and the density of solutions among the real numbers and is an instance of the so-called local-global principle.

To see this, define

$$r(p^K,\lambda) := |\{m \in (\mathbf{Z}/(p^K\mathbf{Z}))^n : P(m) \equiv \lambda \pmod{p^N}\}|,$$

One has

Proposition 28.

$$\sum_{r=0}^{K} K(p^r, 0, \lambda) = p^{-n(K-1)} r(p^K, \lambda).$$

Proof. Note that

$$r(p^K, \lambda) = \sum_{m \ (mod \ p^K)} p^{-K} \sum_{b=1}^{p^K} e^{2\pi i (P(m) - \lambda) \frac{b}{p^K}},$$

since the inner sum is equal to p^K or 0 according to whether $P(m) \equiv \lambda \ (mod \ p^K)$ or not. Next one writes $b = ap^{K-r}$, where $(a, p) = 1, 1 \le a < p^r$ for $r = 0, 1, \ldots, K$, and collects the terms corresponding to a fixed r which turn out to be $K(p^r, 0, \lambda)$. $\qquad\square$

Let us remark that this implies $K_p(\lambda) = \lim_{K \to \infty} p^{-n(K-1)} r(p^K, \lambda)$, which can be viewed as the density of the solutions among the p-adic integers.

To count the number of solutions modulo p^K, one uses the p-adic version of Newton's method.

Lemma 29. *Let p be a prime, λ and let k, l be natural numbers such that $l > 2k$. Suppose there is an $m_0 \in \mathbf{Z}^n$ for which*

$$P(m_0) \equiv \lambda \ (mod \ p^l),$$

moreover suppose, that p^k is the highest power of p which divides all the partial derivatives $\partial_j P(m_0)$.
Then for $K \ge l$, one has $p^{-K(n-1)} r_P(p^K, \lambda) \ge p^{-l(n-1)}$.

Proof. For $K = l$ this is obvious. Suppose it is true for K, and consider all the solutions $m_1 \ (mod \ p^{N+1})$ of the form $m_1 = m + p^{K-k} s$ where $s \ (mod \ p)$. Then

$$P(m + p^{K-k} s) - \lambda = P(m) - \lambda + p^{K-k} P'(m) \cdot s = 0 \ (mod \ p^{K+1}),$$

which yields $a + b \cdot s = 0 \ (mod \ p)$ where $ap^K = P(m) - \lambda$ and $bp^k = P'(m)$. Then $b_j \ne 0 \ (mod \ p)$ for some j hence there are p^{n-1} solutions of this form. All obtained solutions are different $mod \ (p^{K+1})$, and m_1 satisfies the hypothesis of the lemma. $\qquad\square$

We remark that in case of $m = 1, k = 0$ the above argument shows that there are exactly $p^{(K-1)(n-1)}$ solutions m for which $m = m_0 \ (mod \ p)$ and $P(m) = \lambda (mod \ p^K)$. It is not hard to establish now the existence of a set of regular values for the polynomial P.

Lemma 30. *let $P(m)$ be a homogeneous non-singular polynomial of degree $d \geq 2$, then there exists an infinite arithmetic progression Λ and constants $0 < c_P < C_P$, such that for all $\lambda \in \Lambda$*

$$c_P \leq K(\lambda) \leq C_P$$

Proof. Let $\lambda_0 = P(m_0) \neq 0$ for some fixed $m_0 \neq 0$. Let $p_1, \ldots, p_J$ be the set of primes less then R. Let k be an integer s.t. p_j^k does not divide $d\lambda_0$, for all $j \leq J$, where d is degree of $P(m)$. By the homogeneity relation $P'(m_0) \cdot m = d\lambda_0$ it follows that p_j^k does not divide some partial derivative $\partial_i P(m_0)$. Fix l s.t. $l > 2k$ and define the arithmetic progression
$\Lambda = \{\lambda_0 + k \prod_{j=1}^{J} p_j^l : k \geq k_Q\}$. Then we claim that Λ is a set of regular values. Indeed by Proposition 28 one has for $\lambda \in \Lambda$

$$K_{p_j}(\lambda) = \lim_{N \to \infty} p_j^{-n(N-1)} r_Q(p_j^N, \lambda) \geq p_j^{-l(N-1)}.$$

This together with (8.98) ensures that the singular series $K(\lambda)$ remains bounded from below, and the error term becomes negligible by choosing $k = k_P$ large enough. $\qquad\square$

Let us remark that along the same lines it can be shown, that all large numbers are regular values of $P(m)$, if for each prime $p < R$ and each residue class $s \ (mod \ p)$, there is a solution of the equations $P(m) = s \ (mod \ p)$ such that $P'(m) \neq 0 \ (mod \ p)$. This is the case for example for $P(m) = \sum_j m_j^d$.

8.3.2 Upper Bounds for the Discrepancy

We will prove Theorem 12 by extending the arguments given in Sect. 8.2 to the case of a general homogeneous non-singular hypersurface. Our main tool again will be the asymptotic expansion (8.95)

$$\hat{\omega}_\lambda(\xi) = \lambda^{\frac{n}{d}-1} \sum_{q \leq N^{(d-1)\theta}} m_{q,\lambda}(\xi) + E_\lambda(\xi),$$

where

$$m_{q,\lambda}(\xi) = \sum_{l \in \mathbb{Z}^n} K(q, l, \lambda) \, \psi(q\xi - l) \, \tilde{\sigma}(\lambda^{\frac{1}{d}}(\xi - l/q)).$$

Note that $0 < \theta < \frac{1}{2(d-1)}$ and $N = \lambda^{1/d}$. Moreover we will need the decay estimates

$$|\tilde{\sigma}(\xi)| \ll (1 + |\xi|)^{-\frac{n}{D}+1} \tag{8.99}$$

$$|K_p(q, l, \lambda)| \ll_\varepsilon q^{-\frac{n}{D}+1+\varepsilon} \tag{8.100}$$

Recall that the discrepancy of the set $Z'_{P,\lambda} = \{\lambda^{-1/d}m; \ P(m) = \lambda\}$ with respect to caps $C_{a,\xi} = \{x \in S_P : \ |x \cdot \xi \geq a\}$ may be written as

$$D_P(\xi, \lambda) = \sum_{P(m)=\lambda} \chi_a(\lambda^{-1/d}m \cdot \xi) - N_\lambda \int_{S_P} \chi_a(x \cdot \xi) \, d\sigma(x),$$

where N_λ is the number of solutions of the diophantine equation $P(m) = \lambda$, and χ_a is the indicator function of an interval $[a, b]$, b being a fixed constant such that $|x \cdot \xi| \leq b$ for all $x \in S_P$ and $\xi \in S^{n-1}$.

We turn to the proof of Theorem 12. As before, it will be enough to estimate the "smoothed" discrepancy

$$D_P(\phi_{a,\delta}, \xi, \lambda) = \sum_{P(m)=\lambda} \phi_{a,\delta}(\lambda^{-\frac{1}{d}}m \cdot \xi) - N_\lambda \int_{S_P} \phi_{a,\delta}(x \cdot \xi) \, d\sigma(x),$$

for, say $\delta = \lambda^{-n}$. Taking the inverse Fourier transform of the functions $\phi_{a,\delta}$, we have

$$\sum_{P(m)=\lambda} \phi_{a,\delta} (\lambda^{-1/d} \, m \cdot \xi) = \int_{\mathbf{R}} \lambda^{\frac{1}{d}} \tilde{\phi}_{a,\delta}(t\lambda^{\frac{1}{d}}) \, \hat{\omega}_\lambda(t\xi) \, dt \tag{8.101}$$

also

$$\int_{S^{n-1}} \phi_{a,\delta} (x \cdot \xi) \, d\sigma(x) = \int_{\mathbf{R}} \tilde{\phi}_{a,\delta}(t) \, \tilde{\sigma}(t\xi) \, dt. \tag{8.102}$$

Moreover, as in (8.43) and (8.44), set

$$I_{q,\lambda} := \int_{\mathbf{R}} \lambda^{\frac{1}{d}} \tilde{\phi}_{a,\delta}(t\lambda^{\frac{1}{d}}) \, m_{q,\lambda}(t\xi) \, dt,$$

and

$$\mathscr{E}_\lambda := \int_{\mathbf{R}} \lambda^{\frac{1}{2}} \tilde{\phi}_{a,\delta}(t\lambda^{\frac{1}{2}}) \, E_\lambda(t\xi) \, dt.$$

First, we estimate the error term using (8.95)

$$|\mathscr{E}_\lambda| \ll N^{n-d-n\gamma_d} \int_{\mathbf{R}} (1 + |t|)^{-1}(1 + \delta|t|)^{-1} \tag{8.103}$$

$$\ll \lambda^{\frac{n}{d}-1-\frac{n}{d}} (\log \lambda) \ll \lambda^{(\frac{n}{d}-1)(1-\gamma_d)}, \tag{8.104}$$

Next, we decompose the integral $I_{q,\lambda}$ as in (8.46), and observe that for $|t| < 1/8q$

$$m_{q,\lambda}(t\xi) = K(q, 0, \lambda)\, \tilde\sigma(\lambda^{1/d}\, t\xi).$$

Lemma 31. *We have*

$$|\lambda^{\frac{n}{d}-1} \sum_{q \le N^{(d-1)\theta}} I_{q,\lambda}^1 - N_\lambda \int_{S_P} \phi_{a,\delta}\,(x \cdot \xi)\, d\sigma(x)| \ll \lambda^{(\frac{n}{d}-1)(1-\gamma_d)}. \tag{8.105}$$

Proof. By the above observation and a change of variables $t = \lambda^{1/d} t$, we have

$$\sum_{q \le N^{(d-1)\theta}} I_{q,\lambda}^1 = \sum_{q \le N^{(d-1)\theta}} K(q, 0, \lambda) \int_{|t| < N/8q} \tilde\phi_{a,\delta}(t)\, \tilde\sigma(t\xi)\, dt.$$

We extend the integration to the whole real line to exploit (8.102), the error obtained is bounded by

$$\int_{|t| \ge N/8q} |\tilde\phi_{a,\delta}(t)|\, |\tilde\sigma(t\xi)|\, dt \ll \int_{|t| \ge N/8q} (1 + |t|)^{-\frac{n}{D}}\, dt \ll N^{-\frac{n}{D}+1} q^{\frac{n}{D}-1}.$$

Thus

$$\left| \lambda^{\frac{n}{d}-1} \sum_{q \le N^{(d-1)\theta}} I_{q,\lambda}^1 - \sum_{q \le N^{(d-1)\theta}} K(q, 0, \lambda) \int_{S_p} \phi_{a,\delta}\,(x \cdot \xi)\, d\sigma_p(x) \right|$$

$$\ll_\varepsilon N^{-\frac{n}{D}+1} \sum_{q \le N^{(d-1)\theta}} q^{-\frac{n}{D}+1+\varepsilon} q^{\frac{n}{D}-1} \ll N^{-n\gamma_d}, \tag{8.106}$$

using the facts that $(d - 1)\theta < \frac{1}{2}$ and $\frac{n}{D} - 2 > n\gamma_d$, choosing $\varepsilon > 0$ sufficiently small. $\qquad\square$

Lemma 32. *One has*

$$\sum_{q \le N^{(d-1)\theta}} |I_{q,\lambda}^2| \ll N^{-n\gamma_d}. \tag{8.107}$$

Proof. Since $\psi(q\xi - l) = 0$ unless $l = [q\xi]$, the nearest lattice point to the point $q\xi$, we have that

$$m_{q,\lambda}(t\xi) = K(q, [qt\xi], \lambda)\, \psi(\{qt\xi\})\, \tilde\sigma\left(\frac{N}{q}\,\{qt\xi\}\right).$$

By making a change of variables $t := tq$, it follows from (8.99) and (8.100)

$$|I_{q,\lambda}^2| \ll_\varepsilon N^{-\frac{n}{b}+2} q^{-1+\varepsilon} J_\lambda$$

where

$$J_\lambda = \int_{|t| \geq 1/8} |\tilde{\phi}_{a,\delta}(tN/q)| \, \|t\xi\|^{-\frac{n}{b}+1} \, dt.$$

Note that for $q \leq N^{(d-1)\theta} < N^{1/2}$

$$|\tilde{\phi}_{a,\delta}(tN/q)| \ll \frac{q}{N} |t|^{-1}(1 + |\delta t|)^{-1}.$$

By a dyadic decomposition of the range of integration, using (8.36), we have

$$|J_\lambda| \ll_\varepsilon \frac{q}{N} \sum_{j \geq -3} 2^{\varepsilon j} (1 + \delta 2^j)^{-1} \ll q \, N^{-1+\varepsilon'},$$

with $\varepsilon' = nd\varepsilon$. Choosing $\varepsilon > 0$ sufficiently small, this implies

$$\sum_{q \leq N^{(d-1)\theta}} |I_{q,\lambda}^2| \ll_\varepsilon \sum_{q \leq N^{1/2}} q^\varepsilon N^{-\frac{n}{b}+1+\varepsilon'} \ll N^{-\frac{n}{b}+2} \ll N^{-n\gamma_d}. \qquad (8.108)$$

$\square$

Finally, we remark that Theorem 12 follows immediately from estimates (8.103)–(8.107).

$\square$

8.3.3   The Distribution of the Solutions Modulo 1

We will study the distribution of the images of the solutions of a diophantine equation $P(m) = \lambda$ on the flat torus $\mathbf{T}^n = \mathbf{R}^n/\mathbf{Z}^n$, via the map $T_\alpha : (m_1, \ldots, m_n) \to (m_1\alpha_1, \ldots, m_n\alpha_n) \pmod 1$, where $\alpha = (\alpha_1, \ldots, \alpha_n) \in \mathbf{R}^n$ is a given point. We will assume, as before, that P is a positive, homogeneous, non-singular polynomial of degree d, and $n \geq n_d$ is large enough with respect to the degree. Note that if one of the coordinates α_i is rational, say equal to a/q, then $m_i\alpha_i$ can take at most q different values modulo 1, so the images of the solution sets

$$\Omega_{\lambda,\alpha} := \{(m_1\alpha_1, \ldots, m_n\alpha_n) : P(m_1, \ldots, m_n) = \lambda\} \subseteq \mathbf{T}^n \qquad (8.109)$$

cannot become equi-distributed on the torus as $\lambda \to \infty$, even if one restricts to regular values only. In the opposite case, we have

Theorem 33. *Let $\alpha = (\alpha_1, \ldots, \alpha_n)$ be point such that α_i is irrational for all $1 \leq i \leq n$, and ϕ be a smooth function on $\mathbf{T}^n$. If Λ is a set of regular values of the form P, then one has*

$$\lim_{\lambda \to \infty, \lambda \in \Lambda} N_\lambda^{-1} \sum_{P(m) = \lambda} \phi(m_1\alpha_1, \ldots, m_n\alpha_n) = \int_{\mathbf{T}^n} \phi(x)\, dx, \qquad (8.110)$$

where N_λ is the number of solutions of the equation $P(m) = \lambda$.

Proof. For simplicity, let us introduce the notation $m \circ \alpha = (m_1\alpha_1, \ldots, m_n\alpha_n)$. By using the inverse Fourier transform $\phi(\beta) = \sum_{l \in \mathbf{Z}^n} \hat{\phi}(l) e^{2\pi i \beta \cdot l}$, we have

$$\sum_{P(m) = \lambda} \phi(m \circ \alpha) = \sum_{l \in \mathbf{Z}^n} \hat{\phi}(l) \sum_{P(m) = \lambda} e^{2\pi i (m_1 l_1 \alpha_1 + \ldots m_n l_n \alpha_n)}$$

$$= \sum_{l \in \mathbf{Z}^n} \hat{\phi}(l)\, \hat{\omega}_\lambda(l \circ \alpha) = N_\lambda \hat{\phi}(0) + T_\lambda(\alpha), \qquad (8.111)$$

where

$$T_\lambda(\alpha) = \sum_{l \in \mathbf{Z}^n, l \neq 0} \hat{\phi}(l)\, \hat{\omega}_\lambda(l \circ \alpha). \qquad (8.112)$$

Substituting the asymptotic expansion (8.95) into the above expression we have

$$T_\lambda(\alpha) = \sum_{q \leq N^{(d-1)\theta}} \sum_{l \neq 0} m_{q,\lambda}(l \circ \alpha)\hat{\phi}(l) + \sum_{l \neq 0} E_\lambda(l \circ \alpha)\hat{\phi}(l).$$

Using the fact that $\hat{\phi}(l) \leq C_M (1 + |l|)^{-M}$ for all $M \in \mathbf{N}$, estimate (8.97) implies

$$\sum_{l \neq 0} |E_\lambda(l \circ \alpha)\hat{\phi}(l)| \ll N^{n-d-n\gamma_d} \|\hat{\phi}\|_{l^1} \ll N^{n-d-n\gamma_d}, \qquad (8.113)$$

where $N = \lambda^{1/d}$ and $\gamma_d > 0$ is a constant depending on d. Also, by (8.100) one has

$$|m_{q,\lambda}(l \circ \alpha)| \ll_\varepsilon N^{n-d} q^{-\frac{n}{D}+1+\varepsilon} \hat{\sigma}\left(\frac{N}{q} \|ql \circ \alpha\|\right). \qquad (8.114)$$

Since $\alpha \in (\mathbf{R}/\mathbf{Q})^n$ by our assumption and $l \neq 0$ we have that $\|ql \circ \alpha\| > 0$, thus

$$m_{q,\lambda}(l \circ \alpha) \to 0 \quad \text{as} \quad \lambda \to \infty.$$

Let $\varepsilon > 0$ be fixed, then by (8.114) one estimates crudely

$$\sum_{q \geq N^\varepsilon} \sum_{l \neq 0} |m_{q,\lambda}(l \circ \alpha)\,\hat{\phi}(l)| \ll N^{n-d} \sum_{q \geq N^\varepsilon} q^{-\frac{n}{b}+1} \ll N^{n-d-n\varepsilon'}. \qquad (8.115)$$

Also, for a fixed $q \leq N^\varepsilon$

$$\sum_{|l| \geq N^\varepsilon} |m_{q,\lambda}(l \circ \alpha)\,\hat{\phi}(l)| \ll N^{n-d-\varepsilon}, \qquad (8.116)$$

by using the decay estimate $|\hat{\phi}(l)| \ll (1 + |l|)^{-2n}$.

Since for regular values $\lambda \in \Lambda$ the number of solutions is $N_\lambda \approx \lambda^{\frac{n}{d}-1} = N^{n-d}$, (8.110) follows from (8.114)–(8.116). $\qquad \square$

Let $\alpha = (\alpha_1, \dots, \alpha_n)$ be a point such that each of its coordinates α_i is diophantine in the sense that $\|l\alpha_i\| \geq C_\varepsilon |l|^{-1-\varepsilon}$ for $l \in \mathbf{Z}/\{0\}$, for every $\varepsilon > 0$. We will call such points α *diophantine*, and we can extend this definition to points $\alpha \in \mathbf{T}^n$ as α diophantine if and only if $\alpha + m$ is such for any $m \in \mathbf{Z}^n$. Note that this condition on α is different from the notion used in Sects. 8.2–8.3, nevertheless (8.3) implies that the set of diophantine points of the torus has measure 1. Also, it is immediate from the definition that for any $l = (l_1, \dots, l_n) \in \mathbf{Z}^n, l \neq 0$ we have that

$$\|l \circ \alpha\| \geq C_\varepsilon |l|^{-1-\varepsilon}. \qquad (8.117)$$

For diophantine points α we will derive quantitative estimates on the discrepancy of the sets $\Omega_{\lambda,\alpha}$ with respect to both smooth functions and compact, convex bodies. To be more precise, for a smooth function $\phi \in C^\infty(\mathbf{T}^n)$ define the associated discrepancy as

$$D(\phi, \alpha, \lambda) := \sum_{P(m)=\lambda} \phi(m \circ \alpha) - N_\lambda \int_{\mathbf{T}^n} \phi(x)\,dx. \qquad (8.118)$$

Theorem 34. *Let $\alpha \in \mathbf{T}^n$ be a diophantine point, and let $\phi \in C^\infty(\mathbf{T}^n)$. Then for $n > n_d = d(d-1)2^{d+1}$, one has*

$$|D(\phi, \alpha, \lambda)| \ll \lambda^{\frac{n}{d}-1-n\eta_d}, \qquad (8.119)$$

with a constant $\eta_d > 0$ depending only on the degree d.

Proof. We will argue as in the proof of Theorem 33, using condition (8.117) and the decay estimates (8.99) and (8.100). To start, observe that by (8.111)–(8.112)

$$D(\phi, \alpha, \lambda) = T_\lambda(\alpha) \leq \sum_{q \leq N^{(d-1)\theta}} \sum_{l \neq 0} |m_{q,\lambda}|\,|\hat{\phi}(l)| + O(N^{n-d-n\gamma_d}).$$

Since α is assumed to be diophantine we have for all $\varepsilon > 0$

$$|D(\phi, \alpha, \lambda) \ll N^{n-d} \sum_{q \le N^{(d-1)\theta}} \sum_{l \ne 0} q^{-\frac{n}{D}+1} \left(1 + \frac{N}{q} \|ql \circ \alpha\|\right)^{-\frac{n}{D}+1} |\hat{\phi}(l)| \quad (8.120)$$

$$\ll_\varepsilon N^{n-d} \sum_{q \le N^{(d-1)\theta}} \sum_{l \ne 0} q^{-\frac{n}{D}+1} \left(1 + \frac{N}{q^{2+\varepsilon}|l|^{1+\varepsilon}}\right)^{-\frac{n}{D}+1} (1 + |l|)^{-2n}.$$

Now the parameter θ in the asymptotic formula (8.95) was chosen such that $(d - 1)\theta < 1/2$, accordingly we will set $\varepsilon = (1 - 2(d-1)\theta)/4$. This will ensure that

$$\frac{N}{q^{2+\varepsilon}|l|^{1+\varepsilon}} \ge N^\varepsilon,$$

for $1 \le q \le N^{(d-1)\theta}$ and $0 < |l| < N^\varepsilon$, thus by (8.120)

$$\sum_{q \le N^{(d-1)\theta}} \sum_{0 < |l| < N^\varepsilon} |m_{q,\lambda}(l \circ \alpha)| |\hat{\phi}(l)| \ll N^{n-d-\varepsilon(n/D-1)} \ll N^{n-d-n\eta_d},$$

with, say $\eta_d = (1 - 2(d-1)\theta)/8D$. The rest of the sum is estimated crudely by

$$N^{n-d} \sum_{q \le N^{(d-1)\theta}} \sum_{|l| \ge N^\varepsilon} q^{-\frac{n}{D}+1}(1 + |l|)^{-2n} \ll N^{n-d-\varepsilon n}.$$

This finishes the proof of Theorem 34. $\square$

Finally, we will study the discrepancy of the image sets $\Omega_{\alpha,\lambda}$ with respect to compact, convex bodies $K \subseteq (-\frac{1}{2}, \frac{1}{2})^n$, when the flat torus $\mathbf{T}^n$ is identified as a set with $[-\frac{1}{2}, \frac{1}{2})^n$. Let us remark that in this case one cannot hope for better upper bounds than $O(\lambda^{\frac{n}{d}-1-\frac{1}{d}})$. Indeed, consider the discrepancy with respect to the family of cubes $K_c = [-c, c]^n$. The number of solutions of the equation $P(m) = \lambda$ is $\approx \lambda^{n/d-1}$ but (as $P(m) \approx |m|^d$) each coordinate can take $\ll \lambda^{1/d}$ values, thus the number of solutions $m = (m_1, \ldots, m_n)$ with m_1 being fixed is at least $\lambda^{\frac{n}{d}-1-\frac{1}{d}}$, for some value of m_1. Fix such an m_1 and let $c_1 = m_1\alpha_1 \pmod 1$. This means that the boundary of the cube K_{c_1} contains at least $\lambda^{\frac{n}{d}-1-\frac{1}{d}}$ points of the set $\Omega_{\alpha,\lambda}$ so the discrepancy changes by at least this much as c passes through c_1 and thus one cannot have a better uniform upper bound on it. We will prove a similar upper bound, of the form $O(\lambda^{\frac{n}{d}-1-\eta_d})$ with a constant $\eta_d > 0$ depending only on the degree d which as uniform over a large family of convex bodies.

We will use the fact that if $K \subseteq (-\frac{1}{2}, \frac{1}{2})^n$ is a closed convex set with non-empty interior then there exist convex sets K_1 and K_2 such that for sufficiently small $\delta > 0$

$$B(K_1, \delta) \subseteq K \subseteq B(K_2, \delta) \subseteq (-1, 1)^n,$$

where $B(K, \delta)$ is the set of points whose distance to the set K is at most δ. To make our estimates uniform for a large family of convex bodies, define the quantity δ_K as the largest $\delta > 0$ for which there exists a point x such that $x + B_\delta \subseteq K$ and also $K + B_\delta \subseteq [-\frac{1}{2}, \frac{1}{2}]^n$, where B_δ is the closed ball of radius δ centered at the origin.

Lemma 35. *Let $K \subseteq (-\frac{1}{2}, \frac{1}{2})^n$ be a closed convex body, and let x be a point in the interior of K. For given $0 < \delta < \delta_K/10$, $C_0 = 2/\delta_K$, and $\lambda_1 = \lambda_2^{-1} = 1 - C_0\delta$; define the convex bodies $K_1 = x + \lambda_1 K$, $K_2 = x + \lambda_2 K$.*

If $\phi \geq 0$ is a smooth cut-off function supported in $(-1, 1)^n$ such that $\int \phi = 1$, then we have

$$\chi_{K_1} * \phi_\delta \leq \chi_K \leq \chi_{K_2} * \phi_\delta, \tag{8.121}$$

where χ_K stands for the indicator function of a set K, and $\phi_\delta(x) = \delta^{-n}\phi(x/\delta)$.

Proof. From the definition it is immediate that $K_1 \subseteq K \subseteq K_2 \subseteq (-\frac{1}{2}, \frac{1}{2})^n$. By translation invariance we may assume that $x_0 = 0$ and then it is enough to show that $B(K_1, \delta) \subseteq K$ and $B(K, \delta) \subseteq K_2$. Since $K = x_0 + \lambda_1 K_2 = \lambda_1 K_2$ both claim can be shown the same way. Indeed, assume indirect that there is $y \in K_1$ and $z \notin K$ such that $|y - z| \leq \delta$. Then by the Hahn-Banach Theorem there is a unit vector v for which

$$v \cdot y + \delta \geq v \cdot z > \max_{x \in K} v \cdot x \geq \lambda_1^{-1} y \cdot z,$$

since $\lambda_1^{-1} y \in K$. Also, by our assumption $B_{\delta_K} \subseteq K$, hence

$$y \cdot z \geq v \cdot z - \delta > \delta_K - \delta \geq \delta_K/2.$$

This implies

$$\lambda_1 \delta \geq (1 - \lambda_1) y \cdot z \geq C_0 \delta \delta_K/2,$$

which is a contradiction since $\lambda_1 < 1$ and $C_0 \delta_K \geq 2$. The same argument shows that $B(K, \delta) \subseteq K_2$ and (8.121) follows. $\square$

For a closed, convex body $K \subseteq (-\frac{1}{2}, \frac{1}{2})^n$ and a diophantine point α, define the discrepancy

$$D(K, \alpha, \lambda) = \sum_{P(m)=\lambda} \chi_K(m_1\alpha_1, \ldots, m_n\alpha_n) - N_\lambda vol_n(K),$$

where χ_K is the indicator function of K considered as a function on $\mathbf{T}^n$, and $vol_n(K)$ denotes the volume of the body. We have the following uniform estimate on the discrepancy.

Theorem 36. *Let* $n > d(d-1)2^{d+1}$ *and let* P *be a non-singular integral polynomial in* n *variables, and let* $\alpha \in \mathbf{R}^n$ *be diophantine and let* $\delta_0 > 0$. *Then for a closed, convex body* $K \subseteq (-\frac{1}{2}, \frac{1}{2})^n$ *such that* $\delta_K \geq \delta_0$ *we have*

$$|D(K, \alpha, \lambda)| \ll N^{n-d-\eta_d}, \tag{8.122}$$

where $\eta_d > 0$ *is a constant depending only on* d, *and the implicit constant in* (8.122) *depends only on the polynomial* P, *the point* α *and on* δ_0 *and is independent of* K.

Proof. Let us use the notation $\phi_{K,\delta} = \chi_K * \phi_\delta$. By (8.121) we have for $\delta < c\delta_0$ ($c > 0$ being sufficiently small)

$$\sum_{P(m)=\lambda} \phi_{K_1,\delta}(m \circ \alpha) - N_\lambda \int_{\mathbf{T}^n} \phi_{K_2,\delta} \leq D(K, \alpha, \lambda) \leq \sum_{P(m)=\lambda} \phi_{K_2,\delta}(m \circ \alpha) - N_\lambda \int_{\mathbf{T}^n} \phi_{K_1,\delta}.$$

and also

$$\int_{\mathbf{T}^n} (\phi_{K_2,\delta} - \phi_{K_1,\delta}) \leq C\delta \, vol_n(K),$$

with a constant $C \ll \delta_0^{-1}$. Thus

$$|D(K, \alpha, \lambda)| \leq \max_{i=1,2} |D(\phi_{K_i,\delta}, \alpha, \lambda)| + O(N^{n-d}\delta). \tag{8.123}$$

To estimate the discrepancy with respect to the smooth functions $\phi_{K_i,\delta}$ we proceed as before, with exception that now we have the estimates on their Fourier transform

$$|\hat{\phi}_{K_i,\delta}(l)| = |\hat{\phi}_{K_i}(l)\hat{\phi}(\delta l)| \ll (1 + \delta|l|)^{-2n},$$

in particular $\|\hat{\phi}_{K_i,\delta}\|_{l^1} \ll \delta^{-n}$. Thus

$$\left| \sum_{l \neq 0} E_\lambda(l \circ \alpha) \hat{\phi}_{K_i,\delta} \right| \ll N^{n-d-n\gamma_d} \delta^{-n}. \tag{8.124}$$

For the main terms, we have

$$\sum_{0 < |l| < N^\varepsilon} |m_{q,\lambda}(l \circ \alpha) \hat{\phi}_{K_i,\delta}(l)| \ll N^{n-d} q^{-\frac{n}{D}+1} \left(1 + \frac{N}{q^{2+\varepsilon}|l|^{1+\varepsilon}} \right)^{-\frac{n}{D}+1} (1 + \delta|l|)^{-2n}$$

$$\ll N^{n-d-\varepsilon(\frac{n}{D}-1)} q^{-\frac{n}{D}+1} \delta^{-n}, \tag{8.125}$$

for $q \leq N^{(d-1)\theta}$, choosing $\varepsilon = (1 - 2(d - 1)\theta)/4$ as before. Also

$$\sum_{|l| \geq N^\varepsilon} |m_{q,\lambda}(l \circ \alpha) \, \hat{\phi}_{K_i, \delta}(l)| \ll N^{n-d} q^{-\frac{n}{D}+1} \sum_{|l| \geq N^\varepsilon} (1 + \delta |l|)^{-2n}$$

$$\ll N^{n-d} q^{-\frac{n}{D}+1} (1 + \delta N^\varepsilon)^{-2n} N^{\varepsilon n}. \quad (8.126)$$

Let $\delta = N^{-\frac{\varepsilon}{4D}}$ then the right side of both (8.125) and (8.126) is $O(N^{n-d-\frac{\varepsilon}{4D}}$ $q^{-\frac{n}{D}+1})$. Summing for $1 \leq q \leq N^{(d-1)\theta}$ and using (8.123) we obtain the estimate

$$|D(K, \alpha, \lambda)| \ll N^{n-d-\frac{\varepsilon}{4D}}.$$

Finally note that the exponent $\eta_d := \frac{\varepsilon}{4D}$ depend only on the parameter θ and D, hence ultimately only on dimension d, while the implicit constants in our estimates depend on the parameter δ_0 and not on the body K. This finishes the proof of Theorem 36. □

8.3.4 Some Possible Further Directions

Our estimates on the uniformity of the distribution of solutions to diophantine equations in many variables are by no means exhaustive. In fact even in the case of the sphere, it is not clear if our upper bounds are sharp or even what should be the sharp bounds. A closely related problem is to find lower bounds for the mean square average of the discrepancy of the lattice points on spheres over the family of all spherical caps. It is expected that the lattice points are far from optimally distributed and essentially higher lower bounds can be obtained then the uniform lower bounds given in [3, 15] and [12]. To obtain nontrivial lower bounds one may exploit the fact that lattice points in small caps are concentrated on lower dimensional spheres.

For higher degree polynomials it is unrealistic to expect sharp bounds in the generality we have discussed. The special case of the polynomial $P(m) = m_1^d + \ldots + m_n^d$ (d even) deserves special attention as the number of the solutions of the equation $P(m) = \lambda$, the so-called Waring problem, has been studied extensively. In fact much sharper asymptotic formulas have been obtained than the ones which can be derived from the Birch-Davenport method [18]. In general we have only considered positive polynomials of even degree, however there are natural analogues for indefinite forms. Indeed one may identify the solution set of the diophantine equation $P(m_1, \ldots, m_n) = 0$ within the box $|m_i| \leq N$ as the set of lattice points $\mathbf{Z}_P(N) = S \cap \mathbf{Z}^n \cap [-N, N]^n$ where $S = \{P = 0\}$ is the zero surface of P. One can shrink this set by a factor of N and study the discrepancy with respect to caps as $N \to \infty$.

Let us also remark that weaker bounds we have obtained for the distribution of the solutions modulo 1 seemed partly because of we have allowed very rough convex sets K. It might be true that better upper bounds can be given by assuming some smoothness of the boundary of the convex body, however it is not even immediately clear how to improve the bounds on the discrepancy with respect to balls.

Finally, as we mentioned in the introduction, the uniformity of distribution of the solutions modulo 1, is a special case of a more general phenomenon. It can be shown [11] that the images of the solution sets $\{P(m) = \lambda\}$ become equi-distributed when mapped to a probability measure space X via a fully ergodic commuting family of measure preserving transformations. It would be interesting to see if one can get estimates for the rate of equi-distribution for other measure preserving systems than the flat torus with the coordinate shifts.

References

1. V.I. Arnol'd, S.M. Gusejn-Zade, A.N. Varchenko, *Singularities of differentiable maps. Volume I: The classification of critical points, caustics and wave fronts. Transl. from the Russian by Ian Porteous, ed. by V. I. Arnol'd*. Monographs in Mathematics, vol. 82 (Birkhäuser, Boston-Basel-Stuttgart, 1985), p. 382
2. V.I. Arnol'd, S.M. Gusejn-Zade, A.N. Varchenko, *Singularities of differentiable maps. Volume II: Monodromy and asymptotics of integrals. Transl. from the Russian by Hugh Porteous.* Monographs in Mathematics, vol. 83 (Birkhäuser Verlag,Boston, MA, 1988), p. 492
3. J. Beck, Sums of distances between points on a sphere—an application of the theory of irregularities of distribution to discrete geometry. Mathematika **31**, 33–41 (1984). doi:10.1112/S0025579300010639
4. B.J. Birch, Forms in many variables. Proc. Roy. Soc. Lond. Ser. A **265**, 245–263 (1962). doi:10.1098/rspa.1962.0007
5. H. Davenport, Cubic forms in thirty-two variables. Phil. Trans. Roy. Soc. Lond. Ser. A **251**, 193–232 (1959). doi:10.1098/rsta.1959.0002
6. W. Duke, R. Schulze-Pillot, Representation of integers by positive ternary quadratic forms and equidistribution of lattice points on ellipsoids. Invent. Math. **99**(1), 49–57 (1990). doi:10.1007/BF01234411
7. E.P. Golubeva, O.M. Fomenko, Application of spherical functions to a problem of the theory of quadratic forms. Zap. Nauchn. Semin. Leningr. Otd. Mat. Inst. Steklova **144**, 38–45 (1985)
8. G.H. Hardy, A. Wiles, E.M. Wright, *An introduction to the theory of numbers. Edited and revised by D. R. Heath-Brown and J. H. Silverman. With a foreword by Andrew Wiles. 6th ed.* (Oxford University Press, Oxford, 2008), p. 621
9. D.R. Heath-Brown, Cubic forms in 10 variables. Proc. Lond. Math. Soc. III. Ser. **47**, 225–257 (1983). doi:10.1112/plms/s3-47.2.225
10. W. Hurewicz, H. Wallman, *Dimension Theory*. Princeton Mathematical Series, vol. 4 (Princeton University Press, Princeton, NJ, 1941), p. 165
11. A. Magyar, Diophantine equations and ergodic theorems. Am. J. Math. **124**(5), 921–953 (2002). doi:10.1353/ajm.2002.0029
12. A. Magyar, On the discrepancy of point distributions on spheres and hyperbolic spaces. Monatsh. Math. **136**(4), 287–296 (2002). doi:10.1007/s00605-002-0480-5
13. A. Magyar, On the distribution of lattice points on spheres and level surfaces of polynomials. J. Number Theory **122**(1), 69–83 (2007). doi:10.1016/j.jnt.2006.03.006
14. P. Sarnak, *Some Applications of Modular Forms*. Cambridge Tracts in Mathematics, vol. 99 (Cambridge University Press, Cambridge, 1990), p. 111
15. W.M. Schmidt, Irregularities of distribution. IV. Invent. Math. **7**, 55–82 (1969). doi:10.1007/BF01418774

16. W.M. Schmidt, The density of integer points on homogeneous varieties. Acta Math. **154**, 243–296 (1985). doi:10.1007/BF02392473
17. E.M. Stein, *Harmonic analysis: Real-variable methods, orthogonality, and oscillatory integrals. With the assistance of Timothy S. Murphy.* Princeton Mathematical Series, vol. 43 (Princeton University Press, Princeton, NJ, 1993), p. 695
18. T.D. Wooley, On Vinogradov's mean value theorem. Mathematika **39**(2), 379–399 (1992). doi:10.1112/S0025579300015102

Chapter 9
Discrepancy Theory and Quasi-Monte Carlo Integration

Josef Dick and Friedrich Pillichshammer

Abstract In this chapter we show the deep connections between discrepancy theory on the one hand and quasi-Monte Carlo integration on the other. Discrepancy theory was established as an area of research going back to the seminal paper by Weyl [117], whereas Monte Carlo (and later quasi-Monte Carlo) was invented in the 1940s by John von Neumann and Stanislaw Ulam to solve practical problems. The connection between these areas is well understood and will be presented here. We further include state of the art methods for quasi-Monte Carlo integration.

9.1 Introduction: The Connection between Discrepancy Theory and Quasi-Monte Carlo Integration

Let us start with introducing the concepts of discrepancy and quasi-Monte Carlo (QMC) for the domain $[0,1]^s$ and for a point set $\mathscr{P} = \{x_1, \ldots, x_N\}$. To define discrepancy, we define a set of 'test sets' $\mathscr{B}$. For instance, a common choice is the set of all intervals anchored at $\mathbf{0} = (0, \ldots, 0)$, denoted by $[\mathbf{0}, t) = \prod_{i=1}^{s}[0, t_i)$, where $t = (t_1, \ldots, t_s)$. The *local discrepancy* then is

$$\Delta_{\mathscr{P}}(t) = \frac{1}{N} \sum_{n=1}^{N} 1_{[\mathbf{0},t)}(x_n) - \prod_{i=1}^{s} t_i,$$

J. Dick (✉)
School of Mathematics and Statistics, The University of New South Wales, Sydney, NSW 2052, Australia
e-mail: josef.dick@unsw.edu.au

F. Pillichshammer
Institute for Financial Mathematics, University of Linz, Altenbergerstraße 69, 4040 Linz, Austria
e-mail: friedrich.pillichshammer@jku.at

W. Chen et al. (eds.), *A Panorama of Discrepancy Theory*, Lecture Notes in Mathematics 2107, DOI 10.1007/978-3-319-04696-9_9,
© Springer International Publishing Switzerland 2014

Fig. 9.1 The local
discrepancy $\Delta_{\mathscr{P}}(t)$ measures
the difference between the
relative number of points that
belong to the interval $[\mathbf{0}, t)$
and its volume

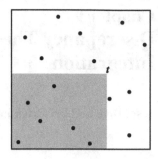

where $1_{[0,t)}$ denotes the *characteristic function* of the interval $[\mathbf{0}, t)$, i.e. $1_{[0,t)}(\mathbf{x})$ is
one if $\mathbf{x}$ belongs to $[\mathbf{0}, t)$ and zero otherwise; see Fig. 9.1. The L_p-*discrepancy* of
$\mathscr{P}$ is then the L_p norm $1 \leq p \leq \infty$ of $\Delta_{\mathscr{P}}$ given by

$$L_p(\mathscr{P}) = \|\Delta_{\mathscr{P}}\|_{L_p} = \left(\int_{[0,1]^s} |\Delta_{\mathscr{P}}(t)|^p \, dt \right)^{1/p}.$$

The L_∞-norm of the discrepancy function is also called the *star-discrepancy*, which
is denoted by $D_N^*(\mathscr{P})$, i.e., $D_N^*(\mathscr{P}) = \|\Delta_{\mathscr{P}}\|_{L_\infty} = \sup |\Delta_{\mathscr{P}}(t)|$, where the
supremum is extended over all $t \in [0, 1]^s$.

A *quasi-Monte Carlo rule* based on a point set $\mathscr{P} = \{\mathbf{x}_1, \ldots, \mathbf{x}_N\}$ is an equal
weight quadrature rule

$$Q_{\mathscr{P}}(f) := \frac{1}{N} \sum_{n=1}^{N} f(\mathbf{x}_n),$$

which can be used to approximate the integral $\int_{[0,1]^s} f(\mathbf{x}) \, d\mathbf{x}$. It is assumed that
the quadrature points are chosen in some deterministic way which yields a small
integration error for certain function classes.

To illustrate the connection between discrepancy and integration error we first
consider discrepancy and numerical integration on the unit interval $[0, 1]$.

9.1.1 An Elementary Approach

Central to showing the connection between the discrepancy of a point set and the
integration error is the characteristic function of an interval. For numbers $x \in \mathbb{R}$ the
characteristic function of an interval I is given by

$$1_I(x) = \begin{cases} 1 \text{ if } x \in I, \\ 0 \text{ otherwise.} \end{cases}$$

To give a glimpse of this connection between discrepancy and integration error, note the following two properties of the characteristic function:

1. Let $\mathscr{P} = \{x_1, \ldots, x_N\} \subseteq [0, 1]$ be a point set. Then

$$\Delta_{\mathscr{P}}(t) := \frac{1}{N} \sum_{n=1}^{N} 1_{[0,t)}(x_n) - t$$

measures the *discrepancy* between the proportion of the points in the interval $[0, t)$ and the length of the interval.

2. Let $f : [0, 1] \to \mathbb{R}$ be continuously differentiable. Then

$$f(x) = f(1) - \int_0^1 f'(t) 1_{(x,1]}(t) \, dt = f(1) - \int_0^1 f'(t) 1_{[0,t)}(x) \, dt. \quad (9.1)$$

These two properties can now be connected naturally in the following way. Let $f : [0, 1] \to \mathbb{R}$ be continuously differentiable. Consider the *integration error* of f using a quasi-Monte Carlo rule $Q_{\mathscr{P}}(f) = \frac{1}{N} \sum_{n=1}^{N} f(x_n)$, given by

$$e(f; \mathscr{P}) = \int_0^1 f(x) \, dx - \frac{1}{N} \sum_{n=1}^{N} f(x_n).$$

Then, using (9.1), we obtain

$$e(f; \mathscr{P})$$

$$= \int_0^1 \left[f(1) - \int_0^1 f'(t) 1_{[0,t)}(x) \, dt \right] dx - \frac{1}{N} \sum_{n=1}^{N} \left[f(1) - \int_0^1 f'(t) 1_{[0,t)}(x_n) \, dt \right]$$

$$= f(1) - \int_0^1 \int_0^1 f'(t) 1_{[0,t)}(x) \, dt \, dx - f(1) + \frac{1}{N} \sum_{n=1}^{N} \int_0^1 f'(t) 1_{[0,t)}(x_n) \, dt$$

$$= \int_0^1 f'(t) \left[\frac{1}{N} \sum_{n=1}^{N} 1_{[0,t)}(x_n) - \int_0^1 1_{[0,t)}(x) \, dx \right] dt$$

$$= \int_0^1 f'(t) \Delta_{\mathscr{P}}(t) \, dt.$$

By using Hölder's inequality we obtain

$$|e(f; \mathscr{P})| \le L_p(\mathscr{P}) \|f'\|_{L_q}, \quad (9.2)$$

where $1 \leq p, q \leq \infty$, $1/p + 1/q = 1$ and $\|g\|_{L_q} = \left(\int_0^1 |g(t)|^q \, dt \right)^{1/q}$ with the obvious modification for $q = \infty$. For $p = \infty$, inequality (9.2) is a simplified version of Koksma's inequality (see Kuipers and Niederreiter [65, Chapter 2, Theorem 5.1]).

Some remarks regarding the last inequality are in order. We have obtained an upper bound on the integration error which is a product of two factors,

- one of which, $\|f'\|_{L_q}$ depends only on the integrand f; it is a semi-norm of f, and
- one of which, the L_p-discrepancy $L_p(\mathscr{P})$ of $\mathscr{P}$, depends only on the point set $\mathscr{P}$.

Thus (9.2) shows that quadrature points with small L_p-discrepancy will yield a small integration error for functions with finite semi-norm $\|f'\|_{L_q}$.

Notice that there is a simple reason why a semi-norm rather than a norm is sufficient in (9.2): for any constant $c \in \mathbb{R}$ we have $e(f + c; \mathscr{P}) = e(f; \mathscr{P})$, i.e., constant functions are integrated exactly by the quasi-Monte Carlo rule $Q_{\mathscr{P}}$.

9.1.2 A Reproducing Kernel Approach

Until now we conveniently assumed that f is continuously differentiable. However, there is a practical framework called reproducing kernel Hilbert spaces by Aronszajn [3], which defines a whole class of functions. On the domain $[0, 1]$, a *reproducing kernel* is a function $K : [0, 1] \times [0, 1] \to \mathbb{C}$ which is

- symmetric: $K(x, y) = \overline{K(y, x)}$ for all $x, y \in [0, 1]$, and
- positive semi-definite, that is,

$$\sum_{k,l=1}^N a_k \bar{a}_l K(x_k, x_l) \geq 0$$

for all $a_1, \ldots, a_N \in \mathbb{C}$ and $x_1, \ldots, x_N \in [0, 1]$. (Here, $\bar{a}_l$ denotes the conjugate complex of a_l.)

A reproducing kernel can naturally be defined using the characteristic function $1_{[0,x)}$ by setting

$$K(x, y) = 1 + \int_0^1 1_{(x,1]}(t) 1_{(y,1]}(t) \, dt = 1 + \min(1 - x, 1 - y).$$

The function K such defined is symmetric and positive definite, and thus a reproducing kernel. Associated with this reproducing kernel is a set $\mathcal{H}(K)$ of functions $f : [0, 1] \to \mathbb{R}$ and an inner product $\langle \cdot, \cdot \rangle_{\mathcal{H}(K)}$ on $\mathcal{H}(K)$ such that

- $K(\cdot, y) \in \mathcal{H}(K)$ for all $y \in [0, 1]$, and
- $f(y) = \langle f, K(\cdot, y) \rangle_{\mathcal{H}(K)}$ for all $y \in [0, 1]$ and $f \in \mathcal{H}(K)$.

From Aronszajn [3] it is known that the function space $\mathcal{H}(K)$ is a Hilbert space with an inner product which is uniquely defined. For functions f, g, which can be represented in the form (9.1), the inner product is given by

$$\langle f, g \rangle_{\mathcal{H}(K)} = f(1)g(1) + \int_0^1 f'(x)g'(x)\,dx.$$

These functions f and g are absolutely continuous and $f', g' \in L_2([0, 1])$, the space of square integrable functions defined on $[0, 1]$.

Let $y \in [0, 1]$ be fixed. Then $k(x) := K(x, y)$ has the representation

$$k(x) = 1 - \int_0^1 1_{(x,1]}(t)[-1_{(y,1]}(t)]\,dt.$$

Thus, by matching it with the pattern from (9.1), $k(1) = 1$ and $k'(x) = -1_{[y,1]}(x)$. Thus

$$\langle f, K(\cdot, y) \rangle_{\mathcal{H}(K)} = f(1)1 - \int_0^1 f'(x)1_{[y,1]}(x)\,dx = f(y).$$

Then the integration error of f using a quasi-Monte Carlo rule based on $\mathscr{P} = \{x_1, \ldots, x_n\}$ is given by

$$
\begin{aligned}
e(f; \mathscr{P}) &= \int_0^1 f(x)\,dx - \frac{1}{N}\sum_{n=1}^N f(x_n) \\
&= \int_0^1 \langle f, K(\cdot, x) \rangle_{\mathcal{H}}\,dx - \frac{1}{N}\sum_{n=1}^N \langle f, K(\cdot, x_n) \rangle_{\mathcal{H}(K)} \\
&= \left\langle f, \int_0^1 K(\cdot, x)\,dx - \frac{1}{N}\sum_{n=1}^N K(\cdot, x_n) \right\rangle_{\mathcal{H}(K)}.
\end{aligned}
$$

We have

$$
\begin{aligned}
h(z) &:= \int_0^1 K(z, x)\,dx - \frac{1}{N}\sum_{n=1}^N K(z, x_n) \\
&= -\int_0^1 1_{(z,1]}(t)\left[\frac{1}{N}\sum_{n=1}^N 1_{(x_n,1]}(t) - \int_0^1 1_{(x,1]}(t)\,dx\right] dt
\end{aligned}
$$

$$= -\int_0^1 1_{(z,1]}(t)\left[\frac{1}{N}\sum_{n=1}^N 1_{[0,t)}(x_n) - t\right]dt$$

$$= -\int_0^1 1_{(z,1]}(t)\Delta_{\mathscr{P}}(t)\,dt.$$

Consequently, matching the representation of h given above with the pattern from (9.1), we obtain $h(1) = 0$ and $h'(x) = \Delta_{\mathscr{P}}(x)$. Hence we have

$$e(f;\mathscr{P}) = \langle f, h\rangle_{\mathscr{H}(K)} = f(1)0 + \int_0^1 f'(x)\Delta_{\mathscr{P}}(x)\,dx = \int_0^1 f'(x)\Delta_{\mathscr{P}}(x)\,dx.$$

Thus, taking the absolute value and using Hölder's inequality we again obtain (9.2).

So far we have considered the integration error for a particular function f. Since we have now a function space $\mathscr{H}(K)$ with inner product $\langle\cdot,\cdot\rangle_{\mathscr{H}(K)}$, we can define the corresponding norm by $\|\cdot\|_{\mathscr{H}(K)} = \langle\cdot,\cdot\rangle_{\mathscr{H}(K)}^{1/2}$. Then it is meaningful to define the worst-case error in the unit ball of $\mathscr{H}(K)$ by

$$e(\mathscr{H}(K);\mathscr{P}) = \sup_{f\in\mathscr{H}(K),\|f\|_{\mathscr{H}(K)}\leq 1}|e(f;\mathscr{P})|.$$

Since $e(f;\mathscr{P}) = \langle f, h\rangle_{\mathscr{H}(K)}$ we obtain

$$e(\mathscr{H}(K);\mathscr{P}) = \sup_{f\in\mathscr{H}(K),\|f\|_{\mathscr{H}(K)}\leq 1}|\langle f, h\rangle_{\mathscr{H}(K)}|$$

$$\leq \sup_{f\in\mathscr{H}(K),\|f\|_{\mathscr{H}(K)}\leq 1}\|f\|_{\mathscr{H}(K)}\|h\|_{\mathscr{H}(K)} = \|h\|_{\mathscr{H}(K)}.$$

On the other hand, we have $h \in \mathscr{H}(K)$ and by choosing $f = h/\|h\|_{\mathscr{H}(K)}$ we obtain that

$$e(\mathscr{H}(K);\mathscr{P}) = \|h\|_{\mathscr{H}(K)}.$$

This yields the formula

$$e^2(\mathscr{H}(K);\mathscr{P}) = \langle h, h\rangle_{\mathscr{H}(K)} \tag{9.3}$$

$$= \int_0^1\int_0^1 K(x,y)\,dx\,dy - \frac{2}{N}\sum_{n=1}^N\int_0^1 K(x,x_n)\,dx + \frac{1}{N^2}\sum_{n,m=1}^N K(x_n,x_m).$$

As $h(1) = 0$ and $h' = \Delta_{\mathscr{P}}$ we have $\|h\|_{\mathscr{H}(K)} = \|\Delta_{\mathscr{P}}\|_{L_2} = L_2(\mathscr{P})$. Thus

$$e(\mathscr{H}(K);\mathscr{P}) = L_2(\mathscr{P})$$

and so (9.3) yields an explicit expression for the L_2-discrepancy (which is the one-dimensional version of a formula that is sometimes attributed to Warnock; see Matoušek [75, Lemma 2.14])

$$(L_2(\mathcal{P}))^2 = \frac{4}{3} - \frac{2}{N} \sum_{n=1}^{N} \frac{3 - x_n^2}{2} + \frac{1}{N^2} \sum_{n,m=1}^{N} [1 + \min(1 - x_n, 1 - x_m)].$$

Since the reproducing kernel function K has a closed form, the worst-case error can be computed for given point sets $\mathcal{P}$.

In the following we also consider another, related, reproducing kernel, namely

$$K(x, y) = \min(1 - x, 1 - y).$$

The corresponding reproducing kernel Hilbert space consists of the same functions f as in the reproducing kernel Hilbert space as above with the restriction that $f(1) = 0$. The corresponding inner product is then simply $\int_0^1 f'(x)g'(x)\,dx$.

9.1.3 Discrepancy and Numerical Integration in Arbitrary Dimension

The step from $[0, 1]$ to $[0, 1]^s$ for some $s \geq 1$ is achieved by considering tensor product function spaces. Let now $\mathcal{P} = \{x_1, \dots, x_N\} \subseteq [0, 1]^s$. The reproducing kernel for functions on $[0, 1]^s$ is simply given by

$$K(x, y) = \int_{[0,1]^s} 1_{(x,1]}(t) 1_{(y,1]}(t)\,dt = \prod_{i=1}^{s} \min(1 - x_i, 1 - y_i), \qquad (9.4)$$

where $x = (x_1, \dots, x_s), y = (y_1, \dots y_s)$, and $(x, 1] = \prod_{i=1}^{s}(x_i, 1]$. The corresponding Hilbert space $\mathcal{H}(K)$ is the s-fold tensor product of the one dimensional reproducing kernel Hilbert spaces with reproducing kernel $K(x, y) = \min(1 - x, 1 - y)$. In particular, if $f \in \mathcal{H}(K)$, then $\|\partial^s f/\partial x\|_{L_2} < \infty$ and $\frac{\partial^{|u|} f}{\partial x_u}(z_u, 1) = 0$ for $u \subseteq \{1, \dots, s\}$, where $\partial x_u = \prod_{i \in u} \partial x_i$ and where $(z_u, 1)$ stands for the vector whose ith component is z_i if $i \in u$ and 1 otherwise. Further, for $u = \emptyset$ we have $\frac{\partial^{|u|} f}{\partial x_u}(z_u, 1) := f(1)$. The inner product is given by

$$\langle f, g \rangle_{\mathcal{H}(K)} = \int_{[0,1]^s} \frac{\partial^s f}{\partial x}(t) \frac{\partial^s g}{\partial x}(t)\,dt.$$

The same steps as in the previous two subsections can be carried out to obtain the discrepancy function

$$\Delta_{\mathscr{P}}(t) = \frac{1}{N} \sum_{n=1}^{N} 1_{[0,t)}(x_n) - \prod_{i=1}^{s} t_i.$$

Again, an analogue of (9.2) holds, namely for the integration error of a function $f \in \mathscr{H}(K)$ using a quasi-Monte Carlo rule based on $\mathscr{P}$ we have

$$|e(f; \mathscr{P})| \le L_p(\mathscr{P}) \|\partial^s f / \partial x\|_{L_q}.$$

Again, for $p = \infty$, this is a simplified version of the Koksma-Hlawka inequality (see Kuipers and Niederreiter [65, Chapter 2, Theorem 5.5]).

The worst-case error is again given by

$$e(\mathscr{H}(K); \mathscr{P}) = \sup_{f \in \mathscr{H}(K), \|f\|_{\mathscr{H}(K)} \le 1} |e(f; \mathscr{P})| = \|h\|_{\mathscr{H}(K)},$$

where $h(z) = \int_{[0,1]^s} K(z, x) \, dx - \frac{1}{N} \sum_{n=1}^{N} K(z, x_n)$. Again we have $\|h\|_{\mathscr{H}(K)} = \|\Delta_{\mathscr{P}}\|_{L_2} = L_2(\mathscr{P})$, therefore we obtain

$$e(\mathscr{H}(K); \mathscr{P}) = L_2(\mathscr{P}).$$

The analogue of (9.3) yields

$$e^2(\mathscr{H}(K); \mathscr{P}) = \langle h, h \rangle_{\mathscr{H}(K)}$$

$$= \int_{[0,1]^s} \int_{[0,1]^s} K(x, y) \, dx \, dy - \frac{2}{N} \sum_{n=1}^{N} \int_{[0,1]^s} K(x, x_n) \, dx$$

$$+ \frac{1}{N^2} \sum_{n,m=1}^{N} K(x_n, x_m).$$

Since there is an explicit expression for the reproducing kernel (9.4) we obtain an explicit expression for the L_2-discrepancy of $\mathscr{P} = \{x_1, \ldots, x_N\}$, sometimes called Warnock's formula

$$(L_2(\mathscr{P}))^2 = \frac{1}{3^s} - \frac{2}{N} \sum_{n=1}^{N} \prod_{i=1}^{s} \frac{1 - x_{n,i}^2}{2} + \frac{1}{N^2} \sum_{n,m=1}^{N} \prod_{i=1}^{s} \min(1 - x_{n,i}, 1 - x_{m,i}),$$

$$(9.5)$$

where $x_{n,i}$ denotes the ith component of the point x_n.

The discrepancy defined this way does not take lower order projections into account. To include also lower dimensional projections we use the reproducing kernel

$$K(x, y) = \int_{[0,1]^s} \prod_{i=1}^{s} \left[1 + 1_{(x_i,1]}(t_i) 1_{(y_i,1]}(t_i) \right] dt$$

$$= \prod_{i=1}^{s} [1 + \min(1 - x_i, 1 - y_i)] .$$

In this case the inner product is given by

$$\langle f, g \rangle_{\mathscr{H}(K)} = \sum_{u \subseteq [s]} \int_{[0,1]^{|u|}} \frac{\partial^{|u|} f}{\partial x_u}(t_u, 1) \frac{\partial^{|u|} g}{\partial x_u}(t_u, 1) \, dt_u,$$

where $[s] = \{1, \ldots, s\}$ and where for $u \subseteq [s]$ and $x = (x_1, \ldots, x_s)$ we write x_u for the $|u|$-dimensional projection of x onto the coordinates given by u and where $(x_u, 1)$ is the s-dimensional vector whose ith component is x_i if $i \in u$ and 1 otherwise. Further we have

$$e^2(\mathscr{H}(K); \mathscr{P}) = \sum_{\emptyset \neq u \subseteq [s]} (L_2(\mathscr{P}_u))^2$$

$$= \frac{4^s}{3^s} - \frac{2}{N} \sum_{n=1}^{N} \prod_{i=1}^{s} \frac{3 - x_{n,i}^2}{2} + \frac{1}{N^2} \sum_{n,m=1}^{N} \prod_{i=1}^{s} [1 + \min(1 - x_{n,i}, 1 - x_{m,i})],$$

where $\mathscr{P}_u$ stands for the projection of the points in $\mathscr{P}$ onto the coordinates in u and $L_2(\mathscr{P}_u)$ stands for the L_2-discrepancy of $\mathscr{P}_u$.

9.1.4 Integration in Weighted Function Spaces

Sloan and Woźniakowski [110] (see also Dick, Sloan, Wang, Woźniakowski [33]) introduced a weighted discrepancy. The idea is that in many applications some projections are more important than others and that this should also be reflected in the quality measure of the point set.

The difference in the importance of projections is usually modelled by introducing so-called weights. Here we restrict ourselves to product-weights. Let $\gamma = (\gamma_i)_{i \geq 1}$ be a sequence of weights in $\mathbb{R}^+$. We use then the reproducing kernel

$$K_\gamma(x, y) = \int_{[0,1]^s} \prod_{i=1}^{s} \left[1 + \gamma_i 1_{(x_i,1]}(t_i) 1_{(y_i,1]}(t_i)\right] \, dt$$

$$= \prod_{i=1}^{s} \left[1 + \gamma_i \min(1 - x_i, 1 - y_i)\right].$$

In this case the inner product is given by

$$\langle f, g \rangle_{\mathcal{H}(K_\gamma)} = \sum_{u \subseteq [s]} \gamma_u^{-1} \int_{[0,1]^{|u|}} \frac{\partial^{|u|} f}{\partial x_u}(t_u, 1) \frac{\partial^{|u|} g}{\partial x_u}(t_u, 1) \, dt_u,$$

where for $u \subseteq [s]$ we write $\gamma_u = \prod_{i \in u} \gamma_i$ and for $u = \emptyset$ we have $\gamma_\emptyset = 1$ and $\frac{\partial^{|u|} f}{\partial x_u}(t_u, 1) := f(1)$.

With

$$h(z) = \int_{[0,1]^s} K_\gamma(z, y) \, dy - \frac{1}{N} \sum_{n=1}^{N} K_\gamma(z, x_n)$$

$$= \prod_{i=1}^{s} \left(1 + \frac{\gamma_i}{2}(1 - z_i^2)\right) - \frac{1}{N} \sum_{n=1}^{N} \prod_{i=1}^{s} [1 + \gamma_i \min(1 - z_i, 1 - x_{n,i})]$$

and

$$\frac{\partial^{|u|} h}{\partial z_u}(t_u, 1) = (-1)^{|u|+1} \gamma_u \Delta_{\mathscr{P}}(t_u)$$

we obtain for the integration error of a function $f \in \mathcal{H}(K_\gamma)$,

$$e(f; \mathscr{P}) = \langle f, h \rangle_{\mathcal{H}(K_\gamma)}$$

$$= \sum_{u \subseteq [s]} \gamma_u^{-1}(-1)^{|u|+1} \gamma_u \int_{[0,1]^{|u|}} \frac{\partial^{|u|} f}{\partial x_u}(t_u, 1) \Delta_{\mathscr{P}}(t_u, 1) \, dt_u.$$

The unweighted version of this formula is due to Hlawka [53] and Zaremba [119] and is called *Hlawka-Zaremba identity*. Applying Hölder's inequality for integrals and sums we obtain

$$|e(f; \mathscr{P})| \leq \|f\|_{\mathcal{H}(K_\gamma),q} L_{p,\gamma}(\mathscr{P}),$$

where $1 \leq p, q \leq \infty$ and $1/p + 1/q = 1$,

$$\|f\|_{\mathcal{H}(K_\gamma),q} = \left(\sum_{u \subseteq [s]} \gamma_u^{-q} \int_{[0,1]^{|u|}} \left|\frac{\partial^{|u|} f}{\partial x_u}(t_u, 1)\right|^q \, dt_u\right)^{1/q}$$

and the so-called *weighted L_p-discrepancy* is given by

$$
L_{p,\gamma}(\mathscr{P}) = \left(\sum_{\emptyset \neq u \subseteq [s]} \gamma_u^p (L_p(\mathscr{P}_u))^p \right)^{1/p}
$$

$$
= \left(\sum_{\emptyset \neq u \subseteq [s]} \gamma_u^p \int_{[0,1]^{|u|}} |\Delta_{\mathscr{P}}((t_u, 1))|^p \, dt_u \right)^{1/p},
$$

where $\Delta_{\mathscr{P}}$ and L_p denote the usual local and L_p-discrepancy, respectively. In the case $p = \infty$ we also write

$$
D_{N,\gamma}^*(\mathscr{P}) = \max_{\emptyset \neq u \subseteq [s]} \gamma_u D_N^*(\mathscr{P}_u).
$$

We call $D_{N,\gamma}^*$ the *weighted star-discrepancy* of $\mathscr{P}$. Note that $D_{N,1}^* = D_N^*$, where $1 = (1)_{i \geq 1}$, the sequence of weights where every weight is equal to one.

We also obtain a weighted discrepancy and Warnock-type formula, given by

$$
e^2(\mathscr{H}(K_\gamma); \mathscr{P}) = (L_{2,\gamma}(\mathscr{P}))^2 = \sum_{\emptyset \neq u \subseteq [s]} \gamma_u^2 (L_2(\mathscr{P}_u))^2
$$

$$
= \prod_{i=1}^s \left(1 + \frac{\gamma_i^2}{3} \right) - \frac{2}{N} \sum_{n=1}^N \prod_{i=1}^s \left(1 + \gamma_i^2 \frac{1 - x_{n,i}^2}{2} \right) \quad (9.6)
$$

$$
+ \frac{1}{N^2} \sum_{n,m=1}^N \prod_{i=1}^s [1 + \gamma_i^2 \min(1 - x_{n,i}, 1 - x_{m,i})].
$$

We remark that the assumption that $s < \infty$ can also be removed. In particular, Gnewuch [44] considered numerical integration in infinite dimensional reproducing kernel Hilbert spaces. Further, integration over $\mathbb{R}^s$ rather than $[0, 1]^s$ has for instance been considered in [23].

9.1.5 Discrepancy and Quasi-Monte Carlo on the Sphere

The above approach can be generalised in various ways, see for instance Gnewuch [45, 88]. In the following we illustrate the above approach on a different domain, see Brauchart and Dick [9]. Consider the sphere $\mathbb{S}^s = \{(x_1, \ldots, x_{s+1}) \in \mathbb{R}^{s+1} : x_1^2 + \cdots + x_{s+1}^2 = 1\}$. As test sets we use spherical caps

$$
C(t, x) = \{z \in \mathbb{S}^s : \langle z, x \rangle \geq t\}, \quad x \in \mathbb{S}^s, -1 \leq t \leq 1,
$$

where $\langle \cdot, \cdot \rangle$ denotes the standard inner product in $\mathbb{R}^{s+1}$. We use the same approach as above. Let now $\mathscr{P} = \{x_1, \ldots, x_N\} \subseteq \mathbb{S}^s$.

We define a reproducing kernel

$$K(x, y) = \int_{-1}^{1} \int_{\mathbb{S}^s} 1_{C(x,t)}(z) 1_{C(y,t)}(z) \, d\mu(z) \, dt,$$

where μ is the Lebesgue measure on the sphere $\mathbb{S}^s$ normalised to a probability measure.

The corresponding reproducing kernel Hilbert space $\mathscr{H}(K)$ then includes functions of the form

$$f(x) = \int_{-1}^{1} \int_{\mathbb{S}^s} 1_{C(x,t)}(z) f_0(z, t) \, d\mu(z) \, dt, \tag{9.7}$$

where $f_0 \in L_2(\mathbb{S}^s \times [-1, 1])$, see Brauchart and Dick [9].

Notice that in this case f_0 is not related to any classical derivative of f. We only assume that there exists a function $f_0 \in L_2(\mathbb{S}^s \times [-1, 1])$ such that (9.7) holds. Notice further that for our purposes it is not necessary to be able to obtain f_0 from some given f (for the cube $[0, 1]^s$ the function f_0 can be obtain via differentiation, but that fact was not used).

For functions $f, g : \mathbb{S}^s \to \mathbb{R}$ with representation of the form (9.7) we can define the inner product

$$\langle f, g \rangle_{\mathscr{H}(K)} = \int_{-1}^{1} \int_{\mathbb{S}^2} f_0(z, t) g_0(z, t) \, d\mu(z) \, dt.$$

Again, going through the same steps as above we obtain the discrepancy function

$$\Delta_{\mathscr{P}}(z, t) = \frac{1}{N} \sum_{n=1}^{N} 1_{C(z,t)}(x_n) - \mu(C(z, t)).$$

Again, we obtain a Koksma-Hlawka type inequality of the form

$$\left| \int_{\mathbb{S}^s} f(x) \, d\mu(x) - \frac{1}{N} \sum_{n=1}^{N} f(x_n) \right| \le \|\Delta_{\mathscr{P}}\|_{L_p} \|f_0\|_{L_q}.$$

We call $\|\Delta_{\mathscr{P}}\|_{L_p}$ the L_p spherical cap discrepancy

$$\|\Delta_{\mathscr{P}}\|_{L_p}^p = \int_{-1}^{1} \int_{\mathbb{S}^s} \left| \frac{1}{N} \sum_{n=1}^{N} 1_{C(z,t)}(x_n) - \mu(C(z, t)) \right|^p \, d\mu(z) \, dt.$$

Again, the L_2-discrepancy is related to the worst-case integration error

$$e(\mathscr{H}(K); \mathscr{P}) = \sup_{f \in \mathscr{H}(K), \|f\|_{\mathscr{H}(K)} \le 1} |e(f; \mathscr{P})| = \|\Delta_{\mathscr{P}}\|_{L_2}.$$

We also obtain

$$e^2(\mathscr{H}(K); \mathscr{P}) = \|\Delta_{\mathscr{P}}\|_{L_2}^2$$

$$= \int_{\mathbb{S}^s} \int_{\mathbb{S}^s} K(x, y) \, d\mu(x) \, d\mu(y) - \frac{2}{N} \sum_{n=1}^{N} \int_{\mathbb{S}^s} K(x, x_n) \, d\mu(x)$$

$$+ \frac{1}{N^2} \sum_{n,m=1}^{N} K(x_n, x_m) \qquad (9.8)$$

$$= \frac{1}{N^2} \sum_{n,m=1}^{N} K(x_n, x_m) - \int_{\mathbb{S}^s} \int_{\mathbb{S}^s} K(x, y) \, d\mu(x) \, d\mu(y). \qquad (9.9)$$

The reproducing kernel for the sphere $\mathbb{S}^s$ even has a concise form, see Brauchart and Dick [9], given by

$$K(x, y) = 1 - \frac{\Gamma(\frac{s+1}{2})}{s\sqrt{\pi}\Gamma(\frac{s}{2})} \|x - y\|, \qquad (9.10)$$

where $\| \cdot \|$ denotes the Euclidean norm in $\mathbb{R}^{s+1}$ and $\Gamma > 0$ is the Gamma function. Thus, using (9.8) and (9.10) we also obtain a Warnock-type formula

$$\|\Delta_{\mathscr{P}}\|_{L_2}^2 = \frac{\Gamma(\frac{s+1}{2})}{s\sqrt{\pi}\Gamma(\frac{s}{2})} \left[\int_{\mathbb{S}^s} \int_{\mathbb{S}^s} \|x - y\| \, d\mu(x) \, d\mu(y) - \frac{1}{N^2} \sum_{n,m=1}^{N} \|x_n - x_m\| \right].$$

This equality is known as *Stolarsky's invariance principle*, see Stolarsky [113]. The value of the distance integral is known explicitly and is given by

$$\int_{\mathbb{S}^s} \int_{\mathbb{S}^s} \|x - y\| \, d\mu(x) \, d\mu(y) = 2^s \frac{\Gamma((s+1)/2)\Gamma((s+1)/2)}{\sqrt{\pi}\Gamma(s-1/2)},$$

where Γ is the Gamma function.

It would be interesting to find generalisations of the geometric discrepancy defined above for other domains (manifolds) where the reproducing kernel also has a concise form to obtain analogues of Stolarsky's invariance principle for other domains.

9.2 Bounds on the Discrepancy

In this section we discuss some bounds on the L_p-discrepancy. For $s, N \in \mathbb{N}$ and $1 \le p \le \infty$ let

$$\mathrm{disc}_p(N, s) = \inf_{\substack{\mathscr{P} \subseteq [0,1)^s \\ |\mathscr{P}| = N}} L_p(\mathscr{P})$$

denote the minimal L_p-discrepancy that can be achieved by point sets consisting of N points in $[0, 1)^s$. Note that for any $1 \le p_1 \le p_2 \le \infty$ we have

$$\mathrm{disc}_{p_1}(N, s) \le \mathrm{disc}_{p_2}(N, s).$$

9.2.1 Asymptotic Bounds

In the case $p = \infty$ it is known that for any fixed $s \in \mathbb{N}$ there exist constants $0 < c_s \le C_s$ such that

$$c_s \frac{(\log N)^{\kappa_s}}{N} \le \mathrm{disc}_\infty(N, s) \le C_s \frac{(\log N)^{s-1}}{N}, \qquad (9.11)$$

where $\kappa_2 = 1$ (see Bejian [7] and Schmidt [101]) and $\kappa_s \ge (s - 1)/2$ for $s \ge 3$, which follows from a result of Roth [98]. For $s \ge 3$ the lower bound on κ_s has recently been improved to $\kappa_s \ge (s - 1)/2 + \delta_s$ for some unknown $0 < \delta_s < 1/2$; see Bilyk, Lacey and Vagharshakyan [8]. The upper bound can even be achieved constructively.

A point set $\mathscr{P}$ whose star-discrepancy satisfies an upper bound of the form $D_N^*(\mathscr{P}) = O((\log N)^{\alpha_s}/N)$ as $N \to \infty$, where $\alpha_s \ge 0$, is sometimes called a *low discrepancy point set*. There are several methods to construct low discrepancy point sets. Examples of such point sets include:

- Hammersley point sets which are based on the infinite van der Corput sequence (see, e.g., [31] and Niederreiter [81]) achieving $\alpha_s = s - 1$.
- Lattice point sets (or, more general, integration lattices) which were introduced independently by Korobov [59] and Hlawka [54] and which are well explained in the books of Niederreiter [81] and of Sloan and Joe [106]. Here it is known that one can achieve $\alpha_2 = 1$ and $\alpha_s = s$ for $s \ge 3$. Lattice point sets will be discussed in Sect. 9.4.
- (t, m, s)-nets in base b which were introduced by Niederreiter [79,81] and which are the main topic of the recent book [31]. Precursors of such nets go back to constructions of Sobol' [112] and Faure [41]. With nets one can achieve $\alpha_s = s - 1$ for all $s \ge 1$. (t, m, s)-nets will be discussed in Sect. 9.3.1.

For $1 < p < \infty$ and for any fixed $s \in \mathbb{N}$ it is known that

$$\text{disc}_p(N, s) \asymp_{s,p} \frac{(\log N)^{(s-1)/2}}{N} \quad \text{as} \quad N \to \infty, \tag{9.12}$$

where $A \asymp_{s,p} B$ means that there are constants $c_{s,p}, C_{s,p} > 0$ depending only on s, p such that $c_{s,p} B \leq A \leq C_{s,p} B$. Here the lower bound is due to Roth [98] for $p \geq 2$ and Schmidt [102] for $1 < p < 2$. The upper bound was shown first for the L_2-discrepancy by Davenport [16] for $s = 2$, by Roth [99, 100] and Frolov [42] for arbitrary dimensions $s \in \mathbb{N}$ and by Chen [11] for the general L_p case. But we know even more. For any $p > 1$, any dimension $s \in \mathbb{N}$ and any integer $N \geq 2$ there is an explicit construction of a point set $\mathscr{P}$ consisting of N points in the s-dimensional unit cube such that

$$L_p(\mathscr{P}) \ll_{s,p} \frac{(\log N)^{(s-1)/2}}{N},$$

where $A \ll_{s,p} B$ means that there is a constant $c'_{s,p} > 0$ depending only on s and p, such that $A \leq c'_{s,p} B$. Such a construction was first given by Davenport for $p = s = 2$ and by Chen and Skriganov [12] for the case $p = 2$ and arbitrary dimension s. Later Skriganov [105] generalised this construction to the L_p case with arbitrary $p > 1$. This construction is also explained in Chen and Skriganov [13] and in [31, Chapter 16]. A different construction was recently presented in [27].

9.2.2 Discrepancy and Tractability

In many applications the dimension s can be rather large. But in this case, the asymptotically almost optimal bounds on the discrepancy given, e.g., in (9.11), are even not useful for a modest number N of points. For example, assume that for every $s, N \in \mathbb{N}$ we have a point set $\mathscr{P}_{s,N}$ in the s-dimensional unit cube of cardinality N with star-discrepancy of at most

$$D_N^*(\mathscr{P}_{s,N}) \ll_s \frac{(\log N)^s}{N}.$$

Hence for any $\varepsilon > 0$ the star-discrepancy behaves asymptotically like $N^{-1+\varepsilon}$, which is the optimal rate of convergence since for dimension $s = 1$ we already have $D_N^*(\mathscr{P}_{1,N}) \geq 1/(2N)$. However, the function $N \to (\log N)^s/N$ decreases to zero not until $N \geq e^s$. For $N \leq e^s$ this function is increasing which means that for cardinality N in this range our discrepancy bounds are useless. But even for moderately large dimension s, the value of e^s is huge, such that point sets with cardinality $N \geq e^s$ cannot be used for practical applications. Therefore, the bound (9.11) is only useful if N is large compared to the dimension s.

Hence we are interested in the discrepancy of point sets with not too large cardinality N (compared to s). To analyse this problem systematically one considers the following quantity. For $\varepsilon > 0$ let

$$N_\infty(s, \varepsilon) = \min\{N \in \mathbb{N} : \mathrm{disc}_\infty(N, s) \leq \varepsilon\},$$

the so-called *inverse of the L_∞-discrepancy*. This is the minimal cardinality N of a point set in $[0, 1)^s$ such that we can achieve a star-discrepancy not larger than ε.

It is known that

$$\mathrm{disc}_\infty(N, s) \leq c \sqrt{\frac{s}{N}} \tag{9.13}$$

for all $N, s \in \mathbb{N}$ from which it follows that

$$N_\infty(s, \varepsilon) \leq C s \varepsilon^{-2} \tag{9.14}$$

for some positive constants c and C. This was shown first by Heinrich, Novak, Wasilkowski and Woźniakowski [48] by using deep results from probability theory. Later, Aistleitner [2] showed by a simplified argument that in (9.13) one can even choose $c = 10$.

Hence, the inverse of star-discrepancy depends only polynomially on s and ε^{-1}. In Information-based Complexity (IBC) theory such a behaviour is called *polynomial tractability*.

Furthermore, it is known that the dependence on the dimension s of the upper bound on the Nth minimal star-discrepancy in (9.14) cannot be improved. It was shown by Hinrichs [51, Theorem 1] that there exist constants $\tilde{c}, \varepsilon_0 > 0$ such that

$$N_\infty(s, \varepsilon) \geq \tilde{c} s / \varepsilon$$

for $0 < \varepsilon < \varepsilon_0$ and $\mathrm{disc}_\infty(N, s) \geq \min(\varepsilon_0, \tilde{c} s / n)$.

The bound (9.13) is only an existence result. Until now no explicit construction of a point set $\mathscr{P}$ of cardinality N in $[0, 1)^s$ for which $D_N^*(\mathscr{P})$ satisfies (9.13) is known. A first constructive approach of such points for which the bound (9.13) is nearly achieved is given in Doerr, Gnewuch and Srivastav [39] which is further improved in Doerr and Gnewuch [38]. There, a deterministic algorithm is presented that constructs point sets $\mathscr{P}_{N,s}$ consisting of N points in $[0, 1)^s$ satisfying

$$D_N^*(\mathscr{P}_{N,s}) = O\left(\frac{s^{1/2}}{N^{1/2}}(\log(N + 1))^{1/2}\right)$$

in run-time $O(s \log(sN)(\sigma N)^s)$, where $\sigma = \sigma(s) = O((\log s)^2/(s \log \log s)) \to 0$ as $s \to \infty$ and where the implied constants in the O-notations are independent of s and N. However, this is still by far too expensive to obtain point sets for high dimensional applications. A small improvement for the run time is presented in Doerr, Gnewuch, Kritzer and Pillichshammer [40]. However, this improvement has to be payed with a worse dependence of the bound for the star-discrepancy on the dimension s.

Let us now turn our attention to the analogue problem for the L_2-discrepancy instead of star-discrepancy. Contrary to the star-discrepancy here it makes little sense to ask for the smallest cardinality of a point set with L_2-discrepancy of at most some $\varepsilon > 0$. The reason for this is that the L_2-discrepancy of the empty point set in the s-dimensional unit cube is exactly $3^{-s/2}$, which follows from (9.5), or in other words, $\mathrm{disc}_2(0, s) = 3^{-s/2}$. Thus for s large enough, the empty set has always L_2-discrepancy smaller than ε. (This is not the case for the star-discrepancy which is always one for the empty set.) This may suggest that for large s, the L_2-discrepancy is not properly scaled. In the following we therefore use the L_2-discrepancy of the empty point set $\mathrm{disc}_2(0, s)$ as a reference.

In general, for $1 \le p \le \infty$, $s \in \mathbb{N}$ and $\varepsilon > 0$ the *inverse of the L_p-discrepancy* is hence defined as

$$N_p(s, \varepsilon) = \min\left\{N \in \mathbb{N} : \mathrm{disc}_p(N, s) \le \varepsilon\, \mathrm{disc}_p(0, s)\right\}.$$

For $N_2(s, \varepsilon)$ the situation is quite different compared to $N_\infty(s, \varepsilon)$. It was shown in Sloan and Woźniakowski [110] and Woźniakowski [118] (in a much more general setting) that for $\varepsilon \in (0, 1)$ we have

$$N_2(s, \varepsilon) \ge (1 - \varepsilon^2) \left(\frac{9}{8}\right)^s. \tag{9.15}$$

Hence $N_2(s, \varepsilon)$ grows exponentially in dimension s. A direct proof of (9.15) is also presented in [31, Proof of Proposition 3.58]. For more general results see Novak and Woźniakowski [88, Chapter 11]. Hence the inverse of the L_2-discrepancy depends at least exponentially on the dimension s. In IBC theory this exponential dependence on the dimension is called *intractability* or the *curse of dimensionality*. For a more detailed discussion of tractability of various notions of discrepancy we refer to the work of Novak and Woźniakowski [85–89].

9.2.3 Weighted Discrepancy and Strong Tractability

One of the reasons for introducing a weighted discrepancy in Sect. 9.1.4 is that with this concept one can overcome the curse of dimensionality for the L_2-discrepancy under suitable conditions on the weights γ. Also for the weighted star-discrepancy one can obtain a weaker dependence on the dimension for suitable choices of weights.

For $s, N \in \mathbb{N}$ and $1 \leq p \leq \infty$ and for a sequence γ of weights we define

$$\text{disc}_{p,\gamma}(N, s) = \inf_{\substack{\mathscr{P} \subseteq [0,1)^s \\ |\mathscr{P}| = N}} L_{p,\gamma}(\mathscr{P}).$$

For $s \in \mathbb{N}$ and $\varepsilon > 0$ the *inverse of the weighted L_p-discrepancy* is defined as

$$N_{p,\gamma}(s, \varepsilon) = \min \left\{ N \in \mathbb{N} : \text{disc}_{p,\gamma}(N, s) \leq \varepsilon \, \text{disc}_{p,\gamma}(0, s) \right\}.$$

In a paper of Hinrichs, Pillichshammer and Schmid [52, Theorem 1] it has been shown that there exists a constant $C > 0$ such that

$$\text{disc}_{\infty,\gamma}(N, s) \leq C \frac{1 + \sqrt{\log s}}{\sqrt{N}} \max_{\emptyset \neq u \subseteq I_s} \gamma_u \sqrt{|u|}. \tag{9.16}$$

Hence, if

$$\sup_{s=1,2,\ldots} \max_{\emptyset \neq u \subseteq [s]} \gamma_u \sqrt{|u|} < \infty, \tag{9.17}$$

then there exists a $C_\gamma > 0$ such that

$$\text{disc}_{\infty,\gamma}(N, s) \leq C_\gamma \frac{1 + \sqrt{\log s}}{\sqrt{N}},$$

and therefore

$$N_{\infty,\gamma}(\varepsilon, s) \leq \left\lceil \tilde{C}_\gamma \left(1 + \sqrt{\log s} \right)^2 \varepsilon^{-2} \right\rceil.$$

for some $\tilde{C}_\gamma > 0$. This means that the weighted star-discrepancy is polynomially tractable whenever the weights satisfy condition (9.17). Compared to the usual star-discrepancy, see (9.14), here we have a much weaker dependence on the dimension s. Note that (9.17) is a very mild condition on the weights. It is enough that the weights γ_i are decreasing and that $\gamma_i < 1$ for an index $i \in \mathbb{N}$. Under a stronger condition on the weights one can even obtain the following property.

If $\sum_{i \geq 1} \gamma_i < \infty$, then for any $\delta > 0$ there exists a $C_{\delta,\gamma} > 0$ such that

$$\text{disc}_{\infty,\gamma}(N, s) \leq \frac{C_{\delta,\gamma}}{N^{1-\delta}} \tag{9.18}$$

and hence

$$N_{\infty,\gamma}(\varepsilon, s) \leq \left\lceil \tilde{C}_{\delta,\gamma} \varepsilon^{-\frac{1}{1-\delta}} \right\rceil \tag{9.19}$$

for some $\tilde{C}_{\delta,\gamma} > 0$. Since this bound is even independent of the dimension one says that the weighted star-discrepancy is *strongly* polynomially tractable. The bound in (9.18) can be achieved with a superposition of polynomial lattice point sets as discussed in Sect. 9.5; see [31, Corollary 10.30].

We know from Sect. 9.2.2 that the classical L_2-discrepancy is subject to the curse of dimensionality. This disadvantage can be overcome when we change to the weighted setting.

Averaging the squared weighted L_2-discrepancy yields

$$\int_{[0,1]^{sN}} \left(L_{2,\gamma}(\{\boldsymbol{\tau}_1,\ldots,\boldsymbol{\tau}_N\})\right)^2 \, \mathrm{d}\boldsymbol{\tau}_1 \cdots \mathrm{d}\boldsymbol{\tau}_N$$

$$= \frac{1}{N} \left(\prod_{i=1}^{s}\left(1 + \frac{\gamma_i^2}{2}\right) - \prod_{i=1}^{s}\left(1 + \frac{\gamma_i^2}{3}\right)\right)$$

and hence

$$\mathrm{disc}_{2,\gamma}(N,s) \le \frac{1}{N^{1/2}} \left(\prod_{i=1}^{s}\left(1 + \frac{\gamma_i^2}{2}\right) - \prod_{i=1}^{s}\left(1 + \frac{\gamma_i^2}{3}\right)\right)^{1/2}.$$

Note that $\mathrm{disc}_{2,\gamma}(0,s) = \left(-1 + \prod_{i=1}^{s}\left(1 + \frac{\gamma_i^2}{3}\right)\right)^{1/2}$. Therefore, we obtain

$$\frac{\mathrm{disc}_{2,\gamma}(N,s)}{\mathrm{disc}_{2,\gamma}(0,s)} \le \frac{1}{N^{1/2}} \exp\left(\frac{1}{6}\sum_{i=1}^{s}\gamma_i^2\right) \tag{9.20}$$

(for details we refer to Sloan and Woźniakowski [110], see also [31, Proof of Theorem 3.64]). Hence if $\sum_{i\ge 1}\gamma_i^2 < \infty$ then there exists a $C_\gamma > 0$ such that

$$N_{2,\gamma}(\varepsilon,s) \le C_\gamma \varepsilon^{-2}.$$

Again this bound is independent of the dimension s and hence the weighted L_2-discrepancy is strongly tractable as long as the squared weights γ_i^2, $i \ge 1$, are summable. On the other hand, this condition is also necessary for strong tractability which follows from (9.15) (see again [31, Proof of Theorem 3.64] for details).

If we only would have $\limsup_{s\to\infty}\sum_{i=1}^{s}\gamma_i^2/(\log s)$, then we still obtain from (9.20) that the weighted L_2-discrepancy is polynomially tractable.

Further results on the tractability of weighted discrepancy can be found in [31], Hinrichs, Pillichshammer and Schmid [52], Leobacher and Pillichshammer [71] and Novak and Woźniakowski [88].

9.2.4   Definition of Tractability for the Worst-Case Integration Error

Let us return to the integration problem for functions from a reproducing kernel Hilbert space $\mathscr{H}(K)$. By $e(\mathscr{H}(K); \mathscr{P})$ we denote the worst-case error of a quasi-Monte Carlo rule based on the point set $\mathscr{P}$. The initial error is defined by

$$e(\mathscr{H}(K); \emptyset) = \sup_{f \in \mathscr{H}(K), \|f\|_{\mathscr{H}(K)} \leq 1} \left| \int_{[0,1]^s} f(x) \, dx \right|.$$

For $\varepsilon > 0$ let $N_{\mathscr{H}(K)}(\varepsilon, s)$ denote the minimal number of nodes that are required to reduce the initial error by a factor of ε, i.e.,

$$N_{\mathscr{H}(K)}(\varepsilon, s)$$
$$= \min\{N \in \mathbb{N} : \exists \mathscr{P} \subseteq [0, 1)^s, |\mathscr{P}| = N \text{ and } e(\mathscr{H}(K); \mathscr{P}) \leq \varepsilon e(\mathscr{H}(K); \emptyset)\}.$$

This number is called the *information complexity* of QMC integration in $\mathscr{H}(K)$.

Now one says that multivariate integration in the space $\mathscr{H}(K)$ is *polynomially (QMC) tractable*, if there exist non-negative C, α, β such that

$$N_{\mathscr{H}(K)}(\varepsilon, s) \leq C s^\alpha \varepsilon^{-\beta}$$

holds for all dimensions $s \in \mathbb{N}$ and for all $\varepsilon > 0$. If this inequality holds with $\alpha = 0$, then one says that multivariate integration in the space $\mathscr{H}(K)$ is *strongly (polynomially) (QMC) tractable*. The infima α and β are called the *s-exponent* and the *ε-exponent* of (strong) polynomial (QMC) tractability.

We remark that there are further notions of tractability such as, e.g., weak tractability or T-tractability. For more information we refer to the books by Novak and Woźniakowski [86, 88, 89].

9.3   Low Discrepancy Point Sets and Sequences

As stated at the beginning, quasi-Monte Carlo rules use deterministic constructions of quadrature points which yield small integration errors. For the reproducing kernel Hilbert spaces on $[0, 1]^s$, we know from Sect. 9.1 that this amounts to constructing point sets with small discrepancy. Explicit constructions of low discrepancy sequences where given by Sobol [112], Faure [41], Niederreiter [79] and Niederreiter-Xing [82]. The following section gives an introduction to the underlying ideas.

9.3.1 Nets and Sequences

The aim is to construct a point set $\mathscr{P}_N = \{x_0, \ldots, x_{N-1}\}$ (in this context it is convenient to index the points from 0 rather than 1) such that the discrepancy $\|\Delta_{\mathscr{P}_N}\|_{L_p}$ converges with the (almost) optimal order. To do so, we discretise the problem by choosing the point set $\mathscr{P}_N$ such that the local discrepancy $\Delta_{\mathscr{P}_N}(z) = 0$ for certain $z \in [0, 1]^s$ (those z in turn are chosen such that the discrepancy of $\mathscr{P}_N$ is small, as we explain below).

It turns out that, if one chooses a base $b \geq 2$ and $N = b^m$, for every dimension $s \geq 1$ there exists a nonnegative integer t such that for all positive integers m there exists a point set $\mathscr{P}_{b^m} = \{x_0, \ldots, x_{b^m-1}\}$ such that $\Delta_{\mathscr{P}_{b^m}}(z) = 0$ for all $z = (z_1, \ldots, z_s)$ of the form

$$z_i = \frac{a_i}{b^{d_i}} \quad \text{for } 1 \leq i \leq s,$$

where $0 < a_i \leq b^{d_i}$ is an integer and $d_1 + \cdots + d_s \leq m - t$ with $d_1, \ldots, d_s \in \mathbb{N}_0$. We stress that the value of t can be chosen independently of m (but has to dependent on s). A point set $\mathscr{P}_{b^m}$ which satisfies this property is called a (t, m, s)-net in base b. An equivalent description of (t, m, s)-nets in base b is given in the following definition.

Definition 1. Let $b \geq 2$, $m, s \geq 1$ and $0 \leq t \leq m$ be integers. A point set $\mathscr{P}_{b^m} = \{x_0, \ldots, x_{b^m-1}\} \subseteq [0, 1)^s$ is called a (t, m, s)-net in base b, if for all $d_1, \ldots, d_s \in \mathbb{N}_0$ with $d_1 + \cdots + d_s = m - t$, the elementary interval

$$\prod_{i=1}^{s} \left[\frac{a_i}{b^{d_i}}, \frac{a_i + 1}{b^{d_i}} \right)$$

contains exactly b^t points of $\mathscr{P}_{b^m}$ for all integers $0 \leq a_i < b^{d_i}$.

A sequence of points $S = (x_0, x_1, \ldots) \subseteq [0, 1)^s$ is called a (t, s)-sequence in base b, if for all $k \geq 1$ and $m > t$ the point set

$$\{x_{(k-1)b^m}, \ldots x_{kb^m-1}\}$$

is a (t, m, s)-net in base b.

Clearly, every point set $\mathscr{P}_{b^m}$ in $[0, 1)^s$ is a (t, m, s)-net in base b with $t = m$. Smaller values of t imply a stronger condition on the point set since elementary intervals of higher resolution are considered. This implies better distribution properties of the point set. However, a necessary condition such that a $(0, m, s)$-net in base b exists is $s \leq b + 1$, and a necessary condition such that a $(0, s)$-sequence in base b exists is $s \leq b$. On the other hand, for fixed base b, for a (t, s)-sequence to exist we must have $t \geq c_b s + d_b$ for some constants $c_b > 0$ and d_b which depend on b but not on s. The parameter t is often referred to as the *quality parameter* of

Fig. 9.2 A $(0, 4, 2)$-net in
base 2

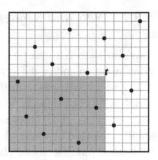

the net. An introduction into the theory of (t, m, s)-nets and (t, s)-sequences can be
found in [31] and in Niederreiter [81, Chapter 4].

As an example, Fig. 9.2 shows a $(0, 4, 2)$-net in base 2 which is a $2^4 = 16$ element
point set in $[0, 1)^2$ where every elementary interval

$$\left[\frac{A}{2^d}, \frac{A+1}{2^d} \right) \times \left[\frac{B}{2^{4-d}}, \frac{B+1}{2^{4-d}} \right)$$

for $d \in \{0, 1, 2, 3, 4\}$, $A \in \{0, \ldots, 2^d - 1\}$ and $B \in \{0, \ldots, 2^{4-d}\}$ contains exactly
one point.

9.3.2 Digital Nets and Sequences

Explicit constructions of (t, m, s)-nets can be obtained using the digital construction
scheme. Such point sets are then called *digital nets* (or *digital (t, m, s)-nets* if the
point set is a (t, m, s)-net).

To describe the digital construction scheme, let b be a prime number and let $\mathbb{Z}_b$
be the finite field of order b (a prime power and the finite field $\mathbb{F}_b$ could be used as
well) and let $d, m \in \mathbb{N}$. Let $C_1, \ldots, C_s \in \mathbb{Z}_b^{dm \times m}$ be s matrices of size $dm \times m$ with
elements in $\mathbb{Z}_b$ (the so-called *generating matrices*). The ith coordinate $x_{n,i}$ of the
nth point $\boldsymbol{x}_n = (x_{n,1}, \ldots, x_{n,s})$, $0 \le n < b^m$ and $1 \le i \le s$, of the digital net is
obtained in the following way.

9.3.2.1 Digital Construction Scheme

- For $0 \le n < b^m$ let $n = n_0 + n_1 b + \cdots + n_{m-1} b^{m-1}$ be the base b representation
  of n and let $\mathbf{n} = (n_0, \ldots, n_{m-1})^\top \in \mathbb{Z}_b^m$ be the digit vector of n.
- Let

$$\mathbf{y}_{n,i} = C_i \mathbf{n}.$$

- For $\mathbf{y}_{n,i} = (y_{n,i,1}, \ldots, y_{n,i,dm})^{\top} \in \mathbb{Z}_b^{dm}$ set

$$x_{n,i} = \frac{y_{n,i,1}}{b} + \cdots + \frac{y_{n,i,dm}}{b^{dm}}.$$

In order to obtain a sequence of points $\mathbf{x}_0, \mathbf{x}_1, \ldots$ one uses generating matrices of size $\infty \times \infty$, that is, $C_1, \ldots, C_s \in \mathbb{Z}_b^{\infty \times \infty}$. Such a sequence is then called a *digital sequence* (or *digital (t, s)-sequence* if the sequence is a (t, s)-sequence).

The classical construction of digital nets proposed by Niederreiter [81] uses $d = 1$.

The search for (t, m, s)-nets and (t, s)-sequences has now been reduced to finding suitable matrices $C_1, \ldots, C_s$. The geometric property of (t, m, s)-nets can also be translated into an algebraic property for the generating matrices.

Definition 2. Let b be prime and $m, s \geq 1$ be integers. Then the point set generated by the matrices $C_1, \ldots, C_s \in \mathbb{Z}_b^{m \times m}$ is called a *digital (t, m, s)-net over $\mathbb{Z}_b$* if for all $d_1, \ldots, d_s \in \mathbb{N}_0$ with $\sum_{i=1}^{s} d_i \leq m - t$ the system of vectors

$$\mathbf{c}_{1,1}, \ldots, \mathbf{c}_{1,d_1}, \ldots, \mathbf{c}_{s,1}, \ldots, \mathbf{c}_{s,d_s} \in \mathbb{Z}_b^m,$$

where $\mathbf{c}_{i,k}$ denotes the kth row of C_i, is linearly independent over $\mathbb{Z}_b$.

The sequence generated by the matrices $C_1, \ldots, C_s \in \mathbb{Z}_b^{\infty \times \infty}$ is called a *digital (t, s)-sequence over $\mathbb{Z}_b$* if for all $m \geq t$ the left-upper $m \times m$ submatrices $C_1^{(m)}, \ldots, C_s^{(m)}$ of $C_1, \ldots, C_s$ generate a digital (t, m, s)-net over $\mathbb{Z}_b$.

In Niederreiter [81] it has been shown that a digital (t, m, s)-net over $\mathbb{Z}_b$ is a (t, m, s)-net in base b and that a digital (t, s)-sequence over $\mathbb{Z}_b$ is a (t, s)-sequence in base b.

Explicit constructions of suitable generating matrices are available, see [31] and Niederreiter [81]. We describe the construction by Niederreiter as an example.

Let $s \in \mathbb{N}$, b be a prime number and let $p_1, \ldots, p_s \in \mathbb{Z}_b[x]$ be distinct monic irreducible polynomials over $\mathbb{Z}_b$. Let $e_i = \deg(p_i)$ for $1 \leq i \leq s$. For $1 \leq i \leq s$, $j \geq 1$ and $0 \leq k < e_i$, consider the expansions

$$\frac{x^{e_i - 1 - k}}{p_i(x)^j} = \sum_{r=0}^{\infty} a^{(i)}(j, k, r) x^{-r-1}$$

over the field $\mathbb{Z}_b((x^{-1}))$ of formal Laurent series. Then we define the matrix $C_i = (c_{j,r}^{(i)})_{j \geq 1, r \geq 0}$ by

$$c_{j,r}^{(i)} = a^{(i)}(Q + 1, k, r) \in \mathbb{Z}_b \quad \text{for } 1 \leq i \leq s, j \geq 1, r \geq 0, \qquad (9.21)$$

where $j - 1 = Qe_i + k$ with integers $Q = Q(i, j)$ and $k = k(i, j)$ satisfying $0 \leq k < e_i$. Digital sequences for which generating matrices are given by (9.21) are called *Niederreiter sequences*. The following result holds:

Theorem 3 (Niederreiter [79, Theorem 1], Dick and Niederreiter [26]). *The digital sequence with generating matrices $C_1, \ldots, C_s \in \mathbb{Z}_b^{\infty \times \infty}$ given by (9.21) is a (t, s)-sequence in base b with*

$$t = \sum_{i=1}^{s} (e_i - 1).$$

If $p_i(x) = x - i - 1 \in \mathbb{Z}_b[x]$ for $1 \leq i \leq s$, then we obtain the digital $(0, s)$-sequence over $\mathbb{Z}_b$ which is known as *Faure sequence* [41]. By setting $b = 2$, $p_1(x) = x \in \mathbb{Z}_2[x]$ and $p_2, \ldots, p_s$ are distinct primitive polynomials over $\mathbb{Z}_2$, then we obtain *Sobol' sequences* [112].

9.3.3 Discrepancy Bounds

We have obtained constructions of (t, m, s)-nets and (t, s)-sequences which yield uniformly distributed point sets and sequences. These nets are designed such that the local discrepancy is 0 for many points. Since the discrepancy can only vary slowly, one can expect that the discrepancy of the net itself is small. This is indeed the case. For instance, the following classical result holds.

Theorem 4 (Niederreiter [81, Theorems 4.5 and 4.6]). *The star-discrepancy of a (t, m, s)-net $\mathscr{P}$ in base b is bounded by*

$$D_{b^m}^*(\mathscr{P}) \leq b^{-(m-t)} \sum_{i=0}^{s-1} \binom{s-1}{i} \binom{m-t}{i} \left\lfloor \frac{b}{2} \right\rfloor^i$$

for $b \geq 3$, and for $b = 2$ we have

$$D_{b^m}^*(\mathscr{P}) \leq 2^{-(m-t)} \sum_{i=0}^{s-1} \binom{m-t}{i}.$$

To illustrate the basic idea for the proof of this discrepancy bound we show the result in the most simple case $s = b = 2$ and $t = 0$. A proof for the general result can be found in [31, Proof of Corollary 5.3].

Proof. For a measurable set C let $A(C)$ denote the number of elements of $\mathscr{P}$ which belong to C.

We consider an interval $B = [0, \alpha) \times [0, \beta)$ where the dyadic digit expansion of α and β is given by

$$\alpha = \frac{a_1}{2} + \frac{a_2}{2^2} + \cdots + \frac{a_m}{2^m} + \cdots,$$

$$\beta = \frac{b_1}{2} + \frac{b_2}{2^2} + \cdots + \frac{b_m}{2^m} + \cdots.$$

The basic idea is to approximate the interval B from the interior and from the exterior with disjoint unions of elementary intervals. Let

$$I_1 := \left[0, \frac{a_1}{2}\right) \times \left[0, \frac{b_1}{2} + \cdots + \frac{b_{m-1}}{2^{m-1}}\right),$$

$$J_1 := \left[0, \frac{a_1}{2}\right) \times \left[0, \frac{b_1}{2} + \cdots + \frac{b_{m-1}}{2^{m-1}} + \frac{1}{2^{m-1}}\right).$$

Then we have $I_1 \subseteq B$ and

$$I_1 = \bigcup_{k=0}^{2^{m-1}b_1 + \cdots + b_{m-1} - 1} \left[0, \frac{a_1}{2}\right) \times \left[\frac{k}{2^{m-1}}, \frac{k+1}{2^{m-1}}\right)$$

is a disjoint union of two-dimensional elementary intervals of area 2^{-m}. By the $(0, m, 2)$-net property we know that each of these intervals contains exactly one element of $\mathscr{P}$. Hence it follows that $A(I_1) = 2^m \lambda(I_1)$. In the same way it follows that $A(J_1) = 2^m \lambda(J_1)$.

Let further

$$I_k := \left[\frac{a_1}{2} + \cdots + \frac{a_{k-1}}{2^{k-1}}, \frac{a_1}{2} + \cdots + \frac{a_k}{2^k}\right) \times \left[0, \frac{b_1}{2} + \cdots + \frac{b_{m-k}}{2^{m-k}}\right),$$

$$J_k := \left[\frac{a_1}{2} + \cdots + \frac{a_{k-1}}{2^{k-1}}, \frac{a_1}{2} + \cdots + \frac{a_k}{2^k}\right) \times \left[0, \frac{b_1}{2} + \cdots + \frac{b_{m-k}}{2^{m-k}} + \frac{1}{2^{m-k}}\right)$$

for $1 \le k \le m - 1$ and put

$$I_m := \left[\frac{a_1}{2} + \cdots + \frac{a_{m-1}}{2^{m-1}}, \frac{a_1}{2} + \cdots + \frac{a_m}{2^m}\right) \times [0, 0) = \emptyset,$$

$$J_m := \left[\frac{a_1}{2} + \cdots + \frac{a_{m-1}}{2^{m-1}}, \frac{a_1}{2} + \cdots + \frac{a_m}{2^m}\right) \times [0, 1).$$

Using the $(0, m, 2)$-net property again, it follows, in the same way as for I_1 and J_1, that $A(I_k) = 2^m \lambda(I_k)$ and $A(J_k) = 2^m \lambda(J_k)$ for all $1 \le k \le m$. Furthermore, note that $\lambda_2(J_k \setminus I_k) \le 2^{-m}$ for all $1 \le k \le m$.

Putting

$$\underline{B} := \bigcup_{k=1}^{m} I_k,$$

$$\overline{B} := \bigcup_{k=1}^{m} J_k \cup \left(\left[\frac{a_1}{2} + \cdots + \frac{a_m}{2^m}, \frac{a_1}{2} + \cdots + \frac{a_m}{2^m} + \frac{1}{2^m}\right) \times [0, 1)\right)$$

we have $\underline{B} \subseteq B \subseteq \overline{B}$, $A(\underline{B}) = 2^m \lambda_2(\underline{B})$ and, by using the $(0, m, 2)$-net property again, $A(\overline{B}) = 2^m \lambda_2(\overline{B})$. Hence

$$\lambda_2(\underline{B}) = 2^{-m} A(\underline{B}) \leq 2^{-m} A(B) \leq 2^{-m} A(\overline{B}) = \lambda_2(\overline{B})$$

and

$$-\lambda_2(\overline{B}) \leq -\lambda_2(B) \leq -\lambda_2(\underline{B}).$$

Therefore, we obtain

$$\lambda_2(\underline{B}) - \lambda_2(\overline{B}) \leq 2^{-m} A(B) - \lambda_2(B) \leq \lambda_2(\overline{B}) - \lambda_2(\underline{B}),$$

and hence

$$|2^{-m} A(B) - \lambda_2(B)| \leq \lambda_2(\overline{B} \setminus \underline{B}) \leq \frac{m}{2^m} + \frac{1}{2^m} = \frac{1}{2^m} \sum_{i=0}^{1} \binom{m}{i},$$

independent of the choice of B. $\square$

The until now best asymptotic result for the star-discrepancy of general (t, m, s)-nets in base b has been shown by Kritzer [61].

Theorem 5 (Kritzer [61]). *The star-discrepancy of a (t, m, s)-net $\mathscr{P}$ in base b with $m > 0$ satisfies*

$$b^m D_{b^m}^*(\mathscr{P}) \leq B(s, b) b^t m^{s-1} + O(b^t m^{s-2}),$$

where the implied O-constant depends only on b and s and where

$$B(s, b) = \left\lfloor \frac{b}{2} \right\rfloor^s \frac{1}{(b + (-1)^b)(s - 1)!(\log b)^{s-1}}.$$

Thus, (t, m, s)-nets achieve a convergence order of $N^{-1}(\log N)^{s-1}$. Notice that since $L_p(\mathscr{P}) \leq D_N^*(\mathscr{P})$ for all $1 \leq p \leq \infty$, this bound also applies to the L_p-discrepancy. Apart from the power in the $\log N$ factor, it is known that this rate of convergence is best possible.

9.3.4 Randomised Quasi-Monte Carlo

So far we have considered deterministic constructions of quadrature points in the unit cube. The advantage of quadrature algorithms based on deterministic constructions is that the convergence rate of the integration error improves for functions with integrable partial mixed derivatives of order up to 1, which is not the case for so-called standard Monte Carlo (MC). Standard MC approximates the integral $\int_{[0,1]^s} f(x)\,dx$ with $\frac{1}{N}\sum_{n=1}^{N} f(z_n)$, where $z_1,\ldots,z_N$ are uniformly i.i.d. in $[0,1]^s$. However, there is also some merit in choosing the quadrature points randomly in $[0,1]^s$ as in the standard MC algorithm. The most obvious case of the usefulness of this choice is if the integrand does not have sufficient smoothness for QMC, in fact, standard MC works for functions in $L_2([0,1]^s)$. Another advantage is that one can obtain a statistical estimation of the variance of the estimator. Let

$$\hat{I}(f;z_1,\ldots,z_N) = \int_{[0,1]^s} f(x)\,dx - \frac{1}{N}\sum_{n=1}^{N} f(z_n).$$

Since the quadrature points $z_1,\ldots,z_N$ are chosen randomly from the uniform distribution, for each given f, the quantity $\hat{I}(f;z_1,\ldots,z_N)$ is a random variable. The variance $\mathrm{Var}(\hat{I}(f;z_1,\ldots,z_N))$ of $\hat{I}(f;z_1,\ldots,z_N)$ satisfies

$$\mathrm{Var}(\hat{I}(f;z_1,\ldots,z_N)) = \mathbb{E}\left(\hat{I}^2(f;z_1,\ldots,z_N)\right)$$

$$= \int_{[0,1]^s}\cdots\int_{[0,1]^s} \hat{I}^2(f;z_1,\ldots,z_N)\,dz_1\ldots dz_N$$

$$= \frac{1}{N}\left(\int_{[0,1]^s} f^2(x)\,dx - \left(\int_{[0,1]^s} f(x)\,dx\right)^2\right)$$

$$= \frac{\mathrm{Var}(f)}{N},$$

which shows the convergence rate of order $N^{-1/2}$ of the standard deviation

$$\mathrm{Std}(\hat{I}(f;z_1,\ldots,z_N)) = \sqrt{\mathrm{Var}(\hat{I}(f;z_1,\ldots,z_N))}$$

to the correct value. For $f \in L_2([0,1]^s)$, the variance decays with order $N^{-1/2}$ for the standard MC method. Functions with higher order smoothness do not yield an improved rate of convergence for standard MC.

The aim of randomised QMC is to construct a hybrid of MC and QMC with 'the best of both worlds'. To define a setting to analyse the variance in this case, one considers the *randomised error*, which one can also call the *worst-case-root-mean-square error*. That is, let $\mathscr{B}$ be some Banach space with norm $\|\cdot\|_{\mathscr{B}}$. Then the randomised error is defined as

$$e_{\mathrm{ran}}(\mathscr{B};\tilde{\mathscr{P}}) = \sup_{f\in\mathscr{B},\|f\|_{\mathscr{B}}\leq 1} \sqrt{\mathrm{Var}(\hat{I}(f))},$$

where $\tilde{\mathscr{P}} = \{z_1, \ldots, z_N\}$ is some randomised point set in $[0, 1]^s$ (concrete examples of such point set are discussed below). In the remainder of this subsection we consider $\mathscr{B}$ to be the reproducing kernel Hilbert space with reproducing kernel $K(x, y) = \prod_{i=1}^{s}(1 + \gamma_i \min(1 - x_i, 1 - y_i))$, i.e., $\mathscr{B} = \mathscr{H}(K)$.

There are several ways of obtaining randomised (t, m, s)-nets and (t, s)-sequences, such that the (t, m, s)-net and (t, s)-sequences structure, respectively, are preserved. A simple way of doing so is by using a digital shift $\sigma \in [0, 1]^s$. Assume that $\{x_0, \ldots, x_{b^m-1}\} \subseteq [0, 1)^s$ forms a (t, m, s)-net in base b, where $x_n = (x_{n,1}, \ldots, x_{n,s})$ and $x_{n,i} = x_{n,i,1}b^{-1} + x_{n,i,2}b^{-2} + \cdots$. Let $\sigma = (\sigma_1, \ldots, \sigma_s) \in [0, 1)^s$, with $\sigma_i = \sigma_{i,1}b^{-1} + \sigma_{i,2}b^{-2} + \cdots$, be i.i.d. uniformly distributed in $[0, 1)^s$. In all the b-adic representations we assume that infinitely many digits are different from $b - 1$.

We now define the randomised point set $\{z_0, \ldots, z_{b^m-1}\}$, where $z_n = (z_{n,1}, \ldots, z_{n,s})$ and $z_{n,i} = z_{n,i,1}b^{-1} + z_{n,i,2}b^{-2} + \cdots$. This is done by defining the digits $z_{n,i,k} \in \{0, \ldots, b - 1\}$ by

$$z_{n,i,k} \equiv x_{n,i,k} + \sigma_{i,k} \pmod{b} \quad \text{for all } 0 \le n < b^m, 1 \le i \le s, k \ge 1. \quad (9.22)$$

The point set $z_0, \ldots, z_{b^m-1}$ is called a *randomly digitally shifted* (t, m, s)-*net*. It can be shown that, with probability 1, the randomly digitally shifted (t, m, s)-nets in base b are again (t, m, s)-nets in base b; see [28, Lemma 3].

There are also variations of this method. For instance, one can use (9.22) for $1 \le k \le m$ and set $z_{n,i,k} = 0$ for $k > m$. Or one can use (9.22) for $1 \le k \le m$ and choose $z_{n,i,k}$ uniformly i.i.d. in $\{0, \ldots, b - 1\}$ for $k > m$. We call this method a *digital shift of depth m*. The convergence rate of the randomised error for functions from the reproducing kernel Hilbert spaces considered in Sect. 9.1.3 is of order $N^{-1}(\log N)^{(s-1)/2}$; see Chen and Skriganov [13], Cristea, Dick and Pillichshammer [15], and [29]. However, it is known that the best possible convergence rate of the randomised error for this function space is of order $N^{-3/2}(\log N)^{c_1(s)}$, again with $c_1(s) \asymp s$; see Bakhvalov [4] and also Novak [84]. We discuss in the following a randomisation method for (t, m, s)-nets and (t, s)-sequences which yields an improvement of the convergence rate of the randomised error for the reproducing kernel Hilbert space with kernel $K(x, y) = \prod_{i=1}^{s}(1 + \gamma_i \min(1 - x_i, 1 - y_i))$. This method goes back to Owen [93–95] and is called *Owen's scrambling*, see also [31, Section 13.5].

Owen's scrambling algorithm is best described for some generic point $x \in [0, 1)^s$, with $x = (x_1, \ldots, x_s)$ and $x_i = \xi_{i,1}b^{-1} + \xi_{i,2}b^{-2} + \cdots$. The scrambled point shall be denoted by $y \in [0, 1)^s$, where $y = (y_1, \ldots, y_s)$ and $y_i = \eta_{i,1}b^{-1} + \eta_{i,2}b^{-2} + \cdots$. The point y is obtained by applying permutations to each digit of each coordinate of x. The permutation applied to $\xi_{i,l}$ depends on $\xi_{i,k}$ for $1 \le k < l$. Specifically, $\eta_{i,1} = \pi_i(\xi_{i,1})$, $\eta_{i,2} = \pi_{i,\xi_{i,1}}(\xi_{i,2})$, $\eta_{i,3} = \pi_{i,\xi_{i,1},\xi_{i,2}}(\xi_{i,3})$, and in general

$$\eta_{i,k} = \pi_{i,\xi_{i,1},\ldots,\xi_{i,k-1}}(\xi_{i,k}), \quad (9.23)$$

where $\pi_{i,\xi_{i,1},...,\xi_{i,k-1}}$ is a random permutation of $\{0,\ldots,b-1\}$. We assume that permutations with different indices are chosen mutually independent from each other and that each permutation is chosen with the same probability.

To describe Owen's scrambling, for $1 \le i \le s$ let

$$\Pi_i = \{\pi_{i,\xi_{i,1},...,\xi_{i,k-1}} : k \in \mathbb{N}, \xi_{i,1}, \ldots, \xi_{i,k-1} \in \{0, \ldots, b-1\}\}$$

be a given set of permutations, where for $k = 1$ we set $\pi_{i,\xi_{i,1},...,\xi_{i,k-1}} = \pi_i$, and let $\Pi = (\Pi_1, \ldots, \Pi_s)$. Then, on applying Owen's scrambling using these permutations to some point $x \in [0,1)^s$, we write $y = \Pi(x)$ and $y_i = \Pi_i(x_i)$ for $1 \le i \le s$, where y is the point obtained by applying Owen's scrambling to x using the set of permutations $\Pi = (\Pi_1, \ldots, \Pi_s)$.

For a (t,s)-sequence $x_0, x_1, \ldots$, the Owen scrambled sequence $y_0, y_1, \ldots$ is then given by $y_n = \Pi(x_n)$ for $n \ge 0$ (for (t,m,s)-nets one just uses $0 \le n < b^m$). The convergence rate of the randomised case error for Owen scrambled (t,m,s)-nets is then

$$e_{\mathrm{ran}}(\mathcal{H}(K); \tilde{\mathcal{P}}) \ll_s \frac{(\log N)^{(s-1)/2}}{N^{3/2}}. \tag{9.24}$$

Further, Loh [72] even proved a central limit theorem for Owen scrambled $(0, m, s)$ nets.

Note that Owen's scrambling is complicated to implement, since all the randomly chosen permutations need to be stored. Therefore, several simplifications have been introduced which simplify the randomisation but still achieve (9.24). The main idea is to design randomisations such that Owen's lemma [94, Lemma 2] still holds. This then implies that also (9.24) still holds.

For instance, the following properties are sufficient for scrambling, see Hong and Hickernell [55] and Matoušek [74]:

A. Each of the sets of permutations Π_i is sampled from the same distribution $\mathcal{D}$ and these sampling are mutually independent.
B. If $x_i \in [0, 1)$ is any real number and Π_i is drawn from the distribution $\mathcal{D}$, then $\Pi_i(x_i)$ is uniformly distributed in $[0, 1)$.
C. Let $a_u = a_{u,1}b^{-1} + a_{u,2}b^{-2} + \cdots$ for $u = 1, 2$ and $c_u = \Pi(a_u) = c_{u,1}b^{-1} + c_{u,2}b^{-2} + \cdots$. Assume that $a_{1,k} = a_{2,k}$ for $1 \le k < r$ and $a_{1,r} \ne a_{2,r}$. Then

   a. $c_{1,k} = c_{2,k}$ for $1 \le k < r$;
   b. $(c_{1,r}, c_{2,r})$ is uniformly distributed on the set $\{(d, e) \in \mathbb{Z}_b^2 : d \ne e\}$;
   c. $c_{u,k}$ are independent for $k > r$ and $u = 1, 2$.

Further simplifications are possible, since often the precise distribution does not have to be known, only their first and second moments, see Matoušek [74].

A scrambling of digital nets which satisfies the conditions above and is easier to implement than Owen's scrambling is the following: Let $C_1, \ldots, C_s \in \mathbb{Z}_b^{\infty \times \infty}$ be generating matrices of a digital (t, s)-sequence over $\mathbb{Z}_b$ and let $L_1, \ldots, L_s \in \mathbb{Z}_b^{\infty \times \infty}$

be non-singular, lower triangular matrices. We choose those matrices randomly such that for $L_i = (\lambda_{u,v})_{u,v \geq 1}$, where $\lambda_{u,v} \in U\{0, \ldots, b-1\}$ i.i.d. for $u > v$, $\lambda_{u,u} \in U\{1, \ldots, b-1\}$ i.i.d. and $\lambda_{u,v} = 0$ for $u < v$. Further, let $e_i \in \mathbb{Z}_b^\infty$ be chosen i.i.d. randomly in $U\{0, \ldots, b-1\}^\infty$. Then we obtain a scrambled sequence by setting

$$\mathbf{y}_{n,i} = L_i C_i \mathbf{n} + e_i \pmod{b} \quad \text{for } 1 \leq i \leq s,$$

and for $\mathbf{y}_{n,i} = (y_{n,i,1}, y_{n,i,2}, \ldots)^\top \in \mathbb{Z}_b^\infty$ we set

$$z_{n,i} = y_{n,i,1} b^{-1} + y_{n,i,2} b^{-2} + \cdots$$

and $z_n = (z_{n,1}, \ldots, z_{n,s}) \in [0,1)^s$.

The sequence $(z_0, z_1, \ldots)$ is again a (t, s)-sequence in base b with probability 1 and also satisfies the properties A, B, C. Therefore, the randomised error for such a sequence is bounded by

$$e_{\mathrm{ran}}(\mathscr{H}(K); \tilde{\mathscr{P}}) \ll_s \frac{(\log N)^{(s-1)/2}}{N^{3/2}}.$$

9.3.5 Higher Order Nets and Sequences

Quasi-Monte Carlo rules using digital nets as quadrature points, as described above, achieve a convergence rate of the integration error of order $N^{-1}(\log N)^s$ for functions of bounded variation. If the integrand has more smoothness, then the above result does not yield a better rate of convergence. In [19, 20] it was shown how to construct QMC rules which can achieve a convergence rate of order $N^{-\alpha}(\log N)^{\alpha s}$ for integrands with square integrable partial mixed derivatives up to order α in each variable (we say that such functions have smoothness α in the following).

Let us now consider digital nets and digital sequences. Above we have seen that an algebraic property of the generating matrices ensures that the corresponding digital net has small discrepancy. These digital nets are therefore useful as quadrature points in a QMC algorithm. Similarly, we now explain the algebraic properties of the generating matrices of the digital nets necessary such that the corresponding QMC rules achieve the almost optimal rate of convergence for integrands of smoothness α. The following definition is a special case of [20, Definition 4.3 and Definition 4.8].

Definition 6. Let $\alpha, m \in \mathbb{N}$. Let $C_1, \ldots, C_s \in \mathbb{Z}_b^{\alpha m \times m}$ be generating matrices of a digital net and let $\mathbf{c}_{i,k}$ denote the kth row of C_i. Then the point set generated by $C_1, \ldots, C_s$ is a *digital $(t, \alpha, \alpha m \times m, s)$-net over $\mathbb{Z}_b$* if for all integers $1 \leq j_{i,1} < \cdots < j_{i,v_i} \leq \alpha m$, $1 \leq i \leq s$, such that

$$\sum_{i=1}^{s} \sum_{k=1}^{\min(\alpha, v_i)} j_{i,k} \leq \alpha m - t,$$

the row vectors

$$\mathbf{c}_{1,j_{1,1}}, \ldots, \mathbf{c}_{1,j_{1,v_1}}, \ldots, \mathbf{c}_{s,j_{s,1}}, \ldots, \mathbf{c}_{s,j_{s,v_s}}$$

are linearly independent over $\mathbb{Z}_b$.

Let $C_1, \ldots, C_s \in \mathbb{Z}_b^{\infty \times \infty}$ be the generating matrices of a digital sequence. If for all $m \geq t/\alpha$ the left-upper $\alpha m \times m$ submatrices $C_1^{(\alpha m, m)}, \ldots, C_m^{(\alpha m, m)}$ of $C_1, \ldots, C_s$ generate a digital $(t, \alpha, \alpha m \times m, s)$-net, then the sequence generated by $C_1, \ldots, C_s$ is a *digital (t, α, s)-sequence over $\mathbb{Z}_b$*.

There is an explicit construction method for such higher order nets and sequences which works the following way.

9.3.5.1 Higher Order Net Construction

- Choose a $(t, m, \alpha s)$-net in base b whose elements are of the form

$$\mathbf{x}_n = (x_{n,1}, \ldots, x_{n,\alpha s}) \in [0, 1]^{\alpha s}$$

and $x_{n,i} = x_{n,i,1} b^{-1} + \cdots + x_{n,i,m} b^{-m} + \cdots$ for $1 \leq i \leq \alpha s$ and $0 \leq n < b^m$.
- For $0 \leq n < b^m$ define $\mathbf{y}_n = (y_{n,1}, \ldots, y_{n,s}) \in [0, 1]^s$ by

$$y_{n,i} = \sum_{j=1}^{m} \sum_{k=1}^{\alpha} x_{n,(i-1)\alpha+k,j} \, b^{-k-(j-1)\alpha} \quad \text{for } 1 \leq i \leq s.$$

The net $\{\mathbf{y}_0, \ldots, \mathbf{y}_{b^m-1}\}$ is called a *higher order net*. This construction can easily be extended to higher order sequences; see [31, Section 15.2]. Furthermore, one can also apply the construction method to the generating matrices directly.

The following explicit construction method of suitable generating matrices was introduced in [19, 20].

Let $\alpha \geq 1$ and let $C_1, \ldots, C_{s\alpha}$ be the generating matrices of a digital $(t', m, \alpha s)$-net. As we will see later, the choice of the underlying digital $(t', m, \alpha s)$-net has a direct impact on the bound of the t-value of the digital $(t, \alpha, \alpha m \times m, s)$-net, which was proven in [19, 20]. Let $C_j = (\mathbf{c}_{j,1}, \ldots, \mathbf{c}_{j,m})^{\top}$ for $1 \leq j \leq \alpha s$; i.e., $\mathbf{c}_{j,l}$ are the row vectors of C_j. Now let the matrix $C_j^{(\alpha)}$ consist of the first rows of the matrices $C_{(j-1)\alpha+1}, \ldots, C_{j\alpha}$, then the second rows of $C_{(j-1)\alpha+1}, \ldots, C_{j\alpha}$, and so on, in the order described in the following: The matrix $C_j^{(\alpha)}$ is a $\alpha m \times m$ matrix; i.e., $C_j^{(\alpha)} = (\mathbf{c}_{j,1}^{(\alpha)}, \ldots, \mathbf{c}_{j,\alpha m}^{(\alpha)})^{\top}$, where $\mathbf{c}_{j,l}^{(\alpha)} = \mathbf{c}_{u,v}$ with $l = (v - j)\alpha + u$, $1 \leq v \leq m$, and $(j - 1)\alpha < u \leq j\alpha$ for $1 \leq l \leq \alpha m$ and $1 \leq j \leq s$.

We remark that this construction can be extended to digital (t, α, s)-sequences by letting $C_j = (\mathbf{c}_{j,1}, \mathbf{c}_{j,2}, \ldots)^{\top}$, for $1 \leq j \leq \alpha s$, denote the generating matrices of a digital $(t', \alpha s)$-sequence; the resulting matrices $C_j^{(\alpha)}$, $1 \leq j \leq s$, are now $\infty \times \infty$

matrices, where again we have $C_j^{(\alpha)} = (\mathbf{c}_{j,1}^{(\alpha)}, \mathbf{c}_{j,2}^{(\alpha)}, \ldots)^\top$, where $\mathbf{c}_{j,l}^{(\alpha)} = \mathbf{c}_{u,v}$ with $l = (v - j)\alpha + u, v \geq 1$, and $(j - 1)\alpha < u \leq j\alpha$ for $l \geq 1$ and $1 \leq j \leq s$. We have the following result on the quality parameter:

Theorem 7 ([20]). *Let $\alpha \in \mathbb{N}$ and let $C_1, \ldots, C_{\alpha s}$ be the generating matrices of a digital $(t', m, \alpha s)$-net over $\mathbb{Z}_b$ of prime order b. Let $C_1^{(\alpha)}, \ldots, C_s^{(\alpha)}$ be defined as above. Then the matrices $C_1^{(\alpha)}, \ldots, C_s^{(\alpha)}$ are the generating matrices of a digital $(t, \alpha, \alpha m \times m, s)$-net over $\mathbb{Z}_b$ with*

$$t = \alpha \min\left(m, t' + \left\lfloor \frac{s(\alpha - 1)}{2} \right\rfloor\right).$$

Furthermore, for $m = \infty$, the matrices $C_1^{(\alpha)}, \ldots, C_s^{(\alpha)}$ obtained from the generating matrices $C_1, \ldots, C_{\alpha s}$ of a $(t', \alpha s)$-sequence over $\mathbb{Z}_b$ are the generating matrices of a digital (t, α, s)-sequence over $\mathbb{Z}_b$ with

$$t = \alpha \left(t' + \left\lfloor \frac{s(\alpha - 1)}{2} \right\rfloor\right).$$

A slight improvement of this result for some cases can be found in Dick and Kritzer [25].

We have the following result on the absolute error for the integration of function with smoothness α with QMC rules based on higher order nets.

Theorem 8 ([20], [31, Chapter 15]). *Let $\{\mathbf{y}_0, \ldots, \mathbf{y}_{b^m-1}\}$ be a higher order net constructed from a digital $(t', m, \alpha s)$-net in base b. Assume that the integrand f has smoothness α. Then the absolute error converges with order*

$$\left| \int_{[0,1]^s} f(\mathbf{x}) \, d\mathbf{x} - \frac{1}{b^m} \sum_{n=0}^{b^m-1} f(\mathbf{y}_n) \right| \ll_{s,b} \frac{m^{\alpha s}}{b^{m\alpha - t}},$$

where $t = \alpha \min(t' + \lfloor s(\alpha - 1)/2 \rfloor, m)$.

Furthermore, it has been shown that, asymptotically, the t-value achieved in the above construction is optimal in the following sense.

Theorem 9 (Dick and Baldeaux [24, Theorem 5]). *Assume that $t, \alpha, s, b \in \mathbb{N}$, b prime, are such that there exists a (t, α, s)-sequence over $\mathbb{Z}_b$. Then*

$$t > s\frac{\alpha(\alpha - 1)}{2} - \alpha.$$

Since explicit constructions of (t, s)-sequences with $t = \mathcal{O}(s)$ are known, see Niederreiter and Xing [83], it follows that the asymptotic behaviour of the t-value of digital (t, α, s)-sequences is

$$t \asymp_b s\alpha^2.$$

Furthermore, explicit constructions can be obtained using the method from [19, 20] introduced above. However, it would be interesting to find other explicit constructions of higher order nets and sequences which can achieve smaller values of t for small values of m and s.

9.3.6 Scrambled Higher Order Nets

In the previous section we have seen how QMC rules can be constructed such that the integration error converges with order $N^{-\alpha}(\log N)^{\alpha s}$. Then, the question arises whether there is also a generalisation of Owen's scrambling such that the randomised error achieves a higher rate of convergence as well. An affirmative answer can be given with the following construction of 'higher order scrambled' nets.

9.3.6.1 Scrambled Higher Order Nets

- Choose a $(t, m, \alpha s)$-net in base b whose elements are of the form $x_n = (x_{n,1}, \ldots, x_{n,\alpha s}) \in [0, 1]^{\alpha s}$ and $x_{n,i} = x_{n,i,1}b^{-1} + \cdots + x_{n,i,m}b^{-m} + \cdots$ for $1 \le i \le \alpha s$ and $0 \le n < b^m$.
- Apply Owen's scrambling or one of its simplifications to the digital net to obtain randomised point set $\{z_0, \ldots, z_{b^m-1}\}$ where $z_n = (z_{n,1}, \ldots, z_{n,\alpha s}) \in [0, 1]^{\alpha s}$ and $z_{n,i} = z_{n,i,1}b^{-1} + \cdots + z_{n,i,m}b^{-m} + \cdots$ for $1 \le i \le \alpha s$ and $0 \le n < b^m$.
- For $0 \le n < b^m$ define $y_n = (y_{n,1}, \ldots, y_{n,s}) \in [0, 1]^s$ by

$$
y_{n,i} = \sum_{j=1}^{\infty} \sum_{k=1}^{\alpha} z_{n,(i-1)\alpha+k,j} b^{-k-(j-1)\alpha} \quad \text{for } 1 \le i \le s.
$$

The net $\{y_0, \ldots, y_{b^m-1}\}$ is called a *scrambled higher order net*. Again, this construction can easily be extended to higher order sequences.

The points $y_0, y_1, \ldots$ can be used in a QMC rule to obtain an improved convergence rate for the randomised error for functions with smoothness α.

Theorem 10 ([22, Theorem 10]). *Let $\{y_0, \ldots, y_{b^m-1}\}$ be a scrambled higher order net constructed from a digital $(t, m, \alpha s)$-net in base b. Assume that the integrand f has smoothness α. Then the randomised error converges with order*

$$
\sqrt{\mathrm{Var}(\hat{I}(f))} \ll_{s,b} \frac{m^{(\alpha+1)s/2}}{b^{(m-t)(\alpha+1/2)}}.
$$

Apart from the power of the $\log_b N(= m)$ factor, this convergence order cannot be improved; see Bakhvalov [4] and also Novak [84].

A few remarks are in order. One cannot change the order in the construction of higher order scrambled nets and sequences. That is, if one applies the higher order construction first, and the scrambling method afterwards, one does not obtain an improved rate of convergence. Further, currently there is no known scrambling method for general higher order digital nets and sequences (only for higher order digital nets or sequences which have been obtain using the higher order construction above). Thus, for instance, the scrambling method above cannot be applied to higher order polynomial lattice rules, which will be defined below.

9.3.7 Digitally Shifted Nets

Digital shifts of depth m of a (t, m, s)-net have already been introduced in Sect. 9.3.4. Here we consider a so-called simplified shift of depth m.

Assume that $\mathscr{P} = \{x_0, \ldots, x_{b^m-1}\}$ forms a (t, m, s)-net in base b where $x_n = (x_{n,1}, \ldots, x_{n,s})$ and $x_{n,i} = x_{n,i,1} b^{-1} + x_{n,i,2} b^{-2} + \cdots$. Choose $\sigma = (\sigma_1, \ldots, \sigma_s)$ where $\sigma_i = \sigma_{i,1} b^{-1} + \cdots + \sigma_{i,m} b^{-m}$ and $\sigma_{i,j}$ are independent and uniformly distributed in $\{0, \ldots, b-1\}$.

We now define the randomised point set $\{z_0, \ldots, z_{b^m-1}\}$, where $z_n = (z_{n,1}, \ldots, z_{n,s})$ and

$$z_{n,i} = \frac{z_{n,i,1}}{b} + \cdots + \frac{z_{n,i,m}}{b^m} + \frac{1}{2b^m},$$

where

$$z_{n,i,k} \equiv x_{n,i,k} + \sigma_{i,k} \pmod{b} \quad \text{for all } 0 \le n < b^m, 1 \le i \le s, 1 \le k \le m.$$

Such a digital shift is called a *simplified digital shift of depth m*. We denote a point set $\mathscr{P}$ that is digitally shifted by a simplified digital shift of depth m by $\hat{\mathscr{P}}_\sigma$.

Note that for the simplified digital shift, we only have b^{sm} possibilities, which means only m digits per dimension need to be selected for performing a simplified digital shift.

A simplified digital shift (of depth m) preserves the (t, m, s)-net structure; see [31, Section 4.4.4].

Theorem 11 (Cristea, Dick and Pillichshammer [15, Theorem 1]). *Let $\mathscr{P}$ be a digital (t, m, s)-net over $\mathbb{Z}_b$ with generating matrices $C_1, \ldots, C_s$. Then the mean square weighted L_2-discrepancy of $\hat{\mathscr{P}}_\sigma$ is given by*

$$\mathbb{E}[L_{2,\gamma}^2(\hat{\mathscr{P}}_\sigma)]$$

$$= \sum_{\emptyset \neq u \subseteq [s]} \gamma_u^2 \left[2\left(\frac{1}{3^{|u|}} - \left(\frac{1}{3} + \frac{1}{24b^{2m}}\right)^{|u|} \right) + \frac{1}{b^m 2^{|u|}} \left(1 - \left(1 - \frac{1}{3b^m}\right)^{|u|} \right) \right]$$

$$+ \sum_{\substack{\emptyset \neq u \subseteq [s] \\ u = \{u_1,\dots,u_e\}}} \frac{\gamma_u^2}{3^{|u|}} \sum_{\substack{k_1,\dots,k_e = 0 \\ (k_1,\dots,k_e) \neq (0,\dots,0) \\ C_{u_1}^\top k_1 + \dots + C_{u_e}^\top k_e = 0}}^{b^m - 1} \prod_{i=1}^{e} \psi(k_i),$$

where $\psi(0) = 1$ and

$$\psi(k) = \frac{3}{2b^{2(r+1)}} \left(\frac{1}{\sin^2(\kappa_r \pi / p)} - \frac{1}{3} \right)$$

if $k = \kappa_0 + \kappa_1 b + \dots + \kappa_r b^r$ with $\kappa_i \in \{0,\dots,b-1\}$ and $\kappa_r \neq 0$.

The proof of this result is based on a Walsh expansion of the Warnock-type formula (9.6) and on the orthogonality properties of Walsh functions. A proof for the unweighted case can also be found in [31, Section 16.5]. An estimate of the sums involved in the above formula yields the following result.

Corollary 12 (Cristea, Dick and Pillichshammer [15, Theorem 2]). *Let $\mathscr{P}$ be a digital (t,m,s)-net over $\mathbb{Z}_b$ with $t < m$. Then the mean square weighted L_2-discrepancy of $\hat{\mathscr{P}}_\sigma$ is bounded by*

$$\mathbb{E}[L_{2,\gamma}^2(\hat{\mathscr{P}}_\sigma)] \leq \frac{1}{b^{2m}} \sum_{\emptyset \neq u \subseteq [s]} \gamma_u^2 \left[\frac{1}{6} + b^{2t} \left(\frac{b^2 - b + 3}{6} \right)^{|u|} (m-t)^{|u|-1} \right].$$

For example, in the unweighted case, i.e., $\gamma = 1 = (1,1,\dots)$ we obtain

$$\mathbb{E}[L_{2,1}^2(\hat{\mathscr{P}}_\sigma)] \ll_{s,b} \frac{(m-t)^{s-1}}{b^{2(m-t)}}.$$

In particular, for every digital (t,m,s)-net $\mathscr{P}$ over $\mathbb{Z}_b$ there exists a simplified digital shift $\sigma^* \in \{0, 1/b^m, \dots, (b^m - 1)/b^m\}^s$ of depth m such that

$$L_{2,1}(\hat{\mathscr{P}}_{\sigma^*}) \ll_{s,b} \frac{(m-t)^{\frac{s-1}{2}}}{b^{m-t}}.$$

According to Roth's lower bound (9.12), for a $(0,m,s)$-net this bound is best possible in the order of magnitude in m.

Let $M_{b,m}$ be the set of all $m \times m$ matrices with entries over $\mathbb{Z}_b$ and let $\mathscr{C}_{s,b} := \{(C_1,\dots,C_s) : C_i \in M_{b,m} \text{ for } 1 \leq i \leq s\}$. Let $1/2 < \lambda \leq 1$. Then consider the average

$$A_{b^m,s,\lambda} := \frac{1}{b^{m^2 s}} \sum_{(C_1,\ldots,C_s) \in \mathscr{C}_{s,b}} \left(\mathbb{E}[L_{2,\gamma}^2(\hat{\mathscr{P}}_\sigma)] \right)^\lambda. \tag{9.25}$$

Using Theorem 11 we have

$$A_{b^m,s,\lambda} \le \frac{1}{b^{2\lambda m}} \sum_{\emptyset \neq u \subseteq [s]} \frac{\gamma_u^{2\lambda}}{3^\lambda 2^{\lambda|u|}} \left(1 + \frac{1}{3b^m} \right)^{\lambda|u|}$$

$$+ \frac{1}{b^{m^2|u|}} \sum_{\substack{\emptyset \neq u \subseteq [s] \\ u = \{u_1,\ldots,u_e\}}} \frac{\gamma_u^{2\lambda}}{3^{\lambda|u|}} \sum_{\substack{k_1,\ldots,k_e=0 \\ (k_1,\ldots,k_e) \neq (0,\ldots,0)}}^{b^m-1} \prod_{i=1}^{e} \psi(k_i)^\lambda \sum_{\substack{C_{u_1},\ldots,C_{u_e} \in M_{b,m} \\ C_{u_1}^\top k_1 + \cdots + C_{u_e}^\top k_e = 0}} 1$$

Let $k_i = \kappa_{i,0} + \kappa_{i,1}b + \ldots + \kappa_{i,m-1}b^{m-1}$ and let $c_{i,j}$ be the jth row vector of the matrix C_i. Since at least one $k_i \neq 0$ it follows that there is a $\kappa_{i,j} \neq 0$. First, assume that $\kappa_{1,0} \neq 0$. Then for any choice of

$$c_{u_1,2}, \ldots, c_{u_1,m}, c_{u_2,1}, \ldots, c_{u_2,m}, \ldots, c_{u_e,1}, \ldots, c_{u_e,m} \in \mathbb{Z}_b^m$$

we can find exactly one vector $c_{u_1,1} \in \mathbb{Z}_b^m$ such that $C_{u_1}^\top k_1 + \cdots + C_{u_e}^\top k_e = 0$ is fulfilled. The same argument holds with $\kappa_{1,0}$ replaced by $\kappa_{i,j}$ and $c_{u_1,1}$ replaced by $c_{u_i,j+1}$. Therefore, we have

$$\sum_{\substack{C_{u_1},\ldots,C_{u_e} \in M_{b,m} \\ C_{u_1}^\top k_1 + \cdots + C_{u_e}^\top k_e = 0}} 1 = b^{m^2|u|-m}$$

and hence, after some elementary algebra, we obtain for $1/2 < \lambda \le 1$,

$$A_{b^m,s} \le \frac{1}{b^{2\lambda m}} \sum_{\emptyset \neq u \subseteq [s]} \frac{\gamma_u^{2\lambda}}{3^\lambda} \left(\frac{2}{3} \right)^{\lambda|u|} + \frac{1}{b^m} \sum_{\emptyset \neq u \subseteq [s]} \frac{\gamma_u^{2\lambda}}{3^{\lambda|u|}} \left(1 + \sum_{k=1}^{b^m-1} \psi(k)^\lambda \right)^{|u|}$$

$$\le \frac{1}{b^m} \sum_{\emptyset \neq u \subseteq [s]} \gamma_u^{2\lambda} c_{b,\lambda}^{|u|}$$

$$\le \frac{1}{b^m} \prod_{i=1}^{s} (1 + c_{b,\lambda} \gamma_i^{2\lambda}),$$

for some $c_{b,\lambda} > 0$.

Thus, for $1/2 < \lambda \le 1$, there exists a sequence $(\hat{\mathscr{P}}_{\sigma^*})_{m \ge 1}$ of simplified digitally shifted digital nets over $\mathbb{Z}_b$ for which we have

$$L_{2,\gamma}(\hat{\mathscr{P}}_{\sigma^*}) \le \frac{1}{b^{m/(2\lambda)}} \prod_{i=1}^{s} (1 + c_{b,\lambda} \gamma_i^{2\lambda})^{1/(2\lambda)} \quad \text{for all } m \in \mathbb{N}.$$

If $\sum_{i \geq 1} \gamma_i^{2\lambda} < \infty$, then we obtain

$$\frac{\text{disc}_{2,\gamma}(b^m, s)}{\text{disc}_{2,\gamma}(0, s)} \leq \frac{L_{2,\gamma}(\mathscr{P}_{\sigma_m^*})}{\text{disc}_{2,\gamma}(0, 1)}$$

$$\leq \frac{1}{b^{m/(2\lambda)}} \exp\left(\frac{1}{2\lambda} \sum_{i=1}^{s} \log(1 + c_{b,\lambda} \gamma_i^{2\lambda})\right) \text{disc}_{2,\gamma}(0, 1)^{-1}$$

$$\leq \frac{1}{b^{m/(2\lambda)}} \exp\left(\frac{c_{b,\lambda}}{2\lambda} \sum_{i=1}^{\infty} \gamma_i^{2\lambda}\right) \text{disc}_{2,\gamma}(0, 1)^{-1} =: \frac{C_{b,\lambda,\gamma}}{b^{m/(2\lambda)}},$$

and this bound is independent of the dimension s. For $\varepsilon > 0$ choose $m \in \mathbb{N}$ such that $b^{m-1} < \lceil (C_{b,\lambda,\gamma}\varepsilon^{-1})^{2\lambda} \rceil =: N \leq b^m$. Then we have

$$\frac{\text{disc}_{2,\gamma}(b^m, s)}{\text{disc}_{2,\gamma}(0, s)} \leq \varepsilon$$

and hence

$$N_{2,\gamma}(\varepsilon, s) \leq b^m < bN = b\lceil (C_{b,\lambda,\gamma}\varepsilon^{-1})^{2\lambda} \rceil.$$

This means that the weighted L_2-discrepancy is strongly tractable with ε-exponent at most 2λ whenever $\sum_{i \geq 1} \gamma_i^{2\lambda} < \infty$ for some $\lambda \in (1/2, 1]$, and the corresponding bounds can be achieved with digitally shifted digital nets.

9.4 Lattice Rules

In this section we present another construction method for low-discrepancy point sets in $[0, 1]^s$ which can be used for QMC algorithms. In the following we write $\{x\} = x - \lfloor x \rfloor$ for the fractional part of a nonnegative real number. For vectors $\boldsymbol{x} = (x_1, \ldots, x_s) \in \mathbb{R}^s$ we set $\{\boldsymbol{x}\} := (\{x_1\}, \ldots, \{x_s\})$.

Definition 13. For an integer $N \geq 2$ and for $\boldsymbol{g} \in \mathbb{Z}^s$ the point set $\mathscr{P}(\boldsymbol{g}, N)$ consisting of the N elements

$$\boldsymbol{x}_n = \left\{\frac{n}{N}\boldsymbol{g}\right\} \quad \text{for all } 0 \leq n < N$$

is called a *lattice point set*. A QMC rule using $\mathscr{P}(\boldsymbol{g}, N)$ as underlying node set is called a *lattice rule*. Hence a lattice rule is of the form

$$Q_{N,\boldsymbol{g}}(f) = \frac{1}{N} \sum_{n=0}^{N-1} f\left(\left\{\frac{n}{N}\boldsymbol{g}\right\}\right).$$

Lattice point sets were introduced independently by Korobov [59] and Hlawka [54]. A detailed treatise can be found in the books of Niederreiter [81] and of Sloan and Joe [106].

An important property of a lattice point set $\mathscr{P}(g, N)$ is that for all $h \in \mathbb{Z}^s$ we have

$$\sum_{n=0}^{N-1} \exp\left(2\pi \mathrm{i} \frac{n}{N} g \cdot h\right) = \begin{cases} N & \text{if } g \cdot h = 0 \pmod{N}, \\ 0 & \text{otherwise.} \end{cases} \tag{9.26}$$

This property motivates the following definition.

Definition 14. The *dual lattice* of the lattice point set $\mathscr{P}(g, N)$ from Definition 13 is defined as

$$\mathscr{L}_{g,N} = \{h \in \mathbb{Z}^s : h \cdot g \equiv 0 \pmod{N}\}.$$

Property (9.26) is the reason why it is most convenient to consider one-periodic functions for the analysis of the integration error of lattice rules. This analysis can again be described in terms of a reproducing kernel as explained in Sect. 9.1.

9.4.1 The Worst-Case Error of Lattice Rules in Weighted Korobov Spaces

We consider a reproducing kernel of the form

$$K_{\mathrm{Kor}}(x, y) = \sum_{h \in \mathbb{Z}^s} \omega_h \exp(2\pi \mathrm{i} h \cdot (x - y))$$

for all x and y in $[0, 1]^s$ with some weights $\omega_h \in \mathbb{R}^+$ for all $h \in \mathbb{Z}^s$ such that $\sum_{h \in \mathbb{Z}^s} \omega_h < \infty$, which may also depend on other parameters. This choice guarantees that the kernel is well defined, since

$$|K_{\mathrm{Kor}}(x, y)| \le K_{\mathrm{Kor}}(x, x) = \sum_{h \in \mathbb{Z}^s} \omega_h < \infty.$$

Obviously, the function $K_{\mathrm{Kor}}(x, y)$ is symmetric in x and y and it is easy to show that it is also positive definite. Therefore, $K_{\mathrm{Kor}}(x, y)$ is indeed a reproducing kernel.

Common examples for ω_h are the following ones:

- $\omega_h = r_{\alpha, \gamma}(h)^{-1}$ where $\alpha > 1$ is a real, $\gamma = (\gamma_1, \gamma_2, \ldots)$ is a sequence of positive reals, for $h = (h_1, \ldots, h_s)$ we put $r_{\alpha, \gamma}(h) = \prod_{i=1}^{s} r_{\alpha, \gamma_i}(h_i)$, and for $h \in \mathbb{Z}$ and $\gamma > 0$ we put

$$r_{\alpha,\gamma}(h) = \begin{cases} 1 & \text{if } h = 0, \\ \gamma^{-1}|h|^{\alpha} & \text{if } h \neq 0. \end{cases}$$

In this case we will write $K_{\mathrm{Kor}} = K_{s,\alpha,\gamma}$. In the unweighted case, i.e., $\gamma = (1, 1, \ldots)$, we simply write r_{α} instead of $r_{\alpha,\gamma}$ and $K_{\mathrm{Kor}} = K_{s,\alpha}$.

- $\omega_{\boldsymbol{h}} = \omega^{\|\boldsymbol{h}\|_1}$ with some $\omega \in (0, 1)$, where $\|\boldsymbol{h}\|_1 := |h_1| + \cdots + |h_s|$ for $\boldsymbol{h} = (h_1, \ldots, h_s)$. In this case we will write $K_{\mathrm{Kor}} = K'_{s,\omega}$.

Associated with this reproducing kernel is now the Hilbert space $\mathscr{H}(K_{\mathrm{Kor}})$ of functions $f : [0, 1]^s \to \mathbb{R}$ which are one-periodic in each variable. The corresponding inner product is given by

$$\langle f, g \rangle_{\mathscr{H}(K_{\mathrm{Kor}})} = \sum_{\boldsymbol{h} \in \mathbb{Z}^s} \omega_{\boldsymbol{h}}^{-1} \hat{f}(\boldsymbol{h}) \overline{\hat{g}(\boldsymbol{h})}, \tag{9.27}$$

where $\hat{f}(\boldsymbol{h}) = \int_{[0,1]^s} f(\boldsymbol{x}) \exp(-2\pi \mathrm{i} \boldsymbol{h} \cdot \boldsymbol{x}) \, \mathrm{d}\boldsymbol{x}$ is the $\boldsymbol{h}$th Fourier coefficient of f. As usual, the norm in $\mathscr{H}(K_{\mathrm{Kor}})$ is defined by

$$\| \cdot \|_{\mathscr{H}(K_{\mathrm{Kor}})}^2 = \langle f, f \rangle_{\mathscr{H}(K_{\mathrm{Kor}})} = \sum_{\boldsymbol{h} \in \mathbb{Z}^s} \omega_{\boldsymbol{h}}^{-1} |\hat{f}(\boldsymbol{h})|^2.$$

The function space $\mathscr{H}(K_{\mathrm{Kor}})$ is called a *Korobov space*.

Using the approach from Sect. 9.1 it follows that the worst-case error for integration in $\mathscr{H}(K_{\mathrm{Kor}})$ using a quasi-Monte Carlo rule based on a point set $\mathscr{P} = \{\boldsymbol{x}_0, \ldots, \boldsymbol{x}_{N-1}\}$ is given by

$$e^2(\mathscr{H}(K_{\mathrm{Kor}}); \mathscr{P}) = \sup_{\substack{f \in \mathscr{H}(K_{\mathrm{Kor}}) \\ \|f\|_{\mathscr{H}(K_{\mathrm{Kor}})} \leq 1}} e^2(f; \mathscr{P})$$

$$= \int_{[0,1]^s} \int_{[0,1]^s} K_{\mathrm{Kor}}(\boldsymbol{x}, \boldsymbol{y}) \, \mathrm{d}\boldsymbol{x} \, \mathrm{d}\boldsymbol{y}$$

$$- \frac{2}{N} \sum_{n=0}^{N-1} \int_{[0,1]^s} K_{\mathrm{Kor}}(\boldsymbol{x}, \boldsymbol{x}_n) \, \mathrm{d}\boldsymbol{x} + \frac{1}{N^2} \sum_{n,m=1}^{N} K_{\mathrm{Kor}}(\boldsymbol{x}_n, \boldsymbol{x}_m)$$

$$= \sum_{\boldsymbol{h} \in \mathbb{Z}^s \setminus \{0\}} \omega_{\boldsymbol{h}} \left| \frac{1}{N} \sum_{n=0}^{N-1} \exp(2\pi \mathrm{i} \boldsymbol{h} \cdot \boldsymbol{x}_n) \right|^2.$$

If the QMC rule is a lattice rule, then, using (9.26) and the notation from Definition 14, we obtain the following simplified expression for the worst-case error.

Theorem 15. *The worst-case error of a lattice rule for integration in the Korobov space $\mathscr{H}(K_{\mathrm{Kor}})$ is given by*

$$e^2(\mathscr{H}(K_{\mathrm{Kor}}); \mathscr{P}(\boldsymbol{g}, N)) = \sum_{\boldsymbol{h} \in \mathscr{L}_{\boldsymbol{g}, N} \setminus \{0\}} \omega_{\boldsymbol{h}}.$$

Remark 16. If $\alpha \geq 2$ is an even integer, then the Bernoulli polynomial B_α of degree α has the Fourier expansion

$$B_\alpha(x) = \frac{(-1)^{(\alpha+2)/2}\alpha!}{(2\pi)^\alpha} \sum_{\substack{h\in\mathbb{Z}\\h\neq0}} \frac{\exp(2\pi\mathrm{i}hx)}{|h|^\alpha} \quad \text{for all } x \in [0,1);$$

see for example Sloan and Joe [106, Appendix C]. Hence in this case we obtain

$$e^2(\mathcal{H}(K_{s,\alpha,\gamma}); \mathscr{P}(g,N))$$

$$= -1 + \frac{1}{N} \sum_{k=0}^{N-1} \prod_{i=1}^{s} \left(1 + \gamma_i \frac{(-1)^{(\alpha+2)/2}(2\pi)^\alpha}{\alpha!} B_\alpha\left(\left\{\frac{kg_i}{N}\right\}\right)\right),$$

so that $e^2(\mathcal{H}(K_{s,\alpha,\gamma}); \mathscr{P}(g,N))$ can be calculated in $O(Ns)$ operations.

We present the following lower bound on the integration error for numerical integration in the Korobov space. We prove this lower bound for quadrature rules of the form

$$Q_{\mathscr{P},w}(f) = \sum_{n=0}^{N-1} w_n f(x_n),$$

where $\mathscr{P} = \{x_0, \ldots, x_{N-1}\} \subseteq [0,1]^s$ and $w = (w_0, \ldots, w_{N-1}) \in \mathbb{R}^N$ are given. The result holds even for more general quadrature rules not considered here, see Bakhvalov [4]. Note that a QMC rule is obtained by choosing $w = (N^{-1}, \ldots, N^{-1}) \in \mathbb{R}^N$. In this case we write $e(\mathcal{H}(K_{\mathrm{Kor}}); \mathscr{P}(g,N); w)$ for the worst-case error of the quadrature rule $Q_{\mathscr{P},w}$ in the Korobov space.

Theorem 17. *Let $\mathscr{P}$ be an arbitrary N-element point set in $[0,1]^s$ and let $w = (w_0, \ldots, w_{N-1}) \in \mathbb{R}^N$.*

1. (Bakhvalov [4], Temlyakov [114]) If $K_{\mathrm{Kor}} = K_{s,\alpha}$ for an $\alpha > 1$, then

$$e(\mathcal{H}(K_{s,\alpha}); \mathscr{P}; w) \geq C(s,\alpha) \frac{(\log N)^{(s-1)/2}}{N^{\alpha/2}},$$

where $C(s,\alpha) > 0$ depends on α and s, but not on N and w.
2. (Šarygin [116]) If $K_{\mathrm{Kor}} = K'_{s,\omega}$ for an $\omega \in (0,1)$, then

$$e(\mathcal{H}(K'_{s,\omega}); \mathscr{P}; w) \geq \left(1 + \frac{2}{1-\omega}\right)^{-s/2} \omega^{(s!(N+1))^{1/s}}.$$

Proof. 1. We follow the proof of Temlyakov [114, Lemma 3.1]. In the same way as was shown above, the worst-case error in the Korobov space for an arbitrary quadrature rule can be written as

$$e^2(\mathcal{H}(K_{s,\alpha}); \mathcal{P}; \boldsymbol{w}) = |1 - \beta|^2 + \sum_{\boldsymbol{h} \in \mathbb{Z}^s \setminus \{0\}} r_\alpha^{-1}(\boldsymbol{h}) \left| \sum_{n=0}^{N-1} w_n \exp(2\pi i \boldsymbol{h} \cdot \boldsymbol{x}_n) \right|^2,$$

where $\beta = \sum_{n=0}^{N-1} w_n$.

Let $f : \mathbb{R} \to \mathbb{R}$ be an infinitely times differentiable function such that $f(x) > 0$ for $x \in (0, 1)$ and $f^{(r)}(x) = 0$ for $x \in \mathbb{R} \setminus (0, 1)$ for all $0 \le r \le a := \lceil \alpha/2 \rceil + 1$. For instance, let

$$f(x) = \begin{cases} x^{a+1}(1-x)^{a+1} & \text{for } x \in (0, 1) \\ 0 & \text{otherwise.} \end{cases} \tag{9.28}$$

For $m \in \mathbb{N}_0$ let $f_m(x) = f(2^{m+2}x)$ and for $\boldsymbol{m} = (m_1, \ldots, m_s) \in \mathbb{N}_0^s$ let

$$f_{\boldsymbol{m}}(\boldsymbol{x}) = \prod_{i=1}^{s} f_{m_i}(x_i),$$

where $\boldsymbol{x} = (x_1, \ldots, x_s)$. Let $\|\boldsymbol{m}\|_1 = m_1 + \cdots + m_s$. We obtain

$$\hat{f}_{\boldsymbol{m}}(\boldsymbol{0}) = \prod_{i=1}^{s} \int_0^1 f(2^{m_i+2}x) \, dx = \prod_{i=1}^{s} \frac{1}{2^{m_i+2}} \int_0^1 f(y) \, dy = \frac{1}{2^{\|\boldsymbol{m}\|_1+2s}} I^s(f),$$

where $I(f) = \int_0^1 f(y) \, dy$. For instance, by choosing f according to (9.28) we obtain

$$I(f) = B(a+2, a+2) = \frac{((a+1)!)^2}{(2a+3)!},$$

where B denotes the beta function.

Let t be such that

$$2N \le 2^t < 4N.$$

Let

$$B_{\boldsymbol{m}} = \left\{ \boldsymbol{y} \in [0, 1]^s : \sum_{n=0}^{N-1} w_n f_{\boldsymbol{m}}(\boldsymbol{x}_n - \boldsymbol{y}) = 0 \right\}.$$

Notice that the support of $f_{\boldsymbol{m}}(\boldsymbol{x}_n - \boldsymbol{y})$ (as a function of $\boldsymbol{y}$) is contained in the interval $\prod_{i=1}^{s}(x_{i,n} - 2^{-m_i-2}, x_{i,n})$ and hence the support of $F(\boldsymbol{y}) = \frac{1}{N} \sum_{n=0}^{N-1} f_{\boldsymbol{m}}(\boldsymbol{x}_n - \boldsymbol{y})$ is contained in $\bigcup_{n=0}^{N-1} \prod_{i=1}^{s}(x_{i,n} - 2^{-m_i-2}, x_{i,n})$. Therefore the area of the support of F is at most $N2^{-\|\boldsymbol{m}\|_1}$. Thus for all $\boldsymbol{m}$ such that $\|\boldsymbol{m}\|_1 = t$ we have

$$\lambda_s(B_{\boldsymbol{m}}) \ge 1 - N2^{-\|\boldsymbol{m}\|_1} = 1 - N2^{-t} > 1/4,$$

where λ_s denotes the s dimensional Lebesgue measure.
Let $\beta = \sum_{n=0}^{N-1} w_n$. Thus we have

$$\lambda_s(B_m)|\hat{f}_m(0)|^2|\beta|^2 = \int_{B_m} |Q_w(f_m(\cdot - y)) - \hat{f}_m(0)\beta|^2 \, dy$$

$$\leq \int_{[0,1]^s} |Q_w(f_m(\cdot - y)) - \hat{f}_m(0)\beta|^2 \, dy$$

$$= \sum_{h \in \mathbb{Z}^s \setminus \{0\}} |\hat{f}_m(h)|^2 \left| \sum_{n=0}^{N-1} w_n \exp(2\pi i h \cdot x_n) \right|^2.$$

We have

$$\hat{f}_m(h) = \int_0^1 f(2^{m+2}x) \exp(-2\pi i h x) \, dx$$

$$= \frac{1}{2^{m+2}} \int_0^1 f(y) \exp(-2\pi i h 2^{-m-2} y) \, dy = \frac{1}{2^{m+2}} \hat{f}(h 2^{-m-2}),$$

where $\hat{f}$ denotes the Fourier transform of f. Since, by assumption, f is infinitely times differentiable, integration by parts shows that for any $m \in \mathbb{N}_0$ we have

$$|\hat{f}_m(h)| \leq 2^{-m-2} |\hat{f}(h 2^{-m-2})| \leq C_a 2^{-m-2} \min(1, (h 2^{-m-2})^{-a}),$$

where the constant $C_a > 0$ depends only on a and f. Then for m with $\|m\|_1 = t$ we have

$$|\hat{f}_m(h)| \leq C(a,s) \prod_{i=1}^{s} (2^{-m_i} \min(1, 2^{a m_i} r_a^{-1}(h_i)))$$

$$= C(a,s) 2^{(\alpha/2-1)t} \prod_{i=1}^{s} (2^{-\alpha m_i/2} \min(1, 2^{a m_i} r_a^{-1}(h_i))).$$

By summing over all choices of m with $\|m\|_1 = t$ we obtain

$$\sum_{\substack{m \in \mathbb{N}_0^s \\ \|m\|_1 = t}} |\hat{f}_m(h)|^2 \leq 2^{(\alpha-2)t} C^2(a,s) \sum_{\substack{m \in \mathbb{N}_0^s \\ \|m\|_1 = t}} \prod_{i=1}^{s} (2^{-\alpha m_i} \min(1, 2^{2 a m_i} r_a^{-2}(h_i)))$$

$$\leq 2^{(\alpha-2)t} C^2(a,s) \prod_{i=1}^{s} \sum_{m=0}^{\infty} 2^{-\alpha m} \min(1, 2^{2 a m} r_a^{-2}(h_i)).$$

The last sum can now be estimated by

$$\sum_{m=0}^{\infty} \frac{1}{2^{\alpha m}} \min(1, 2^{2am} r_a^{-2}(h_i))$$

$$= \sum_{0 \le m \le (\log_2 r_a(h_i))/a} 2^{(2a-\alpha)m} r_a^{-2}(h_i) + \sum_{m > (\log_2 r_a(h_i))/a} 2^{-\alpha m}$$

$$\le \frac{r_{2a-\alpha}(h_i) 2^{2a-\alpha} - 1}{2^{2a-\alpha} - 1} r_{2a}^{-1}(h_i) + \frac{r_\alpha^{-1}(h_i) 2^\alpha}{2^\alpha - 1}$$

$$\le r_\alpha^{-1}(h_i) \left(1 + \frac{2^\alpha}{2^\alpha - 1}\right) \le 3 r_\alpha^{-1}(h_i).$$

Thus we have

$$C_1(a, s) 2^{-(\alpha-2)t} \sum_{\substack{\boldsymbol{m} \in \mathbb{N}_0^s \\ \|\boldsymbol{m}\|_1 = t}} |\hat{f}_{\boldsymbol{m}}(\boldsymbol{h})|^2 \le r_\alpha^{-1}(\boldsymbol{h}).$$

We obtain

$$e^2(\mathcal{H}(K_{s,\alpha}); \mathcal{P}; \boldsymbol{w}) = |1 - \beta|^2 + \sum_{\boldsymbol{h} \in \mathbb{Z}^s \setminus \{\boldsymbol{0}\}} r_\alpha^{-1}(\boldsymbol{h}) \left| \sum_{n=0}^{N-1} w_n \exp(2\pi i \boldsymbol{h} \cdot \boldsymbol{x}_n) \right|^2$$

$$\ge |1 - \beta|^2 + C_1(a, s) \frac{1}{2^{(\alpha-2)t}} \sum_{\substack{\boldsymbol{m} \in \mathbb{N}_0^s \\ \|\boldsymbol{m}\|_1 = t}} \sum_{\boldsymbol{h} \in \mathbb{Z}^s \setminus \{\boldsymbol{0}\}} |\hat{f}_{\boldsymbol{m}}(\boldsymbol{h})|^2 \left| \sum_{n=0}^{N-1} w_n \exp(2\pi i \boldsymbol{h} \cdot \boldsymbol{x}_n) \right|^2$$

$$\ge |1 - \beta|^2 + C_1(a, s) \frac{1}{2^{(\alpha-2)t}} \sum_{\substack{\boldsymbol{m} \in \mathbb{N}_0^s \\ \|\boldsymbol{m}\|_1 = t}} \lambda_s(B_{\boldsymbol{m}}) |\hat{f}_{\boldsymbol{m}}(\boldsymbol{0})|^2 |\beta|^2$$

$$\ge |1 - \beta|^2 + C_2(a, s) |\beta|^2 \frac{2^{2t}}{N^\alpha} \sum_{\substack{\boldsymbol{m} \in \mathbb{N}_0^s \\ \|\boldsymbol{m}\|_1 = t}} 2^{-2t-4s} I^{2s}(f)$$

$$\ge |1 - \beta|^2 + C_3(a, s) |\beta|^2 N^{-\alpha} \binom{t + s - 1}{s - 1}.$$

Set $A = C_3(a, s) N^{-\alpha} \binom{t+s-1}{s-1}$. Then the last expression can be written as $|1 - \beta|^2 + A|\beta|^2$, which satisfies

$$e^2(\mathcal{H}(K_{s,\alpha}); \mathcal{P}; \boldsymbol{w}) \ge |1 - \beta|^2 + A|\beta|^2 \ge \frac{A}{1 + A} \ge C_4(a, s) N^{-\alpha} \binom{t + s - 1}{s - 1},$$

which implies the result, since $t \ge \log_2(N)$.

2. The lower bound in the case that $K_{\mathrm{Kor}} = K'_{s,\omega}$ for some $\omega \in (0,1)$ can be deduced from the following result due to Šarygin [116]: For any point set $\mathscr{P} = \{x_0, \ldots, x_{N-1}\}$ in $[0,1)^s$ one can find a periodic function $f : [0,1]^s \to \mathbb{R}$ with the following properties:

- the Fourier coefficients of f satisfy $|\hat{f}(\boldsymbol{h})| \leq \omega^{|\boldsymbol{h}|_1}$;
- $f(\boldsymbol{x}_n) = 0$ for all $0 \leq n < N$;
- $\int_{[0,1]^s} f(\boldsymbol{x}) \, d\boldsymbol{x} \geq \omega^{(s!(N+1))^{1/s}}$.

Then the function $g(\boldsymbol{x}) = (1 + \frac{2}{1-\omega})^{-s/2} f(\boldsymbol{x})$ is contained in the unit ball of the Korobov space $\mathscr{H}(K'_{s,\omega})$ and hence the result follows from the properties of the function f.□

Remark 18. For the worst-case error of lattice rules for integration in the Korobov spaces $\mathscr{H}(K_{s,\alpha,\boldsymbol{\gamma}})$ and $\mathscr{H}(K'_{s,\omega})$ we have the following lower bounds:

For any integer $N \geq 2$ and $\boldsymbol{g} \in \{0, 1, \ldots, N-1\}^s$ we have

$$e(\mathscr{H}(K_{s,\alpha,\boldsymbol{\gamma}}); \mathscr{P}(\boldsymbol{g}, N)) \geq \left(2\zeta(\alpha) \sum_{i=1}^{s} \gamma_i\right)^{1/2} \frac{1}{N^{\alpha/2}},$$

where $\zeta(\alpha) = \sum_{h \geq 1} h^{-\alpha}$, and

$$e(\mathscr{H}(K'_{s,\omega}); \mathscr{P}(\boldsymbol{g}, N)) \geq \omega^{\frac{1}{2}(s!N)^{1/s}}.$$

Proof. We have

$$e^2(\mathscr{H}(K_{s,\alpha,\boldsymbol{\gamma}}); \mathscr{P}(\boldsymbol{g}, N)) = \sum_{\boldsymbol{h} \in \mathscr{L}_{\boldsymbol{g},N} \setminus \{\boldsymbol{0}\}} \frac{1}{r_{\alpha,\boldsymbol{\gamma}}(\boldsymbol{h})} \geq \sum_{i=1}^{s} \sum_{\substack{h_i \in \mathbb{Z} \setminus \{0\} \\ h_i g_i \equiv 0 \, (\mathrm{mod}\, N)}} \frac{\gamma_i}{|h_i|^\alpha}$$

$$\geq 2 \sum_{i=1}^{s} \gamma_i \sum_{h=1}^{\infty} \frac{1}{(Nh)^\alpha} = 2\zeta(\alpha) \left(\sum_{i=1}^{s} \gamma_i\right) \frac{1}{N^\alpha}.$$

Define $\rho(\boldsymbol{g}) = \min_{\boldsymbol{h} \in \mathscr{L}_{\boldsymbol{g},N} \setminus \{\boldsymbol{0}\}} \|\boldsymbol{h}\|_1$. Then we have

$$e^2(\mathscr{H}(K'_{s,\omega}); \mathscr{P}(\boldsymbol{g}, N)) = \sum_{\boldsymbol{h} \in \mathscr{L}_{\boldsymbol{g},N} \setminus \{\boldsymbol{0}\}} \omega^{\|\boldsymbol{h}\|_1} = \sum_{k=\rho(\boldsymbol{g})}^{\infty} \omega^k \sum_{\substack{\boldsymbol{h} \in \mathscr{L}_{\boldsymbol{g},N} \setminus \{\boldsymbol{0}\} \\ \|\boldsymbol{h}\|_1 = k}} 1 \geq \omega^{\rho(\boldsymbol{h})}.$$

In Lyness [73, Section 5] it is shown that for any $N \in \mathbb{N}$ and any $\boldsymbol{g} \in \{0, 1, \ldots, N-1\}^s$ we have $\rho(\boldsymbol{g}) \leq (s!N)^{1/s}$. Hence we have

$$e^2(\mathscr{H}(K'_{s,\omega}); \mathscr{P}(\boldsymbol{g}, N)) \geq \omega^{(s!N)^{1/s}}.□$$

Remark 19. The quantity $\rho(g)$ used above is the enhanced trigonometric degree of lattice rules, see Cools and Lyness [14] and Lyness [73]. A cubature rule of enhanced trigonometric degree δ is one that integrates all trigonometric polynomials of degree less then δ exactly.

There are also existence results for lattice point sets. In the case of $K_{\mathrm{Kor}} = K_{s,\alpha,\gamma}$ these results are mainly based on averaging arguments over all lattice points $g \in \{0, \ldots, N - 1\}^s$ for given $N \geq 2$. These methods are nowadays quite standard and there are even constructions of such lattice points based on a component-by-component approach. Here one constructs successive the components of g. This approach was introduced by Korobov [60] and later re-invented by Sloan and Reztsov [107]. In the following let $\mathbb{P} = \{2, 3, 5, \ldots\}$ denote the set of prime numbers.

Algorithm 20. Let $N \in \mathbb{P}$ and let $s \geq 2$.

1. Choose $g_1 = 1$.
2. For $d > 1$, assume we have already constructed $g_1, \ldots, g_{d-1}$. Then find $g_d \in \{1, \ldots, N - 1\}$ which minimises $e^2(\mathscr{H}(K_{s,\alpha,\gamma}); \mathscr{P}((g_1, \ldots, g_{d-1}, z), N))$ as a function of $z \in \{1, \ldots, N - 1\}$.

Variations of this algorithm for shifted and randomly shifted lattice rules were analysed by Sloan, Kuo and Joe [108, 109]. The case where N is not necessarily a prime number was first considered in [18].

A straightforward implementation of Algorithm 20 would require $O(N^2 s^2)$ operations for the construction of a lattice point set $\mathscr{P}(g, N)$ in dimension s. Using the so-called *fast component-by-component algorithm* due to Nuyens and Cools the construction costs can be reduced to $O(sN \log N)$ operations; see Nuyens and Cools [90–92] and the references therein for more detailed information.

In the case of $K_{\mathrm{Kor}} = K'_{s,\omega}$ there is, until now, only one existence result for lattice rules. Note, however, that a (modified) regular grid can be used to obtain an exponential rate of convergence with polynomial tractability; see Dick, Larcher, Pillichshammer and Woźniakowski [37]. Since a regular grid can be obtained using the digital construction scheme, it also follows that there are digital nets for which one can obtain an exponential rate of convergence for integrands from $\mathscr{H}(K'_{s,\omega})$.

Theorem 21 (Kuo [66] and Dick, Larcher, Pillichshammer and Woźniakowski [37]).

1. For any prime number N, a vector $g \in \{0, \ldots, N - 1\}^s$ can be found using Algorithm 20 such that

$$e^2(\mathscr{H}(K_{s,\alpha,\gamma}); \mathscr{P}(g, N)) \leq \frac{2^{1/\lambda}}{N^{1/\lambda}} \prod_{i=1}^{s} (1 + 2\gamma_i^\lambda \zeta(\alpha\lambda))^{1/\lambda}$$

for $1/\alpha < \lambda \leq 1$.

2. *For any prime number N, there exists a $g \in \{0, \ldots, N-1\}^s$ such that*

$$e^2(\mathcal{H}(K'_{s,\omega}); \mathcal{P}(g, N)) \leq \omega^{2^{-1}(s! N)^{1/s}} \left(\frac{4e}{\omega - \omega^2}\right)^s N.$$

Proof. 1. A proof of this result can be found in Kuo [66].
2. We have

$$e^2(\mathcal{H}(K'_{s,\omega}); \mathcal{P}(g, N)) = \sum_{h \in \mathscr{L}_{g,N} \setminus \{0\}} \omega^{\|h\|_1}$$

$$= \sum_{k=\rho(g)}^{\infty} \omega^k \sum_{\substack{h \in \mathscr{L}_{g,N} \setminus \{0\} \\ \|h\|_1 = k}} 1 \qquad (9.29)$$

$$\leq \sum_{k=\rho(g)}^{\infty} \omega^k 2^s \binom{k+s-1}{s-1}$$

$$\leq \omega^{\rho(g)} 2^s (1-\omega)^{-s} \binom{\rho(g)+s-1}{s-1}, \qquad (9.30)$$

where we used

$$\sum_{k=\rho}^{\infty} \binom{k+r-1}{r-1} \omega^k \leq \omega^{\rho} \binom{\rho+r-1}{r-1} (1-\omega)^{-r}, \qquad (9.31)$$

which can be shown using the binomial theorem; see Matoušek [74, Lemma 2.18] or [29, Lemma 6].
Now we show that for a prime number N, there exists a $g \in \{0, 1, \ldots, N-1\}^s$ such that

$$\rho(g) \geq \lceil 2^{-1}(s! N)^{1/s} \rceil - s. \qquad (9.32)$$

For a given $h = (h_1, \ldots, h_s) \in \mathbb{Z}^s \setminus \{0\}$ with $|h_i| < N$ for $1 \leq i \leq s$, there are N^{s-1} choices of $g \in \{0, 1, \ldots, N-1\}^s$ such that $g \cdot h \equiv 0 \pmod{N}$. Furthermore,

$$|\{h \in \mathbb{Z}^s : \|h\|_1 = \ell\}| \leq 2^s \binom{\ell+s-1}{s-1}.$$

Let $\rho < N$ be a given positive integer (note that $\rho(g) < N$ always). Then,

$$|\{h \in \mathbb{Z}^s : \|h\|_1 \leq \rho\}| \leq 2^s \sum_{\ell=0}^{\rho} \binom{\ell+s-1}{s-1} = 2^s \binom{\rho+s}{s}.$$

Therefore,

$$|\{g \in \{0, 1, \ldots, N-1\}^s : \rho(g) \leq \rho\}| \leq N^{s-1} 2^s \binom{\rho + s}{s}.$$

Note that the total number of possible generators $g \in \{0, 1, \ldots, N-1\}^s$ is N^s. Thus, if

$$N^{s-1} 2^s \binom{\rho + s}{s} < N^s, \tag{9.33}$$

there exists a $g \in \{0, 1, \ldots, N-1\}^s$ such that $\rho(g) > \rho$. We estimate

$$2^s \binom{\rho + s}{s} \leq 2^s (\rho + s)^s (s!)^{-1}.$$

This means that (9.33) is satisfied if $2^s(\rho + s)^s (s!)^{-1} < N$, i.e., for $\rho = \lceil 2^{-1}(s! \, N)^{1/s} \rceil - s - 1$. Hence (9.32) is shown. Combining (9.30) and (9.32) yields the desired result.□

Let us discuss some tractability issues for the Korobov space $K_{s,\alpha,\gamma}$ where $\alpha > 1$. Assume that $\sum_{i \geq 1} \gamma_i^\lambda < \infty$ for some $\lambda \in (1/\alpha, 1]$. Then we obtain from Theorem 21 that

$$e(\mathscr{H}(K_{s,\alpha,\gamma}); \mathscr{P}(g, N)) \leq \frac{2^{1/(2\lambda)}}{N^{1/(2\lambda)}} \exp\left(\frac{1}{2\lambda} \sum_{i=1}^s \log(1 + 2\gamma_i^\lambda \zeta(\alpha\lambda))\right)$$

$$\leq \frac{2^{1/(2\lambda)}}{N^{1/(2\lambda)}} \exp\left(\frac{\zeta(\alpha\lambda)}{\lambda} \sum_{i \geq 1} \gamma_i^\lambda\right) =: \frac{C_{\alpha,\gamma}}{N^{1/(2\lambda)}}$$

and this bound is independent of the dimension s. Note that for the initial error we have $e(\mathscr{H}(K_{s,\alpha,\gamma}); \emptyset) = 1$; see Sloan and Woźniakowski [111].

For $\varepsilon > 0$ let N be the smallest prime number that is larger or equal $\lceil (C_{\alpha,\gamma} \varepsilon^{-1})^{2\lambda} \rceil =: M$. Then we have $e(\mathscr{H}(K_{s,\alpha,\gamma}); \mathscr{P}(g, N)) \leq \varepsilon$ and hence

$$N_{\mathscr{H}(K_{s,\alpha,\gamma})}(\varepsilon, s) \leq N < 2M = 2\lceil (C_{\alpha,\gamma} \varepsilon^{-1})^{2\lambda} \rceil,$$

where we used Bertrand's postulate which tells us that $M \leq N < 2M$. Hence multivariate integration in $\mathscr{H}(K_{s,\alpha,\gamma})$ is strongly tractable with ε-exponent at most 2λ whenever $\sum_{i \geq 1} \gamma_i^\lambda < \infty$ for some $\lambda \in (1/\alpha, 1]$. The corresponding bounds can be achieved with lattice point sets. Furthermore, in Sloan and Woźniakowski [111, Theorem 5] it was shown that the condition $\sum_{i \geq 1} \gamma_i < \infty$ is also necessary for strong tractability.

Under weaker assumptions on the weights one can still obtain polynomial tractability. For more results in this direction we refer to Sloan and Woźniakowski [111] or to Novak and Woźniakowski [88, Chapter 16].

A discussion of tractability issues for the Korobov space $\mathscr{H}(K'_{s,\omega})$ can be found in Dick, Larcher, Pillichshammer and Woźniakowski [37] and in Kritzer, Pillichshammer, and Woźniakowski [64].

9.4.2 Star-Discrepancy of Lattice Point Sets

For a lattice point set $\mathscr{P}(g, N)$ each point x_n is of the form $x_n = \{y_n/N\}$ with $y_n = ng \in \mathbb{Z}^s$. In particular, the elements of a lattice point set have always rational components. For such point sets there is a variant of the inequality of Erdős-Turán-Koksma due to Niederreiter [81, Theorem 3.10], i.e., a general upper bound for the discrepancy in terms of exponential sums. To formulate this result we need some notation.

For an integer $M \geq 2$, let

$$C(M) = (-M/2, M/2] \cap \mathbb{Z}$$

and let $C_s(M)$ be the Cartesian product of s copies of $C(M)$. Furthermore, let

$$C_s^*(M) = C_s(M) \setminus \{0\}.$$

For $h \in C(M)$ put

$$r(h, M) = \begin{cases} M \sin(\pi|h|/M) & \text{if } h \neq 0, \\ 1 & \text{if } h = 0. \end{cases}$$

For $h = (h_1, \ldots, h_s) \in C_s(M)$, put $r(h, M) = \prod_{i=1}^{s} r(h_i, M)$.

Proposition 22 (Niederreiter [81, Theorem 3.10]). *Let $M \geq 2$ be an integer and let $\mathscr{P} = \{x_0, \ldots, x_{N-1}\}$ be a point set in the s-dimensional unit cube where x_n is of the form $x_n = \{y_n/M\}$ with $y_n \in \mathbb{Z}^s$ for all $0 \leq n < N$. Then we have*

$$D_N^*(\mathscr{P}) \leq 1 - \left(1 - \frac{1}{M}\right)^s + \sum_{h \in C_s^*(M)} \frac{1}{r(h, M)} \left| \frac{1}{N} \sum_{n=0}^{N-1} \exp(2\pi i h \cdot y_n/M) \right|.$$

Applying this result to a lattice point set $\mathscr{P}(g, N)$, i.e., $M = N$, and using (9.26) and the fact that for $h \in C_s^*(N)$ we have $r(h, N) \geq 2r(h)$, where $r(h) := r_{1,1}(h) = \max(1, |h|)$, we obtain the following result.

Theorem 23 (Niederreiter [81, Theorem 5.6]). *For the star-discrepancy of a lattice point set $\mathscr{P}(g, N)$ we have*

$$D_N^*(\mathscr{P}(g, N)) \leq 1 - \left(1 - \frac{1}{N}\right)^s + \frac{1}{2} R(g, N) \leq \frac{s}{N} + \frac{1}{2} R(g, N),$$

where

$$R(g, N) := \sum_{h \in C_s^*(N) \cap \mathscr{L}_{g,N}} \frac{1}{r(h)}.$$

Theorem 23 gives a bound on the star-discrepancy of lattice point sets which is much easier to handle than D_N^* itself (note that the exact computation of the star-discrepancy of a given point set is an NP-hard problem; see Gnewuch, Srivastav, and Winzen [46]). Using (9.26) again, the quantity $R(g, N)$ can be written as

$$R(g, N) = -1 + \frac{1}{N} \sum_{n=0}^{N-1} \prod_{i=1}^{s} \left(1 + \sum_{h \in C^*(N)} \frac{\exp(2\pi i h n g_i / N)}{|h|}\right), \tag{9.34}$$

and hence its calculation requires $O(N^2 s)$ operations. This can be reduced to $O(Ns)$ operations by using an asymptotic expansion; see Joe and Sloan [58].

It has been shown by Larcher [69] that for any dimension $s \geq 2$ there exists some $c_s > 0$ such that for all $N \geq 2$ and all $g \in \mathbb{Z}^s$ we have

$$R(g, N) \geq c_s \frac{(\log N)^s}{N}.$$

On the other hand, it has been shown by Niederreiter [81, Theorem 5.10] that for any integers $s \geq 2$ and $N \geq 2$ we have

$$\frac{1}{|G_s(N)|} \sum_{g \in G_s(N)} R(g, N) = \frac{1}{N} (2 \log N + c)^s - \frac{2s \log N}{N} + O\left(\frac{(\log \log N)^2}{N}\right), \tag{9.35}$$

with $c = 2\gamma - \log 4 + 1 = 0.768 \ldots$, where $\gamma = 0.577 \ldots$ is the Euler constant and where $G_s(N) = \{g = (g_1, \ldots, g_s) \in C_s(N) : \gcd(g_i, N) = 1 \text{ for } 1 \leq i \leq s\}$. In particular, we have the following result.

Theorem 24 (Larcher [69] and Niederreiter [81]). *For any integers $s \geq 2$ and $N \geq 2$ there exist $g \in G_s(N)$ such that*

$$R(g, N) \asymp_s \frac{(\log N)^s}{N}$$

and this order of magnitude is best possible.

If N is a prime we can use the following component-by-component algorithm for the construction of a 'good' lattice point. For a construction for composite N we refer to Sinescu and Joe [103].

Algorithm 25. Let $N \in \mathbb{P}$ and let $s \geq 2$.

1. Choose $g_1 = 1$.
2. For $d > 1$, assume we have already constructed $g_1, \ldots, g_{d-1}$. Then find $g_d \in \{1, \ldots, N-1\}$ which minimises $R_N((g_1, \ldots, g_{d-1}, z))$ as a function of $z \in \{1, \ldots, N-1\}$.

Again, a straightforward implementation of Algorithm 25 would require $O(N^2 s^2)$ operations for the construction of a lattice point set $\mathscr{P}(\boldsymbol{g}, N)$ in dimension s. Using the so-called *fast component-by-component algorithm* due to Nuyens and Cools the construction costs can be reduced to $O(sN \log N)$ operations; see, again, Nuyens and Cools [90–92].

The following result shows that Algorithm 25 provides an optimal lattice point with respect to the order of magnitude of $R(\boldsymbol{g}, N)$.

Theorem 26 (Joe [56]). *Let* $N \in \mathbb{P}$ *and suppose that* $\boldsymbol{g} = (g_1, \ldots, g_s)$ *is constructed according to Algorithm 25. Then for all* $1 \leq d \leq s$ *we have*

$$R(\boldsymbol{g}^{(d)}, N) \leq \frac{1}{N-1}(1 + S_N)^d,$$

where $\boldsymbol{g}^{(d)} = (g_1, \ldots, g_d)$ *and where* $S_N = \sum_{h \in C^*(N)} |h|^{-1}$.

Proof. Since $N \in \mathbb{P}$ it follows that $R(g_1, N) = 0$ for all $g_1 \in \{1, \ldots, N-1\}$. Let $d \geq 1$ and assume that we have

$$R(\boldsymbol{g}, N) \leq \frac{1}{N-1}(1 + S_N)^d,$$

where $\boldsymbol{g} = (g_1, \ldots, g_d)$. Now we consider $(\boldsymbol{g}, g_{d+1}) := (g_1, \ldots, g_d, g_{d+1})$.

As g_{d+1} minimises $R((\boldsymbol{g}, \cdot), N)$ over $\{1, \ldots, N-1\}$ we obtain

$$R((\boldsymbol{g}, g_{d+1}), N) \leq \frac{1}{N-1} \sum_{g_{d+1}=1}^{N-1} \sum_{\substack{(\boldsymbol{h}, h_{d+1}) \in C^*_{d+1}(N) \\ \boldsymbol{h} \cdot \boldsymbol{g} + h_{d+1} g_{d+1} \equiv 0 \pmod N}} \frac{1}{r(\boldsymbol{h})} \frac{1}{r(h_{d+1})}$$

$$= \sum_{(\boldsymbol{h}, h_{d+1}) \in C^*_{d+1}(N)} \frac{1}{r(\boldsymbol{h})} \frac{1}{r(h_{d+1})} \frac{1}{N-1} \sum_{\substack{g_{d+1}=1 \\ h_{d+1} g_{d+1} \equiv -\boldsymbol{h} \cdot \boldsymbol{g} \pmod N}}^{N-1} 1,$$

where we just changed the order of summation. Separating out the term where $h_{d+1} = 0$ we obtain

$$R((\boldsymbol{g}, g_{d+1}), N)$$

$$\leq R(\boldsymbol{g}, N) + \sum_{\boldsymbol{h} \in C_d(N)} \frac{1}{r(\boldsymbol{h})} \sum_{h_{d+1} \in C^*(N)} \frac{1}{r(h_{d+1})} \frac{1}{N-1} \sum_{\substack{g_{d+1}=1 \\ h_{d+1}g_{d+1} \equiv -\boldsymbol{h} \cdot \boldsymbol{g} \pmod{N}}}^{N-1} 1.$$

Since $N \in \mathbb{P}$, the congruence $h_{d+1}g_{d+1} \equiv -\boldsymbol{h} \cdot \boldsymbol{g} \pmod{N}$ has exactly one solution $g_{d+1} \in \{1, \ldots, N-1\}$ if $\boldsymbol{h} \cdot \boldsymbol{g} \not\equiv 0 \pmod{N}$ and no solution in $\{1, \ldots, N-1\}$ if $\boldsymbol{h} \cdot \boldsymbol{g} \equiv 0 \pmod{N}$. From this insight it follows that

$$R((\boldsymbol{g}, g_{d+1}), N) \leq R(\boldsymbol{g}, N) + \frac{1}{N-1} \sum_{\boldsymbol{h} \in C_d(N)} \frac{1}{r(\boldsymbol{h})} \sum_{h_{d+1} \in C^*(N)} \frac{1}{r(h_{d+1})}$$

$$= R(\boldsymbol{g}, N) + \frac{S_N}{N-1} \sum_{\boldsymbol{h} \in C_d(N)} \frac{1}{r(\boldsymbol{h})}$$

$$= R(\boldsymbol{g}, N) + \frac{S_N}{N-1}(1 + S_N)^d$$

$$\leq \frac{1}{N-1}(1 + S_N)^d + \frac{S_N}{N-1}(1 + S_N)^d$$

$$= \frac{1}{N-1}(1 + S_N)^{d+1},$$

where we used the induction hypotheses to bound $R(\boldsymbol{g}, N)$.□

It can be shown that $S_N \leq 2 \log N$ (for a proof of this fact see Niederreiter [76, Lemmas 1 and 2]). Therefore, from Theorems 23 and 26 we obtain the following bound on the star-discrepancy of the lattice point set whose generating vector is constructed with Algorithm 25.

Corollary 27 (Joe [56]). *Let $N \in \mathbb{P}$ and suppose that $\boldsymbol{g} = (g_1, \ldots, g_s)$ is constructed according to Algorithm 25. Then for all $1 \leq d \leq s$ we have*

$$D_N^*(\mathscr{P}(\boldsymbol{g}^{(d)}, N)) \leq \frac{d + (2 \log N)^d}{N},$$

where $\boldsymbol{g}^{(d)} = (g_1, \ldots, g_d)$.

Korobov suggested the use lattice points of the form $\boldsymbol{g} = (1, g, g^2, \ldots, g^{s-1})$ with $g \in \mathbb{Z}$ to restrict the number of candidates that must be inspected. At least for $N \in \mathbb{P}$ there is a result in the vein of (9.35) for such so-called *Korobov lattice points*. Niederreiter [81, Theorem 5.18] showed that for any $N \in \mathbb{P}$ and any integer $s \geq 2$ we have

$$\frac{1}{N} \sum_{g=0}^{N-1} R((1, g, g^2, \ldots, g^{s-1}), N) < \frac{s-1}{N}(2 \log N + 1)^s.$$

Combined with Theorem 23 this leads to the following result.

Corollary 28. *Let $N \in \mathbb{P}$ and let $s \geq 2$ be an integer. For any real $0 < \varepsilon \leq 1$ there exist more than $(1 - \varepsilon)N$ elements $g \in \{0, \ldots, N - 1\}$ such that*

$$D_N^*(\mathscr{P}((1, g, g^2, \ldots, g^{s-1}), N)) \leq \frac{s}{N} + \frac{1}{\varepsilon} \frac{s-1}{2N}(2 \log N + 1)^s.$$

Recently Bykovskii [10] showed the existence of lattice point sets $\mathscr{P}(g, N)$ in dimension s with star discrepancy of order of magnitude $\frac{(\log N)^{s-1} \log \log N}{N}$.

9.4.3 Weighted Star-Discrepancy of Lattice Point Sets

For the weighted star-discrepancy of a lattice point set $\mathscr{P}(g, N)$ we obtain from Theorem 23

$$
\begin{aligned}
D_{N,\gamma}^*(\mathscr{P}(g, N)) &= \max_{\emptyset \neq \mathfrak{u} \subseteq [s]} \gamma_{\mathfrak{u}} D_N^*(\mathscr{P}(g_{\mathfrak{u}}, N)) \\
&\leq \sum_{\emptyset \neq \mathfrak{u} \subseteq [s]} \gamma_{\mathfrak{u}} D_N^*(\mathscr{P}(g_{\mathfrak{u}}, N)) \\
&\leq \sum_{\emptyset \neq \mathfrak{u} \subseteq [s]} \gamma_{\mathfrak{u}} \left(1 - \left(1 - \frac{1}{N}\right)^{|\mathfrak{u}|}\right) + \frac{1}{2} \sum_{\emptyset \neq \mathfrak{u} \subseteq [s]} \gamma_{\mathfrak{u}} R(g_{\mathfrak{u}}, N), \quad (9.36)
\end{aligned}
$$

where $g_{\mathfrak{u}}$ denotes the projection of g onto the components given by $\mathfrak{u}$. Hence $\mathscr{P}(g_{\mathfrak{u}}, N)$ is the $|\mathfrak{u}|$-dimensional lattice point set which is obtained by a projection of the points from $\mathscr{P}(g, N)$ onto the components given by $\mathfrak{u}$.

Set $\tilde{r}(h, \gamma) = 1 + \gamma$ if $h = 0$, and $\gamma r(h)$ if $h \neq 0$ and set $\tilde{r}(h, \gamma) = \prod_{i=1}^{s} \tilde{r}(h_i, \gamma_i)$. Then it follows from (9.34) that

$$
\begin{aligned}
&\sum_{\emptyset \neq \mathfrak{u} \subseteq [s]} \gamma_{\mathfrak{u}} R(g_{\mathfrak{u}}, N) \\
&= -\sum_{\emptyset \neq \mathfrak{u} \subseteq [s]} \gamma_{\mathfrak{u}} + \sum_{\emptyset \neq \mathfrak{u} \subseteq [s]} \frac{1}{N} \sum_{n=0}^{N-1} \prod_{i \in \mathfrak{u}} \gamma_i \left(1 + \sum_{h \in C^*(N)} \frac{\exp(2\pi i h n g_i / N)}{|h|}\right) \\
&= -\prod_{i=1}^{s}(1 + \gamma_i) + \frac{1}{N} \sum_{n=0}^{N-1} \prod_{i=1}^{s} \left(1 + \gamma_i + \gamma_i \sum_{h \in C^*(N)} \frac{\exp(2\pi i h n g_i / N)}{|h|}\right) \\
&= -\prod_{i=1}^{s}(1 + \gamma_i) + \frac{1}{N} \sum_{n=0}^{N-1} \prod_{i=1}^{s} \sum_{h \in C(N)} \tilde{r}(h, \gamma_i) \exp(2\pi i h n g_i / N) \quad (9.37)
\end{aligned}
$$

$$= -\prod_{i=1}^{s}(1 + \gamma_i) + \sum_{\mathbf{h} \in C_s(N)} \tilde{r}(\mathbf{h}, \boldsymbol{\gamma}) \frac{1}{N} \sum_{n=0}^{N-1} \exp(2\pi \mathrm{i} \mathbf{h} \cdot \mathbf{g}n/N)$$

$$= \sum_{\mathbf{h} \in C_s^*(N) \cap \mathscr{L}_{\mathbf{g},N}} \tilde{r}(\mathbf{h}, \boldsymbol{\gamma}) =: \tilde{R}_{\boldsymbol{\gamma}}(\mathbf{g}, N). \tag{9.38}$$

From (9.37) we see that $\tilde{R}_{\boldsymbol{\gamma}}(\mathbf{g}, N)$ can be computed in $O(N^2 s)$ operations. Again this can be reduced to $O(Ns)$ operations by using an asymptotic expansion; for details see Joe [57, Appendix A].

Hence for the weighted star-discrepancy of a lattice point set $\mathscr{P}(\mathbf{g}, N)$ we obtain

$$D_{N,\boldsymbol{\gamma}}^*(\mathscr{P}(\mathbf{g}, N)) \leq \sum_{\emptyset \neq \mathfrak{u} \subseteq [s]} \gamma_{\mathfrak{u}} \left(1 - \left(1 - \frac{1}{N}\right)^{|\mathfrak{u}|}\right) + \frac{1}{2}\tilde{R}_{\boldsymbol{\gamma}}(\mathbf{g}, N).$$

If the weights $\gamma_i, i \geq 1$, are summable, then it has been shown by Joe [57] that

$$\sum_{\emptyset \neq \mathfrak{u} \subseteq [s]} \gamma_{\mathfrak{u}} \left(1 - \left(1 - \frac{1}{N}\right)^{|\mathfrak{u}|}\right) \leq \frac{\max(1, \Gamma) \exp(\sum_{l \geq 1} \gamma_i)}{N} \tag{9.39}$$

where $\Gamma = \sum_{i \geq 1} \gamma_i/(1 + \gamma_i) < \infty$.

If $N \in \mathbb{P}$ one can again use the component-by-component algorithm (Algorithm 25) with R replaced by $\tilde{R}_{\boldsymbol{\gamma}}$ for the construction of a 'good' lattice point.

Theorem 29 (Joe [57]). *Let $N \in \mathbb{P}$ and suppose that $\mathbf{g} = (g_1, \ldots, g_s)$ is constructed according to Algorithm 25 (with R replaced by $\tilde{R}_{\boldsymbol{\gamma}}$). Then for all $1 \leq d \leq s$ we have*

$$\tilde{R}_{\boldsymbol{\gamma}}(\mathbf{g}^{(d)}, N) \leq \frac{1}{N-1} \prod_{i=1}^{d}(1 + \gamma_i(1 + S_N))^d,$$

where $\mathbf{g}^{(d)} = (g_1, \ldots, g_d)$ and $S_N = \sum_{\mathbf{h} \in C^*(N)} |h|^{-1}$.

Using the estimate $S_N \leq 2 \log N$ from the previous section we obtain the following result.

Corollary 30 (Joe [57]). *Let $N \in \mathbb{P}$ and suppose that $\mathbf{g} = (g_1, \ldots, g_s)$ is constructed according to Algorithm 25 (with R replaced by $\tilde{R}_{\boldsymbol{\gamma}}$). Then for all $1 \leq d \leq s$ we have*

$$D_{N,\boldsymbol{\gamma}}^*(\mathscr{P}(\mathbf{g}^{(d)}, N)) \leq \sum_{\emptyset \neq \mathfrak{u} \subseteq [d]} \gamma_{\mathfrak{u}} \left(1 - \left(1 - \frac{1}{N}\right)^{|\mathfrak{u}|}\right) + \frac{1}{N} \prod_{i=1}^{d}(1 + 2\gamma_i \log N),$$

where $\mathbf{g}^{(d)} = (g_1, \ldots, g_d)$ and $[d] = \{1, \ldots, d\}$.

Assume that $\sum_{i\geq 1}\gamma_i < \infty$, then we obtain from Hickernell and Niederreiter [50, Lemma 3] that for any $\delta > 0$ there exists a $c_{\gamma,\delta} > 0$, independent of s and γ, such that

$$\prod_{i=1}^{s}(1 + 2\gamma_i \log N) \leq c_{\gamma,\delta}N^\delta \text{ for any } s \in \mathbb{N}.$$

Using this, (9.39) and Corollary 30 we obtain the following result.

Corollary 31. *Let* $N \in \mathbb{P}$ *and suppose that* g *is constructed according to Algorithm 25 (with R replaced by $\tilde{R}_\gamma$).*

If $\sum_{i\geq 1}\gamma_i < \infty$, *then for any* $\delta > 0$ *there exists a* $c_{\gamma,\delta} > 0$, *independent of s and N, such that the weighted star-discrepancy of* $\mathscr{P}(g, N)$ *satisfies*

$$D^*_{N,\gamma}(\mathscr{P}(g,N)) \leq \frac{c_{\gamma,\delta}}{N^{1-\delta}}.$$

In particular, if $\sum_{i\geq 1}\gamma_i < \infty$, then for any prime number N it follows from Corollary 31 that

$$\text{disc}_{\infty,\gamma}(N, s) \leq \frac{c_{\gamma,\delta}}{N^{1-\delta}}$$

and that the bound can be achieved by a lattice point set.

Note that $\text{disc}_{\infty,\gamma}(0, s) = \max_{\emptyset \neq u \subseteq [s]} \gamma_u \geq \gamma_1 > 0$.

For $\varepsilon > 0$ and $\delta > 0$ let N be the smallest prime number that is larger or equal to $\lceil (c_{\gamma,\delta}\gamma_1^{-1}\varepsilon^{-1})^{1/(1-\delta)}\rceil =: M$. Then we have $\text{disc}_{\infty,\gamma}(N, s) \leq \varepsilon\text{disc}_{\infty,\gamma}(0, s)$ and hence

$$N_{\infty,\gamma}(\varepsilon, s) \leq N < 2M = 2\lceil (c_{\gamma,\delta}\gamma_1^{-1}\varepsilon^{-1})^{1/(1-\delta)}\rceil,$$

where we used Bertrand's postulate which tells us that $M \leq N < 2M$. This bound, which is independent of the dimension s, was already presented in (9.19) and shows that the weighted star-discrepancy is strongly tractable with ε-exponent equal to one whenever the weights γ_i, $i \geq 1$, are summable.

For the weighted L_p-discrepancy of a point set $\mathscr{P}$ we have

$$L_{p,\gamma}(\mathscr{P}) \leq \left(\sum_{\emptyset \neq u \subseteq [s]} \gamma_u^p (D^*_N(\mathscr{P}_u))^p\right)^{1/p} \leq \sum_{\emptyset \neq u \subseteq [s]} \gamma_u D^*_N(\mathscr{P}_u),$$

where we used Jensen's inequality, which states that $\sum_k a_k^p \leq \left(\sum_k a_k\right)^p$ for any $p \geq 1$ and non-negative reals a_k. Hence, for a lattice point set we obtain from (9.36) and (9.38) that

$$L_{p,\gamma}(\mathscr{P}(g, N)) \leq \sum_{\emptyset \neq u \subseteq [s]} \gamma_u \left(1 - \left(1 - \frac{1}{N}\right)^{|u|}\right) + \frac{1}{2}\tilde{R}_\gamma(g, N).$$

This means that the results for the weighted star-discrepancy apply also for the weighted L_p-discrepancy. In particular, if $\sum_{i \geq 1} \gamma_i < \infty$, then the weighted L_p-discrepancy is strongly tractable with ε-exponent equal to one. (Note that $L_{p,\gamma}^p(\emptyset) = -1 + \prod_{i=1}^s (1 + \frac{\gamma_i^p}{p+1})$.)

9.5 Polynomial Lattice Rules

There is also an algebraic analogue of lattice rules which is based on arithmetic of polynomials over finite fields. This construction has first been introduced by Niederreiter [80] as special construction of digital nets over a finite field $\mathbb{F}_q$ where q is a prime power. For simplicity, here we only consider prime bases b and the finite field $\mathbb{Z}_b$ of order b. On the other hand, here we generalize Niederreiter's approach to get a construction for higher order nets. This was first considered in [30].

Let $b \in \mathbb{P}$, let $\mathbb{Z}_b[x]$ denote the set of polynomials in x with coefficients in $\mathbb{Z}_b$ and let $\mathbb{Z}_b((x^{-1}))$ denote the set of formal Laurent series $\sum_{l=w}^\infty u_l x^{-l}$ where $w \in \mathbb{Z}$ and $u_l \in \mathbb{Z}_b$. For $n \in \mathbb{N}$ let $G_{b,n} := \{q \in \mathbb{Z}_b[x] : \deg(q) < n\}$ and $G_{b,n}^* = G_{b,n} \setminus \{0\}$. Note that $|G_n| = b^n$.

Definition 32. Let $\alpha, m, s \in \mathbb{N}$ and choose an irreducible polynomial $p \in \mathbb{Z}_b[x]$ with $\deg(p) = \alpha m$. Further let $q = (q_1, \ldots, q_s) \in G_{b,\alpha m}^s$ and consider the expansions

$$\frac{q_i(x)}{p(x)} = \sum_{l=w_i}^\infty u_l^{(i)} x^{-l} \in \mathbb{Z}_b((x^{-1})),$$

where $w_i \leq 1$. Define the $\alpha m \times m$ matrices $C_1, \ldots, C_s$ over $\mathbb{Z}_b$, $C_i = (c_{k,l}^{(i)})$, by

$$c_{k,l}^{(i)} = u_{k+l-1}^{(i)} \in \mathbb{Z}_b \quad \text{for } 1 \leq i \leq s, 1 \leq k \leq \alpha m, 1 \leq l \leq m.$$

The matrices $C_1, \ldots, C_s$ generate a digital (t, m, s)-net $\mathscr{P}_\alpha(q, p)$ over $\mathbb{Z}_b$ which is called a *polynomial lattice point set*. The quadrature rule $Q_{\mathscr{P}_\alpha(q,p)}$ is called a *polynomial lattice rule*.

Polynomial lattice point sets can also be introduced independent from digital net theory. To this end let v_n, $n \in \mathbb{N}$, be the map from $\mathbb{Z}_b((x^{-1}))$ to the interval $[0, 1)$ defined by

$$v_n\left(\sum_{l=w}^\infty u_l x^{-l}\right) = \sum_{l=\max(1,w)}^n u_l b^{-l}.$$

Then the following construction is equivalent to Definition 32; see [31, Chapter 10]. For a given dimension $s \geq 1$, choose $p \in \mathbb{Z}_b[x]$ with $\deg(p) = \alpha m$ and let $q_1, \ldots, q_s \in \mathbb{Z}_b[x]$. Then $\mathscr{P}_\alpha(\boldsymbol{q}, p)$ is the point set consisting of the b^m points

$$
\boldsymbol{x}_h = \left(v_{\alpha m}\left(\frac{h(x)q_1(x)}{p(x)} \right), \ldots, v_{\alpha m}\left(\frac{h(x)q_s(x)}{p(x)} \right) \right) \quad \text{for } h \in G_{b,m}.
$$

Observe the similarity of this construction with the definition of ordinary lattice point sets from Definition 13, which is the reason for calling $\mathscr{P}_\alpha(\boldsymbol{q}, p)$ a polynomial lattice point set.

Classical polynomial lattice point sets assume that $\alpha = 1$, which also corresponds to the case of classical digital nets; see Niederreiter [80,81], [31, Chapter 10] or [96]. In this case we simply omit the index α and write $\mathscr{P}(\boldsymbol{q}, p)$.

The dual polynomial lattice is of importance for studying polynomial lattice rules.

Definition 33. Let $\alpha, m \in \mathbb{N}$. The *dual polynomial lattice* of a polynomial lattice point set $\mathscr{P}_\alpha(\boldsymbol{q}, p)$ with generating vector $\boldsymbol{q} \in G_{b,\alpha m}^s$ and $p \in \mathbb{Z}_b[x]$ with $\deg(p) = \alpha m$, is given by

$$
\mathscr{D}_\alpha(\boldsymbol{q}, p) = \{ \boldsymbol{k} \in G_{b,\alpha m}^s : \boldsymbol{k} \cdot \boldsymbol{q} \equiv a \pmod{p} \text{ where } \deg(a) < (\alpha - 1)m \}.
$$

For convenience we also write $\mathscr{D}_\alpha^*(\boldsymbol{q}, p) = \mathscr{D}_\alpha(\boldsymbol{q}, p) \setminus \{\boldsymbol{0}\}$. If $\alpha = 1$ we again omit the index α for the sake of simplicity.

For the following we will need the concept of b-adic Walsh functions which we introduce now. For $b \geq 2$ we denote by ω_b the bth primitive root of unity $\exp(2\pi i/b)$.

Definition 34. Let $k \in \mathbb{N}_0$ with b-adic expansion $k = \kappa_0 + \kappa_1 b + \kappa_2 b^2 + \cdots$. The kth b-adic Walsh function ${}_b\mathrm{wal}_k : \mathbb{R} \to \mathbb{C}$, periodic with period one, is defined as

$$
{}_b\mathrm{wal}_k(x) = \omega_b^{\kappa_0 \xi_1 + \kappa_1 \xi_2 + \kappa_2 \xi_3 + \cdots},
$$

for $x \in [0, 1)$ with b-adic expansion $x = \xi_1 b^{-1} + \xi_2 b^{-2} + \xi_3 b^{-3} + \cdots$ (unique in the sense that infinitely many of the digits ξ_i must be different from $b - 1$). We call the system $\{ {}_b\mathrm{wal}_k : k \in \mathbb{N}_0 \}$ the *b-adic Walsh function system*.

Now we generalise the definition of Walsh functions to higher dimensions.

Definition 35. For dimension $s \geq 2$, and $k_1, \ldots, k_s \in \mathbb{N}_0$ we define the s-dimensional b-adic Walsh function ${}_b\mathrm{wal}_{k_1,\ldots,k_s} : \mathbb{R}^s \to \mathbb{C}$ by

$$
{}_b\mathrm{wal}_{k_1,\ldots,k_s}(x_1, \ldots, x_s) := \prod_{j=1}^s {}_b\mathrm{wal}_{k_j}(x_j).
$$

For vectors $\boldsymbol{k} = (k_1, \ldots, k_s) \in \mathbb{N}_0^s$ and $\boldsymbol{x} = (x_1, \ldots, x_s) \in [0, 1)^s$ we write, with some abuse of notation,

$$_b\mathrm{wal}_{\boldsymbol{k}}(\boldsymbol{x}) := {}_b\mathrm{wal}_{k_1, \ldots, k_s}(x_1, \ldots, x_s).$$

The system $\{{}_b\mathrm{wal}_{\boldsymbol{k}} : \boldsymbol{k} \in \mathbb{N}_0^s\}$ is called the *s-dimensional b-adic Walsh function system*.

Basic properties of Walsh functions are summarised in [31, Appendix A].

An important property of polynomial lattice point sets is that

$$\sum_{\boldsymbol{x} \in \mathscr{P}_\alpha(\boldsymbol{q}, p)} {}_b\mathrm{wal}_{\boldsymbol{k}}(\boldsymbol{x}) = \begin{cases} b^m & \text{if } \boldsymbol{k} \in \mathscr{D}_\alpha(\boldsymbol{q}, p), \\ 0 & \text{otherwise.} \end{cases} \tag{9.40}$$

For a proof of this property we refer to [31, Lemmas 4.75 and 10.6] for $\alpha = 1$ and [31, Lemmas 4.75 and 15.25] for $\alpha \geq 1$.

Based on the definition of polynomial lattice point sets in terms of digital nets, the dual polynomial lattice is related to the dual net of a digital net; see [31, Lemma 10.6] or Niederreiter [81, Lemma 4.40]. There is also a concept related to the t-value of digital nets which we introduce in the following.

Definition 36. Let $\alpha, m \in \mathbb{N}$. The *figure of merit* ϱ_α of a polynomial lattice point set $\mathscr{P}_\alpha(\boldsymbol{q}, p)$ with generating vector $\boldsymbol{q} \in G_{b,\alpha m}^s$ and modulus $p \in \mathbb{Z}_b[x]$ with $\deg(p) = \alpha m$ is given by

$$\varrho_\alpha(\boldsymbol{q}, p) = -1 + \min_{\boldsymbol{k} \in \mathscr{D}_\alpha^*(\boldsymbol{q}, p)} \deg_\alpha(\boldsymbol{k}),$$

where $\deg_\alpha(\boldsymbol{k}) = \sum_{i=1}^s \deg_\alpha(k_i)$ for $\boldsymbol{k} = (k_1, \ldots, k_s)$ and

$$\deg_\alpha(k_i) = \begin{cases} \sum_{u=1}^{\min(\alpha, v)} a_u & \text{for } k_i(x) = \kappa_1 x^{a_1-1} + \cdots + \kappa_v x^{a_v-1}, \\ 0 & \text{otherwise,} \end{cases}$$

with $a_1 > a_2 > \cdots > a_v > 0$ and $\kappa_1, \ldots, \kappa_v \in \mathbb{Z}_b \setminus \{0\}$. For $\alpha = 1$ we again omit the index α and write $\varrho(\boldsymbol{q}, p)$.

The following theorem connects the figure of merit of a polynomial lattice point set with the t-value.

Theorem 37 (Niederreiter [81, Theorem 4.42], Dick, Kritzer, Pillichshammer and Schmid [36, Theorem 2]). *Let $\alpha, m \in \mathbb{N}$. Let $\mathscr{P}_\alpha(\boldsymbol{q}, p)$ be a polynomial lattice point set with $\boldsymbol{q} \in G_{b,\alpha m}^s$ and modulus $p \in \mathbb{Z}_b[x]$ with $\deg(p) = \alpha m$. Then $\mathscr{P}_\alpha(\boldsymbol{q}, p)$ is a digital $(t, \alpha, 1, \alpha m \times m, s)$-net in base b with*

$$t = \alpha m - \varrho_\alpha(\boldsymbol{q}, p).$$

This result connects polynomial lattice point sets and digital nets. Hence the bounds on the discrepancy of digital nets, that is, Theorems 4, 5, 8, also apply to polynomial lattice point sets. Further results on the star-discrepancy of polynomial lattice point sets will be presented in Sect. 9.5.3.

The usefulness of polynomial lattice point sets for multivariate integration has been shown by connecting it to digital nets. However, as opposed to digital nets and sequences, no explicit constructions of polynomial lattice point sets are known except for dimension 2, see Niederreiter [81, pp. 87, 88]. For higher dimensions one relies on computer search algorithms to find good polynomial lattice rules. There are several methods of finding good polynomial lattice point sets, each based on a different criterion, which we consider in the following.

9.5.1 The Worst-Case Error of Polynomial Lattice Rules in Weighted Walsh Spaces

Similarly to lattice rules, we introduce a suitable space of functions for which the integration error of polynomial lattice rules for $\alpha = 1$ can be analysed (the case $\alpha > 1$ will be considered later in this section). This function space is based on Walsh functions as introduced in Definitions 34 and 35, respectively.

We define a Walsh function space analogously to the Korobov space, essentially by replacing the exponential function $\exp(2\pi ihx)$ by Walsh functions ${}_b\mathrm{wal}_k$.

Let $s \geq 1$ and $b \geq 2$ be integers, $\delta > 1$ a real and $\boldsymbol{\gamma} = (\gamma_i)_{i \geq 1}$ a sequence of nonnegative reals. The s-dimensional weighted Walsh space is the reproducing kernel Hilbert space of b-adic Walsh series $f(\boldsymbol{x}) = \sum_{\boldsymbol{k} \in \mathbb{N}_0^s} \hat{f}(\boldsymbol{k}) \, {}_b\mathrm{wal}_{\boldsymbol{k}}(\boldsymbol{x})$ with reproducing kernel defined by

$$K_{\mathrm{wal},s,b,\delta,\boldsymbol{\gamma}}(\boldsymbol{x}, \boldsymbol{y}) = \sum_{\boldsymbol{k} \in \mathbb{N}_0^s} r_{\mathrm{wal},b,\delta}(\boldsymbol{k}, \boldsymbol{\gamma}) \, {}_b\mathrm{wal}_{\boldsymbol{k}}(\boldsymbol{x} \ominus \boldsymbol{y}),$$

where for $\boldsymbol{k} = (k_1, \ldots, k_s)$ we put $r_{\mathrm{wal},b,\delta}(\boldsymbol{k}, \boldsymbol{\gamma}) = \prod_{i=1}^{s} r_{\mathrm{wal},b,\delta}(k_i, \gamma_i)$ and for $k \in \mathbb{N}_0$ and $\gamma > 0$ we write

$$r_{\mathrm{wal},b,\delta}(k, \gamma) = \begin{cases} 1 & \text{if } k = 0, \\ \gamma b^{-\delta a} & \text{if } k = \kappa_0 + \kappa_1 b + \cdots + \kappa_a b^a \text{ and } \kappa_a \neq 0. \end{cases}$$

Furthermore, for $x = \sum_{i=w}^{\infty} \xi_i b^{-i}$ and $y = \sum_{i=w}^{\infty} \eta_i b^{-i}$ by $\ominus$ we denote the digit-wise subtraction modulo b, i.e.,

$$x \ominus y := \sum_{i=w}^{\infty} z_i b^{-i} \quad \text{where} \quad z_i := x_i - y_i \pmod{b}.$$

For vectors $\boldsymbol{x}, \boldsymbol{y}$ we apply $\ominus$ component wise.

The inner-product in $\mathcal{H}(K_{\mathrm{wal},s,b,\delta,\gamma})$ is given by

$$\langle f, g \rangle_{\mathcal{H}(K_{\mathrm{wal},s,b,\delta,\gamma})} = \sum_{k \in \mathbb{N}_0^s} r_{\mathrm{wal},b,\delta}(k, \gamma)^{-1} \hat{f}(k) \overline{\hat{g}(k)},$$

where the Walsh coefficients are given by $\hat{f}(k) = \int_{[0,1]^s} f(x) \, _b\mathrm{wal}_k(x) \, \mathrm{d}x$.

Similarly to the Korobov space, the worst-case integration error for a QMC rule in $\mathcal{H}(K_{\mathrm{wal},s,b,\delta,\gamma})$ using a polynomial lattice point set $\mathcal{P}(q, p) = \{x_0, \ldots, x_{N-1}\}$ is given by

$$e^2(\mathcal{H}(K_{\mathrm{wal},s,b,\delta,\gamma}); \mathcal{P}(q, p)) = \sum_{k \in \mathbb{N}_0^s \setminus \{0\}} r_{\mathrm{wal},b,\delta}(k, \gamma) \left| \frac{1}{N} \sum_{n=0}^{N-1} \, _b\mathrm{wal}_k(x_n) \right|^2.$$

There is also an analogue for Theorem 15.

Theorem 38 (Dick, Kuo, Pillichshammer and Sloan [17, Lemma 4.1]). *The worst-case error of a polynomial lattice rule for integration in the Walsh space $\mathcal{H}(K_{\mathrm{wal},s,b,\delta,\gamma})$ is given by*

$$e^2(\mathcal{H}(K_{\mathrm{wal},s,b,\delta,\gamma}); \mathcal{P}(q, p)) = \sum_{h \in \mathcal{D}^*(q,p)} r_{\mathrm{wal},b,\delta}(k, \gamma).$$

In [28, Theorem 2] it was shown that there is a concise formula for the square worst-case error for a QMC rule based on a digital net. Applying this formula for a polynomial lattice point set $\mathcal{P}(q, p) = \{x_0, \ldots, x_{b^m-1}\}$ with $x_n = (x_{n,1}, \ldots, x_{n,s})$, we have

$$e^2(\mathcal{H}(K_{\mathrm{wal},s,b,\delta,\gamma}); \mathcal{P}(q, p)) = -1 + \frac{1}{b^m} \sum_{n=0}^{b^m-1} \prod_{i=1}^{s} (1 + \gamma_i \phi(x_{n,i})),$$

where for $x \in [0, 1)$ we have

$$\phi(x) = \frac{b^\delta(b-1)}{b^\delta - b} - \begin{cases} 0 & \text{if } x = 0, \\ b^{\lfloor \log_b x \rfloor (\delta-1)} \frac{b^{2\delta} - b^\delta}{b^\delta - b} & \text{if } x > 0. \end{cases}$$

This equation can now be used to obtain a construction algorithm for polynomial lattice point sets in the following way.

Algorithm 39. Let $p \in \mathbb{Z}_b[x]$ be a polynomial of degree $m \in \mathbb{N}$ and let $s \geq 2$.

1. Choose $q_1 = 1 \in \mathbb{Z}_b[x]$.
2. For $d > 1$, assume we have already constructed $q_1, \ldots, q_{d-1} \in \mathbb{Z}_b[x]$. Then find $z \in G_{b,m}$ which minimises $e^2(\mathcal{H}(K_{\mathrm{wal},d,b,\delta,\gamma}); \mathcal{P}((q_1, \ldots, q_{d-1}, z), p))$ as a function of $z \in G_{b,m}$.

For polynomial lattice point sets constructed by Algorithm 39 we have the following result.

Theorem 40 (Dick, Kuo, Pillichshammer and Sloan [17, Theorem 4.4]). *Let* $p \in \mathbb{Z}_b[x]$ *be irreducible, with* $\deg(p) = m$. *Suppose* $\boldsymbol{q} = (q_1, \ldots, q_s) \in G_{b,m}^s$ *is constructed by Algorithm 39. Then for all* $1 \leq d \leq s$ *we have*

$$
e(\mathscr{H}(K_{\mathrm{wal},d,b,\delta,\boldsymbol{\gamma}}); \mathscr{P}(\boldsymbol{q}^{(d)}, p)) \leq \frac{2^{1/(2\lambda)}}{b^{m/(2\lambda)}} \prod_{i=1}^{d} (1 + \mu(\delta\lambda)\gamma_i^\lambda)^{1/(2\lambda)},
$$

for all $1/\delta < \lambda \leq 1$, *where* $\boldsymbol{q}^{(d)} = (q_1, \ldots, q_d)$ *and where* $\mu(\delta\lambda) = \frac{b^{\delta\lambda}(b-1)}{b^{\delta\lambda}-b}$.

Assume that $\sum_{i \geq 1} \gamma_i^\lambda < \infty$ for some $\lambda \in (1/\delta, 1]$. Then we obtain from Theorem 40 that

$$
e(\mathscr{H}(K_{\mathrm{wal},s,b,\delta,\boldsymbol{\gamma}}); \mathscr{P}(\boldsymbol{q}, p)) \leq \frac{2^{1/(2\lambda)}}{b^{m/(2\lambda)}} \exp\left(\frac{1}{2\lambda} \sum_{i=1}^{s} \log(1 + \mu(\delta\lambda)\gamma_i^\lambda) \right)
$$

$$
\leq \frac{2^{1/(2\lambda)}}{b^{m/(2\lambda)}} \exp\left(\frac{\mu(\delta\lambda)}{2\lambda} \sum_{i \geq 1} \gamma_i^\lambda \right) =: \frac{C_{b,\delta,\boldsymbol{\gamma}}}{b^{m/(2\lambda)}}
$$

and this bound is independent of the dimension s. Note that for the initial error we have $e(\mathscr{H}(K_{\mathrm{wal},s,b,\delta,\boldsymbol{\gamma}}); \emptyset) = 1$; see [28, p. 162].

For $\varepsilon > 0$ choose $m \in \mathbb{N}$ such that $b^{m-1} < \lceil (C_{b,\delta,\boldsymbol{\gamma}}\varepsilon^{-1})^{2\lambda} \rceil =: M \leq b^m$. Then we have $e(\mathscr{H}(K_{\mathrm{wal},s,b,\delta,\boldsymbol{\gamma}}); \mathscr{P}(\boldsymbol{q}, p)) \leq \varepsilon$ and hence

$$
N_{\mathscr{H}(K_{\mathrm{wal},s,b,\delta,\boldsymbol{\gamma}})}(\varepsilon, s) \leq b^m < bM = b\lceil (C_{b,\delta,\boldsymbol{\gamma}}\varepsilon^{-1})^{2\lambda} \rceil.
$$

Thus multivariate integration in $\mathscr{H}(K_{\mathrm{wal},s,b,\delta,\boldsymbol{\gamma}})$ is strongly tractable with ε-exponent at most 2λ whenever $\sum_{i \geq 1} \gamma_i^\lambda < \infty$ for some $\lambda \in (1/\delta, 1]$. The corresponding bounds can be achieved with polynomial lattice point sets. Furthermore, in [28, Corollary 1] it was shown that the condition $\sum_{i \geq 1} \gamma_i < \infty$ is also necessary for strong tractability.

Under weaker assumption on the weights one can still obtain polynomial tractability. For more results in this direction we refer to [28].

To put the result from Theorem 40 into context, we provide a lower bound on the integration error, which can be viewed as an analogue to Theorem 17. This result shows that the convergence rate in Theorem 40 is almost best possible.

Theorem 41 (Roth [98] and Heinrich, Hickernell, and Yue [47, Theorem 9]). *Let* $\mathscr{P}$ *be an arbitrary* N-*element point set in* $[0,1]^s$ *and let* $\boldsymbol{w} = (w_0, \ldots, w_{N-1}) \in \mathbb{R}^N$. *Then for any* $\delta > 1$

$$e(\mathcal{H}(K_{\mathrm{wal},s,b,\delta,\gamma}); \mathcal{P}; w) \geq C_{s,\delta,\gamma} \frac{(\log N)^{(s-1)/2}}{N^{\delta/2}},$$

where $C_{s,\delta,\gamma} > 0$ depends on δ, s, γ but not on N.

Proof. Choose $t \in \mathbb{N}_0$ such that

$$2N \leq b^t < 4N.$$

Let $\boldsymbol{m} = (m_1, \ldots, m_s) \in \mathbb{N}_0^s$ with $\|\boldsymbol{m}\|_1 = m_1 + \cdots + m_s = t$ and let $\boldsymbol{l} = (l_1, \ldots, l_s) \in \mathbb{N}_0^s$ with $0 \leq l_i < b^{m_i}$ for $1 \leq i \leq s$. Set

$$B_{\boldsymbol{l},\boldsymbol{m}} = \prod_{i=1}^{s} \left[\frac{l_i}{b^{m_i}}, \frac{l_i + 1}{b^{m_i}} \right).$$

For a given $\boldsymbol{x} \in [0, 1)^s$ let $\boldsymbol{l}(\boldsymbol{x})$ be such that $\boldsymbol{x} \in B_{\boldsymbol{l}(\boldsymbol{x}),\boldsymbol{m}}$. Now we define

$$f_{\boldsymbol{m}}(\boldsymbol{x}) = \begin{cases} 0 & \text{if } B_{\boldsymbol{l}(\boldsymbol{x}),\boldsymbol{m}} \cap \mathcal{P} \neq \emptyset, \\ 1 & \text{otherwise.} \end{cases}$$

Then we have

$$\int_{[0,1]^s} f_{\boldsymbol{m}}(\boldsymbol{x}) \, d\boldsymbol{x} \geq 1 - \frac{N}{b^t} \geq \frac{1}{2}.$$

We set

$$F(\boldsymbol{x}) = \sum_{\substack{\boldsymbol{m} \in \mathbb{N}_0^s \\ \|\boldsymbol{m}\|_1 = t}} f_{\boldsymbol{m}}(\boldsymbol{x})$$

and therefore we have

$$\int_{[0,1]^s} F(\boldsymbol{x}) \, d\boldsymbol{x} = \sum_{\substack{\boldsymbol{m} \in \mathbb{N}_0^s \\ \|\boldsymbol{m}\|_1 = t}} \int_{[0,1]^s} f_{\boldsymbol{m}}(\boldsymbol{x}) \, d\boldsymbol{x} \geq \sum_{\substack{\boldsymbol{m} \in \mathbb{N}_0^s \\ \|\boldsymbol{m}\|_1 = t}} \frac{1}{2} \geq \frac{1}{2} \binom{t+s-1}{s-1}.$$

Further, we have $F(\boldsymbol{x}_n) = 0$ for all $1 \leq n \leq N$ and thus

$$\left| \int_{[0,1]^s} F(\boldsymbol{x}) \, d\boldsymbol{x} - \sum_{n=1}^{N} w_n F(\boldsymbol{x}_n) \right| \geq \frac{1}{2} \binom{t+s-1}{s-1}.$$

We now estimate the norm of F. By Parseval's theorem we have

$$\sum_{k \in \mathbb{N}_0^s} |\hat{f}_m(k)|^2 = \int_{[0,1]^s} |f_m(x)|^2 \, dx \leq 1$$

and hence

$$\sum_{k \in \mathbb{N}_0^s} |\hat{f}_m(k) \overline{\hat{f}_{m'}(k)}| \leq \left(\sum_{k \in \mathbb{N}_0^s} |\hat{f}_m(k)|^2 \right)^{1/2} \left(\sum_{k \in \mathbb{N}_0^s} |\hat{f}_{m'}(k)|^2 \right)^{1/2} \leq 1.$$

Further note that $\hat{f}_m(k) = 0$ if there is an $1 \leq i \leq s$ such that $k_i \geq b^{m_i}$. Thus

$$\|F\|_{\mathscr{H}(K_{\mathrm{wal},s,b,\delta,\gamma})}^2$$

$$= \sum_{\substack{m,m' \in \mathbb{N}_0^s \\ \|m\|_1 = \|m'\|_1 = t}} \langle f_m, f_{m'} \rangle_{\mathscr{H}(K_{\mathrm{wal},s,b,\delta,\gamma})}$$

$$\leq \sum_{\substack{m,m' \in \mathbb{N}_0^s \\ \|m\|_1 = \|m'\|_1 = t}} \sum_{u_1=0}^{\min(m_1,m_1')} \cdots \sum_{u_s=0}^{\min(m_s,m_s')}$$

$$\times \sum_{k_1=\lfloor b^{u_1-1} \rfloor}^{b^{u_1}-1} \cdots \sum_{k_s=\lfloor b^{u_s-1} \rfloor}^{b^{u_s}-1} \frac{1}{r_{\mathrm{wal},b,\delta}(k,\gamma)} |\hat{f}_m(k) \overline{\hat{f}_{m'}(k)}|$$

$$\leq \sum_{\substack{m,m' \in \mathbb{N}_0^s \\ \|m\|_1 = \|m'\|_1 = t}} \sum_{u_1=0}^{\min(m_1,m_1')} \cdots \sum_{u_s=0}^{\min(m_s,m_s')} \frac{1}{r_{\mathrm{wal},b,\delta}((b^{u_1}-1,\ldots,b^{u_s}-1),\gamma)}$$

$$\leq \sum_{\substack{m,m' \in \mathbb{N}_0^s \\ \|m\|_1 = \|m'\|_1 = t}} \prod_{i=1}^{s} \left[1 + \gamma_i^{-1} \sum_{u_i=1}^{\min(m_i,m_i')} b^{\delta(u_i-1)} \right]$$

$$= \sum_{\substack{m,m' \in \mathbb{N}_0^s \\ \|m\|_1 = \|m'\|_1 = t}} \prod_{i=1}^{s} \left[1 + \gamma_i^{-1} \frac{b^{\delta \min(m_i,m_i')} - 1}{b^\delta - 1} \right]$$

$$\leq C_{\delta,s,\gamma}' \sum_{\substack{m,m' \in \mathbb{N}_0^s \\ \|m\|_1 = \|m'\|_1 = t}} b^{\delta \sum_{i=1}^{s} \min(m_i,m_i')},$$

where $C_{\delta,s,\gamma}' > 0$ is a constant which depends only on δ, s, γ.

We now estimate the last sum. We have

$$\sum_{\substack{m,m' \in \mathbb{N}_0^s \\ \|m\|_1 = \|m'\|_1 = t}} b^{\delta \sum_{i=1}^s \min(m_i, m_i')} = \sum_{\rho=0}^t b^{\delta\rho} A(\rho, t),$$

where $A(\rho, t)$ is the number of solutions to the system of equations

$$m_1 + \cdots + m_s = t,$$
$$m_1' + \cdots + m_s' = t,$$
$$\min(m_1, m_1') + \cdots + \min(m_s, m_s') = \rho.$$

Thus, for $0 \le \rho \le t$ we have

$$A(\rho, t) \le \sum_{u=1}^{s-1} \binom{s}{u} \binom{\rho + s - 1}{s - 1} \binom{t - \rho + u - 1}{u - 1} \binom{t - \rho + s - u - 1}{s - u - 1}$$
$$\le C_s \rho^{s-1} (t - \rho)^{s-2},$$

for some constant $C_s > 0$ independent of t and ρ. Therefore we have

$$\sum_{\rho=0}^t b^{\delta\rho} A(\rho, t) \le C_s \sum_{\rho=0}^t b^{\delta\rho} \rho^{s-1} (t - \rho)^{s-1}$$
$$\le C_s' \int_0^t b^{\delta\rho} \rho^{s-1} (t - \rho)^{s-1} \, d\rho$$
$$\le C_s'' t^{s-1/2} b^{t\delta/2} I_{s-1/2}(\log b^{t\delta/2}),$$

where we used Prudnikov, Brychkov, and Marichev [97, Subsection 2.3.6, Eq. (1)] and where $I_{s-1/2}$ denotes the modified Bessel function of the first kind. Since $I_{s-1/2}(z) \le c_s \frac{e^z}{\sqrt{2\pi z}}$ (see for instance Abramowitz and Stegun [1, Eq. 9.7.1]), it follows that there is a constant $C > 0$ (depending only on s, δ but not on t) such that

$$\sum_{\rho=0}^t b^{\delta\rho} A(\rho, t) \le C b^{t\delta} t^{s-1}.$$

Thus, there is a constant $C_{\delta,s,\gamma} > 0$, such that

$$\|F\|_{\mathcal{H}(K_{\text{wal},s,b,\delta,\gamma})} \le C_{\delta,s,\gamma} b^{t\delta/2} t^{(s-1)/2}.$$

Let $g(x) = F(x)/\|F\|_{\mathcal{H}(K_{\mathrm{wal},s,b,\delta,\gamma})}$. Then g is in the unit ball of $\mathcal{H}(K_{\mathrm{wal},s,b,\delta,\gamma})$ and we have

$$
e(\mathcal{H}(K_{\mathrm{wal},s,b,\delta,\gamma}); \mathscr{P}; w) \geq \left| \int_{[0,1]^s} g(x)\,\mathrm{d}x - \sum_{n=1}^{N} w_n g(x_n) \right|
$$

$$
\geq \frac{1}{2} \|F\|^{-1} \binom{t+s-1}{s-1}
$$

$$
\geq C N^{-\delta/2} (\log N)^{(s-1)/2}
$$

for some constant $C > 0$ independent of N. $\square$

Notice that in the above constructions it was sufficient to use polynomial lattice point sets $\mathscr{P}(q, p)$. Generally, the space $\mathcal{H}(K_{\mathrm{wal},s,b,\delta,\gamma})$ does not contain smooth functions for $\delta > 1$. Hence, if one wants to consider classical spaces of functions with smoothness α, the above results only work for $\alpha \leq 1$.

To extend the construction of polynomial lattice rules to integrands of smoothness $\alpha > 1$, one needs to use polynomial lattice point sets $\mathscr{P}_\alpha(g, p)$ with $\alpha > 1$. This has been shown in Baldeaux, Dick, Greslehner and Pillichshammer [5] and Baldeaux, Dick, Leobacher, Nuyens and Pillichshammer [6].

We introduce a space of functions of smoothness $\alpha \in \mathbb{N}$ in the following. For such α let $L_{s,\alpha,\gamma} : [0, 1]^s \times [0, 1]^s \to \mathbb{R}$ be the reproducing kernel given by

$$
L_{s,\alpha,\gamma}(x, y) = \prod_{i=1}^{s} \left[1 + \gamma_i \sum_{a=1}^{\alpha} \frac{B_a(x_i) B_a(y_i)}{(a!)^2} - (-1)^\alpha \gamma_i \frac{B_{2\alpha}(|x_i - y_i|)}{(2\alpha)!} \right],
$$

where $x = (x_1, \ldots, x_s)$, $y = (y_1, \ldots, y_s)$, and $B_a(\cdot)$ denotes the Bernoulli polynomial of degree a. For instance we have $B_1(x) = x - 1/2$, $B_2(x) = x^2 - x + 1/6$, $B_3(x) = x^3 - 3x^2/2 + x/2$, and so on. For dimension $s = 1$ the inner product in the reproducing kernel Hilbert space $\mathcal{H}(L_{1,\alpha,\gamma})$ is given by

$$
\langle f, g \rangle_{\mathcal{H}(L_{1,\alpha,\gamma})} = \int_0^1 f(x)\,\mathrm{d}x \int_0^1 \overline{g(x)}\,\mathrm{d}x + \frac{1}{\gamma} \sum_{a=1}^{\alpha-1} \int_0^1 f^{(a)}(x)\,\mathrm{d}x \int_0^1 \overline{g^{(a)}(x)}\,\mathrm{d}x
$$

$$
+ \frac{1}{\gamma} \int_0^1 f^{(\alpha)}(x)\overline{g^{(\alpha)}(x)}\,\mathrm{d}x, \tag{9.41}
$$

where $f^{(a)}$ denotes the ath derivative of f. For $s > 1$ the space $\mathcal{H}(L_{s,\alpha,\gamma})$ is the s-fold tensor product of the one-dimensional spaces $\mathcal{H}(L_{1,\alpha,\gamma_i})$, $1 \leq i \leq s$.

The extension of the construction algorithm in Baldeaux, Dick, Greslehner and Pillichshammer [5] and Baldeaux, Dick, Leobacher, Nuyens and Pillichshammer [6] for $\alpha > 1$ is based on a continuous embedding of $\mathcal{H}(L_{s,\alpha,\gamma})$ into a space of Walsh series [21]. It is then shown that Algorithm 39 can be used with a generalised quality criterion of order $\alpha > 1$ stemming from the Walsh space together with polynomial lattice point sets $\mathscr{P}_\alpha(\boldsymbol{q}, p)$. The convergence rate obtained in this case is of the form

$$e^2(\mathscr{H}(L_{s,\alpha,\gamma}); \mathscr{P}_\alpha(\boldsymbol{q}, p)) \leq b^{-\lambda m} \prod_{i=1}^{s} [1 + \gamma_i^{1/\lambda} C_{b,\alpha,\lambda}]^\lambda,$$

where $C_{b,\alpha,\lambda} > 0$ and $1 \leq \lambda < \alpha$.

Notice that the Korobov space $\mathscr{H}(K_{s,\alpha,\gamma})$ is continuously embedded in the space $\mathscr{H}(L_{s,\alpha,\gamma})$. In fact, the restriction of $\mathscr{H}(L_{s,\alpha,\gamma})$ to functions with one-periodic partial mixed derivatives up to order $\alpha - 1$ in each variable, yields the space $\mathscr{H}(K_{s,\alpha,\gamma})$. This can be seen for instance by considering the one-dimensional inner product for such functions, in which case (9.41) reduces to

$$\langle f, g \rangle_{\mathscr{H}(L_{1,\alpha,\gamma})} = \int_0^1 f(x)\,\mathrm{d}x \int_0^1 \overline{g(x)}\,\mathrm{d}x + \frac{1}{\gamma} \int_0^1 f^{(\alpha)}(x)\overline{g^{(\alpha)}(x)}\,\mathrm{d}x,$$

which is equivalent to the inner product given in (9.27) (which can be shown by substituting the Fourier series for f and g in the inner product above).

Thus, Theorem 17 also applies to multivariate integration in the space $\mathscr{H}(L_{s,\alpha,1})$ and we have

$$e(\mathscr{H}(L_{s,\alpha,1}); \mathscr{P}; \boldsymbol{w}) \geq e(\mathscr{H}(K_{s,\alpha}); \mathscr{P}; \boldsymbol{w}) \geq C(s, \alpha, \beta) \frac{(\log N)^{(s-1)/2}}{N^{\alpha/2}}.$$

Thus, the construction algorithm for higher order polynomial lattice rules yields quadrature rules which are almost best possible in terms of their convergence rate.

In the following we present an alternative approach to constructing higher order polynomial lattice rules using some ideas from Dick, Sloan, Wang and Woźniakowski [34] and Sinescu and L'Ecuyer [104]. First, notice that there is an explicit formula for the worst-case error in $\mathscr{H}(L_{s,\alpha,\gamma})$ using (9.3), given by

$$e^2(\mathscr{H}(L_{s,\alpha,\gamma}); \mathscr{P}) = \int_{[0,1]^s} \int_{[0,1]^s} L_{s,\alpha,\gamma}(\boldsymbol{x}, \boldsymbol{y})\,\mathrm{d}\boldsymbol{x}\,\mathrm{d}\boldsymbol{y}$$

$$- \frac{2}{N} \sum_{n=1}^{N} \int_{[0,1]^s} L_{s,\alpha,\gamma}(\boldsymbol{x}, \boldsymbol{x}_n)\,\mathrm{d}\boldsymbol{x} + \frac{1}{N^2} \sum_{n,n'=1}^{N} L_{s,\alpha,\gamma}(\boldsymbol{x}_n, \boldsymbol{x}_{n'})$$

$$= -1 + \frac{1}{N^2} \sum_{n,n'=1}^{N} L_{s,\alpha,\gamma}(\boldsymbol{x}_n, \boldsymbol{x}_{n'}),$$

where $\mathscr{P} = \{\boldsymbol{x}_1, \ldots, \boldsymbol{x}_N\}$. Thus, for a given point set, the worst-case error can be computed in $O(N^2 s)$ operations. Thus, we can use the following algorithm to find good higher order polynomial lattice rules.

Algorithm 42. Let $\alpha, m, s \geq 2$ be integers, let $p \in \mathbb{Z}_b[x]$ be a polynomial of degree αm.

For $1 \leq d \leq s$, assume we have already constructed $q_1, \ldots, q_{d-1} \in G_{b,\alpha m}$. Then randomly choose c polynomials $h_1, \ldots, h_c \in G_{b,\alpha m}$, where $h_1, \ldots, h_c$ are uniformly i.i.d. Set $q_d = h_u$, where $1 \leq u \leq c$ is the value of w which minimises

$$e^2(\mathcal{H}(L_{d,\alpha,\gamma}); \mathcal{P}_\alpha((q_1, \ldots, q_{d-1}, h_w), p)).$$

The following result stems from Baldeaux, Dick, Greslehner and Pillichshammer [5]. Let $1/\alpha < \lambda \leq 1$. Assume that $\boldsymbol{q} = (q_1, \ldots, q_{d-1}) \in G_{b,\alpha m}^{d-1}$ is such that

$$e^{2\lambda}(\mathcal{H}(L_{d-1,\alpha,\gamma}); \mathcal{P}_\alpha(\boldsymbol{q}, p)) \leq \frac{1}{b^m} \prod_{i=1}^{d-1}(1 + \gamma_i^\lambda C_{b,\alpha,\lambda}), \tag{9.42}$$

where the constant $C_{b,\alpha,\lambda} > 0$ depends only on b, α, λ. Then we have

$$\frac{1}{b^{\alpha m}} \sum_{q \in G_{b,\alpha m}} e^{2\lambda}(\mathcal{H}(L_{d,\alpha,\gamma}); \mathcal{P}_\alpha((\boldsymbol{q}, q), p)) \leq \frac{1}{b^m} \prod_{i=1}^{d}(1 + \gamma_i^\lambda C_{b,\alpha,\lambda}),$$

where $(\boldsymbol{q}, q) = (q_1, \ldots, q_{d-1}, q)$. Using Markov's inequality we obtain, given (9.42) holds, that for all $t \geq 1$ we have

$$\#\left\{ q \in G_{b,\alpha m} : e^{2\lambda}(\mathcal{H}(L_{d,\alpha,\gamma}); \mathcal{P}_\alpha((\boldsymbol{q}, q), p)) \leq \frac{t}{b^m} \prod_{i=1}^{d}(1 + \gamma_i^\lambda C_{b,\alpha,\lambda})\right\}$$

$$> b^{\alpha m}\left(1 - \frac{1}{t}\right),$$

which can be written as

$$\#\left\{ q \in G_{b,\alpha m} : e(\mathcal{H}(L_{d,\alpha,\gamma}); \mathcal{P}_\alpha((\boldsymbol{q}, q), p)) \leq \frac{t}{b^{\frac{m}{2\lambda}}} \prod_{i=1}^{d}(1 + \gamma_i^\lambda C_{b,\alpha,\lambda})^{\frac{1}{2\lambda}}\right\}$$

$$> b^{\alpha m}\left(1 - \frac{1}{t^{2\lambda}}\right).$$

Hence the probability that at least one of $h_1, \ldots, h_c$ satisfies

$$e(\mathcal{H}(L_{d,\alpha,\gamma}); \mathcal{P}_\alpha((\boldsymbol{q}, h_w), p)) \leq \frac{t}{b^{\frac{m}{2\lambda}}} \prod_{i=1}^{d}(1 + \gamma_i^\lambda C_{b,\alpha,\lambda})^{\frac{1}{2\lambda}}$$

is at least $1 - t^{-c/(2\lambda)}$. Thus, we have the following theorem.

Theorem 43. *Let* $1/\alpha < \lambda \leq 1$. *The probability that the vector* $\boldsymbol{q} = (q_1, \ldots, q_s) \in$ $G_{b,\alpha m}^s$ *constructed by Algorithm 42 satisfies*

$$e(\mathcal{H}(L_{d,\alpha,\gamma}); \mathcal{P}_\alpha(\boldsymbol{q}^{(d)}, p)) \leq \frac{t}{b^{\frac{m}{2\lambda}}} \prod_{i=1}^d (1 + \gamma_i^\lambda C_{b,\alpha,\lambda})^{\frac{1}{2\lambda}}$$

for all $1 \leq d \leq s$, *where* $\boldsymbol{q}^{(d)} = (q_1, \ldots, q_d)$, *is at least* $(1 - t^{-c/(2\lambda)})^s \geq 1 - st^{-c/(2\lambda)}$.

9.5.2 The Construction of Polynomial Lattice Rules Based on the Figure of Merit

Another way of constructing polynomial lattice point sets is based on the figure of merit ϱ_α from Definition 36. In Algorithms 39 and 42 we used the worst-case error in some function spaces to compare polynomial lattice rules. In the following we show how Theorems 5, 8 and 37 can also be used to obtain a construction of good polynomial lattice point sets.

The idea is to search for polynomial lattice point sets which maximise the figure of merit.

Algorithm 44. Let $\alpha, b, m, s \in \mathbb{N}$, $b \geq 2$, be given and let $p \in \mathbb{Z}_b[x]$ with $\deg(p) = \alpha m$. Choose $\boldsymbol{q} \in G_{b,\alpha m}^s$ which maximises $\varrho_\alpha(\boldsymbol{q}, p)$.

Since computing the value of $\varrho_\alpha(\boldsymbol{q}, p)$ for given polynomials $\boldsymbol{q}, p$ is computationally expensive, Algorithm 44 can only be used for small values of α, m and s. In order to reduce the size of the search space, one can also consider a simplification due to Korobov [60].

Algorithm 45. Let $\alpha, b, m, s \in \mathbb{N}$, $b \geq 2$, be given and let $p \in \mathbb{Z}_b[x]$ with $\deg(p) = \alpha m$. Choose $q \in G_{b,\alpha m}$ which maximises $\varrho_\alpha((q, q^2, \ldots, q^s), p)$.

If $\alpha = 1$, one can also consider the generating vector $(1, q, \ldots, q^{s-1})$ (as originally proposed by Korobov for lattice rules).

The following result for $\alpha = 1$ was first shown in Larcher, Lauss, Niederreiter, and Schmid [68].

Theorem 46 (Larcher, Lauss, Niederreiter, and Schmid [68]). *Let* $b \in \mathbb{P}$, $m, s \in$ $\mathbb{N}$ *with* $s \geq 2$ *and let* $p \in \mathbb{Z}_b[x]$ *be irreducible over* $\mathbb{Z}_b[x]$ *with* $\deg(p) = m$. *For* $\varrho > 0$ *define*

$$\Delta(s, \varrho) = \sum_{d=0}^{s-1} \binom{s}{d} (b-1)^{s-d} \sum_{l=0}^{\varrho+d} \binom{s-d+l-1}{l} b^l + 1 - b^{\varrho+s}.$$

1. *If $\Delta(s, \varrho) < b^m$, there exists a $q \in G_{b,m}^s$ with*

$$\varrho(q, p) \geq \varrho + s.$$

2. *If $\Delta(s, \varrho) < \frac{b^m}{s-1}$, there exists a polynomial $q \in G_{b,m}$ such that $g \equiv (1, q, \ldots, q^{s-1})$ (mod p) satisfies*

$$\varrho(q, p) \geq \varrho + s.$$

Corollary 47. *Let $b \in \mathbb{P}$, $m, s \in \mathbb{N}$ with $s \geq 2$ and with m sufficiently large. Let $p \in \mathbb{Z}_b[x]$ be irreducible with $\deg(p) = m$.*

1. *There exists a vector $q \in G_{b,m}^s$ with*

$$\varrho(q, p) \geq \left\lfloor m - (s - 1)(\log_b m - 1) + \log_b \frac{(s - 1)!}{(b - 1)^{s-1}} \right\rfloor.$$

2. *There exists a polynomial $q \in G_{b,m}$ such that $q \equiv (1, q, \ldots, q^{s-1})$ (mod p) satisfies*

$$\varrho(q, p) \geq \left\lfloor m - (s - 1)(\log_b m - 1) + \log_b \frac{(s - 2)!}{(b - 1)^{s-1}} \right\rfloor.$$

Together with Theorems 37 and 5 this result shows the existence of polynomial lattice point sets $\mathscr{P}(q, p)$ with star-discrepancy of order

$$D_{b^m}^*(\mathscr{P}(q, p)) \ll_{s,b} \frac{m^{2s-2}}{b^m}.$$

More precise results on the star-discrepancy of polynomial lattice point sets will be presented in Sect. 9.5.3.

For $\alpha > 1$ we have the following result from Dick, Kritzer, Pillichshammer and Schmid [36].

Theorem 48 (Dick, Kritzer, Pillichshammer and Schmid [36, Theorem 3]). *Let $b \in \mathbb{P}$, $m, \alpha, s \in \mathbb{N}$, $\alpha, s \geq 2$, and $p \in \mathbb{Z}_b[x]$ with $\deg(p) = \alpha m$ be irreducible. For $\varrho > 0$ define*

$$\Delta(s, \varrho, \alpha) = \sum_{l=0}^{\varrho} \sum_{i=1}^{s} \binom{s}{i} \sum_{\substack{l_1, \ldots, l_i \geq 1 \\ l_1 + \cdots + l_i = l}} \prod_{z=1}^{i} C(\alpha, l_z),$$

where

$$C(\alpha, l) = \sum_{v=1}^{\alpha-1} (b-1)^v \binom{l - \frac{v(v-1)}{2} - 1}{v - 1}$$

$$+ \sum_{i=1}^{\lfloor l/\alpha \rfloor} (b-1)^{\alpha} b^{i-1} \binom{l - \alpha i - \frac{\alpha(\alpha-3)}{2} - 2}{\alpha - 2}.$$

1. If $\Delta(s, \varrho, \alpha) < b^m$, there exists a $\boldsymbol{q} \in G_{b,\alpha m}^s$ with

$$\varrho_\alpha(\boldsymbol{q}, p) \geq \varrho.$$

2. If $\Delta(s, \varrho, \alpha) < \frac{b^m}{s-1}$, there exists a polynomial $q \in G_{b,\alpha m}$ such that $\boldsymbol{q} \equiv (q, q^2, \ldots, q^s) \pmod{p}$ satisfies

$$\varrho_\alpha(\boldsymbol{q}, p) \geq \varrho.$$

The proofs of Theorems 46 and 48 are based on the following idea applied to codes and going back to Gilbert [43] and Varshamov [115]. We illustrate this idea for Algorithm 44.

First, note that there are $|G_{b,\alpha m}^s| = |G_{b,\alpha m}|^s = b^{\alpha m s}$ vectors $\boldsymbol{q}$ to choose from. The idea is to estimate the number of vectors $\boldsymbol{q} \in G_{b,\alpha m}^s$ for which $\varrho_{\alpha,m}(\boldsymbol{q}, p) < \varrho$ for some chosen $\varrho \geq 0$. If this number is smaller than the total number of possible choices of vectors $\boldsymbol{q} \in G_{b,\alpha m}^s$, it follows that there is at least one vector $\boldsymbol{q}$ with $\varrho_\alpha(\boldsymbol{q}, p) \geq \varrho$. For details we refer to [31, Chapter 10] and [31, Section 15.7.1], or to Dick, Kritzer, Pillichshammer and Schmid [36] and Larcher, Lauss, Niederreiter, and Schmid [68].

In Dick, Kritzer, Pillichshammer and Schmid [36] it was shown that Theorem 48 sometimes yields higher order digital nets with parameters better than the ones obtained using the higher order construction of Sect. 9.3.5.

9.5.3 Star-Discrepancy of Polynomial Lattice Point Sets

For a polynomial lattice point set $\mathscr{P}(\boldsymbol{g}, p)$ with $\alpha = 1$ each point $\boldsymbol{x}_n$ is of the form $\boldsymbol{x}_n = \{\boldsymbol{y}_n/b^m\}$ with $\boldsymbol{y}_n \in \mathbb{Z}^s$. In particular, the elements of a polynomial lattice point set always have a finite b-adic digit expansion. A bound similar to that of Proposition 22 on the star-discrepancy of such point sets was first given by Niederreiter [77, Satz 2] (see also Niederreiter [81, Theorem 3.12]). An approach to this result by means of Walsh functions was described by Hellekalek [49, Theorem 1]. To formulate the result of Hellekalek we again need some notation.

Let $b \geq 2$ be an integer. For a vector $\boldsymbol{k} = (k_1, \ldots, k_s) \in \mathbb{N}_0^s$ we put $\rho_b(\boldsymbol{k}) := \prod_{i=1}^s \rho_b(k_i)$ where $\rho_b(0) = 1$ and where for $k \in \mathbb{N}_0$ we set

$$\rho_b(k) = \frac{1}{b^{r+1} \sin^2(\pi \kappa_r / b)}$$

if $k = \kappa_r b^r + k'$, where $\kappa_r \in \{1, \ldots, b-1\}$ and $0 \leq k' < b^r$.

Proposition 49 (Hellekalek [49] and Niederreiter [77]). *Let $N \geq 1$ and let $\mathscr{P} = \{\boldsymbol{x}_0, \ldots, \boldsymbol{x}_{N-1}\}$ be a point set in the s-dimensional unit cube where $\boldsymbol{x}_n$ is of the form $\boldsymbol{x}_n = \{\boldsymbol{y}_n / b^m\}$ with $\boldsymbol{y}_n \in \mathbb{Z}^s$, and $m, b \in \mathbb{N}$, $b \geq 2$. Then we have*

$$D_N^*(\mathscr{P}) \leq 1 - \left(1 - \frac{1}{b^m}\right)^s + \sum_{\substack{\boldsymbol{k} \in \mathbb{N}_0^s \\ 0 < |\boldsymbol{k}|_\infty < b^m}} \rho_b(\boldsymbol{k}) \left| \frac{1}{N} \sum_{n=0}^{N-1} {}_b\mathrm{wal}_{\boldsymbol{k}}(\boldsymbol{x}_n) \right|.$$

A proof of this result can also be found in [31, Proof of Theorem 3.28].

Applying this result to a polynomial lattice point set $\mathscr{P}(\boldsymbol{q}, p)$, in particular $N = b^m$, and using (9.40), we obtain the following result.

Theorem 50 (Dick, Leobacher and Pillichshammer [32]). *For the star-discrepancy of a polynomial lattice point set $\mathscr{P}(\boldsymbol{q}, p)$ we have*

$$D_{b^m}^*(\mathscr{P}(\boldsymbol{q}, p)) \leq 1 - \left(1 - \frac{1}{b^m}\right)^s + R_b(\boldsymbol{q}, p) \leq \frac{s}{b^m} + R_b(\boldsymbol{q}, p),$$

where

$$R_b(\boldsymbol{q}, p) := \sum_{\boldsymbol{h} \in \mathscr{D}^*(\boldsymbol{q}, p)} \rho_b(\boldsymbol{h}).$$

In the original version of the above results on the star-discrepancy, the squared sine function in the definition of ρ_b can be replaced by the ordinary sine function. Here we deal with the slightly weaker bound since in this case the quantity R_b can be computed efficiently. Assume that $\mathscr{P}(\boldsymbol{q}, p) = \{\boldsymbol{x}_0, \ldots, \boldsymbol{x}_{b^m-1}\}$, where $\boldsymbol{x}_n = (x_{n,1}, \ldots, x_{n,s})$. Then, using (9.40) we can write $R_b(\boldsymbol{q}, p)$ as

$$R_b(\boldsymbol{q}, p) = -1 + \frac{1}{b^m} \sum_{n=0}^{b^m-1} \prod_{i=1}^s \left(1 + \sum_{h=1}^{b^m-1} \rho_b(h) \, {}_b\mathrm{wal}_h(x_{n,i})\right). \tag{9.43}$$

and from this one can deduce that

$$R_b(\boldsymbol{q}, p) = -1 + \frac{1}{b^m} \sum_{n=0}^{b^m-1} \prod_{i=1}^s \phi_{b,m}(x_{n,i}),$$

where for $x = \xi_1 b^{-1} + \cdots + \xi_m b^{-m}$ we have

$$
\phi_{b,m}(x) = \begin{cases} 1 + i_0 \frac{b^2-1}{3b} + \frac{2}{b}\xi_{i_0}(\xi_{i_0} - b) & \text{if } \xi_1 = \cdots = \xi_{i_0-1} = 0 \text{ and} \\ & \xi_{i_0} \neq 0 \text{ with } 1 \leq i_0 \leq m, \\ 1 + m\frac{b^2-1}{3b} & \text{otherwise.} \end{cases}
$$

In particular, $R_b(\boldsymbol{q}, p)$ can be computed in $O(b^m s)$ operations; for a proof we refer to [31, Section 10.2].

It has been shown in Kritzer and Pillichshammer [62] that there exists a $c_{s,b} > 0$ such that for any $p \in \mathbb{Z}_b[x]$ with $\deg(p) = m$ and any $\boldsymbol{q} \in (G^*_{b,m})^s$ we have

$$
R_b(\boldsymbol{q}, p) \geq c_{s,b} b^{\deg(\delta_s)} \frac{(m - \deg(\delta_s))^s}{b^m}, \qquad \text{where } \delta_s := \gcd(q_1, \ldots, q_s, p).
$$

On the other hand, let $b \in \mathbb{P}$, $s, m \in \mathbb{N}$, and let $p \in \mathbb{Z}_b[x]$ be irreducible with $\deg(p) = m$. Then we have

$$
\frac{1}{|G^*_{b,m}|^s} \sum_{\boldsymbol{q} \in (G^*_{b,m})^s} R_b(\boldsymbol{q}, p) = \frac{1}{b^m - 1}\left(\left(1 + m\frac{b^2-1}{3b}\right)^s - 1 - sm\frac{b^2-1}{3b}\right),
$$

see Dick, Leobacher and Pillichshammer [32, Theorem 2.3] or [31, Theorem 10.21] for a proof. See also Niederreiter [81, Theorem 4.43]. In particular, we have the following result.

Theorem 51 (Dick, Leobacher and Pillichshammer [32, Theorem 2.3] and Kritzer and Pillichshammer [62, Theorem 1.1]). *Let $b \in \mathbb{P}$, $s, m \in \mathbb{N}$, $s \geq 2$. For any $p \in \mathbb{Z}_b[x]$ with $\deg(p) = m$ there exists a $\boldsymbol{q} \in (G^*_{b,m})^s$ such that*

$$
R_b(\boldsymbol{q}, p) \asymp_{b,s} \frac{m^s}{b^m}
$$

and this order of magnitude is best possible.

If p is irreducible we can use the following component-by-component algorithm for the construction of $\mathscr{P}(\boldsymbol{g}, p)$.

Algorithm 52. Given $b \in \mathbb{P}$, $s, m \in \mathbb{N}$, and a polynomial $p \in \mathbb{Z}_b[x]$, with $\deg(p) = m$.

1. Choose $q_1 = 1$.
2. For $d > 1$, assume we have already constructed $q_1, \ldots, q_{d-1} \in G^*_{b,m}$. Then find $q_d \in G^*_{b,m}$ which minimises the quantity $R_b((q_1, \ldots, q_{d-1}, z), p)$ as a function of $z \in G^*_{b,m}$.

Since the quantity $R_b(\boldsymbol{q}, p)$ can be calculated in $O(b^m s)$ operations, the cost of Algorithm 52 is of $O(b^{2m} s^2)$ operations. Using the fast component-by-component algorithm due to Nuyens and Cools [91] this can be reduced to $O(smb^m)$ operations; see [31, Section 10.3].

Theorem 53 (Dick, Leobacher and Pillichshammer [32, Theorem 2.7]). *Let* $b \in \mathbb{P}$, $s, m \in \mathbb{N}$, *and let* $p \in \mathbb{Z}_b[x]$ *be irreducible with* $\deg(p) = m$. *Suppose* $q = (q_1, \ldots, q_s) \in (G^*_{b,m})^s$ *is constructed according to Algorithm 52. Then for all* $1 \leq d \leq s$ *we have*

$$R_b(q^{(d)}, p) \leq \frac{1}{b^m - 1} \left(1 + m \frac{b^2 - 1}{3b} \right)^d,$$

where $q^{(d)} = (q_1, \ldots, q_d)$.

The proof of the result relies on similar ideas to those of the corresponding result for lattice rules from Theorem 26. A detailed proof can also be found in [31, Section 10.2.2].

A similar result for not necessarily irreducible polynomials is proven, but with much more technical effort, in Dick, Kritzer, Leobacher and Pillichshammer [35, Theorem 2].

Corollary 54 (Dick, Leobacher and Pillichshammer [32, Corollary 2.8]). *Let* $b \in \mathbb{P}$, $s, m \in \mathbb{N}$, *and let* $p \in \mathbb{Z}_b[x]$ *be irreducible with* $\deg(p) = m$. *Suppose* $q \in (G^*_{b,m})^s$ *is constructed according to Algorithm 52. Then we have*

$$D^*_{b^m}(\mathscr{P}(q, p)) \leq \frac{s}{b^m} + \frac{1}{b^m - 1} \left(1 + m \frac{b^2 - 1}{3b} \right)^s.$$

This result is not quite as good as the best existence result for point sets with low star-discrepancy from (9.11). However, the result is in line with the analogous result from the theory of lattice point sets, cf. Corollary 27.

For polynomial lattice point sets one knows that they also have the digital net structure. Based on this property one can prove the following improved, but still not optimal in the sense of (9.11), existence result.

Theorem 55 (Kritzer and Pillichshammer [63], Larcher [70]). *Let* $b \in \mathbb{P}$ *and* $s \in \mathbb{N}$. *Then for any polynomial* $p \in \mathbb{Z}_b[x]$ *of degree* m *with* $\gcd(p, x) = 1$ *or* $p(x) = x^m$ *there exists a generating vector* $q \in G^s_{b,m}$ *such that*

$$D^*_{b^m}(\mathscr{P}(q, p)) \ll_{s,b} \frac{m^{s-1} \log m}{b^m}.$$

This bound is excellent in an asymptotic sense if $m \to \infty$. The dependence on the dimension s is not known. A construction of polynomial lattice point sets $\mathscr{P}(q, p)$ whose star-discrepancy satisfy the bound from Theorem 55 is not known so far.

In dimension $s = 2$ we have an explicit construction due to Niederreiter of a generating vector q such that $\mathscr{P}(q, x^m)$ has small star-discrepancy. For $s = b = 2$ and for any $m \in \mathbb{N}$ let

$$q_m(x) = \sum_{j=0}^{\lfloor \log_2 m \rfloor + 1} x^{m - \lfloor m/2^j \rfloor} \in \mathbb{Z}_2[x] \quad \text{and} \quad q_m = (1, q_m) \in \mathbb{Z}_2[x]^2.$$

Theorem 56 (Niederreiter [78] and [81]). *For any $m \in \mathbb{N}$ we have*

$$D_{2^m}^*(\mathscr{P}(q_m, x^m)) \leq \left(\frac{m}{3} + \frac{9}{19}\right) \frac{1}{2^m}.$$

Proof. Consider the continued fraction expansion

$$\frac{q_m(x)}{x^m} = [A_1, \ldots, A_h],$$

where $A_i \in \mathbb{Z}_2[x]$ and $\deg(A_i) \geq 1$ for $1 \leq i \leq h$. Niederreiter [78] proved that $K(q_m(x)/x^m) := \max_{1 \leq i \leq h} \deg(A_i) = 1$. From Theorem 37 and from Niederreiter [81, Theorem 4.46] it follows that the two-dimensional polynomial lattice point set $\mathscr{P}(q_m, x^m)$ is a digital $(0, m, 2)$-net over $\mathbb{Z}_2$. In Larcher and P. [67] it has been shown that the star-discrepancy of any digital $(0, m, 2)$-net over $\mathbb{Z}_2$ is at most $\left(\frac{m}{3} + \frac{9}{19}\right) 2^{-m}$ and hence the result follows. $\square$

9.5.4 Weighted Star-Discrepancy of Polynomial Lattice Point Sets

For the weighted star-discrepancy of a polynomial lattice point set we obtain from Theorem 50

$$D_{N, \gamma}^*(\mathscr{P}(q, p)) = \max_{\emptyset \neq u \subseteq [s]} \gamma_u D_{N, \gamma}^*(\mathscr{P}(q_u, p))$$

$$\leq \sum_{\emptyset \neq u \subseteq [s]} \gamma_u \left(1 - \left(1 - \frac{1}{b^m}\right)^{|u|}\right) + \sum_{\emptyset \neq u \subseteq [s]} \gamma_u R_b(q_u, p),$$

where q_u denotes the projection of q onto the components given by u. Hence $\mathscr{P}(q_u, p)$ is the $|u|$-dimensional polynomial lattice point set which is obtained by a projection of the points from $\mathscr{P}(q, p)$ onto the components given by u.

Set $\tilde{\rho}_b(h, \gamma) = 1 + \gamma$ if $h = 0$ and $\gamma \rho_b(h)$ if $h \neq 0$, and set $\tilde{\rho}_b(h, \gamma) = \prod_{i=1}^s \tilde{\rho}_b(h_i, \gamma_i)$. Then it follows from (9.43) in the same way as for the corresponding result for lattice rules that

$$\tilde{R}_{b,\boldsymbol{\gamma}}(\boldsymbol{q},p) := \sum_{\emptyset \neq u \subseteq [s]} \boldsymbol{\gamma}_u R_b(\boldsymbol{q}_u,p)$$

$$= \sum_{\boldsymbol{h} \in \mathscr{D}(\boldsymbol{g},p)^*} \tilde{\rho}_b(\boldsymbol{h},\boldsymbol{\gamma})$$

$$= -\prod_{i=1}^{s}(1+\gamma_i) + \frac{1}{b^m} \sum_{n=0}^{b^m-1} \prod_{i=1}^{s}(1+\gamma_i \phi_{b,m}(x_{n,i})). \qquad (9.44)$$

From (9.44) we see that $\tilde{R}_{b,\boldsymbol{\gamma}}(\boldsymbol{q},p)$ can be computed in $O(b^m s)$ operations.

Hence for the weighted star-discrepancy of a polynomial lattice point set $\mathscr{P}(\boldsymbol{q},p)$ we obtain

$$D^*_{b^m,\boldsymbol{\gamma}}(\mathscr{P}(\boldsymbol{q},p)) \leq \sum_{\emptyset \neq u \subseteq [s]} \boldsymbol{\gamma}_u \left(1 - \left(1 - \frac{1}{b^m}\right)^{|u|}\right) + \tilde{R}_{b,\boldsymbol{\gamma}}(\boldsymbol{q},p).$$

If p is irreducible one can again use the component-by-component algorithm (Algorithm 52) with R_b replaced by $\tilde{R}_{b,\boldsymbol{\gamma}}$ for the construction of a 'good' generating vector.

Theorem 57 (Dick, Leobacher and Pillichshammer [32, Theorem 3.7]). *Let $b \in \mathbb{P}$, $s,m \in \mathbb{N}$, let $p \in \mathbb{Z}_b[x]$ be irreducible with $\deg(p) = m$. Suppose $\boldsymbol{q} = (q_1,\dots,q_s) \in (G^*_{b,m})^s$ is constructed according to Algorithm 52 (with R_b replaced by $\tilde{R}_{b,\boldsymbol{\gamma}}$). Then for all $1 \leq d \leq s$ we have*

$$\tilde{R}_{b,\boldsymbol{\gamma}}(\boldsymbol{q}^{(d)},p) \leq \frac{1}{b^m - 1} \prod_{i=1}^{d}\left(1 + \gamma_i\left(1 + m\frac{b^2 - 1}{3b}\right)\right),$$

where $\boldsymbol{q}^{(d)} = (q_1,\dots,q_d)$.

Corollary 58 (Dick, Leobacher and Pillichshammer [32, Corollary 3.8]). *Let $b \in \mathbb{P}$, $s,m \in \mathbb{N}$, and let $p \in \mathbb{Z}_b[x]$ be irreducible with $\deg(p) = m$. Suppose $\boldsymbol{q} = (q_1,\dots,q_s) \in (G^*_{b,m})^s$ is constructed according to Algorithm 52 (with R_b replaced by $\tilde{R}_{b,\boldsymbol{\gamma}}$). Then for all $1 \leq d \leq s$ we have*

$$D^*_{b^m,\boldsymbol{\gamma}}(\mathscr{P}(\boldsymbol{q}^{(d)},p)) \leq \sum_{\emptyset \neq u \subseteq [d]} \boldsymbol{\gamma}_u \left(1 - \left(1 - \frac{1}{b^m}\right)^{|u|}\right)$$

$$+ \frac{1}{b^m - 1} \prod_{i=1}^{d}\left(1 + \gamma_i\left(1 + m\frac{b^2 - 1}{3b}\right)\right),$$

where $\boldsymbol{q}^{(d)} = (q_1,\dots,q_d)$ and $[d] = \{1,\dots,d\}$.

Similarly as for lattice point sets we can now deduce the following result.

Corollary 59. *Let p be irreducible and suppose that q is constructed according to Algorithm 52 (with R_b replaced by $\tilde{R}_{b,\gamma}$).*

If $\sum_{i \geq 1} \gamma_i < \infty$, then for any $\delta > 0$ there exists a $c_{b,\gamma,\delta} > 0$, independent of s and m, such that the weighted star-discrepancy of $\mathscr{P}(q, p)$ satisfies

$$D^*_{b^m,\gamma}(\mathscr{P}(q, p)) \leq \frac{c_{b,\gamma,\delta}}{b^{m(1-\delta)}}.$$

Assume that $\sum_{i \geq 1} \gamma_i < \infty$. For simplicity, we consider the case $b = 2$ only. Let $\delta > 0$ and let $N \in \mathbb{N}$ with binary representation $N = 2^{m_1} + \cdots + 2^{m_k}$, where $0 \leq m_1 < m_2 < \cdots < m_k$, i.e., $m_k = \lfloor \log_2 N \rfloor$, where $\log_2$ denotes the logarithm in base 2. For each $1 \leq i \leq k$ choose an irreducible polynomial $p_i \in \mathbb{Z}_2[x]$ with $\deg(p_i) = m_i$ and construct a vector q_i according to Algorithm 52 (with $b = 2$ and with R_2 replaced by $\tilde{R}_{2,\gamma}$). Then for the resulting polynomial lattice point sets $\mathscr{P}(q_i, p_i)$ we obtain from Corollary 59 that

$$D^*_{2^{m_i},\gamma}(\mathscr{P}(q_i, p_i)) \leq \frac{c_{\gamma,\delta}}{2^{m_i(1-\delta)}}$$

for all $1 \leq i \leq k$. Let $\mathscr{P}_N = \mathscr{P}(q_1, p_1) \cup \ldots \cup \mathscr{P}(q_k, p_k)$ (here we mean a superposition where the multiplicity of elements matters). Then it follows from the triangle inequality for the star-discrepancy (see [31, Proposition 3.16]) and the definition of the weighted star-discrepancy that

$$D^*_{N,\gamma}(\mathscr{P}_N) \leq \sum_{i=1}^{k} \frac{2^{m_i}}{N} D^*_{2^{m_i},\gamma}(\mathscr{P}(q_i, p_i)) \leq \frac{c_{\gamma,\delta}}{N} \sum_{i=1}^{k} 2^{m_i \delta}$$

$$\leq \frac{c_{\gamma,\delta}}{N} \sum_{j=0}^{\lfloor \log_2 N \rfloor} 2^{j\delta} \leq \frac{\tilde{c}_{\gamma,\delta}}{N^{1-\delta}}.$$

Hence for each $s, N \in \mathbb{N}$ there exists an N-element point set $\mathscr{P}_N$ in $[0, 1)^s$ with $D^*_{N,\gamma}(\mathscr{P}_N) \leq \tilde{c}_{\gamma,\delta} N^{-1+\delta}$ and this point set is a superposition of polynomial lattice point sets.

In particular, if $\sum_{i \geq 1} \gamma_i < \infty$, it follows that for any $s, N \in \mathbb{N}$ we have

$$\text{disc}_{\infty,\gamma}(N, s) \leq \frac{\tilde{c}_{\gamma,\delta}}{N^{1-\delta}}$$

and that the bound can be achieved by a superposition of polynomial lattice point sets.

Recall that $\text{disc}_{\infty,\gamma}(0, s) = \max_{\emptyset \neq u \subseteq [s]} \gamma_u \geq \gamma_1 > 0$.

For $\varepsilon > 0$ and $\delta > 0$ we obtain

$$N_{\infty,\gamma}(\varepsilon, s) \leq \left\lceil (\tilde{c}_{\gamma,\delta} \gamma_1^{-1} \varepsilon^{-1})^{1/(1-\delta)} \right\rceil.$$

This bound, which is independent of the dimension s, was already presented in (9.19) and shows again that the weighted star-discrepancy is strongly tractable with ε-exponent equal to one whenever the weights γ_i, $i \geq 1$, are summable.

As for lattice point sets, for the weighted L_p-discrepancy of a polynomial lattice point set we obtain

$$
L_{p,\gamma}(\mathscr{P}(q, p)) \leq \sum_{\emptyset \neq u \subseteq [s]} \gamma_u \left(1 - \left(1 - \frac{1}{b^m}\right)^{|u|}\right) + \tilde{R}_{b,\gamma}(g, p).
$$

Again, this means that the results for the weighted star-discrepancy apply also for the weighted L_p-discrepancy.

Acknowledgements The first author is supported by a Queen Elizabeth II Fellowship from the Australian Research Council. The second author is partially supported by the Austrian Science Foundation (FWF), Project S9609, that is part of the Austrian National Research Network "Analytic Combinatorics and Probabilistic Number Theory" and Project F5509-N26, that is part of the Special Research Program "Quasi-Monte Carlo Methodes: Theory and Applications".

The authors thank Michaela Szölgyenyi and Henryk Woźniakowski for many helpful suggestions.

References

1. M. Abramowitz, I.A. Stegun (eds.), *Handbook of mathematical functions with formulas, graphs and mathematical tables*. (U.S. Government Printing Office, Washington, DC, 1964)
2. C. Aistleitner, Covering numbers, dyadic chaining and discrepancy. J. Complexity **27**(6), 531–540 (2011). doi:10.1016/j.jco.2011.03.001
3. N. Aronszajn, Theory of reproducing kernels. Trans. Am. Math. Soc. **68**, 337–404 (1950). doi:10.2307/1990404
4. N.S. Bakhvalov, On Approximate Calculation of Multiple Integrals. Vestnik Moskovskogo Universiteta, Seriya Matematiki, Mehaniki, Astronomi, Fiziki, Himii **4**, 3–18 (1959). In Russian
5. J. Baldeaux, J. Dick, J. Greslehner, F. Pillichshammer, Construction algorithms for higher order polynomial lattice rules. J. Complexity **27**(3–4), 281–299 (2011). doi:10.1016/j.jco.2010.06.002
6. J. Baldeaux, J. Dick, G. Leobacher, D. Nuyens, F. Pillichshammer, Efficient calculation of the worst-case error and (fast) component-by-component construction of higher order polynomial lattice rules. Numer. Algorithms **59**(3), 403–431 (2012). doi:10.1007/s11075-011-9497-y
7. R. Béjian, Minoration de la discrépance d'une suite quelconque sur T. Acta Arithmetica **41**, 185–202 (1982)
8. D. Bilyk, M.T. Lacey, A. Vagharshakyan, On the small ball inequality in all dimensions. J. Funct. Anal. **254**(9), 2470–2502 (2008). doi:10.1016/j.jfa.2007.09.010
9. J. Brauchart, J. Dick, A simple proof of stolarsky's invariance principle. Proc. Am. Math. Soc. **141**(6), 2085–2096 (2013)
10. V.A. Bykovskii, The discrepancy of the Korobov lattice points. Izvestiya: Mathematics **76**(3), 446–465 (2012). doi:10.1070/IM2012v076n03ABEH002591
11. W.W.L. Chen, On irregularities of point distribution. Mathematika **27**, 153–170 (1980)

12. W.W.L. Chen, M.M. Skriganov, Explicit constructions in the classical mean squares problem in irregularities of point distribution. J. Reine Angew. Math. **545**, 67–95 (2002). doi:10.1515/crll.2002.037
13. W.W.L. Chen, M.M. Skriganov, Orthogonality and digit shifts in the classical mean squares problem in irregularities of point distribution, in *Diophantine approximation. Festschrift for Wolfgang Schmidt. Based on lectures given at a conference at the Erwin Schrödinger Institute*, ed. by Schlickewei, Hans Peter Developments in Mathematics, vol. 16 (Springer, Wien, 2008), pp. 141–159. doi:10.1007/978-3-211-74280-8_7
14. R. Cools, J.N. Lyness, Three- and four-dimensional K-optimal lattice rules of moderate trigonometric degree. Math. Comput. **70**(236), 1549–1567 (2001). doi:10.1090/S0025-5718-01-01326-6
15. L.L. Cristea, J. Dick, F. Pillichshammer, On the mean square weighted $\mathscr{L}_2$ discrepancy of randomized digital nets in prime base. J. Complexity **22**(5), 605–629 (2006). doi:10.1016/j.jco.2006.03.005
16. H. Davenport, Note on irregularities of distribution. Math. Lond. **3**, 131–135 (1956). doi:10.1112/S0025579300001807
17. J. Dick, F.Y. Kuo, F. Pillichshammer, I.H. Sloan, Construction algorithms for polynomial lattice rules for multivariate integration. Math. Comput. **74**(252), 1895–1921 (2005). doi:10.1090/S0025-5718-05-01742-4
18. J. Dick, On the convergence rate of the component-by-component construction of good lattice rules. J. Complexity **20**(4), 493–522 (2004). doi:10.1016/j.jco.2003.11.008
19. J. Dick, Explicit constructions of quasi-Monte Carlo rules for the numerical integration of high-dimensional periodic functions. SIAM J. Numer. Anal. **45**(5), 2141–2176 (2007). doi:10.1137/060658916
20. J. Dick, Walsh spaces containing smooth functions and quasi-Monte Carlo rules of arbitrary high order. SIAM J. Numer. Anal. **46**(3), 1519–1553 (2008). doi:10.1137/060666639
21. J. Dick, The decay of the Walsh coefficients of smooth functions. Bull. Aust. Math. Soc. **80**(3), 430–453 (2009). doi:10.1017/S0004972709000392
22. J. Dick, Higher order scrambled digital nets achieve the optimal rate of the root mean square error for smooth integrands. Ann. Stat. **39**(3), 1372–1398 (2011). doi:10.1214/11-AOS880
23. J. Dick, Quasi-Monte Carlo numerical integration on $\mathbb{R}^s$: digital nets and worst-case error. SIAM J. Numer. Anal. **49**(4), 1661–1691 (2011). doi:10.1137/100789853
24. J. Dick, J. Baldeaux, Equidistribution properties of generalized nets and sequences., in *Monte Carlo and quasi-Monte Carlo methods 2008. Proceedings of the 8th international conference Monte Carlo and quasi-Monte Carlo methods in scientific computing, Montréal, Canada, July 6–11, 2008.*, ed. by L'Ecuyer, P., Owen, A.B. (Springer, Berlin, 2009), pp. 305–322. doi:10.1007/978-3-642-04107-5_19
25. J. Dick, P. Kritzer, Duality theory and propagation rules for generalized digital nets. Math. Comput. **79**(270), 993–1017 (2010). doi:10.1090/S0025-5718-09-02315-1
26. J. Dick, H. Niederreiter, On the exact t-value of Niederreiter and Sobol' sequences. J. Complexity **24**(5-6), 572–581 (2008). doi:10.1016/j.jco.2008.05.004
27. J. Dick, F. Pillichshammer, Optimal $\mathscr{L}_2$ discrepancy bounds for higher order digital sequences over the finite field $\mathbb{F}_2$. Acta Arithmetica **126**(1), 65–99 (2014)
28. J. Dick, F. Pillichshammer, Multivariate integration in weighted Hilbert spaces based on Walsh functions and weighted Sobolev spaces. J. Complexity **21**(2), 149–195 (2005). doi:10.1016/j.jco.2004.07.003
29. J. Dick, F. Pillichshammer, On the mean square weighted $\mathscr{L}_2$ discrepancy of randomized digital (t, m, s)-nets over $\mathbb{Z}_2$. Acta Arithmetica **117**(4), 371–403 (2005). doi:10.4064/aa117-4-4
30. J. Dick, F. Pillichshammer, Strong tractability of multivariate integration of arbitrary high order using digitally shifted polynomial lattice rules. J. Complexity **23**(4-6), 436–453 (2007). doi:10.1016/j.jco.2007.02.001
31. J. Dick, F. Pillichshammer, *Digital nets and sequences. Discrepancy theory and quasi-Monte Carlo integration.* (Cambridge University Press, Cambridge, 2010)

32. J. Dick, G. Leobacher, F. Pillichshammer, Construction algorithms for digital nets with low weighted star discrepancy. SIAM J. Numer. Anal. **43**(1), 76–95 (2005). doi:10.1137/040604662

33. J. Dick, I.H. Sloan, X. Wang, H. Woźniakowski, Liberating the weights. J. Complexity **20**(5), 593–623 (2004). doi:10.1016/j.jco.2003.06.002

34. J. Dick, I.H. Sloan, X. Wang, H. Woźniakowski, Good lattice rules in weighted Korobov spaces with general weights. Numer. Math. **103**(1), 63–97 (2006). doi:10.1007/s00211-005-0674-6

35. J. Dick, P. Kritzer, G. Leobacher, F. Pillichshammer, Constructions of general polynomial lattice rules based on the weighted star discrepancy. Finite Fields Appl. **13**(4), 1045–1070 (2007). doi:10.1016/j.ffa.2006.09.001

36. J. Dick, P. Kritzer, F. Pillichshammer, W.C. Schmid, On the existence of higher order polynomial lattices based on a generalized figure of merit. J. Complexity **23**(4–6), 581–593 (2007). doi:10.1016/j.jco.2006.12.003

37. J. Dick, G. Larcher, F. Pillichshammer, H. Woźniakowski, Exponential convergence and tractability of multivariate integration for Korobov spaces. Math. Comput. **80**(274), 905–930 (2011). doi:10.1090/S0025-5718-2010-02433-0

38. B. Doerr, M. Gnewuch, Construction of low-discrepancy point sets of small size by bracketing covers and dependent randomized rounding, in *Monte Carlo and Quasi-Monte Carlo Methods 2006* (Springer, Berlin, 2008), pp. 299–312

39. B. Doerr, M. Gnewuch, A. Srivastav, Bounds and constructions for the star-discrepancy via δ-covers. J. Complexity **21**(5), 691–709 (2005). doi:10.1016/j.jco.2005.05.002

40. B. Doerr, M. Gnewuch, P. Kritzer, F. Pillichshammer, Component-by-component construction of low-discrepancy point sets of small size. Monte Carlo Methods Appl. **14**(2), 129–149 (2008). doi:10.1515/MCMA.2008.007

41. H. Faure, Discrépance de suites associées à un système de numération (en dimension s). Acta Arithmetica **41**, 337–351 (1982)

42. K.K. Frolov, An upper estimate of the discrepancy in the L_p-metric, $2 \le p < \infty$. Dokl. Akad. Nauk SSSR **252**, 805–807 (1980)

43. E.N. Gilbert, A comparison of signalling alphabets. Bell Syst. Tech. J. **3**, 504–522 (1952)

44. M. Gnewuch, Infinite-dimensional integration on weighted Hilbert spaces. Math. Comput. **81**(280), 2175–2205 (2012). doi:10.1090/S0025-5718-2012-02583-X

45. M. Gnewuch, Weighted geometric discrepancies and numerical integration on reproducing kernel Hilbert spaces. J. Complexity **28**(1), 2–17 (2012). doi:10.1016/j.jco.2011.02.003

46. M. Gnewuch, A. Srivastav, C. Winzen, Finding optimal volume subintervals with k points and calculating the star discrepancy are NP-hard problems. J. Complexity **25**(2), 115–127 (2009). doi:10.1016/j.jco.2008.10.001

47. S. Heinrich, F.J. Hickernell, R.-X. Yue, Optimal quadrature for Haar wavelet spaces. Math. Comput. **73**(245), 259–277 (2004). doi:10.1090/S0025-5718-03-01531-X

48. S. Heinrich, E. Novak, G.W. Wasilkowski, H. Woźniakowski, The inverse of the star-discrepancy depends linearly on the dimension. Acta Arithmetica **96**(3), 279–302 (2001). doi:10.4064/aa96-3-7

49. P. Hellekalek, General discrepancy estimates: The Walsh function system. Acta Arithmetica **67**(3), 209–218 (1994)

50. F.J. Hickernell, H. Niederreiter, The existence of good extensible rank-1 lattices. J. Complexity **19**(3), 286–300 (2003). doi:10.1016/S0885-064X(02)00026-2

51. A. Hinrichs, Covering numbers, Vapnik-Červonenkis classes and bounds for the star-discrepancy. J. Complexity **20**(4), 477–483 (2004). doi:10.1016/j.jco.2004.01.001

52. A. Hinrichs, F. Pillichshammer, W.C. Schmid, Tractability properties of the weighted star discrepancy. J. Complexity **24**(2), 134–143 (2008). doi:10.1016/j.jco.2007.08.002

53. E. Hlawka, Über die Diskrepanz mehrdimensionaler Folgen mod 1. Math. Z. **77**, 273–284 (1961). doi:10.1007/BF01180179

54. E. Hlawka, Zur angenäherten Berechnung mehrfacher Integrale. Monatsh. Math. **66**, 140–151 (1962). doi:10.1007/BF01387711

55. H.S. Hong, F.J. Hickernell, Algorithm 823: Implementing scrambled digital sequences. ACM Trans. Math. Softw. **29**(2), 95–109 (2003). doi:10.1145/779359.779360
56. S. Joe, Component by component construction of rank-1 lattice rules having $O(n^{-1}(\ln(n))^d)$ star discrepancy, in *Monte Carlo and Quasi-Monte Carlo Methods 2002*, ed. by H. Niederreiter (Springer, Berlin-Heidelberg-New York, 2004)
57. S. Joe, Construction of good rank-1 lattice rules based on the weighted star discrepancy., in *Monte Carlo and Quasi-Monte Carlo Methods 2004*, ed. by H. Niederreiter, D. Talay (Springer, Berlin, 2006), pp. 181–196
58. S. Joe, I.H. Sloan, On computing the lattice rule criterion R. Math. Comput. **59**(200), 557–568 (1992). doi:10.2307/2153074
59. N.M. Korobov, Approximate evolution of repeated integrals. Dokl. Akad. Nauk SSSR **124**, 1207–1210 (1959)
60. N.M. Korobov, *Number-Theoretic Methods in Approximate Analysis* (Gosudarstv. Izdat. Fiz.-Mat. Lit., Moscow, 1963)
61. P. Kritzer, Improved upper bounds on the star discrepancy of (t, m, s)-nets and (t, s)-sequences. J. Complexity **22**(3), 336–347 (2006). doi:10.1016/j.jco.2005.10.004
62. P. Kritzer, F. Pillichshammer, A lower bound on a quantity related to the quality of polynomial lattices. Funct. Approx. Comment. Math. **45**(1), 125–137 (2011)
63. P. Kritzer, F. Pillichshammer, Low discrepancy polynomial lattice point sets. J. Number Theory **132**(11), 2510–2534 (2012). doi:10.1016/j.jnt.2012.05.006
64. P. Kritzer, F. Pillichshammer, H. Woźniakowski, Multivariate integration of infinitely many times differentiable functions in weighted Korobov spaces. Mathematics of Computation **83**(287), 1189–1206 (2014)
65. L. Kuipers, H. Niederreiter, *Uniform Distribution of Sequences*. (Pure and Applied Mathematics. New York etc.: Wiley, a Wiley-Interscience Publication, 390 p., 1974; reprint, Dover Publications, Mineola, NY, 2006)
66. F.Y. Kuo, Component-by-component constructions achieve the optimal rate of convergence for multivariate integration in weighted Korobov and Sobolev spaces. J. Complexity **19**(3), 301–320 (2003). doi:10.1016/S0885-064X(03)00006-2
67. G. Larcher, F. Pillichshammer, Sums of distances to the nearest integer and the discrepancy of digital nets. Acta Arithmetica **106**(4), 379–408 (2003). doi:10.4064/aa106-4-4
68. G. Larcher, A. Lauss, H. Niederreiter, W.C. Schmid, Optimal polynomials for (t, m, s)-nets and numerical integration of multivariate Walsh series. SIAM J. Numer. Anal. **33**(6), 2239–2253 (1996). doi:10.1137/S0036142994264705
69. G. Larcher, A best lower bound for good lattice points. Monatsh. Math. **104**, 45–51 (1987). doi:10.1007/BF01540524
70. G. Larcher, Nets obtained from rational functions over finite fields. Acta Arithmetica **63**(1), 1–13 (1993)
71. G. Leobacher, F. Pillichshammer, Bounds for the weighted L^p discrepancy and tractability of integration. J. Complexity **19**(4), 529–547 (2003). doi:10.1016/S0885-064X(03)00009-8
72. W.-L. Loh, On the asymptotic distribution of scrambled net quadrature. Ann. Statist. **31**(4), 1282–1324 (2003). doi:10.1214/aos/1059655914
73. J.N. Lyness, Notes on lattice rules. Numerical integration and its complexity. J. Complexity **19**(3), 321–331 (2003). doi:10.1016/S0885-064X(03)00005-0
74. J. Matoušek, On the L_2-discrepancy for anchored boxes. J. Complexity **14**(4), 527–556 (1998). doi:10.1006/jcom.1998.0489
75. J. Matoušek, *Geometric discrepancy. An illustrated guide*. Algorithms and Combinatorics, vol. 18 (Springer, Berlin-Heidelberg-New York, 1999), p. 288
76. H. Niederreiter, Existence of good lattice points in the sense of Hlawka. Monatsh. Math. **86**, 203–219 (1978). doi:10.1007/BF01659720
77. H. Niederreiter, Pseudozufallszahlen und die Theorie der Gleichverteilung. (Pseudo-random numbers and the theory of uniform distribution). Sitzungsber. Abt. II Österreich. Akad. Wiss. Math.-Naturwiss. Kl. **195**, 109–138 (1986)

78. H. Niederreiter, Rational functions with partial quotients of small degree in their continued fraction expansion. Monatsh. Math. **103**, 269–288 (1987). doi:10.1007/BF01318069
79. H. Niederreiter, Low-discrepancy and low-dispersion sequences. J. Number Theory **30**(1), 51–70 (1988). doi:10.1016/0022-314X(88)90025-X
80. H. Niederreiter, Low-discrepancy point sets obtained by digital constructions over finite fields. Czechoslovak Math. J. **42**(1), 143–166 (1992)
81. H. Niederreiter, *Random Number Generation and Quasi-Monte Carlo Methods*. CBMS-NSF Regional Conference Series in Applied Mathematics, vol. 63 (SIAM, Society for Industrial and Applied Mathematics, Philadelphia, PA, 1992), p. 241
82. H. Niederreiter, C. Xing, Low-discrepancy sequences and global function fields with many rational places. Finite Fields Appl. **2**(3), 241–273 (1996). doi:10.1006/ffta.1996.0016
83. H. Niederreiter, C. Xing, Algebraic curves over finite fields with many rational points and their applications, in *Number theory*, ed. by R.P. Bambah, (Birkhäuser. Trends in Mathematics, Basel, 2000), pp. 287–300
84. E. Novak, *Deterministic and stochastic error bounds in numerical analysis*. Lecture Notes in Mathematics, vol. 1349 (Springer, Berlin, 1988), p. 113. doi:10.1007/BFb0079792
85. E. Novak, H. Woźniakowski, When are integration and discrepancy tractable?, in *Foundations of computational mathematics. Conference, Oxford, GB, July 18–28, 1999*, ed. by Ronald A. DeVore, Lond. Math. Soc. Lect. Note Ser., vol. 284 (Cambridge University Press, Cambridge, 2001), pp. 211–266
86. E. Novak, H. Woźniakowski, *Tractability of Multivariate Problems. Volume I: Linear Information*. EMS Tracts in Mathematics, vol. 6 (European Mathematical Society (EMS), Zürich, 2008), p. 384
87. E. Novak, H. Woźniakowski, L_2 *Discrepancy and Multivariate Integration.*, ed. by W.W.L. Chen, et al., Analytic number theory. Essays in honour of Klaus Roth on the occasion of his 80th birthday (Cambridge University Press, Cambridge, 2009), pp. 359–388
88. E. Novak, H. Woźniakowski, *Tractability of Multivariate Problems. Volume II: Standard Information for Functionals*. EMS Tracts in Mathematics, vol. 12 (European Mathematical Society (EMS), Zürich, 2010), p. 657
89. E. Novak, H. Woźniakowski, *Tractability of Multivariate Problems. Volume III: Standard Information for Operators* (European Mathematical Society (EMS), Zürich, 2012), p. 586. doi:10.4171/116
90. D. Nuyens, R. Cools, Fast algorithms for component-by-component construction of rank-1 lattice rules in shift-invariant reproducing kernel Hilbert spaces. Math. Comput. **75**(254), 903–920 (2006). doi:10.1090/S0025-5718-06-01785-6
91. D. Nuyens, R. Cools, *Fast Component-by-Component Construction, A Reprise for Different Kernels*, ed. by H. Niederreiter, et al., Monte Carlo and quasi-Monte Carlo methods 2004. Refereed proceedings of the sixth international conference on Monte Carlo and quasi-Monte Carlo methods in scientific computation, Juan-les-Pins, France, June 7–10, 2004 (Springer, Berlin, 2006), pp. 373–387. doi:10.1007/3-540-31186-6
92. D. Nuyens, R. Cools, Fast component-by-component construction of rank-1 lattice rules with a non-prime number of points. J. Complexity **22**(1), 4–28 (2006). doi:10.1016/j.jco.2005.07.002
93. A.B. Owen, *Randomly Permuted (t, m, s)-Nets and (t, s)-Sequences*. ed. by H. Niederreiter, et al., Monte Carlo and quasi-Monte Carlo methods in scientific computing. Proceedings of a conference at the University of Nevada, Las Vegas, Nevada, USA, June 23–25, 1994 (Springer, Berlin, 1995). Lect. Notes Stat. 106, pp. 299–317
94. A.B. Owen, Monte Carlo variance of scrambled net quadrature. SIAM J. Numer. Anal. **34**(5), 1884–1910 (1997). doi:10.1137/S0036142994277468
95. A.B. Owen, Scrambled net variance for intergrals of smooth functions. Ann. Stat. **25**(4), 1541–1562 (1997). doi:10.1214/aos/1031594731
96. F. Pillichshammer, Polynomial Lattice Point Sets., in *Monte Carlo and Quasi-Monte Carlo Methods 2010*, ed. by L. Plaskota, H. Woźniakowski, Springer Proceedings in Mathematics and Statistics, vol. 23 (Springer, Berlin, 2012), pp. 189–210

97. A.P. Prudnikov, Y.A. Brychkov, O.I. Marichev, *Integrals and series. Vol. 1: Elementary functions. Vol. 2: Special functions. Transl. from the Russian by N. M. Queen.* (Gordon & Breach Science Publishers, New York, 1986), p. 798

98. K.F. Roth, On irregularities of distribution. Mathematika **1**, 73–79 (1954). doi:10.1112/ S0025579300000541

99. K.F. Roth, On irregularities of distribution. III. Acta Arithmetica **35**, 373–384 (1979)

100. K.F. Roth, On irregularities of distribution. IV. Acta Arithmetica **37**, 67–75 (1980)

101. W.M. Schmidt, Irregularities of distribution. VII. Acta Arithmetica **21**, 45–50 (1972)

102. W.M. Schmidt, Irregularities of distribution. X., in *Number Theory and Algebra. Collect. Pap. dedic. H. B. Mann, A. E. Ross, O. Taussky-Todd* (Academic Press, New York, 1977), pp. 311–329

103. V. Sinescu, S. Joe, *Good Lattice Rules with a Composite Number of Points Based on the Product Weighted Star Discrepancy*, ed. by A. Keller, et al., Monte Carlo and quasi-Monte Carlo methods 2006. Selected papers based on the presentations at the 7th international conference 'Monte Carlo and quasi-Monte Carlo methods in scientific computing', Ulm, Germany, August 14–18, 2006 (Springer, Berlin, 2008), pp. 645–658

104. V. Sinescu, P. L'Ecuyer, *On the Behavior of the Weighted Star Discrepancy Bounds for Shifted Lattice Rules*, ed. by P. L' Ecuyer, et al., Monte Carlo and quasi-Monte Carlo methods 2008. Proceedings of the 8th international conference Monte Carlo and quasi-Monte Carlo methods in scientific computing, Montréal, Canada, July 6–11, 2008 (Springer, Berlin, 2009), pp. 603– 616. doi:10.1007/978-3-642-04107-5_39

105. M.M. Skriganov, Harmonic analysis on totally disconnected groups and irregularities of point distributions. J. Reine Angew. Math. **600**, 25–49 (2006). doi:10.1515/CRELLE.2006.085

106. I.H. Sloan, S. Joe, *Lattice Methods for Multiple Integration.* (Oxford Science Publications. Clarendon Press, Oxford, 1994), p. 239

107. I.H. Sloan, A.V. Reztsov, Component-by-component construction of good lattice rules. Math. Comput. **71**(237), 263–273 (2002). doi:10.1090/S0025-5718-01-01342-4

108. I.H. Sloan, F.Y. Kuo, S. Joe, Constructing randomly shifted lattice rules in weighted Sobolev spaces. SIAM J. Numer. Anal. **40**(5), 1650–1665 (2002). doi:10.1137/S0036142901393942

109. I.H. Sloan, F.Y. Kuo, S. Joe, On the step-by-step construction of quasi-Monte Carlo integration rules that achieve strong tractability error bounds in weighted Sobolev spaces. Math. Comput. **71**(240), 1609–1640 (2002). doi:10.1090/S0025-5718-02-01420-5

110. I.H. Sloan, H. Woźniakowski, When are quasi-Monte Carlo algorithms efficient for high dimensional integrals? J. Complexity **14**(1), 1–33 (1998). doi:10.1006/jcom.1997.0463

111. I.H. Sloan, H. Woźniakowski, Tractability of multivariate integration for weighted Korobov classes. J. Complexity **17**(4), 697–721 (2001). doi:10.1006/jcom.2001.0599

112. I.M. Sobol, On the distribution of points in a cube and the approximate evaluation of integrals. Z. Vyčisl. Mat. i Mat. Fiz. **7**, 784–802 (1967). doi:10.1016/0041-5553(67)90144-9

113. K.B. Stolarsky, Sums of distances between points on a sphere. II. Proc. Am. Math. Soc. **41**, 575–582 (1973). doi:10.2307/2039137

114. V.N. Temlyakov, Cubature formulas, discrepancy, and nonlinear approximation. Numerical integration and its complexity. J. Complexity **19**(3), 352–391 (2003). doi:10.1016/S0885-064X(02)00025-0

115. R.R. Varshamov, The evaluation of signals in codes with correction of errors. Dokl. Akad. Nauk SSSR **117**, 739–741 (1957)

116. I.F. Šarygin, A lower estimate for the error of quadrature formulas for certain classes of functions. Zh. Vychisl. Mat. i Mat. Fiz. **3**, 370–376 (1963)

117. H. Weyl, Über die Gleichverteilung mod. Eins. Math. Ann. **77**, 313–352 (1916). (German)

118. H. Woźniakowski, *Efficiency of Quasi-Monte Carlo Algorithms for High Dimensional Integrals*, ed. by H. Niederreiter, et al., Monte Carlo and quasi-Monte Carlo methods 1998. Proceedings of a conference held at the Claremont Graduate Univ., Claremont, CA, USA, June 22–26, 1998 (Springer, Berlin, 2000), pp. 114–136

119. S.K. Zaremba, Some applications of multidimensional integration by parts. Ann. Poln. Math. **21**, 85–96 (1968)

Chapter 10
Calculation of Discrepancy Measures and Applications

Carola Doerr, Michael Gnewuch, and Magnus Wahlström

Abstract In this book chapter we survey known approaches and algorithms to compute discrepancy measures of point sets. After providing an introduction which puts the calculation of discrepancy measures in a more general context, we focus on the geometric discrepancy measures for which computation algorithms have been designed. In particular, we explain methods to determine L_2-discrepancies and approaches to tackle the inherently difficult problem to calculate the star discrepancy of given sample sets. We also discuss in more detail three applications of algorithms to approximate discrepancies.

10.1 Introduction and Motivation

In many applications it is of interest to measure the quality of certain point sets, e.g., to test whether successive pseudo-random numbers are statistically independent, see, e.g., [82, 87, 99], or whether certain sample sets are suitable for multivariate numerical integration of certain classes of integrands, see, e.g., [25]. Other areas where the need of such measurements may occur include the generation of low-discrepancy samples, the design of computer experiments, computer graphics, and

C. Doerr
Université Pierre et Marie Curie - Paris 6, LIP6, équipe RO, Paris, France

Max-Planck-Institut für Informatik, Saarbrücken, Germany
e-mail: Carola.Doerr@mpi-inf.mpg.de

M. Gnewuch (✉)
Mathematisches Seminar, Kiel University, Kiel, Germany
e-mail: mig@informatik.uni-kiel.de

M. Wahlström
Max-Planck-Institut für Informatik, Saarbrücken, Germany
e-mail: wahl@mpi-inf.mpg.de

W. Chen et al. (eds.), *A Panorama of Discrepancy Theory*, Lecture Notes
in Mathematics 2107, DOI 10.1007/978-3-319-04696-9_10,
© Springer International Publishing Switzerland 2014

stochastic programming. (We shall describe some of these applications in more detail in Sect. 10.5.) A particularly useful class of quality measures, on which we want to focus in this book chapter, is the class of discrepancy measures. Several different discrepancy measures are known. Some of them allow for an efficient evaluation, others are hard to evaluate in practice. We shall give several examples below, but before doing so, let us provide a rather general definition of a geometric discrepancy measure.

Let (M, Σ) be a measurable space. Now let us consider two measures μ and ν defined on the σ-algebra Σ of M. A typical situation would be that μ is a rather complicated measure, e.g., a continuous measure or a discrete measure supported on a large number of atoms, and ν is a simpler measure, e.g., a discrete probability measure with equal probability weights or a discrete (signed) measure supported on a small number of atoms. We are interested in approximating μ by the simpler object ν in some sense and want to quantify the approximation quality. This can be done with the help of an appropriate discrepancy measure.

Such situations occur, e.g., in numerical integration, where one has to deal with a continuous measure μ to evaluate integrals of the form $\int_M f \, d\mu$ and wants to approximate these integrals with the help of a quadrature formula

$$Qf = \sum_{i=1}^{n} v_i f(x^{(i)}) = \int_M f \, d\nu; \qquad (10.1)$$

here $\nu = \nu_Q$ denotes the discrete signed measure $\nu(A) = \sum_{i=1}^{n} v_i 1_A(x^{(i)})$, with 1_A being the characteristic function of $A \in \Sigma$. Another instructive example is scenario reduction in stochastic programming, which will be discussed in more detail in Sect. 10.5.3.

To quantify the discrepancy of μ and ν one may select a subset $\mathscr{B}$ of the σ-algebra Σ, the class of *test sets*, to define the *local discrepancy* of μ and ν in a test set $B \in \mathscr{B}$ as

$$\Delta(B; \mu, \nu) := \mu(B) - \nu(B),$$

and the *geometric L_∞-discrepancy* of μ and ν with respect to $\mathscr{B}$ as

$$\mathrm{disc}_\infty(\mathscr{B}; \mu, \nu) := \sup_{B \in \mathscr{B}} |\Delta(B; \mu, \nu)|. \qquad (10.2)$$

Instead of considering the geometric L_∞-discrepancy, i.e., the supremum norm of the local discrepancy, one may prefer to consider different norms of the local discrepancy. If, e.g., the class of test sets $\mathscr{B}$ is endowed with a σ-algebra $\Sigma(\mathscr{B})$ and a probability measure ω on $\Sigma(\mathscr{B})$, and the restrictions of μ and ν to $\mathscr{B}$ are measurable functions, then one can consider for $p \in (0, \infty)$ the *geometric L_p-discrepancy* with respect to $\mathscr{B}$, defined by

$$\text{disc}_p(\mathscr{B}; \mu, \nu) := \left(\int_{\mathscr{B}} |\Delta(B; \mu, \nu)|^p \, d\omega(B) \right)^{1/p}.$$

In some cases other norms of the local discrepancy may be of interest, too.

In the remainder of this chapter we restrict ourselves to considering discrete measures of the form

$$\nu(B) = \sum_{i=1}^{n} \nu_i 1_B(x^{(i)}), \quad \text{where } \nu_1, \ldots, \nu_n \in \mathbb{R} \text{ and } x^{(1)}, \ldots, x^{(n)} \in M.$$

(10.3)

In the case where $\nu_i = 1/n$ for all i, the quality of the probability measure $\nu = \nu_X$ is completely determined by the quality of the "sample points" $x^{(1)}, \ldots, x^{(n)}$. The case where not all ν_i are equal to $1/n$ is of considerable interest for numerical integration or stochastic programming (see Sect. 10.5.3). As already mentioned above, in the case of numerical integration it is natural to relate a quadrature rule Qf as in (10.1) to the signed measure $\nu = \nu_Q$ in (10.3). The quality of the quadrature Q is then determined by the sample points $x^{(1)}, \ldots, x^{(n)}$ and the (integration) weights $\nu_1, \ldots, \nu_n$.

Let us provide a list of examples of specifically interesting discrepancy measures.

- *Star discrepancy.* Consider the situation where $M = [0, 1]^d$ for some $d \in \mathbb{N}$, Σ is the usual σ-algebra of Borel sets of M, and μ is the d-dimensional Lebesgue measure λ^d on $[0, 1]^d$. Furthermore, let $\mathscr{C}_d$ be the class of all axis-parallel half-open boxes anchored in zero $[0, y) = [0, y_1) \times \cdots \times [0, y_d)$, $y \in [0, 1]^d$. Then the L_∞-*star discrepancy* of the finite sequence $X = (x^{(i)})_{i=1}^{n}$ in $[0, 1)^d$ is given by

$$d_\infty^*(X) := \text{disc}_\infty(\mathscr{C}_d; \lambda^d, \nu_X) = \sup_{C \in \mathscr{C}_d} |\Delta(C; \lambda^d, \nu_X)|,$$

where

$$\nu_X(C) := \frac{1}{n} \sum_{i=1}^{n} 1_C(x^{(i)}) \quad \text{for all } C \in \Sigma.$$

(10.4)

Thus ν_X is the counting measure that simply counts for given Borel sets C the number of points of X contained in C. The star discrepancy is probably the most extensively studied discrepancy measure. Important results about the star discrepancy and its relation to numerical integration can, e.g., be found in [5, 24, 37, 94, 99] or the book chapters [7, 16, 25].

Since we can identify $\mathscr{C}_d$ with $[0, 1]^d$ via the mapping $[0, y) \mapsto y$, we may choose $\Sigma(\mathscr{C}_d)$ as the σ-algebra of Borel sets of $[0, 1]^d$ and the probability measure ω on $\Sigma(\mathscr{C}_d)$ as λ^d. Then for $1 \le p < \infty$ the L_p-*star discrepancy* is given by

$$d_p^*(X) := \mathrm{disc}_p(\mathscr{C}_d; \lambda^d, \nu_X) = \left(\int_{[0,1]^d} \left| y_1 \cdots y_d - \frac{1}{n} \sum_{i=1}^n 1_{[0,y)}(x^{(i)}) \right|^p \mathrm{d}y \right)^{1/p}.$$

In the last few years also norms of the local discrepancy function different from L_p-norms have been studied in the literature, such as suitable Besov, Triebel-Lizorkin, Orlicz, and BMO[1] norms, see, e.g., [8,68,69,91,92,132] and the book chapter [7].

The star discrepancy is easily generalized to general measures ν. For an application that considers measures ν different from ν_X see Sect. 10.5.3.

Notice that the point 0 plays a distinguished role in the definition of the star discrepancy. That is why 0 is often called the anchor of the star discrepancy. There are discrepancy measures on $[0,1]^d$ similar to the star discrepancy that rely also on axis-parallel boxes and on an anchor different from 0, such as the *centered discrepancy* [65] or *quadrant discrepancies* [66,102]. Such kind of discrepancies are, e.g., discussed in more detail in [101, 103].

- *Extreme discrepancy.* The extreme discrepancy is also known under the names *unanchored discrepancy* and *discrepancy for axis-parallel boxes.* Its definition is analogue to the definition of the star discrepancy, except that we consider the class of test sets $\mathscr{R}_d$ of all half-open axis-parallel boxes $[y, z) = [y_1, z_1) \times \cdots \times [y_d, z_d)$, $y, z \in [0, 1]^d$. We may identify this class with the subset $\{(y, z) \mid y, z \in [0, 1]^d, y \leq z\}$ of $[0, 1]^{2d}$, and endow it with the probability measure $\mathrm{d}\omega(y, z) := 2^d \, \mathrm{d}y \, \mathrm{d}z$. Thus the L_∞-*extreme discrepancy* of the finite sequence $X = (x^{(i)})_{i=1}^n$ in $[0, 1)^d$ is given by

$$d_\infty^e(X) := \mathrm{disc}_\infty(\mathscr{R}_d; \lambda^d, \nu_X) = \sup_{R \in \mathscr{R}_d} |\Delta(R; \lambda^d, \nu_X)|,$$

and for $1 \leq p < \infty$ the L_p-*extreme discrepancy* is given by

$$d_p^e(X) := \mathrm{disc}_p(\mathscr{R}_d; \lambda^d, \nu_X)$$

$$= \left(\int_{[0,1]^d} \int_{[0,z)} \left| \prod_{i=1}^d (z_i - y_i) - \frac{1}{n} \sum_{i=1}^n 1_{[y,z)}(x^{(i)}) \right|^p 2^d \, \mathrm{d}y \, \mathrm{d}z \right)^{1/p}.$$

The L_2-extreme discrepancy was proposed as a quality measure for quasi-random point sets in [96].

To avoid confusion, it should be mentioned that the term "extreme discrepancy" is used by some authors in a different way. Especially in the literature before 1980 the attribute "extreme" often refers to a supremum norm of a local discrepancy function, see, e.g., [83, 138]. Since the beginning of the 1990s several authors

---

[1]BMO stands for "bounded mean oscillation".

used the attribute "extreme" to refer to the set system of unanchored axis-parallel boxes, see, e.g., [96, 99, 103].

- *G-discrepancy.* The G- or G-star discrepancy is defined as the star discrepancy, except that the measure μ is in general not the d-dimensional Lebesgue measure λ^d on $[0, 1]^d$, but some probability measure given by a distribution function G via $\mu([0, x)) = G(x)$ for all $x \in [0, 1]^d$. This is

$$\text{disc}_\infty(\mathscr{C}_d; \mu, \nu_X) = \sup_{C \in \mathscr{C}_d} \left| G(x) - \frac{1}{n} \sum_{i=1}^n 1_C(x^{(i)}) \right|.$$

The G-discrepancy has applications in quasi-Monte Carlo sampling, see, e.g., [105]. Further results on the G-star discrepancy can, e.g., be found in [54].

- *Isotrope discrepancy.* Here we have again $M = [0, 1]^d$ and $\mu = \lambda^d$. As set of test sets we consider $\mathscr{I}_d$, the set of all closed convex subsets of $[0, 1]^d$. Then the *isotrope discrepancy* of a set X is defined as

$$\text{disc}_\infty(\mathscr{I}_d; \lambda^d, \nu_X) := \sup_{R \in \mathscr{I}_d} |\Delta(R; \lambda^d, \nu_X)|. \tag{10.5}$$

This discrepancy was proposed by Hlawka [71]. It has applications in probability theory and statistics and was studied further, e.g., in [4, 97, 113, 124, 143].

- *Hickernell's modified L_p-discrepancy.* For a finite point set $X \subset [0, 1]^d$ and a set of variables $u \subset \{1, \ldots, d\}$ let X_u denote the orthogonal projection of X into the cube $[0, 1]^u$. Then for $1 \le p < \infty$ the *modified L_p-discrepancy* [65] of the point set X is given by

$$D_p^*(X) := \left(\sum_{\emptyset \ne u \subseteq \{1, \ldots, d\}} d_p^*(X_u)^p \right)^{1/p}, \tag{10.6}$$

and for $p = \infty$ by

$$D_\infty^*(X) := \max_{\emptyset \ne u \subseteq \{1, \ldots, d\}} d_\infty^*(X_u). \tag{10.7}$$

In the case where $p = 2$ this discrepancy was already considered by Zaremba in [142]. We will discuss the calculation of the modified L_2-discrepancy in Sect. 10.2 and present an application of it in Sect. 10.5.2.

The modified L_p-discrepancy is an example of a weighted discrepancy, which is the next type of discrepancy we want to present.

- *Weighted discrepancy measures.* In the last years *weighted discrepancy measures* have become very popular, especially in the study of tractability of multivariate and infinite-dimensional integration, see the first paper on this topic, [117], and, e.g., [25, 26, 53, 67, 89, 103] and the literature mentioned therein.

To explain the idea behind the weighted discrepancy let us confine ourselves to the case where $M = [0, 1]^d$ and $\nu = \nu_X$ is a discrete measure as in (10.4). (A more general definition of weighted geometric L_2-discrepancy, which comprises in particular infinite-dimensional discrepancies, can be found in [53].) We assume that there exists a one-dimensional measure μ^1 and a system $\mathcal{B}_1$ of test sets on $[0, 1]$. For $u \subseteq \{1, \ldots, d\}$ we define the product measure $\mu^u := \otimes_{j \in u} \mu^1$ and the system of test sets

$$\mathcal{B}_u := \left\{ B \subseteq [0, 1]^u \;\middle|\; B = \prod_{j \in u} B_j \, , \, B_j \in \mathcal{B}_1 \right\}$$

on $[0, 1]^u$. Again we denote the projection of a set $X \subseteq [0, 1]^d$ to $[0, 1]^u$ by X_u. Put $\mathcal{B} := \mathcal{B}_{\{1,\ldots,d\}}$ and $\mu := \mu^{\{1,\ldots,d\}}$. Let $(\gamma_u)_{u \subseteq \{1,\ldots,d\}}$ be a family of *weights*, i.e., of non-negative numbers. Then the *weighted L_∞-discrepancy* $d_{\infty,\gamma}^*(X)$ is given by

$$d_{\infty,\gamma}^*(X) := \operatorname{disc}_{\infty,\gamma}(\mathcal{B}; \mu, \nu_X) = \max_{\emptyset \neq u \subseteq \{1,\ldots,d\}} \gamma_u \operatorname{disc}_\infty(\mathcal{B}_u; \mu^u, \nu_{X_u}).$$

$$(10.8)$$

If furthermore there exists a probability measure $\omega = \omega^1$ on $\mathcal{B}_1$, put $\omega^u := \otimes_{j \in u} \omega^1$ for $u \subseteq \{1, \ldots, d\}$. The *weighted L_p-discrepancy* $d_{p,\gamma}^*(X)$ is then defined by

$$d_{p,\gamma}^*(X) := \operatorname{disc}_{p,\gamma}(\mathcal{B}; \mu, \nu_X) = \left(\sum_{\emptyset \neq u \subseteq \{1,\ldots,d\}} \gamma_u \operatorname{disc}_p(\mathcal{B}_u; \mu^u, \nu_{X_u})^p \right)^{1/p},$$

where

$$\operatorname{disc}_p(\mathcal{B}_u; \mu^u, \nu_{X_u})^p = \int_{\mathcal{B}_u} |\mu^u(B_u) - \nu_{X_u}(B_u)|^p \, d\omega^u(B_u).$$

Hence weighted discrepancies do not only measure the uniformity of a point set $X \subset [0, 1]^d$ in $[0, 1]^d$, but also take into consideration the uniformity of projections X_u of X in $[0, 1]^u$. Note that Hickernell's modified L_p-discrepancy, see (10.6), is a weighted L_p-star discrepancy for the particular family of weights $(\gamma_u)_{u \subseteq \{1,\ldots,d\}}$ where $\gamma_u = 1$ for all u.

Other interesting discrepancy measures in Euclidean spaces as, e.g., discrepancies with respect to half-spaces, balls, convex polygons or rotated boxes, can be found in [5, 13, 17, 94] and the literature mentioned therein. A discrepancy measure that is defined on a flexible region, i.e., on a certain kind of parameterized variety $M = M(m)$, $m \in (0, \infty)$, of measurable subsets of $[0, 1]^d$, is the *central composite discrepancy* proposed in [19]. For discrepancy measures on manifolds as, e.g., the *spherical cap discrepancy*, we refer to [10, 25, 37] and the literature listed there.

There are further figures of merits known to measure the quality of points sets that are no geometric discrepancies in the sense of our definition. Examples include the classical and the dyadic *diaphony* [62, 145] or the figure of merit $R(z, n)$ [78, 99, 115], which are closely related to numerical integration (see also the comment at the beginning of Sect. 10.3.4). We do not discuss such alternative figures of merit here, but focus solely on geometric discrepancy measures. In fact, we confine ourselves to the discrepancies that can be found in the list above. The reason for this is simple: Although deep theoretical results have been published for other geometric discrepancies, there have been, to the best of our knowledge, no serious attempts to evaluate these geometric discrepancies efficiently. Efficient calculation or approximation algorithms were developed almost exclusively for discrepancies that are based on axis-parallel rectangles, such as the star, the extreme or the centered discrepancy, and weighted versions thereof. We briefly explain in the case of the isotrope discrepancy at the beginning of Sect. 10.3 the typical problem that appears if one wants to approximate other geometric discrepancies than those based on axis-parallel rectangles.

This book chapter is organized as follows: In Sect. 10.2 we consider L_2-discrepancies. In Sect. 10.2.1 we explain why many of these discrepancies can be calculated exactly in a straightforward manner with $O(n^2 d)$ operations, where n denotes (as always) the number of points in X and d the dimension. In Sect. 10.2.2 we discuss some asymptotically faster algorithms which allow for an evaluation of the L_2-star and related L_2-discrepancies in time $O(n \log n)$ (where this time the constant in the big-O-notation depends on d). The problem of calculating L_2-discrepancies is the one for which the fastest algorithms are available. As we will see in the following sections, for $p \neq 2$ there are currently no similarly efficient methods known.

In Sect. 10.3 we discuss the calculation of the L_∞-star discrepancy, which is the most prominent discrepancy measure. To this discrepancy the largest amount of research has been devoted so far, both for theoretical and practical reasons. We remark on known and possible generalizations to other L_∞-discrepancy measures. In Sect. 10.3.1 we present elementary algorithms to calculate the star discrepancy exactly. These algorithms are beneficial in low dimensions, but clearly suffer from the curse of dimensionality. Nevertheless, the ideas used for these algorithms are fundamental for the following subsections of Sect. 10.3. In Sect. 10.3.2 we discuss the more sophisticated algorithm of Dobkin, Eppstein and Mitchell, which clearly improves on the elementary algorithms. In Sect. 10.3.3 we review recent results about the complexity of exactly calculating the star discrepancy. These findings lead us to study approximation algorithms in Sect. 10.3.4. Here we present several different approaches.

In Sect. 10.4 we discuss the calculation of L_p-discrepancy measures for values of p other than 2 and ∞. This Section is the shortest one in this book chapter, due to the relatively small amount of research that has been done on this topic.

In Sect. 10.5 we discuss three applications of discrepancy calculation and approximation algorithms in more detail. These applications are the quality testing of points (Sect. 10.5.1), the generation of low-discrepancy point sets via an optimization approach (Sect. 10.5.2), and scenario reduction in stochastic programming (Sect. 10.5.3). The purpose of this section is to show the reader more recent applications and to give her a feeling of typical instance sizes that can be handled and problems that may occur.

10.2   Calculation of L_2-Discrepancies

L_2-discrepancies are often used as quality measures for sets of sample points. One reason for this is the fact that geometric L_2-discrepancies are equal to the worst-case integration error on corresponding reproducing kernel Hilbert spaces and the average-case integration error on corresponding larger function spaces, see the research articles [43, 53, 65, 67, 102, 117, 141, 142] or the surveys in [25, 103].

An additional advantage of the L_2-star discrepancy and related L_2-discrepancies is that they can be explicitly computed at cost $O(dn^2)$, see Sect. 10.2.1 below. Faster algorithms that are particularly beneficial for lower dimension d and larger number of points n will be presented in Sect. 10.2.2.

10.2.1   Warnock's Formula and Generalizations

It is easily verified by direct calculation that the L_2-star discrepancy of a given n-point set $X = (x^{(i)})_{i=1}^n$ in dimension d can be calculated via Warnock's formula [135]

$$d_2^*(X) = \frac{1}{3^d} - \frac{2^{1-d}}{n} \sum_{i=1}^n \prod_{k=1}^d (1 - (x_k^{(i)})^2) + \frac{1}{n^2} \sum_{i,j=1}^n \prod_{k=1}^d \min\{1 - x_k^{(i)}, 1 - x_k^{(j)}\}$$

(10.9)

with $O(dn^2)$ arithmetic operations. As pointed out in [43, 93], the computation requires a sufficiently high precision, since the three terms in the formula are usually of a considerably larger magnitude than the resulting L_2-star discrepancy. A remedy suggested by T. T. Warnock [136] is to subtract off the expected value of each summand in formula (10.9) (assuming that all coordinate values $x_k^{(i)}$ are uniformly and independently distributed) and to add it back at the end of the computation. This means we write down (10.9) in the equivalent form

$$d_2^*(X) = \frac{1}{n}\left[\frac{1}{2^d} - \frac{1}{3^d}\right] - \frac{2^{1-d}}{n}\sum_{i=1}^{n}\left[\prod_{k=1}^{d}(1-(x_k^{(i)})^2) - \left(\frac{2}{3}\right)^d\right]$$

$$+ \frac{1}{n^2}\left(\sum_{\substack{i,j=1\\i\neq j}}^{n}\left[\prod_{k=1}^{d}\min\{1-x_k^{(i)},1-x_k^{(j)}\} - \frac{1}{3^d}\right] + \sum_{i=1}^{n}\left[\prod_{k=1}^{d}(1-x_k^{(i)}) - \frac{1}{2^d}\right]\right),$$

(10.10)

and calculate first the terms inside the brackets [...] and sum them up afterwards. These terms are, in general, more well-behaved than the terms appearing in the original formula (10.9), and the additional use of Kahan summation [79] helps to further reduce rounding errors [136].

For other L_2-discrepancies similar formulas can easily be deduced by direct calculation. So we have, e.g., that the extreme L_2-discrepancy of X can be written as

$$d_2^e(X) = \frac{1}{12^d} - \frac{2}{6^d n}\sum_{i=1}^{n}\prod_{k=1}^{d}(1-(x_k^{(i)})^3 - (1-x_k^{(i)})^3)$$

(10.11)

$$+ \frac{1}{n^2}\sum_{i,j=1}^{n}\prod_{k=1}^{d}\min\{x_k^{(i)},x_k^{(j)}\}\min\{1-x_k^{(i)},1-x_k^{(j)}\},$$

cf. [60, Section 4], and the weighted L_2-star discrepancy of X for *product weights* $\gamma_u = \prod_{j\in u}\gamma_j$, $\gamma_1 \geq \gamma_2 \geq \cdots \geq \gamma_d \geq 0$, as

$$d_{2,\gamma}^*(X) = \prod_{k=1}^{d}\left(1+\frac{\gamma_k^2}{3}\right) - \frac{2}{n}\sum_{i=1}^{n}\prod_{k=1}^{d}\left(1+\gamma_k^2\frac{1-(x_k^{(i)})^2}{2}\right)$$

(10.12)

$$+ \frac{1}{n^2}\sum_{i,j=1}^{n}\prod_{k=1}^{d}\left(1-\gamma_k^2\min\{1-x_k^{(i)},1-x_k^{(j)}\}\right),$$

cf. also [25]. In particular, the formula holds for the modified L_2-discrepancy (10.6) that corresponds to the case where all weights γ_j, $j = 1,2,\ldots$, are equal to 1. Notice that formulas (10.11) and (10.12) can again be evaluated with $O(dn^2)$ arithmetic operations. In the case of the weighted L_2-star discrepancy this is due to the simple structure of the product weights, whereas for an arbitrary family of weights $(\gamma_u)_{u\subseteq\{1,\ldots,d\}}$ the cost of computing $d_{2,\gamma}^*(X)$ exactly will usually be of order $\Omega(2^d)$.

10.2.2 Asymptotically Faster Methods

For the L_2-star discrepancy S. Heinrich [60] developed an algorithm which is asymptotically faster than the direct calculation of (10.9). For fixed d it uses at most

$O(n \log^d n)$ elementary operations; here the implicit constant in the big-O-notation depends on d. This running time can be further reduced to $O(n \log^{d-1} n)$ by using a modification noted by K. Frank and S. Heinrich in [43].

Let us start with the algorithm from [60]. For a quadrature rule

$$Qf = \sum_{i=1}^{n} v_i f(x^{(i)}), \quad \text{with } v_i \in \mathbb{R} \text{ and } x^{(i)} \in [0,1]^d \text{ for all } i,$$

we define the signed measure ν_Q by

$$\nu_Q(C) := Q(1_C) = \sum_{i=1}^{n} v_i 1_C(x^{(i)}) \quad \text{for arbitrary } C \in \mathscr{C}_d.$$

Then it is straightforward to calculate

$$d_2^*(Q) := \mathrm{disc}_2(\mathscr{C}_d; \lambda^d, \nu_Q)$$

$$= \frac{1}{3^d} - 2^{1-d} \sum_{i=1}^{n} v_i \prod_{k=1}^{d}(1 - (x_k^{(i)})^2) \tag{10.13}$$

$$+ \sum_{i,j=1}^{n} v_i v_j \prod_{k=1}^{d} \min\{1 - x_k^{(i)}, 1 - x_k^{(j)}\}.$$

If we are interested in evaluating this generalized version of (10.9) in time $O(n \log^d n)$ or $O(n \log^{d-1} n)$, it obviously only remains to take care of the efficient calculation of

$$\sum_{i,j=1}^{n} v_i v_j \prod_{k=1}^{d} \min\{y_k^{(i)}, y_k^{(j)}\}, \quad \text{where } y_k^{(i)} := 1 - x_k^{(i)} \text{ for } i = 1, \ldots, n.$$

In the course of the algorithm we have actually to take care of a little bit more general quantities: Let $A = ((v_i, y^{(i)}))_{i=1}^{n}$ and $B = ((w_i, z^{(i)}))_{i=1}^{m}$, where $n, m \in \mathbb{N}$, $v_i, w_i \in \mathbb{R}$ and $y^{(i)}, z^{(i)} \in [0,1]^d$ for all i. Put

$$D(A, B, d) := \sum_{i=1}^{n} \sum_{j=1}^{m} v_i w_j \prod_{k=1}^{d} \min\{y_k^{(i)}, z_k^{(j)}\}.$$

We allow also $d = 0$, in which case we use the convention that the "empty product" is equal to 1.

The algorithm is based on the following observation: If $d \geq 1$ and $y_d^{(i)} \leq z_d^{(j)}$ for all $i \in \{1, \ldots, n\}$, $j \in \{1, \ldots, m\}$, then

$$D(A, B, d) = \sum_{i=1}^{n} \sum_{j=1}^{m} (v_i \, y_d^{(i)}) w_j \prod_{k=1}^{d-1} \min\{y_k^{(i)}, z_k^{(j)}\} = D(\tilde{A}, \overline{B}, d-1),$$

where $\tilde{A} = ((\tilde{v}_i, \overline{y}^{(i)}))_{i=1}^{n}$, $\overline{B} = ((w_i, \tilde{z}^{(i)}))_{i=1}^{m}$ with $\tilde{v}_i = (v_i \, y_d^{(i)})$ and $\overline{y}^{(i)} = (y_k^{(i)})_{k=1}^{d-1}$ and $\tilde{z}^{(i)} = (z_k^{(i)})_{k=1}^{d-1}$. Hence we have reduced the dimension parameter d by 1. But in the case where $d = 0$, we can simply calculate

$$D(A, B, 0) = \left(\sum_{i=1}^{n} v_i \right) \left(\sum_{j=1}^{m} w_i \right) \tag{10.14}$$

with cost of order $O(n + m)$. This observation will be exploited by the algorithm proposed by Heinrich to calculate $D(A, B, d)$ for given $d \geq 1$ and arrays A and B as above.

We describe here the version of the algorithm proposed in [60, Section 2]; see also [93, Section 5]. Let μ denote the median of the dth components of the points $y^{(i)}$, $i = 1, \ldots, n$, from A. Then we split A up into two smaller arrays A_L and A_R with A_L containing $\lfloor n/2 \rfloor$ points $y^{(i)}$ (and corresponding weights v_i) satisfying $y_d^{(i)} \leq \mu$ and A_R containing the remaining $\lceil n/2 \rceil$ points (and corresponding weights) satisfying $y_d^{(i)} \geq \mu$. Similarly, we split up B into the two smaller arrays B_L and B_R that contains the points (and corresponding weights) from B whose dth components are less or equal than μ and greater than μ, respectively.

Since we may determine μ with the help of a *linear-time median-finding algorithm* in time $O(n)$ (see, e.g., [1, Ch. 3]), the whole partitioning procedure can be done in time $O(n+m)$. With the help of this partitioning we can exploit the basic idea of the algorithm to obtain

$$D(A, B, d) = D(A_L, B_L, d) + D(A_R, B_R, d) + D(\tilde{A}_L, \overline{B}_R, d-1)$$

$$+ D(\overline{A}_R, \tilde{B}_L, d-1), \tag{10.15}$$

where, as above, $\tilde{A}_L$ is obtained from A_L by deleting the dth component of the points $y^{(i)}$ and substituting the weights v_i by $v_i \, y_d^{(i)}$, and $\overline{A}_R$ is obtained from A_R by also deleting the dth component of the points $y^{(i)}$, but keeping the weights v_i. In an analogous way we obtain $\tilde{B}_L$ and $\overline{B}_R$, respectively.

The algorithm uses the step (10.15) recursively in a *divide-and-conquer* manner. The "conquer" part consists of three base cases, which are solved directly.

The first base case is $m = 0$; i.e., $B = \emptyset$. Then $D(A, B, d) = 0$.

The second one is the case $d = 0$ already discussed above, where we simply use formula (10.14) for the direct calculation of $D(A, B, 0)$ at cost at most $O(n + m)$.

The third base case is the case where $|A| = 1$. Then we can compute directly

$$D(A, B, d) = v_1 \sum_{j=1}^{m} w_j \prod_{k=1}^{d} \min\{y_k^{(1)}, z_k^{(j)}\}.$$

This computation costs at most $O(m)$.

An inductive cost analysis reveals that the cost of this algorithm to calculate $D(A, B, d)$ is of order $O((n+m) \log^d (n+1))$, see [60, Prop. 1]. As already said, the implicit constant in the big-O-notation depends on d. J. Matoušek provided in [93] a running time analysis of Heinrich's algorithm that also takes care of its dependence on the dimension d and compared it to the cost of the straightforward calculation of (10.13). From this analysis one can conclude that Heinrich's algorithm reasonably outperforms the straightforward method if n is larger than (roughly) 2^{2d}; for details see [93, Section 5]. Moreover, [60, 93] contain modifications of the algorithm and remarks on a practical implementation. Furthermore, Heinrich provides some numerical examples with the number of points ranging from 1,024 to 65,536 in dimensions ranging from 1 to 8 comparing the actual computational effort of his method and the direct calculation of (10.13), see [60, Section 5]. In these examples his method was always more efficient than performing the direct calculation; essentially, the advantage grows if the number of points increases, but shrinks if the dimension increases.

As pointed out by Heinrich in [60, Section 4], his algorithm can be modified easily to calculate L_2-extreme discrepancies instead of L_2-star discrepancies with essentially the same effort. Furthermore, he describes how to generalize his algorithm to calculate "r-smooth" L_2-discrepancies, which were considered in [108, 125]. (Here the smoothness parameter r is a non-negative integer. If $r = 0$, we regain the L_2-star discrepancy. If $r > 0$, then the r-smooth discrepancy is actually not any more a geometric L_2-discrepancy in the sense of our definition given in Sect. 10.1.)

Heinrich's algorithm can be accelerated with the help of the following observation from [43]: Instead of employing (10.14) for the base case $D(A, B, 0)$, it is possible to evaluate already the terms $D(A, B, 1)$ that occur in the course of the algorithm. If we want to calculate $D(A, B, 1)$, we assume that the elements $y^{(1)}, \ldots, y^{(n)} \in [0, 1]$ from A and $z^{(1)}, \ldots, z^{(m)} \in [0, 1]$ from B are already in increasing order. This can be ensured by using a standard sorting algorithm to preprocess the input at cost $O((n + m) \log(n + m))$. Now we determine for each i an index $v(i)$ such that $y^{(i)} \geq z^{(j)}$ for $j = 1, \ldots, v(i)$ and $y^{(i)} < z^{(j)}$ for $j = v(i) + 1, \ldots, m$. If this is done successively, starting with $v(1)$, then this can be done at cost $O(n + m)$. Then

$$D(A, B, 1) = \sum_{i=1}^{n} v_i \left(\sum_{j=1}^{v(i)} w_j z^{(j)} + y^{(i)} \sum_{j=v(i)+1}^{m} w_j \right), \qquad (10.16)$$

and the right hand side can be computed with $O(n+m)$ operations if the inner sums are added up successively. Thus the explicit evaluation of $D(A, B, 1)$ can be done at total cost $O(n + m)$. Using this new base case (10.16) instead of (10.14) reduces the running time of the algorithm to $O((n + m) \log^{d-1}(n + 1))$ (as can easily be checked by adapting the proof of [60, Prop. 1]).

The original intention of the paper [43] is in fact to efficiently calculate the L_2-star discrepancies (10.13) of *Smolyak quadrature rules*. These quadrature rules are also known under different names as, e.g., *sparse grid methods* or *hyperbolic cross points*, see, e.g., [47, 100, 119, 125, 137, 144] and the literature mentioned therein. Frank and Heinrich exploit that a d-dimensional Smolyak quadrature rule is uniquely determined by a sequence of one-dimensional quadratures, and in the special case of composite quadrature rules even by a single one-dimensional quadrature. Their algorithm computes the L_2-star discrepancies of Smolyak quadratures at cost $O(N \log^{2-d} N + d \log^4 N)$ for a general sequence of one-dimensional quadratures and at cost $O(d \log^4 N)$ in the special case of composite quadrature rules; here N denotes the number of quadrature points used by the d-dimensional Smolyak quadrature. This time the implicit constants in the big-O-notation do not depend on the dimension d. With the help of their algorithm Frank and Heinrich are able to calculate the L_2-star discrepancy for extremely large numbers of integration points as, e.g., roughly 10^{35} points in dimension $d = 15$. For the detailed description of the algorithm and numerical experiments we refer to [43].

Notice that both algorithms from [43, 60] use as a starting point formula (10.13). Since the three summands appearing in (10.13) are of similar size, the algorithms should be executed with a sufficiently high precision to avoid cancellation effects.

10.2.3 Notes

Related to the problem of calculating L_2-discrepancies of given point sets is the problem of computing the smallest possible L_2-discrepancy of all point sets of a given size n. For the L_2-star discrepancy and arbitrary dimension d the smallest possible discrepancy value of all point sets of size n was derived in [111] for $n = 1$ and in [85] for $n = 2$.

Regarding the L_2-star discrepancy, one should mention that this discrepancy can be a misleading measure of uniformity for sample sizes n smaller than 2^d. For instance, Matoušek pointed out that for small n the pathological point set that consists of n copies of the point $(1, \ldots, 1)$ in $[0, 1]^d$ has almost the best possible L_2-star discrepancy, see [93, Section 2]. A possible remedy is to consider a weighted version of the L_2-star discrepancy instead, as, e.g., the modified L_2-discrepancy.

Matoušek's observation may also be interpreted in the context of numerical integration. The L_2-star discrepancy is equal to the worst-case error of quasi-Monte Carlo (QMC) integration on the unanchored Sobolev space. More precisely, we have for a finite sequence $X = (x^{(i)})_{i=1}^n$ in $[0, 1]^d$ that

$$d_2^*(X) = \sup_{f \in B} \left| \int_{[0,1]^d} f(x) \, dx - Q(f) \right|,$$

where B is the norm unit ball of the unanchored Sobolev space and Q is the QMC algorithm

$$Q(f) = \frac{1}{n} \sum_{i=1}^{n} f(x^{(i)});$$

see, e.g., [103, Chapter 9]. Now Matoušek's observation indicates that if for given n smaller than 2^d one is interested in minimizing the worst-case integration error with the help of a general quadrature rule of the form (10.1), then one should not use QMC algorithms with equal integration weights $1/n$. In fact, already normalized QMC algorithms with suitably chosen equal integration weights $a = a(n, d) < 1/n$ as stated in [103, (10.12)] improve over conventional QMC algorithms with weights $1/n$; for a detailed discussion see [103, Section 10.7.6]. This suggests that for n smaller than 2^d the L_2-star discrepancy modified by substituting the factor $1/n$ by $a < 1/n$ from [103, (10.12)] may be a better measure of uniformity than the L_2-star discrepancy itself.

10.3 Calculation of L_∞-Discrepancies

In this section we survey algorithms which can be used to calculate or approximate the L_∞-star discrepancy. Most of these algorithms have a straightforward extension to other "L_∞-rectangle discrepancies", as, e.g., to the extreme discrepancy discussed above, the centered discrepancy [65], or other quadrant discrepancies [66, 103]. Algorithms for the L_∞-star discrepancy are also necessary to compute or estimate weighted L_∞-star discrepancies. Let us, e.g., assume that we are interested in finding tight upper or lower bounds for the weighted L_∞-discrepancy $d^*_{\infty,\gamma}(X)$, as defined in (10.8). Then we may divide the family of weights $(\gamma_u)_{u \subseteq \{1,...,d\}}$ into a set S of suitably small weights and a set L of larger weights and use the fact that the star discrepancy has the following monotonicity behavior with respect to the dimension: If $u \subseteq v$, then $d^*_\infty(X_u) \leq d^*_\infty(X_v)$. We can use the algorithms discussed below to calculate or bound the discrepancies $d^*_\infty(X_u)$, $u \in L$. The remaining discrepancies $d^*_\infty(X_v)$, $v \in S$, corresponding to the less important weights can be upper-bounded simply by 1 and lower-bounded by $\max_{u \in L ; u \subset v} d^*_\infty(X_u)$ (or even by 0 if the weights are negligible small).

In general, it is not easy to calculate L_∞-discrepancies as defined in (10.2); the cardinality of the system $\mathscr{B}$ of test sets is typically infinite. Since we obviously cannot compute the local discrepancies for an infinite number of test boxes, we usually have to find a finite subset $\mathscr{B}_\delta \subset \mathscr{B}$ such that $\mathrm{disc}_\infty(\mathscr{B}; \mu, \nu) = \mathrm{disc}_\infty(\mathscr{B}_\delta; \mu, \nu)$ or at least $\mathrm{disc}_\infty(\mathscr{B}; \mu, \nu) \leq \mathrm{disc}_\infty(\mathscr{B}_\delta; \mu, \nu) + \delta$ for sufficiently small δ. (This "discretization method" is also important for finding upper bounds for the best possible discrepancy behavior with the help of *probabilistic proofs*, see, e.g., [3] or the book chapter [52].) For most systems $\mathscr{B}$ of test sets this is not a trivial task. If one is, for instance, interested in the isotrope discrepancy, see (10.5), it is not completely obvious to see how the system $\mathscr{I}_d$ can be substituted by a finite set system that leads to a (arbitrarily) close approximation of $\mathrm{disc}_\infty(\mathscr{I}_d; \lambda^d, \nu_X)$. In [97] H. Niederreiter pointed out that it is sufficient to consider the smaller system of test sets $\mathscr{E}_d$ of all open and closed polytopes P contained in $[0,1]^d$ with the property that each face of P is lying entirely on the boundary of $[0,1]^d$ or contains a point of X. Note that $\mathscr{E}_d$ still consists of infinitely many test sets and that further work has to be done before this observation can be used for a concrete algorithm to approximate the isotrope discrepancy.

For the star discrepancy it is easier to find useful discretizations, as we will show below.

10.3.1   Calculating the Star Discrepancy in Low Dimension

Let us have a closer look at the problem of calculating the L_∞-star discrepancy: Let $X = (x^{(i)})_{i=1}^n$ be some fixed finite sequence in $[0,1)^d$. For convenience we introduce for an arbitrary point $y \in [0,1]^d$ the short-hands

$$V_y := \prod_{i=1}^d y_i,$$

and

$$A(y, X) := \sum_{i=1}^n 1_{[0,y)}(x^{(i)}), \quad \text{as well as} \quad \overline{A}(y, X) := \sum_{i=1}^n 1_{[0,y]}(x^{(i)}),$$

i.e., V_y is the volume of the test box $[0, y)$, $A(y, X)$ the number of points of X lying inside the half-open box $[0, y)$, and $\overline{A}(y, X)$ the number of points of X lying in the closed box $[0, y]$. Let us furthermore set

$$\delta(y, X) := V_y - \frac{1}{n} A(y, X) \quad \text{and} \quad \overline{\delta}(y, X) := \frac{1}{n} \overline{A}(y, X) - V_y.$$

Putting $\delta^*(y, X) := \max\{\delta(y, X), \overline{\delta}(y, X)\}$, we have

$$d_\infty^*(X) = \sup_{y \in [0,1]^d} \delta^*(y, X).$$

We define for $j \in \{1, \ldots, d\}$

$$\Gamma_j(X) := \{x_j^{(i)} \mid i \in \{1, \ldots, n\}\} \quad \text{and} \quad \overline{\Gamma}_j(X) := \Gamma_j(X) \cup \{1\},$$

Fig. 10.1 Some set
$X = (x^{(i)})_{i=1}^5$ in $[0, 1)^2$, a
test box $[0, y)$, and
$x \in \Gamma(X), z \in \overline{\Gamma}(X)$ with
$x \leq y \leq z$

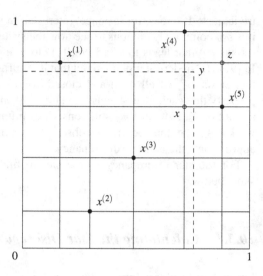

and put

$$\Gamma(X) := \Gamma_1(X) \times \cdots \times \Gamma_d(X) \quad \text{and} \quad \overline{\Gamma}(X) := \overline{\Gamma}_1(X) \times \cdots \times \overline{\Gamma}_d(X).$$

We refer to $\Gamma(X)$ and to $\overline{\Gamma}(X)$ as *grids induced by X*.

Lemma 1. *Let $X = (x^{(i)})_{i=1}^n$ be a sequence in $[0, 1)^d$. Then*

$$d_\infty^*(X) = \max \left\{ \max_{y \in \overline{\Gamma}(X)} \delta(y, X), \max_{y \in \Gamma(X)} \overline{\delta}(y, X) \right\}. \tag{10.17}$$

Formulas similar to (10.17) can be found in several places in the literature—the first reference we are aware of is [97, Thm. 2].

Proof. Consider an arbitrary test box $[0, y)$, $y \in [0, 1]^d$, see Fig. 10.1.
Then for every $j \in \{1, \ldots, d\}$ we find a maximal $x_j \in \Gamma_j(X) \cup \{0\}$ and a minimal $z_j \in \overline{\Gamma}(X)$ satisfying $x_j \leq y_j \leq z_j$. Put $x = (x_1, \ldots, x_d)$ and $z = (z_1, \ldots, z_d)$. We get the inequalities

$$V_y - \frac{1}{n} A(y, X) = V_y - \frac{1}{n} A(z, X) \leq V_z - \frac{1}{n} A(z, X),$$

and

$$\frac{1}{n} A(y, X) - V_y = \frac{1}{n} \overline{A}(x, X) - V_y \leq \frac{1}{n} \overline{A}(x, X) - V_x.$$

Observing that $\overline{A}(x, X) = 0$ if $x_j = 0 \notin \Gamma_j(X)$ for some $j \in \{1, \ldots, d\}$, we see that the right hand side of (10.17) is at least as large as $d^*_\infty(X)$.

Let us now show that it cannot be larger than $d^*_\infty(X)$. So let $y \in \Gamma(X)$ be given. Then we may consider for a small $\varepsilon > 0$ the vector $y(\varepsilon)$, defined by $y(\varepsilon)_j = \min\{y_j + \varepsilon, 1\}$ for $j = 1, \ldots, d$. Clearly, $y(\varepsilon) \in [0, 1]^d$ and

$$\lim_{\varepsilon \to 0} \left(\frac{1}{n} A(y(\varepsilon), X) - V_{y(\varepsilon)} \right) = \frac{1}{n} \overline{A}(y, X) - V_y.$$

These arguments show that (10.17) is valid. $\square$

Lemma 1 shows that an enumeration algorithm would provide us with the exact value of $d^*_\infty(X)$. But since the cardinality of $\Gamma(X)$ for almost all X is n^d, such an algorithm would be infeasible for large values of n and d. Indeed, for a random n-point set X we have almost surely $|\Gamma(X)| = n^d$, resulting in $\Omega(n^d)$ test boxes that we have to take into account to calculate (10.17). This underlines that (10.17) is in general impractical if n and d are large. There are some more sophisticated methods known to calculate the star discrepancy, which are especially helpful in low dimensions. In the one-dimensional case H. Niederreiter derived the following formula, see [97] or [98].

Theorem 2 ([97, Thm. 1]). *Let* $X = (x^{(i)})_{i=1}^n$ *be a sequence in* $[0, 1)$. *If* $x^{(1)} \le x^{(2)} \le \cdots \le x^{(n)}$, *then*

$$d^*_\infty(X) = \max_{i=1}^n \max \left\{ \frac{i}{n} - x^{(i)}, x^{(i)} - \frac{i-1}{n} \right\} = \frac{1}{2n} + \max_{i=1}^n \left| x^{(i)} - \frac{2i-1}{2n} \right|.$$
$$(10.18)$$

Proof. The first identity follows directly from (10.17), since for $y = 1 \in \overline{\Gamma}(X)$ we have $V_y - \frac{1}{n} A(y, X) = 0$. The second identity follows from

$$\max \left\{ \frac{i}{n} - x^{(i)}, x^{(i)} - \frac{i-1}{n} \right\} = \left| x^{(i)} - \frac{2i-1}{2n} \right| + \frac{1}{2n}, \quad i = 1, \ldots, n.$$

$\square$

Notice that (10.18) implies immediately that for $d = 1$ the set $\{\frac{1}{2n}, \frac{3}{2n}, \ldots, \frac{2n-1}{2n}\}$ is the uniquely determined n-point set that achieves the minimal star discrepancy $1/2n$.

In higher dimension the calculation of the star discrepancy unfortunately becomes more complicated.

In dimension $d = 2$ a reduction of the number of steps to calculate (10.17) was achieved by L. De Clerk [22]. In [11] her formula was slightly extended and simplified by P. Bundschuh and Y. Zhu.

Theorem 3 ([22, Section II],[11, Thm. 1]). *Let* $X = (x^{(i)})_{i=1}^n$ *be a sequence in* $[0, 1)^2$. *Assume that* $x_1^{(1)} \le x_1^{(2)} \le \cdots \le x_1^{(n)}$ *and rearrange for each* $i \in \{1, \ldots, n\}$

the numbers $0, x_2^{(1)}, \ldots, x_2^{(i)}, 1$ *in increasing order and rewrite them as* $0 = \xi_{i,0} \le \xi_{i,1} \le \cdots \le \xi_{i,i} \le \xi_{i,i+1} = 1$. *Then*

$$d_\infty^*(X) = \max_{i=0}^{n} \max_{k=0}^{i} \max \left\{ \frac{k}{n} - x_1^{(i)} \xi_{i,k}, \; x_1^{(i+1)} \xi_{i,k+1} - \frac{k}{n} \right\}. \tag{10.19}$$

The derivation of this formula is mainly based on the observation that in dimension $d \ge 2$ the discrepancy functions $\delta(\cdot, X)$ and $\bar{\delta}(\cdot, X)$ can attain their maxima only in some of the points $y \in \overline{\Gamma}(X)$ and $y \in \Gamma(X)$, respectively, which we shall call *critical points*. The formal definition we state here is equivalent to the one given in [56, Section 4.1].

Definition 4. Let $y \in [0, 1]^d$. The point y and the test box $[0, y)$ are $\delta(X)$-*critical*, if we have for all $0 \ne \varepsilon \in [0, 1 - y_1] \times \cdots \times [0, 1 - y_d]$ that $A(y + \varepsilon, X) > A(y, X)$. The point y and the test box $[0, y]$ are $\bar{\delta}(X)$-*critical*, if we have for all $\varepsilon \in [0, y) \setminus \{0\}$ that $\overline{A}(y - \varepsilon, X) < \overline{A}(y, X)$. We denote the set of $\delta(X)$-critical points by $\mathscr{C}(X)$ and the set of $\bar{\delta}(X)$-critical points by $\overline{\mathscr{C}}(X)$, and we put $\mathscr{C}^*(X) := \mathscr{C}(X) \cup \overline{\mathscr{C}}(X)$.

In Fig. 10.1 we have, e.g., that $y = (x_1^{(5)}, x_2^{(4)})$ and $y' = (x_1^{(5)}, 1)$ are $\delta(X)$-critical points, while $y'' = (x_1^{(5)}, x_2^{(3)})$ is not. Furthermore, $\bar{y} = (x_1^{(3)}, x_2^{(3)})$ and $\bar{y}' = (x_1^{(5)}, x_2^{(1)})$ are $\bar{\delta}(X)$-critical points, while $\bar{y}'' = (x_1^{(4)}, x_2^{(2)})$ is not. This shows that, in contrast to the one-dimensional situation, for $d \ge 2$ not all points in $\Gamma(X)$ and $\overline{\Gamma}(X)$ are critical points.

With a similar argument as in the proof of Lemma 1, the following lemma can be established.

Lemma 5. *Let* $X = (x^{(i)})_{i=1}^{n}$ *be a sequence in* $[0, 1]^d$. *Then* $\mathscr{C}(X) \subseteq \overline{\Gamma}(X)$ *and* $\overline{\mathscr{C}}(X) \subseteq \Gamma(X)$, *as well as*

$$d_\infty^*(X) = \max \left\{ \max_{y \in \mathscr{C}(X)} \delta(y, X), \; \max_{y \in \overline{\mathscr{C}}(X)} \bar{\delta}(y, X) \right\}. \tag{10.20}$$

For a rigorous proof see [22, Section II] or [56, Lemma 4.1].

The set of critical test boxes may be subdivided further. For $j = 0, 1, \ldots, n$ put

$$\mathscr{C}^k(X) := \{ y \in \mathscr{C}(X) \mid A(y, X) = k \},$$

and

$$\overline{\mathscr{C}}^k(X) := \left\{ y \in \overline{\mathscr{C}}(X) \mid \overline{A}(y, X) = k \right\}.$$

Then

$$d_\infty^*(X) = \max_{k=1}^{n} \max \left\{ \max_{y \in \mathscr{C}^{k-1}(X)} \delta(y, X), \; \max_{y \in \overline{\mathscr{C}}^k(X)} \bar{\delta}(y, X) \right\}, \tag{10.21}$$

see [22, Section II]. For $d = 2$ De Clerk characterized the points in $\mathscr{C}^k(X)$ and $\overline{\mathscr{C}}^k(X)$ and derived (10.19) under the assumption that $|\Gamma_j(X)| = n$ for $j = 1, \ldots, d$ [22, Section II]. Bundschuh and Zhu got rid of this assumption, which resulted in a technically more involved proof [11, Section 2].

De Clerck used her formula (10.19) to provide explicit formulas for the discrepancies of *Hammersley point sets* [59] for arbitrary basis b and $n = b^m$, $m \in \mathbb{N}$, see [22, Section III]. Her results generalize the formulas for the star discrepancy of two-dimensional Hammersley sets in basis 2 provided by Halton and Zaremba [58]. To establish the explicit formula, she used a recursion property of two-dimensional Hammersley point sets X of size b^m [39] and the facts that these sets are symmetric with respect to the main diagonal of the unit square [109] and that their discrepancy function $\delta(\cdot, X)$ is never positive [45], which implies that for the calculation of the discrepancy of these sets one only has to consider $\bar{\delta}(X)$-critical test boxes.

In dimension $d = 3$ Bundschuh and Zhu provided a formula similar to (10.19).

Theorem 6 ([11, Thm. 2]). *Let* $X = (x^{(i)})_{i=1}^n$ *be a sequence in* $[0, 1)^3$. *Put* $x^{(0)} := (0, 0, 0)$ *and* $x^{(n+1)} := (1, 1, 1)$, *and assume that* $x_1^{(1)} \leq x_1^{(2)} \leq \cdots \leq x_1^{(n)}$. *For* $i \in \{1, \ldots, n\}$ *rearrange the second components* $x_2^{(0)}, x_2^{(1)}, \ldots, x_2^{(i)}, x_2^{(n+1)}$ *in increasing order and rewrite them as* $0 = \xi_{i,0} \leq \xi_{i,1} \leq \cdots \leq \xi_{i,i} \leq \xi_{i,i+1} = 1$ *and denote the corresponding third components* $x_3^{(i)}$, $i = 0, 1, \ldots, i, n + 1$ *by* $\tilde{\xi}_{i,0}, \tilde{\xi}_{i,1}, \ldots, \tilde{\xi}_{i,i+1}$. *Now for fixed* i *and* $k = 0, 1, \ldots, i$ *rearrange* $\tilde{\xi}_{i,0}, \tilde{\xi}_{i,1}, \ldots, \tilde{\xi}_{i,k}, \tilde{\xi}_{i,i+1}$ *and rewrite them as* $0 = \eta_{i,k,0} \leq \eta_{i,k,1} \leq \cdots \leq \eta_{i,k,k} \leq \eta_{i,k,k+1} = 1$. *Then*

$$d_\infty^*(X) = \max_{i=0}^{n} \max_{k=0}^{i} \max_{\ell=0}^{k} \max \left\{ \frac{k}{n} - x_1^{(i)} \xi_{i,k} \eta_{i,k,\ell} , \; x_1^{(i+1)} \xi_{i,k+1} \eta_{i,k,\ell+1} - \frac{k}{n} \right\}. \tag{10.22}$$

The method can be generalized to arbitrary dimension d and requires for generic point sets roughly $O(n^d/d!)$ elementary operations. This method was, e.g., used in [139] to calculate the exact discrepancy of particular point sets, so-called rank-1 lattice rules (cf. [24, 99, 115]), up to size $n = 236$ in dimension $d = 5$ and to $n = 92$ in dimension $d = 6$. But as pointed out by P. Winker and K.-T. Fang, for this method instances like, e.g., sets of size $n \geq 2,000$ in $d = 6$ are infeasible. This method can thus only be used in a very limited number of dimensions.

A method that calculates the exact star discrepancy of a point set in a running time with a more favorable dependence on the dimension d, namely time $O(n^{1+d/2})$, was proposed by Dobkin, Eppstein, and Mitchell in [27]. We discuss this more elaborate algorithm in the next subsection.

10.3.2   The Algorithm of Dobkin, Eppstein, and Mitchell

In order to describe the algorithm of Dobkin, Eppstein, and Mitchell [27], we begin with a problem from computational geometry. In *Klee's measure problem*, the input is a set of n axis-parallel rectangles in $\mathbb{R}^d$, and the problem is to compute the volume of the union of the rectangles. The question of the best possible running time started with V. Klee [81], who gave an $O(n \log n)$-time algorithm for the one-dimensional case, and asked whether this was optimal; it was later found that this is the case [44]. The general case was considered by J. L. Bentley [6], who gave an $O(n^{d-1} \log n)$-time algorithm for $d \geq 2$, thus (surprisingly) giving an algorithm for the $d = 2$ case with an asymptotic running time matching that of the $d = 1$ case. By the lower bound for $d = 1$, Bentley's algorithm is tight for $d = 2$, but as we shall see, not for $d \geq 3$.

The essential breakthrough, which also lies behind the result of [27], was given by Overmars and Yap [107]. They showed that $\mathbb{R}^d$ can be partitioned (in an input-dependent way) into $O(n^{d/2})$ regions, where each region is an axis-parallel box, such that the intersection of the region with the rectangles of the input behaves in a particular regular way. Let us fix some terminology. Let $C = [a, b]$ be a region in the decomposition. A *slab* in C is an axis-parallel box contained in C which has full length in all but at most one dimension, i.e., a box $\prod_{j=1}^{d} I_j$ where $I_j = [a_j, b_j]$ for all but at most one j. A finite union of slabs is called a *trellis*. Overmars and Yap show the following result.

Theorem 7 ([107]). *Let d be a fixed dimension. Given a set of n rectangles in $\mathbb{R}^d$, there is a partitioning of the space into $O(n^{d/2})$ regions where for every region, the intersection of the region with the union of all rectangles of the input forms a trellis.*

An algorithm to enumerate this partition, and a polynomial-time algorithm for computing the volume of a trellis, would now combine into an $O(n^{d/2+c})$-time algorithm for Klee's measure problem, for some constant $c > 0$; Overmars and Yap further improve it to $O(n^{d/2} \log n)$ time (and $O(n)$ space) by using dynamic data structures and careful partial evaluations of the decomposition. Recently, T. M. Chan [12] gave a slightly improved running time of $n^{d/2} 2^{O(\log^* n)}$, where $\log^* n$ denotes the iterated logarithm of n, i.e., the number of times the logarithm function must be iteratively applied before the result is less than or equal to 1.

While no direct reductions between the measure problem and discrepancy computation are known, the principle behind the above decomposition is still useful. Dobkin et al. [27] apply it via a dualization, turning each point x into an orthant (x, ∞), and each box $B = [0, y)$ into a point y, so that x is contained in B if and only if the point y is contained in the orthant. The problem of star discrepancy computation can then be solved by finding, for each $i \in \{1, \ldots, n\}$, the "largest" and "smallest" point $y \in \overline{\Gamma}(X)$ contained in at most respectively at least i orthants; here largest and smallest refer to the value V_y defined previously, i.e., the volume of the box $[0, y)$. Note that this is the dual problem to (10.21). As Dobkin et al. [27] show, the partitioning of $\mathbb{R}^d$ of Overmars and Yap can be used for this purpose. In particular, the base case of a region B such that the intersections of the input rectangles with B form a trellis, corresponds for us to a case where for every point x there is at most one coordinate $j \in \{1, \ldots, d\}$ such that for any $y \in B$ only the value of y_j determines whether $x \in [0, y)$. Given such a base case, it is not difficult to compute the maximum discrepancy relative to points $y \in B$ in polynomial time.

We now sketch the algorithm in some more detail, circumventing the dualization step for a more direct presentation. For simplicity of presentation, we are assuming that for each fixed $j \in \{1, \ldots, d\}$, the coordinates $x_j^{(1)}, \ldots, x_j^{(n)}$ are pairwise different. We will also ignore slight issues with points that lie exactly on a region boundary.

We will subdivide the space $[0, 1]^d$ recursively into regions of the form $[a_1, b_1] \times \ldots \times [a_i, b_i] \times [0, 1]^{d-i}$; call this a *region at level i*. We identify a region with $[a, b]$; if the region is at level $i < d$, then for $j > i$ we have $a_j = 0$ and $b_j = 1$. For $j \in \{1, \ldots, i\}$, we say that a point x is *internal in dimension j*, relative to a region $[a, b]$ at level i, if x is contained in $[0, b)$ and $a_j < x_j < b_j$. We will maintain two invariants as follows.

1. For every point x contained in the box $[0, b)$, there is at most one coordinate $j \in \{1, \ldots, i\}$ such that x is internal in dimension j.
2. For any region at level $i > 0$ and any $j \in \{1, \ldots, i\}$ there are only $O(\sqrt{n})$ points internal in dimension j.

Once we reach a region at level d, called a *cell*, the first condition ensures that we reach the above-described base case, i.e., that every point contained in the box $[0, b)$ is internal in at most one dimension. The second condition, as we will see, ensures that the decomposition can be performed while creating only $O(n^{d/2})$ cells.

The process goes as follows. We will describe a recursive procedure, subdividing each region at level $i < d$ into $O(\sqrt{n})$ regions at level $i + 1$, ensuring $O(\sqrt{n}^d) = O(n^{d/2})$ cells in the final decomposition. In the process, if $[a, b]$ is the region currently being subdivided, we let X_b be the set of points contained in the box $[0, b)$, and X_I the subset of those points which are internal in some dimension $j \le i$. We initialize the process with the region $[0, 1]^d$ at level 0, with X_b being the full set of points of the input, and $X_I = \emptyset$.

Given a region $[a, b]$ at level $i < d$, with X_b and X_I as described, we will partition along dimension $i+1$ into segments $[\xi_j, \xi_{j+1}]$, for some ξ_j, $j \in \{1, \ldots, \ell\}$, where ℓ is the number of subdivisions. Concretely, for each $j \in \{1, \ldots, \ell - 1\}$ we generate a region $[a^{(j)}, b^{(j)}] = [a_1, b_1] \times \ldots \times [a_i, b_i] \times [\xi_j, \xi_{j+1}] \times [0, 1]^{d-i-1}$ at level $i + 1$, set $X_b^{(j)} = X_b \cap [0, b^{(j)})$ and $X_I^{(j)} = (X_I \cap [0, b^{(j)})) \cup \{x \in X_b^{(j)} : \xi_j < x_{i+1} < \xi_{j+1}\}$, and recursively process this region, with point set $X_b^{(j)}$ and internal points $X_I^{(j)}$. Observe that the points added to $X_I^{(j)}$ indeed are internal points, in dimension $i + 1$. The coordinates ξ_j are chosen to fulfill the following conditions.

For all $x \in X_I$, we have $x_{i+1} \in \{\xi_1, \ldots, \xi_\ell\}$.   (10.23)

For all $j \in \{1, \ldots, \ell - 1\}$, we have $|\{x \in X_b : \xi_j < x_{i+1} < \xi_{j+1}\}| = O(\sqrt{n})$.

(10.24)

We briefly argue that this is necessary and sufficient to maintain our two invariants while ensuring that $\ell = O(\sqrt{n})$ at every level. Indeed, if condition (10.23) is not fulfilled, then the point x is internal in two dimensions in some region $[a^{(j)}, b^{(j)}]$, and if condition (10.24) is not fulfilled for some j, then $|X_I^{(j)}|$ is larger than $O(\sqrt{n})$ in the same region. To create a set of coordinates ξ_j fulfilling these conditions is not difficult; simply begin with the coordinate set $\{x_{i+1} : x \in X_I\}$ to fulfill (10.23), and insert additional coordinates as needed to satisfy (10.24). This requires at most $|X_I| + (n/\sqrt{n})$ coordinates; thus $\ell = O(\sqrt{n})$ as required, and one finds inductively that both invariants are maintained at every region created at level $i + 1$.

To finally sketch the procedure for computing the discrepancy inside a given cell, let $B = [a, b]$ be the cell, X_b the points contained in $[0, b)$, and X' the set of points in $[0, b)$ which are not also contained in $[0, a]$. The points X' must thus be internal in at least one dimension, so by invariant 1, the points X' are internal in exactly one dimension $j \in \{1, \ldots, d\}$. Note that if a point $x \in X'$ is internal in dimension j, and hence has $x_i \le a_i$ for every $i \in \{1, \ldots, d\}$, $i \ne j$, then for any $y \in (a, b)$ we have that $x \in [0, y)$ if and only if $y_j > x_j$; that is, if x is internal in dimension j, then the membership of x in $[0, y)$ is determined only by y_j. Now, for each $j \in \{1, \ldots, d\}$, let $X'_j = \{x \in X' : a_j < x_j < b_j\}$ be the points internal in dimension j; note that this partitions X'. We can then determine $X \cap [0, y)$ for any $y \in (a, b)$ by looking at the coordinates y_j independently, that is,

$$[0, y) \cap X = ([0, a] \cap X_b) \cup \bigcup_{j=1}^{d} \{x \in X'_j : x_j < y_j\}.$$

This independence makes the problem suitable for the algorithmic technique of *dynamic programming* (see, e.g., [20]). Briefly, let $f_j(y) = |X_b \cap [0, a]| + \sum_{k=1}^{j} |\{x \in X'_k : x_k < y_k\}|$. For $i \in \{1, \ldots, n\}$ and $j \in \{1, \ldots, d\}$, let $p(i, j)$ be the minimum value of $\prod_{k=1}^{j} y_k$ such that $f_j(y) \ge i$, and let $q(i, j)$ be the maximum value of $\prod_{k=1}^{j} y_k$ such that $f_j(y) \le i$ and $y \in (a, b)$. By sorting the coordinates x_j of every set X'_j, it is easy both to compute the values of $p(\cdot, 1)$ and $q(\cdot, 1)$, and to use the values of $p(\cdot, j)$ and $q(\cdot, j)$ to compute the values of $p(\cdot, j + 1)$ and $q(\cdot, j + 1)$. The maximum discrepancy attained for a box $[0, y)$ for $y \in (a, b)$ can then be computed from $p(\cdot, d)$ and $q(\cdot, d)$; note in particular that for $y \in (a, b)$, we have $f_d(y) = |X \cap [0, y)|$. For details, we refer to [27]. Discrepancy for boxes $[0, y]$ with $y_j = 1$ for one or several $j \in \{1, \ldots, d\}$ can be handled in a similar way.

Some slight extra care in the analysis of the dynamic programming, and application of a more intricate form of the decomposition of Overmars and Yap, will lead to a running time of $O(n^{d/2+1})$ and $O(n)$ space, as shown in [27].

10.3.3 Complexity Results

We saw in the previous sections that for fixed dimension d the most efficient algorithm for calculating the star discrepancy of arbitrary n-point sets in $[0, 1)^d$, the algorithm of Dobkin, Eppstein, and Mitchell, has a running time of order $O(n^{1+d/2})$.

So the obvious question is if it is possible to construct a faster algorithm for the discrepancy calculation whose running time does *not* depend exponentially on the dimension d.

There are two recent complexity theoretical results that suggest that such an algorithm does not exist—at least not if the famous hypotheses from complexity theory, namely that P $\neq$ NP and the stronger *exponential time hypothesis*[74], are true.

10.3.3.1 Calculating the Star Discrepancy Is NP-Hard

Looking at identity (10.17), we see that the calculation of the star discrepancy of a given point set X is in fact a *discrete optimization problem*, namely the problem to find an $y \in \overline{\Gamma}$ that maximizes the function value

$$\delta^*(y, X) = \max \left\{ \delta(y, X), \overline{\delta}(y, X) \right\}.$$

In [55] it was proved that the calculation of the star discrepancy is in fact an NP-hard optimization problem. Actually, a stronger statement was proved in [55], namely that the calculation of the star discrepancy is NP-hard even if we restrict ourselves to the easier sub-problem where all the coordinates of the input have finite binary expansion, i.e., are of the form $k2^{-\kappa}$ for some $\kappa \in \mathbb{N}$ and some integer $0 \leq k \leq 2^{\kappa}$. To explain this result in more detail let us start by defining the *coding length* of a real number from the interval $[0, 1)$ to be the number of digits in its binary expansion.

Informally speaking, the class NP is the class of all *decision problems*, i.e., problems with a true-or-false answer, for which the instances with an affirmative answer can be decided in polynomial time by a *non-deterministic Turing machine*[2]; here "polynomial" means polynomial in the coding length of the input. Such a non-deterministic Turing machine can be described as consisting of a non-deterministic part, which generates for a given instance a polynomial-length candidate ("certificate") for the solution, and a deterministic part, which verifies in polynomial time whether this candidate leads to a valid solution.

In general, the NP-hardness of an optimization problem U is proved by verifying that deciding the so-called *threshold language* of U is an NP-hard decision problem (see, e.g., [73, Section 2.3.3] or, for a more informal explanation, [46, Section 2.1]). Thus it was actually shown in [55] that the following decision problem is NP-hard:

Decision Problem. STAR DISCREPANCY

Instance: Natural numbers $n, d \in \mathbb{N}$, sequence $X = (x^{(i)})_{i=1}^{n}$ in $[0, 1)^d, \varepsilon \in (0, 1]$

Question: Is $d_{\infty}^*(X) \geq \varepsilon$?

---

[2]NP stands for "non-deterministic polynomial time".

Notice that the input of STAR DISCREPANCY has only finite coding length if the binary expansion of all coordinates and of ε is finite. The standard approach to prove NP-hardness of some decision problem is to show that another decision problem that is known to be NP-hard can be reduced to it in polynomial time. This approach was also used in [55], where the graph theoretical problem DOMINATING SET was reduced to STAR DISCREPANCY. The decision problem DOMINATING SET is defined as follows.

Definition 8. Let $G = (V, E)$ be a graph, where V is the finite set of vertices and $E \subseteq \{\{v, w\} \mid v, w \in V, v \neq w\}$ the set of edges of G. Let M be a subset of V. Then M is called a *dominating set of G* if for all $v \in V \setminus M$ there exists a $w \in M$ such that $\{v, w\}$ is contained in E.

Decision Problem. DOMINATING SET

> *Instance:* Graph $G = (V, E), m \in \{1, \ldots, |V|\}$
> *Question:* Is there a dominating set $M \subseteq V$ of cardinality at most m?

The decision problem DOMINATING SET is well studied in the literature and known to be NP-complete, see, e.g., [46].

Before we explain how to reduce DOMINATING SET to STAR DISCREPANCY, let us introduce another closely related decision problem, which will also be important in Sect. 10.3.4.2.

Decision Problem. EMPTY HALF-OPEN BOX

> *Instance:* Natural numbers $n, d \in \mathbb{N}$, sequence $X = (x^{(i)})_{i=1}^n$ in $[0, 1)^d$, $\varepsilon \in (0, 1]$
> *Question:* Is there a $y \in \overline{\Gamma}(X)$ with $A(y, X) = 0$ and $V_y \geq \varepsilon$?

As in the problem STAR DISCREPANCY the coding length of the input of EMPTY HALF-OPEN BOX is only finite if the binary expansion of all coordinates and of ε is finite. For the reduction of DOMINATING SET it is sufficient to consider these instances with finite coding length. We now explain how to reduce DOMINATING SET to EMPTY HALF-OPEN BOX in polynomial time; this proof step can afterwards be re-used to establish that DOMINATING SET can indeed be reduced to STAR DISCREPANCY.

Theorem 9 ([55, Thm. 2.7]). *The decision problem* EMPTY HALF-OPEN BOX *is NP-hard.*

Proof. Let $G = (V, E), m \in \{1, \ldots, |V|\}$ be an instance of DOMINATING SET. We may assume that $V = \{1, \ldots, n\}$ for $n := |V|$. Let $\alpha, \beta \in [0, 1)$ have finite binary expansions and satisfy $\beta < \alpha^n$; for instance, we may choose $\alpha = 1/2$ and $\beta = 0$. For $i, j \in \{1, \ldots, n\}$ put

$$x_j^{(i)} := \begin{cases} \alpha, & \text{if } \{i, j\} \in E \text{ or } i = j, \\ \beta, & \text{otherwise,} \end{cases}$$

and put $x^{(i)} := (x_j^{(i)})_{j=1}^n \in [0,1)^n$ and $X := (x^{(i)})_{i=1}^n$. We shall show that there is a dominating set $M \subseteq V$ of cardinality at most m if and only if there is a $y \in \overline{\Gamma}(X)$ such that $A(y, X) = 0$ and $V_y \geq \alpha^m$.

Firstly, assume that there is a dominating set $M \subseteq V$ of cardinality at most m. Put

$$y_j := \begin{cases} \alpha, & \text{if } j \in M, \\ 1, & \text{otherwise,} \end{cases}$$

and $y := (y_j)_{j=1}^n$. Then $y \in \overline{\Gamma}(X)$ and $V_y = \alpha^{|M|} \geq \alpha^m$. Hence it suffices to prove that $[0, y) \cap X = \emptyset$. Now for each $i \in M$ we have $x_i^{(i)} = \alpha$, i.e., $x^{(i)} \notin [0, y)$. For every $i \in V \setminus M$ there is, by definition of a dominating set, a $v \in M$ such that $\{i, v\} \in E$, implying $x_v^{(i)} = \alpha$ which in turn yields $x^{(i)} \notin [0, y)$. Therefore $A(y, X) = 0$.

Secondly, assume the existence of a $y \in \overline{\Gamma}(X)$ such that $A(y, X) = 0$ and $V_y \geq \alpha^m$. Recall that $y \in \overline{\Gamma}(X)$ implies that $y_j \in \{\beta, \alpha, 1\}$ for all j. Since $\beta < \alpha^n \leq V_y$, we have $|\{j \in \{1, \ldots, n\} \mid y_j \geq \alpha\}| = n$. Putting $M := \{i \in \{1, \ldots, n\} \mid y_i = \alpha\}$, we have $|M| \leq m$. Since $A(y, X) = 0$, we obtain $|M| \geq 1$, and for each $i \in \{1, \ldots, n\}$ there exists a $v \in M$ such that $\{i, v\} \in E$ or $i \in M$. Hence M is a dominating set of G with size at most m.   □

In [55] further decision problems of the type "maximal half-open box for k points" and "minimal closed box for k points" were studied; these problems are relevant for an algorithm of E. Thiémard that is based on integer linear programming, see Sect. 10.3.4.2.

Theorem 10 ([55, Thm. 3.1]). STAR DISCREPANCY *is NP-hard.*

Proof (Sketch). Due to identity (10.17) the decision problem STAR DISCREPANCY can be formulated in an equivalent way: Is there a $y \in \overline{\Gamma}(X)$ such that $\delta(y, X) \geq \varepsilon$ or $\bar{\delta}(y, X) \geq \varepsilon$? The NP-hardness of this equivalent formulation can again be shown by polynomial time reduction from DOMINATING SET. So let $V = \{1, \ldots, n\}$, and let $G = (V, E), m \in \{1, \ldots, n\}$ be an instance of DOMINATING SET. We may assume without loss of generality $n \geq 2$ and $m < n$. Put $\alpha := 1 - 2^{-(n+1)}, \beta := 0$, and

$$x_j^{(i)} := \begin{cases} \alpha, & \text{if } \{i, j\} \in E \text{ or } i = j, \\ \beta, & \text{otherwise.} \end{cases}$$

The main idea is now to prove for $X := (x^{(i)})_{i=1}^n$ that $d_\infty^*(X) \geq \alpha^m =: \varepsilon$ if and only if there is a dominating set $M \subseteq V$ for G with $|M| \leq m$.

From the proof of Theorem 9 we know that the existence of such a dominating set M is equivalent to the existence of a $y \in \overline{\Gamma}(X)$ with $A(y, X) = 0$ and $V_y \geq \alpha^m$. Since the existence of such a y implies $d_\infty^*(X) \geq \alpha^m$, it remains only to show that

$d^*_\infty(X) \geq \alpha^m$ implies in turn the existence of such a y. This can be checked with the help of Bernoulli's inequality, for the technical details we refer to [55].

10.3.3.2 Calculating the Star Discrepancy Is W[1]-Hard

Although NP-hardness (assuming $P \neq NP$) excludes a running time of $(n + d)^{O(1)}$ for computing the star discrepancy of an input of n points in d dimensions, this still does not completely address our running time concerns. In a nutshell, we know that the problem can be solved in polynomial time for every fixed d (e.g., by the algorithm of Dobkin, Eppstein, and Mitchell), and that it is NP-hard for arbitrary inputs, but we have no way of separating a running time of $n^{\Theta(d)}$ from, say, $O(2^d n^2)$, which would of course be a huge breakthrough for computing low-dimensional star discrepancy.

The usual framework for addressing such questions is *parameterized complexity*. Without going into too much technical detail, a parameterized problem is a decision problem whose inputs are given with a *parameter* k. Such a problem is *fixed-parameter tractable* (FPT) if instances of total length n and with parameter k can be solved in time $O(f(k)n^c)$, for some constant c and an arbitrary function $f(k)$. Observe that this is equivalent to being solvable in $O(n^c)$ time for every fixed k, as contrasted to the previous notion of polynomial time for every fixed k, which also includes running times such as $O(n^k)$. We shall see in this section that, unfortunately, under standard complexity theoretical assumptions, no such algorithm is possible (in fact, under a stronger assumption, not even a running time of $O(f(d)n^{o(d)})$ is possible).

Complementing the notion of FPT is a notion of parameterized hardness. A *parameterized reduction* from a parameterized problem $\mathcal{Q}$ to a parameterized problem $\mathcal{Q}'$ is a reduction which maps an instance I with parameter k of $\mathcal{Q}$ to an instance I' with parameter k' of $\mathcal{Q}'$, such that

1. (I, k) is a true instance of $\mathcal{Q}$ if and only if (I', k') is a true instance of $\mathcal{Q}'$,
2. $k' \leq f(k)$ for some function $f(k)$, and
3. the total running time of the reduction is FPT (i.e., bounded by $O(g(k)\|I\|^c)$, for some function $g(k)$ and constant c, where $\|I\|$ denotes the total coding length of the input).

It can be verified that such a reduction and an FPT-algorithm for $\mathcal{Q}'$ imply an FPT-algorithm for $\mathcal{Q}$.

The basic hardness class of parameterized complexity, analogous to the class NP of classical computational complexity, is known as W[1], and can be defined as follows. Given a graph $G = (V, E)$, a *clique* is a set $X \subseteq V$ such that for any $u, v \in X, u \neq v$, we have $\{u, v\} \in E$. Let k-CLIQUE be the following parameterized problem.

Parameterized Problem. k-CLIQUE

Instance: A graph $G = (V, E)$; an integer k.
Parameter: k
Question: Is there a clique of cardinality k in G?

The class W[1] is then the class of problems reducible to k-CLIQUE under parameterized reductions, and a parameterized problem is W[1]-hard if there is a parameterized reduction to it from k-CLIQUE. (Note that, similarly, the complexity class NP can be defined as the closure of, e.g., CLIQUE or 3-SAT under standard polynomial-time reductions.) The basic complexity assumption of parameterized complexity is that FPT $\neq$ W[1], or equivalently, that the k-CLIQUE problem is not fixed-parameter tractable; this is analogous to the assumption in classical complexity that P $\neq$ NP.

For more details, see the books of R. G. Downey and M. R. Fellows [36] or J. Flum and M. Grohe [42]. In particular, we remark that there is a different definition of W[1] in terms of problems solvable by a class of restricted circuits (in fact, W[1] is just the lowest level of a hierarchy of such classes, the so-called *W-hierarchy*), which arguably makes the class definition more natural, and the conjecture FPT $\neq$ W[1] more believable.

P. Giannopoulus et al. [48] showed the following.

Theorem 11 ([48]). *There is a polynomial-time parameterized reduction from k-*CLIQUE *with parameter k to* STAR DISCREPANCY *with parameter $d = 2k$.*

Thus, the results of [48] imply that STAR DISCREPANCY has no algorithm with a running time of $O(f(d)n^c)$ for any function $f(d)$ and constant c, unless FPT $=$ W[1].

A stronger consequence can be found by considering the so-called exponential-time hypothesis (ETH). This hypothesis, formalized by Impagliazzo and Paturi [74], states that 3-SAT on n variables cannot be solved in time $O(2^{o(n)})$; in a related paper [75], this was shown to be equivalent to similar statements about several other problems, including that k-SAT on n variables cannot be solved in time $O(2^{o(n)})$ for $k \geq 3$ and that 3-SAT cannot be solved in time $O(2^{o(m)})$, where m equals the total number of clauses in an instance (the latter form is particularly useful in reductions).

It has been shown by J. Chen et al. [14, 15] that k-CLIQUE cannot be solved in $f(k)n^{o(k)}$ time, for any function $f(k)$, unless ETH is false. We thus get the following corollary.

Corollary 12 ([48]). STAR DISCREPANCY *for n points in d dimensions cannot be solved exactly in $f(d)n^{o(d)}$ time for any function $f(d)$, unless ETH is false.*

In fact, it seems that even the constant factor in the exponent of the running time of the algorithm of Dobkin, Eppstein and Mitchell [27] would be very difficult to improve; e.g., a running time of $O(n^{d/3+c})$ for some constant c would imply a new faster algorithm for k-CLIQUE. (See [48] for more details, and for a description of the reduction.)

10.3.4 Approximation of the Star Discrepancy

As seen in Sects. 10.3.3.1 and 10.3.3.2, complexity theory tells us that the exact calculation of the star discrepancy of large point sets in high dimensions is infeasible.

The theoretical bounds for the star discrepancy of low-discrepancy point sets that are available in the literature describe the asymptotic behavior of the star discrepancy well only if the number of points n tends to infinity. These bounds are typically useful for point sets of size $n \gg e^d$, but give no helpful information for moderate values of n. To give concrete examples, we restate here some numerical results provided in [76] for bounds based on inequalities of *Erdős-Turán-Koksma-type* (see, e.g., [24, 99]).

If the point set $P_n(z)$ is an n-point rank-1 lattice in $[0, 1]^d$ with generating vector $z \in \mathbb{Z}^d$ (see, e.g., [99, 115]), then its discrepancy can be bounded by

$$d_\infty^*(P_n(z)) \leq 1 - \left(1 - \frac{1}{n}\right)^d + T(z, n) \leq 1 - \left(1 - \frac{1}{n}\right)^d + W(z, n)$$

$$\leq 1 - \left(1 - \frac{1}{n}\right)^d + R(z, n)/2; \tag{10.25}$$

here the quantities $W(z, n)$ and $R(z, n)$ can be calculated to a fixed precision in $O(nd)$ operations [76, 78], while the calculation of $T(z, n)$ requires $O(n^2 d)$ operations (at least there is so far no faster algorithm known).

S. Joe presented in [76] numerical examples where he calculated the values of $T(z, n)$, $W(z, n)$, and $R(z, n)$ for generators z provided by a *component-by-component algorithm* (see, e.g., [84, 104, 116, 118]) from [76, Section 4]. For dimension $d = 2$ and 3 he was able to compare the quantities with the exact values

$$d_\infty^*(P_n(z)) - \left[1 - \left(1 - \frac{1}{n}\right)^d\right] =: E(z, n). \tag{10.26}$$

In $d = 2$ for point sets ranging from $n = 157$ to 10,007, the smallest of the three quantities, $T(z, n)$, was 8–10 times larger than $E(z, n)$. In dimension $d = 3$ for point sets ranging from $n = 157$ to 619 it was even more than 18 times larger. (Joe used for the computation of $E(z, n)$ the algorithm proposed in [11] and was therefore limited to this range of examples.)

Computations for $d = 10$ and 20 and $n = 320,009$ led to $T(z, n) = 1.29 \cdot 10^4$ and $5.29 \cdot 10^{13}$, respectively. Recall that the star discrepancy and $E(z, n)$ are always bounded by 1. The more efficiently computable quantities $W(z, n)$ and $R(z, n)$ led obviously to worse results, but $W(z, n)$ was at least very close to $T(z, n)$.

This example demonstrates that for good estimates of the star discrepancy for point sets of practicable size we can unfortunately not rely on theoretical bounds.

Since the exact calculation of the star discrepancy is infeasible in high dimensions, the only remaining alternative is to consider approximation algorithms. In the following we present the known approaches.

10.3.4.1 An Approach Based on Bracketing Covers

An approach that approximates the star discrepancy of a given set X up to a user-specified error δ was presented by E. Thiémard [128, 130]. It is in principle based on the generation of suitable δ-*bracketing covers* (which were not named in this way in [128, 130]). Let us describe Thiémard's approach in detail.

The first step is to "discretize" the star discrepancy at the cost of an approximation error of at most δ. The corresponding discretization is different from the one described in Sect. 10.3.1; in particular, it is completely independent of the input set X. The discretization is done by choosing a suitable finite set of test boxes anchored in zero whose upper right corners form a so-called δ-*cover*. We repeat here the definition from [30].

Definition 13. A finite subset Γ of $[0, 1]^d$ is called a δ-*cover* of the class $\mathscr{C}_d$ of all axis-parallel half-open boxes anchored in zero (or of $[0, 1]^d$) if for all $y \in [0, 1]^d$ there exist $x, z \in \Gamma \setminus \{0\}$ such that

$$x \leq y \leq z \quad \text{and} \quad V_z - V_x \leq \delta.$$

Put

$$N(\mathscr{C}_d, \delta) := \min\{|\Gamma| \mid \Gamma \ \delta\text{-cover of } \mathscr{C}_d.\}$$

Any δ-cover Γ of $\mathscr{C}_d$ satisfies the following approximation property:

Lemma 14. *Let Γ be a δ-cover of $\mathscr{C}_d$. For all finite sequences X in $[0, 1]^d$ we have*

$$d_\infty^*(X) \leq d_\Gamma^*(X) + \delta, \tag{10.27}$$

where

$$d_\Gamma^*(X) := \max_{y \in \Gamma} |V_y - A(y, X)|$$

can be seen as a discretized version of the star discrepancy.

A proof of Lemma 14 is straightforward. (Nevertheless, it is, e.g., contained in [30].)

In [50] the notion of δ-covers was related to the concept of bracketing entropy, which is well known in the theory of empirical processes. We state here the definition for the set system $\mathscr{C}_d$ of anchored axis-parallel boxes (a general definition can, e.g., be found in [50, Section 1]):

Definition 15. A closed axis-parallel box $[x, z] \subset [0, 1]^d$ is a δ-*bracket* of $\mathscr{C}_d$ if $x \leq z$ and $V_z - V_x \leq \delta$. A δ-*bracketing cover* of $\mathscr{C}_d$ is a set of δ-brackets whose union is $[0, 1]^d$. By $N_{[]}(\mathscr{C}_d, \delta)$ we denote the *bracketing number* of $\mathscr{C}_d$ (or of $[0, 1]^d$), i.e., the smallest number of δ-brackets whose union is $[0, 1]^d$. The quantity $\ln N_{[]}(\mathscr{C}_d, \delta)$ is called the *bracketing entropy*.

The bracketing number and the quantity $N(d, \delta)$ are related to the *covering* and the L_1-*packing number*, see, e.g., [30, Rem. 2.10].

It is not hard to verify that

$$N(\mathscr{C}_d, \delta) \leq 2N_{[]}(\mathscr{C}_d, \delta) \leq N(\mathscr{C}_d, \delta)(N(\mathscr{C}_d, \delta) + 1) \tag{10.28}$$

holds. Indeed, if $\mathscr{B}$ is a δ-bracketing cover, then it is easy to see that

$$\Gamma_{\mathscr{B}} := \{x \in [0, 1]^d \setminus \{0\} \,|\, \exists y \in [0, 1]^d : [x, y] \in \mathscr{B} \text{ or } [y, x] \in \mathscr{B}\} \tag{10.29}$$

is a δ-cover. If Γ is a δ-cover, then

$$\mathscr{B}_\Gamma := \{[x, y] \,|\, x, y \in \Gamma \cup \{0\}, [x, y] \text{ is a } \delta\text{-bracket}, x \neq y\}$$

is a δ-bracketing cover. These two observations imply (10.28).

In [50] it is shown that

$$\delta^{-d}(1 - O_d(\delta)) \leq N_{[]}(\mathscr{C}_d, \delta) \leq 2^{d-1}(2\pi d)^{-1/2} e^d (\delta^{-1} + 1)^d, \tag{10.30}$$

see [50, Thm. 1.5 and 1.15]. The construction that leads to the upper bound in (10.30) implies also

$$N_{[]}(\mathscr{C}_d, \delta) \leq (2\pi d)^{-1/2} e^d \delta^{-d} + O_d(\delta^{-d+1}) \tag{10.31}$$

(see [50, Remark 1.16]) and

$$N(\mathscr{C}_d, \delta) \leq 2^d (2\pi d)^{-1/2} e^d (\delta^{-1} + 1)^d. \tag{10.32}$$

For more information about δ-covers and δ-bracketing covers we refer to the original articles [30, 50, 51] and the survey article [52]; the articles [51] and [52] contain also several figures showing explicit two-dimensional constructions.

The essential idea of Thiémard's algorithm from [128, 130] is to generate for a given point set X and a user-specified error δ a small δ-bracketing cover $\mathscr{B} = \mathscr{B}_\delta$ of $[0, 1]^d$ and to approximate $d_\infty^*(X)$ by $d_\Gamma^*(X)$, where $\Gamma = \Gamma_{\mathscr{B}}$ as in (10.29), up to an error of at most δ, see Lemma 14.

The costs of generating $\mathscr{B}_\delta$ are of order $\Theta(d|\mathscr{B}_\delta|)$. If we count the number of points in $[0, y)$ for each $y \in \Gamma_{\mathscr{B}}$ in a naive way, this results in an overall running time of $\Theta(dn|\mathscr{B}_\delta|)$ for the whole algorithm. As Thiémard pointed out in [130], this *orthogonal range counting* can be done in moderate dimension d more effectively by employing data structures based on *range trees*, see, e.g., [21,95]. This approach reduces in moderate dimension d the time $O(dn)$ per test box that is needed for the naive counting to $O(\log^d n)$. Since a range tree for n points can be generated in $O(C^d n \log^d n)$ time, $C > 1$ some constant, this results in an overall running time of

$$O((d + \log^d n)|\mathscr{B}_\delta| + C^d n \log^d n).$$

As this approach of orthogonal range counting is obviously not very beneficial in higher dimension (say, $d > 5$), we do not further explain it here, but refer for the details to [130].

The smallest bracketing covers $\mathscr{T}_\delta$ used by Thiémard can be found in [130]; they differ from the constructions provided in [50, 51], see [51, 53]. He proved for his best constructions the upper bound

$$|\mathscr{T}_\delta| \le e^d \left(\frac{\ln \delta^{-1}}{\delta} + 1 \right)^d,$$

a weaker bound than (10.31) and (10.32), which both hold for the δ-bracketing covers constructed in [50]. Concrete comparisons of Thiémard's bracketing covers with other constructions in dimension $d = 2$ can be found in [51], where also optimal two-dimensional bracketing covers are provided.

The lower bound in (10.30) proved in [50] immediately implies a lower bound for the running time of Thiémard's algorithm, regardless how cleverly the δ-bracketing covers are chosen. That is because the dominating factor in the running time is the construction of the δ-bracketing cover $|\mathscr{B}_\delta|$, which is of order $\Theta(d|\mathscr{B}_\delta|)$. Thus (10.30) shows that the running time of the algorithm is exponential in d. (Nevertheless smaller δ-bracketing covers, which may, e.g., be generated by extending the ideas from [51] to arbitrary dimension d, would widen the range of applicability of Thiémard's algorithm.) Despite its limitations, Thiémard's algorithm is a helpful tool in moderate dimensions, as was reported, e.g., in [32, 128, 130] or [106], see also Sect. 10.5.1.

For more specific details we refer to [110, 128, 130]. For a modification of Thiémard's approach to approximate L_∞-extreme discrepancies see [50, Section 2.2].

10.3.4.2 An Approach Based on Integer Linear Programming

Since the large scale enumeration problem (10.17) is infeasible in high dimensions, a number of algorithms have been developed that are based on heuristic approaches. One such approach was suggested by E. Thiémard in [131] (a more detailed description of his algorithm can be found in his PhD thesis [129]). Thiémard's algorithm is based on *integer linear programming*, a concept that we shall describe below. His approach is interesting in that, despite being based on heuristics in the initialization phase, it allows for an arbitrarily good approximation of the star discrepancy of any given point set. Furthermore, the user can decide on the fly which approximation error he is willing to tolerate. This is possible because the algorithm outputs, during the optimization of the star discrepancy approximation, upper and lower bounds for the exact $d_\infty^*(X)$-value. The user can abort the optimization procedure once the difference between the lower and upper bound are small enough for his needs, or he may wait until the optimization procedure is finished, and the exact star discrepancy value $d_\infty^*(X)$ is computed.

Before we describe a few details of the algorithm, let us mention that numerical tests in [130, 131] suggest that the algorithm from [131] outperforms the one from [130] (see Sect. 10.3.4.1 for a description of the latter algorithm). In particular, instances that are infeasible for the algorithm from [130] can be solved using the integer linear programming approach described below, see also the discussion in Sect. 10.5.1.

The basic idea of Thiémard's algorithm is to split optimization problem (10.17) into $2n$ optimization problems similarly as done in Eq. (10.21), and to transform these problems into integer linear programs. To be more precise, he considers for each value $k \in \{0, 1, 2, \ldots, n\}$ the volume of the smallest and the largest box containing exactly k points of X. These values are denoted $V_{\min}^k$ and $V_{\max}^k$, respectively. It is easily verified, using similar arguments as in Lemma 1, that these values (if they exist) are obtained by the grid points of X, i.e., there exist grid points $y_{\min}^k \in \Gamma(X)$, $y_{\max}^k \in \overline{\Gamma}(X)$ such that

$$\overline{A}(y_{\min}^k, X) = k \text{ and } V_{y_{\min}^k} = V_{\min}^k,$$

$$A(y_{\max}^k, X) = k \text{ and } V_{y_{\max}^k} = V_{\max}^k.$$

It follows from (10.21) that

$$d_\infty^*(X) = \max_{k \in \{0,1,2,\ldots,n\}} \max \left\{ \frac{k}{n} - V_{\min}^k, V_{\max}^k - \frac{k}{n} \right\}. \tag{10.33}$$

As noted in [55], boxes containing *exactly* k points may not exist, even if the n points are pairwise different. However, they do exist if for at least one dimension $j \in \{1, \ldots, d\}$ the coordinates $(x_j)_{x \in X}$ are pairwise different. If they are pairwise different for all $j \in \{1, \ldots, d\}$, in addition we have

$$V_{\min}^1 \leq \cdots \leq V_{\min}^n \quad \text{and} \quad V_{\max}^0 \leq \cdots \leq V_{\max}^{n-1}.$$

Note that we obviously have $V_{\min}^0 = 0$ and $V_{\max}^n = 1$. We may therefore disregard these two values in Eq. (10.33).

As mentioned in Sect. 10.3.3.1 (cf. Theorem 9 and the text thereafter), already the related problem "Is $V_{\max}^0 \geq \epsilon$?" is an NP-hard one. By adding "dummy points" at or close to the origin $(0, \ldots, 0)$ one can easily generalize this result to all questions of the type "Is $V_{\max}^k \geq \epsilon$?". Likewise, it is shown in [55] that the following decision problem is NP-hard.

Decision Problem. $V_{\min}^k$-BOX

Instance: Natural numbers $n, d \in \mathbb{N}$, $k \in \{0, 1, \ldots, n\}$, sequence $X = (x^{(i)})_{i=1}^n$ in $[0, 1)^d$, $\varepsilon \in (0, 1]$
Question: Is there a point $y \in \Gamma(X)$ such that $\overline{A}(y, X) \geq k$ and $V_y \leq \varepsilon$?

This suggests that the $V_{\max}^k$- and $V_{\min}^k$-problems are difficult to solve to optimality. As we shall see below, in Thiémard's integer linear programming ansatz, we will not have to solve all $2n$ optimization problems in (10.33) to optimality. Instead it turns out that for most practical applications of his algorithms only very few of them need to be solved exactly, whereas for most of the problems it suffices to find good upper and lower bounds. We shall discuss this in more detail below.

For each of the $2n$ subproblems of computing $V_{\min}^k$ and $V_{\max}^k$, respectively, Thiémard formulates, by taking the logarithm of the volumes, an integer linear program with $n + d(n - k)$ binary variables. The size of the linear program is linear in the size nd of the input X. We present here the integer linear program (ILP) for the $V_{\min}^k$-problems. The ones for $V_{\max}^k$-problems are similar. However, before we are ready to formulate the ILPs, we need to fix some notation.

For every $n \in \mathbb{N}$ we abbreviate by S_n the set of permutations of $\{1, \ldots, n\}$. For all $j \in \{1, \ldots, d\}$ put $x_j^{(n+1)} := 1$ and let $\sigma_j \in S_{n+1}$ such that

$$x_j^{(\sigma_j(1))} \leq \ldots \leq x_j^{(\sigma_j(n))} \leq x_j^{(\sigma_j(n+1))} = 1.$$

With σ_j at hand, we can define, for every index $\delta = (\delta_1, \ldots, \delta_d) \in \{1, \ldots, n + 1\}^d$ the closed and half-open boxes *induced by* δ,

$$[0, \delta] := \prod_{j=1}^d [0, x_j^{(\sigma_j(\delta_j))}] \quad \text{and} \quad [0, \delta) = \prod_{j=1}^d [0, x_j^{(\sigma_j(\delta_j))}).$$

One of the crucial observations for the formulation of the ILPs is the fact that $x^{(i)} \in [0, \delta]$ (resp. $x^{(i)} \in [0, \delta)$), if and only if for all $j \in \{1, \ldots, d\}$ it holds that $\sigma_j^{-1}(i) \leq \delta_j$ (resp. $\sigma_j^{-1}(i) \leq \delta_j - 1$). We set

$$z_j^i(\delta) := \begin{cases} 1, & \text{if } \sigma_j^{-1}(i) \leq \delta_j, \\ 0, & \text{otherwise.} \end{cases}$$

Every δ induces exactly one sequence $z = ((z_j^{(i)}(\delta)))_{j=1}^d)_{i=1}^n$ in $(\{0, 1\}^d)^n$, and, likewise, for every *feasible* sequence z there is exactly one $\delta(z) \in \{1, \ldots, n + 1\}^d$ with $z = z(\delta(z))$. In the following linear program formulation we introduce also the variables $y^{(1)}, \ldots, y^{(n)}$, and we shall have $y^{(i)} = 1$ if and only if $x^{(i)} \in [0, \delta(z)]$.

The integer linear program for the $V_{\min}^k$-problem can now be defined as follows.

$$\ln(V_{\min}^k) = \min \sum_{j=1}^{d}[\ln(x_j^{(\sigma_j(1))}) + \sum_{i=2}^{n} z_j^{(\sigma_j(i))}(\ln(x_j^{(\sigma_j(i))}) - \ln(x_j^{(\sigma_j(i-1))}))]$$

(10.34)

subject to

(i) $\quad 1 = z_j^{(\sigma_j(1))} = \ldots = z_j^{(\sigma_j(k))} \geq \ldots \geq z_j^{(\sigma_j(n))} \quad \forall j \in \{1,\ldots,d\}$

(ii) $\quad z_j^{(\sigma_j(i))} = z_j^{(\sigma_j(i+1))} \qquad\qquad\qquad \forall j \in \{1,\ldots,d\} \forall i \in \{1,\ldots,n\}:$
$$\qquad\qquad\qquad\qquad\qquad\qquad\qquad\qquad x_j^{(\sigma_j(i))} = x_j^{(\sigma_j(i+1))}$$

(iii) $\quad y^{(i)} \leq z_j^{(i)} \qquad\qquad\qquad\qquad\quad \forall j \in \{1,\ldots,d\} \forall i \in \{1,\ldots,n\}$

(iv) $\quad y^{(i)} \geq 1 - d + \sum_{j=1}^{d} z_j^{(i)} \qquad\quad \forall i \in \{1,\ldots,n\}$

(v) $\quad \sum_{i=1}^{n} y^{(i)} \geq k$

(vi) $\quad y^{(i)} \in \{0,1\} \qquad\qquad\qquad\qquad \forall i \in \{1,\ldots,n\}$

(vii) $\quad z_j^{(i)} \in \{0,1\} \qquad\qquad\qquad\qquad \forall j \in \{1,\ldots,d\} \forall i \in \{1,\ldots,n\}$

We briefly discuss the constraints of the integer linear program (10.34).

- Since we request at least k points to lie in the box $[0, \delta(z)]$, the inequality $x_j^{(\sigma_j(\delta(z)_j))} \geq x_j^{(\sigma_j(k))}$ must hold for all $j \in \{1,\ldots,d\}$. We may thus fix the values $1 = z_j^{(\sigma_j(1))} = \ldots = z_j^{(\sigma_j(k))}$.
- The second constraint expresses that for two points with the same coordinate $x_j^{(\sigma_j(i))} = x_j^{(\sigma_j(i+1))}$ in the jth dimension, we must satisfy $z_j^{(\sigma_j(i))} = z_j^{(\sigma_j(i+1))}$.
- The third and fourth condition say that $y^{(i)} = 1$ if and only if $x^{(i)} \in [0, \delta(z)]$. For $x^{(i)} \in [0, \delta(z)]$ we have $\sigma_j^{-1}(i) \leq \delta(z)_j$ and thus $z_j^{(i)} = 1$, $j \in \{1,\ldots,d\}$. According to condition (iv) this implies $y^{(i)} \geq 1 - d + \sum_{j=1}^{d} z_j^{(i)} = 1$, and thus $y^{(i)} = 1$. If, on the other hand, $x^{(i)} \notin [0, \delta(z)]$, there exists a coordinate $j \in \{1,\ldots,d\}$ with $x_j^{(i)} > \delta(z)_j$. Thus, $z_j^{(i)} = 0$ and condition (iii) implies $y^{(i)} \leq z_j^{(i)} = 0$.
- Condition (v) ensures the existence of at least k points inside $[0, \delta(z)]$.
- Conditions (vi) and (vii) are called the *integer* or *binary constraints*. Since only integer (binary) values are allowed, the linear program (10.34) is called a *(binary) integer linear program*. We shall see below that by changing these conditions to $y^{(i)} \in [0,1]$ and $z_j^{(i)} \in [0,1]$, we get the *linear relaxation* of the integer linear program (10.34). The solution of this linear relaxation is a lower bound for the true $V_{\min}^k$-solution.

Using the $V_{\min}^k$- and $V_{\max}^k$-integer linear programs, Thiémard computes the star discrepancy of a set X in a sequence of optimization steps, each of which possibly deals with a different k-box problem. Before the optimization phase kicks in, there is an *initialization phase*, in which for each k an upper bound $\overline{V}^k_{\min}$ for $V^k_{\min}$ and a lower bound $\underline{V}^k_{\max}$ for $V^k_{\max}$ is computed. This is done by a simple greedy strategy followed by a local optimization procedure that helps to improve the initial value of the greedy strategy. Thiémard reports that the estimates obtained for the $V^k_{\max}$-problems are usually quite good already, whereas the estimates for the $V^k_{\min}$-problems are usually too pessimistic. A lower bound $\underline{V}^k_{\min}$ for the $V^k_{\min}$-problem is also computed, using the simple observation that for each dimension $j \in \{1, \ldots, d\}$, the jth coordinate of the smallest $V^k_{\min}$-box must be at least as large as the kth smallest coordinate in $(x_j)_{x \in X}$. That is, we have $V^k_{\min} \geq \prod_{j=1}^{d} x_j^{(\sigma_j(k))}$. We initialize the lower bound $\underline{V}^k_{\min}$ by setting it equal to this expression. Similarly, $\prod_{j=1}^{d} x_j^{(\sigma_j(k+1))}$ is a lower bound for the $V^k_{\max}$-problem, but this bound is usually much worse than the one provided by the heuristics. For an initial upper bound of $V^k_{\max}$, Thiémard observes that the $V^{n-1}_{\max}$-problem can be solved easily. In fact, we have $V^{n-1}_{\max} = \max\{x_j^{(\sigma_j(n))} \mid j \in \{1, \ldots, d\}\}$. As mentioned in the introduction to this section, if for all $j \in \{1, \ldots, d\}$ the jth coordinates $x_j^{(1)}, \ldots, x_j^{(n)}$ of the points in X are pairwise different, we have $V^0_{\max} \leq \ldots \leq V^{n-1}_{\max}$. Thus, in this case, we may initialize $\overline{V}^k_{\max}$, $k = 1, \ldots, n-1$, by $V^{n-1}_{\max}$.

From these values (we neglect a few minor steps in Thiémard's computation) we compute the following estimates

$$\underline{D}^k_{\min}(X) := \tfrac{k}{n} - \overline{V}^k_{\min} \text{ and } \overline{D}^k_{\min}(X) := \tfrac{k}{n} - \underline{V}^k_{\min},$$

$$\underline{D}^k_{\max}(X) := \underline{V}^k_{\max} - \tfrac{k}{n} \text{ and } \overline{D}^k_{\max}(X) := \overline{V}^k_{\max} - \tfrac{k}{n}.$$

Clearly, $\underline{D}^k_{\min}(X) \leq \tfrac{k}{n} - V^k_{\min} \leq \overline{D}^k_{\min}(X)$ and $\underline{D}^k_{\max}(X) \leq V^k_{\max} - \tfrac{k}{n} \leq \overline{D}^k_{\max}(X)$.

After this initialization phase, the *optimization phase* begins. It proceeds in rounds. In each round, the k-box problem yielding the largest estimate

$$\overline{D}^*_{\infty}(X) := \max\left\{ \max_{k \in \{1,2,\ldots,n\}} \overline{D}^k_{\min}(X), \max_{k \in \{0,1,\ldots,n-1\}} \overline{D}^k_{\max}(X) \right\}$$

is investigated further.

If we consider a $V^k_{\min}$- or a $V^k_{\max}$-problem for the first time, we regard the *linear relaxation* of the integer linear program for $\ln(V^k_{\min})$ or $\ln(V^k_{\max})$, respectively. That is—cf. the comments below the formulation of the ILP for $V^k_{\min}$-problem above— instead of requiring the variables $y^{(i)}$ and $z_j^{(i)}$ to be either 0 or 1, we only require them to be in the interval $[0, 1]$. This turns the integer linear program into a linear program. Although it may seem that this relaxation does not change much, linear programs are known to be polynomial time solvable, and many fast readily available solving procedures, e.g., commercial tools such as CPLEX, are available. Integer linear programs and binary integer linear programs such as ours, on the other hand, are known to be NP-hard in general, and are usually solvable only with considerable computational effort. Relaxing the binary constraints (vi) and (vii) in (10.34) can thus be seen as a heuristic to get an initial approximation of the $V_{\min}^k$- and $V_{\max}^k$-problems, respectively.

In case of a $V_{\min}^k$-problem the value of this relaxed program is a lower bound for $\ln(V_{\min}^k)$—we thus obtain new estimates for the two values $\underline{V}_{\min}^k$ and $\overline{D}_{\min}^k$. If, on the other hand, we regard a $V_{\max}^k$-problem, the solution of the relaxed linear program establishes an upper bound for $\ln(V_{\max}^k)$; and we thus get new estimates for $\overline{V}_{\max}^k$ and $\overline{D}_{\max}^k$. We may be lucky that we get an integral solution, in which case we have determined $V_{\min}^k$ or $V_{\max}^k$, respectively, and do not need to consider this problem in any further iteration of the algorithm.

If we consider a $V_{\min}^k$- or $V_{\max}^k$-problem for the second time, we solve the integer linear program itself, using a standard *branch and bound* technique. Branch and bound resembles a divide and conquer approach: the problem is divided into smaller subproblems, for each of which upper and lower bounds are computed.

Let us assume that we are, for now, considering a fixed $V_{\min}^k$-problem ($V_{\max}^k$-problems are treated the same way). As mentioned above, we divide this problem into several subproblems, and we compute upper and lower bounds for these subproblems. We then investigate the most "promising" subproblems (i.e., the ones with the largest upper bound and smallest lower bound for the value of $V_{\min}^k$) further, until the original $V_{\min}^k$-problem at hand has been solved to optimality or until the bounds for $V_{\min}^k$ are good enough to infer that this k-box problem does not cause the maximal discrepancy value in (10.33).

A key success factor of the branch and bound step is a further strengthening of the integer linear program at hand. Thiémard introduces further constraints to the ILP, some of which are based on straightforward combinatorial properties of the k-box problems and others which are based on more sophisticated techniques such as *cutting planes* and *variable forcing* (cf. Thiémard's PhD thesis [129] for details). These additional constraints and techniques strengthen the ILP in the sense that the solution to the linear relaxation is closer to that of the integer program. Thiémard provides some numerical results indicating that these methods frequently yield solutions based on which we can exclude the k-box problem at hand from our considerations for optimizing (10.33). That is, only few of the $2n$ many k-box problems in (10.33) need to be solved to optimality, cf. [131] for the details.

As explained above, Thiémard's approach computes upper and lower bounds for the star discrepancy of a given point set at the same time. Numerical experiments indicate that the lower bounds are usually quite strong from the beginning, whereas the initial upper bounds are typically too large, and decrease only slowly during the optimization phase, cf. [131, Section 4.2] for a representative graph of the convergence behavior. Typical running times of the algorithms can be found in [131] and in [56]. The latter report contains also a comparison to the alternative approach described in the next section.

10.3.4.3 Approaches Based on Threshold-Accepting

In the next two sections we describe three heuristic approaches to compute lower bounds for the star discrepancy of a given point set X. All three algorithms are based on randomized local search heuristics; two of them on a so-called *threshold accepting approach*, see this section, and one of them on a *genetic algorithm*, see Sect. 10.3.4.4.

Randomized local search heuristics are problem-independent algorithms that can be used as frameworks for the optimization of inherently difficult problems, such as combinatorial problems, graph problems, etc. We distinguish between Monte-Carlo algorithms and Las Vegas algorithms. Las Vegas algorithms are known to converge to the optimal solution, but their exact running time cannot be determined in advance. Monte-Carlo algorithms, on the other hand, have a fixed running time (usually measured by the number of iterations or the number of function evaluations performed), but we usually do not know the quality of the final output. The two threshold accepting algorithms presented next are Monte-Carlo algorithms for which the user may specify the number of iterations he is willing to invest for a good approximation of the star discrepancy value. The genetic algorithm presented in Sect. 10.3.4.4, on the other hand, is a Monte-Carlo algorithm with unpredictable running time (as we shall see below, in this algorithm, unconventionally, the computation is aborted when no improvement has happened for some t iterations in a row).

This said, it is clear that the lower bounds computed by both the threshold accepting algorithms as well as the one computed by the genetic algorithm may be arbitrarily bad. However, as all reported numerical experiments suggest, they are usually quite good approximations of the true discrepancy value—in almost all cases for which the correct discrepancy value can be computed the same value was also reported by the improved threshold accepting heuristic [56] described below. We note that these heuristic approaches allow the computation of lower bounds for the star discrepancy also in those settings where the running time of exact algorithms like the one of Dobkin, Eppstein, and Mitchell described in Sect. 10.3.2 are not feasible.

Threshold accepting is based on a similar idea as the well-known simulated annealing algorithm [80]. In fact, it can be seen as a simulated annealing algorithm in which the selection step is derandomized, cf. Algorithm 1 for the general scheme of a threshold accepting algorithm. In our application of computing star discrepancy values, we accept a new candidate solution z if its local discrepancy $d_\infty^*(z, X)$ is not much worse than that of the previous step, and we discard z otherwise. More precisely, we accept z if and only if the difference $d_\infty^*(z, X) - d_\infty^*(y, X)$ is at least as large as some *threshold value* T. The threshold value is a parameter to be specified by the user. We typically have $T < 0$. $T < 0$ is a reasonable choice as it prevents the algorithm from getting stuck in some local maximum of the local discrepancy function. In the two threshold accepting algorithms presented below, T will be updated frequently during the run of the algorithm (details follow).

Algorithm 1: Simplified scheme of a threshold accepting algorithm for the computation of star discrepancy values. I is the runtime of the algorithm (number of iterations), and T is the threshold value for the acceptance of a new candidate solution z

---

1 Initialization: Select $y \in \overline{\Gamma}(X)$ uniformly at random and compute $d_\infty^*(y, X)$;
2 for $i = 1, 2, \ldots, I$ **do**
3   **Mutation Step:** Select a random neighbor z of y and compute $d_\infty^*(z, X)$;
4   **Selection step: if** $d_\infty^*(z, X) - d_\infty^*(y, X) \geq T$ **then** $y \leftarrow z$;
5 Output $d_\infty^*(y, X)$;

The first to apply threshold accepting to the computation of star discrepancies were P. Winker and K.-T. Fang [139]. Their algorithm was later improved in [56]. In this section, we briefly present the original algorithm from [139], followed by a short discussion of the modifications made in [56].

The algorithm of Winker and Fang uses the grid structure $\overline{\Gamma}(X)$. As in line 1 of Algorithm 1, they initialize the algorithm by selecting a grid point $y \in \overline{\Gamma}(X)$ uniformly at random. In the mutation step (line 3), a point z is sampled uniformly at random from the neighborhood $\mathcal{N}_k^{mc}(y)$ of y. For the definition of $\mathcal{N}_k^{mc}(y)$ let us first introduce the functions φ_j, $j \in \{1, \ldots, d\}$, which order the elements in $\overline{\Gamma}_j(X)$; i.e., for $n_j := |\overline{\Gamma}_j(X)|$ the function φ_j is a permutation of $\{1, \ldots, n_j\}$ with $x_j^{(\varphi_j(1))} \leq \ldots \leq x_j^{(\varphi_j(n_j))} = 1$. For sampling a neighbor z of y we first draw mc coordinates $j_1, \ldots, j_{mc}$ from $\{1, \ldots, d\}$ uniformly at random. We then select, independently and uniformly at random, for each j_i, $i = 1, \ldots, mc$, a value $k_i \in \{-k, \ldots, -1, 0, 1, \ldots, k\}$. Finally, we let $z = (z_1, \ldots, z_d)$ with

$$z_j := \begin{cases} y_j, & \text{for } j \notin \{j_1, \ldots, j_{mc}\}, \\ y_j + k_j, & \text{for } j \in \{j_1, \ldots, j_{mc}\}. \end{cases}$$

Both the values $mc \in \{1, \ldots, d\}$ and $k \in \{1, \ldots, n/2\}$ are inputs of the algorithm to be specified by the user. For example, if we choose $mc = 3$ and $k = 50$, then in the mutation step we change up to three coordinates of y, and for each such coordinate we allow to do up to 50 steps on the grid $\overline{\Gamma}_j(X)$, either to the "right" or to the "left".

In the selection step (line 4), the search point z is accepted if its discrepancy value is better than that of y or if it is at least not worse than $d_\infty^*(y, X) + T$, for some threshold $T \leq 0$ that is determined in a precomputation step of the algorithm. Winker and Fang decided to keep the same threshold value for $\sqrt{I}$ iterations, and to replace it every $\sqrt{I}$ iterations with a new value $0 \geq T' > T$. The increasing sequence of threshold values guarantees that the algorithm has enough flexibility in the beginning to explore the search space, and enough stability towards its end so that it finally converges to a local maximum. This is achieved by letting T be very close to zero towards the end of the algorithm.

The algorithm by Winker and Fang performs well in numerical tests on rank-1 lattice rules, and it frequently computes the correct star discrepancy values in cases where this can be checked. However, as pointed out in [56], their algorithm does not perform very well in dimensions 10 and larger. For this reason, a number of modifications have been introduced in [56]. These modification also improve the performance of the algorithm in small dimensions.

The main differences of the algorithm presented in [56] include a refined neighborhood structure that takes into account the topological structure of the point set X and the usage of the concept of critical boxes as introduced in Definition 4. Besides this, there are few minor changes such as a variable size of the neighborhood structures and splitting the optimization process of $d_\infty^*(\cdot, X)$ into two separate processes for $\delta(\cdot, X)$ and $\bar{\delta}(\cdot, X)$, respectively. Extensive numerical experiments are presented in [56]. As mentioned above, in particular for large dimension this refined algorithm seems to compute better lower bounds for $d_\infty^*(\cdot, X)$ than the basic one from [139].

We briefly describe the refined neighborhoods used in [56]. To this end, we first note that the neighborhoods used in Winker and Fang's algorithm do not take into account the absolute size of the gaps $x_j^{(\varphi(i+1))} - x_j^{(\varphi(i))}$ between two successive coordinates of grid points. This is unsatisfactory since large gaps usually indicate large differences in the local discrepancy function. Furthermore, for a grid cell $[y, z]$ in $\overline{\Gamma}(X)$ (i.e., $y, z \in \overline{\Gamma}(X)$ and $(z_j = x_j^{(\varphi_j(i_j))}) \Rightarrow (y_j = x_j^{(\varphi_j(i_j+1))})$ for all $j \in [d]$) with large volume, we would expect that $\bar{\delta}(y, X)$ or $\delta(z, X)$ are also rather large. For this reason, the following continuous neighborhood is considered. As in the algorithm by Winker and Fang we sample mc coordinates $j_1, \ldots, j_{mc}$ from $\{1, \ldots, d\}$ uniformly at random. The neighborhood of y is the set $\mathcal{N}_k^{mc}(y) := [\ell_1, u_1] \times \ldots \times [\ell_d, u_d]$ with

$$[\ell_j, u_j] := \begin{cases} \{y_j\}, & \text{for } j \notin \{j_1, \ldots, j_{mc}\}, \\ [x^{(\varphi_j(\varphi_j^{-1}(y_j)-k \vee 1))}, x^{(\varphi_j(\varphi_j^{-1}(y_j)+k \wedge n_j))}], & \text{for } j \in \{j_1, \ldots, j_{mc}\}, \end{cases}$$

where we abbreviate $\varphi_j^{-1}(y_j) - k \vee 1 := \max\{\varphi_j^{-1}(y_j) - k, 1\}$ and, likewise, $\varphi_j^{-1}(y_j) + k \wedge n_j := \min\{\varphi_j^{-1}(y_j) + k, n_j\}$. That is, for each of the coordinates $j \in \{j_1, \ldots, j_{mc}\}$ we do k steps to the "left" and k steps to the "right". We sample a point $\tilde{z} \in \mathcal{N}_k^{mc}(y)$ (not uniformly, but according to some probability function described below) and we round $\tilde{z}$ once up and once down to the nearest critical box. For both these points $\tilde{z}^-$ and $\tilde{z}^+$ we compute the local discrepancy value, and we set as neighbor of y the point $z \in \arg\max\{\bar{\delta}(\tilde{z}^-, X), \delta(\tilde{z}^+, X)\}$. The rounded grid points $\tilde{z}^+$ and $\tilde{z}^-$ are obtained by the *snapping procedure* described in [56, Section 4.1]. We omit the details but mention briefly that rounding down to $\tilde{z}^-$ can be done deterministically ($\tilde{z}^-$ is unique), whereas for the upward rounding to $\tilde{z}^+$ there are several choices. The strategy proposed in [56] is based on a randomized greedy approach.

We owe the reader the explanation of how to sample the point $\tilde{z}$. To this end, we need to define the functions

$$\Psi_j : [\ell_j, u_j] \to [0, 1], r \mapsto \frac{r^d - (\ell_j)^d}{(u_j)^d - (\ell_j)^d}, j \in \{j_1, \dots, j_{mc}\}$$

whose inverse functions are

$$\Psi_j^{-1} : [0, 1] \to [\ell_j, u_j], s \mapsto \left(((u_j)^d - (\ell_j)^d)s + (\ell_j)^d\right)^{1/d}.$$

To sample $\tilde{z}$, we first sample values $s_1, \dots, s_{mc} \in [0, 1]$ independently and uniformly at random. We set $\tilde{z}_j := \Psi_j^{-1}(s_j)$ for $j \in \{j_1, \dots, j_{mc}\}$ and we set $\tilde{z}_j := y_j$ for $j \notin \{j_1, \dots, j_{mc}\}$. The intuition behind this probability measure is the fact that it favors larger coordinates than the uniform distribution. To make this precise, observe that in the case where $mc = d$, the probability measure on $\mathcal{N}_k^{mc}(y)$ is induced by the affine transformation from $\mathcal{N}_k^{mc}(y)$ to $[0, 1]^d$ and the polynomial product measure

$$\pi^d(dx) = \otimes_{j=1}^d f(x_j) \lambda(dx_j) \text{ with density function } f : [0, 1] \to \mathbb{R}, r \mapsto dr^{d-1}$$

on $[0, 1]^d$. The expected value of a point selected according to π^d is $d/(d+1)$, whereas the expected value of a point selected according to the uniform measure (which implicitly is the one employed by Winker and Fang) has expected value $1/2$. Some theoretical and experimental justifications for the choice of this probability measure are given in [56, Section 5.1]. The paper also contains numerical results for the computation of rank-1 lattice rules, Sobol' sequences, Faure sequences, and Halton sequences up to dimension 50. The new algorithm based on threshold accepting outperforms all other algorithms that we are aware of. For more recent applications of this algorithm we refer the reader to Sect. 10.5.2, where we present one example that indicates the future potential of this algorithm.

10.3.4.4 An Approach Based on Genetic Algorithms

A different randomized algorithm to calculate lower bounds for the star discrepancy of a given point set was proposed by M. Shah in [114]. His algorithm is a *genetic algorithm*. Genetic algorithms are a class of local search heuristics that have been introduced in the sixties and seventies of the last century, cf. [72] for the seminal work on evolutionary and genetic algorithms. In the context of geometric discrepancies, genetic algorithms have also been successfully applied to the design of low-discrepancy sequences (cf. Sect. 10.5.2 for more details). In this section, we provide a very brief introduction into this class of algorithms, and we outline its future potential in the analysis of discrepancies.

Algorithm 2: Simplified scheme of a $(\mu + \lambda)$ evolutionary algorithm for the computation of star discrepancy values. I is the runtime of the algorithm (i.e., the number of iterations), C is the number of crossover steps per generation, and M is the number of mutation steps

---

1 Initialization: Select $y^{(1)}, \ldots, y^{(\mu)} \in \overline{\Gamma}(X)$ uniformly at random and compute
  $d_\infty^*(y^{(1)}, X), \ldots, d_\infty^*(y^{(\mu)}, X)$;
2 for $i = 1, 2, \ldots, I$ **do**
3   **Crossover Step: for** $j = 1, 2, \ldots, C$ **do**
4   Select two individuals $y, y' \in \{y^{(1)}, \ldots, y^{(\mu)}\}$ at random and create from y and y'
             a new individual $z^{(j)}$ by recombination;
5   Compute $d_\infty^*(z^{(j)}, X)$;

6   **Mutation Steps: for** $j = 1, 2, \ldots, M$ **do**
7   Select an individual $y \in \{y^{(1)}, \ldots, y^{(\mu)}, z^{(1)}, \ldots, z^{(C)}\}$ at random;
8   Sample a neighbor $n^{(j)}$ from y and compute $d_\infty^*(n^{(j)}, X)$;

9   **Selection step:**
10   From $\{y^{(1)}, \ldots, y^{(\mu)}, z^{(1)}, \ldots, z^{(C)}, n^{(1)}, \ldots, n^{(M)}\}$ select—based on their local
             discrepancy values $d_\infty^*(\cdot, X)$—a subset of size μ;
11   Rename these individuals $y^{(1)}, \ldots, y^{(\mu)}$;
12 Output $d_\infty^*(y, X)$;

While threshold accepting algorithms take their inspiration from physics, genetic algorithms are inspired by biology. Unlike the algorithms presented in the previous section, in genetic algorithms, we typically do not keep only one solution candidate at a time, but we maintain a whole set of candidate solutions instead. This set is referred to as a *population* in the genetic algorithms literature. Algorithm 2 provides a high-level pseudo-code for genetic algorithms, adjusted again to the problem of computing lower bounds for the star discrepancy of a given point configuration. More precisely, this algorithm is a so-called $(\mu + \lambda)$ evolutionary algorithm (with $\lambda = C + M$ in this case). Evolutionary Algorithms are genetic algorithms that are based on Darwinian evolution principles. We discuss the features of such algorithms further below.

As mentioned above, the nomenclature used in the genetic algorithms literature deviates from the standard one used in introductory books to algorithms. We briefly name a few differences. In a high-level overview, a genetic algorithm runs in several *generations* (steps, iterations), in which the solution candidates (*individuals*) from the current population are being *recombined* and *mutated*.

We initialize such an algorithm by selecting μ individuals at random. They form the *parent population* (line 1). To this population we first apply a series of *crossover* steps (line 5), through which two (or more, depending on the implementation) individuals from the parent population are recombined. A very popular recombination operator is the so-called *uniform crossover* through which two search points $y, y' \in \overline{\Gamma}(X)$ are recombined to some search point z by setting $z_j := y_j$ with probability $1/2$, and by setting $z_j = y'_j$ otherwise. Several other recombination operators exist, and they are often adjusted to the problem at hand. The random choice of the parents to be recombined must not be uniform, and it may very well depend on the local discrepancy values $d_\infty^*(y^{(1)}, X), \ldots, d_\infty^*(y^{(\mu)}, X)$, which are also referred to as the *fitness* of these individuals.

Once C such recombined individuals $z^{(1)}, \ldots, z^{(C)}$ have been created and evaluated, we enter the *mutation* step, in which we compute for a number M of search points one neighbor each, cf. line 8. Similarly as in the threshold accepting algorithm, Algorithm 1, it is crucial here to find a meaningful notion of neighborhood. This again depends on the particular application. For our problem of computing lower bounds for the star discrepancy value of a given point configuration, we have presented two possible neighborhood definitions in Sect. 10.3.4.3. The newly sampled search points are evaluated, and from the set of old and new search points a new population is selected in the *selection* step, line 9. The selection typically depends again on the fitness values of the individuals. If always the μ search points of largest local discrepancy value are selected, we speak of an *elitist selection scheme*. This is the most commonly used selection operator in practice. However, to maintain more diversity in the population, it may also be reasonable to use other selection schemes, or to randomize the decision.

In the scheme of Algorithm 2, the algorithm runs for a fixed number I of iterations. However, as we mentioned in the beginning of Sect. 10.3.4.3, Shah's algorithm works slightly different. His algorithm stops when no improvement has happened for some t iterations in a row, where t is a parameter to be set by the user.

The details of Shah's implementation can be found in [114]. His algorithm was used in [106] for the approximation of the star discrepancy value of ten-dimensional permuted Halton sequences. Further numerical results are presented in [114]. The problem instances considered in [114], however, are not demanding enough to make a proper comparison between his algorithm and the ones presented in the previous section. On the few instances where a comparison seems meaningful, the results based on the threshold accepting algorithms outperform the ones of the genetic algorithm, cf. [56, Section 6.4] for the numerical results. Nevertheless, it seems that the computation of star discrepancy values with genetic and evolutionary algorithms is a promising direction and further research would be of interest.

10.3.5 Notes

In the literature one can find some attempts to compute for L_∞-discrepancies the smallest possible discrepancy value of all n-point configurations. For the star discrepancy B. White determined in [138] the smallest possible discrepancy values for $n = 1, 2, \ldots, 6$ in dimension $d = 2$, and T. Pillards, B. Vandewoestyne, and R. Cools in [111] for $n = 1$ in arbitrary dimension d. G. Larcher and F. Pillichshammer provided in [85] for the star and the extreme discrepancy the smallest discrepancy values for $n = 2$ in arbitrary dimension d. Furthermore, they derived for the isotrope discrepancy the smallest value for $n = 3$ in dimension

$d = 2$ and presented good bounds for the smallest value for $n = d + 1$ in arbitrary dimension $d \geq 3$. (Note that the isotrope discrepancy of $n < d + 1$ points in dimension d is necessarily the worst possible discrepancy 1.)

10.4 Calculation of L_p-Discrepancies for $p \notin \{2, \infty\}$

This section is the shortest section in this book chapter. The reason for this is not that the computation of L_p-discrepancies, $p \notin \{2, \infty\}$, is an easy task which is quickly explained, but rather that not much work has been done so far and that therefore, unfortunately, not much is known to date. We present here a generalization of Warnock's formula for even p.

Let $\gamma_1 \geq \gamma_2 \geq \cdots \geq \gamma_d \geq 0$, and let $(\gamma_u)_{u \subseteq \{1,\ldots,d\}}$ be the corresponding product weights; i.e., $\gamma_u = \prod_{j \in u} \gamma_j$ for all u. For this type of weights G. Leobacher and F. Pillichshammer derived a formula for the weighted L_p-star discrepancy $d_{p,\gamma}^*$ for arbitrary even positive integers p that generalizes the formula (10.12):

$$(d_{p,\gamma}^*(X))^p =$$

$$\sum_{\ell=0}^{p} \binom{p}{\ell} \left(-\frac{1}{n}\right)^{\ell} \sum_{(i_1,\ldots,i_\ell) \in \{1,\ldots,n\}^\ell} \prod_{j=1}^{d} \left(1 + \gamma_j \frac{1 - \max_{1 \leq k \leq \ell}(x_j^{(i_k)})^{p-\ell+1}}{p - \ell + 1}\right),$$

$$\tag{10.35}$$

see [89, Thm. 2.1] (notice that in their definition of the weighted discrepancy they replaced the weights γ_u appearing in our definition (10.8) by $\gamma_u^{p/2}$). Recall that in the special case where $\gamma_1 = \gamma_2 = \cdots = \gamma_d = 1$, the weighted L_p-star discrepancy $d_{p,\gamma}^*(X)$ coincides with Hickernell's modified L_p-discrepancy. Using the approach from [89] one may derive analogous formulas for the L_p-star and L_p-extreme discrepancy. The formula (10.35) can be evaluated directly at cost $O(dn^p)$, where the implicit constant in the big-O-notation depends on p. Obviously, the computational burden will become infeasible even for moderate values of n if p is very large.

Apart from the result in [89] we are not aware of any further results that are helpful for the calculation or approximation of weighted L_p-discrepancies.

10.4.1 Notes

The calculation of *average L_p-discrepancies* of Monte Carlo point sets attracted reasonable attention in the literature. One reason for this is that L_p-discrepancies such as, e.g., the L_p-star or L_p-extreme discrepancy, converge to the corresponding L_∞-discrepancy if p tends to infinity, see, e.g., [49,61]. Thus the L_p-discrepancies can be used to derive results for the corresponding L_∞-discrepancy. In the literature one can find explicit representations of average L_p-discrepancies in terms of sums involving Stirling numbers of the first and second kind as well as upper bounds and formulas for their asymptotic behavior, see, e.g., [49, 61, 70, 89, 123].

10.5   Some Applications

We present some applications of algorithms that approximate discrepancy measures. The aim is to show here some more recent examples of how the algorithms are used in practice, what typical instances are, and what kind of problems occur. Of course, the selection of topics reflects the interest of the authors and is far from being complete. Further applications can, e.g., be found in the design of computer experiments ("experimental design"), see [38, 88], the generation of pseudo-random numbers, see [82, 87, 99, 126], or in computer graphics, see [27].

10.5.1   Quality Testing of Point Sets

A rather obvious application of discrepancy approximation algorithms is to estimate the quality of low-discrepancy point sets or, more generally, deterministic or randomized quasi-Monte Carlo point configurations.

Thiémard, e.g., used his algorithm from [130], which we described in Sect. 10.3.4.1, to provide upper and lower bounds for the star discrepancy of Faure $(0, m, s)$-nets [40] with sample sizes varying from 1,048,576 points in the smallest dimension $d = 2$ to 101 points in the largest dimension 100 (where, not very surprisingly, the resulting discrepancy is almost 1). He also uses his algorithm to compare the performance of two sequences of pseudo-random numbers, generated by Rand() and MRG32k3a [86], and Faure, Halton [57], and Sobol' [120] sequences by calculating bounds for their star discrepancy for sample sizes between 30 and 250 points in dimension 7.

For the same instances Thiémard was able to calculate the exact star discrepancy of the Faure, Halton and Sobol' sequences by using his algorithm from [131], which we described in Sect. 10.3.4.2, see [131, Section 4.3].

In the same paper he provided the exact star discrepancy of Faure $(0, m, s)$-nets ranging from sample sizes of 625 points in dimension 4 to 169 points in dimension 12. These results complement the computational results he achieved for (less demanding) instances in [128] with the help of the algorithm presented there.

Algorithms to approximate discrepancy measures were also used to judge the quality of different types of generalized Halton sequences. Since these sequences are also important for our explanation in Sect. 10.5.2, we give a definition here.

Halton sequences are a generalization of the one-dimensional van der Corput sequences. For a prime base p and a positive integer $i \in \mathbb{N}$ let $i = d_k d_{k-1} \ldots d_2 d_1$

Fig. 10.2 The first 200 points of the 20-dimensional Halton sequence, projected to dimensions 19 and 20

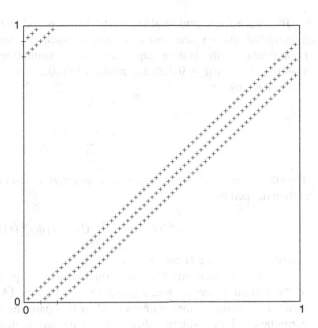

be the digital expansion of i in base p. That is, let $d_1, \ldots, d_k$ be such that $i = \sum_{\ell=1}^{k} d_\ell p^{\ell-1}$. Define the *radical inverse function* ϕ_p in base p by

$$\phi_p(i) := \sum_{\ell=1}^{k} d_\ell p^{-\ell}. \qquad (10.36)$$

Let p_j denote the jth prime number. The ith element of the d-dimensional *Halton sequence* is defined as

$$x^{(i)} := (\phi_{p_1}(i), \ldots, \phi_{p_d}(i)).$$

The Halton sequence is a low-discrepancy sequence, i.e., its first n points $X = (x^{(i)})_{i=1}^{n}$ in dimension d satisfy the star discrepancy bound

$$d_\infty^*(X) = O(n^{-1} \ln(n)^d). \qquad (10.37)$$

In fact, the Halton sequence was the first construction for which (10.37) was verified for any dimension d [57], and up to now there is no sequence known that exhibits a better asymptotical behavior than (10.37).

Nevertheless, for higher dimension d the use of larger prime p_j bases leads to some irregularity phenomenon, which is often referred to as *high correlation between higher bases* [9]. This phenomenon can easily be visualized by looking at two-dimensional projections of Halton sequences over higher coordinates, see Fig. 10.2.

To reduce these undesirable effects, Braaten and Weller [9] suggested to use generalized Halton sequences. To obtain such a sequence, one applies digit permutations to the Halton sequence: For a permutation π_p of $\{0, 1, \ldots, p - 1\}$ with fixpoint $\pi_p(0) = 0$ define in analogy to (10.36) the *scrambled radical inverse function* ϕ_p^π by

$$\phi_p^\pi(i) := \sum_{\ell=1}^{k} \pi_p(d_\ell) p^{-\ell}.$$

The ith element of the *generalized* (or *scrambled*) *Halton sequence* in d dimensions is then defined by

$$x^{(i)}(\Pi) := (\phi_{p_1}^{\pi_{p_1}}(i), \ldots, \phi_{p_d}^{\pi_{p_d}}(i)), \tag{10.38}$$

where we abbreviate $\Pi := (\pi_{p_1}, \ldots, \pi_{p_d})$.

In several publications different permutations were proposed to shuffle the digits of the Halton sequence, see, e.g., [2, 18, 23, 41, 106, 133, 134] and the literature mentioned therein. Many authors tried to compare the quality of some of these permutations by considering theoretical or numerical discrepancy bounds or other numerical tests.

Vandewoestyne and Cools [133], e.g., calculated the L_2-star and L_2-extreme discrepancy of several generalized Halton sequences by using Warnock's formula (10.9) and the corresponding modification (10.11), respectively. The instances they studied ranged from 10,000 sample points in dimension 8 to 1,000 points in dimension 64. They reported that those generalized Halton sequences performed best which are induced by the simple *reverse permutations* $\pi_p(0) = 0$ and $\pi_p(j) = p - j$ for $j = 1, \ldots, p - 1$. These sequences showed usually a smaller L_2-star and L_2-extreme discrepancy as, e.g., the original Halton sequences, the generalized Halton sequences proposed in [2,9] or the randomized Halton sequences from [134]; for more details see [133].

Ökten, Shah and Goncharov [106] tried to compare different generalized Halton sequences by calculating bounds for their L_∞-star discrepancy. For the calculations of the upper bounds for the star discrepancy they used Thiémard's algorithm presented in [130]. For the calculation of the lower bounds they used Shah's algorithm [114], which gave consistently better lower bounds than Thiémard's algorithm. They did the calculation for instances of 100 points in dimension 5 and 10 and studied different cases of prime bases. So they considered, e.g., different generalized Halton sequences in $d = 5$ where the prime bases are the 46th to 50th prime numbers $p_{46}, \ldots, p_{50}$ (which corresponds to the study of the projections over the last 5 coordinates of 50-dimensional sequences induced by the first 50 primes). Ökten et al. found that, apart from the original Halton sequence, the reverse permutations led to the highest discrepancy bounds, in contrast to their very low L_2-star and L_2-extreme discrepancy values. (This indicates again that the conventional L_2-star and L_2-extreme discrepancy are of limited use for judging the uniformity of point sets, cf. also the notes at the end of Sect. 10.2.) Furthermore, they reported that the average star discrepancy bounds of generalized Halton sequences induced by random permutations were rather low. For more details we refer to [106].

Algorithms to approximate star discrepancies were used in [32] to compare the quality of small samples of classical low-discrepancy points and Monte Carlo points with new point sets generated in a randomized and in a deterministic fashion. Here, "small samples" has to be understood as point sets whose size is small compared to the dimension, say, bounded by a constant times d or $d^{3/2}$. Notice that for this sample sizes asymptotic bounds like (10.37) do not provide any helpful information. The classical point sets were Halton-Hammersley points [24, 57, 99], Sobol' points [77, 120, 121], and Faure points shuffled by a Gray code [40, 126, 127]. The algorithms that generated the new point sets rely on certain random experiments based on randomized rounding with hard constraints [28, 122] and large deviation bounds that guarantee small discrepancies with high probability. The deterministic versions of these algorithms make use of derandomization techniques from [28, 34]. The concrete algorithms are randomized versions of the component-by-component (CBC) construction proposed in [33] and implemented in [31] and a randomized and derandomized version of an algorithm proposed in [29, 32]. The CBC construction has the advantage that it is faster and can be used to extend given low-discrepancy point sets in the dimension, but its theoretical error bound is worse than the one for the latter algorithm.

In the numerical experiments different methods to approximate the discrepancy were used, depending on the instance sizes. For instances of 145–155 points in dimension 7 and 85–95 points in dimension 9 exact discrepancy calculations were performed by using the method of Bundschuh and Zhu [11]. For larger instances this was infeasible, so for instances ranging from 145–155 points in dimension 9 to 65–75 points in dimension 12 a variant of the algorithm of Dobkin, Eppstein and Mitchell [27] was used that gained speed by allowing for an imprecision of order d/n. It should be noted that since the publication of these experiments, the implementation of this algorithm has been improved; a version that attains the same speed without making the d/n-order sacrifice of precision is available at the third author's homepage at http://www.mpi-inf.mpg.de/~wahl/. For the final set of instances, ranging from 145–155 points in dimension 12 to 95–105 points in dimension 21, the authors relied on upper bounds from Thiémard's algorithm from [130] and on lower bounds from a randomized algorithm based on threshold accepting from [140], which is a precursor of the algorithm from [56]. Similarly as in [106], the randomized algorithm led consistently to better lower bounds than Thiémard's algorithm. For the final instances of 95–105 points in dimension 21 the gaps between the upper and lower discrepancy bounds were roughly of the size 0.3, thus these results can only be taken as very coarse indicators for the quality of the considered point sets.

As expected, the experiments indicate that for small dimension and relatively large sample sizes the Faure and Sobol' points are superior, but the derandomized algorithms (and also their randomized versions) performed better than Monte Carlo, Halton-Hammersley, and Faure points in higher dimension for relatively small sample sizes. From the classical low-discrepancy point sets the Sobol' points performed best and were competitive for all instances. For more details see [32, Section 4].

10.5.2 Generating Low-Discrepancy Points via an Optimization Approach

In Sect. 10.3.4.4 we have seen that genetic algorithms can be used for an approximation of the star discrepancy values of a given point configuration. Here in this section we present another interesting application of biology-inspired algorithms in the field of geometric discrepancies.

Whereas the works presented in Sect. 10.3.4 focus mainly on the computation of star discrepancy values, the authors of [23], F.-M. de Rainville, C. Gagné, O. Teytaud, and D. Laurendeau, apply evolutionary algorithms (cf. Sect. 10.3.4.4 for a brief introduction) to generate *low discrepancy point configurations*. Since the fastest known algorithms to compute the exact star discrepancy values have running time exponential in the dimension (cf. Sect. 10.3.2 for a discussion of this algorithm and Sect. 10.3.3 for related complexity-theoretic results), and the authors were—naturally—interested in a fast computation of the results (cf. [23, Page 3]), the point configurations considered in [23] are optimized for Hickernell's modified L_2-star discrepancies, see Eq. (10.6) in the introduction to this chapter. As mentioned in Sect. 10.2, L_2-star discrepancies can be computed efficiently via Warnock's formula, cf. Eq. (10.9). Similarly, Hickernell's modified L_2-discrepancy can be computed efficiently with $O(dn^2)$ arithmetic operations.

The point sets generated by the algorithm of de Rainville, Gagné, Teytaud, and Laurendeau are generalized Halton sequences. As explained in Sect. 10.5.1, generalized Halton sequences are digit-permuted versions of the Halton sequence, cf. Eq. (10.38). The aim of their work is thus to find permutations $\pi_{p_1}, \ldots, \pi_{p_d}$ such that the induced generalized Halton sequence has small modified L_2-discrepancy. They present numerical results for sequences with 2,500 points in dimensions 20, 50, and 100 and they compare the result of their evolutionary algorithm with that of standard algorithms from the literature. Both for the modified L_2-discrepancy as well as for the L_2-star discrepancy the results indicate that the evolutionary algorithm is at least comparable, if not superior, to classic approaches.

The evolutionary algorithm of de Rainville, Gagné, Teytaud, and Laurendeau uses both the concept of crossover and mutation (as mentioned above, see Sect. 10.3.4.4 for a brief introduction into the notations of genetic and evolutionary algorithms). The search points are permutations, or, to be more precise configurations of permutations. The mutation operator shuffles the values of the permutation by deciding independently for each value in the domain of the permutation if its current value shall be swapped, and if so, with which value to swap it. The crossover operator recombines two permutations by iteratively swapping pairs of values in them.

We recall that in the definition of the generalized Halton sequence, Eq. (10.38), we need a vector (a *configuration*) $\Pi = (\pi_{p_1}, \ldots, \pi_{p_d})$ of permutations of different length. The length of permutation π_{p_i} is determined by the ith prime number p_i. These configurations are extended component-by-component. That is, the evolutionary algorithm first computes an optimal permutation π_2 on $\{0, 1\}$ and sets $\Pi := (\pi_2)$. It then adds to Π, one after the other, the optimal permutation π_3 for $\{0, 1, 2\}$, the one π_5 on $\{0, 1, 2, 3, 4\}$, and so on. Here, "optimal" is measured with respect to the fitness function, which is simply the modified (Hickernell) L_2-discrepancy of the point set induced by Π. We recall from Eq. (10.38) that a vector Π of permutations fully determines the resulting generalized Halton sequence.

The iterative construction of Π allows the user to use, for any two positive integers $d < D$, the same permutations $\pi_{p_1}, \ldots, \pi_{p_d}$ in the first d dimensions of the two sequences in dimensions d and D, respectively.

The population sizes vary from 500 individuals for the 20-dimensional point sets to 750 individuals for the 50- and 100-dimensional point configurations. Similarly, the number of generations that the author allow the algorithm for finding the best suitable permutation ranges from 500 in the 20-dimensional case to 750 for the 50-dimensional one, and 1,000 for the 100-dimensional configuration.

It seems evident that combining the evolutionary algorithm of de Rainville, Gagné, Teytaud, and Laurendeau with the approximation algorithms presented in Sect. 10.3.4 is a promising idea to generate low star discrepancy point sequences. This is ongoing work of Doerr and de Rainville [35]. In their work, Doerr and de Rainville use the threshold accepting algorithm from [56] (cf. Sect. 10.3.4.3) for the intermediate evaluation of the candidate permutations. Only the final evaluation of the resulting point set is done by the exact algorithm of Dobkin, Eppstein, and Mitchell [27] (cf. Sect. 10.3.2). This allows the computation of low star discrepancy sequences in dimensions where an exact evaluation of the intermediate permutations of the generalized Halton sequences would be far too costly. Using this approach, one can, at the moment, not hope to get results for such dimensions where the algorithm of Dobkin, Eppstein, and Mitchell does not allow for a final evaluation of the point configurations. However, due to the running time savings during the optimization process, one may hope to get good results for moderate dimensions for up to, say, 12 or 13. Furthermore, it seems possible to get a reasonable indication of good generating permutations in (10.38) for dimensions much beyond this, if one uses only the threshold accepting algorithm to guide the search. As is the case for all applications of the approximation algorithms presented in Sect. 10.3.4, in such dimensions, however, we do not have a *proof* that the computed lower bounds are close to the exact discrepancy values.

We conclude this section by emphasizing that, in contrast to the previous subsection, where the discrepancy approximation algorithms were only used to *compare* the quality of different point sets, here in this application the calculation of the discrepancy is an integral part of the optimization process to *generate* good low discrepancy point sets.

10.5.2.1 Notes

Also other researchers used algorithms to approximate discrepancy measures in combination with optimization approaches to generate low discrepancy point sets. A common approach described in [27] is to apply a multi-dimensional optimization algorithm repeatedly to randomly jiggled versions of low discrepancy point sets to search for smaller local minima of the discrepancy function. In [38, 90] the authors used the optimization heuristic threshold accepting to generate experimental designs with small discrepancy. The discrepancy measures considered in [38] are the L_∞- and the L_2-star discrepancy, as well as the centered, the symmetric, and the modified L_2-discrepancy. The discrepancy measure considered in [90] is the central composite discrepancy for a flexible region as defined in [19].

10.5.3 Scenario Reduction in Stochastic Programming

Another interesting application of discrepancy theory is in the area of *scenario reduction*. We briefly describe the underlying problem, and we illustrate the importance of discrepancy computation in this context.

Many real-world optimization problems, e.g., of financial nature, are subject to stochastic uncertainty. *Stochastic programming* is a suitable tool to model such problems. That is, in stochastic programming we aim at minimizing or maximizing the expected value of a random process, typically taking into account several constraints. These problems, of both continuous and discrete nature, are often infeasible to solve optimally. Hence, to deal with such problems in practice, one often resorts to approximating the underlying (continuous or discrete) probability distribution by a discrete one of much smaller support. In the literature, this approach is often referred to as *scenario reduction*, cf. [64, 112] and the numerous references therein.

For most real-world situations, the size of the support of the approximating probability distribution has a very strong influence on (i.e., "determines") the complexity of solving the resulting stochastic program. On the other hand, it is also true for all relevant similarity measures between the original and the approximating probability distribution that a larger support of the approximating distribution allows for a better approximation of the original one. Therefore, we have a fundamental trade-off between the running time of the stochastic program and the quality of the approximation. This trade-off has to be taken care of by the practitioner, and his decision typically depends on his time constraints and the availability of computational resources.

To explain the scenario reduction problem more formally, we need to define a suitable measure of similarity between two distributions. For simplicity, we regard here only distance measures for two discrete distributions whose support is a subset of $[0, 1]^d$. Unsurprisingly, a measure often regarded in the scenario reduction literature is based on discrepancies, cf. again [64, 112] and references therein. Formally, let P and Q be two discrete Borel distributions on $[0, 1]^d$ and let $\mathcal{B}$ be a system of Borel sets of $[0, 1]^d$. The $\mathcal{B}$-discrepancy between P and Q is

$$\text{disc}_\infty(\mathcal{B}; P, Q) = \sup_{B \in \mathcal{B}} |P(B) - Q(B)|.$$

The right choice of the set $\mathcal{B}$ of test sets depends on the particular application. Common choices for $\mathcal{B}$ are

- $\mathcal{C}_d$, the class of all axis-parallel half-open boxes,
- $\mathcal{R}_d$, the class of all half-open axis-parallel boxes,
- $\mathcal{P}_{d,k}$, the class of all polyhedra having at most k vertices, and
- $\mathcal{I}_d$, the set of all closed, convex subsets,

which were introduced in Sect. 10.1.

In the scenario reduction literature, the discrepancy measure associated with $\mathcal{C}_d$ is referred to as star discrepancy, *uniform*, or *Kolmogorov metric*; the one associated with $\mathcal{I}_d$ is called *isotrope discrepancy*, whereas the distance measure induced by $\mathcal{R}_d$ is simply called the *rectangle discrepancy* measure, and the one induced by $\mathcal{P}_{d,k}$ as *polyhedral discrepancy*.

With these distance measures at hand, we can now describe the resulting approximation problem. For a given distribution P of support $\{x^{(1)}, \ldots, x^{(N)}\}$ we aim at finding, for a given integer $n < N$, a distribution Q such that (i) the support $\{y^{(1)}, \ldots, y^{(n)}\}$ of Q is a subset of the support of P of size n and that (ii) Q minimizes the distance between P and Q.

Letting $\delta(z)$ denote the Dirac measure placing mass one at point z, $p^{(i)} := P(x^{(i)})$, $1 \leq i \leq N$, and $q^{(i)} := P(x^{(i)})$, $1 \leq i \leq n$, we can formulate this minimization problem as

$$\text{minimize} \quad \text{disc}_\infty(\mathcal{B}; P, Q) = \text{disc}_\infty(\mathcal{B}; \sum_{i=1}^{N} p^{(i)}\delta(x^{(i)}), \sum_{i=1}^{n} q^{(i)}\delta(y^{(i)}))$$

$$\tag{10.39}$$

$$\text{subject to} \quad \{y^{(1)}, \ldots, y^{(n)}\} \subseteq \{x^{(1)}, \ldots, x^{(N)}\},$$

$$q^{(i)} > 0, \quad 1 \leq i \leq n,$$

$$\sum_{i=1}^{n} q^{(i)} = 1.$$

This optimization problem can be decomposed into an *outer* optimization problem of finding the support $\{y^{(1)}, \ldots, y^{(n)}\}$ and an *inner* optimization problem of finding—for fixed support $\{y^{(1)}, \ldots, y^{(n)}\}$—the optimal probabilities $q^{(1)}, \ldots, q^{(n)}$.

Some heuristics for solving both the inner and the outer optimization problems have been suggested in [64]. In that paper, Henrion, Küchler, and Römisch mainly regard the star discrepancy measure, but results for other distance measures are provided as well; see also the paper [63] by the same set of authors for results on minimizing the distance with respect to polyhedral discrepancies.

We present here a few details about the algorithms that Henrion, Küchler, and Römisch developed for the optimization problem (10.39) with respect to the star discrepancy. A more detailed description can be found in their paper [64].

Two simple heuristics for the outer optimization problem are *forward* and *backward selection*. In forward selection we start with an empty support set Y, and we add to Y, one after the other, the element from $X = \{x^{(1)}, \ldots, x^{(N)}\}$ that minimizes the star discrepancy between P and Q, for an optimal allocation Q of probabilities $q^{(1)}, \ldots, q^{(|Y|)}$ to the points in Y. We stop this forward selection when Y has reached its desired size, i.e., when $|Y| = n$.

Backward selection follows an orthogonal idea. We start with the full support set $Y = X$ and we remove from Y, one after the other, the element such that an optimal probability distribution Q on the remaining points of Y minimizes the star discrepancy $\text{disc}_\infty(\mathscr{C}_d; P, Q)$. Again we stop once $|Y| = n$. It seems natural that forward selection is favorable for values n that are much smaller than N, and, likewise, backward selection is more efficient when the difference $N - n$ is small.

For the inner optimization problem of determining the probability distribution Q for a fixed support Y, Henrion, Küchler, and Römisch formulate a linear optimization problem. Interestingly, independently of the discrepancy community, the authors develop to this end the concept of *supporting* boxes—a concept that coincides with the critical boxes introduced in Definition 4, Sect. 10.3.1. Using these supporting boxes, they obtain a linear program that has much less constraints than the natural straightforward formulation resulting from problem (10.39). However, the authors remark that not the solution to the reduced linear program itself is computationally challenging, but the computation of the supporting (i.e., critical) boxes. Thus, despite significantly reducing the size of the original problem (10.39), introducing the concept of critical boxes alone is not sufficient to considerably reduce the computational effort required to solve problem (10.39). The problems considered in [64] are thus only of moderate dimension and moderate values of N and n. More precisely, results for four and eight dimensions with $N = 100$, $N = 200$, and $N = 300$ points are computed. The reduced scenarios have $n = 10$, 20, and 30 points.

Since most of the running time is caused by the computation of the star discrepancy values, it seems thus promising to follow a similar approach as in Sect. 10.5.2 and to use one of the heuristic approaches for star discrepancy estimation presented in Sect. 10.3.4 for an intermediate evaluation of candidate probability distributions Q. This would allow us to compute the exact distance of P and Q only for the resulting approximative distribution Q.

We conclude this section by mentioning that similar approaches could be useful also for the other discrepancy measures, e.g., the rectangle or the isotrope discrepancy of P and Q. However, for this to materialize, more research is needed to develop good approximations of such discrepancy values. (This is particularly the case for the isotrope discrepancy—see the comment regarding the isotrope discrepancy at the beginning of Sect. 10.3.)

Acknowledgements The authors would like to thank Sergei Kucherenko, Shu Tezuka, Tony Warnock, Greg Wasilkowski, Peter Winker, and an anonymous referee for their valuable comments.

Carola Doerr is supported by a Feodor Lynen postdoctoral research fellowship of the Alexander von Humboldt Foundation and by the Agence Nationale de la Recherche under the project ANR-09-JCJC-0067-01.

The work of Michael Gnewuch was supported by the German Science Foundation DFG under grant GN-91/3 and the Australian Research Council ARC.

The work of Magnus Wahlström was supported by the German Science Foundation DFG via its priority program "SPP 1307: Algorithm Engineering" under grant DO 749/4-1.

References

1. A.V. Aho, J.E. Hopcroft, J.D. Ullman, *The Design and Analysis of Computer Algorithms* (Addison-Wesley, Reading, 1974)
2. E.I. Atanassov, On the discrepancy of the Halton sequences. Math. Balkanica (N. S.) **18**, 15–32 (2004)
3. J. Beck, Some upper bounds in the theory of irregulatities of distribution. Acta Arithmetica **43**, 115–130 (1984)
4. J. Beck, On the discrepancy of convex plane sets. Monatsh. Math., 91–106 (1988)
5. J. Beck, W.W.L. Chen, *Irregularities of Distribution* (Cambridge University Press, Cambridge, 1987)
6. J.L. Bentley, 1977, Algorithms for Klee's rectangle problem.Unpublished notes, Dept. of Computer Science, Carnegie Mellon University
7. D. Bilyk, Roth's orthogonal function method in discrepancy theory and some new connections, in *A Panorama of Discrepancy Theory*, ed. by W.W.L. Chen, A. Srivastav, G. Travaglini (Springer, Berlin, 2012)
8. D. Bilyk, M.T. Lacey, I. Parissis, A. Vagharshakyan, Exponential sqared integrability of the discrepancy function in two dimensions. Mathematika **55**, 1–27 (2009)
9. E. Braaten, G. Weller, An improved low-discrepancy sequence for multidimensional quasi-Monte Carlo integration. J. Comput. Phys. **33**, 249–258 (1979)
10. L. Brandolini, C. Choirat, L. Colzani, G. Gigante, R. Seri, G. Travaglini, *Quadrature rules and distribution of points on manifolds*, 2011. To appear in: Ann. Scuola Norm. Sup. Pisa Cl. Sci.
11. P. Bundschuh, Y.C. Zhu, A method for exact calculation of the discrepancy of low-dimensional point sets I. Abh. Math. Sem. Univ. Hamburg **63**, 115–133 (1993)
12. T.M. Chan, A (slightly) faster algorithm for Klee's measure problem. Comput. Geom. **43**(3), 243–250 (2010)
13. B. Chazelle, *The Discrepancy Method* (Cambridge University Press, Cambridge, 2000)
14. J. Chen, B. Chor, M. Fellows, X. Huang, D.W. Juedes, I.A. Kanj, G. Xia, Tight lower bounds for certain parameterized NP-hard problems. Inf. Comput. **201**(2), 216–231 (2005)
15. J. Chen, X. Huang, I.A. Kanj, G. Xia, Strong computational lower bounds via parameterized complexity. J. Comput. Syst. Sci. **72**(8), 1346–1367 (2006)
16. W.W.L. Chen, M.M. Skriganov, Upper bounds in irregularities of point distribution, in *A Panorama of Discrepancy Theory*, ed. by W.W.L. Chen, A. Srivastav, G. Travaglini (Springer, Berlin, 2012)
17. W.W.L. Chen, G. Travaglini, Discrepancy with respect to convex polygons. J. Complexity **23**, 662–672 (2007)
18. H. Chi, M. Mascagni, T. Warnock, On the optimal Halton sequence. Math. Comput. Simul. **70**, 9–21 (2005)

19. S.C. Chuang, Y.C. Hung, Uniform design over general input domains with applications to target region estimation in computer experiments. Comput. Stat. Data Anal. **54**, 219–232 (2010). http://dx.doi.org/10.1016/j.csda.2010.01.032

20. T.H. Cormen, C.E. Leiserson, R.L. Rivest, C. Stein, *Introduction to Algorithms (3. ed.)*(MIT Press, Cambridge, 2009)

21. M. de Berg, O. Cheong, M. van Kreveld, M.H. Overmars, *Computational Geometry. Algorithms and Applications* (Springer, Berlin, 2008)

22. L. de Clerck, A method for exact calculation of the star-discrepancy of plane sets applied to the sequence of Hammersley. Monatsh. Math. **101**, 261–278 (1986)

23. F.-M. De Rainville, C. Gagné, O. Teytaud, D. Laurendeau, Evolutionary optimization of low-discrepancy sequences. ACM Trans. Model. Comput. Simul. **22** (2012). Article 9

24. J. Dick, F. Pillichshammer, *Digital Nets and Sequences* (Cambridge University Press, Cambridge, 2010)

25. J. Dick, F. Pillichshammer, Discrepancy theory and quasi-Monte Carlo integration, in *A Panorama of Discrepancy Theory*, ed. by W.W.L. Chen, A. Srivastav, G. Travaglini (Springer, Berlin, 2012)

26. J. Dick, I.H. Sloan, X. Wang, H. Woźniakowski, Liberating the weights. J. Complexity **20**(5), 593–623 (2004). http://dx.doi.org/10.1016/j.jco.2003.06.002

27. D.P. Dobkin, D. Eppstein, D.P. Mitchell, Computing the discrepancy with applications to supersampling patterns. ACM Trans. Graph. **15**, 354–376 (1996)

28. B. Doerr, Generating randomized roundings with cardinality constraints and derandomizations, in *Proceedings of the 23rd Annual Symposium on Theoretical Aspects of Computer Science (STACS'06)*, ed. by B. Durand, W. Thomas LNiCS, vol. 3884 (Springer, Berlin and Heidelberg, 2006)

29. B. Doerr, M. Gnewuch, Construction of low-discrepancy point sets of small size by bracketing covers and dependent randomized rounding, in *Monte Carlo and Quasi-Monte Carlo Methods 2006*, ed. by A. Keller, S. Heinrich, H. Niederreiter (Springer, Berlin and Heidelberg, 2008)

30. B. Doerr, M. Gnewuch, A. Srivastav, Bounds and constructions for the star discrepancy via δ-covers. J. Complexity **21**, 691–709 (2005)

31. B. Doerr, M. Gnewuch, M. Wahlström, Implementation of a component-by-component algorithm to generate low-discrepancy samples, in *Monte Carlo and Quasi-Monte Carlo Methods 2008*, ed. by P. L'Ecuyer, A.B. Owen (Springer, Berlin and Heidelberg, 2009)

32. B. Doerr, M. Gnewuch, M. Wahlström, Algorithmic construction of low-discrepancy point sets via dependent randomized rounding. J. Complexity **26**, 490–507 (2010)

33. B. Doerr, M. Gnewuch, P. Kritzer, F. Pillichshammer, Component-by-component construction of low-discrepancy point sets of small size. Monte Carlo Methods Appl. **14**, 129–149 (2008)

34. B. Doerr, M. Wahlström, Randomized rounding in the presence of a cardinality constraint, in *Proceedings of the Workshop on Algorithm Engineering and Experiments (ALENEX 2009)*, ed. by I. Finocchi, J. Hershberger (SIAM, Philadelphia, 2009)

35. C. Doerr (nee Winzen), F.-M. De Rainville, Constructing Low Star Discrepancy Point Sets with Genetic Algorithms, in *Proc. of Genetic and Evolutionary Computation Conference (GECCO'13)* (ACM, New York, 2013)

36. R.G. Downey, M.R. Fellows, *Parameterized Complexity* (Springer, Berlin, 1999)

37. M. Drmota, R.F. Tichy, *Sequences, Discrepancies and Applications*. Lecture Notes in Mathematics, vol. 1651 (Springer, Berlin and Heidelberg, 1997)

38. K.-T. Fang, D.K.J. Lin, P. Winker, Y. Zhang, Uniform design: theory and application. Technometrics **42**, 237–248 (2000)

39. H. Faure, Discrépance de suites associées à un systéme de numération (en dimension un). Bull. Soc. Math. France **109**, 143–182 (1981)

40. H. Faure, Discrépance de suites associées à un systéme de numération (en dimension *s*). Acta Arithmetica **41**, 338–351 (1982)

41. H. Faure, C. Lemieux, Generalized Halton sequences in 2008: A comparative study. ACM Trans. Model. Comput. Simul. **19**, 15–131 (2009)

42. J. Flum, M. Grohe, *Parameterized Complexity Theory* (Springer, New York, 2006)

43. K. Frank, S. Heinrich, Computing discrepancies of Smolyak quadrature rules. J. Complexity **12**, 287–314 (1996)
44. M.L. Fredman, B.W. Weide, On the complexity of computing the measure of $\bigcup[a_i, b_i]$. Commun. ACM **21**(7), 540–544 (1978)
45. H. Gabai, On the discrepancy of certain sequences mod 1. Indaq. Math. **25**, 603–605 (1963)
46. M.R. Garey, D.S. Johnson, *Computers and Intractability: A Guide to the Theory of NP-Completeness* (W. H. Freeman and Co., San Francisco, 1979)
47. T. Gerstner, M. Griebel, Numerical integration using sparse grids. Numer. Algorithms, 209–232 (1998)
48. P. Giannopoulus, C. Knauer, M. Wahlström, D. Werner, Hardness of discrepancy computation and epsilon-net verification in high dimensions. J. Complexity **28**, 162–176 (2012)
49. M. Gnewuch, Bounds for the average L^p-extreme and the L^∞-extreme discrepancy. Electron. J. Combin., 1–11 (2005). Research Paper 54
50. M. Gnewuch, Bracketing numbers for axis-parallel boxes and applications to geometric discrepancy. J. Complexity **24**, 154–172 (2008)
51. M. Gnewuch, Construction of minimal bracketing covers for rectangles. Electron. J. Combin. **15** (2008). Research Paper 95
52. M. Gnewuch, Entropy, Randomization, Derandomization, and Discrepancy, in *Monte Carlo and Quasi-Monte Carlo Methods 2010*, ed. by L. Plaskota, H. Woźniakowski (Springer, Berlin and Heidelberg, 2012)
53. M. Gnewuch, Weighted geometric discrepancies and numerical integration on reproducing kernel Hilbert spaces. J. Complexity **28**, 2–17 (2012)
54. M. Gnewuch, A. Roşca, On G-discrepancy and mixed Monte Carlo and quasi-Monte Carlo sequences. Acta Univ. Apul. Math. Inform. **18**, 97–110 (2009)
55. M. Gnewuch, A. Srivastav, C. Winzen, Finding optimal volume subintervals with k points and calculating the star discrepancy are NP-hard problems. J. Complexity **25**, 115–127 (2009)
56. M. Gnewuch, M. Wahlström, C. Winzen, A new randomized algorithm to approximate the star discrepancy based on threshold accepting. SIAM J. Numer. Anal. **50**, 781–807 (2012)
57. J.H. Halton, On the efficiency of certain quasi-random sequences of points in evaluating multidimensional integrals. Numer. Math. **2**, 84–90 (1960)
58. J.H. Halton, S.K. Zaremba, The extreme and L^2-discrepancies of some plane sets. Monatshefte Math. **73**, 316–328 (1969)
59. J.M. Hammersley, Monte Carlo methods for solving multivariate problems. Ann. New York Acad. Sci. **86**, 844–874 (1960)
60. S. Heinrich, Efficient algorithms for computing the L_2 discrepancy. Math. Comp. **65**, 1621–1633 (1996)
61. S. Heinrich, E. Novak, G.W. Wasilkowski, H. Woźniakowski, The inverse of the star-discrepancy depends linearly on the dimension. Acta Arithmetica **96**, 279–302 (2001)
62. P. Hellekalek, H. Leeb, Dyadic diaphony. Acta Arithmetica **80**, 187–196 (1997)
63. R. Henrion, C. Küchler, W. Römisch, Discrepancy distances and scenario reduction in two-stage stochastic mixed-integer programming. J. Ind. Manag. Optim. **4**, 363–384 (2008)
64. R. Henrion, C. Küchler, W. Römisch, Scenario reduction in stochastic programming with respect to discrepancy distances. Comput. Optim. Appl. **43**, 67–93 (2009)
65. F.J. Hickernell, A generalized discrepancy and quadrature error bound. Math. Comp. **67**, 299–322 (1998)
66. F.J. Hickernell, I.H. Sloan, G.W. Wasilkowski, On tractability of weighted integration over bounded and unbounded regions in $\mathbb{R}^s$. Math. Comp. **73**, 1885–1901 (2004)
67. F.J. Hickernell, X. Wang, The error bounds and tractability of quasi-Monte Carlo algorithms in infinite dimensions. Math. Comp. **71**, 1641–1661 (2001)
68. A. Hinrichs, Discrepancy of Hammersley points in Besov spaces of dominated mixed smoothness. Math. Nachr. **283**, 477–483 (2010)
69. A. Hinrichs, Discrepancy, Integration and Tractability, in *Monte Carlo and Quasi-Monte Carlo Methods 2012*, ed. by J. Dick, F.Y. Kuo, G. Peters, I.H. Sloan (Springer, Berlin and Heidelberg, 2013). pp. 129–172

70. A. Hinrichs, H. Weyhausen, Asymptotic behavior of average L_p-discrepancies. J. Complexity **28**, 425–439 (2012)
71. E. Hlawka, Discrepancy and uniform distribution of sequences. Compositio Math. **16**, 83–91 (1964)
72. J.H. Holland, *Adaptation In Natural And Artificial Systems* (University of Michigan Press, Ann Arbor, 1975)
73. J. Hromković, *Algorithms for Hard Problems*, 2nd edn (Springer, Berlin and Heidelberg, 2003)
74. R. Impagliazzo, R. Paturi, The complexity of k-SAT, in *Proc. 14th IEEE Conf. on Computational Complexity* (1999)
75. R. Impagliazzo, R. Paturi, F. Zane, Which problems have strongly exponential complexity? J. Comput. Syst. Sci. **63**(4), 512–530 (2001)
76. S. Joe, An intermediate bound on the star discrepancy, in *Monte Carlo and Quasi-Monte Carlo Methods 2010*, ed. by L. Plaskota, H. Woźniakowski (Springer, Berlin and Heidelberg, 2012)
77. S. Joe, F.Y. Kuo, Constructing Sobol' sequences with better two-dimensional projections. SIAM J. Sci. Comput. **30**, 2635–2654 (2008)
78. S. Joe, I.H. Sloan, On computing the lattice rule criterion R. Math. Comp. **59**, 557–568 (1992)
79. W. Kahan, Further remarks on reducing truncation errors. Commun. ACM **8**, 40 (1965)
80. S. Kirckpatrick, C. Gelatt, M. Vecchi, Optimization by simulated annealing. Science **20**, 671–680 (1983)
81. V. Klee, Can the measure of $\bigcup_1^n [a_i, b_i]$ be computed in less than $O(n \log n)$ steps? The American Mathematical Monthly **84**(4), 284–285 (1977). http://www.jstor.org/stable/2318871
82. D.E. Knuth, *The Art of Computer Programming. Vol. 2. Seminumerical Algorithms*, 3rd edn. Addison-Wesley Series in Computer Science and Information Processing (Addison-Wesley, Reading, 1997)
83. L. Kuipers, H. Niederreiter, *Uniform Distribution of Sequences* (Wiley-Interscience [Wiley], New York, 1974). Pure and Applied Mathematics
84. F.Y. Kuo, Component-by-component constructions achieve the optimal rate of convergence for multivariate integration in weighted Korobov and Sobolev spaces. J. Complexity **19**, 301–320 (2003)
85. G. Larcher, F. Pillichshammer, A note on optimal point distributions in $[0, 1)^s$. J. Comput. Appl. Math. **206**(2), 977–985 (2007). http://dx.doi.org/10.1016/j.cam.2006.09.004
86. P. L'Ecuyer, Good parameter sets for combined multiple recursive random number generators. Oper. Res. **47**, 159–164 (1999)
87. P. L'Ecuyer, P. Hellekalek, Random number generators: selection criteria and testing, in *Random and quasi-random point sets*. Lecture Notes in Statist., vol. 138 (Springer, New York, 1998). http://dx.doi.org/10.1007/978-1-4612-1702-2_5
88. C. Lemieux, *Monte Carlo and Quasi-Monte Carlo Sampling* (Springer, New York, 2009)
89. G. Leobacher, F. Pillichshammer, Bounds for the weighted L_p discrepancy and tractability of integration. J. Complexity **529**–547 (2003)
90. D.K.J. Lin, C. Sharpe, P. Winker, Optimized U-type designs on flexible regions. Comput. Stat. Data Anal. **54**(6), 1505–1515 (2010). http://dx.doi.org/10.1016/j.csda.2010.01.032
91. L. Markhasin, Discrepancy of generalized Hammersley type point sets in Besov spaces with dominating mixed smoothness. Uniform Distribution Theory **8**, 135–164 (2013)
92. L. Markhasin, Quasi-Monte Carlo methods for integration of functions with dominating mixed smoothness in arbitrary dimension. J. Complexity **29**(5), 370–388 (2013)
93. J. Matoušek, On the L_2-discrepancy for anchored boxes. J. Complexity **14**, 527–556 (1998)
94. J. Matoušek, *Geometric Discrepancy*, 2nd edn. (Springer, Berlin, 2010)
95. K. Mehlhorn, *Multi-dimensional Searching and Computational Geometry, Data Structures and Algorithms 3* (Springer, Berlin/New York, 1984)

96. W. Morokoff, R. Caflisch, Quasi-random sequences and their discrepancies. SIAM J. Sci. Comput. **15**, 1251–1279 (1994)

97. H. Niederreiter, Discrepancy and convex programming. Ann. Mat. Pura Appl. **93**, 89–97 (1972)

98. H. Niederreiter, Methods for estimating discrepancy, in *Applications of Number Theory to Numerical Analysis*, ed. by S.K. Zaremba (Academic Press, New York, 1972)

99. H. Niederreiter, *Random Number Generation and Quasi-Monte Carlo Methods*. SIAM CBMS-NSF Regional Conference Series in Applied Mathematics, vol. 63 (SIAM, Philadelphia, 1992)

100. E. Novak, K. Ritter, High dimensional integration of smooth functions over cubes. Numer. Math. **75**, 79–97 (1996)

101. E. Novak, H. Woźniakowski, *Tractability of Multivariate Problems. Vol. 1, Linear Information*. EMS Tracts in Mathematics (European Mathematical Society (EMS), Zürich, 2008)

102. E. Novak, H. Woźniakowski, L_2 discrepancy and multivariate integration, in *Analytic number theory. Essays in honour of Klaus Roth.*, ed. by W.W.L. Chen, W.T. Gowers, H. Halberstam, W.M. Schmidt, R.C. Vaughan (Cambridge University Press, Cambridge, 2009)

103. E. Novak, H. Woźniakowski, *Tractability of Multivariate Problems. Vol. 2, Standard Information for Functionals*. EMS Tracts in Mathematics (European Mathematical Society (EMS), Zürich, 2010)

104. D. Nuyens, R. Cools, Fast algorithms for component-by-component construction of rank-1 lattice rules in shift-invariant reproducing kernel Hilbert spaces. Math. Comp. **75**, 903–920 (2006)

105. G. Ökten, Error reduction techniques in quasi-Monte Carlo integration. Math. Comput. Modelling **30**, 61–69 (1999)

106. G. Ökten, M. Shah, Y. Goncharov, Random and deterministic digit permutations of the Halton sequence, in *Monte Carlo and Quasi-Monte Carlo Methods 2010*, ed. by L. Plaskota, H. Woźniakowski (Springer, Berlin and Heidelberg, 2012)

107. M.H. Overmars, C.-K. Yap, New upper bounds in Klee's measure problem. SIAM J. Comput. **20**(6), 1034–1045 (1991)

108. S.H. Paskov, Average case complexity of multivariate integration for smooth functions. J. Complexity **9**, 291–312 (1993)

109. P. Peart, The dispersion of the Hammersley sequence in the unit square. Monatshefte Math. **94**, 249–261 (1982)

110. T. Pillards, R. Cools, A note on E. Thiémard's algorithm to compute bounds for the star discrepancy. J. Complexity **21**, 320–323 (2005)

111. T. Pillards, B. Vandewoestyne, R. Cools, Minimizing the L_2 and L_∞ star discrepancies of a single point in the unit hypercube. J. Comput. Appl. Math. **197**(1), 282–285 (2006). http://dx.doi.org/10.1016/j.cam.2005.11.005

112. W. Römisch, Scenario reduction techniques in stochastic programming, in *SAGA 2009*, ed. by O. Watanabe, T. Zeugmann LNCS, vol. 43 (Springer, Berlin-Heidelberg, 2009)

113. W.M. Schmidt, Irregularities of distribution IX. Acta Arithmetica, 385–396 (1975)

114. M. Shah, A genetic algorithm approach to estimate lower bounds of the star discrepancy. Monte Carlo Methods Appl. **16**, 379–398 (2010)

115. I.H. Sloan, S. Joe, *Lattice Methods for Multiple Integration* (Clarendon Press, Oxford, 1994)

116. I.H. Sloan, A.V. Reztsov, Component-by-component construction of good lattice rules. Math. Comp. **71**, 263–273 (2002)

117. I.H. Sloan, H. Woźniakowski, When are quasi-Monte Carlo algorithms efficient for high dimensional integrals? J. Complexity **14**, 1–33 (1998)

118. I.H. Sloan, F.Y. Kuo, S. Joe, On the step-by-step construction of quasi-Monte Carlo integration rules that achieve strong tractability error bounds in weighted Sobolev spaces. Math. Comp. **71**, 1609–1640 (2002)

119. S.A. Smolyak, Quadrature and interpolation formulas for tensor products of certain classes of functions. Dokl. Akad. Nauk. SSSR **4**, 1042–1045 (1963)

120. I.M. Sobol', The distribution of points in a cube and the approximate evaluation of integrals. Zh. Vychisl. Mat. i. Mat. Fiz. **7**, 784–802 (1967)
121. I.M. Sobol', D. Asotsky, A. Kreinin, S. Kucherenko, Construction and comparison of high-dimensional Sobol' generators. Wilmott Mag. **Nov. 2011**, 64–79 (2011)
122. A. Srinivasan, Distributions on level-sets with applications to approximation algorithms, in *Proceedings of FOCS'01* (2001)
123. S. Steinerberger, The asymptotic behavior of the average L_p-discrepancies and a randomized discrepancy. Electron. J. Combin., 1–18 (2010). Research Paper 106
124. W. Stute, Convergence rates for the isotrope discrepancy. Ann. Probab., 707–723 (1977)
125. V.N. Temlyakov, On a way of obtaining lower estimates for the errors of quadrature formulas. Math. USSR Sbornik **71**, 247–257 (1992)
126. S. Tezuka, *Uniform Random Numbers: Theory and Practice* (Kluwer Academic, Boston, 1995)
127. E. Thiémard, 1998, Economic generation of low-discrepancy sequences with a b-ary Gray code. EPFL-DMA-ROSO, RO981201, http://rosowww.epfl.ch/papers/grayfaure/
128. E. Thiémard, Computing bounds for the star discrepancy. Computing **65**, 169–186 (2000)
129. E. Thiémard, *Sur le calcul et la majoration de la discrépance à l'origine* (PhD thesis École polytechnique fédérale de Lausanne EPFL, nbr 2259, Lausanne, 2000). Available from http://infoscience.epfl.ch/record/32735
130. E. Thiémard, An algorithm to compute bounds for the star discrepancy. J. Complexity **17**, 850–880 (2001)
131. E. Thiémard, Optimal volume subintervals with k points and star discrepancy via integer programming. Math. Meth. Oper. Res. **54**, 21–45 (2001)
132. H. Triebel, *Bases in Function Spaces, Sampling, Discrepancy, Numerical Integration*. EMS Tracts in Mathematics 11 (European Mathematical Society, Zürich, 2010)
133. B. Vandewoestyne, R. Cools, Good permutations for deterministic scrambled Halton sequences in terms of L_2-discrepancy. J. Comput. Appl. Math. **189**, 341–361 (2006)
134. X. Wang, F.J. Hickernell, Randomized Halton sequences. Math. Comput. Modell. **32**, 887–899 (2000)
135. T.T. Warnock, Computational investigations of low-discrepancy point sets, in *Applications of number theory to numerical analysis*, ed. by S.K. Zaremba (Academic Press, New York, 1972)
136. T.T. Warnock, 2013. Personal communication
137. G.W. Wasilkowski, H. Woźniakowski, Explicit cost bounds of algorithms for multivariate tensor product problems. J. Complexity **11**, 1–56 (1995)
138. B.E. White, On optimal extreme-discrepancy point sets in the square. Numer. Math. **27**, 157–164 (1976/77)
139. P. Winker, K.T. Fang, Applications of threshold-accepting to the evaluation of the discrepancy of a set of points. SIAM J. Numer. Anal. **34**, 2028–2042 (1997)
140. C. Winzen, *Approximative Berechnung der Sterndiskrepanz* (Christian-Albrechts-Universität zu Kiel, Kiel, 2007)
141. H. Woźniakowski, Average case complexity of multivariate integration. Bull. Am. Math. Soc. (N. S.) **24**, 185–191 (1991)
142. S.K. Zaremba, Some applications of multidimensional integration by parts. Ann. Polon. Math. **21**, 85–96 (1968)
143. S.K. Zaremba, La discrépance isotrope et l'intégration numérique. Ann. Mat. Pura Appl. **37**, 125–136 (1970)
144. C. Zenger, Sparse grids, in *Parallel Algorithms for Partial Differential Equations*, ed. by W. Hackbusch (Vieweg, Braunschweig, 1991)
145. P. Zinterhof, Über einige Abschätzungen bei der Approximation von Funktionen mit Gleich-verteilungsmethoden. Sitzungsber. Österr. Akad. Wiss. Math.-Naturwiss. Kl. II **185**, 121–132 (1976)

Author Index

Aho, A. V. 631, 673
Aistleitner, C. 554, 614
Alon, N. 319, 334, 335, 345, 421, 429, 456, 464, 483
Anderson, T. 121, 153
Appel, K. 322, 421
Arnol'd, V. 521, 537
Aronszajn, N. 542, 543, 614
Asotsky, D. 667, 678
Atanassov, E. I. 666, 673

Babai, L. 331, 421
Babu, G. S. S. 474, 483
Bak, J.-G. 163, 217
Bakhvalov, N. S. 566, 572, 578, 614
Baldeaux, J. 570, 602, 604, 614, 615
Baluja, S. 471, 483
Banaszczyk, W. 327, 328, 421, 430, 443, 456
Bansal, N. viii–x, 326, 327, 421, 430, 431, 456, 461, 483
Bárány, I. 327, 328, 351, 421
Bass, R. F. 119, 153
Beck, J. vi–viii, x, xi, 5–7, 20, 68, 74, 76, 80, 81, 89, 134, 136–138, 153, 160, 211, 213, 217, 227, 230, 234, 243, 315, 316, 319–321, 325, 327, 329, 330, 346, 351, 355, 375–377, 383, 385, 407, 421, 422, 429, 432, 456, 490, 536, 537, 623, 625, 626, 634, 673
Béjian, R. 75, 108, 153, 552, 614
Bellhouse, D. 213, 217
Bentley, J. L. 640, 673
Bernard, A. 103, 153
Bilyk, D. 18, 68, 76, 80, 89, 99, 106, 107, 110, 118, 119, 121, 134, 135, 137–140, 145, 148–151, 153, 154, 552, 614, 623, 624, 673
Birch, B. 488, 491, 510, 511, 516, 524, 525, 537
Bogachev, V. I. 127, 154
Bohman, T. 325, 422
Bourgain, J. 432, 456
Boyd, S. 437, 454, 457
Braaten, E. 665, 666, 673
Brandolini, L. vii, xi, 160, 161, 163, 164, 166, 172, 176, 178, 186, 187, 198, 211, 213, 217, 218, 626, 673
Brauchart, J. 549–551, 614
Bruna, J. 163, 218
Brychkov, Y. 601, 619
Bundschuh, P. 637, 639, 648, 667, 673
Burkholder, D. L. 96, 154
Bykovskii, V. A. 590, 614

Caflisch, R. 624, 625, 676
Caruana, R. 471, 483
Cassels, J. 207, 218, 237, 316
Chan, T. M. 640, 673
Chandrasekaran, K. 432, 456
Chang, S.-Y. A. 98, 103, 106, 154
Charikar, M. 335, 422, 427, 456
Chazelle, B. vi, viii, xi, 80, 81, 87, 154, 316, 319, 385, 422, 426, 430, 457, 626, 673
Chen, J. 647, 673
Chen, W. W. L. vi, vii, x, xi, 7, 15, 20, 35, 48, 52, 55, 59, 67, 68, 74, 76, 77, 80, 149–151, 153, 154, 160, 192, 210, 211, 213, 217, 218, 316, 320, 421, 553, 566, 614, 615, 623, 626, 673
Cheong, O. 651, 674

W. Chen et al. (eds.), *A Panorama of Discrepancy Theory*, Lecture Notes in Mathematics 2107, DOI 10.1007/978-3-319-04696-9,
© Springer International Publishing Switzerland 2014

Chi, H. 666, 673
Choirat, C. 626, 673
Chor, B. 647, 673
Chuang, S. C. 626, 670, 674
Cilleruelo, J. 378, 388, 422
Colin de Verdière, Y. 198, 218
Colzani, L. vii, xi, 160, 161, 163, 172, 178,
       188, 198, 218, 626, 673
Cools, R. ix, xii, 583, 588, 610, 615, 618, 633,
       648, 651, 662, 666, 677, 678
Cormen, T. H. 642, 674
Cover, T. M. 434, 457
Cristea, L. L. 566, 572, 573, 615
Csaki, E. 119, 154

Das, D. B. 474, 483
Daubechies, I. 84, 154
Davenport, H. 18, 20, 67, 68, 74, 77, 150, 154,
       160, 218, 491, 510, 514, 537, 553, 615
de Berg, M. 651, 674
de Clerck, L. 637–639, 674
De Rainville, F.-M. 666, 668, 669, 674
Dick, J. ix–xi, 77, 80, 91, 95, 149, 151, 155,
       547, 549–553, 555, 557, 560–562, 566,
       568–573, 583, 584, 586, 593–595, 597,
       598, 602–604, 606–610, 612–616, 621,
       623, 625, 626, 628, 629, 639, 648, 667,
       674
Dobkin, D. P. 639, 640, 642, 647, 664, 667,
       669, 670, 674
Doerr, B. viii, xi, 325, 326, 330, 331, 333,
       334, 339, 341, 346, 349, 351, 354–356,
       364–367, 377, 405, 421, 422, 428, 457,
       554, 555, 616, 649–651, 667, 668, 674
Doerr (nee Winzen), C. 669, 674
Downey, R. G. 647, 674
Drmota, M. vi, xi, 426, 457, 623, 626, 674
Duke, W. 494, 537
Dunker, T. and Kühn, T. and Lifshits, M. and
       Linde, W. 119, 124, 155

Eisenbrand, F. 432, 457
Eppstein, D. 639, 640, 642, 647, 664, 667, 669,
       670, 674
Erdős, P. 319, 330, 334, 335, 345, 421, 422,
       465, 467, 483

Fang, K.-T. 664, 670, 674
Faure, H. ix, xi, 48, 69, 149, 151, 155, 552,
       558, 562, 616, 639, 664, 666, 667, 674
Fefferman, R. 99, 103, 106, 154, 155
Feller, W. 119, 155, 237, 316

Fellows, M. 647, 673
Fellows, M. R. 647, 674
Fiala, T. viii, xi, 321, 327, 346, 422, 429, 432,
       456
Fine, N. 44, 66, 69
Flum, J. 647, 674
Fomenko, O. 494, 495, 537
Frank, K. 628, 630, 632, 633, 674
Fredman, M. L. 640, 675
Freiman, G. A. 378, 422
Frolov, K. K. 77, 155, 553, 616

Gabai, H. 639, 675
Gagné, C. 666, 668, 674
Garey, M. R. 643, 644, 675
Gelatt, C. 657, 676
Gerstner, T. 633, 675
Giannopoulos, A. 326, 422
Giannopoulus, P. 647, 675
Gigante, G. 163, 211, 213, 218, 626, 673
Gilbert, E. N. 126, 155, 607, 616
Gioev, D. 172, 218
Gluskin, E. D. 326, 422
Gnewuch, M. ix, xi, 366, 367, 422, 549, 554,
       555, 587, 616, 625, 626, 628, 634, 638,
       643–646, 649–652, 656–660, 662–664,
       667–669, 674, 675
Goldberg, D. E. 472, 483, 484
Golubeva, E. 494, 495, 537
Goncharov, Y. 651, 662, 666, 667, 677
Gonthier, G. 322, 422
Grafakos, L. 95, 116, 155
Graham, R. L. 316, 319, 356, 369, 422
Greenleaf, A. 186, 218
Greslehner, J. 602, 604, 614
Griebel, M. 633, 675
Grinberg, V. S. 327, 328, 421
Grohe, M. 647, 674
Gruber, P. M. 186, 198, 219
Guruswami, V. 451, 457
Gusejn-Zade, S. 521, 537

Haar, A. 83, 155
Haken, W. 322, 421
Halász, G. 75, 77, 89, 104, 108, 111, 113, 115,
       155, 259, 316
Halton, J. H. ix, xi, 18, 29, 48, 69, 76, 149–151,
       155, 639, 664, 665, 667, 675
Hammersley, J. M. 76, 149, 155, 639, 675
Han, K.-H. 461, 472, 474, 478, 483
Hardy, G. H. 19, 69, 212, 219, 507, 537
Harik, G. R. 472, 483

Hauschild, M. 474, 483
Haussler, D. 345, 422
Hayes, T. P. 331, 421
Heath-Brown, D. 496, 506, 537
Hebbinghaus, N. viii, xi, 366, 367, 371, 378, 379, 388, 393, 397, 402, 405, 412, 413, 422, 423
Heinrich, S. 82, 155, 554, 598, 616, 628–633, 663, 664, 674, 675
Hellekalek, P. 607, 608, 616, 621, 627, 664, 675, 676
Helm, M. 327, 423
Henrion, R. 670–672, 675
Herz, C. 163, 219
Hickernell, F. J. 219, 567, 592, 598, 616, 624, 625, 628, 634, 666, 675, 678
Hinrichs, A. 85, 90, 105, 149–152, 155, 554, 556, 557, 616, 624, 664, 675
Hlawka, E. 163, 219, 548, 552, 576, 616, 625, 676
Hoeffding, W. 110, 155
Hofmann, S. vii, xi, 161, 164, 166, 217
Holland, J. H. 660, 676
Holzman, R. 325, 422
Hong, H. S. 567, 616
Hopcroft, J. E. 631, 673
Hromković, J. 643, 676
Huang, X. 647, 673
Hung, Y. C. 626, 670, 674
Hurewicz, W. 516, 537
Huxley, M. N. 186, 219

Impagliazzo, R. 643, 647, 676
Iosevich, A. vii, xi, 161, 164, 166, 178, 186, 198, 217, 218

Joe, S. ix, xii, 552, 576, 578, 583, 587–589, 591, 617, 619, 627, 639, 648, 667, 676, 677
Johnson, D. S. 643, 644, 675
Juedes, D. W. 647, 673

Kac, M. 316
Kahan, W. 629, 676
Kanj, I. A. 647, 673
Kasabov, N. 473, 479, 484
Katznelson, Y. 116, 117, 155
Keller, A. 211, 213, 219
Kendall, D. G. 160, 187, 219
Khintchine, A. 235, 236, 316
Kim, J.-H. 461, 472, 474, 478, 483

Kimmel, P. G. 331, 421
Kirckpatrick, S. 657, 676
Klee, V. 640, 676
Kleitman, D. 434, 457
Kliemann, L. 354, 423, 462, 469, 474, 478, 480–483
Kliemann, O. 354, 423, 462, 469, 474, 478, 480–483
Knauer, C. 647, 675
Knieper, P. 359, 375, 423
Knuth, D. E. 621, 664, 676
Koch, J. 322, 421
Kollig, T. 211, 213, 219
Kolmogorov, A. 237, 316
Kolountzakis, M. N. 174, 219
Konyagin, S. V. 16, 69, 189, 219
Korobov, N. M. 552, 576, 583, 605, 617
Krätzel, Ekkehard 186, 219
Kreinin, A. 667, 678
Kritzer, P. 149, 151, 155, 555, 564, 570, 586, 595, 606, 607, 609, 610, 615–617, 667, 674
Kucherenko, S. 667, 678
Küchler, C. 670–672, 675
Kuelbs, J. 124, 127, 155
Kuipers, L. 73, 80, 155, 542, 546, 617, 624, 676
Kuo, F. Y. ix, xi, xii, 583, 584, 597, 598, 615, 617, 619, 648, 667, 676, 677

Lacey, M. T. 18, 68, 76, 80, 89, 99, 103, 104, 106, 107, 110, 118, 119, 121, 134, 135, 137–140, 145, 148–151, 153–155, 552, 614, 624, 673
Lang, S. 229, 316
Larcher, G. 149, 151, 155, 583, 586, 587, 605, 607, 610, 611, 616, 617, 633, 662, 676
Laurendeau, D. 666, 668, 674
Lauss, A. 605, 607, 617
L'Ecuyer, P. 603, 619, 621, 664, 676
Ledoux, M. 127, 155
Leeb, H. 627, 675
Leiserson, C. E. 642, 674
Lekkerkerker, C. G. 186, 198, 219
Lemieux, C. 664, 666, 674, 676
Leobacher, G. 557, 602, 608–610, 612, 614–617, 625, 663, 664, 676
Lerch, M. 19, 69, 75, 155
Lévy, P. 120, 156
Li, W. V. 119, 124, 127, 155, 156
Liardet, P. 75, 108, 156
Lidl, R. 42, 69, 410, 423
Lifshits, M. A. 124, 127, 156

Lin, D. K. J. 664, 670, 674, 676
Lindenstrauss, J. 104, 156
Littlewood, J. E. 19, 69
Llorà, X. 472, 484
Lobo, F. G. 472, 483
Loh, W.-L. 567, 617
Lovász, L. 428, 430, 457
Lovett, S. 326, 423, 431, 445, 447–449, 457
Lvov, A. viii, xi
Lyness, J. N. 582, 583, 615, 617

MacWilliams, F. 42, 69
Magyar, A. 489–491, 494, 510, 536, 537
Mani, A. 474, 483
Marichev, O. 601, 619
Markhasin, L. 85, 90, 105, 152, 155, 156, 624, 676
Mascagni, M. 666, 673
Matoušek, J. vi, viii–xii, 80, 156, 192, 219, 316, 319–321, 326, 329, 330, 345, 354, 356, 363, 423, 426, 430–433, 435, 453, 454, 457, 461, 464, 483, 545, 567, 584, 617, 623, 626, 628, 631–633, 676
McMichael, D. 163, 217
Mehlhorn, K. 651, 676
Meka, R. ix, xii, 326, 423, 431, 445, 447–449, 457
Mitchell, D. P. 211, 213, 219, 639, 640, 642, 647, 664, 667, 669, 670, 674
Montgomery, H. L. vii, xii, 80, 156, 160, 207, 209, 219
Morokoff, W. 624, 625, 676
Muckenhoupt, B. 106, 156
Mühlenbein, H. 471, 483
Mulmuley, K. 334, 423

Nagel, A. 163, 218
Nathanson, M. B. 320, 378, 423
Newman, A. 335, 422, 427, 456
Niederreiter, H. ix, x, xii, 42, 69, 73, 80, 155, 410, 423, 542, 546, 552, 558, 560–562, 570, 576, 586, 587, 589, 592–596, 605, 607–609, 611, 615–618, 621, 623–625, 627, 635–637, 639, 648, 664, 667, 676, 677
Nikolov, A. 328, 329, 335, 422, 423, 427, 432, 456, 457
Novak, E. ix, xii, 91, 95, 156, 549, 554, 555, 557, 558, 566, 572, 586, 616, 618, 624, 625, 628, 633, 634, 663, 664, 675, 677
Nuyens, D. ix, xii, 583, 588, 602, 610, 614, 618, 648, 677

Ökten, G. 651, 662, 666, 667, 677
Ostrowski, A. 316
Ou, W. 106, 156
Overmars, M. H. 640, 651, 674, 677
Owen, A. B. ix, xii, 566, 567, 618

Paaß, G. 471, 483
Pálvölgyi, D. 431, 432, 453, 457
Parissis, I. 89, 106, 107, 118, 148–151, 154, 624, 673
Parnovski, L. 189, 200, 219
Paskov, S. H. 632, 677
Paturi, R. 643, 647, 676
Patvardhan, C. 354, 423, 462, 469, 474, 478, 480–483
Peart, P. 639, 677
Pelikan, M. 474, 483
Pereyra, M. C. 95, 156
Pillards, T. 633, 651, 662, 677
Pillichshammer, F. x, xi, 77, 80, 91, 95, 149, 151, 155, 552, 553, 555–557, 560–562, 566, 569, 570, 572, 573, 583, 584, 586, 593–595, 597, 598, 602, 604, 606–618, 621, 623, 625, 626, 628, 629, 633, 639, 648, 662–664, 667, 674, 676
Pipher, J. 99, 155, 156
Platel, M. D. 473, 479, 484
Podkorytov, A. 164, 176, 178, 198, 218, 219
Pollard, D. 9, 69
Pólya, G. 74, 156
Prakash, P. 474, 483
Price, J. 66, 69
Přívětivý, A. viii, xii, 378, 383, 424
Prudnikov, A. 601, 619

Rado, R. 356, 424
Ramaré, O. 163, 219
Randol, B. 163, 194, 198, 219
Reztsov, A. V. 583, 619, 648, 677
Ricci, F. 163, 219
Riesz, F. 115, 118, 156
Rigoli, M. 160, 161, 176, 186, 187, 218
Riordan, J. 367, 424
Ritter, K. 633, 677
Rivest, R. L. 642, 674
Robertson, N. 322, 424
Roşca, A. 625, 675
Römisch, W. 670–672, 675, 677
Roth, K. F. v, vi, viii, ix, xii, 4, 17, 18, 24, 48, 52, 67, 69, 72–74, 77, 81, 83, 84, 89, 149, 150, 156, 160, 219, 258, 259, 316,

320, 330, 356, 359, 360, 394, 424, 461, 484, 552, 553, 598, 619
Rothschild, B. L. 316, 319, 356, 369, 422
Rothvoß, T. 431, 432, 457
Rudin, W. 357, 358, 360, 424

Sanders, D. P. 322, 424
Šarygin, I. 578, 582, 619
Sarnak, P. 506, 509, 537
Sastry, K. 472, 484
Sauerland, V. 354, 423, 424, 460–462, 465, 469, 474, 478, 480–484
Schliebs, S. 473, 479, 484
Schmid, W. C. 556, 557, 595, 605–607, 616, 617
Schmidt, W. M. vii, xii, 18, 69, 74, 75, 87, 89, 95, 108, 156, 157, 159, 220, 259, 316, 490, 510, 536, 537, 552, 553, 619, 625, 677
Schneider, R. 220
Schoen, T. viii, xi, 405, 412, 413, 423
Schulze-Pillot, R. 494, 537
Seri, R. 626, 673
Seymour, P. 322, 424
Shachar Lovett, S. ix, xii
Shah, M. 651, 660, 662, 666, 667, 677
Shao, Q.-M. 119, 124, 156
Sharpe, C. 670, 676
Sidon, S. 115, 116, 157
Sidorova, N. 189, 219
Sinescu, V. 588, 603, 619
Sipser, M. 427, 457
Skriganov, M. M. vi, xi, 16, 48, 64, 67–69, 77, 149–151, 154, 157, 189, 199, 217–220, 553, 566, 615, 619, 623, 673
Sloan, I. H. ix, xii, 547, 552, 555, 557, 576, 578, 583, 585–587, 597, 598, 603, 615–617, 619, 624, 625, 627, 628, 634, 639, 648, 674–677
Sloane, N. 42, 69
Smolyak, S. A. 633, 677
Sobol', I. M. 664, 667, 678
Sobolev, A. V. 16, 69, 189, 200, 219
Sogge, C. D. 194, 220
Sós, V. T. vi, xi, 319, 327, 329, 351, 355, 376, 383, 385, 407, 422
Spencer, J. viii, ix, xii
Spencer, J. H. viii–xi, 316, 319–321, 325, 326, 328–330, 334, 335, 343, 345, 354, 356, 363, 369, 377, 421–424, 427–430, 456, 457, 461, 464, 465, 467, 483
Srinivasan, A. 327, 424, 430, 431, 435, 457, 667, 678

Srivastav, A. viii, xi, 330, 339, 341, 345, 346, 349, 354–356, 364, 365, 377, 393, 397, 402, 405, 412, 413, 422–424, 462, 469, 474, 478, 480–483, 554, 587, 616, 643–646, 649, 650, 652, 674, 675
Stangier, P. 345, 424
Stein, C. 642, 674
Stein, E. M. 95, 96, 99, 103, 157, 161–163, 166, 175, 182, 192, 193, 199, 220, 503, 538
Steinerberger, S. 664, 678
Stolarsky, K. B. 551, 619
Stute, W. 625, 678

Taibleson, M. H. 193, 220
Talagrand, M. 111, 119, 121, 124, 127, 155, 157
Talwar, K. 329, 423, 432, 457
Tarnopolska-Weiss, M. 194, 220
Temlyakov, V. N. 74, 77, 80, 82, 91, 95, 110, 111, 124, 132, 133, 148, 157, 578, 619, 632, 633, 678
Teytaud, O. 666, 668, 674
Tezuka, S. 664, 667, 678
Thangavelu, S. 163, 218
Thiémard, E. 649–652, 656, 664, 666, 667, 678
Thomas, J. A. 434, 457
Thomas, R. 322, 424
Tichy, R. F. vi, xi, 426, 457, 623, 626, 674
Travaglini, G. vii, xi, 15, 20, 68, 80, 154, 160, 161, 163, 172, 176, 178, 186, 187, 192, 194, 198, 211, 213, 217–220, 626, 673
Triebel, H. 80, 105, 152, 157, 624, 678
Tzafriri, L. 104, 156, 432, 456

Ullman, J. D. 631, 673

Vagharshakyan, A. 18, 68, 76, 89, 106, 107, 118, 119, 121, 134, 135, 137–140, 145, 148–151, 154, 552, 614, 624, 673
van Aardenne-Ehrenfest, T. v, xii, 4, 69, 73, 157
van der Corput, J. G. 4, 69, 73, 75, 149, 157, 316
van der Waerden, B. L. 330, 424
van Kreveld, M. 651, 674
Vance, J. 163, 217
Vandenberghe, L. 437, 454, 457
Vandewoestyne, B. 633, 662, 666, 677, 678
Varchenko, A. 521, 537

Varshamov, R. R. 126, 157, 607, 619
Vecchi, M. 657, 676
Vempala, S. 432, 456
Vershynin, R. 432, 457
Vesztergombi, K. viii, xi, 325, 329, 423, 428,
    430, 457
Vondrák, J. 464, 483
Vu, V. H. 432, 457

Wahlström, M. 638, 647, 651, 656–660, 662,
    667–669, 674, 675
Wainger, S. 163, 217, 218
Wallman, H. 516, 537
Wang, G. 96, 157
Wang, X. 547, 603, 616, 625, 628, 666, 674,
    675, 678
Warnock, T. 666, 673
Warnock, T. T. 628, 629, 678
Wasilkowski, G. W. 554, 616, 624, 633, 634,
    663, 664, 675, 678
Watson, G. N. 162, 163, 220
Wehr (nee Knieper), P. viii, xi
Wehr, P. 355, 356, 364, 365, 377, 422
Weide, B. W. 640, 675
Weiss, G. 162, 175, 182, 192, 193, 220
Weiss, N. J. 199, 220
Weller, G. 665, 666, 673
Welzl, E. viii, xii, 321, 345, 422, 423
Werner, D. 647, 675
Wernisch, L. viii, xii, 321, 345, 423
Werth, S. 405, 422
Weyhausen, H. 664, 675
Weyl, H. v, xii, 316

Wheeden, R. L. 162, 179, 220
White, B. E. 624, 662, 678
Wiles, A. 507, 537
Wilson, J. 98, 154
Winker, P. 639, 658, 659, 664, 670, 674, 676,
    678
Winzen, C. 587, 616, 638, 643–646, 652,
    656–660, 662, 667, 669, 675, 678
Wolff, T. 98, 154, 174, 219
Wooley, T. D. 536, 538
Woźniakowski, H. 625, 628, 663, 664, 675,
    677
Wright, E. M. 212, 219, 507, 537

Xia, G. 647, 673
Xing, C. 558, 570, 618

Yap, C.-K. 640, 677
Yue, R.-X. 598, 616

Zagier, D. B. 229, 316
Zane, F. 647, 676
Zaremba, S. K. 29, 69, 149–151, 155, 548,
    619, 625, 628, 639, 675, 678
Zenger, C. 633, 678
Zhang, L. 329, 423, 432, 457
Zhang, Y. 664, 670, 674
Zhu, Y. C. 637, 639, 648, 667, 673
Zinterhof, P. 627, 678
Zygmund, A. 115, 116, 118, 158, 162, 179,
    220

Index

Symbols

1-set   59
2-color discrepancy   314, 323, 461
2-coloring   314
3-SAT
   decision problem   647

A

Abelian group
   locally compact   357
accepting a sample   472
algorithm
   compact genetic (cGA)   472
   component-by-component   588, 591, 609,
      648, 667, 669
   de Rainville–Gagné–Teytaud–Laurendeau
      668
   divide-and-conquer   631, 656
   Dobkin–Eppstein–Mitchell   627, 639
   FPT   646
   genetic (GA)   657, 660
   Heinrich   629
   Las Vegas   657
   Monte-Carlo (MC)   565, 657
   $(\mu + \lambda)$ evolutionary   661
   Nuyens–Cools fast component-by-
      component   583, 588
   quadrature   564
   quantum-inspired evolutionary (QEA)
      472
   quasi-Monte Carlo (QMC)   634
   Thiémard's bracketing cover   650

Thiémard's ILP-based   651
threshold accepting   657, 658
univariate marginal distribution (UMDA)
   471
Winker– Fang   658
alternating mirror technique   469
Anderson's lemma   121, 129, 130
Area Principle
   for $\sqrt{2}$ 239
   general   240
   Naive   224
arithmetic progressions (AP)   330, 460, 488,
   527
   Cartesian product of   355
   centered   394, 399
   d-dimensional   355
   in $\mathbb{Z}_p$ 393, 467
   sum of   320, 378
   symmetric Cartesian product of   364
   with common difference   371
attractor   472
attractor population QEA (apQEA)   474

B

backward selection   672
balanced coloring   465
Beck gain   137
Bernoulli polynomial   578, 602
Bertrand's postulate   585
Besov norm   105, 624
Besov space   79, 105
Bessel function   16, 162, 601
Bessel's inequality   79, 90
best hyperbolic cross approximation   132

Birch–Davenport method   494, 509, 510
Bochner integral   99
Borel function   357
Borel's theorem on normal numbers   236
Borell's inequality   129, 130
bounded mean oscillation (BMO)
    norm   624
    space   79, 106
bracketing entropy   649, 650
bracketing number   650
branch and bound   656
Brownian sheet   80, 118, 119
brute force   113

C

Cantor set   247
Cantor set construction   247
cap   489
    direction   489, 494
    discrepancy   489
cell   641
centered discrepancy   624
central composite discrepancy   626
central limit theorem   234, 567
Chang–Wilson–Wolff inequality   98
character (of a locally compact Abelian group)
        357
Chebyshev inequality   122, 192
Chernoff bound   335
Chernoff inequality   345
Chernoff–Hoeffding inequality   345
Chinese remainder theorem   49, 509
Cholesky decomposition   437
Chrestenson–Levy function   62
circle problem   235
circulant matrix   355
coding length   643
coloring   425
    balanced   465
    fair   339
    generating   465
    modulo   465
    of a graph   139
    of edges   139
    partial   431
    random   429, 464
combinatorial 2-color discrepancy   323, 461
combinatorial multicolor discrepancy   320
compact genetic algorithm (cGA)   472
complexity class
    FPT   646

NP   322, 643
    W[1]   646, 647
component-by-component algorithm   588,
        591, 609, 648, 667, 669
conditional expectation   147, 148
continued fractions   223, 300
convolution   357
coprime lattice point   254
covariance kernel   128
covering number   123
critical points   638
critical test box   638
cubature formula   74, 91
    Kolmogorov's method   74, 90
cubature rule   583
curse of dimensionality   82, 555, 627
cutoff function   164, 165, 492
cutting plane   656

D

Davenport reflection   19
decision problem   643
    3-SAT   647
    discrepancy   334
    dominating set   644
    empty half-open box   644
    star discrepancy   643
δ-bracket   650
δ-bracketing cover   649, 650
δ-cover   649
diaphony   627
digital $(t, \alpha, \alpha m \times m, s)$-net over $\mathbb{Z}_b$ 568
digital (t, s)-sequence
    over $\mathbb{Z}_b$ 561
digital construction scheme   560
    generating matrices   560
digital net   560
    dual   595
digital sequence   560, 561
    Faure   562
    Niederreiter   561
    Sobol'   562
digital shift   566
    simplified   572
digital (t, α, s)-sequence over $\mathbb{Z}_b$ 569
digital (t, m, s)-net   560
    over $\mathbb{Z}_b$ 561
digital (t, s)-sequence   561
dilation   187
diophantine
    approximation   19, 226, 239

coordinate 490
direction 490
equation 222, 487
inequality 227
point 489, 532
Dirac measure 13, 215
Dirichlet's principle 495
discrepancy 426
  2-color 314, 323, 461
  axis-parallel rectangle based 627
  cap 489
  centered 624
  central composite 626
  combinatorial 2-color 323, 461
  combinatorial multicolor 320
  decision problem 334
  determinant lower bound 430
  function 73, 187
  extreme 624
  for axis-parallel boxes (*see* extreme
      discrepancy)
  G-star 625
  geometric 72
  half space 192
  hereditary 325, 427
  Hickernell's modified L_p 625
  isotrope 625, 671
  L^2-square 383
  L_2 545, 546, 627, 628
  L_2-extreme 624, 629
  L_2-star 627
  linear 324
  local 539
  L_p 540, 663
  L_p spherical cap 550
  L_p-extreme 624
  L_p-star 623
  L_∞ 622
  L_∞-extreme 624
  L_∞-star 623
  measure 622
  modified L_2 625
  modified L_p 625
  multicolor 320
  of a point set 4
  of a vector sequence 327
  of finite point sets 74
  of infinite sequences 74
  of matrices 324
  one-sided 405
  phenomena of large 5
  phenomena of small 5
  polyhedral 671
  problem of large 5

problem of small 17
quadrant 624
r-smooth L_2 632
rectangle 671
smoothed 503
spherical cap 626
star 73, 540, 623
unanchored (*see* discrepancy, extreme)
weighted 325, 549, 555, 625
weighted combinatorial multicolor 332
weighted L_2 557
weighted L_p 549, 626
weighted L_∞ 626
weighted star 549
discrete optimization problem 643
divergence theorem 164
divide-and-conquer 631, 656
Dobkin–Eppstein–Mitchell algorithm 627,
    639
dominating point 267
dominating set 644
  decision problem 644
dual lattice 576
dual polynomial lattice 594
dyadic box 92, 93
dyadic interval 26, 28, 83, 94, 506
dyadic rectangle 74, 84, 85
dyadic square function 96
  product 101
dynamic programming 642

E

EDA
  incremental 471
  population-based 471
edge coloring 139
ellipsoid method 437
empty half-open box
  decision problem 644
entropy 434
  method 429, 433
  metric 127, 130
  number 82, 123
equation
  diophantine 222, 487
  Parseval-type 412
  Pell 222
Erdős–Turán–Koksma inequality 586,
    648
Euclidean Fourier transform 492
Euler's formula 251

evolutionary algorithm (EA)
   $(\mu + \lambda)$ 661
exponential time hypothesis (ETH)   647
exponentially square integrable function   98
extra large deviation   253, 257
extreme discrepancy   624
   L_∞ 624

F

facet   162
fair coloring   339
Farey dissection   496
Farey fraction   507
fast component-by-component algorithm
   583, 588
Faure construction   48
Faure sequence   562, 660
Faure set   55, 56, 667
Fejér kernel   208
Fine formula   44
Fine–Price formula   65
fitness function   471
flat torus   490, 530
flatline termination criterion   476
formula
   cubature   74, 91
   Euler   251
   Fine   44
   Fine–Price   65
   inclusion-exclusion   143
   Parseval   234
   Poisson summation   196, 234
   quadrature   622
   Stirling   126
   Warnock   545, 546, 549, 551, 628
   Warnock, generalization   663
forward selection   672
Fourier series
   lacunary   80, 115, 116
Fourier transform decay   161, 180, 186, 523
Fourier–Walsh analysis   40
   p-adic   62
Fourier–Walsh coefficients   40, 63
Fourier–Walsh series   40, 63
FPT (compexity class)   646
FPT-algorithm   646
function
   exponentially square integrable   98
   subgaussian   98
function space   80
   Besov   79, 105

   BMO   79, 106
   exponential Orlicz   79, 98, 106
   Hardy   79, 103
   Korobov   576, 577
   L_p 79
   product BMO   106
   reproducing kernel Hilbert   124, 128, 542,
      543, 545, 628
   Schwartz   500
   Sobolev   105, 128, 633
   Walsh   596
   weighted Walsh   596

G

G-star discrepancy   625
Gallai's theorem   356
gap condition
   strong   250
   weak   251
Gauss' circle problem   198
Gaussian curvature   163, 164, 180
Gaussian measure   326
Gaussian sum   497
Gelfand–Leray form   521
Gelfand–Leray measure   523
generating coloring   465
generation
   in EDA   471
genetic algorithm (GA)   657, 660
   (parent) population   661
   crossover   661
   elitist selection scheme   662
   fitness (of an individual)   662
   generation   661
   individual   661
   mutation   661, 662
   population   661
   recombination   661
   selection   662
   uniform crossover   661
geometric discrepancy   72
Gilbert–Varshamov bound   126, 130
graph coloring   139
graph matching   139
great open problem   19, 76

H

Hölder function   91
   product   92

Hölder inequality 191, 541
Haar basis 74, 83
  product 84
Haar coefficient 74
Haar expansion 81, 96, 105
Haar function 74, 83
Haar function method of Roth 75
Haar measure 357
Haar wavelet 83, 259
Hadamard matrix 326
Hahn–Banach theorem 534
half-space discrepancy 192
Halton construction 48
Halton sequence 49, 660, 665
  generalized 666
  scrambled 666
Halton set 49
  extended 52
  translated 53
Halton–Hammersley set 667
Hammersley set 149, 552, 639
Hamming distance 125
Hamming weight 125
Hardy norm 103
Hardy space 79, 103
Hardy–Littlewood maximal function theorem
    179
Hardy–Littlewood method 491
harmonic analysis 80, 95
Heinrich's algorithm 629
hereditary discrepancy 325, 427
  multicolor 333
Hickernell's modified L_p-discrepancy 625
high correlation between higher bases 665
higher order net 568, 569
  scrambled 571
higher order sequence 568, 569
  scrambled 571
Hilbert space technique 73
Hlawka–Zaremba identity 548
Hoeffding inequality 9
Hurwitz's theorem 227
Hyperbola Problem 235
hyperbolic cross 131, 132
  best approximation 132
  dyadic 132
hyperbolic cross points (see Smolyak
      quadrature rule)
hyperbolic needle 224, 226, 243
hypergraph 322
  of arithmetic progressions 330
  of arithmetic progressions in $\mathbb{Z}_p$ 467
  of arithmetic progressions in the first n
      integers 460

of linear hyperplanes 405
  wrapped 385
  $\mathbb{Z}_p$-invariant 394
hyperplane 171
hypersurface 509

I

identity
  Hlawka–Zaremba 548
  Parseval 14, 96, 117, 173, 188
  Plancherel 516
inclusion-exclusion formula 143
individual 471
induced box 653
induced grid 636
inequality
  Bessel 79, 90
  Borell 129
  Chang–Wilson–Wolff 98
  Chebyshev 122, 192
  Chernoff 345
  Chernoff–Hoeffding 345
  diophantine 227
  Erdős–Turán–Koksma 586, 648
  Hölder 191, 541
  Hoeffding 9
  isoperimetric 129, 434
  Jensen 111, 592
  Khintchine 79, 97, 99
  Koksma 542
  Koksma–Hlawka 546, 550
  large deviation 9, 140
  Littlewood–Paley 96
  Littlewood–Paley type 75, 79, 101, 132
  Markov 604
  Minkowski (integral) 169
  Pell 223, 224
  small ball 75, 79, 107
information complexity 558
Information-based Complexity (IBC) 554
integer linear programming 651
integral weight 340
integration
  error 541
  numerical 84, 90, 91
  operator 91, 120
  polynomially (QMC) tractable 558
  quasi-Monte Carlo (QMC) 539
  strongly polynomially (QMC) tractable
      558
internal point dimension 641

intractability   555
irregularity of distribution   73, 74
isoperimetric inequality   129, 434
isotrope discrepancy   625, 671

J

Jacobi symbol   509
Jensen's inequality   111, 592

K

k-clique
    parameterized problem   647
Kahan summation   629
Khintchine inequality   79, 97, 99
Khintchine's theorem   236
Klee's measure problem   639
Kloosterman refinement   495, 496, 506
Kloosterman sum   506
Koksma inequality   542
Koksma–Hlawka inequality   546, 550
Kolmogorov
    cubature error bound method   74
Kolmogorov metric   671
Kolmogorov width   82
Kolmogorov's cubature error bound method
    90
Korobov lattice points   589
Korobov space   576, 577
Kuelbs–Li equivalence   124
Kuelbs–Li theorem   124

L

L^2-square discrepancy   383
L_2-discrepancy   545, 546, 627, 628
    r-smooth   632
    modified   625
L_2-extreme discrepancy   624, 629
L_2-star discrepancy   627
lacunarity   116
lacunary Fourier series   115, 116
lacunary sequence   116
Lagrange's theorem   11
large deviation   119
    inequality   9, 140
    technique   8

large discrepancy phenomena   5
large discrepancy problem   5
Las Vegas algorithm   657
Latin square   362
lattice point set   552, 575
lattice rule   575, 639
    polynomial   593
Laurent series   561, 593
Law of the Iterated Logarithm   230, 236
    Cassels's form   237
    Khintchine's form   235, 236
    Kolmogorov–Erdős form   237
LCA group   357
learning rate   472
learning resolution   473
Lebesgue measure   3, 17, 82, 238, 550, 580,
    623, 625
Legendre symbol   467
lemma
    Anderson   121
    Owen   567
    partial coloring   433, 447
level surface   488 (*see also* hypersurface)
linear discrepancy   324
    multicolor   333
linear learning   473
linear relaxation   654
Littlewood–Paley inequality   75, 79, 96, 101,
    132
    dyadic   79
Littlewood–Paley product square function
    90
Littlewood–Paley square function   97
Littlewood–Paley theory   79, 95, 96, 101
local discrepancy   539, 623
    of measures   622
local-global principle   525
low discrepancy point set   552, 558
low discrepancy sequence   558
L_p space   79
L_p spherical cap discrepancy   550
L_p-discrepancy   540, 663
    inverse of   555
    modified   625
    of measures   622
L_p-extreme discrepancy   624
L_p-star discrepancy   623
L_∞-discrepancy
    inverse of   554
    of measures   622
L_∞-extreme discrepancy   624
L_∞-star discrepancy   623

M

major arc   511
Markov's inequality   604
martingale   97
martingale difference square function   97
matching (in a graph)   139
mean value theorem   179
measure
   Dirac   13, 215
   Gaussian   326
   Gelfand–Leray   523
   Haar   357
   Lebesgue   3, 17, 82, 238, 550, 580, 623,
     625
   of discrepancy   622
measure-preserving transformation   491
method
   Birch–Davenport   494, 509, 510
   branch and bound   656
   divide-and-conquer   631, 656
   ellipsoid   437
   entropy   429, 433
   Hardy–Littlewood   491
   Kolmogorov's cubature error bound   74
   nested intervals   247
   partial coloring   429
   probabilistic   335
   Roth (orthogonal function)   75–77, 80, 83
   sparse grid   633
metric entropy   80, 123, 124, 127, 130
migration   473
   global   473
   local   473
Minkowski inequality (integral)   169
Minkowski's theorem   227
minor arc   511
mirror construction   465
modulo coloring   465
Monte-Carlo (MC)   565
Monte-Carlo (MC) algorithm   565, 657
Monte-Carlo point set   667
$(\mu + \lambda)$ evolutionary algorithm (EA)   661
multicolor discrepancy   320

N

Naive Area Principle   224
nested interval method   247
Niederreiter sequence   561
norm

Besov   105, 624
BMO   624
Hardy   103
Orlicz   624
Triebel–Lizorkin   624
NP (complexity class)   322, 643
NP-completeness   322, 644
NP-hard problem   322
NP-hardness   322, 643
numerical integration   84, 90, 91

O

Orlicz function   98
Orlicz norm   98, 624
Orlicz space
   exponential   79, 98, 106
orthogonal basis   84, 96, 105
orthogonal decomposition   73
orthogonal function method of Roth   75–77,
   80, 83
orthogonal range counting   650
orthogonal transformation   6
Owen scrambled sequence   567
Owen's lemma   567
Owen's scrambling   566

P

p-adic integer   525
parallel section function   162
parameterized complexity   646
parameterized problem   646
parameterized reduction   646
Parseval formula   234
Parseval identity   14, 96, 117, 173, 188
Parseval's theorem   600
Parseval-type equation   412
partial coloring   431
   lemma   433, 447
   method   429
   SDP   446
partition tree
   for a color set   336
   for a positive integer   338
Pell inequality   223, 224
Pell's equation   222
Plancherel identity   516
Plancherel theorem   150, 358
point set discrepancy   4
Poisson sum   496

Poisson summation formula　196, 234
polar coordinates　162
polygon　181
polyhedral discrepancy　671
polyhedron　180
polynomial
　　Bernoulli　578, 602
　　non-singular　488
　　regular value of　490
　　regular value set of　488, 510
　　trigonometric　117, 131
polynomial lattice point set　557, 593
　　figure of merit　595
polynomial lattice rule　593
polynomial tractability　554
population-based incremental learning (PBIL)
　　471
premature convergence　473
principle
　　Dirichlet　495
　　local-global　525
　　restricted invertibility　432
　　Roth　83, 85
　　Stolarsky invariance　551
probabilistic method　335
probabilistic proof　634
problem
　　fixed-parameter tractable (FPT)　646
　　Gauss' circle　198
　　great open　19, 76
　　Klee's measure　639
　　large discrepancy　5
　　NP-complete　644
　　NP-hard　322, 643
　　parameterized　646
　　small ball　118, 128
　　small deviation　118, 121
　　small discrepancy　17
　　W[1]-hard　646, 647
　　Waring　536
product BMO space　106
product Hölder function　92
product Haar basis　84
product rule　112
property B　334
pseudo-random numbers　621

Q

QEA
　　attractor population (apQEA)　474
　　standard (sQEA)　472

　　versatile (vQEA)　473
quadrant discrepancy　624
quadrature algorithm　564
quadrature formula　622
quadrature points　540
quadrature rule　540, 578, 623
　　Smolyak　633
"quantum-inspired" attribute　473
quantum-inspired evolutionary algorithm
　　(QEA)　472
quasi-Monte Carlo (QMC)　539
　　algorithm　634
　　integration　539, 558, 633
　　randomised　564, 565
　　rule　540
　　sampling　625

R

r-function　86, 112, 113
Rademacher function　86, 97, 235
　　generalized　86
　　modified　259, 260
radical inverse function　665
Ramsey theory　323
random coloring　429, 464
random walk　235, 440, 445
　　gaussian　430, 431, 447
　　infinite　235
randomised error　565
range tree　651
rectangle discrepancy　671
rectangle property　253, 254
region level　641
regular value　490
regular value set　488, 510
reproducing kernel　129, 542
reproducing kernel Hilbert space　124, 128,
　　542, 543, 545, 628
restricted invertibility principle　432
Riesz product　75, 112, 115, 116, 136, 249,
　　251, 259
Rosenblum–Tsfasman weight　67
rotation learning　473
Roth's principle　83, 85

S

Salie sum　509
sawtooth function　21

Schwartz space   500
scrambled radical inverse function   666
semidefinite programming (SDP)   430, 436
   for partial coloring   446
   vector program view   437
separation constant   309
sequence
   digital   560, 561
   digital (t, s) 561
   digital (t, s) over $\mathbb{Z}_b$ 561
   Faure   660
   Halton   660, 665
   higher order   568, 569
   lacunary   116
   low discrepancy   558
   Niederreiter   561
   Owen scrambled   567
   Sobol'   660
   (t, s) in base b 559
   (t, s) over $\mathbb{Z}_b$ 570
   uniformly distributed   72
   van der Corput   25, 552, 664
set
   Cantor   247
   dominating   644
   dual polynomial lattice   594
   extended Halton   52
   extended van der Corput   31
   Faure   55, 56, 667
   Halton   49, 50
   Halton–Hammersley   667
   Hammersley   552, 639
   lattice point   552, 575
   low discrepancy   552, 558
   Monte-Carlo   667
   of shattered hyperedges   334
   polynomial lattice point   557, 593
   regular value   488, 510
   shifted van der Corput   35
   Sobol'   667
   translated Halton   53
   translated van der Corput   31
   van der Corput   25, 26, 80, 149, 151
shatter function
   dual   345
   primal   345
shattered hyperedge set   334
shortcutting computation   462
Sidon's theorem   80, 116, 251
σ-separation   309
simplex   182, 367
single dominant term rule   264
single term domination   264, 272
singular series   491, 524

slab   640
small ball conjecture   76, 108
   generic signed   110
   signed   109
   trigonometric   133
small ball inequality   75, 79, 107
   signed   145
small ball probability   119, 122, 124, 127,
      130, 134
small ball problem   118, 128
small deviation probability   80, 118, 119
small deviation problem   118, 121
small discrepancy phenomena   5
small discrepancy problem   17
Smolyak quadrature rule   633
smoothed discrepancy   503
snapping procedure   659
Sobol' sequence   562, 660
Sobol' set   667
Sobolev space   105, 128, 633
sparse grid method   (see Smolyak quadrature
      rule)
spherical average   163
spherical cap   164
spherical cap discrepancy   626
spherical mean   163, 172
square function
   martingale difference   79
standard QEA (sQEA)   472
star discrepancy   73, 107, 540, 623
   decision problem   643
   inverse of   554
   L_p 623
   L_∞ 623
star discrepancy metric   671
Stirling number   366
Stirling's formula   126
stochastic program   670
stochastic programming   622, 670
   scenario reduction   622, 670
Stolarsky's invariance principle   551
strong tractability   555, 557
strongly distinct vectors   137
subgaussian function   98
sum
   Gaussian   497
   Kloosterman   506
   of arithmetic progressions   320, 378
   Poisson   496
   Salie   509
summation
   Kahan   629
   Poisson   196, 234
super-orthogonality   259, 260

superirregularity   233
support   165
supporting box   672

T

T-tractability   558
tensor product   121, 545
tensor product technique   321
theorem
    Beck–Fiala   321
    Borel's normal number   236
    Bárány–Grinberg   321
    central limit   234, 567
    Chinese remainder   49
    divergence   164
    Gallai   356
    Hahn–Banach   534
    Hardy–Littlewood (maximal function)
        179
    Hurwitz   227
    Khintchine   236
    Kuelbs–Li   124
    Lagrange   11
    mean value   179
    Minkowski   227
    Minkowski lattice point   326
    Parseval   600
    Plancherel   150, 358
    Sidon   80, 116, 251
    six standard deviations   321
    van der Waerden   330
theory
    Littlewood–Paley   79, 95, 96,
        101
    Ramsey   323
Thiémard's algorithm
    bracketing cover approach   650
    ILP-based   651
threshold accepting   657, 658
threshold language   643
(t, m, s)-net
    digital   560
    in base b 552, 559
    Owen scrambled   567
    quality parameter   559
    randomly digitally shifted   566
total unimodular matrix   325
tractability
    polynomial   554
    strong   557
trellis   640

Triebel–Lizorkin norm   624
trigonometric polynomial   117, 131
trivial error   261, 264
(t, s)-sequence
    in base b 559
    over $\mathbb{Z}_b$ 570
Turing machine (non-deterministic)   643
type ε point   501

U

unanchored discrepancy   (*see* extreme
        discrepancy)
uniform metric   671
uniformly distributed sequence   72
unit level surface   488
univariate marginal distribution algorithm
        (UMDA)   471

V

van der Corput conjecture   4
van der Corput sequence   25, 552, 664
    generalization by Faure   (*see* Faure
        construction)
    generalization by Halton   (*see* Halton
        construction)
van der Corput set   25, 26, 80, 149, 151
    extended   31
    Fourier–Walsh approach   38
    shifted   35
    translated   31
Vandermonde matrix   58
variable forcing   656
VC-dimension   334
versatile QEA (vQEA)   473
vertical translation   266

W

W[1] (complexity class)   646, 647
W[1]-hardness   646, 647
Walsh coefficients   597
Walsh expansion   573
Walsh function   39, 62, 77, 573
    b-adic   594
    b-adic system   594
    s-dimensional b-adic   595

Walsh series   602
Walsh space   596
Waring problem   536
Warnock formula   545, 546, 549, 551, 628
    generalization   663
weak tractability   558
weighted discrepancy   325, 549, 555, 625
weighted L_2-discrepancy   557
weighted L_p-discrepancy   549, 626
    inverse of   556
weighted L_∞-discrepancy   626
weighted star-discrepancy   549

weighted Walsh space   596
Weil's estimate   506
Weyl-type estimate   511
Winker–Fang algorithm   658
worst-case-root-mean-square error   565
wrapped hypergraph   385

Z

$\mathbb{Z}_p$-invariant hypergraph   394

LECTURE NOTES IN MATHEMATICS  Springer

Edited by J.-M. Morel, B. Teissier; P.K. Maini

Editorial Policy (for Multi-Author Publications: Summer Schools / Intensive Courses)

1. Lecture Notes aim to report new developments in all areas of mathematics and their applications - quickly, informally and at a high level. Mathematical texts analysing new developments in modelling and numerical simulation are welcome. Manuscripts should be reasonably selfcontained and rounded off. Thus they may, and often will, present not only results of the author but also related work by other people. They should provide sufficient motivation, examples and applications. There should also be an introduction making the text comprehensible to a wider audience. This clearly distinguishes Lecture Notes from journal articles or technical reports which normally are very concise. Articles intended for a journal but too long to be accepted by most journals, usually do not have this "lecture notes" character.

2. In general SUMMER SCHOOLS and other similar INTENSIVE COURSES are held to present mathematical topics that are close to the frontiers of recent research to an audience at the beginning or intermediate graduate level, who may want to continue with this area of work, for a thesis or later. This makes demands on the didactic aspects of the presentation. Because the subjects of such schools are advanced, there often exists no textbook, and so ideally, the publication resulting from such a school could be a first approximation to such a textbook. Usually several authors are involved in the writing, so it is not always simple to obtain a unified approach to the presentation.

   For prospective publication in LNM, the resulting manuscript should not be just a collection of course notes, each of which has been developed by an individual author with little or no coordination with the others, and with little or no common concept. The subject matter should dictate the structure of the book, and the authorship of each part or chapter should take secondary importance. Of course the choice of authors is crucial to the quality of the material at the school and in the book, and the intention here is not to belittle their impact, but simply to say that the book should be planned to be written by these authors jointly, and not just assembled as a result of what these authors happen to submit.

   This represents considerable preparatory work (as it is imperative to ensure that the authors know these criteria before they invest work on a manuscript), and also considerable editing work afterwards, to get the book into final shape. Still it is the form that holds the most promise of a successful book that will be used by its intended audience, rather than yet another volume of proceedings for the library shelf.

3. Manuscripts should be submitted either online at www.editorialmanager.com/lnm/ to Springer's mathematics editorial, or to one of the series editors. Volume editors are expected to arrange for the refereeing, to the usual scientific standards, of the individual contributions. If the resulting reports can be forwarded to us (series editors or Springer) this is very helpful. If no reports are forwarded or if other questions remain unclear in respect of homogeneity etc, the series editors may wish to consult external referees for an overall evaluation of the volume. A final decision to publish can be made only on the basis of the complete manuscript; however a preliminary decision can be based on a pre-final or incomplete manuscript. The strict minimum amount of material that will be considered should include a detailed outline describing the planned contents of each chapter.

   Volume editors and authors should be aware that incomplete or insufficiently close to final manuscripts almost always result in longer evaluation times. They should also be aware that parallel submission of their manuscript to another publisher while under consideration for LNM will in general lead to immediate rejection.

4. Manuscripts should in general be submitted in English. Final manuscripts should contain at least 100 pages of mathematical text and should always include

   - a general table of contents;
   - an informative introduction, with adequate motivation and perhaps some historical remarks: it should be accessible to a reader not intimately familiar with the topic treated;
   - a global subject index: as a rule this is genuinely helpful for the reader.

   Lecture Notes volumes are, as a rule, printed digitally from the authors' files. We strongly recommend that all contributions in a volume be written in the same LaTeX version, preferably LaTeX2e. To ensure best results, authors are asked to use the LaTeX2e style files available from Springer's web-server at
   ftp://ftp.springer.de/pub/tex/latex/svmonot1/ (for monographs) and
   ftp://ftp.springer.de/pub/tex/latex/svmultt1/ (for summer schools/tutorials).
   Additional technical instructions, if necessary, are available on request from: lnm@springer.com.

5. Careful preparation of the manuscripts will help keep production time short besides ensuring satisfactory appearance of the finished book in print and online. After acceptance of the manuscript authors will be asked to prepare the final LaTeX source files and also the corresponding dvi-, pdf- or zipped ps-file. The LaTeX source files are essential for producing the full-text online version of the book. For the existing online volumes of LNM see:
   http://www.springerlink.com/openurl.asp?genre=journal&issn=0075-8434.
   The actual production of a Lecture Notes volume takes approximately 12 weeks.

6. Volume editors receive a total of 50 free copies of their volume to be shared with the authors, but no royalties. They and the authors are entitled to a discount of 33.3 % on the price of Springer books purchased for their personal use, if ordering directly from Springer.

7. Commitment to publish is made by letter of intent rather than by signing a formal contract. Springer-Verlag secures the copyright for each volume. Authors are free to reuse material contained in their LNM volumes in later publications: a brief written (or e-mail) request for formal permission is sufficient.

Addresses:
Professor J.-M. Morel, CMLA,
École Normale Supérieure de Cachan,
61 Avenue du Président Wilson, 94235 Cachan Cedex, France
E-mail: morel@cmla.ens-cachan.fr

Professor B. Teissier, Institut Mathématique de Jussieu,
UMR 7586 du CNRS, Équipe "Géométrie et Dynamique",
175 rue du Chevaleret,
75013 Paris, France
E-mail: teissier@math.jussieu.fr

For the "Mathematical Biosciences Subseries" of LNM:

Professor P. K. Maini, Center for Mathematical Biology,
Mathematical Institute, 24-29 St Giles,
Oxford OX1 3LP, UK
E-mail: maini@maths.ox.ac.uk

Springer, Mathematics Editorial I,
Tiergartenstr. 17,
69121 Heidelberg, Germany,
Tel.: +49 (6221) 4876-8259
Fax: +49 (6221) 4876-8259
E-mail: lnm@springer.com